Multimessenger Astronomy in Practice

AAS AMERICAN ASTRONOMICAL SOCIETY 

AAS Editor in Chief

About the program:

AAS-IOP Astronomy ebooks is the official book program of the American Astronomical Society (AAS) and aims to share in depth the most fascinating areas of astronomy, astrophysics, solar physics, and planetary science. The program includes publications in the following topics:

GALAXIES AND COSMOLOGY

INTERSTELLAR MATTER AND THE LOCAL UNIVERSE

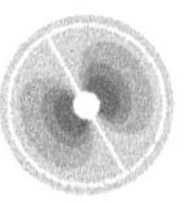
STARS AND STELLAR PHYSICS

EDUCATION, OUTREACH, AND HERITAGE

HIGH-ENERGY PHENOMENA AND FUNDAMENTAL PHYSICS

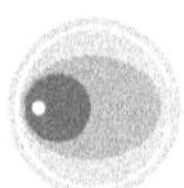
THE SUN AND THE HELIOSPHERE

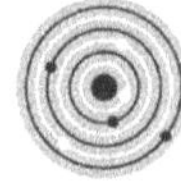
THE SOLAR SYSTEM, EXOPLANETS, AND ASTROBIOLOGY

LABORATORY ASTROPHYSICS, INSTRUMENTATION, SOFTWARE, AND DATA

Books in the program range in level from short introductory texts on fast-moving areas, graduate and upper-level undergraduate textbooks, research monographs, and practical handbooks.

For a complete list of published and forthcoming titles, please visit iopscience.org/books/aas.

About the American Astronomical Society

The American Astronomical Society (aas.org), established in 1899, is the major organization of professional astronomers in North America. The membership (~7,000) also includes physicists, mathematicians, geologists, engineers, and others whose research interests lie within the broad spectrum of subjects now comprising the contemporary astronomical sciences. The mission of the Society is to enhance and share humanity's scientific understanding of the universe.

Editorial Advisory Board

Multimessenger Astronomy in Practice

Edited by
Miroslav D Filipović and Nicholas F H Tothill
Western Sydney University, Penrith, NSW, Australia

IOP Publishing, Bristol, UK

ISBN 978-0-7503-2344-4 (ebook)
ISBN 978-0-7503-2342-0 (print)
ISBN 978-0-7503-2345-1 (myPrint)
ISBN 978-0-7503-2343-7 (mobi)

DOI 10.1088/2514-3433/ac2256

Version: 20220101

AAS–IOP Astronomy
ISSN 2514-3433 (online)
ISSN 2515-141X (print)

British Library Cataloguing-in-Publication Data: A catalogue record for this book is available from the British Library.

Published by IOP Publishing, wholly owned by The Institute of Physics, London

IOP Publishing, Temple Circus, Temple Way, Bristol, BS1 6HG, UK

US Office: IOP Publishing, Inc., 190 North Independence Mall West, Suite 601, Philadelphia, PA 19106, USA

Image credit: X-ray: NASA/CXC/CfA/M.Markevitch et al.; Optical: NASA/STScI; Magellan/U.Arizona/D.Clowe et al.; Lensing Map: NASA/STScI; ESO WFI; Magellan/U.Arizona/D.Clowe et al.

Miroslav D. Filipović: I dedicate this book to all my students who graduated under my supervision. Well done, all! I am really proud of your fantastic work. At the same time, I would like to dedicate this book to my supervisors, Professor Graeme L. White and Professor Raymond F. Haynes, who I call my academic fathers.

Nicholas F. H. Tothill: To Megan, Aidan, and Callum, who put up with the writing of this book; and to the many colleagues, teachers, and students who have given up their time to explain its contents to me over the years.

Ray Norris: Many of the telescopes around the world described in this volume are located on lands traditionally owned by Indigenous people. We acknowledge the sacrifice made by these people in allowing our telescopes to be built on their land, and in particular, we acknowledge the Wajarri Yamatji, Wiradjuri, and Gamilaraay people as the traditional owners of the Australian radio observatories discussed in this book.

Jeffrey Payne: To my family, Brandon, Peter, Lora, and Mark who have supported my love of astronomy over the years, and especially to my supervisors, Miroslav D. Filipović and Graeme L. White, who opened up the academic doors of astronomy for me.

Clancy W. James: To my wonderful daughter Diana, the people in the world battling to keep Earth a beautiful place for her to grow up in.

Pierre Maggi: To Hsin-Yin, the star in my sky, for her tremendous support over the years; to my family, for their support in my upbringing and education; to colleagues, friends, and students that make working in astronomy worth it.

Sebastian Gurovich: To the teachers of the world: most of whom work for little pay, and Satoshi Nakamoto: whose actions are stimulating Open Development.

Contents

Preface

Our aim is to prepare a newcomer to research in some particular part of multimessenger astronomy, in order that they may be able to understand what their new colleagues are talking about, get an overview of their new field, and make substantial progress toward being able to contribute themselves. The newcomer might be a new graduate, an experienced researcher who wishes to work outside their own research mainstream, or anyone in between.

The aim is ambitious, and it has led us to produce two books, the products of orthogonal approaches to the task, of which this is the second. In our first book, *Principles of Multimessenger Astronomy*, our priority was to give a unified view of the emerging multimessenger program. We tried to pick out the common concerns and methods used by scientists who try to decode these messengers, over the wide range of natures and energies now available to us. But these fundamental principles give rise to an almost-fractal proliferation of disciplines, areas, topics, groups, and so on. In order to contribute at the frontiers, a newcomer must become familiar with the domain in which they wish to operate. Fundamental principles become less important than knowledge of how a particular group of scientists does things. A detail-oriented, bottom-up approach must complement the bird's-eye view—and that is the domain of this book.

As both editors and contributors, we have asked our colleagues to contribute chapters in their own particular field. They share our basic aim, but each takes their own approach. This diversity is deliberate; imposing a uniform structure would undercut the design—which is to introduce the reader to a community, with its own habits of thought and ways of seeing the universe. Our chapter authors understand the project and have risen to the challenge enthusiastically.

Whether you, the reader, are a student trying to get to grips with the habits, assumptions, and "common knowledge" of a field you aspire to join; or a researcher who, in the busyness of producing their output, has missed out on some elements of their own field; or an established worker in one field who suddenly needs or wants to get to grips with another—we hope you will find this work useful, both as an introduction and as a reference.

Foreword

These are thrilling times to be launching a career studying the universe—are you ready? Is your toolkit well equipped? Is your knowledge base broad enough so that when something new and exciting comes along, you'll be ready to pounce on it, finding unexpected connections and novel insights of your own? Will this book on multimessenger astronomy be part of your toolkit?

The 1960s and 1970s, when my science journey was starting, were a similarly exciting time. It's hard now to picture the world without even the cosmic microwave background, without quasars and pulsars, without the X-ray sky. But my toolkit had been building and I was eager to jump in. When I graduated high school, my father took me to a fine-goods shop in downtown Philadelphia to purchase a high-quality slide rule, suitably ensconced in its own polished green leather case. The adepts among us could manage 2.5 significant figures! Logs—natural and base 10, sines, cosines, squares, and cubes. For more accurate work, I acquired the 2.3 kg *Handbook of Chemistry and Physics*, with its long mathematical tables and interpolation schemes. Radio interferometry was in its infancy, and the Fourier transforms were done most efficiently with mechanical calculators that clicked and whirred or nixie-tube devices (Google it!) that flashed in iridescent colors. Sine and cosine coefficients were written on 3 × 5 cards, stacked in long drawers for easy retrieval. And my college graduation gift was a miraculous calculator, the HP45. It had registers! It could do not only those sines and cosines but arcsines and arccosines, means and rms's, and even convert to polar coordinates! And it worked at top speed without clicks or flashes.

So here we are with a modern toolkit in book form, covering the dizzying array of techniques and findings that today's astronomers/physicists need to know. Figure 1.1 sets one stage—the phenomena we study contain unique signatures across the landscape, particles like cosmic rays and neutrinos, to electromagnetic radiation spanning radio to gamma rays, and to the latest space-and-time warping effects of gravitational waves. And any bets on when the laboratory experiments described in the dark matter chapter will turn our world on its head? A second stage represents discipline-spanning techniques such as machine learning and mining of large data sets; these tools are as fundamental and essential today as the HP45 that gave me so much power and pleasure. The third stage is more abstract—how do the experts in each of these technical or domain areas think about their science? What kinds of questions do they ask? For me, this came through most clearly in the excitement of establishing entirely new areas of astronomy, e.g., in gravitational waves and neutrinos, and in the SETI chapter, with the mind-bending aspects of thinking about how to detect Dyson spheres.

My Ph.D. was in Physics, working under the inspiring Dave Wilkinson (of WMAP renown) at Princeton. Graduate school was minimalist in terms of courses; there were no required classes, only the mandate to pass the preliminary examinations by the end of two years. And so, with only one undergraduate and no graduate courses in astrophysics, it was up to me to identify and study whatever I needed to know. I wish I had a book like this one. Today, for early-career astronomers, the situation is much the same as it was for me; no matter what you did in your formal education, it's not enough today, and certainly won't be enough tomorrow. So enjoy, dip in and out of these chapters, build your toolkit, and be ready to make those fresh connections. And try to picture, if you dare, how you will write the next chapters.

Lawrence Rudnick Professor Emeritus Minnesota Institute for Astrophysics School of Physics and Astronomy, University of Minnesota (http://umn.edu/~larry; email: larry@umn.edu).

Editor biographies

Miroslav D. Filipović

Professor Miroslav D. Filipović is a scientist, philosopher, and philanthropist with over 30 years of experience in astronomy. Astronomy, science, philosophy, and computing are his profession, hobby, interest, and passion. Research in astronomy has been a source of fascination for him since the early 1980s. Since May 2002, Professor Filipović has been affiliated with Western Sydney University (WSU) and has been responsible for the development of astronomy at WSU. Between 1997 and 2000, he held full-time research positions at the Max-Planck-Institut für extraterrestrische Physik (MPE; Germany). He is Chair of the largest public observatory in Australia (the WSU's Penrith Observatory) and has over 200 refereed publications. His research interests center on supernovae, high-energy astrophysics, planetary nebulae, Milky Way structure and mass extinctions, HII regions, X-ray binaries, active galactic nuclei, deep fields (SPT and Pavo), and stellar content (Wolf–Rayet, O, and B stars) in nearby galaxies including the Magellanic Clouds. All of this research is closely related to further our understanding of the interactions between galaxies and the processes of stellar formation and star evolution as they affect galaxy evolution. One of the scientific highlights of professor Filipović's scientific career is flying on board NASA's SOFIA science mission in 2018 June. He is a fellow of the Astronomical Society of Australia, Australian Institute of Physics, editorial board member of the Serbian Astronomical Journal, and a member of the International Astronomical Union. He lives in Sydney with his family.

Nicholas F. H. Tothill

Nick Tothill was born in the south of England, read Physics at Corpus Christi College, Cambridge, and received his M.Sc. from the University of Manchester, studying at Jodrell Bank. He graduated with a Ph.D. in Astrophysics from the University of London, after research in star formation and the interstellar medium at Queen Mary & Westfield College. He has held research appointments in Heidelberg, Nova Scotia, Boston, Exeter, and Sydney, and worked on telescopes in Spain, Chile, Hawaii, Arizona, Australia, and Antarctica. In 2004, he was a winterover scientist and Station Science Leader at the Amundsen-Scott South Pole Research Station. In 2011, He joined Western Sydney University, where he is now Senior Lecturer in the School of Science and Director of the Penrith Observatory. He is a member of the Astronomical Society of Australia and the International Astronomical Union. His research still centers on the interstellar medium of the Milky Way Galaxy but includes topics as diverse as high-redshift galaxy surveys, Antarctic astronomy, and cosmic-ray astrophysics. He lives in Sydney with his family.

Author biographies

Natasha Hurley-Walker

Dr. Natasha Hurley-Walker is an experienced radio astronomer who has conducted several radio-frequency surveys of the sky and believes the best science comes from well-calibrated data covering large areas in exquisite detail. She has helped to commission two radio telescopes, published over 100 refereed publications, and released 10 large astronomical data sets. Supporting open science and repeatable work, she uploads all of her code to a public github repository. Dr. Hurley-Walker studies supernova remnants, galaxy clusters, radio galaxies, and radio astronomical techniques. One of her driving motivations is to bring her work to the public, and she created a free public online viewer and mobile phone app so that everyone in the world could view the sky in radio color. She has spoken on the radio and TV multiple times, including on the world's longest-running TV show, the BBC's "Sky at Night," and conducts local outreach events on a monthly basis. She was named the WA Tall Poppies Scientist of the Year for 2017 and one of the ABC's Top 5 Scientists for 2018. Her TED talk on low-frequency radio astronomy has been viewed over 1 million times. She is a keen advocate for diversity in STEM and is one of Science and Technology Australia's Superstars of STEM for 2019–2020, empowering young women to study in STEM by acting as a positive role model. In her spare time, she is a keen advocate for cycling for transport, reads a plethora of speculative and non-fiction, and has a discerning taste for loose-leaf tea. She lives in Perth with her family.

Michelle Cluver

Dr. Michelle Cluver is an Australian Research Council (ARC) Future Fellow and Senior Lecturer at Swinburne University of Technology. She obtained a Ph.D. in Astronomy from the University of Cape Town (2009) and was awarded the Ph.D. Medal for Best Thesis in the Faculty of Science. Her first postdoctoral position was at the Spitzer Science Center, at the California Institute for Technology, followed by an ARC Super Science Fellowship at the Australian Astronomical Observatory where she worked on the Galaxy and Mass Assembly (GAMA) survey. She is an experienced user of space-based mid-infrared telescopes, such as Spitzer and WISE, as well as several optical and radio ground-based telescopes. Her current research focus is on the evolution of galaxies within groups, a key environment in the transformation of galaxies, using multiwavelength photometry combined with radio data from the SKA Pathfinders. She lives in Melbourne.

Thomas H. Jarrett

Professor Thomas H. Jarrett is the SARChI Chair in Astrophysics and Space Science hosted in the Department of Astronomy at the University of Cape Town (UCT). While he is world renowned for his expertise in the near and mid-infrared, he also has hands-on experience in all other bands of the electromagnetic spectrum, from radio to X-ray, both Earth and space based. He is a passionate researcher whose interest and expertise lie in extragalactic large-scale structure—and visualization thereof—of the nearby universe, the Zone of Avoidance, interacting galaxies, star formation processes, and galaxy evolution. He is now the lead scientist of the Visualisation Lab at the UCT Department of Astronomy as well as of the Visualisation Research group at the Iziko Planetarium. He lives in Cape Town.

Sebastian Gurovich

Professor Sebastian Gurovich grew up in the eastern suburbs of Sydney. At 11 years of age, he saw Halley's comet over the South Bondi cliffs. He completed school at Dover Heights High and then a B.Sc. (applied physics, Honors) at the University of Western Sydney (Nepean). He obtained his Ph.D. in 2007 at Mount Stromlo Observatory (ANU) studying dark matter in disk galaxies followed by a postdoc at the Observatorio Astronómico de Córdoba, Argentina (UNC-OAC-CONICET), where he is currently a CONICET researcher and assistant professor. His research interests include galaxy formation and evolution, active galactic nuclei, stellar astrophysics, machine learning abd data mining, and open development of science. Sebastian has ample observational experience using optical telescopes around the world and is part of the Vista Variables in the Vía Láctea survey (VVV/VVVx) and the Transient Robotic Survey of the South (TOROS) that follows up gravitational-wave sources. Sebastian currently represents Argentina at the International Virtual Observatory Alliance (IVOA) on the executive commission. He lives in Córdoba with his family.

Denis Leahy

Professor Denis Leahy grew up on a wheat farm in Alberta, Canada, where the clear night skies stimulated his interest in mathematics and astronomy. He completed a B.Sc. in Physics at the University of Waterloo. His M.Sc. and Ph.D. in Physics, with theses on pulsar magnetospheres and black hole evaporation, were completed at the University of British Columbia. His postdoctoral fellowship at NASA's Marshall Space Flight Center, working for Martin Weisskopf, was in X-ray astronomy. He started at the University of Calgary, in Canada, with a postdoctoral position before becoming a faculty member. His research interests are in multiwavelength (radio, ultraviolet, X-ray, and gamma-ray) analysis of X-ray binaries, supernova remnants, neutron stars, gamma-ray bursts, and stars. He has published more than 250 refereed papers. Outside of work, he enjoys hiking and skiing in the Canadian Rockies, swimming, cycling, and spending time with his family in Calgary.

Pierre Maggi

Dr. Pierre Maggi was trained at Strasbourg University (France), where he specialized in astrophysics at Strasbourg Astronomical Observatory. He obtained his Ph.D. at the Max-Planck Institute for extraterrestrial Physics in Germany, focusing on the supernova remnant population of the Large Magellanic Cloud. After post-doctoral studies at CEA Saclay near Paris, he came back in 2018 to his alma mater as an astronomer at Strasbourg Observatory. When he is not teaching to the next generation of astrophysicists, his main research focuses on the high-energy cosmic phenomena occurring at the end of stars' life. In parallel, he is actively involved in the development and preparation of future X-ray space observatories. He lives in Strasbourg.

Gavin Rowell

Professor Gavin Rowell has nearly 30 years of experience in gamma-ray astronomy and high-energy astrophysics. He specializes in ground-based gamma-ray astronomy at Tera-electron-volt energies and the associated high-frequency radio astronomy to map the interstellar gas in our Milky Way. Following postdoctoral positions in Japan and Germany, he returned to Australia in 2006 via a QE-II Fellowship at The University of Adelaide. With this he brought together the fields of gamma-ray and high-frequency radio astronomy in Australia. He has discovered several gamma-ray sources including the first extended source, associated with a star cluster, and the first observational evidence of cosmic-ray-generated gamma-ray emission from a nearby supernova remnant. His focus is on the particle acceleration in extreme objects like supernova remnants, pulsar-powered nebulae, and transient sources such as gamma-ray bursts. He now leads Australia's efforts in the multinational next-generation TeV gamma-ray facility—the Cherenkov Telescope Array.

Clancy William James

Dr. Clancy William James is a Research Fellow at the International Centre for Radio Astronomy Research (ICRAR), Curtin University, Western Australia. His research focuses on neutrino astronomy, the radio detection of cosmic rays, and radio transients. He is a member of the ANTARES and KM3NeT neutrino telescope collaborations and has served on the publication committees of both. At his previous appointment at the Erlangen Centre of Astroparticle Physics, he headed the KM3NeT Simulations

Working Group and was a coeditor of the KM3NeT Letter of Intent. His research in neutrino astronomy has focused on neutrino fluxes from blazars—in conjunction with the TANAMI VLBI and multiwavelength collaboration—and on neutrino interactions and their light signatures in water. Dr. James also currently works with the Australian Square Kilometre Array Pathfinder (ASKAP) and Murchison Widefield Array (MWA) radio telescopes to perform searches for fast radio bursts (FRBs) as part of the CRAFT collaboration and is involved in projects to follow up gravitational-wave and neutrino transient events. Dr. James has previously worked on the theory of radio emission from cosmic-ray extensive air showers, at Radboud University Nijmegen, and its detection by the LOFAR radio telescope. He is coleading a project to implement this detection mode with the MWA and the Square Kilometre Array. His Ph.D., obtained from the University of Adelaide in 2009, was on the radio detection of high-energy neutrinos, for which he was awarded the 2010 Bragg Gold Medal by the Australian Institute of Physics. He lives in Perth.

Paul Lasky

Dr. Paul Lasky has dedicated his career to further our understanding of the most exotic regions of the universe. He is an active member of the LIGO Scientific Collaboration that, in 2016, transformed the very foundations of astrophysics by announcing the first detection of gravitational waves—tiny ripples in the fabric of spacetime—coming from two colliding black holes over one billion light-years from Earth. He has identified new ways of studying the interiors of neutron stars using their gravitational-wave signatures, as well as new ways of testing Einstein's theory of gravity in regions of the universe where new physics is most likely to occur at the surfaces of black holes. Lasky is currently an Australian Research Council Future Fellow and Senior Lecturer in the School of Physics and Astronomy at Monash University.

Csaba Balázs

Professor Csaba Balázs is a particle astrophysicist at the School of Physics and Astronomy at Monash University in Melbourne. He grew up and completed his undergraduate studies in Hungary. Following two years of military service, he worked at the Central Research Institute for Physics in Budapest, logged trees in British Columbia, and programmed robots in California. After completing a Master of Arts degree at Temple University in Philadelphia, he decided to study theoretical physics. He obtained a Ph.D. in physics at Michigan State University and went on doing postdoctoral work at the University of Hawaii, Florida State University, and at Argonne National Lab, the research laboratory of the University of Chicago. From there, he moved to Monash University to establish a particle physics group where, during the last decade, he was the Director of the Monash Node of the ARC Centre of Excellence for Particle Physics. Professor Balázs is the coauthor of more than 10 books and scientific reports, over a hundred journal publications, and several publicly available software packages. His research topics include dark matter, cosmic matter–antimatter asymmetry, primordial gravitational waves, Higgs portals to new physics, supersymmetry phenomenology, and the unification of fundamental forces. He lives in Melbourne.

Ray P. Norris

Professor Ray P. Norris is a Research Professor in Data Science at Western Sydney University and an Emeritus Fellow at CSIRO Astronomy & Space Science. He leads the EMU (Evolutionary Map of the Universe) project, which uses CSIRO's new ASKAP telescope to study the evolution of galaxies and their black holes since the dawn of the universe. EMU expects to detect 70 million galaxies, compared to the 2.5 million currently known. Ray mainly works on radio continuum surveys, galaxy evolution, machine learning, and the process of astronomical discovery. He also studies the astronomy of Aboriginal Australians, has given hundreds of public talks, and frequently appears in the media. Previous career highlights include ATNF Deputy Director, ATNF Head of Astrophysics, and Director of the Australian Astronomy Major National Research Facility, which (among other things) funded the development that led to ASKAP, and funded the ATCA CABB correlator. He returned to hands-on research in 2006. He has also written a novel, *Graven Images*; many popular science articles; and his work was chosen for the anthology of *Best Science Writing in 2017*. He lives in Sydney with his family.

Branislav Vukotić

Dr. Branislav Vukotić is an associate research professor at the Astronomical Observatory in Belgrade, Serbia. He obtained his M.Sc. and Ph.D. from the University of Belgrade. As part of the European Union–funded Erasmus+ ASTROMUNDUS program scholar exchange, he spent 3 months at the University of Innsbruck, Austria, in 2013, where he practiced the astrobiological history of the Milky Way and N-body simulations. Within the series of numerical experiments, he developed probabilistic cellular automata code used for testing the astrobiological evolutionary timescales in order to fit our current perception of life outside Earth. The rest of his research is in data statistics and supernovae remnants. He is a member of the Editorial Board of the Publications of the Astronomical Observatory of Belgrade and a vice chair of the Management Board of the Astronomical Observatory in Belgrade. He lives in Niš with his family.

Milan M. Ćirković

Professor Milan M. Ćirković (b. 1971) is a research professor at the Astronomical Observatory of Belgrade (Serbia) and a research associate of the Future of Humanity Institute at Oxford University (UK). His primary research interests are in the fields of astrobiology (Galactic habitable zone, SETI studies, and catastrophic episodes in the history of life), philosophy of science (future studies, science in pop culture, and philosophy of physics), and risk analysis (global catastrophes, observation selection effects, and epistemology of risk). He coedited the anthology on Global Catastrophic Risks (Oxford University Press, 2008), wrote three monographs (the latest being *The Great Silence*, Oxford University Press, 2018) and two anthologies of essays, and authored about 200 research and professional papers. In addition, he translated several science books, including titles by Richard P. Feynman, Paul C. W. Davies, and Sir Roger Penrose. He lives in Belgrade with his family.

Jeffrey L. Payne

Dr. Jeffrey L. Payne is a physician M.D. and Ph.D. astronomer who studies supernova remnants, particularly those in the Magellanic Clouds. He has authored or coauthored over 40 refereed publications and has also worked with predictive modeling and big data. He has taught numerous astronomy and medicine classes in the past. He lives near Evansville, Indiana (USA).

Contributors

Csaba Balazs School of Physics and Astronomy, Monash University, VIC 3800, Australia

Milan M. Ćirković Astronomical Observatory, Volgina 7, PO Box 74 11060 Belgrade, Serbia

Michelle E. Cluver Centre for Astrophysics and Supercomputing, Swinburne University of Technology, John Street, Hawthorn, 3122, Australia

Miroslav D. Filipović Western Sydney University, Locked Bag 1797, Penrith South DC, NSW 2751, Australia

Sebastian Gurovich Observatorio Astronómico de Córdoba, X5000BGQ, Córdoba, Argentina

Natasha Hurley-Walker International Centre for Radio Astronomy Research, Curtin University, Bentley, WA 6102, Australia

Clancy James International Centre for Radio Astronomy Research, Curtin University, Bentley, WA 6102, Australia

Thomas H. Jarrett Astrophysics, Cosmology and Gravity Centre (ACGC), Astronomy Department, University of Cape Town, Private Bag X3, Rondebosch 7701, South Africa

Paul D. Lasky School of Physics and Astronomy, Monash University, VIC 3800, Australia

Denis Leahy Department of Physics and Astronomy, University of Calgary, Calgary, Alberta, T2N 1N4, Canada

Pierre Maggi Observatoire Astronomique de Strasbourg, Université de Strasbourg, CNRS, 11 rue de l'Université F-67000 Strasbourg, France

Raymond P. Norris Western Sydney University, Locked Bag 1797, Penrith South DC, NSW 2751, Australia

Jeffrey L. Payne School of Science, Western Sydney University, Locked Bag 1797, Penrith South DC, NSW 2751, Australia

Gavin Rowell School of Physical Sciences, The University of Adelaide, Adelaide 5005, Australia

Nicholas F. H. Tothill Western Sydney University, Locked Bag 1797, Penrith South DC, NSW 2751, Australia

Branislav Vukotić Astronomical Observatory, Volgina 7, PO Box 74, 11060 Belgrade, Serbia

Chapter 1

Multimessenger Astronomy in Practice: Celestial Sources in Action

Miroslav D Filipović, Jeffrey L Payne and Nicholas F H Tothill

1.1 Introduction

Although the first nonelectromagnetic messengers from space—cosmic rays—were discovered in the early 20th century,[1] it is only now that multimessenger astronomy is coming into its own. Neutrino and gravitational-wave detections are being combined with cosmic-ray and electromagnetic messengers to illuminate our view of the cosmos, especially parts with high energy density and fast variation. Gravitational-wave detections of energetic mergers and their aftermaths involving neutron stars and stellar-mass black holes are opening doors of understanding to events including gamma-ray bursts (GRBs), fast radio bursts (FRBs), and high-energy neutrino detection. Supernovae (and their remnants), pulsars, magnetars, and active galactic nuclei (AGNs) are natural particle accelerators that can be used to analyze very high-energy processes.

As discussed in the companion volume to this book (*Principles of Multimessenger Astronomy*, hereafter "Book 1"), the invention of the telescope and its subsequent refinement and use by Galileo marked the birth of the modern scientific method, setting the stage for a dramatic reassessment of our place in the cosmos. This technological breakthrough demonstrated that there is much more to the universe than is available to our unaided senses. These revelations, in time, have established the unforeseen vastness of our dynamic, expanding universe; shown that our Galaxy is but one among countless others; and introduced us to a wealth of exotic astrophysical structures. There are now telescopes to cover the entire multimessenger spectrum, located on Earth, in the sky and in space.

[1] Meteorites could be regarded as messengers, but are out of our scope.

doi:10.1088/2514-3433/ac2256ch1

Galileo's *Letters on Sunspots*[2] can also be thought of as a foundation of time-domain astronomy—the study of how astronomical objects change with time, which is a fundamental aspect of today's multimessenger astronomy. Gravitational waves, for example, arise from changes in physical structure. Advanced Laser Interferometer Gravitational Wave Observatory (LIGO) detected the first gravitational wave (GW 150914) in 2015—even before the final test run was complete. Since then, several binary black hole (BBH) mergers as well as at least one binary neutron star coalescence have been discovered (see Book 1, Chapter 8, and Chapters 9 and 10 of this work).

1.1.1 Observational Breakthroughs

New and intriguing observations have been performed at scales from compact stellar-mass objects to galaxy clusters. The fact that most of these observations can be classified into a rough size scale is itself an achievement of multimessenger observations.

Stellar Scale:

- The binary neutron star merger GW 170817, detected in gravitational waves and as a GRB (Chapter 9);
- TeV gamma-ray observations of several GRBs, probing prompt and afterglow phases (Chapter 7);
- The F-type star KIC 8462852 (a.k.a. Boyajian's star) at a distance of only 450 pc, hypothesized to have a swarm of artificial objects in order to explain its peculiar light curve (Chapter 11);
- The ultraviolet (UV) spectra of the hot star ξ Per (Chapter 5);
- Detection of FRB 200428 in the same direction as the magnetar SGR 1935+2154 (Section 1.2.5).

Galactic Scale:

- The revelation of giant gamma-ray structures—the Fermi bubbles—emanating from the center of the Milky Way Galaxy, extending ~7700 pc (north and south) of the Galactic Plane (Chapter 7);
- The detection of neutrino emission from the blazar TXS 0506+056 (Chapter 8); IceCube's detection of "astrophysical" neutrinos is a benchmark discovery with great implications, prompting questions such as where are the neutrinos produced—in the interstellar medium (ISM), in the intergalactic medium, or as a superposition of individual sources, starburst galaxies, AGN, etc.?
- Discovery of the first low-frequency radio counterpart to an unidentified TeV gamma-ray source (Chapters 2 and 7);
- Discovery of mysterious odd radio circles (ORCs) (Chapter 12).

[2] *Istoria e Dimostrazioni intorno alle Macchie Solari.*

Cluster Scale:

- Unequivocal evidence for the extragalactic origin of the highest-energy cosmic rays (Chapter 7);
- The Bullet Cluster (1E 0657–56) and dark matter (Chapter 10).

1.1.2 New Instruments

A new generation of detection instruments, together with upgrades to existing ones, are being designed and are coming online at a rapid pace, including both ground and space-based detectors.

The US-based Cosmic Explorer will be similar to the LIGO detector but with 40 km arms. The Einstein telescope will have three 10 km long arms in a triangular configuration and will be located partially underground. These detectors have a minimum frequency limited to 5 Hz. Space-based observatories will probe much lower frequencies to detect massive black hole binaries and gravitational-wave backgrounds. They include LISA, TianQin, and DECIGO. In addition to laser interferometers, pulsar timing arrays (PTAs) are being designed for gravitational-wave detection of frequencies down to nanohertz (Kembhavi & Khare 2020).

Telescopes based on the detection of Čerenkov radiation (see Book 1, Chapter 7, and Chapter 7 of this work) can detect high-energy messengers such as cosmic rays, gamma rays, and neutrinos using the particles they create, such as muons. These detectors are found on the ground, underground, and deep in oceans. Example instruments include the Cherenkov Telescope Array (CTA) for TeV gamma-ray detection and IceCube, the Cubic KiloMetre (km3) Neutrino Telescope (KM3NeT), and the Pierre Auger Observatory for neutrino and cosmic-ray detection (see Book 1, Chapter 7 and Chapter 8).

An essential complement to these new instruments is the coordinating organization that allows the information from one detector to be disseminated quickly to the others, which can then search for counterparts. This is vital to the transition from multiwavelength to multimessenger astronomy.

1.1.3 Theoretical Synergies

In order to create a realistic picture of the universe, theoretical research is essential to understand multimessenger astrophysics in a consistent and coherent way. The most crucial requirements to enable theoretical contributions are free access to data and open communication. Computational requirements range from minimal to high-performance computing facilities (Chapter 12).

Example applications of theoretical studies in high-energy multimessenger astrophysics include:

- Understanding the astrophysical processes that lead to the generation of electromagnetic radiation, neutrinos, and cosmic rays. Such processes include cosmic-ray/proton collisions, inverse-Compton scattering, bremsstrahlung, matter/antimatter production and annihilation, photopion production, curvature radiation, synchrotron radiation, and radioactive decay (see Book 1); they arise in supernova remnants (SNRs), pulsar environments, accreting

objects (AGNs, X-ray binaries, microquasars), starburst galaxies, cataclysmic events (supernovae, hypernovae, kilonovae, compact mergers), massive stellar winds and clusters, and ISM clouds.
- Jet simulations from X-ray to radio data, allowing us to understand the ejecta from GW 170817, probe microquasar physics, and simulate blazar neutrino production.
- Simulations of Galactic cosmic-ray and electron propagation using magnetic fields, infrared (IR) photon, and ISM gas distributions, reproducing the cosmic-ray flux at Earth, diffuse GeV gamma-ray emission, and diffuse Galactic neutrino fluxes. Such simulations[3] can be used to provide improved predictions of the diffuse TeV gamma-ray emission that is expected to be detected by the CTA.
- Prediction of TeV gamma-ray morphology from the propagation of cosmic rays and electrons into ISM clouds, predicting the emission seen by HESS (High Energy Stereoscopic System) and to be explored more deeply by CTA.
- Central Milky Way outflow modeling from accretion or stellar winds.
- Simulations of low-mass galaxies, in which cosmic-ray pressure may open magnetic field lines, allowing gas escape that stops star formation.
- High-energy gamma-ray and neutrino observations to understand the dark matter that dominates the formation of cosmic structure, the first stars, and galaxies.
- Using gamma-ray observations from GW 170817 and AGN flares to place strong limits on Lorentz invariance violations.

1.2 The Multimessenger Event Zoo

The story of multimessenger astronomy so far is one of particular sources and classes of sources (Figure 1.1), for which multimessenger observations yield unique insights into their nature. Ultimately, however, the goal is to understand the astrophysical processes that lead to the emission of each messenger and that affect the messengers as they traverse intergalactic and interstellar space to reach us (Figure 1.2). More details of specific sources or events can be found throughout various chapters of this book.

1.2.1 Gamma-Ray Bursts

GRBs are a diverse group of energetic explosions lasting from milliseconds to hours and associated with "afterglows" at wavelengths longer than gamma rays (Figure 1.1; also see Chapter 7). These transient gamma-ray events were first discovered in the 1960s by the Vela satellites.[4] For decades, all that was known about GRBs was their gamma-ray emission, that they were isotropically distributed over the sky, that they did not repeat, and that they were clearly extraterrestrial.

[3] Run with the GALPROP code, at the time of writing.

[4] Designed to verify treaties that banned atmospheric nuclear detonations, they detected no illicit nuclear tests—except possibly in 1979.

Event Class		G.W.	C.R.	ν	Electromagnetic Radiation						Example	Ref.
					γ	X	UV	O	IR	R		
GRB	*short*	✓	✓	✓	✓	✓	✓	✓	✓	✓	GRB 170817A	[1]
	long	✓	✓	✓	✓	✓	✓	✓	✓	✓	GRB 030329	[2]
SN	*Ia*	?	✓	✓	✓	✓	✓	✓	✓	✓	N103B	[3]
	CC	✓	✓	✓	✓	✓	✓	✓	✓	✓	SN 1987A	[4]
AGN		✓	✓	✓	✓	✓	✓	?	✓	✓	TXS 0506+056	[5]
PBH		?	?	✓	✓						GW190814?	[6]
FRB		?	?	?	✓					✓	FRB 200428	[7]
BBH		✓				✓		✓		✓	GW 150914	[8]
BNS		✓	✓	✓	✓	✓		✓	✓	✓	GW 170817	[8]
PULSAR		✓	✓	✓	✓	✓		✓		✓	Crab Pulsar (M1)	[9]
UHECR			✓	✓							Oh-My-God	[10]
TDE		?	?	✓	✓	✓	✓	✓			ASASSN-19bt	[11]
SGR		✓	✓	✓	✓	✓	✓	✓	✓	✓	SGR 0525-66	[12]

Figure 1.1. Examples of multimessenger events. Abbreviations: G.W.—gravitational waves, C.R.—cosmic rays, ν—neutrinos, γ—gamma rays, X—X-rays, O—optical, R—radio, IR—infrared, UV—ultraviolet, GRB—gamma-ray burst, SN—supernova, CC—core collapse, AGN—active galactic nucleus, PBH—primordial black hole, FRB—fast radio burst, BBH—binary black holes, BNS—binary neutron stars, UHECR—ultra-high-energy cosmic rays, TDE—tidal disruptive event, and SGR—soft gamma repeaters. References: [1] Abbott et al. (2017); [2] Stanek et al. (2003); [3] Li et al. (2017); [4] Fryer et al. (2019); [5] Britzen et al. (2019); [6] Scholtz & Unwin (2020); [7] Zhang (2020); [8] Kembhavi & Khare (2020); [9] Hewish et al. (1968); [10] Anchordoqui (2019); [11] Holoien et al. (2019); [12] Moskvitch (2020).

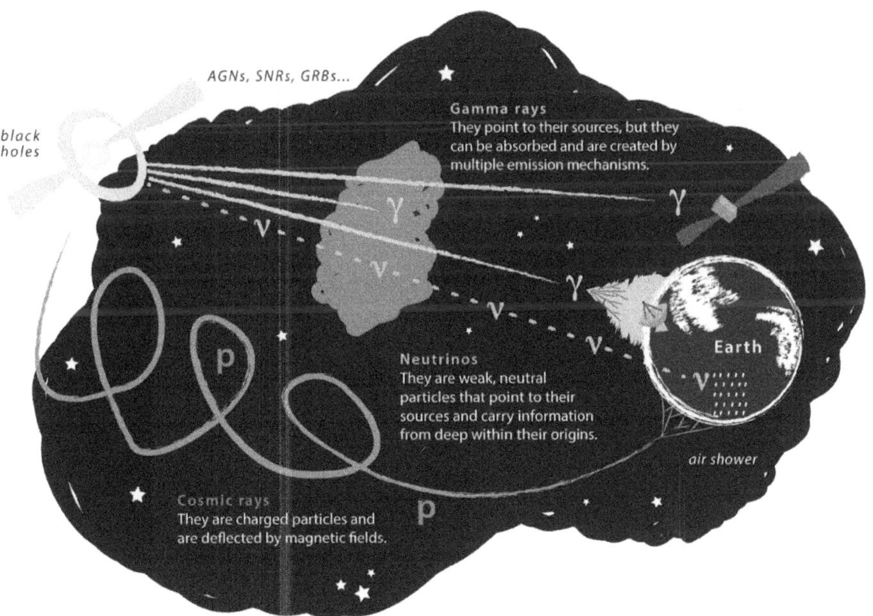

Figure 1.2. High-energy sources emit multiple messengers, including gamma rays, neutrinos, and cosmic rays. Cosmic rays are deflected by magnetic fields, making it hard to trace their origin at lower energies. Image credit: IceCube Collaboration/WIPAC, Juan Antonio Aguilar, and Jamie Yang.

The debate over the nature of GRBs was intense and ongoing. Over 30 models for their origin had been proposed by the end of the 1970s,[5] generally falling into three classes: accretion, stellar activity, and stellar destruction. Some of these models had GRBs as nearby Galactic phenomena, while some models placed them at cosmological distances, requiring very high luminosity.

In order to understand GRBs, it was necessary to observe more than just the gamma rays. The key to unraveling the mystery was a combination of specialist monitoring instruments to spot new GRBs and organization to observe them as soon as they were detected—all within the minutes-to-hours timeframe. The BATSE[6] instrument on the Compton Gamma Ray Observatory satellite[7] detected about one GRB per day over the course of its 9 year mission, at sub-MeV energies. The rapid response required for facilities at other wavelengths drove significant development in telescope automation, operation and scheduling software, and human organization.[8] Detection and follow-up were brought together by the Swift satellite,[9] which combines a wide-field gamma-ray telescope to detect the bursts with x-ray and UV/optical telescopes to observe the burst position.

These coordinated observing campaigns showed that the afterglows of GRBs were to be found in distant galaxies with redshifts of one or more, settling the GRB debate in favor of cosmological models. With reasonably well-known distances, the energy release of a GRB can be estimated; that of a long burst ranges from 10^{52} to 10^{54} erg.[10] Energies of this magnitude are equivalent to the conversion of about a solar mass into energy, with an efficiency of about 10%. The difficulty of explaining such a high energy leads to models that invoke geometric beaming to boost the brightness of the burst.

The basic gamma-ray emission mechanism for GRBs is thought to be an inverse-Compton scattering, where preexisting lower-energy photons are scattered by relativistic electrons within an explosion, gaining energy from the scattering event and becoming gamma rays. The longer-wavelength afterglow emission is thought to be the result of the explosion moving outward at close to the speed of light, colliding with surrounding interstellar gas and creating a shock wave, with a possible reverse shock propagating back into the ejecta.

For an excellent "historical primer" about GRBs, the reader is directed to Andrew Levan's "Gamma-Ray Bursts" publication (Levan 2018).

[5] And up to 118 models by the early 1990s.

[6] Burst and Transient Source Experiment.

[7] CGRO, one of NASA's Great Observatories, operated in 1991–2000.

[8] Astronomers already had systems in place to handle these problems in the hours-to-days timeframe—the Central Bureau for Astronomical Telegrams was founded in the 19th century. The challenge was to set up systems for follow-up observations in minutes. The *Hotwiring the Transient Universe* conference series brought many of these elements together.

[9] The Neil Gehrels Swift Observatory, a NASA medium-class Explorer satellite, launched in 2004, still operational at the time of writing.

[10] The most commonly used energy unit in astronomy, equivalent to 10^{-7} J; shorter bursts (under about 2 s) are less energetic by approximately two orders of magnitude.

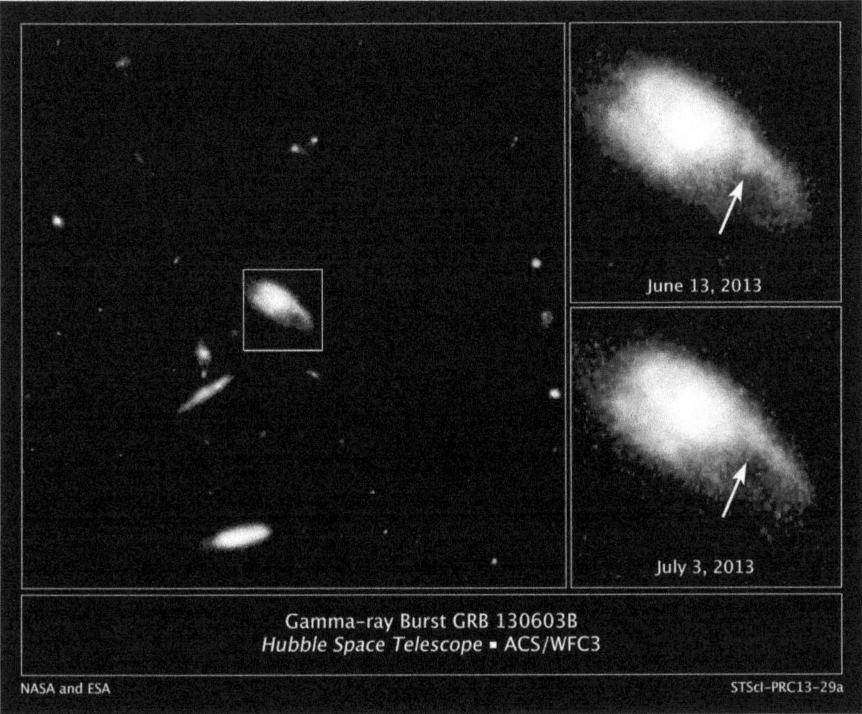

Figure 1.3. Kilonova or glow created during the short GRB 130603B event as seen by the HST. The glow was clearly seen on 2013 June 13 but had faded when observed on 2013 July 3. Image credit: National Aeronautics and Space Administration (NASA), European Space Agency (ESA), N. Tanvir (University of Leicester), A. Fruchter (STSci), and A. Levan (University of Warwick).

1.2.1.1 Short GRBs

Short-duration GRBs are events with a duration less than 2 s and account for 30% of all GRBs. The nature of these events was initially unknown but one clue was their short mean duration of 0.2 s, suggesting the physical diameter of their progenitors was less than 0.2 light-second (about four times Earth's diameter).

These events are most likely associated with binary mergers, specifically, a neutron star merging with another neutron star or black hole. This produces a kilonova[11] (Figure 1.3). This identification of the origin of short GRBs with kilonovae (Section 4.4.1) has become more secure with the detection of GRB 170817A associated with gravitational wave GW 170817 that signaled the merger of two neutron stars (Abbott et al. 2017).

1.2.1.2 Long GRBs

Long GRBs (duration >2 s) comprise the majority of events, last longer, and have the brightest afterglows, so they have been studied more extensively. Most long events are associated with star-forming galaxies, and specifically with core-collapse

[11] Although a small number may be produced by giant flares from soft gamma repeaters in nearby galaxies.

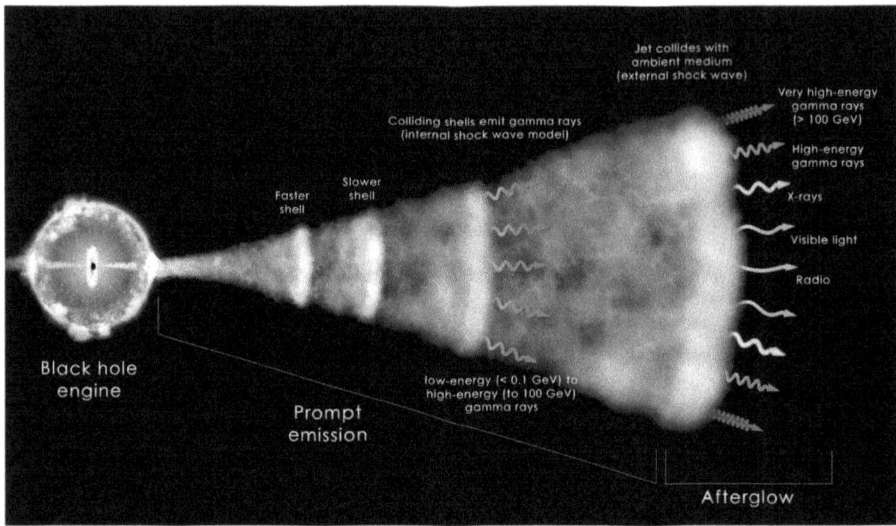

Figure 1.4. General picture of the shocks and related particle acceleration regions within a GRB jet originating from either a core-collapse supernova (LGRB) or a compact object merger (SGRB). The prompt and afterglow regions are also defined. Image credit: NASA/Goddard Space Flight Center/ICRAR.

supernovae (CCSNe; Woosley & Bloom 2006). This identification arose from sources such as GRB 030329, the first burst whose afterglow showed the characteristics of a supernova (Stanek et al. 2003). As with any supernova event, high-energy neutrinos may also be produced in these explosions. These neutrinos would be in the TeV range, distinguishable from lower-energy (MeV) neutrinos from supernovae or the Sun; however, they are yet to be found. The production of cosmic rays is also likely to arise, due to the acceleration of charged particles by supernova shocks (Figure 1.4). These protons and heavier nuclei could be accelerated to relativistic velocities, possibly yielding ultra-high-energy cosmic rays.

A few GRB events have lasted more than 10^4 s and have been proposed as a separate class (ultralong gamma-ray bursts). Proposed progenitors include the collapse of a blue supergiant (Gendre et al. 2013), a tidal disruption (Greiner & Mazzali 2015), or a newborn magnetar[12] (Greiner & Mazzali 2015).

1.2.2 Supernovae

Supernova (SN) events can be divided into two major categories: Type Ia events, in which a white dwarf star undergoes thermonuclear detonation and core-collapse events, in which the core of a high-mass star undergoes gravitational collapse after exhausting its supplies of fuel for nuclear fusion. Both types of explosion drive matter out into the interstellar medium around the star, and the latter type leaves a compact object remnant (a neutron star or a black hole). Understanding the physics

[12] A type of neutron star with an extremely powerful magnetic field, 10^{13}–10^{15} G, 10^{9}–10^{11} T.

of these events requires extensive use of data from all available messengers (see Chapters 2, 3, 4, 5, 6, 7, and 8).

1.2.2.1 Type Ia Detonation

Type Ia SNe are thought to be thermonuclear detonations of carbon–oxygen white dwarfs (WDs). Although the events leading up to the detonation are not perfectly understood, the outline is fairly clear: These stars cannot have mass $>1.44\ M_{\odot}$ (the Chandrasekhar limit), but if they are in binary systems they will gradually accrete mass until they reach the limit. As they reach the limit, the carbon and oxygen undergo runaway nuclear fusion, which drives the explosion. The nature of the progenitor is often unclear: WD–WD binary systems can give rise to a double-degenerate supernova, while systems made up of a WD and a main-sequence or giant star generate single-degenerate supernovae. Examples of a Type Ia SN in which the nature of the progenitor system is yet to be confirmed are J0509–66731 and N 103B located in the Large Magellanic Cloud (LMC; Bozzetto et al. 2014; Li et al. 2017; Roper et al. 2018, Sano et al. 2018; Alsaberi et al. 2019). However, a single-degenerate scenario is preferred for SNR N103B but double degenerate for SNR J0509-6731. Double-degenerate supernovae from binary WDs would produce gravitational-wave signals from the inspiral, leading to the merger and explosion (in addition to electromagnetic, neutrino, and cosmic-ray messengers), so space-based gravitational-wave detectors sensitive in the decihertz range such as DECIGO could observe such WD–WD mergers directly (Kinugawa et al. 2019).

Neutrinos detected from CCSNe can reveal important information about the dynamics of the explosion, and neutrinos from thermonuclear SNe also have similar potential. Type Ia SNe are dimmer neutrino sources, but detectors such as Hyper-K may be able to detect their neutrinos out to 10 kpc. For a relatively near Type Ia SN at 1 kpc, JUNO, Super-K, and DUNE could find a few events, while IceCube, KM3NeT, and Hyper-K could find several tens of events (Wright et al. 2016; Aiello et al. 2019; Aiello & Albert 2021).

1.2.2.2 Core Collapse

Multimessenger observations of CCSNe could be considered to date back a half-century to the study of dust grains that probed the products of a CCSN, but for our purposes, multimessenger observations of CCSNe started with the concurrent neutrino and electromagnetic observations of SN 1987A in the LMC (Figure 1.5). Most of the energy output of a CCSN is carried by MeV neutrinos, and these should be detectable out to a few tens of kiloparsecs by next-generation detectors (Fryer et al. 2019; Aiello & Albert 2021), such as Super-Kamiokande, DUNE, JUNO, KM3NeT, and IceCube. Because the neutrino pulse is generated by the nuclear reactions in the core of the star, it can be used in conjunction with detailed modeling to probe the physics of supernova explosions. In addition, observation of the diffuse neutrino background has the potential to place limits on populations of CCSNe.

Observations of electromagnetic emission span the progenitor star, the explosion itself, and the remnants. If the star that underwent the supernova was already known, cataloged, and hopefully even classified spectroscopically, the supernova

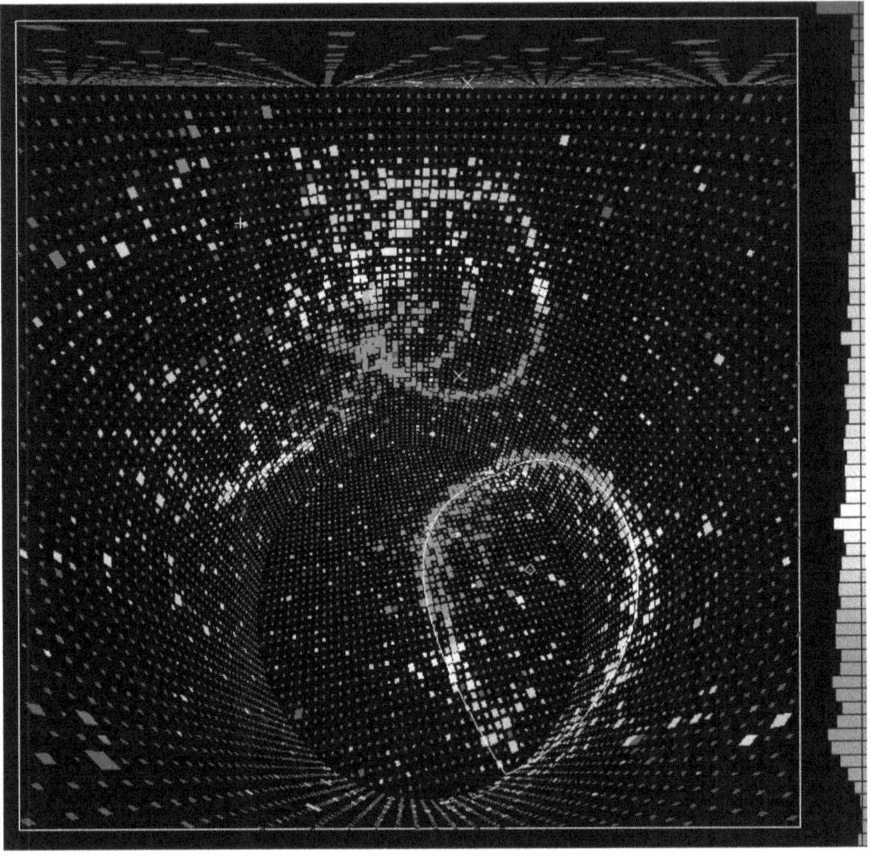

Figure 1.5. Multiple neutrino events, reconstructed from separate neutrino detectors. In 1987, three independent detectors that were sensitive to energetic neutrinos and antineutrinos detected a total of 25 particles in a single burst spanning 13 s. A few hours later, the light arrived as well. Image credit: Super-Kamiokande Collaboration/Tomasz Barszczak.

will deliver far more astrophysical insight, as was the case for SN 1987A. In the early stages of the explosion, the electromagnetic radiation is trapped until shock breakout, accompanied by a burst of UV and X-ray photons. The modern fleet of transient-focused facilities (particularly Swift) have made it possible to observe shock breakout, as they can get to a supernova in time to see it.

The nucleosynthesis that occurs in CCSN is an inherently multimessenger phenomenon (Fryer et al. 2019), producing the neutrino pulse and delivering new elements into the remnant. Electromagnetic observations can then probe the nucleosynthetic yields of the explosion, via IR, optical, and UV spectra (especially in the nebular phase) and in gamma-ray decay lines.

The shocks that CCSN drive into their surrounding medium are engines for cosmic-ray generation, making cosmic rays a probe of the aftermath of the supernova, while gravitational waves are expected to deliver information from the "engine" of the supernova explosion. In particular, strong gravitational-wave signals should arise from rapidly rotating stars, probing stellar rotation, asymmetry, and the convective engine. CCSNe may be able to produce gravitational waves detectable with advanced LIGO to 10 Mpc.

Distance is the key limitation to what we can learn about CCSNe. Detailed analysis is often limited to Galactic objects (Hurley-Walker et al. 2019a, 2019b) or at best the nearby Magellanic Clouds (Maggi et al. 2016, 2019; Bozzetto et al. 2017), while most SNe happen farther out in the universe. Validation of our models with nearby events, however, may eventually allow us to use the array of messengers to study more distant CCSNe, even back to the early universe.

1.2.3 Active Galaxies

The centers of most (and maybe all) large galaxies contain a supermassive black hole (SMBH) with mass of order $10^8\ M_{\odot}$. These black holes accrete interstellar gas episodically, and during accretion are observed as AGNs. The accretion process releases vast amounts of electromagnetic radiation, and some of the matter from the accretion disk is ejected perpendicular to the disk as bipolar plasma jets at relativistic speeds which in turn drive high-speed gas outflows.

The initial signatures of AGN emission were found in the first half of the 20th century; radio astronomy was a major catalyst to understanding them (for example see Figure 1.6). The process of understanding them has been long, not only because they are often very distant, but because they seem to have a complex structure consisting of an SMBH surrounded by an accretion disk, a complex torus of material around the disk, and bipolar jets. The high magnetic fields of these jets and

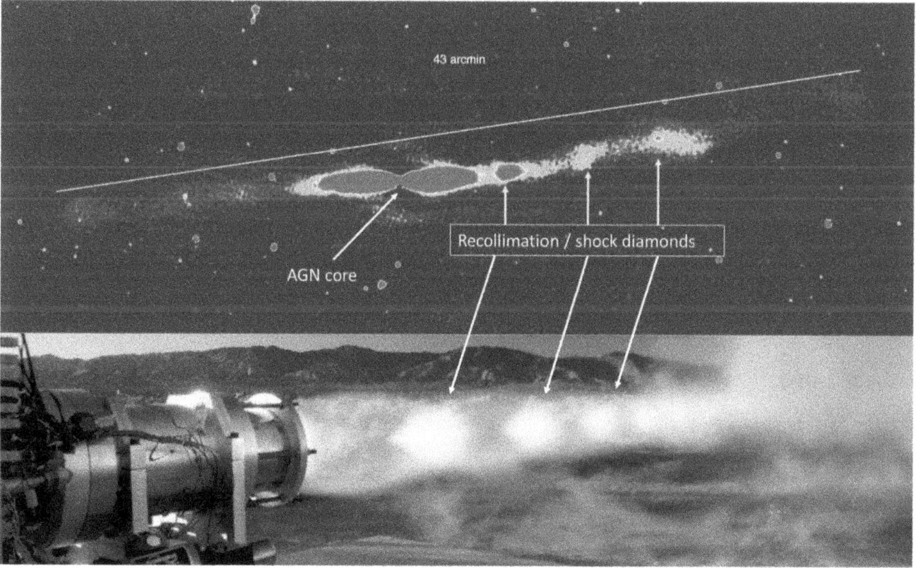

Figure 1.6. (Top) Australian Square Kilometre Array Pathfinder (ASKAP) 1.5 GHz image of the AGN NGC 2663, showing recollimation knots in the southern jet. (Bottom) "Shock diamonds" in jet engine exhaust, resembling the recollimation knots in NGC 2663, but on a much smaller scale. Image credit: (top) V. Velović and M. D. Filipović; (bottom) Mike Massee/XCOR.

the shocks that are found in them are efficient particle accelerators and the best candidates for the production of cosmic rays beyond the "knee" (10^{15} eV)—particularly quasar jets (Chapter 8).

AGNs can be divided into radio-quiet and radio-loud objects—the latter due to synchrotron emission from both the jets and the radio lobes of plasma that the jets inflate. They can also have radically different appearances, depending on the direction from which they are viewed (Figure 1.7). Based on characteristics like activity, emission lines (narrow or broad), variability, jets, and the presence of X-ray, UV, and far-IR radiation, AGNs have been classified as normal, LINER (low-ionization nuclear emission-line region), Seyfert I, Seyfert II, quasar, blazar, BL Lacertae, OVV (optically violent variables), radio galaxies, and more. Although these classifications are still used, it is accepted that they refer to similar objects—an idea known as unification.

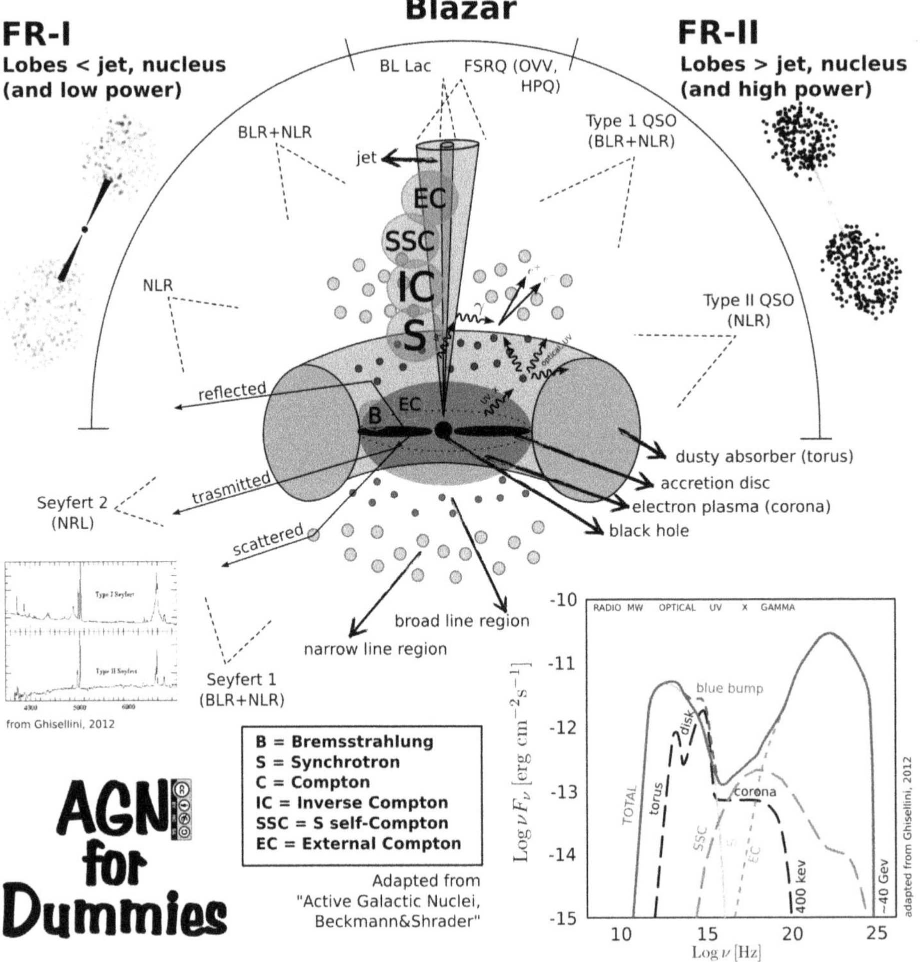

Figure 1.7. The unified AGN Model. Image credit: Brunetto M. Ziosi (http://brunettoziosi.com/posts/agn-for-dummies/).

Blazars, for example, are AGNs viewed from a position illuminated by the cone of a relativistic jet. Within this jet, particles are accelerated by Fermi shocks in matter blobs traveling with a Lorentz factor (γ) of 10 or higher. This environment allows the production of high-energy particles including pions and neutrinos (Zuber 2020), making blazars potential neutrino sources. The neutrino event IceCube 170922A detected at the South Pole was reported to originate from TXS 0506 +056, of BL Lac type, at a distance of order a gigaparsec ($z = 0.34$) from Earth (Chapter 8) with a suspected precessing jet–jet interaction (Britzen et al. 2019).

Quasars, the most energetic AGNs, are triggered during mergers of gas-rich spiral galaxies. These mergers not only trigger massive star formation but also result in SMBH binaries at the center of the newly merged galaxy. When SMBH binaries coalesce, enormous outbursts of gravitational waves and electromagnetic radiation are expected, including radio, microwave, IR, optical, UV, X-ray, and gamma-ray wavebands. Gravitational-wave-producing inspirals could also be produced by the formation in the AGN region of massive stars that would collapse into compact objects and merge into the central SMBH. A population of these could contribute to the gravitational-wave background detectable by LISA above a few millihertz at a rate of 10–100 per year (Schnittman et al. 2006).

Our own Milky Way Galaxy—which contains a central SMBH—has probably undergone AGN episodes in the past, leaving the Fermi Bubbles (Figure 1.8) as relics of bipolar outflow (see Section 7.5.5).

1.2.4 Primordial Black Holes

PBHs are a hypothetical population of black holes formed by the gravitational collapse of overdense regions of space when the universe was less than a second old. They were proposed by Zeldovich[13] and Novikov[14] in the mid-1960s, and studied in depth by Hawking.[15] Because PBHs arise from dense regions rather than stars, they do not have the same mass limitations as stellar black holes; there is no known stellar evolutionary pathway that can deliver a black hole with mass less than a few solar masses—but PBHs can have masses as little as 10^{-8} kg. The idea has even been entertained that the proposed Planet 9 could be a PBH with mass several times that of Earth (Figure 1.9), but which is almost undetectable (Scholtz & Unwin 2020).

As gravitational-wave detectors start to find more black holes in binary systems, more candidates are found whose masses are inconsistent with stellar evolutionary theory, and these may be relic PBHs from the early universe, which have grown by accretion and/or merger. The 2.5–2.7 $M_\odot$ component of GW 190814 is an example (Clesse and Garcia-Bellido 2020; and see Chapter 9), as are the gravitational-wave-detected black holes of a few tens of solar masses. Attributing these binary merger components to PBHs, however, requires fine-tuning the amount of time that elapses between formation and merger of such objects to explain the merger rate (Vattis et al. 2020).

[13] Yakov Borisovich Zeldovich (1914–1987), Russian astrophysicist.
[14] Igor Dmitriyevich Novikov (1935–), Russian astrophysicist.
[15] Stephen Hawking (1942–2018), British theoretical physicist and cosmologist.

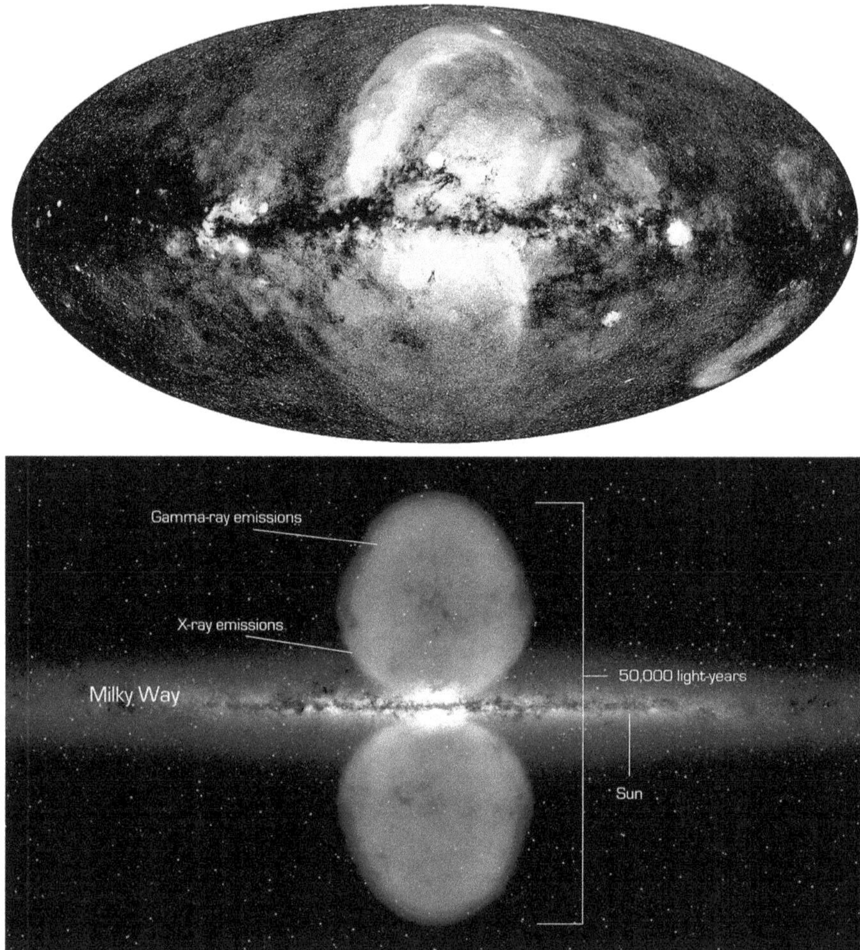

Figure 1.8. Upper: the SRG/eROSITA all-sky map of the Milky Way as a false-color image, with energies from 6 $\times 10^6$ K (red) to 15 $\times$ 10^6 K (blue). Bottom: artist impression of large-scale Fermi bubbles in the Milky Way halo. Credit: Wikipedia: JohannesBuchner (CC BY-SA 4.0)/Predehl et al. (2020) and NASA (NASA's Goddard Space Flight Center).

PBHs could also be part of the population of massive compact halo objects (MACHOs) and thus nonbaryonic dark matter candidates. However, the PBH contribution to dark matter (as well as ultra-high-energy cosmic rays) has been constrained by several observations. PBHs would be expected to generate observable phenomena, including lensing of GRBs, capture by neutron stars with rapid star destruction, capture with rapid detonation by white dwarfs, stellar microlensing, Type Ia SN microlensing, and temperature anisotropies of the CMB. Further constraints on the PBH population may come from the next-generation Square Kilometre Array (SKA) radio telescopes probing the effects on the reionization history of the universe due to energy injection into the intergalactic medium by accretion of matter onto PBHs.

The search for PBH encompasses many different messengers and techniques. Because PBHs can take on very low masses, they may be able to evaporate on

Figure 1.9. Exact scale (1:1) illustration of a 5 Earth-mass primordial black hole. Is this the mysterious Planet 9? Image credit: Reprinted with permission from Scholtz & Unwin (2020). Copyright 2020 by the American Physical Society. (CC BY 4.0).

timescales similar to the life of the universe, in which case we might expect to be able to observe the burst of gamma-ray Hawking radiation and high-energy neutrinos that are produced. The next generation of gamma-ray telescopes such as CTA will search for evaporating PBHs during their last second to year of existence, with bursts detectable out to distances of 10^{-3} to 0.1 pc. At the same time, the upgraded KM3NeT and IceCube will also be able to detect the expected neutrino emission from PBH. Gravitational-wave detectors such as LIGO could discover PBHs by mass reconstruction, finding low-mass black holes, and it may also be possible to measure large orbital eccentricities produced by PBH binaries. PTAs and LISA may find a stochastic background of gravitational waves from PBH binaries.

Observable influences of PBHs may come from observations of faint dwarf galaxy central star clusters and even positions and velocities of stars within the Milky Way. Small PBHs may pass unharmed through Earth producing acoustic signals or leave a seismic signature when passing through a star. Monitoring for microlensing of quasars by PBHs may also be possible.

1.2.5 Fast Radio Bursts

FRBs are radio pulses lasting a fraction to a few milliseconds, some of which, repeating FRBs, recur on a regular basis. Although the signals are broad band, they are often detected at frequencies around a gigahertz. The first detected FRB was found in archival data from the Parkes telescope in 2007 by Duncan Lorimer—so this initial detection is often called the Lorimer burst.

Searching for more FRBs so as to find a population has proven to be hard. FRBs can occur anywhere on the sky at any time, so finding them requires continuous monitoring of a large area of sky. The Parkes telescope, with a multiplexed receiver, was an initial leader in the field, and the ASKAP telescope, with its wide-field phased array feeds, has also found FRBs. Most FRBs are currently being discovered by more specialized instruments such as UTMOST in Australia and CHIME in Canada.

Based purely on the radio pulses, not much can be inferred about FRBs. The dispersion in their signals and their isotropic distribution on the sky are consistent with an extragalactic origin. The short duration of the pulses suggests that they arise in a compact region, and the observed polarization may constrain our models. But in order to make much more progress, it is necessary to localize them—that is, to estimate their position on the sky sufficiently accurately to allow comparison with maps from other wavebands and messengers. Wide-field radio observations generally have quite poor instantaneous angular resolution, and so it is not easy to generate a precise location for the signal. Technical developments at ASKAP allowed the first localization of a nonrepeating FRB (Bannister et al. 2019) to a position outside the center of a distant galaxy (redshift 0.3). These developments, along with those at UTMOST, should lead to greater rates of localized detection.

Repeating FRBs are much easier to localize, as the localization need not be instantaneous—however, there are only very few known repeating FRBs. The first known repeating FRB, FRB 121102, was not localized until 2017 (Chatterjee et al. 2017) and found to be associated with a star-forming dwarf galaxy (Tendulkar et al. 2017), whereas the second to be localized, FRB 180916.J0158+65, was found outside the center of a massive spiral galaxy (Marcote et al. 2020), and the two environments have little in common. With more repeating FRBs being found, this part of the puzzle may soon be solved.

There is no consensus as to the origin of these bursts, but explanations range from the collision of black holes or neutron stars to extraterrestrials. Current candidates could include any high-energy phenomenon. For example, one hypothesis is that they originate from flares emanating from magnetars (Figure 1.10): the magnetar SGR 1935+2154 is responsible for the repeating FRB 200428 (Kirsten et al. 2021),

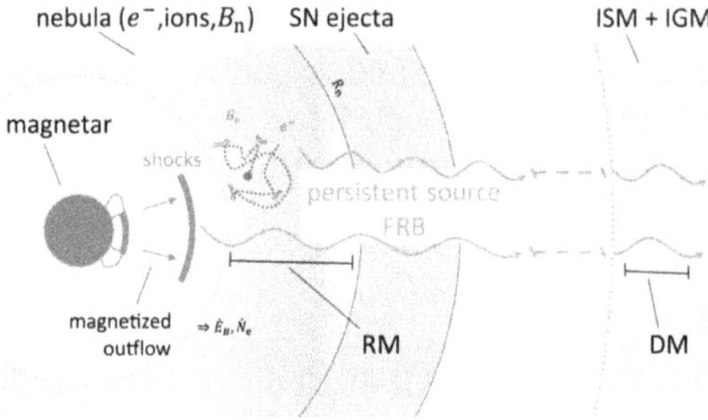

Figure 1.10. A hypothetical emission model for FRBs arising from a magnetar (highly magnetized neutron star). Image credit: Margalit & Metzger (2018).

and localizations to noncentral regions of galaxies would be consistent with this idea. Other possible FRB progenitors include young pulsars, black-hole-related outbursts, blitzars (pulsars that can rapidly collapse into black holes), black hole evaporation, stellar-mass black hole coalescence, or superradiance from PBHs (Burke-Spolaor 2018). It is not even known whether repeating and nonrepeating FRBs are the same kind of source, though current opinion is leaning toward the idea that all FRBs repeat in some way, though possibly with varying brightness and period.

Electromagnetic messengers are the only ones detected from FRBs so far, but the likelihood that they come from a region of very high energy density (implied by the short pulse duration and high power) suggests that multimessenger observations may play a role in our understanding of the phenomenon (Burke-Spolaor 2018). Neutrino detection from these events could tell us about energetic atomic decay processes and hadronic accelerations while gravitational waves can tell us about the relativistic or explosive motion of mass charge.

1.2.6 Binary Events

The merging of the components of a binary system generates gravitational waves in the frequency band most easily detected by current instrumentation, and so these events are at the forefront of gravitational-wave research. They can emerge from many different types of binary systems: black hole–black hole, neutron star–neutron star, neutron star–black hole, WD–neutron star, giant star–neutron star, and so forth. This variety of progenitors allows our detections of these events to advance our understanding of many different fields: the origin of GRBs, the physics of black holes, the origin of heavy elements, the composition of neutron star cores, and the neutron star equation of state.

1.2.6.1 Binary Black Holes

Binary black holes are a preeminent example of a phenomenon known chiefly from a nonelectromagnetic messenger—gravitational waves. Most of what we know about these systems come from the analysis of the gravitational waves emitted when they merge. This analysis is carried out by building numerical models of the system using the general theory of relativity, but simple "back of the envelope" arguments can give some insight (for fuller details, see Chapter 9 and Kembhavi & Khare 2020).

On 2015 September 14, at 09:50:45 UTC, the advanced LIGO detectors (at Hanford, WA and Livingston, LA) recorded an event lasting about 0.2 s, now known as GW 150914.[16] The LIGO arms went through about eight strain cycles, with increasing amplitude and frequency to give a "chirp" signal. The maximum amplitude was at a frequency of 150 Hz, after which the amplitude decreased in the "ringdown" phase. The signal is characteristic of a binary "inspiral," with the bodies orbiting each other faster and faster until they merge at maximum amplitude.

[16] The first direct detection of gravitational waves, this resulted in the award of the 2017 Nobel Prize in Physics to Rainer Weiss, Barry Barish, and Kip Thorne.

The frequency of gravitational waves emitted by a binary system is twice the orbital frequency, so the two components were orbiting each other 75 times a second at merger. The rate of frequency increase allows the chirp mass (Chapter 9, Equation (9.3)) of 30 $M_\odot$ to be calculated. The distance between them at merger can be estimated to be 350 km.

These basic parameters imply that the system must be a binary black hole: It cannot be a binary neutron star because neutron stars cannot exceed about 3 $M_\odot$; nor can it be a black hole–neutron star binary, because the 75 Hz orbital frequency would then imply a black hole mass of $\sim 1000 M_\odot$ with Schwarzschild radius[17] of ~3000 km—and the neutron star would have merged into the black hole before reaching the orbital separation implied by the signal.

The far more sophisticated modeling carried out with numerical relativity calculations is able to give more details: The two progenitor black holes had masses of 36 $M_\odot$ and 29 $M_\odot$; at merger, they were traveling at 60% of the speed of light; they merged into a Kerr–Newman black hole of mass 62 $M_\odot$ rotating 100 times a second. The mass deficit of 3 $M_\odot$ represents the energy released in gravitational radiation in a fraction of a second.[18] Signal modeling also gives a distance of 440 Mpc (redshift 0.093)—this distance estimate does not need electromagnetic radiation, and merger signals like this can be used to constrain cosmological models.

From the number of subsequent detections of black hole–black hole mergers, it appears that binary black holes are fairly common, which was unexpected. It is also possible that there are binary supermassive black holes. We know of no binary SMBH in the nearby universe, but an SMBH–SMBH merger at 250 Mpc would likely be detectable with multiple messengers.

1.2.6.2 Binary Neutron Stars

Several BBH mergers had already been detected by 2017 August 17, when advanced LIGO and advanced Virgo in Italy detected a 100 s signal denoted GW 170817 (Chapter 9); because all three detectors were operational, the gravitational-wave source could be localized to an area of 31 deg^2 on the sky. A GRB was also detected in the same part of the sky 1.7 s after the peak of the gravitational-wave signal, and this generated a GCN[19] alert in under a minute, using the well-established procedures established by the GRB community. The gravitational-wave data analysis proved challenging, so the gravitational-wave event was not sent out for some hours. It was the GRB that allowed telescopes to study the source within minutes of the signal arrival.

The combination of gravitational-wave data, gamma-ray burst, and multiwavelength electromagnetic follow-up shows this event to have been the merger of a neutron star–neutron star binary system (or BNS). Such a merger is very different from the BBH merger outlined above. While the BBH cannot be seen except in

---

[17] Named after Karl Schwarzschild (1873–1916), German astrophysicist.

[18] Because the progenitors were black holes, we expect that very little of this energy would have been released in other messengers.

[19] Gamma-ray burst Coordinates Network.

gravitational waves just before the merger, the BNS merger results in a very bright transient phenomenon seen throughout the electromagnetic spectrum. If one of the progenitor neutron stars is a pulsar, it may even be detected long before the event.

The differences arise because the neutron stars are extended objects, so the gravitational field of each produces strong tides on the other. These tides tear the stars apart so that some of the matter is ejected from the merger. This will form a cloud or disk of ejecta, and this is the source of the electromagnetic radiation.

One part of the ejected matter is likely to be a jet, and this is thought to be the location of the short GRB. The rate of gamma-ray emission from GW 170817 was 10^4 times smaller than other such known bursts. It has been suggested that the weaker emission is due to the jet being seen at an angle. Optical and near-IR spectra of the ejecta of GW 170817 show large amounts of heavy elements created by the r-process[20] after the merger. The decay of these unstable isotopes powers a kilonova, and it is the kilonova that is observable by its electromagnetic radiation. These observations are consistent with kilonova models, including the production of large amounts of gold and platinum—BNS mergers may be the principal source of these elements in the universe.

Modeling of the gravitational waves indicates that the progenitors of GW 170817 had masses of 0.86 to 1.36 $M_{\odot}$ and 1.36 to 2.26 $M_{\odot}$, with a total combined mass between 2.73 and 2.82 $M_{\odot}$ before the merger.[21] The product of the merger is modeled to be a hypermassive neutron star[22] with mass $\sim 2.8 M_{\odot}$, larger than many estimates of the upper mass limit for a neutron star, so it is likely that it very quickly collapsed into a black hole. This would be consistent with the lack of gravitational-wave emission after the initial signal; but there are signs of such emission in later analyses, which would be consistent with the merger product being a hypermassive magnetar. Continued modeling of this event is yielding new insights into the structure, composition, and merger of neutron stars (Figure 1.11).

Gravitational-wave and electromagnetic observations agree on the location of the merger—the galaxy NGC 4993 (Figure 9.8), about 40 Mpc away (Section 9.7.1), much closer than the previously detected BBH. GW 170817 is one of the prototypical multimessenger events, in which the combination of gravitational-wave detection and electromagnetic-wave observations reveal an extraordinary breadth and depth of detail about this event. As such, it demonstrates some of the promise of multimessenger astrophysics. BNS mergers like it may go on to produce neutrino or cosmic-ray detections as well.

1.2.7 Pulsars

The idea of a neutron star—a stellar-mass object composed almost entirely of neutrons at high density—was first suggested by Zwicky[23] and Baade[24] in 1934,

[20] Rapid neutron capture, leading to high-mass-number isotopes of heavy elements.
[21] Uncertainty in the masses results from not knowing the amount of spin the objects have.
[22] If a black hole were immediately formed by the merger, there would be less ejecta than observed.
[23] Fritz Zwicky (1898–1974), Swiss-American astrophysicist.
[24] Walter Baade (1893–1960), German-American astronomer.

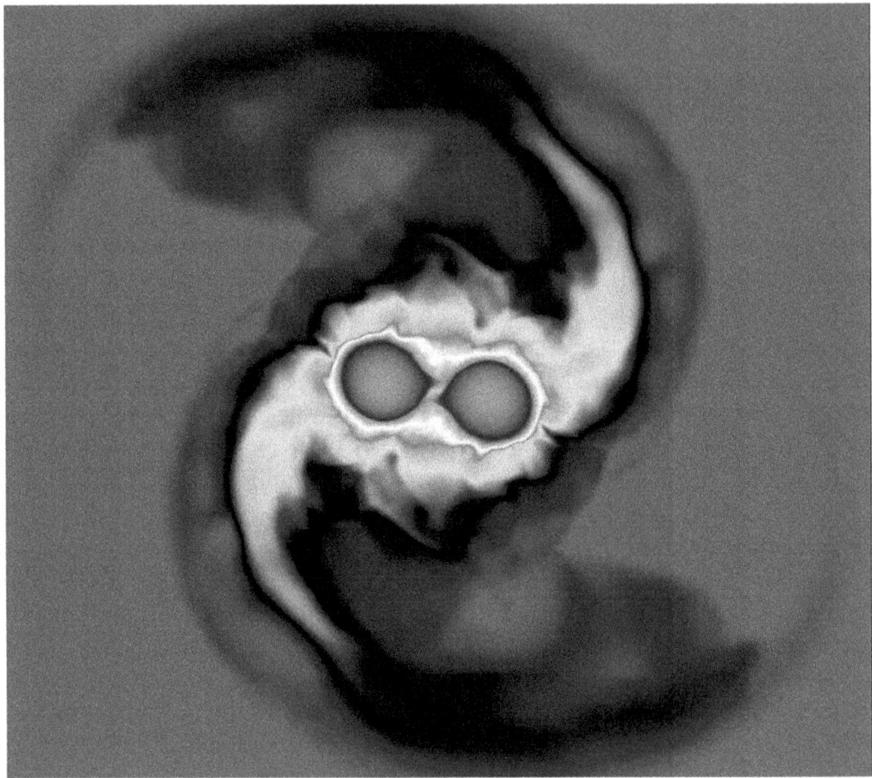

Figure 1.11. Numerical relativity simulation of two merging neutron stars similar to GW 170817. Image credit: Patricia Schmidt and Geraint Pratten (University of Birmingham); the Einstein Toolkit (https://einsteintoolkit.org/).

shortly after the discovery of the neutron itself by Chadwick.[25] Zwicky and Baade even predicted that the collapse of a star into a neutron star with the conversion of significant mass into energy would explain supernovae; however, at this time, the idea was purely hypothetical, and it seemed unlikely that such an object could ever be detected—they would be so small and so cold that there would be no significant thermal emission from them.

It took the development of radio astronomy, with its sensitivity to nonthermal emission, for neutron stars to be observationally confirmed, in the form of pulsars. Hewish[26] and Bell[27] designed and built a radio telescope (the Interplanetary Scintillation Array), which they were using to study the newly discovered quasars. The telescope was designed to work in the time domain (using chart recorders), so it recorded the regular arrival of radio pulses from one point in the sky. Bell's persistence in tracking down this periodic signal led to the

[25] James Chadwick (1891–1974), British physicist.

[26] Antony Hewish (1924–2021), British astrophysicist, winner of the Nobel Prize in Physics 1974.

[27] Jocelyn Bell Burnell (1943–), British astrophysicist.

discovery of pulsars (Hewish et al. 1968).[28] By the time of the discovery, theoretical models of radio pulses from neutron stars had been put forward independently by Pacini[29] and by Gold.[30] These rotating neutron star models were the only ones that could account for the very short rotation period of the Crab Pulsar (33 ms), measured in 1968.

While the astronomical heritage of pulsar astronomy lies in the radio part of the electromagnetic spectrum, pulsars play a role in multimessenger astronomy (see Section 7.5.3). As an early case of the importance of time-domain observations, they served as an early laboratory for the time-domain techniques that are key to many of the multimessenger observations we make now. Pulsars are extremely precise clocks, so the combination of signals from a pulsar timing array (PTA) may be able to detect the passing of gravitational waves in space—a very different approach to their detection. Gravitational waves may also be emitted from pulsars: Fast-spinning highly magnetized young neutron stars can have nonsymmetric rotational perturbations and may emit gravitational waves with frequencies of a few hundred Hertz—within the frequency range of ground-based detectors. Gravitational-wave detection would then probe the rotation and deformation of these compact objects (Chapter 9).

Neutron stars may also emit neutrinos, either as thermal emission in the MeV range (Chapter 8) or due to hadron acceleration processes in the magnetized wind or jet yielding GeV to EeV neutrinos. The high-energy neutrinos may therefore probe the highly magnetized winds (Kashiyama et al. 2016).

1.2.7.1 Pulsar Wind Nebulae

As the product of a supernova explosion, pulsars are often found within a cavity, surrounded by the shell of matter ejected by the explosion. Some pulsars, however, are surrounded by centrally concentrated pulsar wind nebulae (PWNs) or plerions; the prototype of these is the Crab pulsar.

PWNs are made up of highly relativistic leptons (electrons and positrons) from the neutron star wind; they are generally found around young pulsars with fairly high magnetic fields. The relativistic leptons emit synchrotron radiation across a very broad spectrum from radio to X-rays, and gamma rays (up to TeV energies) by inverse-Compton emission. PWNs usually appear within a few hundred years after a pulsar's creation and last about 100,000 years. They can be seen in visible light, but most observations are carried out in the radio continuum (Chapter 2), soft X-rays, hard X-rays (Chapter 6), and gamma rays (Chapter 7; Section 7.5.2). Although observed in electromagnetic radiation, PWNs are sources of high-energy leptonic cosmic rays.

[28] The discovery of at least four such sources in different parts of the sky made it unlikely that these were signals from alien civilizations.

[29] Franco Pacini (1939–2012), Italian astrophysicist.

[30] Thomas Gold (1920–2004), Austrian–British–American astrophysicist.

1.2.8 Ultra-high-energy Cosmic Rays

An ultra-high-energy cosmic ray (UHECR) has energy greater than 1 EeV (10^{18} eV), placing it among the highest-energy—and rarest—cosmic rays. These cosmic rays have energies comparable to macroscopic phenomena; an EeV is equivalent to 0.16 J.

Cosmic rays with energies $>5 \times 10^{19}$ eV (the Greisen–Zatsepin–Kuzmin limit or GZK limit) will lose energy by interaction with the photons of the cosmic microwave background, so this should be a practical upper limit to the cosmic-ray spectrum. However, particles with energy of order 10^{20} eV have been detected.[31] Such particles, known as extreme energy cosmic rays (EECRs), should have an effective range of about 50 Mpc due to the GZK limit, implying an origin within or near the Milky Way. The GZK limit is derived for protons, which make up the majority of cosmic rays. There is some evidence that the highest-energy cosmic rays are dominated by heavier nuclei up to iron (Anchordoqui 2019) but these have their own equivalents to the GZK limit, which are fairly similar in energy.

A fundamental question about UHECRs is simply how they are accelerated to such high kinetic energy. Acceleration theories fall into two basic categories: "One-shot" processes assume direct acceleration to high speed by an extended electric field, which could be produced by the rapid rotation of compact highly magnetized objects such as white dwarfs, neutron stars (pulsars), or black holes. The second category involves gradual energy gain through multiple stochastic encounters with moving magnetized plasmas, found on scales throughout the universe from local (the interplanetary medium) to galactic (e.g., SNRs, the Galactic disk and halo, and microquasar systems) and intergalactic (e.g., AGNs, jets and lobes of giant radio galaxies, blazars, GRBs, starburst superwinds, and clusters of galaxies). Stochastic methods, which by definition are probabilistic and random, are relatively slow and inefficient. Other hypothetical sources of these cosmic rays include hypernovae, relativistic SNe, the decay of supermassive particles from the early universe, dark matter particle decay, and even compact stars composed of conjectured subcomponents of quarks and leptons called preons.

The major detector experiment focused on UHECRs is currently the Pierre Auger Observatory in Argentina. Other projects include the Telescope Array in Utah (successor to the Fly's Eye experiment and the Japanese AGASA) and TAIGA in Siberia. Innovative detection methods using radar and even distributed mobile phones have also been proposed to find these very rare events. Space-based detection, by monitoring of Earth's atmosphere, has also been proposed, such as the Probe Of Extreme Multi Messenger Astrophysics (POEMMA) concept, designed to find the air showers produced by UHECRs with energy >20 EeV.

UHECRs can also produce high- and ultra-high-energy cosmic neutrinos (Anchordoqui 2019), as well as the neutrinos produced when UHECRs enter

[31] The "Oh-My-God particle," for example, was found on 1991 October 15 over Utah by the "Fly's Eye" experiment, with energy 3.2×10^{20} eV.

Earth's atmosphere. A multimessenger (cosmic-ray/neutrino) approach may be fruitful in identifying cosmic accelerators throughout the universe (Chapter 8).

1.2.9 Tidal Disruptive Events

Tidal disruptive events (TDEs) occur when a star becomes close enough to a supermassive black hole so that the star is ripped apart by the black hole's tidal forces[32] (Chapter 9). All black holes have a tidal disruption radius, but for stellar-mass black holes, it lies within the Schwarzschild radius. For supermassive black holes, the tidal disruption radius lies outside the Schwarzschild radius, so a star can be tidally disrupted without disappearing completely into the black hole.

There are a few dozen TDE candidates identified from wide-field optical and UV transient surveys, as well as X-ray telescope observations (Dai 2018). Interestingly, some distinct classes of TDEs have been observed; some radiate in the near UV and optical while others have prominent X-rays. Some also have relativistic jets. Unified models have been proposed for these different classes of TDEs in which the spectral properties depend on the viewing angle of the observer to the disk.

For example, ASASSN-19bt was detected by the All Sky Automated Survey for SNe project on 2019 January 21. The star was destroyed by a black hole in 2MASX J07001137–6602251, a galaxy 115 Mpc away (Holoien et al. 2019). UV observations with the Swift satellite showed a drop from 40,000 to 20,000 K over a period of a few days near the same time.

Although the disk associated with these SMBHs should obey a maximum accretion rate known as the Eddington limit due to light pressure, models have predicted emission beyond soft X-rays. This gives rise to the concept of super-Eddington accretion and may yield energies up to gamma rays. The mechanism of Super-Eddington accretion is not understood at this time.

Therefore, given the high-energy environments of SMBHs, the tidal disruption and accretion of stellar matter could produce many different messengers in the electromagnetic realm from radio to gamma rays as well as neutrinos and possibly cosmic rays.

1.2.10 Soft Gamma Repeaters

On 1979 March 5, a powerful wave of gamma radiation was detected by multiple satellites—100 times stronger than the GRB detected by the Vela satellites in 1967. This burst was only a fraction of a second long but had as much energy as the Sun emits in 10,000 years. This first peak was followed by a 100 s tail with periodic repeating peaks.

Several of these events have now been detected. They are termed soft gamma repeaters (SGRs), which are defined to emit large bursts of gamma rays and X-rays at irregular intervals and are associated with a type of neutron star called a magnetar (Section 1.2.10.1).

[32] The tidal force elongates the material, stretching it in one dimension and compressing it in the others, hence the term "spaghettification."

1.2.10.1 Magnetars

Magnetars are a type of neutron star with extremely powerful magnetic fields ($\approx 10^{13}$–10^{15} G) that can decay and give rise to high-energy emission of X-rays and gamma rays. Magnetars rotate relatively slowly every 2–10 s. While the magnetic field can emit strong bursts of X-rays and gamma rays, their fields decay after 10^4 years. Gamma-ray flares may occur also, which may be the result of starquakes. It is important to note that magnetars may not be pulsars[33] (Figure 1.12).

The origins of magnetars are theorized to result from a magnetohydrodynamic dynamo process inside the neutron star. It is hypothesized that the inside of these young neutron stars remained an ordinary fluid at 30 billion K for a short time. This neutron fluid could bob up and down at thousands of kilometers per second. If the initial magnetic field is strong enough and the rotation is more than 200 per second, a few seconds of the dynamo effect will amplify the magnetic field to greater than 10^{15} G.[34]

According to models, the stronger the initial magnetic field, the faster the pulsar will die. The high magnetic field causes the pulsar's rotation period to slow down dramatically and rapidly. Thus, magnetars are born spinning faster than a typical pulsar, but the stars' rotational energy is quickly lost. For a field of 10^{15} G, a magnetar might slow to a period of about 8 s.

Eight seconds was the periodicity of the SGR event observed on 1979 March 5 (SGR 0525–66), suggesting a connection between magnetars and SGRs. The origin of this event was traced to the LMC and associated with a 5000 year old remnant known as N49 in the LMC. This particular SGR was intensely bright with a tail of regular pulses that was thought to occur because the magnetic fields and the remainder of the outburst were dragged out as the neutron star rotated. Each time it faced Earth, a pulse was detected.

The huge energy associated with SGR 0525–66 required the field to be stronger than 10^{14} G. It is suspected that when an extremely strong field drifts through the solid, relatively stable crust of a neutron star, the twisting and turning creates a starquake. It is the twisting of the magnetic field outside the star energizing electrons and positrons that trigger hard gamma rays. The pulsating tail was powered by the residue of a dispersing shrinking hot cloud of electron–positron pairs trapped by the stars' magnetic field. This residue cools down as escaping X-rays also. In these strong magnetic fields, X-ray photons may split in two and atoms can become long and thin; the polarization of hard X-rays may also occur (Moskvitch 2020).

It has also been proposed that gravitational-wave bursts may be detected at the time of SGR flares, as well as long-lived quasiperiodic[35] waves after the

---

[33] An excellent reference list about magnetars with article links can be found at https://en.wikipedia.org/wiki/Magnetar.

[34] For reference, a refrigerator magnet is about 100 G.

[35] Quasiperiodic behavior is a pattern of recurrence with a component of unpredictability that does not lend itself to precise measurement.

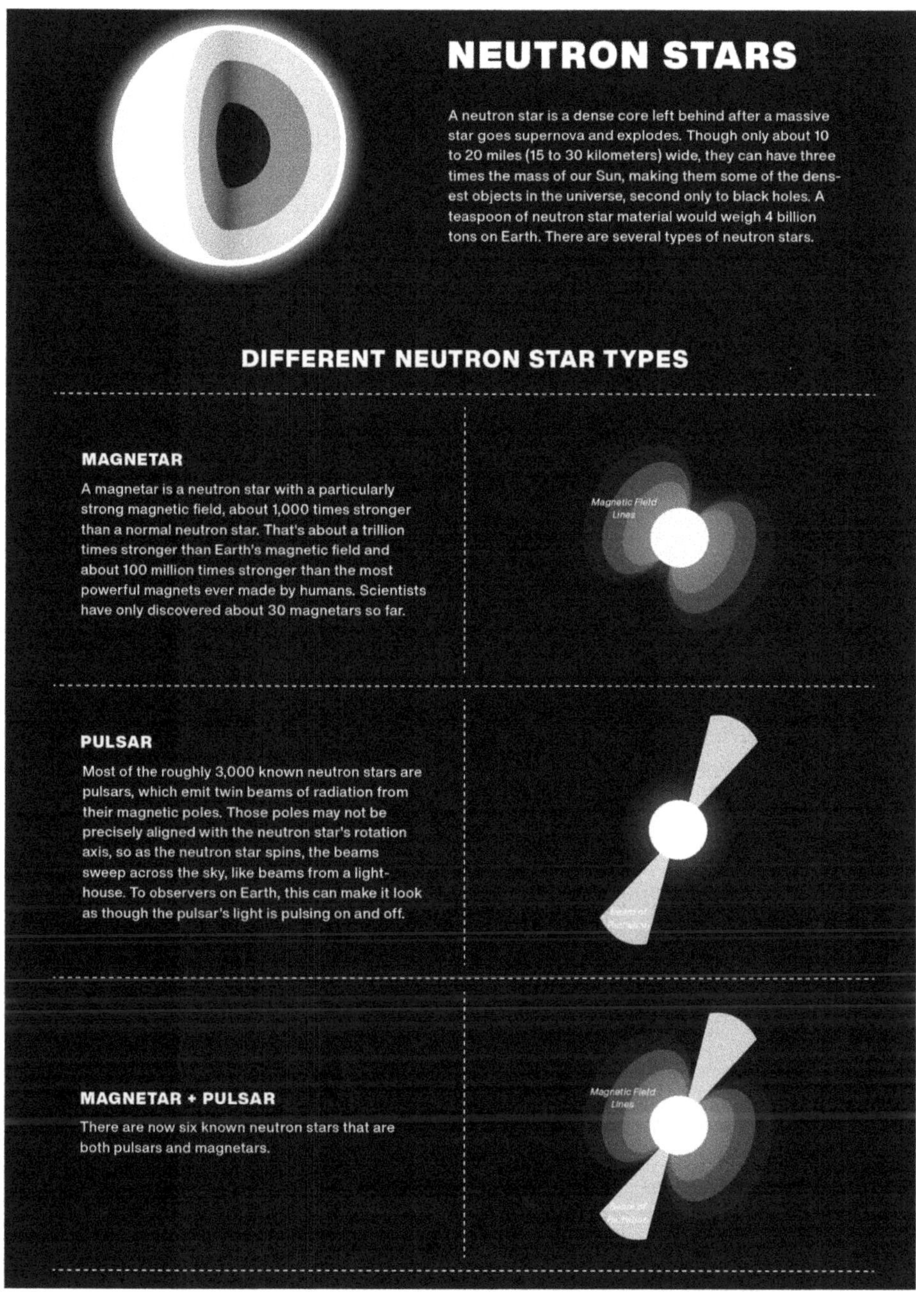

Figure 1.12. Different neutron star types. Image credit: NASA/JPL.

giant flare. Also, given that these magnetars have huge magnetic fields, they may well be cosmic-ray accelerators or produce gamma rays up to TeV energies; even high-energy neutrinos could also be produced (Halzen et al. 2005).

1.3 Conclusion

In this chapter, we have introduced the more practical aspect of multimessenger astronomy by discussing a multitude of astrophysical events; including some history of the evolution of our understanding. As was the case for the concept of multiwavelength astronomy, multimessenger astronomy will give us another order of magnitude of insight into astrophysical processes in the near and distant universe.

As discussed in this chapter, three major confirmations of multimessenger events beyond the Milky Way include (Moskvitch 2020):

- The detection of SN 1987A using both optical telescopes and neutrino observatories.
- The location of the origin of a cosmic neutrino, 170922A, to a blazar, TXS 0506+056, 3.8 billion lt-yr away using optical telescopes and the IceCube neutrino observatory.
- The detection and observation of GW 170817 and EM 170817 in multiple wavelength domains.

Finally, we also mention the Astrophysical Multimessenger Observatory Network (AMON; https://www.amon.psu.edu/) as a currently unique virtual observatory that has been built with the purpose of enabling near real-time coincidence searches using data from leading multimessenger observatories and astronomical facilities (Ayala Solares et al. 2020). Given the astrophysical processes in the zoo of events outlined here, it is obvious that in practice, multimessenger astronomy is rapidly improving our insight into high-energy events within the universe.

References

Abbott, B. P., Abbott, R., Abbott, T. D., et al. 2017, PhRvL, 119, 161101
Aiello, S., Akrame, S. E., Ameli, F., et al. 2019, APh, 111, 100
Aiello, S., & Albert, A. 2021, EPJC, 81, 445
Alsaberi, R. Z. E., Barnes, L. A., Filipović, M. D., et al. 2019, Ap&SS, 364, 204
Anchordoqui, L. A. 2019, PhR, 801, 1
Ayala Solares, H. A., Coutu, S., Cowen, D. F., et al. 2020, APh, 114, 68
Bannister, K. W., Deller, A. T., Phillips, C., et al. 2019, Sci, 365, 565
Bozzetto, L. M., Filipović, M. D., Urošević, D., Kothes, R., & Crawford, E. J. 2014, MNRAS, 440, 3220
Bozzetto, L. M., Filipović, M. D., Vukotić, B., et al. 2017, ApJS, 230, 2
Britzen, S., Fendt, C., Böttcher, M., et al. 2019, A&A, 630, A103
Burke-Spolaor, S. 2018, NatAs, 2, 845
Chatterjee, S., Law, C. J., Wharton, R. S., et al. 2017, Natur, 541, 58
Clesse, S., & Garcia-Bellido, J. 2020, arXiv:2007.06481
Dai, L. 2018, ApJ, 859, L20
Fryer, C. L., Burns, E., Roming, P., et al. 2019, BAAS, 51, 122
Gendre, B., Stratta, G., Atteia, J. L., et al. 2013, ApJ, 766, 30
Greiner, J., & Mazzali, P. A. 2015, Natur, 523, 189
Halzen, F., Landsman, H., & Montaruli, T. 2005, arXiv:astro-ph/0503348

Hewish, A., Bell, S. J., Pilkington, J. D. H., Scott, P. F., & Collins, R. A. 1968, Natur, 217, 709
Holoien, T. W. S., Vallely, P. J., Auchettl, K., et al. 2019, ApJ, 883, 111
Hurley-Walker, N., Filipović, M. D., Gaensler, B. M., et al. 2019a, PASA, 36, e045
Hurley-Walker, N., Gaensler, B. M., Leahy, D. A., et al. 2019b, PASA, 36, e048
Kashiyama, K., Murase, K., Bartos, I., Kiuchi, K., & Margutti, R. 2016, ApJ, 818, 94
Kembhavi, A., & Khare, P. 2020, Gravitational Waves: A New Window to the Universe (Singapore: Springer)
Kinugawa, T., Takeda, H., & Yamaguchi, H. 2019, arXiv:1910.01063
Kirsten, F., Snelders, M. P., Jenkins, M., et al. 2021, NatAs, 5, 414
Levan, A. 2018, Gamma-Ray Bursts (Bristol: IOP Publishing)
Li, C.-J., Chu, Y.-H., Gruendl, R. A., et al. 2017, ApJ, 836, 85
Maggi, P., Haberl, F., Kavanagh, P. J., et al. 2016, A&A, 585, A162
Maggi, P., Filipović, M. D., Vukotić, B., et al. 2019, A&A, 631, A127
Marcote, B., Nimmo, K., Hessels, J. W. T., et al. 2020, Natur, 577, 190
Margalit, B., & Metzger, B. D. 2018, ApJ, 868, L4
Moskvitch, K. 2020, Neutron Stars; The Quest to Understand the Zombies of the Cosmos (Cambridge, MA: Harvard Univ. Press)
Predehl, P., Sunyaev, R. A., Becker, W., et al. 2020, Natur, 588, 227
Roper, Q., Filipovic, M., Allen, G. E., et al. 2018, MNRAS, 479, 1800
Sano, H., Yamane, Y., Tokuda, K., et al. 2018, ApJ, 867, 7
Schnittman, J., Sigl, G., & Buonanno, A. 2006, in AIP Conf. Ser. 873, Laser Interferometer Space Antenna, ed. S. M. Merkovitz, & J. C. Livas (Melville, NY: AIP), 437
Scholtz, J., & Unwin, J. 2020, PhRvL, 125, 051103
Stanek, K. Z., Matheson, T., Garnavich, P. M., et al. 2003, ApJ, 591, L17
Tendulkar, S. P., Bassa, C. G., Cordes, J. M., et al. 2017, ApJ, 834, L7
Vattis, K., Goldstein, I. S., & Koushiappas, S. M. 2020, PhRvD, 102, 061301
Woosley, S. E., & Bloom, J. S. 2006, ARA&A, 44, 507
Wright, W. P., Nagaraj, G., Kneller, J. P., Scholberg, K., & Seitenzahl, I. R. 2016, PhRvD, 94, 025026
Zhang, B 2020, Natur, 587, 45
Zuber, K. 2020, Neutrino Physics (3rd ed. Boca Raton, FL: CRC Press)

Chapter 2

Long-wave (Radio) Astronomy with Coherent Detection from Radio to THz

Natasha Hurley-Walker, Jeffrey L Payne, Miroslav D Filipović and Nicholas Tothill

Radio wavelengths are the longest discussed in this book, and therefore have the lowest energy. The very largest radio telescopes that astronomers have constructed collect barely enough energy to power a single light bulb. Despite this, radio observations give us insight into some of the most energetic phenomena in the universe. Radio is the first astronomical technique that allowed astronomers to view beyond the visible spectrum and as such is one of the oldest multimessenger techniques available. And yet at the time of writing, the Ph.D. students of the first radio astronomers are still alive and well and documenting the beginnings of the field (see *Principles of Multimessenger Astronomy* (Book 1), Section 1.3.3). Such has been the progress of expanding multiwavelength capability in the last century.

You may have heard that radio data are difficult to work with, and most astronomers wouldn't disagree! But the best way of understanding whether your data really support your astrophysical conclusions is to understand the instrument and processing chain or even do the data reduction yourself. The latter may be a step that will be eliminated with the rise of massive radio arrays with "Big Data" challenges (Chapter 12), but nevertheless, a robust understanding of the challenges of radio astronomy will allow you to propose the right kind of observations and understand the final images.

This chapter brings the enthusiastic reader up to speed on the utility of radio techniques for exploring our universe, including the astrophysical insights it brings us, the plethora of telescopes and observing modes available, and notes on the kinds of data processing necessary to transform radio data into useful images that we can study.

2.1 Radio Continuum

Continuum emission is simply that emission that varies slowly in brightness with respect to frequency. It arises through several fundamental physical processes and is

doi:10.1088/2514-3433/ac2256ch2

typically sorted into two loose categories: "thermal" (consistent with thermal blackbody radiation) and "nonthermal" (inconsistent with such an interpretation). These two emission mechanisms were presented in our Book 1. Here, we briefly emphasize their importance for radio astronomy.

2.1.1 Synchrotron

The first emission mechanism we will discuss is the most common in radio astronomy. Synchrotron emission is a special case of "bremsstrahlung" or "braking radiation" (Book 1; Chapter 6), where charged particles change their velocity and thus emit electromagnetic waves. In the case of synchrotron, the particles are usually electrons, and the acceleration is caused by magnetic fields, and thus it is also known as "magnetobremsstrahlung."[1] The electrons describe spiral paths around the magnetic field lines, and the photons emitted are relativistically beamed in the electron's instantaneous direction of motion, with an energy (and thus frequency) determined by the speed of the electron at that moment. The energy spectrum for a single electron is a narrow peak centered about this frequency (Book 1; Figure 6.2).

In radio sources there typically arises a large population of electrons of various energies. If the electron energy spectrum has power-law index p, the spectral index (α) of the synchrotron emission of these electrons, defined by

$$J(\nu) \propto \nu^{\alpha}, \tag{2.1}$$

is

$$\alpha = \frac{-(p-1)}{2}. \tag{2.2}$$

Thus, observations of the brightness of a source with respect to frequency give direct insight into the distribution of electron energies in that source (Book 1; Figure 6.3).

Clearly, the spectrum cannot rise forever toward lower and lower frequencies, and indeed a new process kicks in to limit the number of low-frequency photons emitted by a synchrotron source: synchrotron self-absorption. At low-enough frequencies, the brightness temperature of the source may approach the kinetic temperature of the radiating electrons. Thermodynamically, the source cannot emit radiation of brightness temperature greater than its kinetic temperature. Thus, the photons begin to be reabsorbed by the electrons, and the source becomes opaque at that frequency. As the frequency of observation is lowered, the source becomes more and more opaque and the brightness decreases. A full derivation yields that the "optically thick" spectrum of a synchrotron source has $\alpha = +2.5$ (Longair 2011). The position of the turnover between the optically thick and optically thin parts of the spectrum, combined with an estimate of the source's true size, allows one to derive the kinetic temperature of the source. Early radio astronomers determined that synchrotron sources could have "absurd" temperatures of $\approx 10^{11}$ K, which provided direct evidence for relativistic electrons in these sources.

[1] If the particle is nonrelativistic, then the emission is called cyclotron emission.

We can also examine the properties of the magnetic field of distant synchrotron sources. The emitted photons are linearly polarized in the plane perpendicular to the magnetic field. Thus, a measurement of the direction of polarization yields a direct measurement of the orientation of the magnetic field with respect to the observer. The fraction of polarization (linearly polarized flux density over total flux density) gives an idea of the order of the magnetic field: very regular and ordered fields will produce strong linear polarization, while messy and disordered magnetic fields will produce low levels of measured polarization. In Book 1 (Section 6.1.2), we show a list of the most common radio sources that produce synchrotron emission, including their spectral indices. More discussion on these sources and their properties are presented in Sections 1.2 and 2.7.

2.1.2 Bremsstrahlung

The second most common emission mechanism in radio sources is standard bremsstrahlung: particles (usually electrons) emitting radiation due to acceleration, but in this case, the acceleration is caused by other charged particles (usually protons). Because both particles are not in a bound state, this emission is also sometimes called "free–free." This typically occurs in extremely hot, ionized gases, such as HII regions ($T \approx 10^4$ K) and the centers of galaxy clusters ($T \approx 10^9$ K).

The acceleration of the electron perpendicular to the line of flight is a very brief impulse, unlike a synchrotron-emitting electron, which essentially radiates continuously. The effect of this impulsive emission is to produce a flat spectrum ($\alpha \approx 0$) with equal brightness at a broad range of frequencies. The energy of the photon can never exceed the kinetic energy of the electron, so at frequencies corresponding to this energy, the spectrum falls off exponentially with increasing frequency. The location of the turnover corresponds roughly to the frequency which is the inverse of the duration of the collision; hotter gases will produce collisions with shorter durations, and thus a higher cutoff.

At the lower-frequency end, self-absorption once again causes the optical depth to rise with decreasing frequency. Depending on the conditions, it is possible for the spectral index of this optically thick part of the spectrum to be steeper than +2.5; thus, a measurement of $\alpha > +2.5$ can often indicate the presence of hot ionized gas causing bremsstrahlung.

The aggregate emission from all of the randomly oriented interactions in a hot plasma produces unpolarized emission, so this can be a useful discriminant between synchrotron and bremsstrahlung if only a single frequency of observation is used.

Plasmas that emit bremsstrahlung can also be seen in absorption against a continuum source; for instance, the low-frequency synchrotron glow of the Milky Way is increasingly absorbed by foreground HII regions at low frequencies. Measurements across a wide low-frequency band can thus detect these regions as "holes" against the brighter background, and comparing the on- and off-source brightnesses can trace the line-of-sight synchrotron emission, making it possible to create three-dimensional maps of electrons in our Galaxy ("cosmic-ray tomography").

2.1.3 Inverse-Compton Scattering

Inverse-Compton (iC) scattering involves the scattering of low-energy photons to high energies by ultrarelativistic electrons (Blumenthal & Gould 1970; also see Chapter 7 and Book 1, Section 7.3.1). The process is the inverse of standard Compton scattering because the electrons, rather than the photons, lose energy. The maximum energy that the photon can acquire corresponds to a head-on collision in which the photon is sent back along its original path. The ratio of the upscattered to original frequency is dependent on the square of the Lorentz factor: the more relativistic the electrons, the more dramatic the amount by which photons are upscattered. For instance, radio waves of gigahertz frequencies can be upscattered by relativistic gases into the ultraviolet!

For a single electron energy, the spectrum of radiation due to iC scattering is a linear rise with frequency and then a sharp dropoff at the maximum energy that can possibly be imparted. Similarly to synchrotron radiation, a power-law distribution of electron energies will produce a power-law distribution of flux densities from iC, with the same spectral index (Equation (2.2)).

Note that iC does not create new photons, only upscatters existing photons. Thus, if light with some incident distribution of energies encounters a relativistic plasma, its total spectrum may be steepened upward by iC. This is demonstrated dramatically by the scattering of cosmic microwave background (CMB) photons in the centers of galaxy clusters—the Sunyaev–Zeldovich effect. Relative to the CMB spectrum, observers at high radio frequencies see an excess of photons, while observers at low radio frequencies see a dearth of photons. This has been used successfully by various teams to measure the mass of relativistic plasma in galaxy clusters.

2.1.4 Dust Emission

Blackbodies also emit in the radio, although due to the shape of the blackbody spectrum this emission is quite faint below gigahertz frequencies. The most common form of thermal continuum emission observed in the radio is that of dust at temperatures of 10–100 K, which absorb optical and UV light and reradiate it as part of their blackbody emission.

Some dust grains are charged, and their rotation can produce microwave emission, which is often called "anomalous microwave emission," as it was detected as an excess in wideband measurements of the CMB (also see Book 1, Chapter 7).

2.2 Radio Spectral Lines

In contrast to continuum, spectral lines consist of abrupt changes in brightness with respect to frequency. All atoms and molecules have quantum transitions, and many of these produce lines in the radio part of the spectrum. Common features of radio spectral lines include:

- Small intrinsic line widths; movement of the emitting medium causes the red- and blueshift of the lines. The Doppler-broadened line profile and position can therefore be used to determine the line-of-sight velocity distribution,
- Stimulated emission is important because $h\nu < <kT$. This causes line opacities to vary as T^{-1} and favors the formation of natural masers,
- The ability to penetrate dust in our Galaxy and in other galaxies allows the detection of line emission emerging from dusty molecular clouds, protostars, and molecular disks orbiting active galactic nuclei (AGNs), and
- Frequency (inverse time) can be measured with much higher precision than wavelength (length), so very sensitive searches for small changes in fundamental physical constants over cosmic timescales can be made.

The International Astronomical Union (IAU) maintains a list of "important spectral lines": https://www.craf.eu/iau-list-of-important-spectral-lines/ as well as The Cologne Database for Molecular Spectroscopy (CDMS): https://cdms.astro.uni-koeln.de/.

2.2.1 21 cm Line

Special attention should be paid by budding radio astronomers to the "21 cm" line, which traces the hyperfine transition in hydrogen. Radiation of wavelength 21 cm is emitted when the electron in a hydrogen atom flips its spin from aligned with the proton spin to the opposite direction. The wavelength of 21 cm translates to 1.4 GHz, a very convenient frequency for observing with radio telescopes (see Section 2.4). If the hydrogen has a velocity with respect to the observer, the line will be blueshifted to higher frequency or redshifted to lower frequency. Therefore, spectral line observations around this frequency allow us to trace the cosmic movement of vast quantities of neutral hydrogen, which is an important feedstock for star formation, and outweighs the stars in galaxies by an order of magnitude.

The 21 cm line can be observed both in emission and absorption. In the first case, clouds of hydrogen can be self-absorbing, depending on their density. This means that an opacity correction must usually be made to determine the mass of hydrogen observed in emission. For absorption lines, no correction needs to be made, so as long as a bright continuum source can be found behind the source of interest, absorption can give a more accurate mass estimate.

2.2.2 Radio Recombination Lines

A common low-energy transition is the radio recombination line. The electrons in atoms can exist in higher- or lower-energy quantized orbits; electron transitions down these orbits produce long-wavelength photons. Astronomers label each recombination line by the name of the element, the lower level number n, and successive letters in the Greek alphabet to denote the level change δn: e.g., α for $\delta n = 1$, β for $\delta n = 2$, γ for $\delta n = 3$, etc. For example, Konovalenko & Sodin (1980) observed a low-frequency (26.13 MHz) absorption line toward the supernova remnant (SNR) Cassiopeia A (Cas A; G111.7–02.1), which turned out to be the transition of carbon between the 631st to 632nd energy state, i.e., C631α.

Radio recombination lines often trace the outskirts of sources of ionizing emission, where there is enough incident energy to "pump up" the electrons to higher-energy states or free them, but where the temperatures are low enough that the particles can recombine. Hydrogen radio recombination lines require electrons of energy >13.6 eV, and so are most commonly found near HII regions, while carbon radio recombination lines tend to be found in cooler environments, such as the surfaces of molecular clouds, where the carbon atoms may be ionized by background far-UV radiation.

2.2.3 Rotational Transition Lines

Molecules in the interstellar medium emit and absorb radio waves through transitions between quantized states of rotational angular momentum (denoted J); this emission is usually found at high frequencies, tens to hundreds of gigahertz. For one of these spectral lines to be observable:

1. The transition must be fairly strong, which requires that the molecule has a significant electric dipole moment and is reasonably abundant;
2. the local volume density[2] must be comparable to or higher than a critical density, n_{crit};
3. and the local kinetic temperature must be high enough that enough molecules are excited into the higher-energy state.

Because the brightness of a spectral line depends on both density and temperature, these molecular lines can be used to probe both of these conditions in molecular gas. The critical density is the density required to ensure that collisions between the emitting molecules and other molecules happen often enough that the distribution of emitting molecules over rotational quantum states is characterized by a Maxwell–Boltzmann distribution with excitation temperature $T_x = T_k$, where T_k is the kinetic temperature. This condition is known as local thermodynamic equilibrium, LTE, and the spectral line is said to be thermalized. Even then, however, the Maxwell–Boltzmann distribution at that temperature must put enough molecules into the higher level to yield significant line emission; if the energy of the upper state is much higher than kT_k, the line will become much fainter. Each individual spectral line has its own critical density and upper-state energy, and so is effectively a separate probe of the physical conditions of the gas. For example, the discovery of ammonia (NH_3) in the direction of the Galactic Center (Cheung et al. 1968) gave astronomers insight that there are much denser cooler regions of the interstellar medium (ISM) than they had expected, because NH_3 has such a high critical density but must be in a cool region to even form.

The interstellar medium is composed of about 90% hydrogen and about 10% helium; the most abundant trace elements ("metals"), in decreasing order of

[2] Measured in particles per unit volume, generally cm^{-3}.

abundance, are oxygen and carbon at a few hundredths of a percent, neon and nitrogen at about a tenth that of oxygen, and sulfur at a few percent of oxygen. The most abundant component of molecular gas is molecular hydrogen, H_2, which has no dipole moment and whose transitions have such high energy that they are not excited in cold dense clouds. Helium and neon are monatomic and have no low-energy transitions. The most abundant observable components of molecular gas are therefore molecules of O, C, N, and S, often with hydrogen atoms attached.[3]

In molecular clouds, the majority of gas-phase carbon atoms are bound to (more abundant) oxygen atoms, making carbon monoxide, CO, the most abundant molecule. Its lowest transition, $J = 1 \rightarrow 0$ (often written CO 1–0), has a frequency of 115 GHz (wavelength 2.6 mm), corresponding to a temperature of 5.5 K and a critical density of a few hundred cm^{-3}. This makes it an excellent tracer of molecular hydrogen: molecular clouds are dense enough to thermalize it and warm enough to give a bright transition. It is such a good tracer that the mass of molecular gas is often derived simply by multiplying the velocity-integrated line brightness of CO 1–0 by an "X factor."

Observation of comparatively low-frequency molecular transitions such as CO 1–0 can be carried out from most observatory sites by most radio telescopes, such as the Mopra radio telescope in Australia, the Green Bank Telescope in the USA, and many others. Higher-frequency lines at submillimeter wavelengths are usually the province of more specialized telescopes, designed for the purpose, such as the ALMA array in Chile or the James Clerk Maxwell Telescope on Maunakea.

2.2.4 Masers

Most people with an interest in physics are familiar with the concept of a laser, which is based on "light amplification by stimulated emission of radiation." Stimulated emission is the quantum mechanical effect where an incident electromagnetic field on a system with an electron with a potential energy transition will affect the quantum mechanical state of the system without being absorbed. As the electron makes a transition between the two levels, it enters a transition state, which has a dipole field, and this dipole oscillates at a characteristic frequency. If the incident field has the same frequency, the probability of the electron entering this transition state is greatly increased. Thus, the rate of transitions between two stationary states is increased beyond that of spontaneous emission. A transition from the higher to a lower-energy state produces an additional photon with the same phase and direction as the incident photon; this is the process of stimulated emission.

Masers were discovered first, and the M indicates "microwave," which for our purposes is considered a radio wavelength. Molecular clouds experiencing incident radiation will mase at the frequencies of the stimulated transition, creating masers. The radiation needs to be of high-enough frequency to "pump" a population inversion, where the higher-energy levels are more populated, but not so high energy that the molecules dissociate. Maser emission is exponentially sensitive to how

[3] For example, CO, CS, HCN, and HCO^+.

coherently the cloud is moving, how far it can propagate, and the size of the population inversion (up to a saturation limit). This means that small variations in these quantities can have a huge difference on the observability of the maser. Thus, masers tend to be point-like sources with enormous (10^9–10^{14} K) brightness temperatures and rapid variability. Masers are often very highly polarized, sometimes 100% (in the case of some OH masers) in a circular fashion and to a lesser degree in a linear fashion. This polarization is due to some combination of the Zeeman effect (see Section 2.2.5), magnetic beaming of the maser radiation, and anisotropic pumping, which favors certain magnetic-state transitions.

At the time of writing, 13 different molecular species had been detected via their maser emission, in comets, moons, star-forming regions, SNRs, and AGN. The detection of masers shows the presence of specific molecules like OH or water, and an anisotropic radiation field required to pump the population inversion (also see Section 2.7.2.4). Polarization gives insights into the magnetic field strengths and directions.

2.2.5 Zeeman Effect

In Section 2.1.1 (also see Book 1, Section 6.1.2) we discussed the way the local magnetic field affects the polarization of any emitted synchrotron radiation. Magnetic fields can also affect the polarization of spectral line emission via the Zeeman effect: the spectral transitions of molecules and atoms split into magnetic sublevels. As a consequence, there are two processes that polarize the spectral line emission.

2.2.5.1 The Goldreich–Kylafis Effect

Anisotropic electromagnetic radiation fields (extremely common in the ISM) will cause the magnetic sublevels of the molecular rotational energy levels to have unequal populations; through a subtle quantum-mechanical effect, this causes the transition radiation to become linearly polarized (Goldreich & Kylafis 1981). This is rarely observed in practice because the optical depths must be exactly correct, and the excitation of the transition needs to be by something other than collisions. The Goldreich–Kylafis effect (Goldreich & Kylafis 1981) is sensitive to the direction of the magnetic field projected in the plane of the sky, but it does not provide information of the magnetic field strength.

2.2.5.2 Magnetic Sub-Level Polarisation

The different magnetic sublevels have slightly different energies. This causes the resulting atomic and molecular transitions to be partially circularly polarized, and the degree of this polarization depends on the magnetic field strength along the line of sight and on the molecule's magnetic dipole moment. Under typical ISM magnetic field (B) strengths, only HI and molecules with an unpaired electron in the outer layer (i.e., radicals such as OH, CN, C_2S, C_2H) have magnetic moments high enough to be detectable at radio wavelengths. Maser emission is so strong that the circular polarization can be relatively easily detected.

2.3 Polarization

Polarization is the behavior of the electric field with time. Astrophysical processes like synchrotron radiation can emit partially polarized emission, but never fully polarized. Interstellar matter can polarize random background emission or depolarize polarized background emission.

As we have mentioned in passing in previous sections, electromagnetic radiation can be polarized, i.e., the wave has a particular angle at which the electric and magnetic fields oscillate. To describe a wave with nonzero and nonunity polarization, one needs four parameters, which can be described in various reference frames. Radio astronomers typically care about the power of the received waves, so we often use the Stokes[4] parameters[5] to transform the electromagnetic wave amplitudes into powers (Figure 2.1):

- Stokes $I = E_{0^\circ}^2 + E_{90^\circ}^2$
- Stokes $Q = E_{0^\circ}^2 - E_{90^\circ}^2$
- Stokes $U = E_{45^\circ}^2 - E_{-45^\circ}^2$
- Stokes $V = E_{\rm RCP}^2 - E_{\rm LCP}^2$,

where I is total intensity and sum of any two orthogonal polarizations, Q and U completely specify linear polarization, and V completely specifies circular polarization. Here, the E parameters describe the electromagnetic waves at various angles, as well as the right circular polarization (RCP) and left circular polarization (LCP). Any linear polarization can be decomposed into the superposition of two equal-amplitude circularly polarized components of opposite handedness and different phase. Conversely a circularly polarized wave can be decomposed into two orthogonal linear waves. The sum of two circular waves of unequal amplitude is

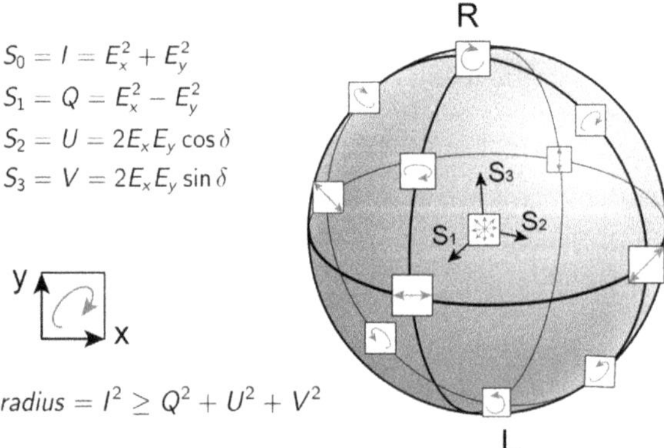

Figure 2.1. Depiction of the polarization states on a Poincaré sphere. Image credit: Wikipedia: 2128506 (CC0 1.0).

[4] Sir George Gabriel Stokes, 1st Baronet (1819–1903).

[5] Adopted for astronomy by Chandrasekhar in 1947.

elliptical. The sum of two orthogonal linearly-polarized waves with a phase difference of between 0 and $\pi/2$ is also elliptical.

These are equivalents to these transformations of the power responses from crossed linear orthogonal dipoles (receiving voltages X and Y), an oft-used design for radio astronomy feeds:

- Stokes $I = XX + YY$
- Stokes $Q = XX - YY$
- Stokes $U = XY + YX$
- Stokes $V = i(YX - XY)$.

Stokes I is the unpolarized power and can only be positive, while Q, U, and V are differences of similar values and are just as likely to be negative as positive. When unconcerned with the detail of the orientation of the polarization, it can be helpful to calculate the polarized fraction p:

$$p = \frac{\sqrt{Q^2 + U^2 + V^2}}{I} \tag{2.3}$$

and should the angle be important, e.g., for the determination of the magnetic field orientation, the polarization angle χ can be obtained from

$$\chi = \frac{1}{2} \arctan U, Q. \tag{2.4}$$

For a synchrotron source, the intrinsic position angle of the electromagnetic field is perpendicular to the projection of the magnetic field onto the sky, so maps of χ can be transformed via a 90° rotation into maps of magnetic fields. Unfortunately, there are two remaining subtleties we must first understand before we can do this.

2.3.1 Faraday Rotation

Faraday rotation (also see Book 1, Chapter 2) is caused by left and right circularly polarized waves propagating at slightly different speeds through an ionized magnetic medium, a property known as circular birefringence. This induces a relative phase shift, the effect of which is to rotate the orientation of a wave's linear polarization:

$$\chi = \chi_0 + \text{RM} \times \lambda^2, \tag{2.5}$$

where χ and χ_0 are the measured and intrinsic polarization angles, respectively, λ is the observing wavelength, and RM (rotation measure) is a constant of proportionality that traces conditions in the intervening medium:

$$\text{RM} \propto \int n_e \vec{B} \cdot \vec{dl}. \tag{2.6}$$

Thus, observations over multiple frequencies can be used to determine the electron density n_e and magnetic field value along the line of sight, $\vec{B}$. Figure 2.2 shows such measurements, taken by Park et al. (2019) along the jet of M87.

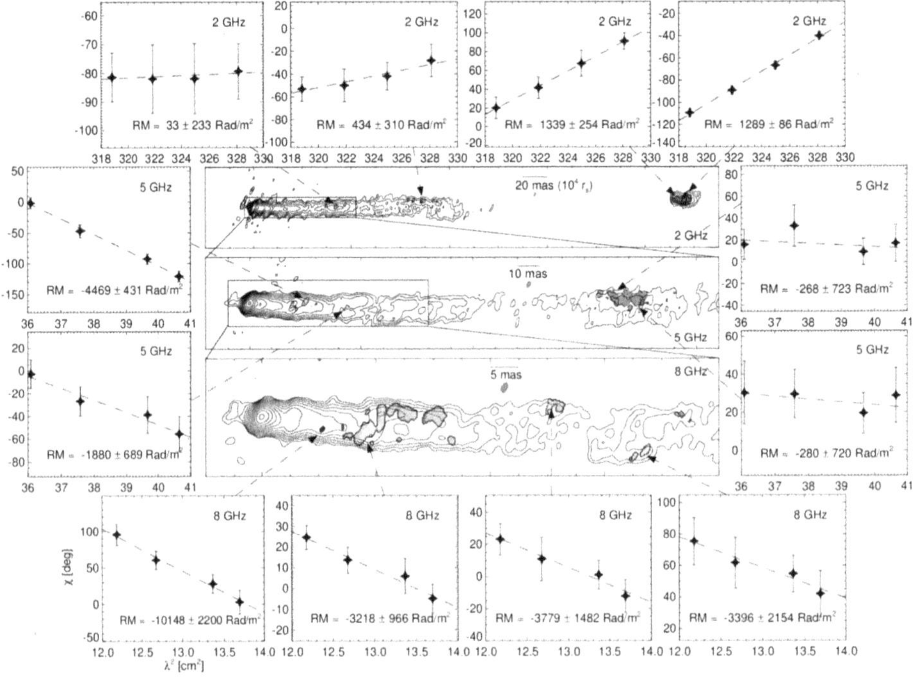

Figure 2.2. The frequency dependence of the polarization position angle at three locations along the jet of M87. Different gradients of the lines allow for different values of n_e and $\overrightarrow{B}$ (Equations (2.5) and (2.6)). Image credit: Park et al. (2019).

2.3.2 Depolarization

Various astrophysical and instrumental effects may reduce the measured polarization of radio waves (Gardner & Whiteoak 1966). An astrophysical source is in truth an extended object emitting over a broad band, and we make measurements with instruments that do not truly resolve the objects nor measure their spectral behavior with infinite resolution. Thus, if there are variations of polarization angle over the source and our beam does not resolve these variations on their true scale, they are averaged together and the overall measured polarization is reduced ("beam depolarization"). Similarly, the angle of polarization will vary across the bandwidth of the spectral channel ($\Delta\nu$) by the fractional bandwidth ($\Delta\nu/\nu$) multiplied by the Faraday rotation (Equation (2.5)). Thus, a high spectral resolution is needed to adequately sample sources at large RM.

There is also an effect analogous to the apparent spectral steepening of sources that are partly resolved out by longer baselines (Section 2.5.1.3): If the polarization angle varies over the face of an emission region as large as or larger than the synthesized beam, the changing resolution with frequency will cause an apparent change in the degree of polarization. The solution is to perform all measurements at all frequencies at a common resolution. Note also that in real radio galaxies, high frequencies tend to sample the cores and jets, in which the magnetic fields are quite ordered; at low frequencies, the disordered and turbulent lobes become dominant. Thus, it is important to disentangle the instrumental causes of apparent depolarization from the astrophysical.

2.4 Radio Astronomy Instrumentation

A comprehensive review of radio astronomy instrumentation is presented in Book 1 (Chapter 1). Here, we focus on a modern-day radio telescopes and their use for astronomical observations.

2.4.1 Radio Dishes

Aside from the visible wavelengths, radio is the only other part of the electromagnetic spectrum that travels freely through our atmosphere, unlike the many wavelengths that require the use of space telescopes (Book 1, Section 1.3.3). The first radio telescope was inadvertently built in 1932 by the American physicist and radio engineer Karl G. Janksy, who was examining the radio spectrum at 20.5 MHz for the Bell Telephone company, for the purpose of building a transatlantic radio telephone service. His discovery of a sidereally periodic radio hiss showed that the astronomical sky was a potent source of radio waves. Gröte Reber, a hobbyist radio enthusiast, built the world's first single-dish radio telescope in his mother's back garden and used it six days a week to map the sky,[6] discovering the nonthermal emission of the Milky Way. The excess of radar equipment available after World War II allowed radio astronomy to grow into a mainstream field after 1945. For a thorough and excellent review of the early history of radio astronomy, we highly recommend Sullivan (2009).

Reber's first telescope hit upon many of the principles that were used for decades to design single-dish radio telescopes, which are simply telescopes optimized to receive radio waves from space (Figure 2.3). They have three main components:

- An antenna to collect radio waves (usually following the parabolic design Reber used) because it causes the radio waves to converge a single focal point; some antenna designs place a secondary reflector at this point, while others (including Reber's original design) place the receiver elsewhere,
- A receiver and amplifier to boost the very weak radio signal to a measurable level. The thermal noise of the receiver dominates the system temperature at high frequencies (Equation (2.7)) so usually receivers above the L band ($\nu = 1$–2 GHz or $\lambda = 30$–15 cm) are cooled,
- A recorder to measure the signal; early telescopes connected the receiver to a pen and spool of chart paper; nowadays, recordings are performed digitally.

Some astronomers are interested in the brightness of the sky as a function of time (e.g., pulsars; Roy & Gangadhara 2019) but many prefer to obtain the brightness as a function of frequency, because this allows us to obtain information about the physics of the source under observation. Figure 2.4 shows the transformation of

[6] The telescope was used by his mother as a washing line on Mondays.

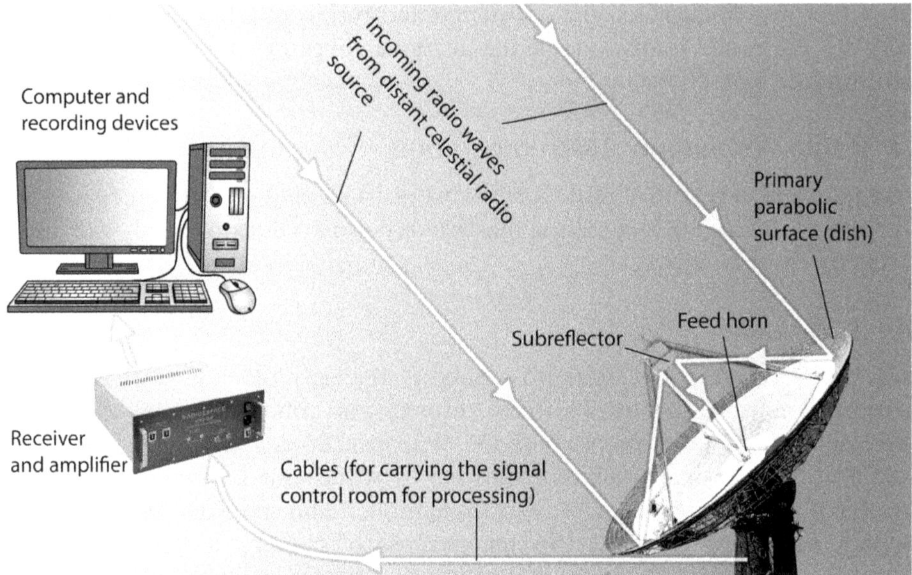

Figure 2.3. Basic components of a radio telescope. Image credit: Aleksandar Zorkić.

time-domain recordings into frequency-domain signals via a heterodyne recording circuit, which can be created using analog electronics as follows:

- The radio-frequency (RF) low-noise amplifier (LNA) boosts the signal from the antenna while trying not to add extra noise; it may also include a filter to exclude certain frequencies (e.g., those known to contain more interference than astronomical signal): The range of frequencies allowed through is shown on the right panel as $\Delta\nu_{\mathrm{RF}}$;
- The mixer takes a signal from a local oscillator (LO) and mixes (multiplies) it into the astronomical signal, reducing it to an intermediate frequency (IF) (shifting the center of ν_{RF} down to ν_{IF}). This is done for two reasons: electronics that operate at these low frequencies are cheaper and easier to build (although this is changing as technology improves), and operations performed on the new signal that might leak back toward the antenna will not contaminate the incoming signal, because they are now at different frequencies;
- The IF amplifier boosts the IF signal further, and more bandpass filtering can be performed at this stage (where the electronics are cheaper and easier to build);
- A quadratic detector (usually a diode) squares the incoming signal, changing the oscillating positive and negative voltages into a positive signal, which will be proportional to the power incoming to the receiver. In the frequency domain, this is equivalent to recording the information from zero frequency up to the bandwidth of the observation;

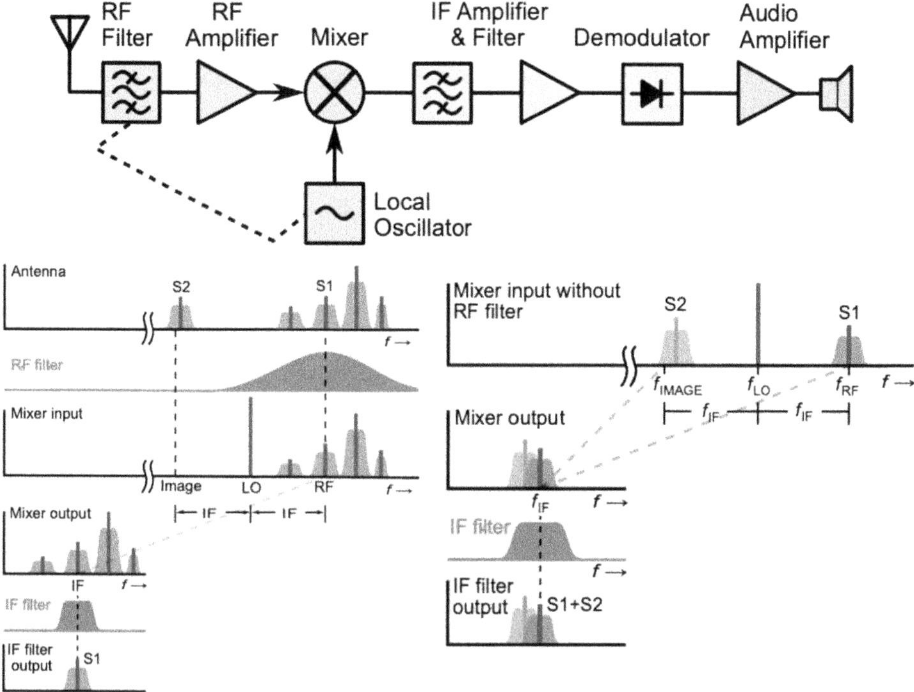

Figure 2.4. A heterodyne receiver circuit. Image credit: Wikipedia: Chetvorno (CC0 1.0).

- An integrator accumulates this signal over time τ_0, which gives it a frequency channel width of τ_0^{-1}.

The sensitivity of a simple single-dish system like this can be described by the radiometer equation:

$$\Delta T = \frac{T_{\text{sys}}}{\sqrt{\Delta\nu\,\tau_0}}. \tag{2.7}$$

Analogous to optical lenses, the angular resolution θ of a single-dish telescope is defined by the diffraction limit: $\theta = \lambda/D$, where λ is the receiving wavelength and D is the dish diameter, both in the same units (usually meters). Thus, while radio wavelengths from 1 mm to 10 m can propagate through the atmosphere, to obtain resolution equivalent to optical telescopes ($\approx 1''$), we require dishes of 200 m to 2000 km in diameter.

Fully steerable dishes can only be built up to $D \approx 100$ m; such engineering marvels include Effelsberg (100 m), the Lovell Telescope (76 m), and Parkes (64 m). After 27 years of operation, the original Green Bank Radio Telescope (100 m) collapsed in 1988, reducing to 7000 tons of crumpled steel girders; it was replaced with a sturdier design that remains in operation today. Larger dishes cannot be steered; the now-collapsed 300 m diameter Arecibo dish was built into an extinct volcanic caldera in Puerto Rico and the Five-hundred-meter Aperture Spherical radio Telescope (FAST) into a concave hill in China. In both cases, the receiver is suspended high above the dish and tilted to produce steering over a zenith angle range of tens of degrees.

Because single-dish telescopes measure total power at a given sky location, they do not produce an image in a single scan. Instead, the power must be recorded at multiple sky locations and aggregated into a map. The field of view is the same as the resolution, so high-resolution surveys can take a long time to build up. High-frequency measurements require dish surfaces with aberrations less than the wavelength of interest, so millimeter-wave astronomy becomes infeasible with large dishes, ultimately limiting the resolution of the single-dish approach. Thus, astronomers have turned to interferometry to more efficiently explore the radio sky at high resolution.

2.4.2 Interferometry

One of the simplest ways of understanding interferometry is to imagine splitting up your large single-dish telescope into a multitude of smaller pieces and distributing them over flat ground. Instead of using the parabolic dish shape to combine the signals together, you add an electrical delay to the signal received by each piece, which mimics the physical delay that would have been experienced were they still arranged in a dish shape. Now you are free to move the pieces as far apart as you like, as long as you insert the correct delay.

If you took the signal received by all these pieces, and squared and integrated it as per the single-dish case (Section 2.4.1), you would retrieve the summed power of the array. The direction of interest can be changed by modifying the electrical delays by routing the signals through different lengths of electric circuits or cables. This is called a "phased array"; these were originally invented as a means of boosting the resolution of military radar systems. This improved resolution can be approximated by

$$\theta = \frac{\lambda}{d}, \tag{2.8}$$

where d is the distance between the two most separated array elements, or the "longest baseline." Similarly, the telescope field of view can be estimated as

$$\theta_{\mathrm{FoV}} = 1.22\frac{\lambda}{d}, \tag{2.9}$$

where θ_{FoV} is in (radians), λ is the wavelength (in meters), and d is the diameter of the telescope's aperture (in meters).

A phased array combines the square of the voltages, losing the phase information. Shortly after the end of World War II, Australian physicists Ruby Payne-Scott and Joseph Pawsey realized that instead, the voltages could be multiplied together, retaining the phase information. They repurposed radar equipment to build a 200 MHz "sea-cliff interferometer," which they used to measure the Sun. They multiplied

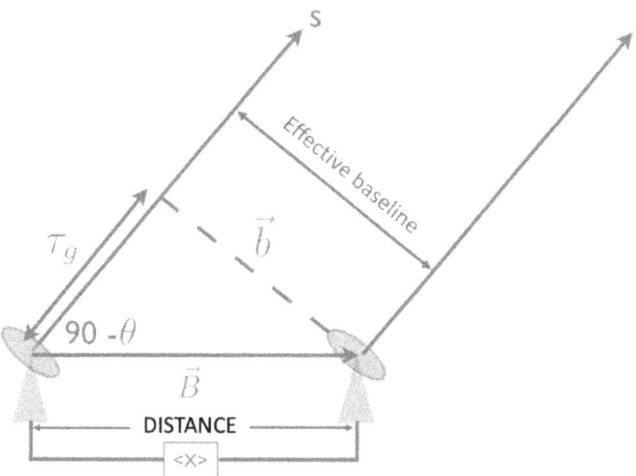

Figure 2.5. Two-element radio interferometer observing a source at infinite distance.

(or "correlated") the signals of the Sun and the reflection of the Sun off the sea to measure the interference pattern, creating an effective baseline of 200 m, yielding a resolution of 0.5°, smaller than the disk of the Sun. With this they were able to identify that sunspots were responsible for bursts of solar radiation and published this exciting result along with the principles of radio interferometry in a 1947 paper.

An interferometer measures the coherence of the electric field between the receiving elements. Figure 2.5 shows a simple two-element interferometer observing a source at infinite distance; the concepts introduced here can be scaled to interferometers consisting of any number of elements. $\vec{B}$ is the baseline vector, $\vec{b}$ is the projected baseline for a source at elevation θ, and τ_g is the geometrical delay caused by the difference in light travel time between the two rays from source S. The complex voltages at telescopes 1 and 2 are given by V_1 and V_2, and each is composed of the source signal V_S and the receiver noise V_R. When these signals are multiplied and averaged at the correlator,

$$\langle V_1 . V_2 \rangle = \langle (V_{R1} V_S + V_{R1} V_{R2} + V_S V_{R2} + V_S^2) \rangle. \tag{2.10}$$

The receiver noise at each telescope is uncorrelated, and is also not correlated with the source signal, so all terms except V_S^2 average to zero over time. As a result, many potential sources of noise, such as fluctuations in the receiver gains or outside interference, drop out of the signal. The amplitudes V_1 and V_2 are proportional to the electric field generated from the signal from each telescope, i.e., the gains of the two telescopes. The power of the final signal is proportional to the autocorrelation of the incident fields, which themselves are proportional to the square roots of the antenna areas, A_1 and A_2. Thus, the effective collecting area is proportional to $\sqrt{A_1 A_2}$.

In the 1950s, Martin Ryle (1918–1984) and colleagues at the University of Cambridge further developed the technique by laying out antennas in an east–west arrangement, pointing them a target, and allowing Earth's rotation to rotate the row of antennas 180° in 12 hours, synthesizing a complete aperture from a smaller set of components: the aperture synthesis technique. A completely sampled aperture could be inverted via a Fourier transform to yield a perfect image of the sky, and for some time the focus was on building east–west arrays that could, over time, build up such a sampling.

In the 1960s and 1970s, advances in computing allowed radio astronomers to build arrays with less complete spatial coverage, using computers to handle the resulting images (see Section 2.6.2.6). The number of baselines of an interferometer of N antennas is given by $\frac{N \times (N-1)}{2}$, and this forms an upper bound of the number of spatial scales that can be sampled instantaneously (antennas can also be arranged to produce near-identical, i.e., "redundant" baselines, which can be useful for some science cases). As computers have become more powerful, it has become possible to construct radio interferometers with larger N, which make more completely sampled images of the sky in less integration time, allowing them to carry out large-scale surveys.[7] Also, we can now simulate interferometric observations with computers (Figure 2.6).

Combining this idea with the radiometer equation (Equation (2.7)) and the effective area of the interferometer, we can calculate the interferometric radiometer equation:

$$\Delta T = \frac{T_{\rm sys}}{\sqrt{\Delta\nu N(N-1)\tau_0}}. \tag{2.11}$$

A further development has been the ability to create radio interferometers with extremely long baselines, by correlating data from antennas separated by hundreds or even thousands of kilometers. This technique is called Very Long Baseline Interferometry (VLBI) and requires exquisite timing and position precision in order that the incoming voltages can be combined coherently. Until the advent of extremely high-speed internet, it was necessary to record the voltages in real time from each antenna on to high-speed large-capacity tapes, then bring them to a central location for correlation. The East-Asian VLBI Network and European VLBI Networks can now perform VLBI measurements in real time.

At the time of writing, the most impressive achievement in radio interferometry was the direct imaging of the lensed accretion disk around the central supermassive black hole of M87 (Figure 2.7). Composed of antennas distributed across an entire hemisphere of Earth, the Event Horizon Telescope (EHT) recorded data at 230 GHz, yielding a resolution of 40 μas—the equivalent of being able to see the footprints left by the Apollo 11 astronauts on the Moon. This required exquisitely cold, dry weather at all sites, in order to make calibration possible.

[7] The best example of modern radio interferometers are Atacama Large Millimeter/submillimetre Array (ALMA), ASKAP, and MeerKAT.

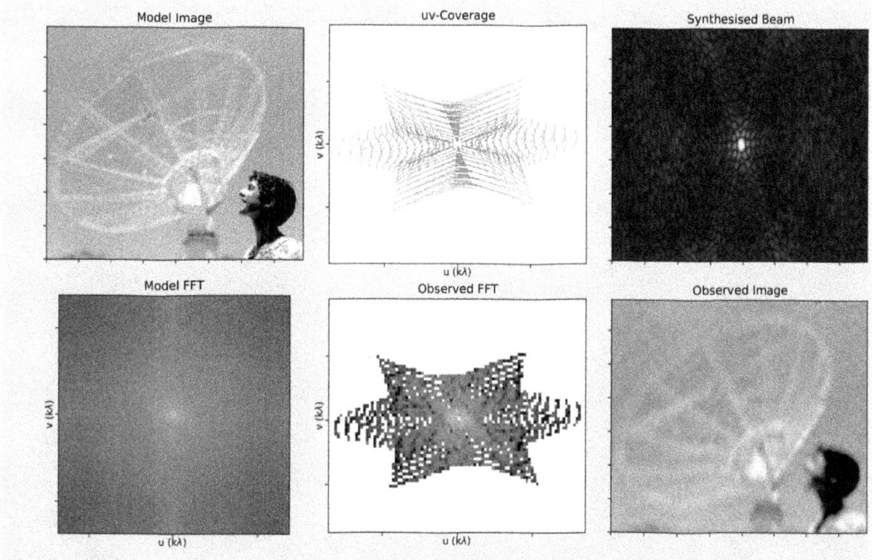

Figure 2.6. A simulated "observation" of the photograph shown in the top left, using the Karl G. Jansky Very Large Array (VLA) in D configuration (a compact layout). The top-middle image shows the (u, v) coverage of this configuration for a two-hour observation 45° from the zenith; this shortens the projected baselines in the north–south direction, causing the (u, v) coverage to have a "squashed" appearance. The top-right panel shows the Fourier transform of the (u, v) coverage, the synthesized beam. The bottom-left panel shows the Fourier transform of the image, and the middle-left panel shows the observed (u, v) samples that result from multiplying the (u, v) coverage by this transformed model (this is what a real radio telescope would measure—we would not know the top-left image a priori). The bottom-right image shows the inverse Fourier transform of the "observation": a low-resolution image that can still be interpreted by the observer to recover the original information, i.e., the author standing by the Giant Metrewave Radio Telescope near Pune, India.

2.5 Observational Considerations

Now armed with an understanding of what we can see with radio telescopes, as well as an idea of how they work, perhaps you want to obtain some radio data to explore an interesting target (or ten, or a hundred!). Most radio telescopes allow individuals or groups to apply for time, sometimes with priority or allocation given to members of funding countries or consortia. Telescopes often have some amounts of time set aside for dedicated observatory projects, such as large surveys. Calls for proposals are usually issued every 6 or 12 months, with a 1–2 month window in which to submit your application, after which it will be considered by a time allocation committee, a decision made, and the observations scheduled.

When applying for time you will usually need to write a scientific justification, giving the wider context of the field, stating the research problem, and then showing how your observations will help you solve it. It is also important to check and show that there are not existing archive observations (or even publications) that already solve the problem. A technical justification is then usually written separately to explain how you have considered the specifications of the observation to best meet

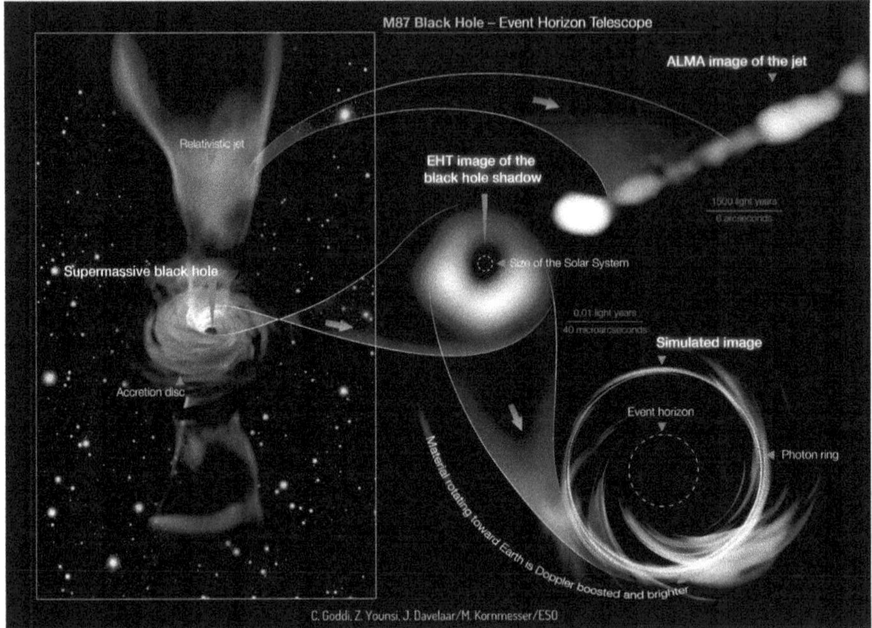

Figure 2.7. First M87 Event Horizon Telescope results. Image Credit: C. Goddi, Z. Younsi, J. Davelaar/M. Kornmesser/European Southern Observatory (ESO)/EHT, Goddi et al. (2019) reproduced with permission © ESO.

your science goals. In this section we'll go through the kinds of parameters that you need to pay attention to when designing your observations.

2.5.1 Experimental Goal

Most observing goals can be broken into two general categories: a detection experiment, where the aim is to find or set limits on the brightness of sources, or an imaging experiment, where the aim is to map sources in detail and determine their morphology. There is more of a distinction between the two in radio astronomy because it is possible to modify the spatial scales recovered by the instrument, either by changing the array layout (e.g., the VLA), changing the receiver frequency, or modifying the imaging (e.g., tweaking the weighting between natural and uniform, see Section 2.6.2.6). Surveys are the widest-area detection experiments and generally are designed to find a large number of sources down to a trustworthy and uniform brightness limit (high "completeness") with low numbers of false positives and negatives (high "reliability").

In general, detection experiments favor shorter baselines, because these are sensitive to sources of all angular scales below the spatial frequency they sample; baselines that produce a spatial frequency sensitivity that is of a smaller scale than that of typical sources will "resolve out" sources that are diffuse and do not have detail on those scales. A classic example of this is the National Radio Astronomy Observatory (NRAO) VLA Sky Survey (NVSS; Condon et al. 1998), which was

Table 2.1. Important Radio Observational Parameters

Consideration	Detection	Imaging
Frequency	Low for $\alpha < 0$, high for $\alpha > 0$	At the highest frequency possible
Array configuration	Baseline distribution peaks at expected maximum scale of emission	Baseline distribution covering all scales of emission
Image weighting	Robust $\rightarrow 2$	Robust $\rightarrow -2$

Notes. One of the key parameters is the robust parameter, which sets R in the Briggs equations. The scaling of R is such that $R = 0$ gives a good tradeoff between resolution and sensitivity. The robust R takes values between −2.0 (close to uniform weighting) to 2.0 (close to natural).

performed at the L band in the most compact configuration of the VLA: The aim was to detect as many sources as possible, with high completeness and reliability. The details of the sources could be worked out later by following them up with higher-resolution observations. In contrast, the Faint Images of the Radio Sky at Twenty-Centimeters (FIRST; Beck et al. 1994; Becker et al. 1995) has nearly an order of magnitude improvement in resolution over NVSS, which makes it better matched to optical surveys, but many arcminute-scale radio galaxies are completely invisible in FIRST, making it much less complete than NVSS. Table 2.1 suggests parameters to tweak between detection and imaging experiments.

2.5.1.1 Sky Position

Radio telescopes are ground-based instruments and typically have elevation limits of at best 10°, limiting their observable declination range to their latitude ± usable zenith angle. Sources at the edge of the declination range will also appear only for short times, limiting the (u, v) coverage and thus the image quality. East–west arrays are unable to create images of sources close to the equator, because Earth's rotation does not change the (u, v) coordinates of the source, and so the point-spread function (PSF) remains a fan beam. To simplify engineering and computation, many 20th century telescopes (e.g., VLA, Giant Metrewave Radio Telescope, GMRT) were built with antennas arranged in long straight "arms" from a central core. When a source is at a low elevation and lying "along" the arm, a large number of antennas can shadow each other, making it impossible to observe the source.

Any radio telescope with moving dishes will have a maximum slew speed, which dictates how quickly it can point from one source to another. Widely scattered targets are less efficient to observe. Most observing campaigns are therefore broken up into "scheduling blocks," which aggregate nearby sources together, along with any convenient bandpass, polarization, and temporal gain calibrators. Many telescopes provide users with software to help design these blocks in advance of observations, while others provide support staff who will design the schedule for approved observers.

The position of the source in the sky, the longitude of the telescope, and the time of year when the observations are scheduled will also determine whether the source is up during the day or during the night. The effect of day or night changes with observing frequency:

- <1 **GHz**: ionospheric layer separation at sunset and sunrise can lead to greater ionospheric effects on the data. Avoid sunrise and sunset for high-quality data. The quiet Sun is a blackbody and so is dimmer at lower frequencies, and the ionosphere may often be more consistent during the day than during the night. However, the Sun is also a variable, extended source, which occasionally produces steep-spectrum cyclotron emission and care should be taken to keep it out of the side- or main lobes.
- **1–10 GHz**: avoid observing during daytime (including sunrise and sunset) at times of high solar activity.
- **10–100 GHz**: atmospheric temperature has an increasingly large effect on the total system temperature. Phase varies due to turbulence in the troposphere. Amplitude is mostly dependent on variations in the integrated precipitable water vapor (PWV) column (i.e., atmospheric opacity; Book 1, Section 5.3.1). PWV increases as the secant of the zenith angle, and changes as clouds with varying water content move across the sky. Ideally observe in cold, dry, constant conditions, and avoid warm, wet, and rapidly-changing conditions. This is one of the reasons there are few radio telescopes built in sub- and tropical regions! An injected noise signal, or "rain gauge," can sometimes be used to measure and compensate for the incident water column.
- >100 **GHz** Telescopes that operate in this regime are located in cold, high, dry sites such as the Atacama Desert (Chile) and the South Pole to minimize the huge impact of the troposphere on observing. Per-antenna water vapor radiometers provide an additional calibration stage to account for the attenuation and phase differences over the array.

2.5.1.2 Sensitivity and Integration Time

In the radiometer equation (Equation (2.11)), the brightness fluctuation to which the radio telescope is sensitive is inversely proportional to the square root of the integration time. Thus, every doubling of sensitivity requires a commensurate quadrupling of integration time.

The theoretical thermal noise expected for an image using natural weighting of the visibility data is given by

$$\Delta I_m = \frac{\mathrm{SEFD}}{\eta_c\sqrt{n_{\mathrm{pol}}N(N-1)t_{\mathrm{int}}}}, \tag{2.12}$$

where η_c is the correlator efficiency, n_{pol} is the number of polarization products included in the image $n_{\mathrm{pol}} = 2$ for images in Stokes I, Q, U, or V and $n_{\mathrm{pol}} = 1$ for images in RCP or LCP; N is the number of antennas; t_{int} is the total on-source integration time in seconds; and $\Delta\nu$ is the bandwidth in Hertz. Radio telescope sensitivity is given by the system equivalent flux density (SEFD, in Jansky), which is the flux density of a radio source that doubles the system temperature. Lower values of the SEFD indicate more sensitive performance. SEFD can be decreased by building more or larger antennas, lowering the system temperature, or making the antennas more efficient, all of which cost money! Because there are always overheads to observing (setting a lower limit on the observation length), and one tends to be competing with other observers (setting an upper limit on the observation length), it is the SEFD that typically defines the noise level achievable with a radio telescope within reasonable amounts of observing time.

A natural conclusion from reading Section 2.5.1 and examining Equation (2.12) might be that detection experiments are best served by observing with the lowest-resolution telescopes for the largest amount of time. However, in this regime, we will start to encounter confusion, which will limit the observation's sensitivity.

There are two kinds of confusion:

- *Classical Confusion:* well-known to observers across any wavelength who are blessed with a multitude of photons, e.g., infrared (Chapter 3). This effect arises when multiple faint sources are confused together in a low-resolution image. The noise reaches an upper limit and no further integration time will lower it. Most of the radio sky is dominated by extragalactic radio sources, with ever greater numbers at lower flux densities. If the signal-to-noise ratio (S/N) limit of a survey is σ, and the cumulative number counts of sources $N(>S) \propto S^{-\gamma}$, $\frac{S}{b} = \frac{3-\gamma}{\sigma^2}$ (Condon 1974). Longer baselines decrease the confusion floor as the square of the length of the baseline, so most radio telescopes are designed to keep the baselines long enough that confusion is not important for typical integration times of hours to days. Lower-frequency instruments have historically been more prone to confusion due to their greater difficulty in accessing finer spatial scales.
- *Sidelobe Confusion:* the noise floor that ensues when the sidelobes of sources are not properly subtracted from the observations, e.g., by a CLEAN process (see Section 2.6.2.6). Calibration errors, sensitivity to sources outside of the field of view, and less complete (u, v) coverage can all lead to increased sidelobe confusion. These factors can be difficult to simulate accurately, so often the sidelobe confusion level of an instrument is determined empirically by accumulating observing experience with a given instrument. Observing guides for the instrument in question will usually mention sidelobe confusion if it is expected to be a strong constraint on observing. One method of dealing with sidelobe confusion is to directly subtract a source model from the visibilities.

2.5.1.3 Use of Bandwidth

From Equation (2.11) we see that we can obtain greater sensitivity by integrating over a larger bandwidth. This also has positive effects on the image quality (see

Section 2.6.2.6). Most 20th century radio telescopes had small bandwidths available for each receiver because of the limitations of analog circuitry and digital computation. In the 21st century, most radio facilities have upgraded their receivers to operate over wider bandwidths, using improved digital signal processing. This has vastly improved the sensitivity of such instruments, while increasing the complexity and computational cost of calibrating and imaging the data, as well as massively increasing the data storage requirements.

Many radio facilities offer different frequency bands of observing, which usually correspond to different physical receivers. These are usually embedded in a "feed horn" at the focus which rotates the different receivers in for different observations. Often, a further digital window is set, capturing some subset of the available receiver bandwidth. Sometimes there is the option of restricting the frequency window in order to boost the time resolution or digital sampling depth of the observations (Section 2.5.1.2). In other circumstances, it may be useful to change the observing window to reduce contamination from radio-frequency interference (RFI) or the Sun, or to capture a specific spectral line.

A further consideration is the changing brightness of sources within the frequency band. Imaging such sources over a wide bandwidth can cause artifacts unless mitigating techniques are used (see Section 2.6.2.6). More usefully, the imaging strategy can aim to divide the observed bandwidth into multiple sections and examine the changing brightness of the source to measure its spectral index. "Resolved spectral index mapping" is a powerful tool, e.g., for investigating the age distribution of electrons in astrophysical plasmas. However, care must be taken to match the sampled spatial scales both while planning and imaging observations. For instance, a source whose largest extent is only just captured by the shortest baseline at the lowest frequency in the band will have a reduced apparent flux density at the highest frequency in the band, causing an artificially steep spectral index.

Finally, the field of view of the telescope is proportional to the observing wavelength (Equations (2.8) and (2.9)). Ideally, the source of interest fits inside the field of view. In the case where it is larger, special techniques (such as mosaicking) must be used to reconstruct the very largest spatial scales; for comprehensive review see Mason (2020) and https://science.nrao.edu/facilities/vla/docs/manuals/obsguide/modes/mosaicking.

2.5.2 Spectral Line Observing

In order to observe spectral lines, one would need to plan observations around four key features:

- The central frequency,
- The total velocity range,
- The optimum velocity resolution,
- The expected line intensity.

2.5.2.1 Central Frequency

The central frequency is that of the line you are interested in observing, Doppler-shifted by the systemic velocity of your source. There are two conventions for this shift:

1. The "optical" convention: $V^{\text{optical}} = \frac{\lambda - \lambda_0}{\lambda_0} c = cz$, and
2. The "radio" convention: $V^{\text{radio}} = \frac{\nu_0 - \nu}{\nu_0} c = \frac{\lambda - \lambda_0}{\lambda} c \neq V^{\text{optical}}$

where z is the redshift of the source, and λ/ν and λ_0/ν_0 are the corresponding observed and rest wavelengths/frequency, respectively.

The radio convention is more commonly used for radio observatories because their natural axis of measurement is frequency, not velocity, so radio velocities can be transformed by a simple multiplication into frequency channels.

A further velocity correction that must be made is the velocity of the observatory relative to the rest frame of the source. Depending on the source, a different rest frame should be used. For instance, the local standard of rest (LSR) is generally used as the rest frame for Galactic astronomy, whereas the barycentric frame is generally used for extragalactic work (Table 2.2).

Some observatories are able to Doppler track, shifting the observing band over the course of a scheduling block or an observing campaign depending on the calculated Earth velocity. In other observatories (e.g., Australia Telescope Compact

Table 2.2. Velocity Rest-frame Designations (Credit: https://science.nrao.edu/facilities/vla/docs/manuals/obsguide/modes/line)

Rest-frame Name	Rest Frame	Correct for:	Max. Amplitude ($km\ s^{-1}$)
Topocentric	Telescope	Nothing	0
Geocentric	Earth Center	Earth rotation	0.5
Earth-Moon Barycentric	Earth+Moon center of mass	Motion around Earth +Moon center of mass	0.013
Heliocentric	Center of the Sun	Earth orbital motion	30
Barycentric	Earth+Sun center of mass	Earth+Sun center of mass	0.012
LSR	Center of Mass of local stars	Solar motion relative to nearby stars	20
Galactocentric	Center of Milky Way	Milky Way Rotation	230
Local Group Barycentric	Local Group center of mass	Milky Way Motion	100
Virgocentric	Center of the Local Virgo supercluster	Local Group Motion	300
CMB	CMB	Local Supercluster Motion	600

Array, ATCA) you must perform the calculation yourself when setting up the observations. With the advent of wider band receivers, correctly estimating the Doppler shift in advance is less important, as the line is very unlikely to be shifted outside of the observing window by the $\approx$0.1–5 MHz shifts induced by Earth's annual and diurnal movement.

Observatories typically make available calculators to determine these values in advance of your observations. Additionally, astropy's SpectralCoord Class enables these transformations to be performed straightforwardly in Python (https://docs.astropy.org/en/stable/coordinates/spectralcoord.html).

2.5.2.2 Bandwidth

The expected velocity range for your source should be determined from existing work in the literature; for instance, HI in other galaxies can orbit with velocities of $\pm$1000 km s^{-1}, while water masers in our own Galaxy could be expected to move with velocities of $\pm$100 km s^{-1}. From this, we calculate the required bandwidth Δf via

$$\Delta f \approx \frac{f \Delta v}{c} \tag{2.13}$$

where f is the rest frequency, Δv is the required velocity range in km s^{-1}, and speed of light is $c = 3 \times 10^5$ km s^{-1}.

2.5.2.3 Frequency Resolution

The finer the frequency resolution ($\Delta \nu$), the more detail you will obtain on the line-of-sight velocity distribution of your line-emitting source. Just as with image-plane PSFs (Section 2.5.1.1), it is important to have a velocity resolution with at least three to four spectral channels over the FWHM of the typical spectral line. Undersampling the source velocity distribution will reduce the intensity of the detected line. Oversampling has little penalty: if the S/N is low in a given channel, you can always average channels together. Most correlators have a maximum number of channels or throughput determined by the number of channels multiplied by the bandwidth. You will need to work out the compromise between a high-enough frequency (velocity) resolution to sample the sources, but retaining enough bandwidth to cover the possible central frequencies (velocities) at which those sources are located. Equation (2.8) can be used to calculate the resolution in the same way. For instance, individual water masers at 22.2 GHz in our Galaxy have typical velocity distributions of $\pm$1 km s^{-1}, so a resolution of $\Delta f = \frac{\frac{1}{3}}{3 \times 10^5} \times 22.2 \times 10^9 = 25$ kHz would be required to adequately sample their individual lines.

2.5.2.4 Sensitivity

The SEFD (Equation (2.12)) is even more important to spectral line observing, because observations are vanishingly unlikely to be confusion limited in the spatial plane, and the more raw sensitivity—the finer the channels can be divided up, yielding greater velocity resolution, and reducing confusion along the spectral axis. The predicted rms (or σ) in a spectral channel i is given by

$$\sigma_i = \frac{\text{SEFD}}{k(\tau\Delta\nu)^{\frac{1}{2}}}, \tag{2.14}$$

where τ is the integration time and k is a constant determined by the instrumental sampling (Section 2.5.1.2).

2.5.3 Overheads and Calibration

The final data must be calibrated to be usable. To do this, you must include known reference sources in the observing program, interspersed with the main field(s) of interest. Section 2.6.2.1 describes the possible calibration tasks required, and different telescopes have guides as to which calibrators to use and how often to observe them. Be aware that telescopes can take time to switch targets (the "slew" time), so a larger number of targets means including a larger observing slewing overhead. Using multiple frequencies necessitates observing the calibrators at those same frequencies, again increasing overheads. And of course, all the above considerations of sensitivity, frequency resolution, and sky location apply to scheduling calibrator observations.

2.6 Processing Radio Data

2.6.1 Single Dish

When an optical image of an object is formed by reflecting light rays from a mirror onto a focus, quite sharp images of extended objects can be made. A single-dish radio telescope is not nearly as effective at detecting the fine detail in the objects it focuses on, and the radio image of an object from a single-dish radio telescope is very blurry. The blurring stems from a basic principle that affects all types of telescopes. In order to "resolve" images (i.e., make sharp images), the diameter of a telescope's collecting area must be many times greater than the wavelength of the radiation it detects. Light waves have wavelengths of less than one-millionth of a meter, and so collecting mirrors are large enough compared to a wavelength of light that they can resolve the details of objects observed. However, radio waves have wavelengths of roughly 1/10 of a meter and so even large radio telescope dishes produce blurry images. A single radio dish would have to be many kilometers across to achieve a sharp image at radio wavelengths and such telescopes have been too difficult to build.

Due to the fact the radio waves interact with matter differently than visible light, the telescopes and instruments used to detect radio waves are necessarily different. Typically, large parabolic dishes reflect and focus the radio energy onto an antenna and receiver (or "feed") sensitive to a particular radio frequency band. The antenna and receiver are similar to the antenna and receiver for a television set or a radio in a car. It is simply an instrument where the radio waves induce electrons to accelerate and thus create an alternating current that is then amplified and recorded.

One advantage of working at such long wavelengths is that parabolic surfaces with tolerable deviations are easier to obtain. Typically these surfaces (whether for optical or radio telescopes) need to be figured to within some small fraction of a wavelength. For a radio telescope operating at 21 cm, this means that variations of up to 1 cm are tolerable. For optical telescopes, imperfections must be below 25 nm.

Contemporary computing facilities offer interactive usage and high-resolution graphical displays, fast data processing, and large data storage. A number of software packages that can deal with the single-dish data exist. From the early NOD2 to the upgraded NOD3 (Müller et al. 2017); Australia Telescope National Facility (ATNF) Spectral Analysis Package (ASAP; https://www.atnf.csiro.au/computing/software/asap/userguide/), Astronomical Image Processing System (AIPS++; https://casa.nrao.edu/aips2_docs/aips++.html), Common Astronomy Software Applications (CASA package; McMullin et al. 2007; https://casa.nrao.edu/docs/UserMan/casa_cookbook009.html), etc.

ASAP is a single-dish spectral line processing package currently being developed by the ATNF. It is intended to process data from all ATNF antennas and can probably be used for other antennas if they can produce "Single-Dish FITS" format. It is based on the AIPS++ package. The CASA single-dish tool suite is built separately to the interferometer suite, having an early development history based on the ASAP, which operated on a "scantable" format. Work has been ongoing toward transiting the "scantable" format to the "Measurement Set" format in use by the interferometer CASA. In doing so, both software and development effort of CASA are shared more openly between CASA single-dish and interferometer developers, and the longer-term future development is more straightforward, without requiring very specialized knowledge.

In principle, there are four major and very simple steps when it comes to single-dish data processing (reduction). Namely, they are

1. Initial Data Inspection and Editing. Filtering out RFI in the so-called raw continuum data stream from the output channels of the polarimeter,
2. Calibration (of the Stokes parameters),
3. Imaging, and
4. Analysis.

As suggested by Müller et al. (2017), it is essential to initially correct the base levels and suppress the scanning effects including noise reduction. Here, we point to a single-beam procedure technique called basket-weaving (Figure 2.8). Basket-weaving is a technique that is used to remove scan-line patterns from single-dish radio maps; by scanning the telescope over the map area in two orthogonal directions, the effect of scan-line artefacts is diluted.

The image restoration is the final step and can additionally reduce scanning effects such as weather, atmospheric or ground radiation, and receiver instabilities. If the horns are mounted parallel to the azimuth direction and the telescope covers the field of the sky with azimuth scans, the assumption is that the horns detect the same clouds but at different positions on the sky. One of the initial methods to restore multiple-horn observations obtained with a single-dish radio telescope was introduced by Emerson et al. (1979). The method uses the difference of two maps

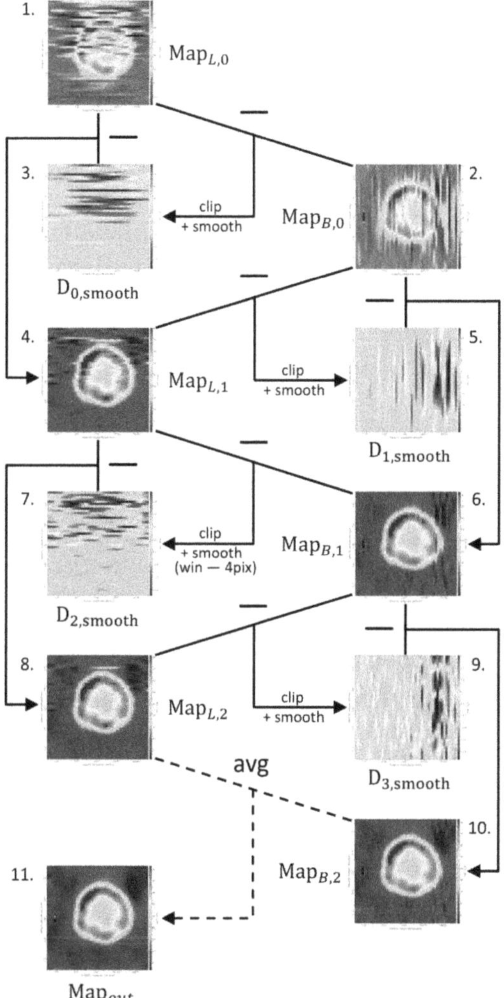

Figure 2.8. The workflow of the basket-weaving procedure, based on the Effelsberg 100 m telescope observations of the Tycho supernova remnant (at 4.85 GHz). Image credit: Müller et al. (2017) reproduced with permission © ESO.

covered by different horns, with the underlying idea that the weather effects disappear while the astronomical signal remains in difference, shifted by the horn separation (for more details, see Müller et al. 2017).

Finally, all modern single-dish software packages offer plot routines and several methods for map analysis such as an integration in sectors of elliptical rings in maps of total intensity and polarization (e.g., in the projected plane of an inclined galaxy) and box integration along strips including scale-height determinations (e.g., in edge-on galaxies).

2.6.2 Interferometry

Interferometry is a technique used to overcome the blurring and produce sharper radio images. Astronomers are able to electronically simulate the effect of a very large dish by using the combined signal from many smaller single-dish radio telescopes. The technique they use is called interferometry.

Interferometry exploits one advantage of the much larger wavelength of radio waves—use of detectors that can measure the phase as well as the intensity of the radio waves received. If a pair of single-dish radio telescopes, separated on the ground, observes a single small source, the combined output oscillates with time. This is because, as Earth rotates, the relative phase of the signals received by the two dishes varies as the distances from the source to each dish vary. This means that sometimes the radio wave from the object received by the two different dishes will add up in phase and produce a large signal and sometimes be out of phase and produce a small signal, depending on the slightly different distances from the object to each dish.

Interferometry uses the constructive and destructive addition of the radiation to determine information about the intensity and size of the object being observed. The spacing between the radio telescope dishes in an interferometer determines the size of the objects that can be resolved by the interferometer. If a source being observed covers an angle in the sky much smaller than the ratio (detected wavelength)/(twice the separation between two radio telescope dishes), then the combined output from the pair of dishes observing the source oscillates strongly with time. The ratio above is called the resolution of the telescope (see Equation (2.8)).

As the separation between the dishes increases, the angular size of the source that the telescope can detect decreases. For sources of angular size much larger than the resolution, the output ceases to vary and less information can be obtained about the sources. To study large-scale structure, astronomers need to use a lower resolution, i.e., a more closely spaced pair of dishes. An interferometer works most efficiently, in the sense of returning the most information about the intensity and size of the source being observed, for a source whose size is comparable with the telescope's resolution. Using many dishes together in an interferometer array allows us to form more complete images of objects.

2.6.2.1 Calibration

The visibilities produced by an interferometer are the raw correlations of voltages x between multiple antennas, stored as complex values associated with coordinates in (u, v), as well as time t and frequency ν. The sky and instrument can introduce effects that can change the values of these visibilities. We use the process of calibration to remove these effects and so create visibilities that are ready to transform into images (Section 2.6.2.6).

We typically express this process via the measurement equation,[8] where for a pair of antennas i and j, the visibilities are

$$V_{ij}(u_{ij}, vij) = \langle x_i(t) \cdot x_j^*(t) \rangle_{\Delta t} = J_{ij} V(u_{ij}, v_{ij}), \tag{2.15}$$

[8] A full discussion of the measurement equation is beyond the scope of this handbook, but we refer an interested reader to Smirnov (2011) for a detailed description.

Table 2.3. A Nonexhaustive List of Effects that May Need to Be Calibrated Out of Radio Observations, along with the Additional Observations Required to Make This Possible

Effect	Time Dependence	Frequency Dependence	Required Observations
Antenna positions	After array reconfiguration	Linear	Geodesy (handled by observatory)
Clock errors	Telescope dependent: ≈ never—hourly	Linear	Reference signal (handled by observatory)
Flux density scale	Constant over observation	Telescope dependent	Primary calibrator source
Delays	Constant over observation	Linear	Primary calibrator source
Antenna gains	Days—hours	Telescope dependent	Secondary calibrator source
Pointing	Telescope dependent: days—hours	Linear	Secondary calibrator on/off pointings
Polarization	Telescope dependent: days—hours	Linear	Polarization calibrator
Ionosphere	Hours—seconds	Quadratic (negligible at high ν)	Varies; field-based calibration
Atmospheric phase	Hours—seconds	Quadratic (negligible at low ν)	Water-vapor radiometer

where J_{ij} is an operator that characterizes the net effect of the observing process on this baseline, and it is this that we must solve for and remove. The deleterious effects can have different behaviors with respect to time and frequency, and whether they affect antennas individually, or apply to each baseline. Table 2.3 shows some of the possible calibration that can be performed for radio interferometers, and below we cover the typical steps for most common dish arrays such as the VLA or the ATCA.

2.6.2.2 Bandpass and Delay

The output of a correlator is at an arbitrary scale, with the input gains tuned such that the inputs do not saturate nor register only as zero. We must first set the overall scale of the visibilities such that the units are at least approximately in Janskys rather than arbitrary correlator units. As shown in Table 2.3, observations of a primary calibrator source with a well-known flux density are required. This is placed at the center of the field of view during the observation, so its visibilities when corrected have a fixed amplitude at each frequency channel and zero phase (e.g., the black points in Figure 2.9).

The corrections made at this stage are to the complex antenna gains, i.e., using our formalism defined above, $J_{ij} = J_i J_j$. To determine the values of these for antennas i, j... N, we use the fact that there are $N \times (N - 1)$ independent measurements

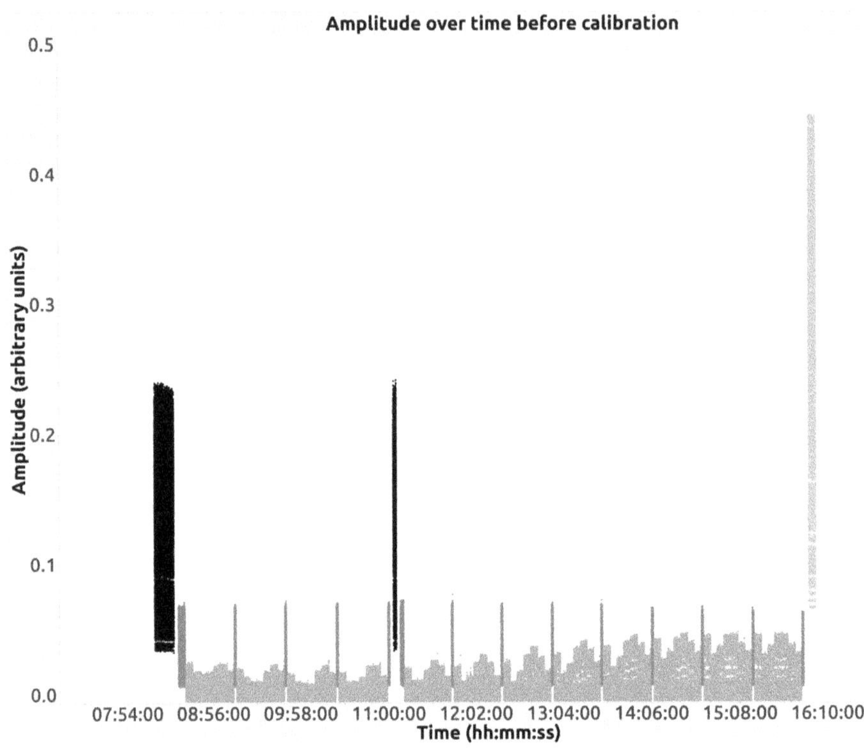

Figure 2.9. The amplitude of the visibilities with respect to time for a typical observation with the VLA. The colors indicate the planned utility of each part of the observation: the black points are calibrators used to set the overall flux density scale and delays (Section 2.6.2.2), the pink points are calibrators used to set the time-varying complex gains (Section 2.6.2.3), the green points show the polarization calibrator (Section 2.6.2.5), and the orange points show the field of interest.

(visibilities) while we only need to constrain N complex gains. We therefore reframe the problem as a minimization fit, typically performed via a least-squares algorithm. The target ("model") values of the visibilities are constant amplitude and zero phase, and the antenna gains are iteratively adjusted until the data visibilities most closely resemble this model.

The amplitude part of these complex gains typically shows the bandpass of the instrument, while the phase part of these complex gains typically shows the residual delays in the system, e.g., slow temperature-dependent changes to the electrical path length of cables connecting the antenna to the correlator. Figure 2.10 shows a single baseline before and after delay calibration.

2.6.2.3 Antenna Gains

At low frequencies (<1 GHz) a single bandpass and delay calibration is often correct for up to hours of observing, because these systems tend to have stable and simple electronics, and the local temperature does not make up a very large fraction of the total system noise, which is dominated by the sky temperature. At higher

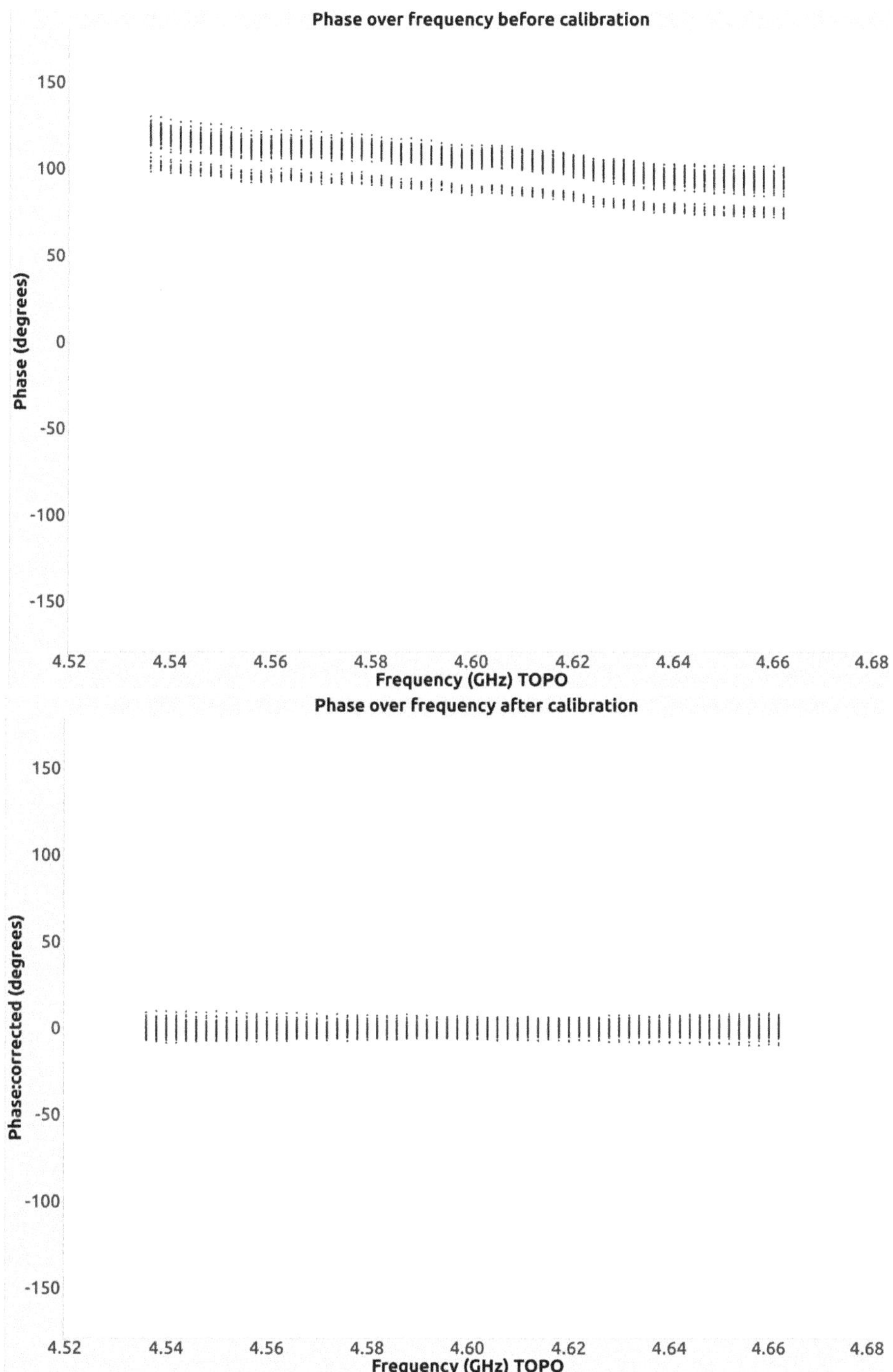

Figure 2.10. The phase of the visibilities with respect to frequency for a single baseline and single-polarization measurement of the gain calibrator of a typical observation with the VLA. The top panel shows pre-calibration data, where uncalibrated delay causes a linear slope in phase with time, and the bottom panel shows post-calibration data, where the delay has been calculated and removed.

frequencies, the properties of the instrument-sky system can vary more quickly with time, and thus calibration must be recalculated more frequently. The time between required calibration is often called the coherence time: the typical time in which external effects cause the visibility phase to change by a radian. Additional repeated calibration scans on a sufficiently bright nearby source (a "secondary" calibrator) prevent the observation from becoming incoherent and are often referred to as antenna gain or complex gain calibration.

The software process used to calculate these gains is very similar to that of calculating bandpass and delay (Section 2.6.2.2). The difference is usually that that the flux density of the secondary calibrator is not known. Fortunately, amplitude variations tend to be much slower than phase variations, and it is usually sufficient to calculate the predicted flux density of the secondary from the primary calibrator and then use this throughout the rest of the observation.

2.6.2.4 Pointing Calibration

The antennas of most radio telescopes are placed on rotating and tilting mounts which allow their antennas to be directed toward a target on the sky. Software typically transforms the RA and Dec of the source into an azimuth and elevation, which are fed to the different mount drives. However, mechanical differences in drive response, changes in temperature, and deformation of the antennas by gravity can cause the pointing to be inaccurate. Because primary beams reduce in size and resolution increases as a function of increasing frequency, pointing errors become more and more critical at higher frequencies.

These pointing errors can be removed by using a nearby reference source, which can be the same as other calibrators. Typically the observing software allows you to label the source as a pointing calibrator, and the scheduler will automatically perform a "five-point" calibration observation. In this procedure, the antennas are directed at the calibrator, then at four locations that form a cross around the target. Differences in the response away from symmetry allow the measurement of the pointing error, which can then be applied to the rest of the data. Different telescopes will have different guides to how often this procedure needs to be repeated.

2.6.2.5 Polarization Calibration

Polarization calibration has two aims: to remove any instrumental polarization and to fix the polarization angle. The procedure differs strongly from instrument to instrument, especially whether they have opposing circular feeds (RR, LL) or orthogonal linear feeds (XX, YY). In general, it is easier to use circular feeds to precisely measure linear polarization and to use linear feeds to precisely measure circular polarization.

In general, polarization calibration follows three main steps:

- (if applicable) determine and remove any instrumental delay that differs between the two polarization outputs: as per Section 2.6.2.3;
- solve for the instrumental polarization, otherwise known as the leakage or "D-terms";
- calibrate the absolute polarization angle.

There are typically two approaches to determine the leakage terms: observe at least one bright polarized calibrator source over a wide range of parallactic angles, or observe a bright unpolarized ($P<<1\%$) calibrator source in a single scan. To calibrate the absolute polarization angle, the calibrator itself must have a well-known polarization angle. Every observatory has different suggestions for their preferred polarization calibrators, but in general the calibrator S/N must be such that the calibration can converge and solve for spectral variations in instrumental polarization.

The angle describing the difference between the sky frame and the celestial source is called the parallactic angle. Some telescopes, such as ASKAP, rotate their feeds so that this angle remains constant or zero. Otherwise, data reduction software can usually calculate this value and insert it where necessary, for instance during polarization calibration. Different software packages and telescopes use different routines to perform this calibration, such as `polcal` in CASA and `gpcal` in MIRIAD. For an unpolarized source, any observed polarization is instrumental and can therefore be placed into the gain calibration. For a polarized source, the degeneracy between the instrumental and astrophysical polarization can be solved by including the parallactic angle correction: The polarized source terms will remain fixed while the instrumental terms will change. The final adjustment of the polarization angle must be made using a polarized source and can be calculated by comparing the remaining polarization angle to that of the known source.

2.6.2.6 Imaging

The (u, v) coverage of a radio interferometer is essentially continuous. In order to perform a Fast Fourier Transform (FFT) on a computer we need to turn the (u, v) points into a 2D grid. We can split the (u, v) plane up into a discrete grid and lay down the points in each cell. Then, we can average together the values in each cell and perform the FFT. This gives us a "dirty" image. Because we have Fourier-transformed, the (sky distribution × instrument function), the resulting image is the (image of the sky convolved with the image-plane instrument function, or synthesized beam).

Dirty images can be useful in some cases, e.g., if only the approximate positions of sources are required, or in the case of a detection experiment where the expected signal will have high S/N. However, most astronomers would prefer the removal of the effects of the instrument, because then it is easier to interpret the image. Note that even with perfect removal, the radio interferometric image is never a true representation of the sky.

The most common method to remove the effects of the instrument convolution, i.e., deconvolve the image, is the CLEAN algorithm (Högbom 1974). In this method, the instrument function is modeled and then removed by an iterative image-based approach. Almost all modern methods are derived from this approach.

The initial step in the CLEAN algorithm is to determine what the sky image has been convolved with: We call this the synthesized beam. In the case of a perfect instrument, i.e., a completely filled aperture, i.e., a dish, you are Fourier-transforming a wide

Gaussian, so your resulting synthesized beam is a narrow Gaussian. In the case of interferometers, it is as if parts of the dish have been removed, so the resulting synthesized beam becomes less and less like a Gaussian. The shape of the synthesized beam is given by the FT of the original weighted (u, v) distribution, with values of amplitude = 1 and phase = 0 in each cell.

The top middle panel of Figure 2.6 shows the (u, v) coverage of the VLA in D configuration (a compact layout) after performing a short observation; the resulting synthesized beam is shown in the top-right panel. The center of the synthesized beam is approximately Gaussian and peaks at unity. The positive and negative features close to the center are called the "near-in sidelobes" of the synthesized beam and are typically responsible for the strongest corrupting artifacts around bright sources. The less bright radial streaks farther out in the image are called the "far sidelobes" of the synthesized beam. While they may not be as obviously problematic as the near-in sidelobes, the influence of the far sidelobes of all of the sources in an image can combine together to produce a large amount of noise. For wide-field instruments, which are sensitive to many sources, this "sidelobe confusion" can be dominant over thermal noise unless careful deconvolution is performed. The image at the bottom right is a "dirty" image—a Fourier transform of the observed (u, v) data (bottom-middle panel), and it is this that we wish to CLEAN.

In the CLEAN algorithm, the following steps are performed iteratively until a stopping condition is reached:

1. The pixel in the image with the greatest absolute value of brightness is identified;
2. At this location, an image of the synthesized beam is subtracted, multiplied by the brightness of the pixel and the "minor loop gain," which is typically about 10%;
3. The value of the brightness multiplied by the minor loop gain is added to an (initially blank) model at that pixel location.

This iterative process results in the removal of all structure from the dirty image, eventually producing a noise-like image called the residual image. Stopping conditions for CLEAN include (roughly in order of increasing utility):

- The number of iterations performed ("niter");
- A cutoff value in brightness of pixels selected ("threshold");
- Selecting a pixel with negative brightness ("-stop-negative");
- The residual image reaching some prescribed level of noise;
- The pixel brightness reaching some level of S/N compared to the residual image.

The model image and residual image are of limited interpretability: the model image has the same resolution as the pixel grid, which is higher than the spatial scales that are truly sampled by the interferometer, and is in the generally unhelpful units of Janskys per pixel. The residual image should contain mostly noise and any faint extended structures below the deconvolution threshold, where their peak brightness was not enough to trigger a CLEAN iteration.

To make our images easier to interpret, we wish to form an image that combines the faint extended structures in the residual image, with the bright detections made in the model image, with a representation as close to possible of the true spatial frequency sensitivity of our instrument. This is the process of "restoring" our image. To do this, we fit a 2D elliptical Gaussian to the center of the synthesized beam image. For well-behaved synthesized beams where the (u, v) coverage of the observation has been selected to be sensitive to a good range of spatial scales, this Gaussian fit produces a useful "restoring" beam.[9]

To restore the image, we convolve the model image with the restoring beam and add it to the residual image. The resolution of the resulting "clean" image is the FWHM of the restoring beam. Typically, the parameters of the Gaussian fit made to the restoring beam are added to the FITS header of the image with three keywords: BMAJ, BMIN, and BPA, and are the size of the restoring beam in sky coordinates at the phase center of the image. Figure 2.11 shows a flow diagram of the CLEAN process in action.

Another phrase sometimes used to describe the resolution of a radio image is the PSF, which is derived from its equivalent use in optical astronomy, where it describes how much the light from a point (e.g., a star) is spread over the CCD. Its shape is determined by the instrumental characteristics, and for ground-based instruments, the atmosphere. However, in radio astronomy, "PSF" can be used to describe either the synthesized beam, the restoring beam, or the shape of point sources after many images have been stacked together. This can make it somewhat confusing to new users, so for the purposes of this guide, we use it only to describe the final situation, where it is most analogous to the optical case.

We can change the weighting of the cells in the gridded (u, v) data in order to produce different images that may be more useful for different science goals. Natural weighting means giving more weight to the cells with more points in them. This retains the full sensitivity of the instrument, and the resulting image will have the lowest possible noise, ignoring other effects. It will also be most sensitive to the largest-scale structures in the sky sampled by the interferometer: by geometry, there are always more short baselines than long, so natural weighting may help detect larger and fainter objects. However, the synthesized beam will typically have the most dramatic sidelobes, dictated entirely by the (u, v) coverage of the array. This may make deconvolution more difficult, increasing sidelobe confusion, complicating the calculation of the restoring beam, and potentially limiting final image quality.

Filling the (u, v) plane with cells of identical weight gives the most well-behaved synthesized beam. To that end, a uniform weighting scheme can be used, in which, no matter how many (u, v) points were added to a cell, it retains the same weight during the FT. This of course reduces the sensitivity of the final image by quite large amounts for heterogeneous (u, v) distributions. However, it typically does produce a synthesized beam with less bright sidelobes, which immediately reduces the sidelobe noise in the dirty image, makes the center of the beam more Gaussian and thus

[9] In the case of poor (u, v) coverage, e.g., VLBI short observations on an E–W array with a high concentration of points in very different parts of the (u, v) plane, this process may not work optimally.

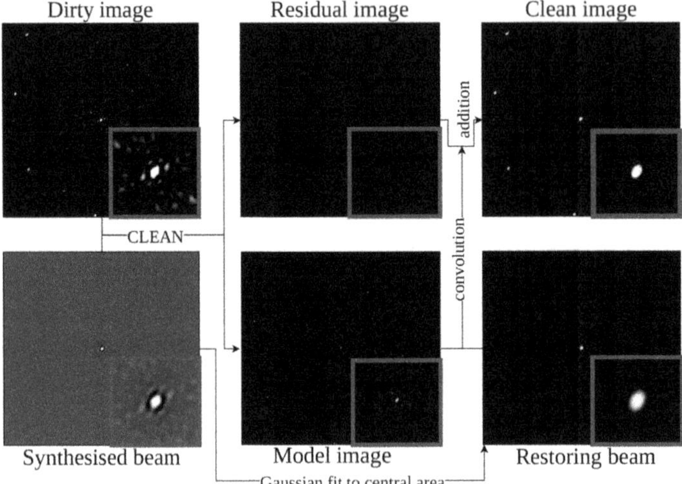

Figure 2.11. A 3° × 3° section of sky as observed by the Murchison Widefield Array at 150 MHz in a two-minute snapshot, going through the CLEAN deconvolution and restoration process. The inset boxes at the bottom right of each panel show a 10′ × 10′ zoom on the center to illustrate the process for a single source. To enhance details, the images have different color scales; the top row is set to −0.1–1 Jy beam^{-1}, the bottom left to −0.2–0.5 Jy beam^{-1}, the bottom middle to 0–0.1 Jy pixel^{-1}, and the bottom right to 0–1 Jy beam^{-1}. A Fourier transform of the (calibrated) (u, v) data yields the dirty image at the top left, and a Fourier transform of the (u, v) coverage (i.e., amplitudes set to unity and phases set to zero) yields the synthesized beam at the bottom left. The CLEAN process iteratively removes the synthesized beam from the dirty image, creating a set of model components (bottom middle) and noise-like residuals (top middle). To produce the final clean image (top right), we convolve the model image with the restoring beam (bottom right), then coadd this with the residuals.

better represented by the restoring beam, and often improves deconvolution. It typically also increases the resolution of the final image: due to geometry, there are always fewer long baselines than short, and in a natural scheme, these contribute less weight to the final image. In a uniform scheme, those sparsely filled (u, v) cells are given the same weight as those more filled by short baselines, and so the final resolution is higher.

Most commonly used in radio astronomy today is the Briggs weighting scheme, which forms a balance between natural and uniform weighting (http://www.aoc.nrao.edu/dissertations/dbriggs/). The actual weighting scheme used is

$$W_i = \frac{\omega_i}{1 + W_k f^2}, \tag{2.16}$$

$$f^2 = \frac{(5 \times 10^{-R})^2}{\dfrac{\Sigma_k W_k^2}{\Sigma_i \omega_i}}, \tag{2.17}$$

where W_k is defined as in uniform and superuniform weighting, and a robustness R of +2 is equivalent to natural weighting, while −2 is equivalent to uniform weighting. Depending on the (u, v) coverage of the array, and the science case being performed, users may choose their own robustness parameter, with the typical drivers of the compromise being raw sensitivity and resolution.

The original Högbom CLEAN is rarely used today: the full synthesized beam must be shifted and calculated many tens of thousands of times in order to deconvolve an image, which becomes very expensive when the images are large. The most important advance was made by Clark (1980) who implemented two major changes: calculating and subtracting a spatially truncated synthesized beam instead of the entire thing, which dramatically increases the speed of the process, and implementing major cycles, in which the clean components found in a set of minor cycles are removed in a single step from the visibilities, which makes up for the poorer identification of the correct components using only a truncated synthesized beam.

Major cycles have their own gain factor, which can be much higher than the minor cycle gain. It is typically set between 80% and 90%, although if the PSF has high sidelobes, lower values may be a better approach and sometimes is increased to 95% or even 100% if speed is more important than image quality.

A further advance enabled by improved computing was the advent of multiscale CLEAN (Cornwell 2008), where instead of using a delta function to represent each cleaned component, a set of different scale functions (typically Gaussians) could be used. This vastly improved the handling of diffuse structure, which could be fitted with a few large functions rather than tens of thousands of delta functions. However, multiscale clean can be prone to develop nonlinear oscillations, for instance choosing to clean an increasingly large negative component on one scale and an increasingly large positive component on another scale, such that the cleaning never converges and the noise approaches infinity. Reducing the major gain is often the solution.

Up to this point, we have considered only the narrowband case, where all the visibilities and resulting images are at a single frequency. Especially in modern radio astronomy, receivers are rarely narrow band, and the visibilities typically include a spectral dimension. The effect on the (u, v) coverage is to improve it if the information can be combined in a satisfactory way.

Using the simple techniques previously introduced, the spectral information can be used in two different ways:

1. To make a measurement of the observed region at two or more different frequencies,
2. To combine together the information to improve the (u, v) coverage of the data, and therefore the sensitivity of the final image (multifrequency synthesis or MFS).

Which approach is taken is a function of the science case and the instrument: are the sources of interest spectrally smooth across the bandwidth studied? Does the instrument have significant chromatic effects (e.g., primary beam shape and sensitivity) over that bandwidth? The need to disentangle spectral structure or to account for instrumental chromatic effects may push the user toward forming more spectral channels.

2.6.3 Combination of Single-dish with Interferometer Observations

One of the major and a well-known problem that large holes in the (u, v) plane of interferometric data causes "missing flux" (a.k.a. zero spacing) because telescope arrays do not cover all short baselines. This missing flux can be added from a large single dish such as the Parkes or Effelsberg radio telescopes. Ideally, the receiver frequencies should cover the same range at both observations. Instead of combining both maps in the Fourier Plane as in AIPS or CASA, Müller et al. (2017) proposed a straightforward combination method in the image plane. The resulting map is similar to the result of "feathering" in CASA.

2.6.4 Single Dish versus Interferometer

It is certainly true that both have their advantages and disadvantages. In essence, it is up to us, the observers, to choose the right "tools" for the job. Often, a combination of both tools may be required in order to do good science.

The fundamental characteristics of single-dish observations are good potential sensitivity to large-scale structure and lack of sensitivity to fine structure, or high spatial frequencies. Single dishes have a high spatial frequency cutoff in resolution set by the antenna diameter. On the other hand, the interferometer has a low spatial frequency cutoff set by the minimum antenna separation. However, sometimes the interferometer's low-frequency cutoff is advantageous.

Usually, single-dish maps are analyzed in a way that removes the lowest spatial frequencies too. The relative flux in low spatial frequencies is typically far greater than that at higher spatial frequencies

According to Emerson (2015), practical advantages of single-dish observing are:

- Spatial frequency response,
- Sensitivity,
    1. Sensitivity in Jansky (point source) depends on collecting area, single-dish or interferometer,
    2. Sensitivity in brightness temperature in Kelvin (extended emission) gets worse as (Max. Baseline) squared, for the same collecting area—i.e., roughly as $(d/\mathrm{D})^2$
        - 100 m single dish: $\sim$2 K Jy^{-1}
        - 1 mile max baseline aperture synthesis telescope: $\sim$1600 K Jy^{-1}

- Ability to map very extended areas quickly,
- May provide large collecting area with manageable electronic complexity,
- Simplicity: one receiver, not N receivers, nor $N(N-1)/2$ correlations,
- Flexibility,
- Relative ease of upgrading, customizing hardware to an experiment,
- Relative ease of implementing imaging arrays, including bolometers,
- Relatively modest investment to use multifrequency receivers,
- A single large dish can add significant sensitivity to (e.g.) VLBI arrays, and
- Software can be simpler: "conceptually" easier to understand for novice astronomers.

And, practical disadvantages of single-dish observing (Emerson 2015) are:
- Spatial frequency response,
- Mechanical complexity replaces electronic complexity,
- Susceptibility to instrumental drifts in gain and noise—don't have the correlation advantage of interferometers,
- Interferometers can in principle give high sensitivity and high total collecting area, and
- Aperture synthesis imaging is a form of multibeaming—arguably obtaining more information from the radiation falling on a telescope than is possible with a single dish.

The overall key parameters are spatial frequency response and relative complexity.

2.6.5 Analogies and Terminology

Here, we present some of the analogies (Table 2.4) and terminologies (Table 2.5) in radio astronomy that may correspond to optical astronomy according to Ekers (2015).

2.7 Radio Astronomy of Astrophysical Sources

We examine the astrophysical sources that can be studied in the radio domain. The brief outline presented here follows Pannuti (2020); Pannuti's highly accessible volume goes into much greater detail with copious example calculations. It is recommended for both students and professionals in other fields who want to learn more about radio astronomy.

2.7.1 Solar System Radio Astronomy

Solar system radio astronomy includes all of the objects the solar system contains: Sun, planets, comets, asteroids, and dust. Multimessenger astronomy might not only

Table 2.4. Analogies Between Radio and Optical Astronomy (Table Credit: Ekers 2015)

Radio		Optical
Grating responses	$\Longleftrightarrow$	Aliased orders
Primary beam direction	$\Longleftrightarrow$	Grating blaze angle
Visibility (u, v) plane	$\Longleftrightarrow$	Hologram
Bandwidth smearing	$\Longleftrightarrow$	Chromatic aberration
Local oscillator	$\Longleftrightarrow$	Reference beam

Table 2.5. Terminology Used in Radio Astronomy that Corresponds to Optical Astronomy (Table Credit: Ekers 2015)

Radio		Optical
Map	⟺	Image
Source	⟺	Object
Image plane	⟺	Image plane
Aperture plane	⟺	Pupil plane
(u, v) plane	⟺	Fourier plane
Aperture	⟺	Entrance pupil
(u, v) coverage	⟺	Modulation transfer function
(u, v) visibility plane	⟺	Hologram
Dynamic range	⟺	Contrast
Phased array	⟺	Beam combiner
Correlator	⟺	No analog
No analog	⟺	Correlator
Receiver	⟺	Detector
Taper	⟺	Apodize
Self-calibration	⟺	Wavefront sensing (Adaptive optics)

include the study of electromagnetic emission from these objects but may also include augmented studies from space probes that can directly analyze them.

Due to the relatively small distances in the solar system, radar studies are also a feasible means of study. Objects studied by radar might include mountains on Venus, water deposits on Mercury and the Moon as well as more exact measurements of distances to these astrophysical objects. We also note a number of attempts to detect nearby comets as described in Jones et al. (2006). Radar observation may be monostatic where the same antenna is used or bistatic whereby the emitting and receiving antenna are different.

Radio astronomy began its advancement after Jansky and Reber during and after the second world war when radar first came into use (Book 1, Chapter 1). Initial observations of the Sun occurred when radar operators in England were confused by solar emissions that interfered with their equipment. The evolution of radar and surplus radio equipment after the war allowed researchers to explore and expand the field of radio astronomy, especially in Great Britain, Australia, and the USA. Continuum radio emission over their extent allowed measurements of flux densities from blackbody, synchrotron, or bremsstrahlung radiation found within the solar system.

2.7.1.1 The Sun

Main-sequence stars are relatively weak sources for radio detection due to the combination of blackbody surface temperatures near 10^4 K and small angular extents. The Sun and a small number of nearby stars are exceptions because they are much closer.

Components of radio emission from the Sun have both a thermal blackbody and nonthermal synchrotron component. The effective blackbody temperature of the Sun is about 5700 K, corresponding to its photosphere. The photosphere is considered the surface of the Sun while layers above it, including the chromosphere and corona are considered the solar atmosphere.

The thin corona in the Sun's atmosphere is by contrast a source of synchrotron emission. This layer is composed of ionized plasma at a few million Kelvin and a density of order 10^{15} particles m^{-3}, currently believed to be energized by solar magnetic fields. Radio observation frequencies may be as high as 30 GHz and vary over time, tied to the 11 year sunspot cycle.

2.7.1.2 Planets

In addition to radar observations of the terrestrial planets and moon, the observed radio emission for these objects correspond to their equivalent blackbody temperatures and angular extent. Radar observations of planets such as Venus have allowed the determination of properties such as its retrograde rotation, compared to Earth (Goldstein & Carpenter 1963).

There is a thermal component to the radio emission of the gas giant planets, also modeled as blackbodies with temperatures of about 100 K. These giant planets have a nonthermal cyclotron component that dominates longer wavelengths and extend beyond their apparent angular extent. These bodies, including Earth, have a dynamo effect that produces large magnetic fields. This is due to an electrically conducting liquid that has both circular and convective motion in the planet's core. For Saturn and Jupiter, this liquid is metallic hydrogen. For Uranus and Neptune, this liquid may be a mixture of water, ammonia and methane. The nonthermal emission from Jupiter may be very complex in comparison to other planets, due to charged particles created from the ejection of sulfur dioxide from volcanoes on Io (see Figures 2.12 and 2.13).

Radio emission from smaller bodies in the solar system may be modest at best but can yield information about their evolution. For example, thermal emission from comets may be small, but rotational molecular transitions (e.g., hydroxyl radicals or cyanide compounds) may tell us about their composition.

2.7.2 Galactic Radio Astronomy

Some aspects of stellar evolution from the ISM to end-stage interstellar enrichment can be studied in radio. Both discrete and diffuse sources can be found in our Milky Way.

2.7.2.1 HI

The ISM can have such a low temperature and density that hydrogen can exist in its atomic state. The electron in this lowest energy level ($n = 1$) of atomic hydrogen doesn't encounter enough collisions to promote it to higher levels. However, the state of the electron's spin can be parallel or antiparallel to the spin of the proton. The difference between the lower energy antiparallel state and the higher energy parallel state is very low ($\approx 10^{-24}$J), corresponding to a wavelength of 21 cm. The rate

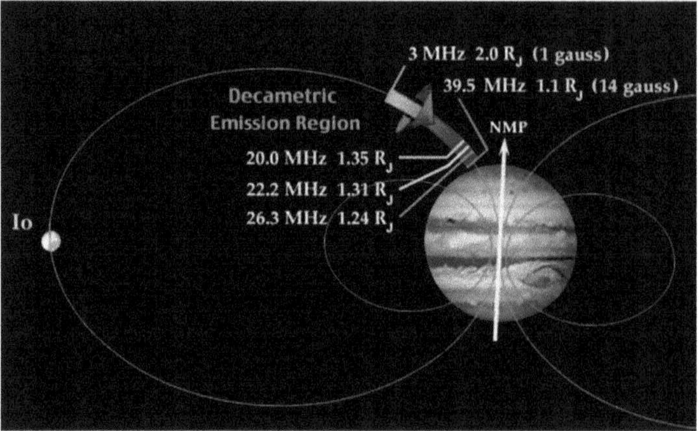

Figure 2.12. Electrons spiraling in Jupiter's magnetic field are thought to be the cause of decametric radio waves with frequencies between 10 and 40 MHz. Because frequency depends on the strength of a magnetic field, we can estimate the strength of Jupiter's field. Image credit: Wikipedia: Gge002, NASA Radio JOVE.

of this spin–flip transition is only 10^{-15} s^{-1} with a corresponding excited lifetime of 10^7 years.

These gigantic clouds of cold interstellar hydrogen are so large that we can detect and study the 21 cm radio signal they emit (see Figure 2.14). This is because the excitation temperature necessary for the spin–flip transition is quite small, 0.069 K. It is easy to see that the CMB with a temperature of 3 K can easily excite large fractions of these atoms. The ability to use the 21 cm emission from atomic hydrogen thus provides a powerful tool to estimate the motion, temperature, mass, and density of these clouds.

2.7.2.2 Star Formation Sites

Star formation occurs within molecular clouds found within the ISM when shocks, often from supernova explosions, trigger gravitational collapse.[10] Because the resulting protostars are contained in dense material opaque to shorter wavelengths, longer wavelengths such as infrared and radio are necessary to study them. Such radio telescope arrays as ALMA, operating at millimeter wavelengths, are able to study these regions.

Molecular clouds have densities over 10^8 m^{-3} and a temperature of about 15 K. Because molecular hydrogen produce few spectral lines, CO (carbon monoxide), the second most common molecule in the universe, produces molecular rotational transitions that can generate several spectral lines, the lowest at 115 GHz. Using an

[10] The virial theorem can be applied to stable systems, relating the total energy E, kinetic energy K and potential energy U of the system. If initially all of the particles are separated at large distances ($U = 0$) at rest ($K = 0$), as the cloud collapses from gravity, the kinetic energy increases while the potential energy decreases. At equilibrium, $E = K - 2K$, the system radiates away half of its potential energy, leaving a bound system such that $E = -K$.

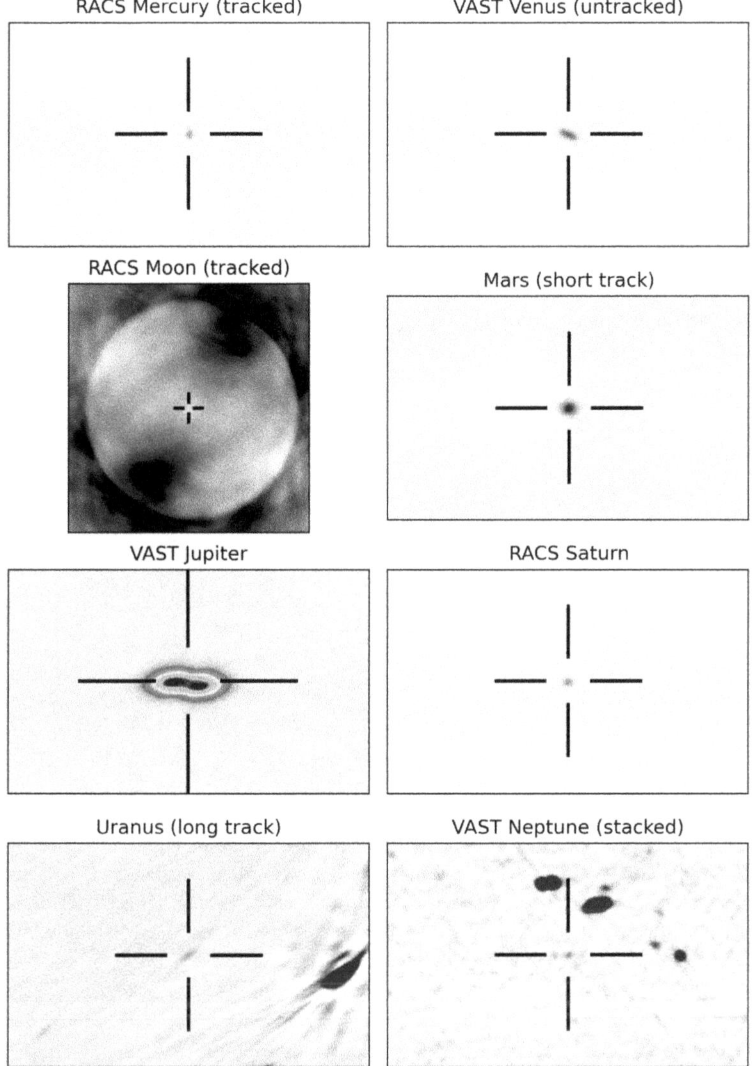

Figure 2.13. ASKAP images of the solar system planets and the Moon. Image credit: Emil Lenc (CSIRO).

assumed ratio between CO and molecular hydrogen can allow molecular clouds to be quantified.

Observations of molecular clouds reveal discrete areas of emission from hydroxyl radicals (OH) or water (H_2O). These are areas of microwave amplification by the stimulated emission of radiation, or simply masers. In these situations, molecules can be excited or pumped up and then stimulated to deexcite in a coordinated fashion, creating very intense, narrow, singular-frequency emissions.

2.7.2.3 HII Regions

In HII regions, the majority of hydrogen atoms are in the ionized state creating a plasma of electrons and protons. These are found in stellar nurseries where the most

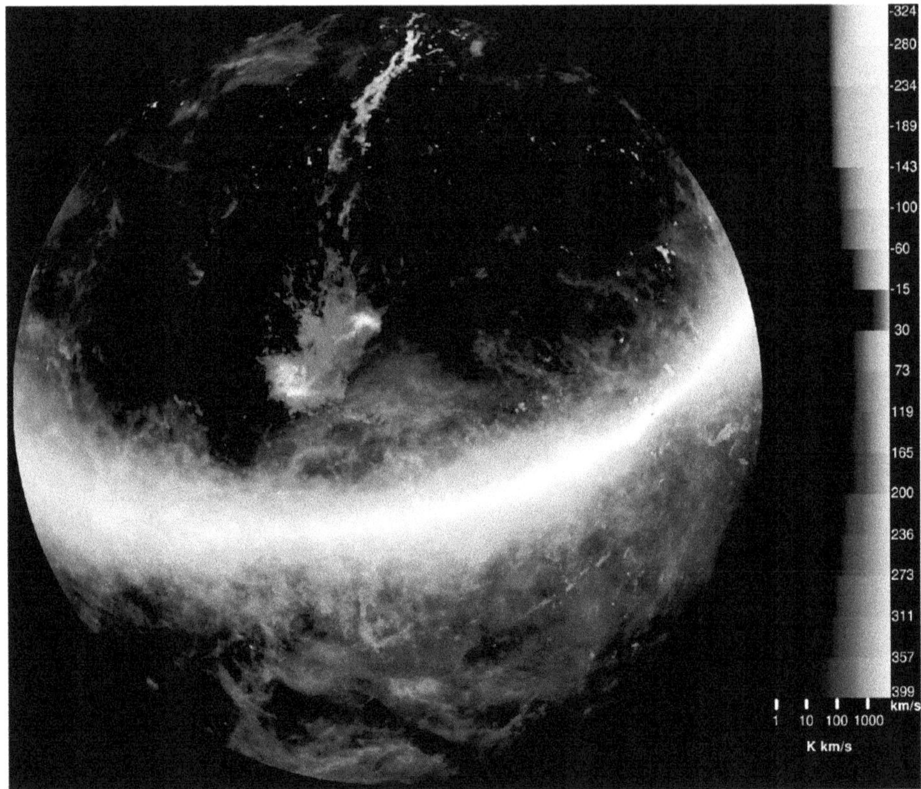

Figure 2.14. The Parkes Galactic All-Sky Survey (GASS) is a survey of Galactic atomic hydrogen (HI) emission in the southern sky. The map is in zenithal equal-area coordinates, centered on the South Celestial Pole. The Galactic Plane stretches across the middle of the map, with the Magellanic Clouds above the center, and the Magellanic Stream stretching upwards. Brightness corresponds to the HI line intensity, while color denotes the radial velocity. Image credit: McClure-Griffiths et al. (2009).

massive stars (O and B type) are created. According to Wien's displacement law for blackbodies (Book 1, Chapter 2), O and B stars emit large amounts of UV light because of their high surface temperatures. The high-energy UV photons can ionize any hydrogen atoms nearby, creating these HII regions.

Within these HII regions, radio detections include those from line and continuum emission (bremsstrahlung). When electrons in these regions recombine with protons, certain transitions are allowed (high probability) and others are forbidden (low probability). Photons detected in the radio domain tend to be higher-level transitions; transitions to low levels can produce visible or UV light (e.g., Balmer and Lyman series). For example, a transition between the 110th and 109th level produces the H109α line commonly studied in radio at 5.03 GHz. Densities can be calculated by comparing the ratios of fluxes from two closely spaced transitions.

Continuum radio radiation from these regions comes in the form of bremsstrahlung emission. The spectrum from this radiation is described as thermal and is flat when flux density is compared to frequency. This can be contrasted to the steep synchrotron radiation found in SNRs which can be described as nonthermal. In fact, this is a common tool used to help distinguish these two types of sources in radio.

The thermal bremsstrahlung spectra of HII regions have two parts separated by a turnover frequency. At low frequency, the HII region becomes opaque with self-absorption decreasing the emission detected. At higher frequencies, the HII region is transparent, allowing emission to be detected. The turnover frequency is useful in that it allows the calculation of the number density of particles in the HII region. In Figure 2.15, we show the largest and brightest HII region complex in our Local Group of galaxies—30 Doradus in the LMC.

2.7.2.4 Maser Emission from Evolved Stars

Stars depleted of their core hydrogen leave the main sequence to become red giants or supergiants depending on their mass. These evolved stars undergo significant mass loss that is not well understood. The outer atmospheres of these stars contain silicon monoxide (SiO) masers that allow detailed studies of thermal motions and magnetic fields using high-resolution radio observations. The masing transition of SiO at 43 GHz facilitates a better understanding of proper motion, redshift, and blueshift within the stellar atmosphere. Polarization studies also allow insights into the orientation of magnetic fields. These fields serve to split energy levels via the Zeeman effect; polarized spectral-line imaging gives insight about the magnitude of these magnetic fields. Using these ideas, it was found that mass loss seems greatest in regions where the magnetic field is disrupted (Kemball & Diamond 1997). In Figure 2.16, we show Very Long Baseline Array (VLBA) high-resolution observations of water vapor and silicon monoxide masers in the protoplanetary nebula OH 231.8+4.2 (Desmurs et al. 2007).

2.7.2.5 Planetary Nebulae

Planetary nebulae (PNe) are emission nebulae consisting of an expanding glowing shell of ionized gas ejected from Red Giant stars nearing the end of their stellar life cycle. This end stage for stars with masses between 1 and 8 $M_{\odot}$, consists of a core PN nucleus that emits UV light that then energizes the ejected shell and reradiates that light into magnificent optical colors.

These sources consist of ionized, neutral, atomic, molecular and solid states of matter in diverse regions with different densities and morphological structure having temperatures ranging from 10^2 to more than 10^6 K (Filipovic et al. 2009). They radiate from the X-ray to the radio domain with detection influenced by selection effects from intervening dust and gas, instrument sensitivity and distance.

Most known PNe are weak thermal radio sources and although morphologies of these objects are similar to their optical counterparts, radio interferometric observations allow us to image the structure of their ionized component. For example, a

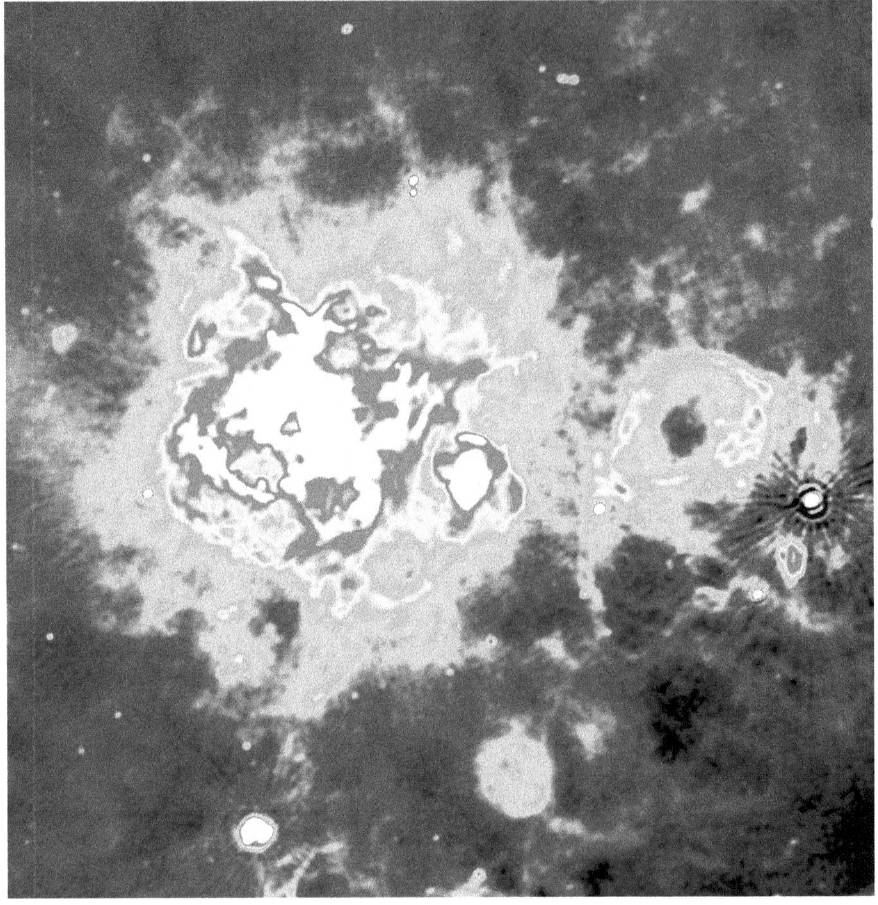

Figure 2.15. The ASKAP 888 MHz image of the largest H II region in the Local Group of Galaxies—30 Doradus (LMC). The bright (strong) point source toward the edge of the right hand side is SN 1987A.

spherically symmetric uniform-density PNe has an ionized mass, M_i, that can be expressed as

$$M_i = 282(D_{\mathrm{kpc}})^2 F_5 (n_e)^{-1} M_\odot, \tag{2.18}$$

where D_{kpc} is distance (kiloparsecs), F_5 is the radio flux density at 5 GHz (Jansky), and n_e represents the electron density (cm^{-3}) derived from forbidden-line ratios.

The current number of known Galactic PNe is ~3.5×10^3 as compiled in the HASH database and research platform (Parker et al. 2016; Bojičić et al. 2017), while estimates of the total number of Galactic PNe vary from 10,000 (Jacoby 1980) to 46,000 (Moe & De Marco 2006). Although the majority of known Galactic PNe (~90%; Frew & Parker 2010) are detected and successfully observed in optical (narrow) bands, this method is probably reaching its limits. Our current understanding of the field predicts that the majority of not-yet-detected Galactic PNe are either too distant, too evolved (and so of very low surface brightness), or lie at low

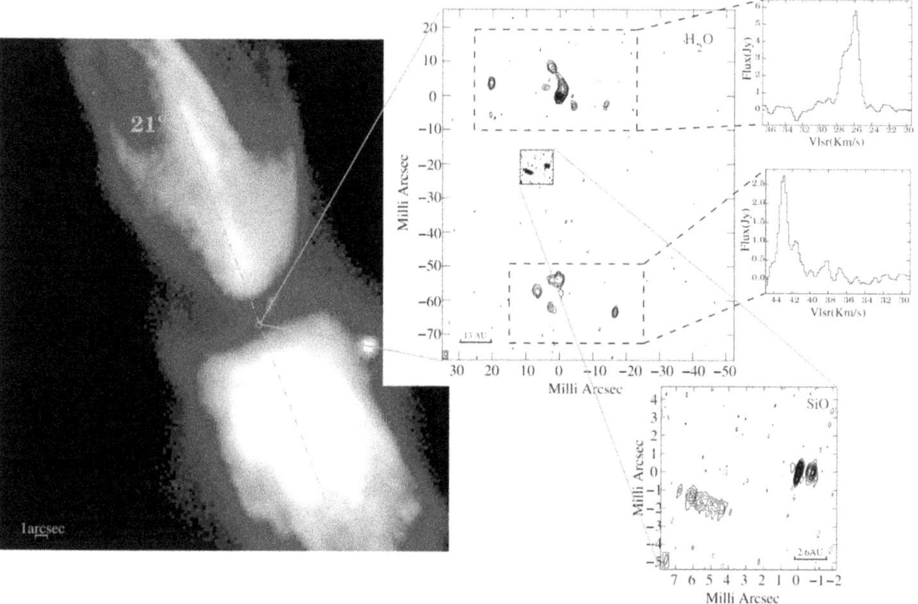

Figure 2.16. Masers at the heart of PNe OH231.8+4.2. Image credit: Desmurs et al. (2007) reproduced with permission © ESO.

galactic latitudes and are heavily obscured in optical bands by interstellar dust, or a combination of all factors.

New detection and confirmation methods have been employed in recent years (Gledhill et al. 2018; Fragkou et al. 2018; Irabor et al. 2018; Anderson et al. 2012; Parker et al. 2012a). Most, if not all, of these methods are based on radio and mid-infrared properties of PNe (Figure 2.17). Radio continuum based measurements in particular are expected to become a critical tool in future studies as they are effectively insensitive to the presence of intervening dust. Hence, the next generation of high-resolution interferometric radio continuum all-sky surveys will likely become the preeminent tool to estimate angular diameters of PNe. The key here will not be just the ability to peer through the dust and detect new, obscured PNe but the inherent higher angular resolutions that these radio surveys provide compared to ground-based optical data—this is a critical factor. Finally, we also mention our very successful search for radio PNe in the nearby Small and Large Magellanic Clouds (MCs; Filipovic et al. 2009; Leverenz et al. 2016, Joseph et al. 2019).

2.7.2.6 Microquasars

Binary systems with stellar-mass black holes and a less-evolved stellar companion can form accretion disks heated by friction to more than a million Kelvin, producing blackbody X-ray emission. These objects can be thought of as scaled-down quasars in which an accretion disk forms around supermassive black holes found at the center of most galaxies. A combination of magnetic fields and particle acceleration in jets associated with these objects gives rise to synchrotron emission in the radio domain.

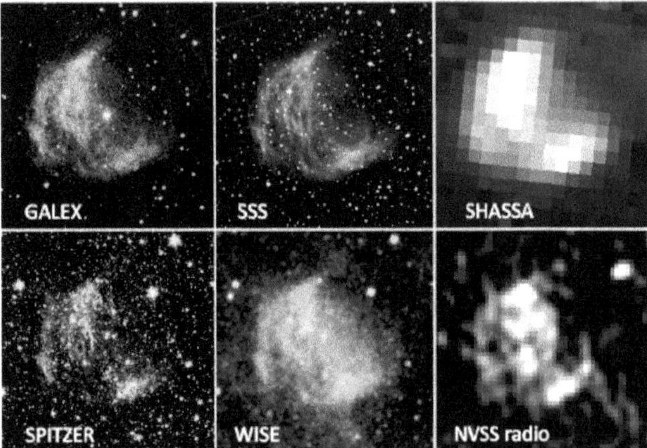

Figure 2.17. Multiwavelength six-image montage of PNe Abell 21 a.k.a. the Medusa Nebula. Image credit: Parker et al. (2012b) with permission of Cambridge University Press.

These jets are even seen to precess around the axis perpendicular to the accretion disk creating a corkscrew pattern. An example of a microquasar is shown in Figure 2.18.

2.7.2.7 Supernova Remnants

As explained in Section 1.2.2, supernovae represent the violent end of stars or star systems when the Chandrasekhar limit of 1.4 solar masses is exceeded. Type Ia supernovae occur in binary systems when an evolved white dwarf accretes enough material from a companion star, creating a single degenerate supernova, or when two evolved stars in the form of white dwarfs collide creating what is known as a double degenerate supernova.[11]

Supernovae may also occur when high-mass stars greater than 8 solar masses can no longer generate enough energy in their cores to prevent collapse. Depending on mass, the end result of these explosions may either be a neutron star or black hole. Based on how these stars evolve and release mass prior to the explosion, these supernovae may be classified as Type Ib, Type Ic, or Type II. There are about a dozen historical supernovae observed in our Galaxy over the past 2000 years and almost 350 Galactic SNRs known to exist.

The expanding shock of stellar ejecta from these explosions sweep up and enrich the surrounding ISM. This expanding shock front with swept-up material is known as an SNR and is a strong source of synchrotron emission at radio frequencies. The synchrotron radiation is emitted because energetic electrons interact with compressed and amplified magnetic fields. The similarity between the spectrum of cosmic-ray electrons and this synchrotron radiation suggests SNRs are the origin of cosmic rays, at least to the knee (10^{15} eV) of its spectrum. The mechanism for this is diffusive shock acceleration (first-order Fermi acceleration) at the expanding shock

[11] A kilonova is thought to occur when two neutron stars collide and may be a source of high-mass elements such as gold (see Section 4.4.1).

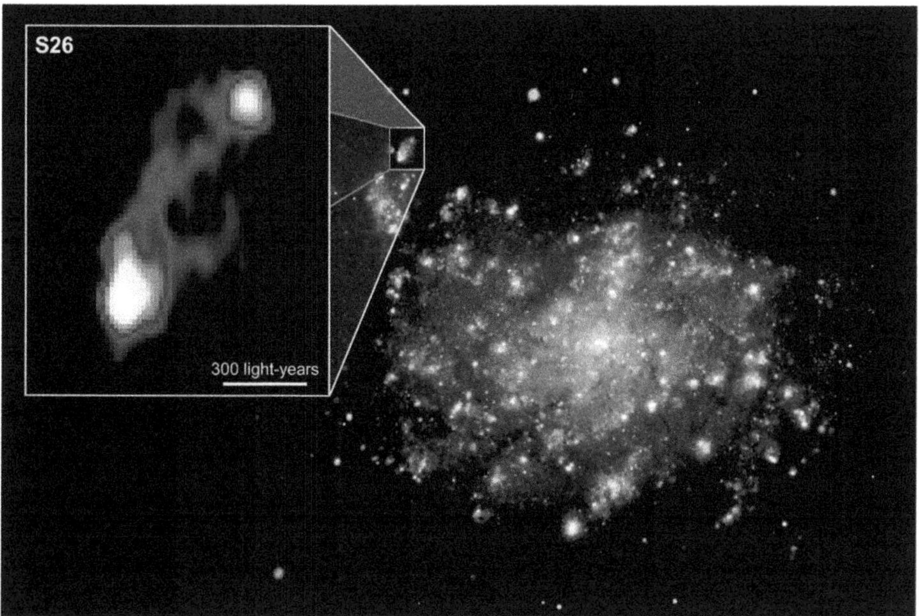

Figure 2.18. A composite image showing the position of the microquasar S26 in the Galaxy NGC 7793. The image of S26 is a radio image, made with a CSIRO ATCA radio telescope while the image of the Galaxy is made from combined X-ray and optical data. Image credit: Roberto Soria.

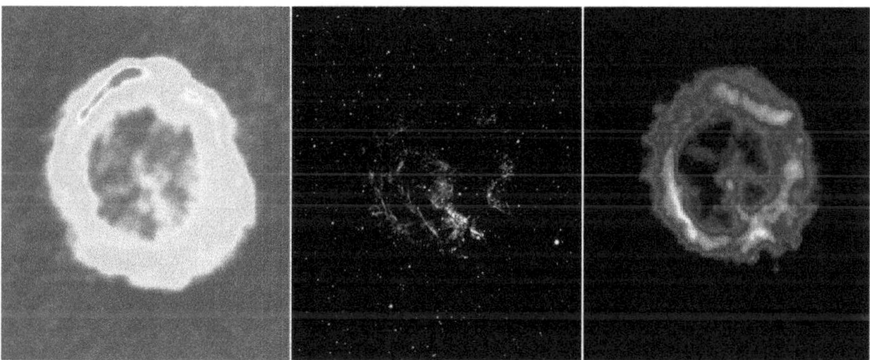

Figure 2.19. ATCA images of the Small Magellanic Cloud (SMC) SNR 1E0102–72 at 5.5 GHz (left), HST (middle), and Chandra (right).

of the remnant whereby particles trapped by magnetic fields cross over the shock multiple times until they gain enough energy to escape at relativistic speeds.

In summary, SNRs represent the mechanism by which the products of stellar evolution enrich the ISM, trigger star formation by interactions with molecular clouds and produce cosmic rays. The physics of the four stages of SNRs, including free expansion, Sedov or adiabatic, radiative, and dissipative, are the subject of intense study, especially in the radio domain but also in other messengers (e.g., X-rays and gamma rays; Sections 6.5.1 and 7.5.1).

2.7.2.8 Neutron Stars and Pulsars

Neutron stars are the degenerate core remains of massive supernovae with a radius of only 10 km and a mass between 1.4 and 3 $M_{\odot}$. Conservation of angular momentum and magnetic flux gives them short rotation periods and powerful magnetic fields that create magnetic beams that follow closely to the rotational axis.

Pulsars are neutron stars from which we can observe pulses; they occur from charged particles following beams of magnetic flux directed near the neutron star's axis of rotation. Electrons can emit synchrotron radiation in the radio domain that sweep off axis as the neutron star rotates, giving a lighthouse effect. If this momentary sweeping of radiation occurs in our field of view, it gives us a pulse with the time between pulses representing the period. Every pulsar is a neutron star but not all neutron stars line up in such a way that we can detect them as pulsars.

Measurable parameters of pulsars include their period and spin-down time; the spin rate of pulsars is expected to decrease as rotational energy is converted into the radiation that is emitted. The period of a pulsar is inversely proportional to the square root of its mass density.

Another parameter is the minimum neutron star magnetic field strength, proportional to the square root of the product of its period and spin-down rate. Assuming the pulsar is rotation powered and its period monotonically decreases with time, it is also possible to estimate the age of the pulsar.[12]

Of course, there are exceptions to a smooth pulsar period that decreases with time in a constant fashion. An example are glitches where periods decreases suddenly by one part in a million for a short time. This may be due to a decoupling between the core and the surface of the neutron star. There are also millisecond pulsars with very short periods created by accretion from a companion star.

2.7.2.9 Pulsar Rotation Measure and Dispersion Measure

The plane of polarization of a light wave rotates as it passes through a magnetic field along the direction of propagation. This phenomenon is known as Faraday rotation (Book 1, Section 2.2.4). The Rotation Measure (RM) measures the amount by which the plane of polarization is rotated, with the single wavelength-squared dependence omitted. The RM may alternatively be found through observations of separate Faraday rotations made at two different wavelengths.

By convention, if the RM is positive, the parallel component of the magnetic field points toward the observer. With certain assumptions, the usefulness of the RM is that the distance to a pulsar may be determined. Or if the distance to the pulsar is known, the electron number density can be found.

The dispersion measure (DM) is another way to measure distance based on the idea that the velocity of light through a medium is a function of the light's frequency.

---

[12] The braking index is another parameter that describes how a neutron star loses its rotational energy (see Pannuti 2020).

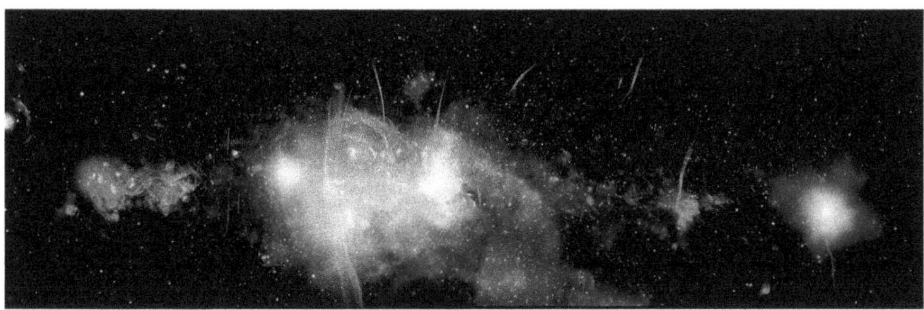

Figure 2.20. The region around Sagittarius A* is shown in this new composite image with Chandra data (green and blue) combined with radio data (red) from the MeerKAT telescope in South Africa, which will eventually become part of the SKA. Image credit: X-Ray: NASA/CXC/UMass/D. Wang et al.; radio: NRF/SARAO/MeerKAT.

When light at radio frequencies travels through the ISM, the lower the frequency, the lower the velocity of light. The measurement of this delay can provide an estimate of the pulsar's distance with certain assumptions (again a full discussion of this is found, with equations and worked examples, in Pannuti 2020 and Section 1.2.7).

RM and DM can also be combined in order to measure the magnetic field, $B_{||}$, along the line of sight toward a pulsar. This has been used to estimate the ambient magnetic field strength of the interstellar medium. Using this method, the ambient magnetic field in the interstellar medium has been found to range between 0.3μG and 3μG.

2.7.2.10 Galactic Center

Although the Galactic Center is not very accessible to visible light, radio observations of it and its surrounding environment have yielded many crucial insights. In fact, the center was the first astronomical radio source found by Karl Jansky. The central object is denoted SGR A* and is thought to be a supermassive black hole. A RGB map is shown in Figure 2.20. There are a wide diversity of radio sources including SNRs and HII regions in this image. This emphasizes that in addition to destroying stars that come too close to the supermassive black hole, this is also an area of active star formation and birth.

2.7.3 Extragalactic Radio Astronomy

Edwin Hubble is responsible for confirming that the Milky Way was only one of an uncountable number of galaxies in the universe. His classification included spiral galaxies with spiral arms and a disk, the tightest wound ones classed as Sa to looser ones designated as Sc (see Figure 2.21). Those with a bar in their disk were designated as SBa to SBc. The second major category of galaxies is the ellipticals that lack an arm structure and have a more spherical morphology from E0 (the most spherical) to E7, the most oblate. The third category is the lenticular galaxies, which are hybrids of the previous two; like spirals, these have disks and copious dust, but lack gas and spiral arms similar to ellipticals. Irregular galaxies, the fourth category,

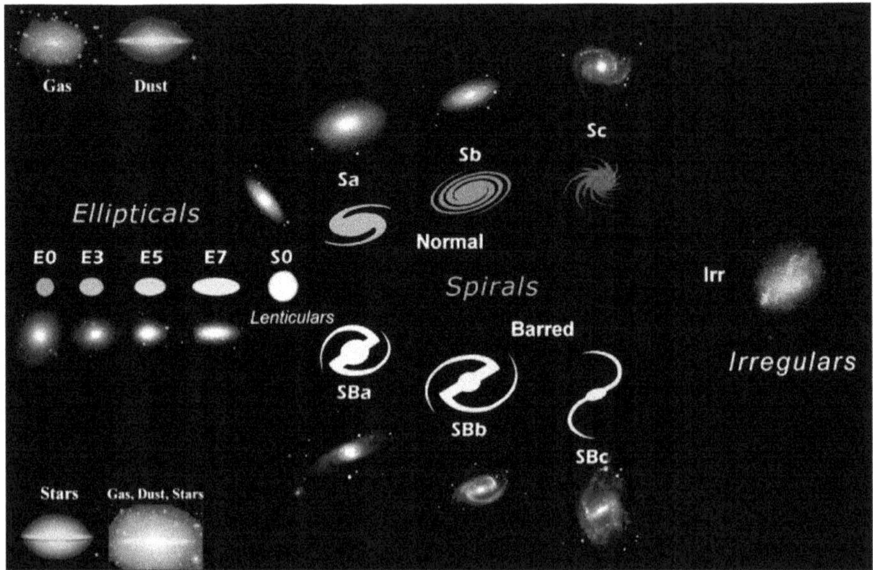

Figure 2.21. An example of Hubble's tuning fork classification of galaxies. Image credit: Wikipedia: Cosmo0.

are unlike the other three, having no regular structure. An example of the last category are the LMC and SMC satellite galaxies of the Milky Way.

2.7.3.1 Radio Observations of Normal Galaxies

Galaxy radio continuum observations can detect synchrotron emission from cosmic-ray electrons accelerated within the shocks of SNRs. The amount of this emission can be proportional to the Galaxy's star formation rate, as would be expected from the number of supernovae found. HI can reveal atomic hydrogen gas and CO observations can detect the amount of molecular hydrogen gas within a galaxy (see the example of the nearby Andromeda galaxy, a.k.a. M31, in Figures 2.22 and 2.23).

The identification of HII regions and SNRs can also be accomplished by examining their radio morphology if the Galaxy is near enough. A classic example is the sample of Magellanic Cloud SNRs (Bojičić et al. 2007; Cajko et al. 2009; Crawford et al. 2008, 2010, 2014; Bozzetto et al. 2010; Grondin et al. 2012; Bozzetto et al. 2017; De Horta et al. 2012, 2014; Brantseg et al. 2014; Kavanagh et al. 2013, 2015a, 2015b, 2015c, 2016; Reid et al. 2015; Maggi et al. 2014; Warth et al. 2014; Bozzetto & Filipovíc 2014; Bozzetto et al. 2012a, 2012b, 2012c, 2012d, 2013, 2014a, 2014b, 2015; Leahy 2017; Ghavamian et al. 2017; Dopita et al. 2018; Maitra et al. 2019; Alsaberi et al. 2019; Warth et al. 2014; Leahy et al. 2019; Pennock et al. 2004). Nearby galaxies may also have their nucleus resolved by radio telescopes, telling us more about the dynamics of the supermassive black hole that powers them as well as their interactions with the surrounding environment.

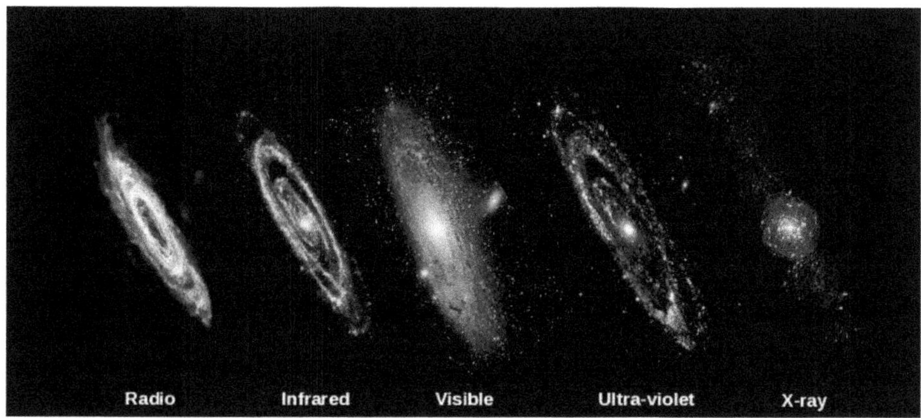

Figure 2.22. This multiwavelength view of the nearby Andromeda (M31) galaxy shows what is revealed in radio continuum, infrared, visible, ultraviolet, and X-ray light. Gas, dust, stars, and stellar remnants that emit light in different energies and at different temperatures can all be highlighted, depending on which wavelength is chosen. Image credit: Lee (2019); radio: WSRT/R. Braun; infrared: NASA/Spitzer/K. Gordon; visible: Robert Gendler; ultraviolet: NASA/GALEX; X-ray: ESA/XMM/W. Pietsch.

21 cm HI line observations not only detect the mass of galaxies by integrating over the HI flux profile, but also trace the morphology, such as warping of the disks due to tidal interactions with other galaxies. These can tell us how nearby galaxies interact with each other, information not available in optical images. HI emission also may extend out twice as far as visible light from a face-on viewed galaxy. By studying HI redshifts, it is not only possible to find the radial or systemic velocity for a spiral galaxy as a whole, but it is also possible to better understand the dynamics of their rotation. HI observations can also determine the maximum rotational velocity of a spiral galaxy. Using this, it is possible to relate the Galaxy to its luminosity distance using the Tully–Fisher relation (Tully & Fisher 1977).

As previously noted, low-angular-resolution radio observations can tell us about star formation rates of a galaxy based on a direct correlation to its synchrotron emission from relativistic electrons generated from the shocks of SNRs.[13]

High-resolution interferometric radio observations can reveal these discrete sources such as SNRs and HII regions. Radio classification of these sources can be made based on their spectral index measurements.[14] HII regions emit in radio through thermal bremsstrahlung processes and have flatter indices while SNRs, neutron stars, and background galaxies emit synchrotron radiation with steeper indices.

Use of coincident sources at other wavelength domains can further aid in classification. Background galaxies may not have a clear optical counterpart, SNRs and neutron stars are distinguished in X-ray because X-rays from SNRs

[13] The greater the number of supernovae from massive stars, the more progenitor stars that had to be created, and these were created in star-forming regions.

[14] The index, α, expresses how the flux density, S, of a radio source varies with the frequency, ν, such that $S \propto \nu^{\alpha}$.

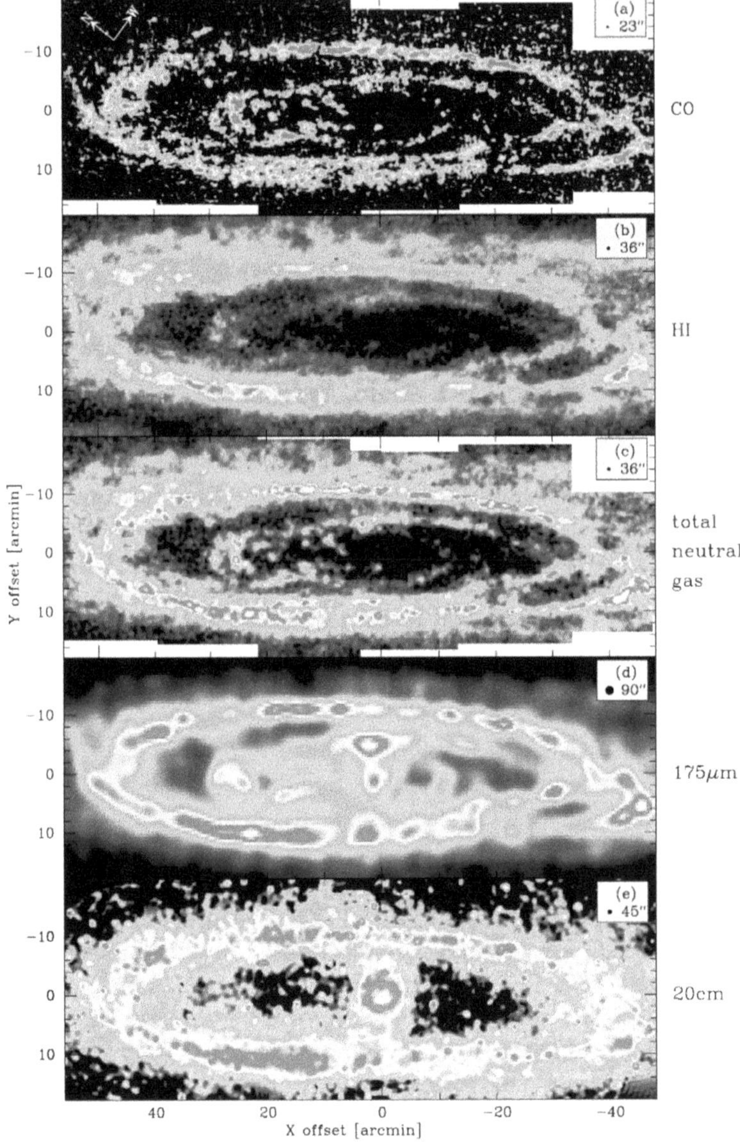

Figure 2.23. Distributions of neutral gas, cold dust, and radio continuum in the nearby neighboring spiral M31. From top to bottom: (a) emission observed in the ^{12}CO(1–0) line (b) emission observed in the HI line (Brinks & Shane 1984), (c) emission from the total neutral gas, N(HI) + 2N(H_2), with X_{CO} = 1.9 $\times 10^{20}$ mol cm^{-2} (K km s^{-1})$^{-1}$, (d) emission from cold dust at λ175 μm (Haas et al. 1998), (e) radio continuum emission at λ20 cm (Beck et al. 1998) (also see Galvin et al. 2012; Galvin & Filipovic 2014). The half-power beamwidth is indicated in the upper right-hand corner of each map. Image credit: Nieten et al. (2006) reproduced with permission © ESO.

are "softer," detected as having lower energy photons. This is because the mechanism for X-ray emission in SNRs is thermal bremsstrahlung, while X-ray emission in neutron stars arises from the synchrotron emission of very high-energy particles accelerating within magnetic fields.

2.7.3.2 Radio Galaxies and Active Galactic Nuclei

Carl Seyfert was studying spiral galaxies with unusually bright optical nuclei in the 1940s when he found their spectra contained Doppler-broadened emission lines. Unexpectedly, two of the galaxies in Seyfert's sample were later found to contain strong radio sources. This gave rise to a class of galaxies named after Seyfert, based on Doppler-broadened permitted and narrow forbidden lines in their spectra (Type I) or only narrow permitted and forbidden lines (Type II). The correlation of both bright optical nuclei with radio emission became associated with these sources.

The first sky radio surveys revealed these sources associated with point-like optical sources resembling ordinary stars. If "stars," these would have normally been expected to be weak radio sources. Optical spectra also revealed them to have high redshifts, and using Hubble's law, this implied they were much farther away than expected. This in turn meant they must be intrinsically very luminous. These objects, such as 3C 48, were termed quasars or quasi-stellar objects, because while they resembled a star in the optical, detailed optical and radio observations proved they were extragalactic at very large distances.

Later, high angular-resolution optical observations showed that although these objects are galaxies, their central nuclei are so luminous that the surrounding emission is lost by lower-resolution telescopes, making them appear point-like. We now know these galaxies contain a supermassive black hole in their center that can become active, consuming large amounts of matter and emitting copious amounts of radiation. They are termed "active" because of this activity and hence belong to a general class of objects known as AGNs.

For some time after their discovery, the structure and processes within AGNs were not understood. It was very similar to a group of blind people who studied an elephant using their sense of touch. Everyone reported something different, and it took some time to put all the pieces together.

One interesting part of the elephant were blazars with very bright radio-loud (RL) cores. The name came from one of the first discovered, BL Lac. It was later found these AGNs were emission jets from accretion disks of supermassive black holes pointed in the direction of the observer. These jets appear quite different when the AGNs are viewed face on; AGNs with jets emptying into radio lobes render the lobes transparent when looking down the jet's throat.

Blazars can be detected as gamma-ray sources by synchrotron self-Compton emission and are highly variable at all wavelengths (Section 7.6.1). Spectral energy distributions change wavelengths such that they can be divided into many subclasses based on properties that change over time:

- Optically violent variables (OVVs) have fluxes that change by 50% in just a day and have broad emission lines in their optical spectra absent from some other types of blazars.
- Flat-spectrum radio-loud quasars (FSRQs) have very broad emission lines but radio spectra with a flat spectral index, unlike many whose index approaches $\alpha \approx -0.8$. These are observed at higher redshifts than other blazars.

- Some are known as low- or high-frequency peaked blazars (LBLs and HBLs), the former with spectral energy distribution spectral peaks in the infrared while the latter with such peaks in the ultraviolet or X-ray.

As the reader may suspect, these are but a few examples.

2.7.3.3 Gigahertz Peak Spectrum and Compact Steep Spectrum Sources

Gigahertz peak spectrum (GPS) and compact steep spectrum (CSS) Sources sources are thought to contribute ~30%–40% of radio sources (An & Baan 2012; Lonsdale & Whittle 2016). However, their true nature is still under discussion. This debate, to uncover the absorption mechanism that contributes to the unique spectral features of this population, has continued for over 30 years. The most extensive study of GPS and CSS sources is presented by O'Dea (1998), who hypothesized regarding their origin. He additionally used these sources as constraints on AGN physical evolution as it is believed that these sources represent the early stage of the AGN phase (Fanti et al. 1995; Polatidis & Conway 2003; Fanti 2009; Randall et al. 2011; Callingham et al. 2017; Collier et al. 2018).

GPS and CSS sources have similar characteristics in that they are both compact with steep spectra ($\alpha < -0.8$). GPS sources have a much higher turnover frequency apart from their steep spectrum. This observable property allows their physical size to be estimated using the relation shown by Orienti & Dallacasa (2014), who agree with previous work by O'Dea & Baum (1997) and O'Dea (1998). They define the relation as

$$\log\ \nu_m = -0.21(\pm 0.05) - 0.65(\pm 0.05)\ \log\ \mathrm{LLS}, \tag{2.19}$$

where ν_m is the turnover frequency (in gigahertz) and LLS is the largest linear size (in kiloparsecs). This suggests a simple physical process between GPS and CSS populations that have been separately defined due to an arbitrary selection of turnover frequencies. Generally, it is believed that these objects are young, whose growth was frustrated by dense gas and dust or are simply young radio sources moving along their evolutionary path.

Fanti et al. (1995) presented the "young source" hypothesis, postulating that these sources typically have ages of 10^6 years and tend to decrease their radio luminosity as they move along their evolutionary path and grow in size from a few kiloparsecs to a few hundred kiloparsecs.

CSS sources are classified as such if they show an α of <-0.8, while GPS sources have an average α of $+0.56$ below the turnover frequency and -0.77 above the turnover.

2.7.3.4 High-frequency Peakers

We follow the convention given by Dallacasa et al. (2000) in naming radio continuum sources with convex spectra that peak above 5 GHz, high-frequency peaker (HFPs). These high-frequency peaking sources are thought to be very young and compact, following the anticorrelation shown in Equation (2.19). Additionally, these sources are thought to be rapidly changing in radio flux density, size, and flux density peak frequency on timescales of a few decades. These sources are thought to then evolve into GPS sources, then become CSS sources and finally evolve into large-scale objects with a significantly flatter spectrum (Fanti et al. 1995; Readhead et al. 1996; Alexander 2000; Marecki et al. 2003; Orienti & Dallacasa 2008; An & Baan 2012).

Dallacasa et al. (2000) presented a catalog of 55 HFP sources above 300 mJy, with the majority of these being unresolved sources. Additionally, they defined a sample of true HFPs as sources with an $\alpha < -0.56$ above the turnover frequency. Understanding and modeling these sources are of vital importance for three main reasons:

1. Their evolutionary path is currently not very well understood,
2. These sources are highly variable on short timescales, with this change likely driven by significant events in their host galaxy,
3. Due to the high frequency of their emission peak, these sources can strongly contribute to measurements of the CMB radiation as observed with the Wilkinson Microwave Anisotropy Probe and Planck telescopes, and as such these sources need to be accurately subtracted (De Zotti et al. 2000). Dallacasa & Orienti (2016) show that a common behavior of HFP sources is a spectral profile change correlated with an increase in optically thick emission and a decrease in optically thin emission.

2.7.3.5 High-redshift Radio Galaxies

Finding high-redshift radio galaxies is important in understanding the formation and evolution of galaxies at higher redshifts and in dense environments. By far, radio sources with an utrasteep spectrum (USS; $\alpha < -1.3$) have been the most efficient tracers of HzRGs at $z > 2$. This technique is based on the observed steepening of radio spectrum with both redshift and frequency.

Almost all known HzRGs including the most distant one at $z = 5.72$ (Saxena et al. 2018) have been identified using the USS selection method only. However, Yamashita et al. (2020) demonstrated that the USS selection criterion is not fully efficient, presenting the discovery of a radio galaxy, HSC J083 913.17+011 308.1, at $z = 4.72$ with a flatter spectral index of -1.1 between 325 MHz and 1400 MHz.

HzRGs are known to be extremely rare, and less than about 400 are known in the entire sky. In contrast, simulations demonstrate that the number density of radio galaxies in the distant universe ($z > 3$) is of the order of thousands, which indicates a significant deviation from what is observed (Raccanelli et al. 2012).

Infrared-faint radio sources (IFRSs) are a class of high-redshift RL-AGN originally identified as bright radio sources with no cospatial IR counterpart when cross-matching the 1.4 GHz ATLAS radio survey with the deep SWIRE IR survey between 3.6 μm and 24 μm (Norris et al. 2006). As radio sources, the nature of IFRSs was unknown at the time. It was suggested that they were either star-forming galaxies or AGNs as needed to produce IR emission. Numerous follow-up studies have been conducted because of their discovery to select more IFRSs and to understand what they are. Zinn et al. (2011) redefined the original selection criterion from Norris et al. (2006) and proposed a set of survey-independent criteria, which enable the selection of a brighter population of IFRSs with optical and IR counterparts. They are

- Radio-to-IR flux density ratio, $S_{20\ cm}/S_{3.6\ \mu m} > 500$, and
- 3.6 μm flux density, $S_{3.6\ \mu m} < 30\ \mu$Jy.

To date, a total of ~1400 IFRSs have because been identified from various radio surveys using the criteria of Zinn et al. (2011). All studies so far suggested that most of the observed IFRSs, if not all, are RL-AGNs at $z > 2$. Additionally, some studies presented evidence that the IFRS population not only encompasses USS sources but nonsteep and flat-spectrum sources too. Thus, IFRSs serve as a potential candidate to discover HzRGs.

2.7.3.6 Fanaroff and Riley AGN Classification

The Fanaroff and Riley classification system divides these galaxies into class I and II (FR-I and FR-II). FR-I galaxies (e.g., 3C 296) have a lower radio luminosity and are brighter in their cores than their lobes with a decreasing surface brightness near their edges. The radio lobes are connected to the core by defined jets and they have steep radio spectral indices. Beyond a transition luminosity at 1.4 GHz of 10^{32} erg s^{-1} Hz^{-1}, FR-II galaxies (e.g., 3C 334) are more radio luminous with brighter lobes that have hot spots and limb brightening.

Radio jets are composed of subatomic particles accelerated to relativistic velocities that by least resistance travel out of the Galaxy plane from the central black hole. The detected radio emission is synchrotron in origin due to the interaction of magnetic fields and charged particles such as electrons and are associated with the accretion disk of the black hole.

RL galaxies (mostly elliptical) comprise 10% of AGNs, defined by the ratio of the radio flux density at 5 GHz to the flux at a wavelength 440 nm of 10 or greater. The jets are powered by the rotational energy of the accretion disk, the black hole, or a combination of both. Magnetic fields are essential to this process. A current model for how energy from the accretion disk leads to the formation of the jet is known as the Blandford–Znajek mechanism; this is explained in detail in Pannuti (2020).

Relativistic particles from the jets collect in giant radio lobes that extend over tens of kiloparsecs. The lobe is formed when particles at the jet tip decelerate when encountering the active galaxy and the ISM, causing the formation of a shock. Radio lobes contain vast amounts of energy (10^{54} J), radio spectral indices of −0.5, and have high levels of linear polarization. It is thought that as particles travel in the lobe, they radiate away their energy through diffusive shock acceleration.

The spectral energy distribution (SED) of a radio galaxy can be quite complex, having a steep radio slope due to synchrotron emission and a turnover with a peak known as the infrared peak. The slope of this emission is subject to synchrotron self-absorption processes.[15] There is also a big blue bump in the SED, found in the X-ray

[15] In blazars, this peak may instead occur in the X-ray domain.

and possibly gamma-ray domain. The process for this bump is inverse-Compton scattering (the transfer of energy to a photon from an electron through collisions).

Blazars are detected as gamma-ray sources from synchrotron self-Compton emission. Here, photons emitted by electrons act as seed photons that electrons can scatter to increase photon energies. These upscattered photons may then Compton scatter off electrons, which can then scatter off photons, creating a chain reaction. For this process the jet medium is required to be optically thick to the seed photons so that the initial upscattering may occur. Eventually, the most energetic gamma-ray photons will perceive the medium as optically thin, escaping and thereby acting to cool the jet.

In some active galaxies, the radio double lobe may be visible but not the jet, due to the process of Doppler boosting that can either enhance or reduce the emission detected due to tightness of the jet and the angle of its emission.

Most of the parts of the elephant have now been assembled in the unified AGN model as shown in Figure 1.7. The type of emission and the structure of these objects

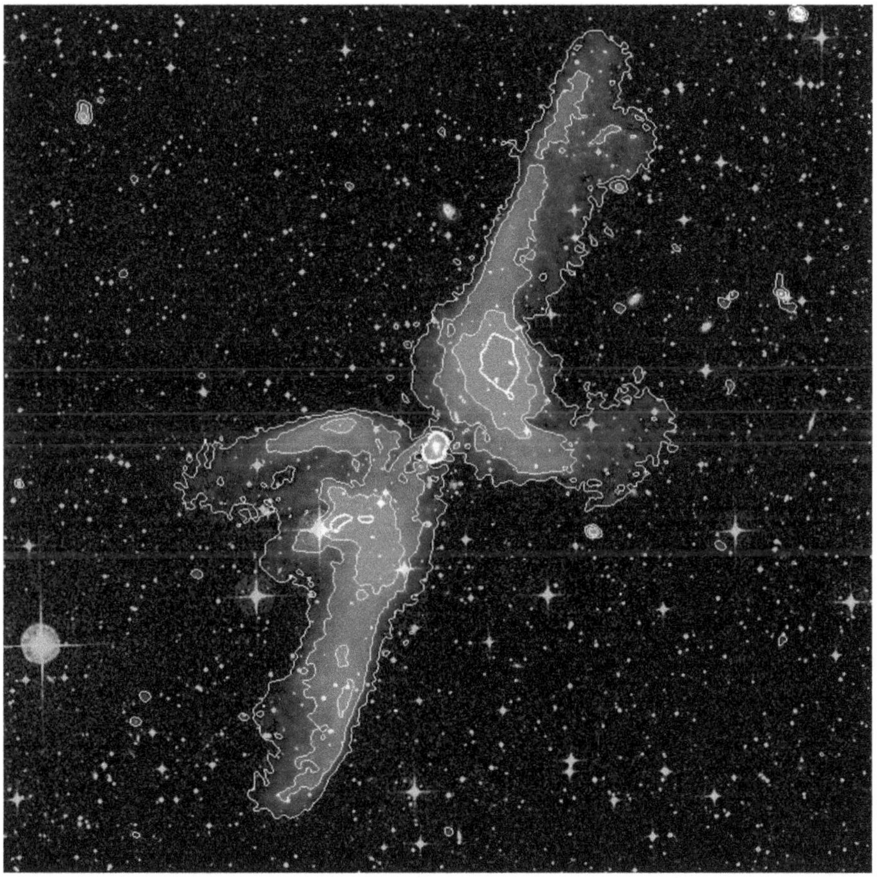

Figure 2.24. ASKAP image of the giant X-shaped radio galaxy PKS 2014–55. Image credit: Bärbel Koribalski.

depend on both the orientation as well as the astrophysical processes occurring at a particular time in the supermassive black hole's history. An exercise for the reader might be to assemble multiwavelength observations expected from observing an AGN from a particular orientation. Indeed, these objects alone contain most if not all of the astrophysics currently known.

2.7.3.7 Galaxy Clusters and Associated Diffuse Radio Emission

The structure of the universe includes galaxies that are bound by gravity into clusters, which in turn assemble into superclusters. One idea is that this is the result of billions of years of mergers into larger and larger assemblies. These cluster mergers release around 10^{57} J of energy over a billion years. Shocks driven into the intercluster medium coupled with the magnetic field create radio synchrotron emission that can be observed.

The diffuse radio structures within clusters include a radio halo toward the center having smooth morphology and matching spatially the emission of thermal X-ray radiation from hot intracluster gas. These areas are not polarized.

A second type of diffuse cluster structure comprises radio relics with irregular morphologies at the cluster perimeter with high levels of polarization. Both haloes and relics have similar spectral indices ($\alpha \approx -1.0$) and are associated with disturbed systems or mergers.

An important method used to study these clusters is by the determination of the RM. The RM from background galaxies as seen through the intracluster medium can probe its magnetic field strength and be used to derive its electron number density.

2.7.3.8 The Cosmic Microwave Background

Initially, about 13.7 billion years ago, according to the Big Bang theory, the universe was filled with a soup of particles and very high-energy gamma rays that constantly photodissociated nuclei. Due to the expansion of the universe, these photons lost energy until the recombination epoch began where bound atoms could form. The continuing expansion of the universe has stretched the energy and wavelengths of these photons to the point they act as a thermal blackbody with a peak wavelength of 1.06 mm. This blackbody spectrum is known as the CMB, well within the scope of radio astronomy (Sections 7.4 and 10.1.5 as well as Book 1, Section 7.3.1.). Several maps of the CMB across the sky have been made in order to understand its remarkable smoothness and yet minute fluctuations are thought to be responsible for the structure of the universe (see Figure 10.6 and 10.5).

The Greisen[16]–Zatsepin[17]–Kuzmin[18] limit (the GZK limit) is a theoretical upper limit on the energy of cosmic rays from distant sources (4.10×10^{19} eV), calculated (independently) in 1966. It provides the maximum energy a cosmic-ray proton can have between galaxies. As cosmic-ray protons with energies exceeding this limit (8 J) travel though space, they interact with the cosmic microwave photons. To the

[16] Kenneth Greisen (1918–2007), American physicist and astronomer.
[17] Georgiy Zatsepin (1917–2010), Russian astrophysicist.
[18] Vadim Kuzmin (1937–2015), Russian theoretical physicist.

protons, the photons appear blueshifted to much larger energies. These energetic protons interact with these photons to produce pions that bleed energy from the protons until the GZK limit of about 8 J is reached. Because the length scale of this interaction is about 50 Mpc (the GZK horizon), it is unlikely protons that exceed the limit will be observed on Earth. But how is it that there are high-energy particles with even higher energies that are observed? It is thought that these cosmic rays are most likely atomic nuclei and not high-energy isolated protons.

Between galaxies in galaxy clusters, space is filled with a hot X-ray-emitting ionized plasma among the sea of CMB photons that occur everywhere. Free electrons in the ionized plasma interact with CMB photons through iC scattering, imparting energy to the photons and fractionally decreasing the background CMB temperature, normally 2.73 K. This is known as the Sunyaev–Zeldovich effect (Section 2.1.3). Using this effect, measurement of the fractional temperature decrease ($\Delta T/T$) allows the distance to the cluster to be determined. When combined with the recessional velocity of the cluster, it is possible to give an estimate for the Hubble constant, H_0.[19] One value of the Hubble constant using this effect was reported to be, $H_0 \approx 61$ km s^{-1} Mpc^{-1} (Reese 2004).

2.8 Concluding Statement

The discussion in this chapter emphasizes how important the detection and measurement of radio emission is to understanding the astrophysics of both multiwavelength and multimessenger astronomy. This can be seen at all distance scales from our own star, the Sun and its satellites, to the farthest reaches of the universe.

References

Alexander, P. 2000, MNRAS, 319, 8
Alsaberi, R. Z. E., Barnes, L. A., Filipović, M. D., et al. 2019, Ap&SS, 364, 204
An, T., & Baan, W. A. 2012, ApJ, 760, 77
Anderson, L. D., Zavagno, A., Barlow, M. J., García-Lario, P., & Noriega-Crespo, A. 2012, A&A, 537, A1
Beck, R., Berkhuijsen, E. M., & Hoernes, P. 1998, A&AS, 129, 329
Becker, R. H., White, R. L., & Helfand, D. J. 1994, in ASP Conf. Ser. 61, ed. D. R. Crabtree, R. J. Hanisch, & J. Barnes (San Francisco, CA: ASP), 165
Becker, R. H., White, R. L., & Helfand, D. J. 1995, ApJ, 450, 559
Blumenthal, G. R., & Gould, R. J. 1970, RvMP, 42, 237
Bojičić, I. S., Filipović, M. D., Parker, Q. A., et al. 2007, MNRAS, 378, 1237
Bojičić, I. S., Parker, Q. A., & Frew, D. J. 2017, in Planetary Nebulae: Multi-Wavelength Probes of Stellar and Galactic Evolution Vol. 323, ed. X. Liu, L. Stanghellini, & A. Karakas (Cambridge: Cambridge Univ. Press), 327
Bozzetto, L. M., & Filipović, M. D. 2014, Ap&SS, 351, 207
Bozzetto, L. M., Filipović, M. D., Crawford, E. J., et al. 2010, SerAJ, 181, 43

[19] See Pannuti (2020) for details.

Bozzetto, L. M., Filipović, M. D., Crawford, E. J., De Horta, A. Y., & Stupar, M. 2012d, SerAJ, 184, 69
Bozzetto, L. M., Filipović, M. D., Crawford, E. J., et al. 2012a, MNRAS, 420, 2588
Bozzetto, L. M., Filipović, M. D., Crawford, E. J., et al. 2012b, RMxAA, 48, 41
Bozzetto, L. M., Filipović, M. D., Crawford, E. J., et al. 2013, MNRAS, 432, 2177
Bozzetto, L. M., Filipović, M. D., Urosevic, D., & Crawford, E. J. 2012c, SerAJ, 185, 25
Bozzetto, L. M., Filipović, M. D., Urošević, D., Kothes, R., & Crawford, E. J. 2014a, MNRAS, 440, 3220
Bozzetto, L. M., Kavanagh, P. J., Maggi, P., et al. 2014b, MNRAS, 439, 1110
Bozzetto, L. M., Filipović, M. D., Haberl, F., et al. 2015, PKAS, 30, 149
Bozzetto, L. M., Filipović, M. D., Vukotić, B., et al. 2017, ApJS, 230, 2
Brantseg, T., McEntaffer, R. L., Bozzetto, L. M., Filipović, M., & Grieves, N. 2014, ApJ, 780, 50
Brinks, E., & Shane, W. W. 1984, A&AS, 55, 179
Cajko, K. O., Crawford, E. J., & Filipović, M. D. 2009, SerAJ, 179, 55
Callingham, J. R., Ekers, R. D., Gaensler, B. M., et al. 2017, ApJ, 836, 174
Cheung, A. C., Rank, D. M., Townes, C. H., Thornton, D. D., & Welch, W. J. 1968, PhRvL, 21, 1701
Clark, B. G. 1980, A&A, 89, 377
Collier, J. D., Tingay, S. J., Callingham, J. R., et al. 2018, MNRAS, 477, 578
Condon, J. J. 1974, ApJ, 188, 279
Condon, J. J., Cotton, W. D., Greisen, E. W., et al. 1998, AJ, 115, 1693
Cornwell, T. J. 2008, ISTSP, 2, 793
Crawford, E. J., Filipović, M. D., De Horta, A. Y., Stootman, F. H., & Payne, J. L. 2008, SerAJ, 177, 61
Crawford, E. J., Filipović, M. D., Haberl, F., et al. 2010, A&A, 518, A35
Crawford, E. J., Filipović, M. D., McEntaffer, R. L., et al. 2014, AJ, 148, 99
Dallacasa, D., & Orienti, M. 2016, AN, 337, 120
Dallacasa, D., Stanghellini, C., Centonza, M., & Fanti, R. 2000, A&A, 363, 887
De Horta, A. Y., Filipović, M. D., Bozzetto, L. M., et al. 2012, A&A, 540, A25
De Horta, A. Y., Sommer, E. R., Filipović, M. D., et al. 2014, AJ, 147, 162
De Zotti, G., Granato, G. L., Silva, L., Maino, D., & Danese, L. 2000, A&A, 354, 467
Desmurs, J. F., Alcolea, J., Bujarrabal, V., Sánchez Contreras, C., & Colomer, F. 2007, A&A, 468, 189
Dopita, M. A., Vogt, F. P. A., Sutherland, R. S., et al. 2018, ApJS, 237, 10
Ekers, R. D. 2012, Principles of Interferometry I (CASS Radio Astronomy School)
Emerson, D. 2002, in ASP Conf. Proc. 278, Single-Dish Radio Astronomy: Techniques and Applications, ed. S. Stanimirovic, D. Altschuler, P. Goldsmith, & C. Salter (San Francisco, CA: ASP), 27
Emerson, D. T., Klein, U., & Haslam, C. G. T. 1979, A&A, 76, 92
Fanti, C. 2009, AN, 330, 120
Fanti, C., Fanti, R., Dallacasa, D., et al. 1995, A&A, 302, 317
Filipovic, M. D., Cohen, M., Reid, W. A., et al. 2009, MNRAS, 399, 769
Fragkou, V., Parker, Q. A., Bojičić, I. S., & Aksaker, N. 2018, MNRAS, 480, 2916
Frew, D. J., & Parker, Q. A. 2010, PASA, 27, 129
Galvin, T. J., & Filipovic, M. D. 2014, SerAJ, 189, 15
Galvin, T. J., Filipovic, M. D., Crawford, E. J., et al. 2012, SerAJ, 184, 41

Gardner, F. F., & Whiteoak, J. B. 1966, ARA&A, 4, 245
Ghavamian, P., Seitenzahl, I. R., Vogt, F. P. A., et al. 2017, ApJ, 847, 122
Gledhill, T. M., Froebrich, D., Campbell-White, J., & Jones, A. M. 2018, MNRAS, 479, 3759
Goddi, C., Crew, G., Impellizzeri, V., et al. 2019, Msngr, 177, 25
Goldreich, P., & Kylafis, N. D. 1981, ApJ, 243, L75
Goldstein, R. M., & Carpenter, R. L. 1963, Sci, 139, 910
Grondin, M.-H., Sasaki, M., Haberl, F., et al. 2012, A&A, 539, A15
Haas, M., Lemke, D., Stickel, M., et al. 1998, A&A, 338, L33
Högbom, J. A. 1974, A&AS, 15, 417
Irabor, T., Hoare, M. G., Oudmaijer, R. D., et al. 2018, MNRAS, 480, 2423
Jacoby, G. H. 1980, ApJS, 42, 1
Jones, P. A., Sarkissian, J. M., Burton, M. G., Voronkov, M. A., & Filipović, M. D. 2006, MNRAS, 369, 1995
Joseph, T. D., Filipović, M. D., Crawford, E. J., et al. 2019, MNRAS, 490, 1202
Kavanagh, P. J., Sasaki, M., Bozzetto, L. M., et al. 2015a, A&A, 573, A73
Kavanagh, P. J., Sasaki, M., Bozzetto, L. M., et al. 2016, A&A, 586, A4
Kavanagh, P. J., Sasaki, M., Bozzetto, L. M., et al. 2015b, A&A, 583, A121
Kavanagh, P. J., Sasaki, M., Points, S. D., et al. 2013, A&A, 549, A99
Kavanagh, P. J., Sasaki, M., Whelan, E. T., et al. 2015c, A&A, 579, A63
Kemball, A. J., & Diamond, P. J. 1997, ApJ, 481, L111
Konovalenko, A. A., & Sodin, L. G. 1980, Natur, 283, 360
Leahy, D. A. 2017, ApJ, 837, 36
Leahy, D., Wang, Y., Lawton, B., Ranasinghe, S., & Filipović, M. 2019, AJ, 158, 149
Lee, C.-H. 2019, in Time-Domain Studies of the Andromeda Galaxy (Beograd: Institute of Physics), 5–1
Leverenz, H., Filipović, M. D., Bojičić, I. S., et al. 2016, Ap&SS, 361, 108
Leverenz, H., Filipović, M. D., Vukotić, B., & Uro, D. 2017, MNRAS, 468, 1794
Longair, M. S. 2011, High Energy Astrophysics (3rd ed.; Cambridge: Cambridge University Press)
Lonsdale, C. J., Whittle, M., Trapp, A., et al. 2016, AN, 337, 194
Maggi, P., Haberl, F., Kavanagh, P. J., et al. 2014, A&A, 561, A76
Maitra, C., Haberl, F., Filipović, M. D., et al. 2019, MNRAS, 490, 5494
Marecki, A., Spencer, R. E., & Kunert, M. 2003, PASA, 20, 46
Mason, B. S. 2020, arXiv:2006.06549
McClure-Griffiths, N. M., Pisano, D. J., Calabretta, M. R., et al. 2009, ApJS, 181, 398
McMullin, J. P., Waters, B., Schiebel, D., Young, W., & Golap, K. 2007, in ASP Conf. Ser. 376, ed. R. A. Shaw, F. Hill, & D. J. Bell (San Francisco, CA: ASP), 127
Moe, M., & De Marco, O. 2006, ApJ, 650, 916
Müller, P., Krause, M., Beck, R., & Schmidt, P. 2017, A&A, 606, A41
Nieten, C., Neininger, N., Guélin, M., et al. 2006, A&A, 453, 459
Norris, R. P., Afonso, J., Appleton, P. N., et al. 2006, AJ, 132, 2409
O'Dea, C. P. 1998, PASP, 110, 493
O'Dea, C. P., & Baum, S. A. 1997, AJ, 113, 148
Orienti, M., & Dallacasa, D. 2008, A&A, 477, 807
Orienti, M., & Dallacasa, D. 2014, MNRAS, 438, 463

Pannuti, T. G. 2020, The Physical Processes and Observing Techniques of Radio Astronomy; An Introduction (1st ed.; Berlin: Springer)
Park, J., Hada, K., Kino, M., et al. 2019, ApJ, 871, 257
Parker, Q. A., Bojičić, I. S., & Frew, D. J. 2016, JPhCS, 728, 032008
Parker, Q. A., Cohen, M., Stupar, M., et al. 2012a, MNRAS, 427, 3016
Parker, Q. A., Frew, D. J., Acker, A., & Miszalski, B. 2012b, in IAU Symp. 283, Planetary Nebulae: An Eye to the Future, 9
Pennock, C. M., van Loon, J. T., Bell, C., et al. 2004, arXiv:2004.04531
Polatidis, A. G., & Conway, J. E. 2003, PASA, 20, 69
Raccanelli, A., Zhao, G.-B., Bacon, D. J., et al. 2012, MNRAS, 424, 801
Randall, K. E., Hopkins, A. M., Norris, R. P., & Edwards, P. G. 2011, MNRAS, 416, 1135
Readhead, A. C. S., Taylor, G. B., Pearson, T. J., & Wilkinson, P. N. 1996, ApJ, 460, 634
Reese, E. D. 2004, in Measuring and Modeling the Universe, ed. W. L. Freedman (Cambridge: Cambridge Univ. Press), 138
Reid, W. A., Stupar, M., Bozzetto, L. M., Parker, Q. A., & Filipović, M. D. 2015, MNRAS, 454, 991
Roy, T., & Gangadhara, R. T. 2019, ApJ, 878, 148
Saxena, A., Marinello, M., Overzier, R. A., et al. 2018, MNRAS, 480, 2733
Seitenzahl, I. R., Ghavamian, P., Laming, J. M., & Vogt, F. P. A. 2019, PhRvL, 123, 041101
Smirnov, O. M. 2011, A&A, 527, A106
Sullivan, W. T. 2009, Cosmic Noise: A History of Early Radio Astronomy (Cambridge: Cambridge University Press)
Tully, R. B., & Fisher, J. R. 1977, A&A, 500, 105
Warth, G., Sasaki, M., Kavanagh, P. J., et al. 2014, A&A, 567, A136
Yamashita, T., Nagao, T., Ikeda, H., et al. 2020, AJ, 160, 60
Zinn, P. C., Middelberg, E., & Ibar, E. 2011, A&A, 531, A14

Chapter 3

Long-wave Astronomy with Incoherent Detection

Michelle Cluver and Thomas Jarrett

Stars and star formation underpin almost every aspect of baryon evolution in our universe. With the advent of large-format arrays and sensitive bolometers some four decades ago, infrared and submillimeter astronomy have revolutionized our understanding of the lifecycle of stars, from the cradle to the grave. On physical scales that range from single stars (subparsec), to Giant Molecular Clouds (GMCs at 50 pc), to kiloparsec scales in the extragalactic realm of galaxies, the wavelengths from 1 to 1000 μm have been used to penetrate and reveal the dusty and gas-dense stellar nurseries where stars are born in the Milky Way. Not only the tenuous gas and dust components of the universe but the infrared is also used to probe the ancient, cool, and luminous stars that form the gravitational backbones of the largest conglomerations of baryons we see in the observable universe—the mighty galaxies. In this chapter, we review the origins of the great infrared astronomy missions carried out since the 1980s and up to today, highlighting some of the most important discoveries of the—previously—hidden universe, as well as introduce concepts and methods that are used to analyze the all-important data from these special instruments and telescopes that explore the infrared window to the universe.

3.1 Overview

The infrared to submillimeter band can be subdivided into smaller windows based on the physical processes that give rise to the electromagnetic radiation we observe, in combination with whether observations are predominantly ground or space based. The infrared and submillimeter bands are also home to numerous rotational and vibrational transitions of various molecules forming the basis of astrochemistry. However, this extends to the water and CO_2 in our atmosphere, and as a result, only the near-infrared (NIR) from 1 to 3 μm is accessible from ground observatories, while mid-infrared to far-infrared studies are generally only feasible from above the

doi:10.1088/2514-3433/ac2256ch3

atmosphere, or well above the troposphere. Although challenging, submillimeter observatories (e.g., SCUBA and SCUBA-2 on the JCMT, the ALMA interferometer) located high atop dry mountains have been incredibly successful in studying the star formation of distant galaxies.

As a consequence of the blanketing atmosphere, the most powerful and sensitive instruments in this regime have been space telescopes. Launched in 1983, IRAS represents the day astrophysics was changed forever. Imaging the whole sky (Figure 3.1) at 12, 25, 60, and 100 μm bands, IRAS revealed for the first time protoplanetary disks, young stellar objects (YSOs), and a new class of powerful object in the universe, the luminous infrared galaxies (LIRGs; Sanders & Mirabel 1996), among many other discoveries and milestones in modern astrophysics. IRAS detected about 350,000 infrared sources, many for the first time (e.g., the core of our Galaxy), which represented an increase in the number of total cataloged astronomical sources by approximately 70%.

The success of IRAS led to more missions in the 1990s and 2000s; taking advantage of the rapid technological advances in this relatively new field meant space telescopes with increasingly higher resolution and better sensitivity. From the ISO, Spitzer, Herschel Space Observatory (HSO) with far-IR and submillimeter capability, to AKARI, and the most recent, the WISE, launched in 2009. Soon, the next generation will be led by the JWST, whose mid-infrared instrumentation will be notably focused on exoplanets and the most distant galaxies—the first galaxies—in the observable universe.

All of these telescopes deployed infrared-light sensitive large-format arrays and bolometers, enabling the efficient mapping and spectroscopy of stars, the interstellar

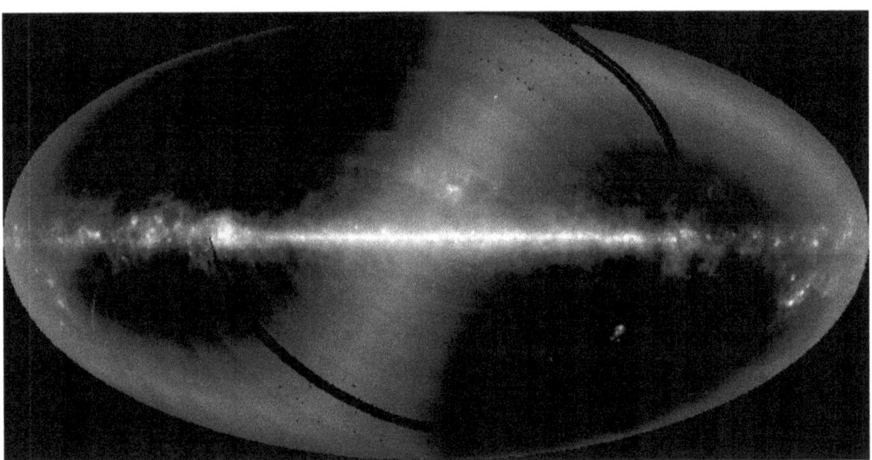

Figure 3.1. IRAS view of the whole sky with an equal-area projection. The Milky Way is front and center, its band of light is the Galactic Plane, composed of star formation regions—the stellar nurseries that continue to grow our home galaxy. The primary colors represent infrared emission detected in three of four IRAS bands: blue is 12 μm; yellow-green is 60 μm, and red is 100 μm. The ghostly blue S-shape stripe that runs across the sky is the zodiacal light from thermal-heated small dust grain in our solar system. Black stripes are regions of the sky that were not scanned by the telescope. Image credit: NASA/JPL-Caltech.

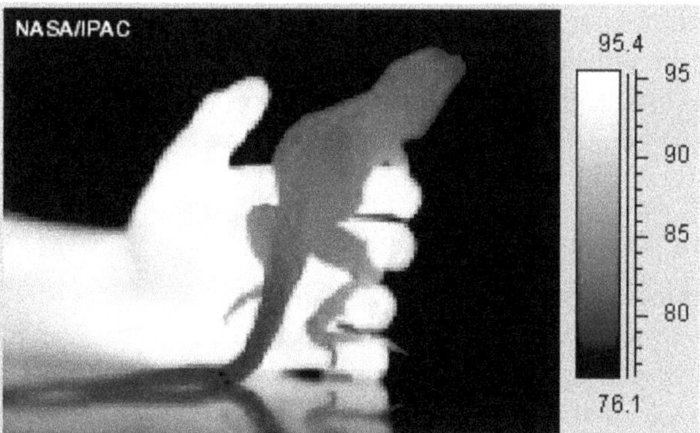

Figure 3.2. Thermal infrared imaging of a human and lizard, demonstrating the temperature differences between warm- and cold-blooded animals. Image credit: NASA/IPAC.

medium, and galaxies. In parallel, new methods for data reduction and analysis were developed to optimally exploit the capabilities of these telescopes and their instrumentation.

Earth-based observations with NIR detectors largely preceded the space telescopes, and hence the methods developed for working with large-format arrays from ground-based instruments—such as the Wide Field Infrared Camera (WIRC) on board the Hale 5 m telescope, and the 2MASS, which used telescopes in the southern and northern hemispheres to cover the whole sky—were largely adapted for use on the space-based missions.

In this chapter, we will introduce some examples of the major instrumentation, data reduction, and analysis methods that have been developed for infrared–submillimeter astrophysics. We will discuss the underlying physics driving the radiation mechanisms that we observe in both the continuum and in emission lines (Figure 3.2).

3.2 Introduction to Infrared and Submillimeter Astronomy

The existence of radiation at wavelengths longer than those of the optical spectrum was first discovered by William Herschel in 1800. Placing a thermometer beyond the red portion of the spectrum generated from sunlight through a glass prism revealed that although no radiation was visible, the temperature was in fact higher than for any of the colors in the spectrum. And so "infrared," i.e., "below red" radiation was born.

Although we can feel infrared radiation as thermal heat, e.g., when we stand close to a brick wall that has been in the Sun all day, the technical challenge of building sensitive receivers that work efficiently in this regime (i.e., between radio-type coherent and optical-like incoherent detection techniques) was not overcome until the 1960s, leaving this part of the electromagnetic spectrum largely inaccessible.

Since then, however, thermal imaging at infrared wavelengths has found widespread application, ranging from meteorology to fire fighting. In astronomy, the divisions between near-, mid-, and far-infrared are a combination of atmospheric transmission windows and detector technology, as well as the distinctive physical mechanisms that drive the emission in those windows. For the purposes of this chapter, we adopt near-infrared (NIR) to be 1–5 μm, mid-infrared (mid-IR) to be 5–50 μm, far-infrared (far-IR) to be 50–200 μm, and submillimeter from 200 μm to 1 mm.

3.2.1 Atmospheric Transmission Windows

In contrast to optical electromagnetic radiation, which passes relatively unimpeded through Earth's atmosphere (albeit with some distortion), the infrared and submillimeter parts of the spectrum are subject to large swathes of obscuration (see Figure 3.3). The lion's share of this obscuration is due to water (H_20) and carbon dioxide (CO_2) molecules (but also O_2 and CH_4) in the atmosphere absorbing incoming radiation and producing "absorption bands" (Rudolf et al. 2016). Because the water vapor content of the atmosphere varies, this telluric contamination can vary on seasonal and even hourly timescales. As we see from the diagram, there are small confined windows in the NIR, defining the filter bandpasses for ground-based astronomy, while the mid- and far-IR are completely opaque from the ground. The transmission steadily improves at millimeter wavelengths, allowing ground-based (high mountaintop) observations, even if challenging due to thermal (environmental) noise; and finally, the radio regime is fully open to ground observations, even from relatively low elevations.

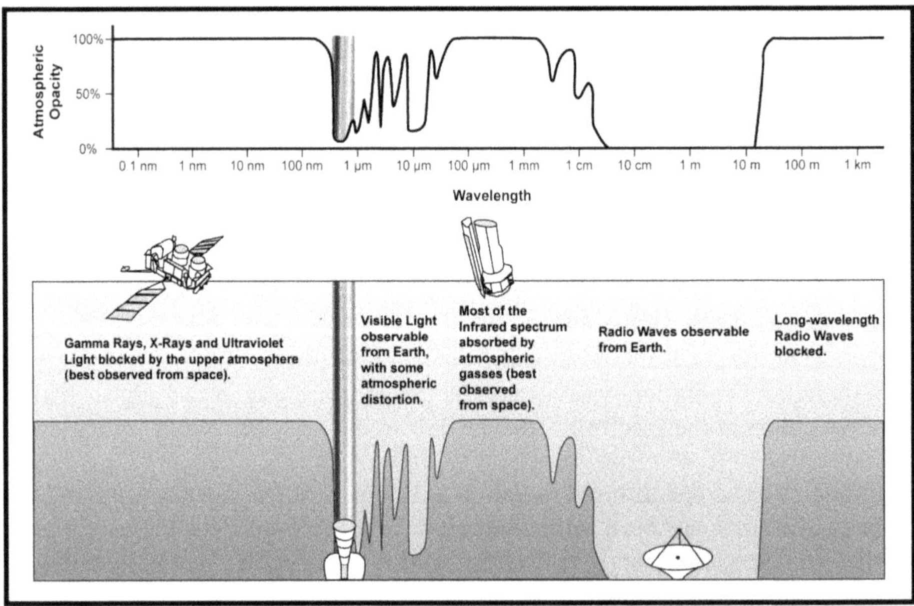

Figure 3.3. Atmospheric transmission of electromagnetic radiation. NIR can be done from the ground using the *J* (1.2 μm), *H* (1.6 μm) and *K* (2.2 μm) bands, and even 10 μm in the mid-IR is possible (although the thermal background is generally just too high even from the highest mountaintops), whereas far-IR to submillimeter is best done from space. Image credit: NASA/IPAC.

3.2.2 Detector Technologies

While optical astronomy relies largely on the CCD, these are incompatible with detecting radiation at longer, infrared wavelengths. NIR detectors typically use HgCdTe (mercury–cadmium–telluride) arrays (e.g., the 2MASS detectors, NIRCAM on JWST), sensitive to wavelengths between 0.5 and 5 μm. The array layers absorb light and convert it into voltages in individual pixels. The short-wavelength (3.6 μm and 4.5 μm) detectors of Spitzer IRAC, however, used indium antimonide technology. Other hybrid technologies have been developed for the NIR window, which are increasingly more sensitive and rivaling CCD silicon technology in the optical.

For the longer wavelengths of the mid-infrared, silicon doped with arsenic, abbreviated as Si:As, has been widely used (e.g., IRAC 5.8 μm and 8 μm, and MIPS 24 μm arrays on Spitzer, MIRI on JWST). The IRS spectrograph on Spitzer used Si: As arrays for its short-wavelength bands (e.g., short-low, 5–14 μm) and silicon doped with antimony (Si:Sb) at longer wavelengths (e.g., long-low 14–40 μm). Indium is used in both detector types to connect pixels in the absorber layer to a silicon readout circuit.

Radiation in the far-infrared can be detected using either photoconductor technology, i.e., by capturing photons using, for example, gallium-doped germanium (Ge:Ga) arrays (e.g., MIPS 70 μm and 160 μm arrays on Spitzer, and for PACS spectroscopy on Herschel) or bolometers (e.g., PACS imaging at 70 μm and 160 μm on Herschel). A bolometer is a thermal detector and measures the power of incident electromagnetic radiation through the heating of the array material, which has a temperature-dependent electrical resistance. For a general introduction to infrared astronomy detectors, see Rieke (2007).

Moving into the submillimeter regime (200 μm to 1 mm), incoherent detection is achieved almost exclusively through bolometers (e.g., SCUBA and SCUBA-2 on the JCMT, MAMBO on IRAM, and the Planck satellite). For example, the SPIRE instrument on Herschel used bolometer detector arrays for both its imaging photometer (250, 360, and 520 μm) and its Fourier transform spectrometer (300–670 μm). The detector arrays were feedhorn-coupled, spider-web bolometers with neutron transmutation doped (NTD) germanium temperature sensors.

We note that at these long wavelengths, coherent detection can be achieved using entirely different technology. For example, HIFI on Herschel was a heterodyne spectrometer in the wavelength range 157–625 μm. The principle of heterodyne detection involves the frequency range of the astronomical signal being translated to an equivalent signal at a lower frequency where standard electronics can be used for signal processing. The advantage is significantly higher spectral resolution (10^7 to 10^8) and hence kinematic fidelity. This process requires superconductor–insulator–superconductor (SIS) and hot electron bolometer (HEB) mixers. SIS mixers are also used by ALMA for bands 3 through 10. The complexity of the receivers generally means that only a single receiver is installed at each telescope focal array plane, which is why HIFI could only observe a single pixel on the sky at a time.

An example of light interacting with dust grains is when a star illuminates a dust cloud giving rise to a reflection nebula. The optical wavelengths of the star's light are comparable to the size of the dust grains and as a result, the grains scatter, i.e., reflect back, the radiation. However, blue wavelengths are scattered more efficiently, which is why these clouds usually appear blue (Figure 3.4).

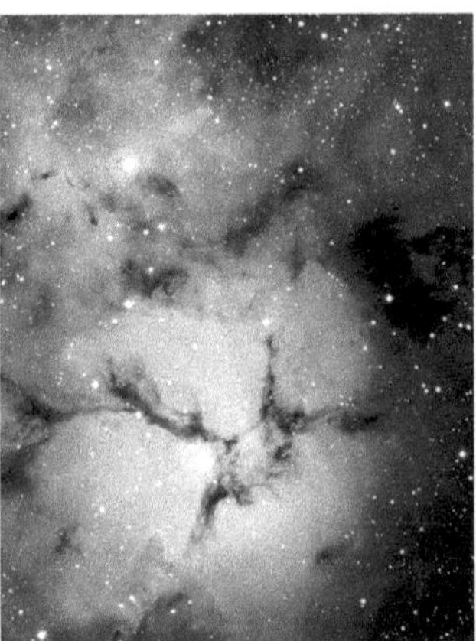

Figure 3.4. The famous Trifid Nebula. Image credit: ESO (CC BY 3.0).

This is incidentally the same physical principle that produces our blue sky, with the difference that molecules in our atmosphere are responsible for the scattering. The smoke from fires can result in excess dust particles in our sky—this can give the sky a reddish hue. This is because with all the additional particles in the air, all the blue light coming from the Sun through our atmosphere has been scattered out by the time it reaches us, leaving only the red wavelengths of optical light. This is the same principle that causes sunsets to be red—when we look toward the horizon, light passes through more of the atmosphere, which can result in all the blue light having been removed by the time it reaches us. In this image of the Trifid Nebula, we see dark dust features obscuring some of the light from the glowing emission nebula (red), creating its distinctive appearance. The red color indicates the presence of hydrogen being excited by hot, young stars. The cloud lying above is a reflection nebula, where the scattering of light by dust grains produces the bluish color we see here.

3.2.3 Near-infrared Astronomy

NIR astronomy was transformed by the advent of the large-format NICMOS and Hawai'i detectors, with their large field of view and high-resolution imaging capabilities. Combined with adaptive optics, these are workhorse instruments even today (e.g., Figure 3.5), with wide-ranging scientific applications covering all scales in astrophysics, from Earth and planetary sciences, to Galactic (Milky Way) and extragalactic studies covering stars, the interstellar medium, and even the largest structures in the universe. Major ground-based NIR observatories include those in

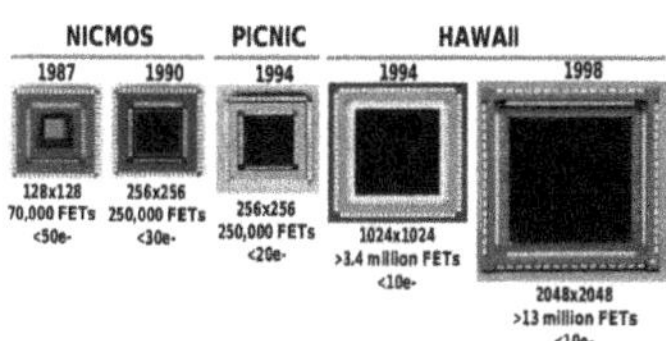

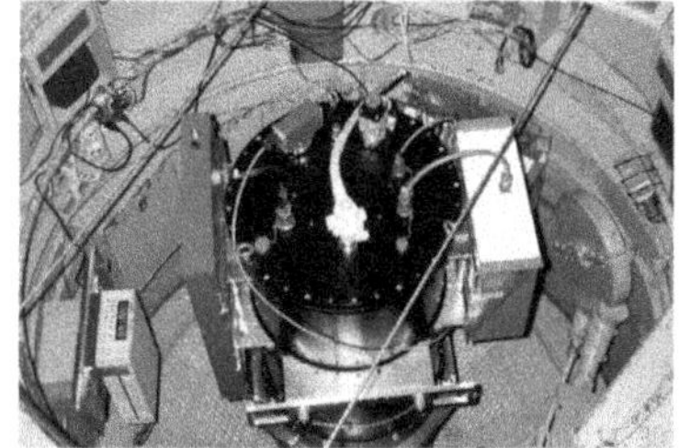

Figure 3.5. Instrumentation for ground-based near-infrared astronomy has greatly advanced since the 1990s, due in large part to detector development (upper left). The Palomar observatory and the Hale 5 m telescope (right) have hosted a number of instruments that evolved from the NICMOS to the Hawaii-II capability. Mounted at the prime focus of the Hale 5 m telescope, one of the most powerful near-infrared cameras is the Wide Field Infrared Camera (WIRC; lower left) with its 2048 ×2048 Hawaii-II HgCdTe. Image credits: Rockwell Scientific, T. Jarrett, & J.C. Wilson, respectively.

California (e.g., Palomar), Hawai'i (e.g., IRST, Subaru, Keck Observatory), and Chile (ESO-VLT). In space, the most notable has been HST (with the NICMOS camera), but future imaging and spectroscopic missions are on the horizon. Large-area surveys include 2MASS (carried out in the late 1990s) and the VISTA Hemisphere Survey (from ESO's Paranal Observatory), which is the most recent large-area mapping successor to 2MASS.

The 2MASS survey (Figure 3.6) was a natural follow-up to IRAS. Using ground-based telescopes and the new large-format detectors, it covered the whole sky with fidelity to angular scales of several arcseconds, revealing the Milky Way and the greater universe. It was in large part designed to detect the lowest-mass stars, those that are too small to ignite their cores—the so-called L and T dwarfs. The T dwarf companion to Gliese 570ABC is so cool that its atmospheric methane absorbs the 2 μm light, making it prominent in the H and J bands, ideal for 2MASS colors.

Compared to the optical window, the NIR can penetrate the thick layer of dust that obscures the Galactic Plane, revealing the stellar nurseries, YSOs (baby stars), star clusters, and supernova remnants (SNRs) that inhabit our sublime Galaxy (Figure 3.7).

Peering beyond the Milky Way, the background universe comes into high definition, namely, the large-scale structure of the Cosmic Web. This was the second keystone goal of 2MASS: pairing NIR imaging with spectroscopic redshift surveys, creating for the first time a 3D roadmap of the universe (Figure 3.8), changing our view of the local and high-redshift universe forever.

In the local universe, wavelengths between $1 < \lambda < 5$ μm are sensitive to the continuum arising from evolved stellar populations, forming the tail of the

Figure 3.6. The whole near-IR sky as viewed in an equal-area Galactic-coordinate projection, from 100 million 2MASS stars. At the center is the Milky Way, whose plane stretches across the sky. Also evident are the satellite Magellanic Clouds (to the lower right). Deep within are asteroids, comets, planets, brown dwarfs, star formation regions, stellar clusters, exploding stars, dying stars, and everything in between. Peer even deeper and you will see what is behind the Milky Way—galaxies! Image credit: NASA/IPAC.

Rayleigh–Jeans emission (where the flux is proportional to T/λ^4) from the aggregate stellar population of a galaxy. In the NIR, nearly all galaxies appear very similar in color (e.g., $J - K$), dominated by these luminous old stars. This uniform property can be used effectively as a direct measure of the total stellar mass by adopting an appropriate mass-to-light ratio (M/L) based on galaxy color or Hubble Type. It has the distinct advantage that dust extinction is minimal at these wavelengths, due to the mismatch in wavelength and the typical grain size of dust particles. At optical wavelengths (0.5 to 0.7 μm), the size distribution of interstellar dust grains means that those of similar size to the wavelength of incident light interact with and scatter (as well as absorb) this radiation—this contributes directly to the poor visibility that results from smoke-filled air (which is why firefighters often rely on infrared cameras that can "see" through the particles).

Another "homogeneous" feature of the NIR window is the stellar "bump" peaking at 1.6 μm, which is associated with the H^- ion found in the atmospheres of cool stars and ubiquitous to virtually all stellar populations (Jarrett 2000). The photometric signature of this feature can be used to estimate the redshifts of galaxies (Sawicki 2002); this will be heavily exploited by the Spherex mission.[1]

Although the NIR window is largely considered to be a relatively dust-free probe of old stars, emission from hot dust can be present between 2 and 5 μm. Dust absorbs radiation at shorter wavelengths and reradiates this energy as thermal emission. In the case of NGC 1068 (a.k.a. Messier 77, Figure 3.9), the source of the emission is the dust torus near a central active galactic nucleus (AGN) being heated to a high temperature ($T \gtrsim 1000$ K), causing the dust to radiate into the NIR.

[1] https://spherex.caltech.edu/Science.html

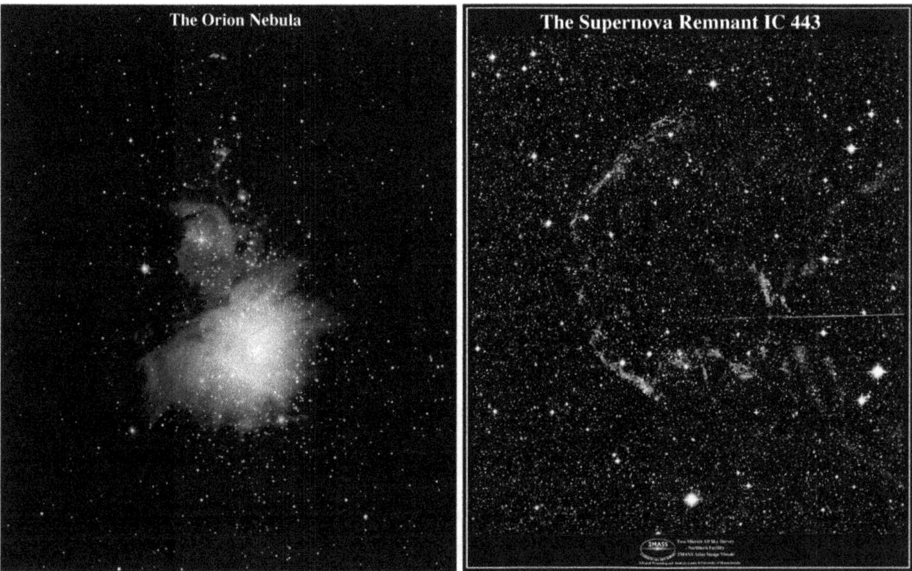

Figure 3.7. 2MASS near-IR view of star formation and star death. The Orion Nebulae (left) features the spectacular Trapezium Cluster of newborn, hot stars. But there are many thousands of stars that are newly forming in the giant molecular cloud of Orion; the NIR can penetrate the thick gas and dust to reveal the action-packed inner murky cores. (Right) At the other end of the circle of stellar life is the IC 443 SNR, showing the end result of a massive explosion thousands of years ago. The bluish arc to the northeast arises from shock-excited iron in the J and H bands, notably [Fe II] at 1.64 μm, whereas to the south, the ejecta has slammed into the molecular cloud, creating a shock front that is revealed by molecular hydrogen at 2.12 μm, one of the few (and rare!) ways to directly observe and measure molecular hydrogen. Image credit: NASA/IPAC and Rho et al. (2001).

The rest-frame NIR spectrum is also home to several prominent emission lines, tracing a variety of processes such as heating from star formation, gas composition, and shock excitation. Notable are the Paschen and Brackett series of hydrogen transitions ranging from the J to the K bands. For example, Paα (1.87 μm) observed by HST is widely used as a star formation tracer—under Case B recombination, this line is one-ninth the strength of Hα, but twice as strong as Paβ (at 1.28 μm in the J band), 12 times stronger than Brγ (2.16 μm in the K band), the two next best tracers in this window.

High-excitation lines in this band include the AGN diagnostic line [Si VI] λ 1.96 and the shock tracers [Fe II] (1.257 and 1.64 μm), [Si I] (1.64 μm), and [He I] 1.08 μm. First observed in the Orion Nebula (Gautier et al. 1976), excited molecular hydrogen serves as a tracer of star formation, as well as interstellar and intergalactic shocks (e.g., SNRs, the galaxy–galaxy collision in Stephan's Quintet, etc.), and is observed in the NIR as rovibrational transitions (e.g., H_2 1–0 S(1) λ2.122 and H_2 2–1 S(1) λ2.248). In addition, the CO molecule has rovibrational transitions in the K band, creating band heads at 2.32, 2.35, and 2.38 μm, respectively, which can be used to assess Galactic gas abundances and excitation temperatures. A beautiful example of NIR emission lines used to study the shock and gas environment of the ISM is shown in Figure 3.7(b), the SNR IC 443.

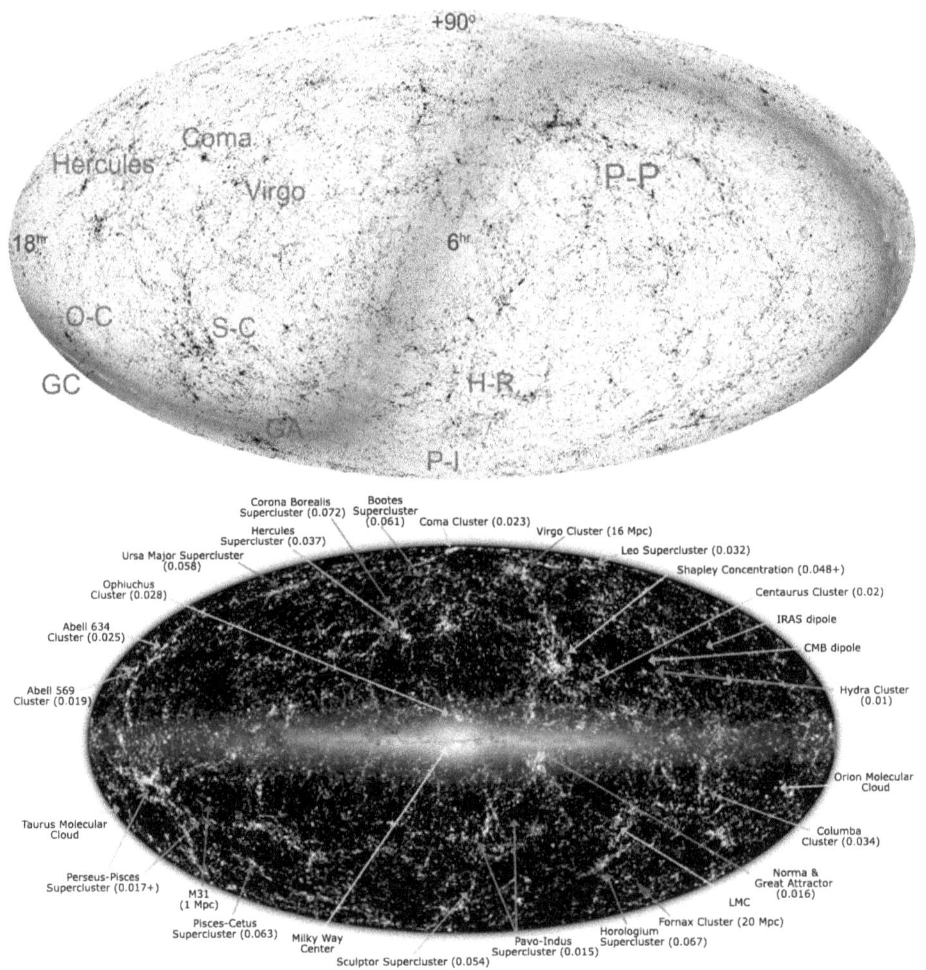

Figure 3.8. Panoramic views—equatorial (top) and galactic (bottom)—of the local universe reveal the distribution of galaxies beyond the Milky Way. The images are derived from the 2MASS Extended Source Catalog (XSC)—more than 1.5 million galaxies—and the Point Source Catalog (PSC)—nearly 0.5 billion Milky Way stars. (Bottom) The galaxies are color-coded by redshift, blue are the nearest sources ($z < 0.01$), green are at moderate distances ($0.01 < z < 0.04$), and red are the most distant sources that 2MASS resolves ($0.04 < z < 0.1$). Image credit: NASA/IPAC/Jarrett.

Owing to its longer-wavelength window, the NIR has also been exploited to study galaxies and large scale structure (the "Cosmic Web"; see Figure 3.8) and extreme objects at high redshifts, where prominent emission lines (e.g., Hα) are redshifted into the NIR window. The example shown here (Figure 3.10) is an NIR spectrum from the Gemini telescope, revealing the (rest) optical window, with continuum, absorption, and emission lines of a distant compact galaxy, located at a redshift of 2.1865. The most prominent emission lines are the [O II] and Hα, revealing how incredibly distant—and hence, early epoch—this galaxy lives.

Figure 3.9. The 2MASS view of NGC 1068. Image Credit: 2MASS.

3.2.4 Mid-infrared Astronomy

Space telescopes opened the window to the "cool" cosmos, starting with IRAS and its short-wavelength imaging bands of 12 and 25 μm, followed by ever more complex and capable machines: ISO, Spitzer, AKARI, and WISE. These missions combine to fully cover the mid-IR window between 5 and 50 μm, and with angular and spectral fidelity to reveal hidden star formation regions in the Milky Way and nearby galaxies, the engines that grow and evolve our universe.

Whereas NIR wavelengths are particularly useful for ameliorating the impact of dust extinction, shifting to the longer wavelengths of the mid-infrared (5–50 μm) allows us to see the emission from dust and the interstellar medium. This is illustrated in Figure 3.11, which is of the inner Milky Way, normally hopelessly obscured by gas and dust in the optical bands, yet unveiled to be fully active and alive with star formation in the mid-IR.

The unique capabilities of the mid-IR window are even better demonstrated in Figure 3.12, where M51 is used to contrast the differences between dust obscuration (in the optical/visible) and dust emission (in the infrared). In this band, there are several sources of emission: continuum emission from starlight, thermal emission from the ionized interstellar medium, and most prominently, continuum emission arising from dust "reprocessing" radiation. Dust absorbs ultraviolet radiation from massive stars, is heated, and then reradiates this energy as thermal emission at mid-infrared wavelengths.

In particular, very small dust grains are transiently heated by nonionizing UV photons, where these small grains can be heated to high temperatures by single UV photons. This is therefore a key window to study dusty star-forming regions where YSOs are being produced, both in the Milky (e.g., the nearby Rho Ophiuchus giant molecular cloud) and well beyond (e.g., the LMC at 50 kpc distance; see Figure 3.13).

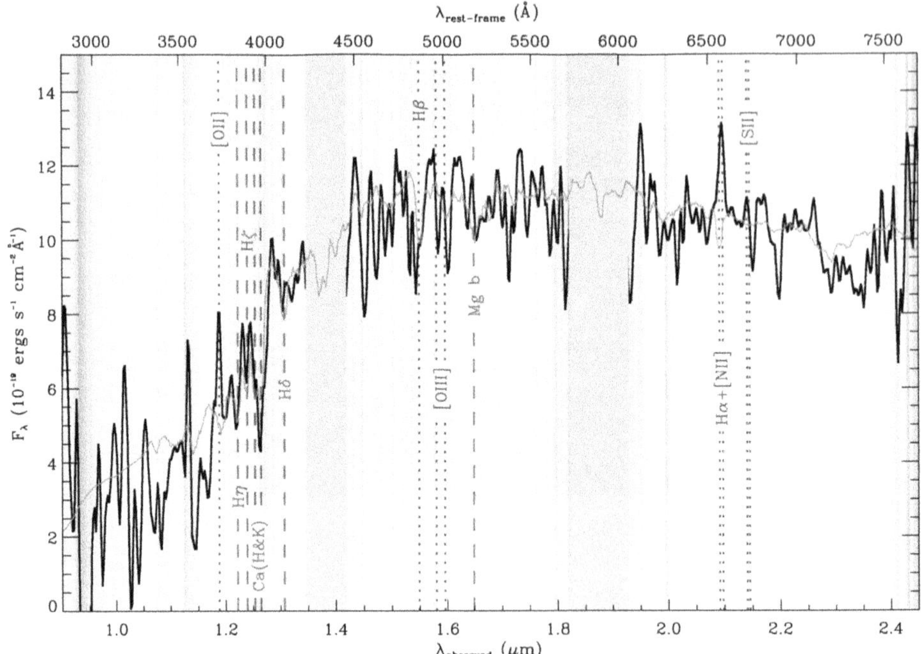

Figure 3.10. Emission-line galaxy at a redshift of 2.1865, acquired with the Gemini Near-Infrared Spectrograph. Note the Hα line at 2.1 μm. Image credit: Kriek et al. (2009).

Astrophysical $\lambda > 5$ μm emission is sensitive to warm ($T > 100$ K) dust, and emission from PAHs is found between 6 and 17 μm. PAHs produce distinctive broad emission bands in the MIR and are linked to ongoing or recent star formation (see text box). However, PAH molecules are destroyed by intense radiation such as those found in HII regions and close to AGNs (e.g., Genzel et al. 1998) and therefore provide a sensitive diagnostic for determining the dominant processes contributing to observed mid-IR emission (see Li 2020 for an excellent review article).

Both ISO and Spitzer have highlighted the immense value that mid-infrared spectroscopy holds for decoding the internal physics that drives galaxy evolution. It probes ionized gas (which produces fine-structure lines), molecular gas (e.g., warm molecular hydrogen), and dust grains in the interstellar medium. Warm (small graphite) dust grains produce the mid-infrared continuum longwards of 10 μm, while large dust grains (silicates) give rise to absorption features (Chiar & Tielens 2006).

Large molecules produce emission and absorption features, in particular, the bending and stretching modes of PAHs already discussed. They also reveal the composition of the interstellar medium. Mid-infrared diagnostics based on continuum and emission-line measures can be used to ascertain key properties. For example, one can separate star formation vs AGN activity by using [Ne II] λ12.81 and [Ne III] λ15.56 (which in combination act as excitation measures), [S III] λ33.48 μm (tracing HII regions), and [Si II] λ34.82, which traces dense PDRs (photodissociation regions) and

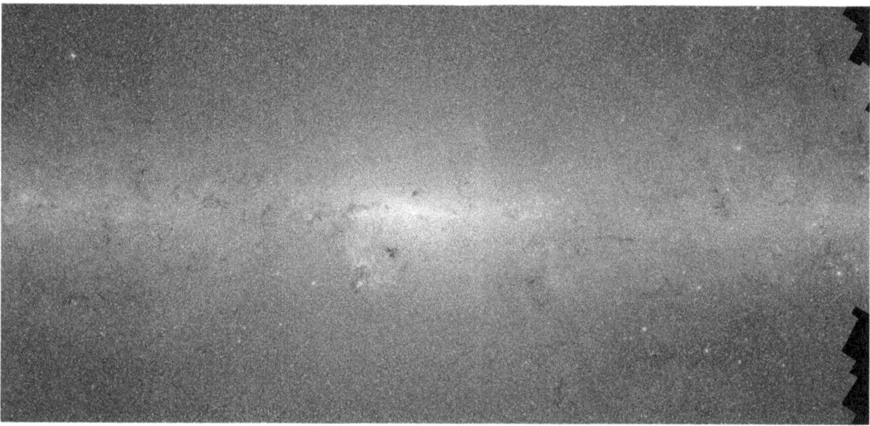

Figure 3.11. Center of the Milky Way as seen by the Spitzer. This panorama is the Galactic Legacy Infrared Mid-Plane Survey Extraordinaire (GLIMPSE) that surveyed the plane of the inner Galaxy. The image colors are gauging the temperature of galactic objects, from stars (white to blue) to the interstellar medium and dusty star formation regions (oranges to reds). Image credit: NASA/Spitzer.

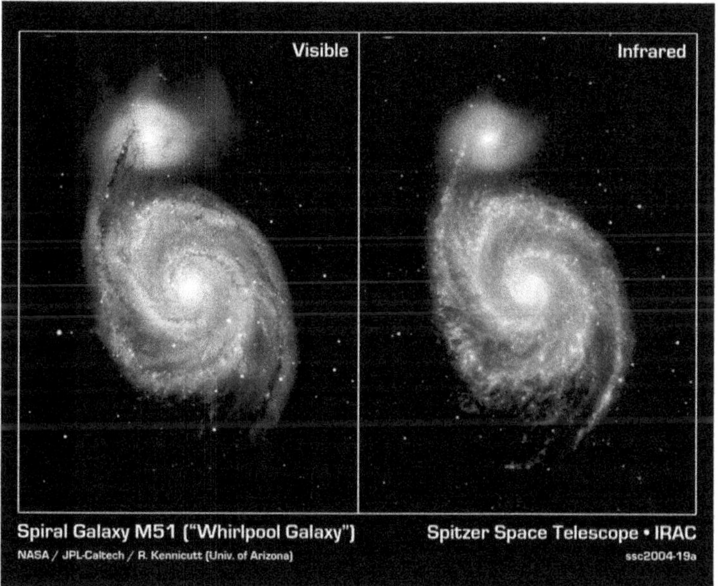

Figure 3.12. The magnificent Whirlpool Galaxy (M51), comparing optical and mid-IR views. This is an example of a closely interacting galaxy pair—M51a is the central grand-design spiral, and M51b is the old and compact ellipsoidal companion, whose tidal dance induces a density wave that compresses the gas and dust (most evident in the visual image) into tight spiral arms that burst with the stellar nurseries (revealed in the mid-IR) that build the galaxy. The visual/optical window is limited by the dust absorption and reflection, whereas the mid-IR "shines" as the dust reemits the absorbed light from newly born stars. It also reveals the underlying stellar structure from the older generation of stars. Image credit: NASA/Spitzer.

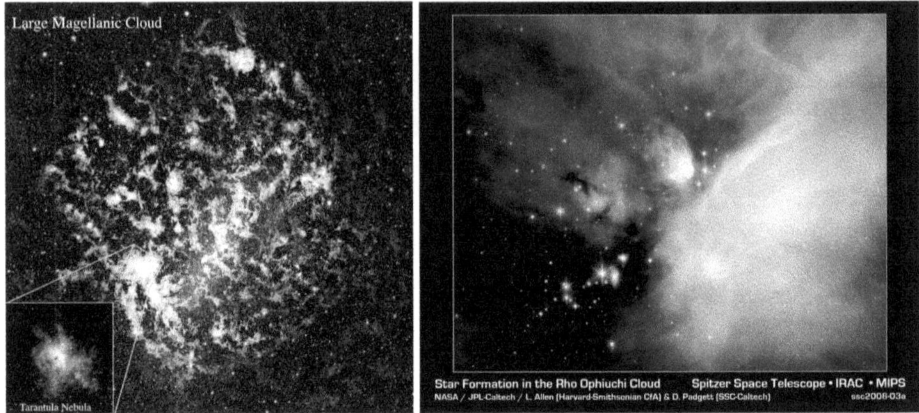

Figure 3.13. Star formation revealed in the mid-IR. (Left) The Milky Way's largest satellite, the LMC, and its luminous "Tarantula Nebula" star formation complex as viewed by WISE. (Right) The ρ Ophiuchi giant molecular cloud as imaged by NASA's Spitzer. One of the nearest of its kind (a mere 400 lt-yr from the Sun), the dark cloud is a baby star factory, with its YSO well under a million years in age. Image credit: NASA/Spitzer and Jarrett et al. (2019).

X-ray-dominated regions powered by AGNs (Dale et al. 2006). Similarly, combinations of continuum and PAH emission can be used to establish the relative contributions of AGN versus PDR versus HII excitation (Peeters et al. 2004; Armus et al. 2007). The high-excitation lines of [Ne V] λ14.3 and [O IV] λ25.89 are particularly useful for identifying AGNs. In combination, therefore, the mid-infrared can be used to determine the likelihood of an AGN being present and its connection to its galaxy host (Lacy & Sajina 2020).

Further to this, electron density can be determined from the ratio of [S III] λ18.71 and [S III] λ33.48, while [Ne II] λ12.81 and [Ne III] λ15.56 can be used as a star formation rate (SFR) indicator (Ho & Keto 2007). The mid-infrared also boasts access to the pure rotational transitions of warm (excited) H_2—the only direct measure of molecular hydrogen. Although H_2 is typically excited in PDRs, cosmic-ray and X-ray heating are additional excitation mechanisms. In fact, the ratio of H_2/PAH emission has been shown to be a powerful discriminator of shock heating (Ogle et al. 2007). Several of these key lines can be seen in Figure 3.14, which shows the Spitzer low-resolution spectrum of the main shock region in Stephan's Quintet, dominated by H_2 and other tracers (Cluver et al. 2010a).

In short, mid-infrared observations enable the study of the composition and physical processes that drive the evolution of the ISM, including stellar feedback mechanisms.

Stellar Mass and Star Formation: The mid-IR has the distinct advantages of (1) a considerably lower dust extinction in the 3–5 μm range, (2) its sensitivity to the 1–5 μm emission of luminous, evolved stars that typically dominate the underlying (aggregate) mass of a galaxy, and (3) its sensitivity to star formation activity through the emission of molecules and dust grains that have been excited by powerful stellar emission at shorter (UV/optical) wavelengths.

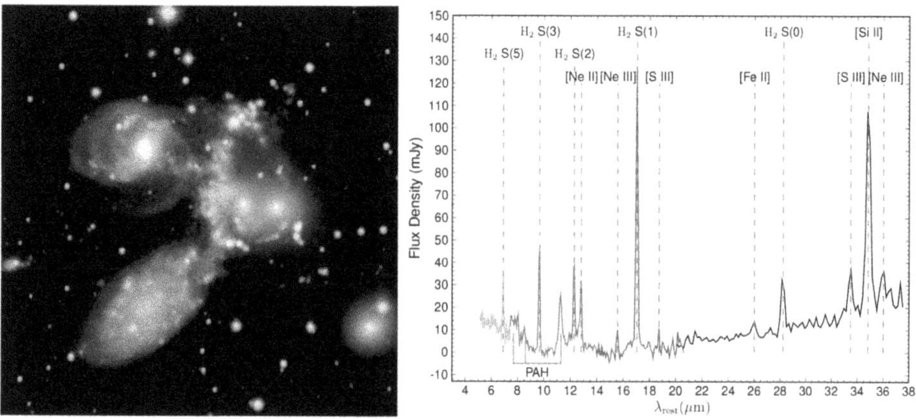

Figure 3.14. Stephan's Quintet of galaxies (left), revealed by Spitzer to be undergoing a massive tidal shock. (Right) Spitzer mid-IR spectrum of the main shock excitation region (long green ridge in image)—dominated by strong emission lines—demonstrating one of the few ways to directly measure molecular hydrogen in the universe. Image credit: NASA/Spitzer and Cluver et al. (2010a).

The Spitzer and the WISE mid-IR all-sky survey were designed with these advantages in mind, simultaneously imaging the NIR (i.e., stellar-dominated) and mid-IR (ISM-dominated) windows. Hence, a powerful new method for studying galaxy evolution has come from these missions, most notably, through the calibration of the global (a) stellar mass and (b) star formation rate (SFR) metrics fundamental to understanding galaxy growth; Figure 3.15 illustrates the evolutionary history of the spiral galaxy M83.

An example of how astronomers study ensembles of galaxies using these metrics is given here, where the largest galaxies in the sky have been measured using WISE imaging and large-aperture photometry (Jarrett et al. 2019). The stellar mass, M_{stellar}, of each galaxy is estimated using the W1 3.4 μm integrated flux and the W1 – W2 color, whose value for typical galaxies was calibrated to the mass-to-light ratio (M/L) by Cluver et al. (2014). In the NIR, from 1 to 5 μm, the M/L ranges from unity at short wavelengths, to a value of about 0.3 for the longer wavelengths, with a color dependency (likely related to metallicity and population morphology). In the diagram, Figure 3.16, the WISE colors show a clear separation between galaxies whose light is dominated by old stars (these are the spheroidal types) and those dominated by young, disky populations (i.e., star-forming), which also tracks with the M/L: early-type spheroidals have the highest M/L ratios while the disk galaxies have the lowest.

To gauge the star formation activity, Cluver et al. (2017) calibrated the WISE W3 and W4 bands to the total infrared luminosity, which closely traces the dust-obscured star formation in galaxies, dominated by the far-IR component (see next section). Hence, the W3 spectral luminosity can be used to estimate the SFR in solar masses per year:

$$\text{Log}_{10}\,\text{SFR}_{\text{TIR}}(M_{\odot}/\text{yr}) = 0.889(\pm 0.018) \times \text{Log}_{10}\,\nu L_{12\,\mu\text{m}}(L_{\odot}) - 7.76(\pm 0.15), \quad (3.1)$$

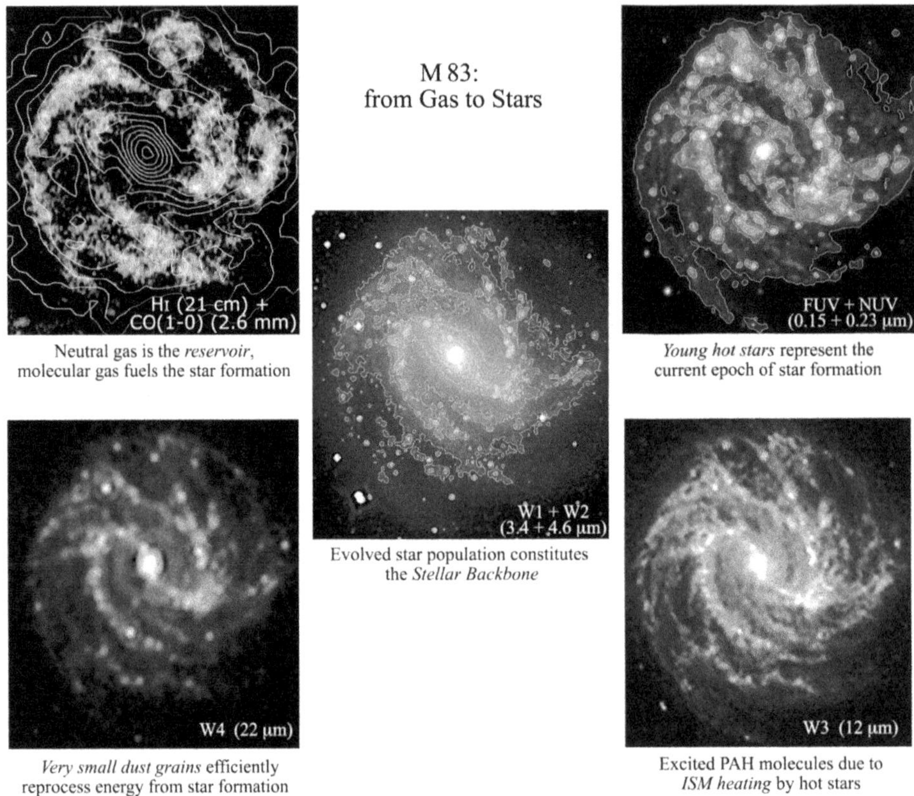

Figure 3.15. Star formation history of the M83 galaxy. Clockwise from upper left: starting with the hydrogen fuel reservoir, characterized by spiral arms and a core of molecular gas, which collapses to form stars, where the most massive OB stars (seen in this GALEX image) energize the ISM, gas, and dust reradiating in the mid-IR as seen in the WISE W3 and W4 images. Finally, over time, the stellar content builds, forming disks and bulges, and the past history of M83 is revealed with the near-IR WISE W1 and W2 images (center). Image credit: Jarrett et al. (2013).

where $\nu L_{12\,\mu\mathrm{m}}$ is the spectral luminosity, normalized by the solar luminosity ($L_\odot$), and the stellar continuum removed using the W1 luminosity as a proxy for the stellar continuum (see Cluver et al. 2017 for details). Figure 3.16 shows the resulting SFR as a function of the integrated stellar mass, perhaps the most important global property for galaxies. We see a clear "sequence" from low to high stellar mass: As galaxies grow larger, their SFR increases. This secular evolution continues for Gyrs, with occasional starburst phases where the Galaxy suddenly moves well above the "main sequence" (e.g., M82) until the burst phase ends (50–100 Myrs), and then it settles back down on the sequence with a higher stellar mass. In this way, galaxies continue to grow until they reach a phase in their evolution when their fuel is consumed, or the SF disk becomes too hot, or the star formation itself is disrupted due to powerful events, such as gravitational tidal interactions or strong AGN activity phases. The nature of these "quenching" mechanisms is less understood and is a source of vigorous research in the field of galaxies and cosmology.

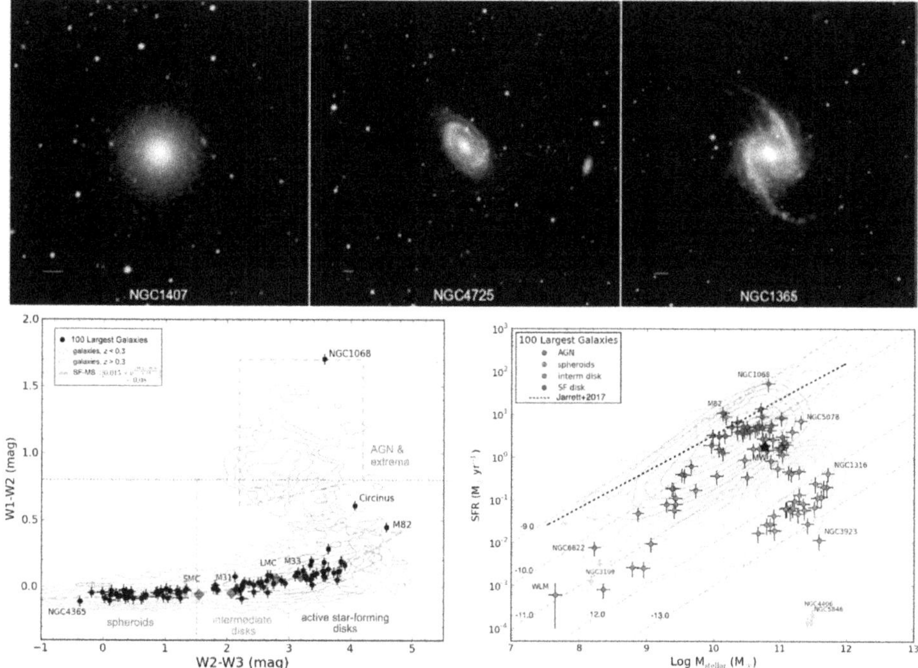

Figure 3.16. Fundamental galaxy properties as measured in the mid-IR. The top panels show three examples of galaxies that are spheroidal (left), intermediate disk (middle), and star-forming spirals (right). The bottom panels are statistical metrics: (left) shows the WISE color–color diagram for the largest of galaxies, whose colors separate the different classes of galaxies, as well as help identify those with AGNs. (Right) SFR as a function of the aggregate stellar mass; this is known as the star-forming "main sequence" (SFMS). It is clear that the morphological type of the Galaxy correlates with positions on the SFMS, from the extrema of starburst galaxies (e.g., M82) with unusually high SFRs, to the quenched "red and dead" elliptical/spheroidal galaxies that are pouring down from the SFMS. Image credit: Jarrett et al. (2019).

3.2.5 Far-infrared and Submillimeter Astronomy

Probing the cold interstellar medium, the bulk of the bolometric luminosity from dust-enshrouded star formation emits in this window, from 70 to 1000 μm. Blackbody continuum emission arises from a range of dust-grain compositions and sizes, although predominantly from larger grains at thermal equilibrium (compared to the "warm" dust continuum of the mid-infrared). Moreover, the submillimeter to millimeter window contains a multitude of atomic and molecular (emission-line) transitions, most notably those of carbon monoxide (CO), which may be used to probe the densest and coldest cores of star formation regions, hence providing insight into the chemistry of star formation.

Tracing this cold gas and dust in the diffuse and dense ISM, with temperatures of 10–50 K, requires capturing emission at very low frequencies using special bolometer and heterodyne technology in combination with space telescopes (or telescopes located on the most remote mountaintop locations on Earth). As IRAS first showed, far-IR emission is ubiquitous throughout the Milky Way and across to

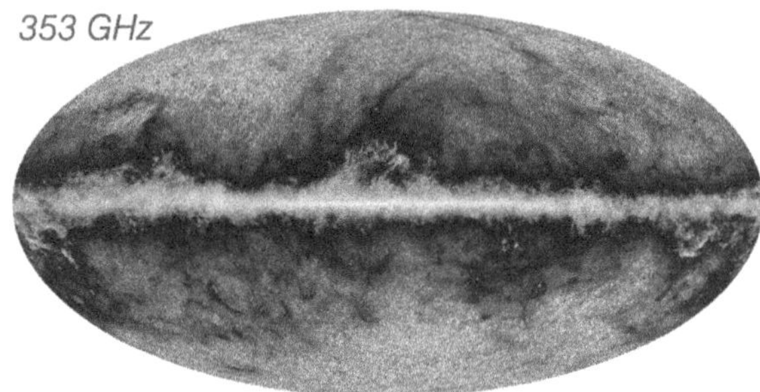

Figure 3.17. Planck view of the observable universe at 850 μm (353 GHz), dominated by the "foreground" emission from the Milky Way itself. Image credit: ESA/Planck.

the farthest reaches that our telescopes can peer into the early universe. More recent international missions, most notably from the Herschel and Planck Space Telescopes (see Figure 3.17), have spectacularly shown that the cold dust component of the ISM is by far the most important mechanism for absorbing starlight, which is then reemitted at these longer wavelengths, filling the entire sky with a blaze of (predominantly) Milky Way light.

In order to study the extragalactic sky and the early universe with its distant echo of the Big Bang, this "foreground" light must be identified and removed to clearly see what is beyond. Planck measured and characterized this foreground light in several bands across the far-infrared to the submillimeter, including at 850 μm (Figure 3.17), allowing for robust subtraction of the Milky Way and local universe (nearby and distant galaxies) components, and thereby reveal with unprecedented fidelity the structure in the cosmic microwave background. For the most recent results, see Planck Collaboration et al. (2020).

The far-IR to submillimeter window is the most effective measure of current star formation in dust-rich systems, as the ultraviolet and optical-blue radiation from massive stars is absorbed by the interstellar grains and reradiated as thermal continuum (Cram et al. 1998). Some important emission lines are also found in this window, including the fine-structure lines of [O I] at 63 μm and ionized [C II] at 158 μm, which are powerful coolants of the SF-excited gas of photodissociation regions, with temperatures of 20–2000 K. Molecular CO (3—2) at 869 μm is an important tracer of dense gas ($>10^4$ cm^{-3}) associated with newly forming stars (Figure 3.18).

The far-IR–submillimeter window is a direct probe of the physical conditions in the ISM of the Milky Way and galaxies, which in turn is energized by newly forming stars, dying stars through atmospheric shedding, winds and explosions, and large-scale (>1 pc) jets from a variety of sources including black holes and YSOs. Astronomers attempt to disentangle these physical (and feedback) mechanisms by gathering as much multiwavelength data that complements the far-IR and submillimeter, including imaging maps, and spectroscopic and even kinematic information.

For example, Herschel spectroscopy has been used to discover large-scale outflows from galaxies where superwinds have punched through their flattened disk, spouting a kiloparsec-sized fountain of dense material (gas and dust)—starbursts are shedding into the intergalactic medium. The example here (Fischer et al. 2010)

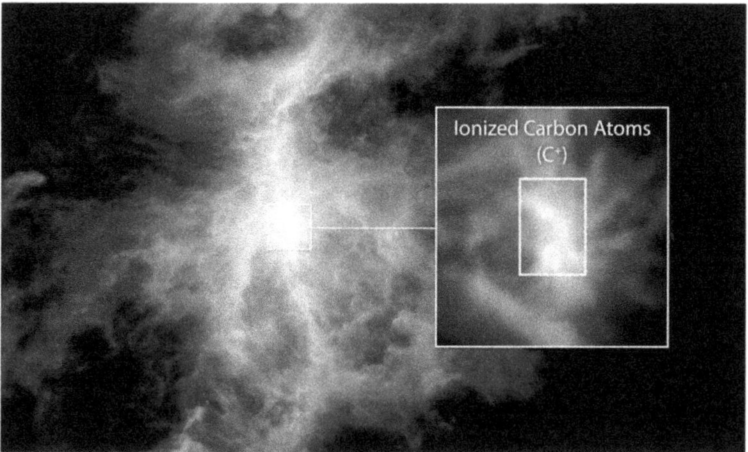

Figure 3.18. Into the Sword of Orion: radiation from cold gas and dust emanates from one of the most famous stellar nurseries in the Milky Way. This image is a composite of far-IR continuum from Herschel-PACS at 100 μm (blue) and 160 μm (green), and submillimeter from Herschel-SPIRE at 250 μm. At the center, shown in yellow, [C II] (ionized carbon) line emission at 158 μm from Herschel-HIFI is one of the most important coolants of the SF-excited interstellar medium. Image credit: ESA/Herschel/IPAC.

shows the OH (hydroxyl) double emission line at 120 μm, seen toward the remarkable AGN–starburst hybrid galaxy Mrk 231, exhibiting a classic P Cygni profile characteristic of gas outflow (Figure 3.19).

Simple models can be fit to SEDs using modified blackbody emission curves corresponding to a range of grain sizes and compositions (translating to emissivity) in order to characterize the temperature, density, and aggregate mass of the dust grains that are producing the bulk of the luminosity. An example of characterizing the infrared radiation from the most luminous objects in the universe—ULIRGS and HyLIRGs[2]—is shown in Figure 3.20.

First discovered by IRAS, Arp 220 may be extraordinarily luminous, but it is also invisible at shorter wavelengths because it is so dense with gas and dust (Soifer et al. 1987). The object remained a mystery until quite recently when higher resolution NIR observations showed it to be a merging system, likely two gas-rich spirals that merged (known as a "wet" merger), compressing the cold molecular gas (see the dense core of Arp 220 in Figure 3.20, as traced by CO (3–2) at 346 GHz) and triggering a starburst phase. The galaxy is consuming the gas reservoir at a rate that would completely exhaust the fuel in well less than a gigayear, although it will likely quench before all gas is consumed. Star formation creates metals and dust, copious amounts of it, which quickly enshroud the source in UV-absorbing materials that reradiate at much longer wavelengths from the mid-IR (warm dust component), peaking in the far-IR (cold dust) and diminishing in the submillimeter (very cold dust; see Figure 3.20). This massive thermal signature of starbursting galaxies gives way to synchrotron emission in the radio (centimeter-wave) window.

[2] Ultraluminous infrared galaxies and hyperluminous infrared galaxies.

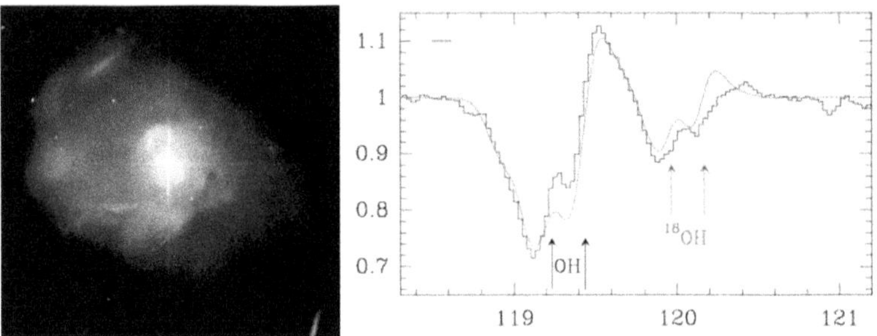

Figure 3.19. AGN superwind! Outflows from the core of the powerful Mrk 231 galaxy (left, HST optical composite), as revealed by the classic P Cygni profile of the OH doublet at 120 μm (right, from the Herschel Space Telescope). Image Credit: NASA/ESA and Fischer et al. (2010) reproduced with permission © ESO.

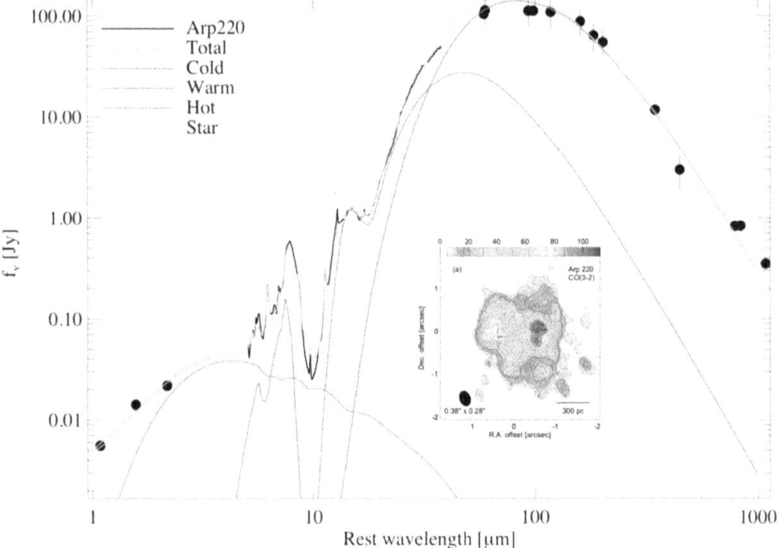

Figure 3.20. Infrared energy distribution for the ultraluminous infrared galaxy (ULIRG) Arp 220. Spanning the near-IR to the submillimeter, the flux density, measured from a combination of space and ground-based telescopes, of this remarkable starburst galaxy is dominated by the cold dust component (red line), peaking at ~100 μm, corresponding to temperatures of 20 to 30 K. The warm dust component (green) line arises from smaller and much warmer grains, ranging from 300 to 700 K. The inset shows Submillimeter Array (SMA) mapping of the cold molecular gas in Arp 220, traced from the CO (3–2) emission line at 869 μm, which reveals dense ($>10^4$ cm^{-3}) cores where stars are born. Image credit: Armus et al (2007) and Sakamoto et al. (2008).

Galaxies forming, growing, and evolving at much earlier epochs are of great interest in our quest to understand galaxies and their evolution within the Cosmic Web. The far-IR–submillimeter window is notably powerful to this end, as redshifted light from high-z galaxies may be detected at the longest wavelengths. The reason is that the thermal peak we observe in the Milky Way and nearby galaxies (such as Arp 220) moves to higher wavelengths, shifting the light from the far-IR to the submillimeter. Hence, this "negative" k-correction works in favor of the submillimeter. This is similar to how the mid-IR WISE bands get brighter as starlight that peaks in the NIR H band at 1.6 μm redshifts into these longer bands.

In addition to space telescopes, from the driest locations on Earth (the Andes, Maunakea, Antarctica) it is possible to detect submillimeter photons from distant galaxies. Pioneering work in the late 1990s from these sites (and well before the launch of Herschel), astronomers were building bolometer arrays (e.g., SCUBA, MAMBO) that did small surveys, detecting sources that were invisible (or nearly so) at all other wavelengths. These were the submillimeter galaxies (SMGs; Blain et al 2002). They are characterized as being compact, dusty and luminous, and invisible to other observations.

Many studies later, including those from the powerful Herschel Space Telescope revealed that SMGs are indeed high-redshift (typically $z > 3$), gas-rich, high-SFR (typically $>500\ M_{\odot}/yr$), and massive (~Milky Way) systems that are likely the early-epoch counterparts to the ULIRGS that we observe in the local universe (e.g., Arp 220), but have a higher volume density and were therefore more important in the early universe. Sensitive radio interferometric surveys are able to detect SMGs, which is due to the thermal infrared-to-radio synchrotron connection in star-forming galaxies. Finally, while high-z SMGs may harbor AGNs, the starburst emission will usually dominate the far-IR–submillimeter light we see today.

We highlight one final interesting case of note, discovered by Herschel and probed in detail by ALMA. Consider an SMG lensed by an invisible foreground galaxy, and thus appearing unusually (if not, unphysically) luminous given the apparent redshift. Because Herschel is unable to see the foreground galaxy—it is simply too faint compared to the dusty and luminous background galaxy—the lensing is in fact undetected. Follow-up observations, or archival deep optical–IR observations, may reveal the interloper and hence the lens. One such outstanding example is SMG SDP 81 (Figure 3.21).

3.2.6 Remarkable Infrared and Submillimeter Telescopes

In Figure 3.22 we show images of the remarkable instruments used to study the infrared and submillimeter universe.

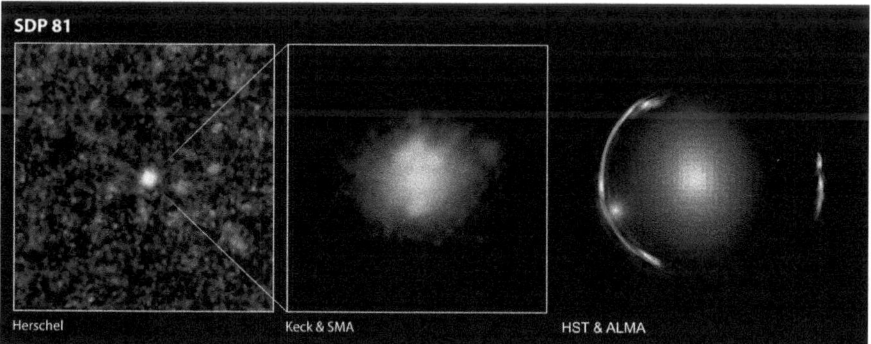

Figure 3.21. SMG SDP 81, discovered by Herschel, is more than meets the eye: high-resolution imaging from Keck and HST, and long-baseline submillimeter/millimeter interferometry from ALMA reveal a foreground "lensing" galaxy and a stunning, nearly perfect "Einstein Ring." The high-redshift (background) dusty galaxy is lens distorted and amplified in the far-IR–submillimeter window, easily detected by Herschel as an unresolved point source. Image credit: ESA/NASA/JPL-Caltech/Keck/SMA, Y. Hezaveh/Stanford University/ALMA (NRAO/ESO/NAOJ), NASA/ESA Hubble Space Telescope (CC BY 3.0).

Figure 3.22. Beyond what our eyes can see: Since the early 1980s astronomers have peered into the cosmos using these incredible telescopes and instruments that can sense and discern the universe in the cool window between 1 and 1000 μm. Image credit: (top left to bottom right) (1) Infrared Processing and Analysis Center, Caltech/JPL; (2) ESA; (3) Infrared Processing and Analysis Center, Caltech/JPL; (4) NASA/Caltech/JPL/Spitzer Space Telescope Science Center; (5) F. Kerschbaum/ESO; (6) Courtesy of NASA; (7) Courtesy of NASA/Jim Ross; (8) This Scale model of the Akari (ASTRO-F, 2006-005A) exhibited at Noshiro City Children's Center.jpg image has been obtained by the authors from the Wikimedia website where it was made available by User:掬茶 under a CC BY-SA 4.0 licence. It is included within this article on that basis. It is attributed to User:掬茶; (9) This Model of James Clerk Maxwell Telescope (JCMT).jpg image has been obtained by the authors from the Wikimedia website where it was made available Mike Peel under a CC BY-SA 4.0 licence. It is included within this article on that basis. It is attributed to Mike Peel (www.mikepeel.net); (10) NASA/Univ. of Hawaii; (11) Gemini Observatory, NoirLab; (12) Infrared Processing and Analysis Center, Caltech/JPL; (13) Reprinted with permission from the Karin Öberg/ Center for Astrophysics | Harvard & Smithsonian; (14) Caltech Submillimeter Observatory (CSO); (15) Kaunana, University of Hawaii. Reprinted with permission from Klaus Hodapp; (16) ESA/NASA/JPL-Caltech.

What are polycyclic aromatic hydrocarbons (PAHs), and why do they matter? PAHs are ubiquitous. If you've ever burned your toast, you have created PAHs! And you've definitely seen them as air pollution. In astronomy, carbon-rich evolved stars are believed to be the main source of PAHs, formed during their mass loss phase. They are found in regions ranging from planetary nebulae, protoplanetary disks, H II regions, and reflection nebulae, and as an essential ingredient in the interstellar medium of galaxies.

PAHs are hydrocarbon molecules, whose easily (UV and visual) photon-excited bending and stretching mode "vibrational" bands pepper the mid-IR window. IRAS first encountered them in its observations of clouds within our Milky Way. However, mostly a mystery at the time, they were called unidentified infrared emission bands (UIBs) and later identified in laboratories as belonging to the vast and varied PAH family of ringed hydrocarbons. ISO and Spitzer spectroscopy revealed their emission bands, notably at 3.3, 6.2, 7.7, 8.6, 11.3, and 12.7 μm, but they appear elsewhere in the mid-IR window and perhaps remain unidentified to this day.

They are indeed ubiquitous and important in the ISM—perhaps 10% of all carbon is locked up in PAHs. Even high-redshift galaxies have their distinctive emission signature, revealing to us the very same processes of gas and very small dust grains heated by star formation occurring in earlier epochs.

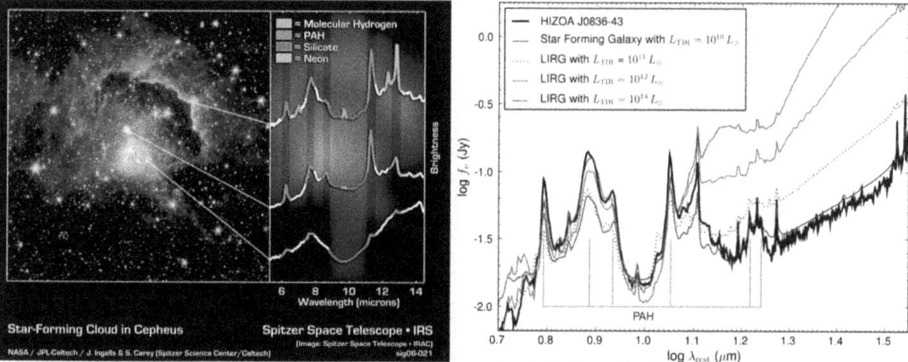

PAHs in a Milky Way star-forming region (left) and in a starburst galaxy (right). Note the broad emission features at 6, 8, and 11 μm from PAH bands, as well as an underlying continuum from stochastically heated small dust grains, and the broad silicate absorption feature at 10 μm. The strength of the emission features depends on the heating sources, the stellar radiation field they are bathed in, and the density and composition of carbon (metallicity). If the radiation field is too strong, PAHs may be destroyed, as with close proximity to AGNs. In contrast, the PAH bands may be exceptionally strong in starbursting galaxies; the example here (right) is one of the most powerful PAH emitters ever discovered, the Galaxy HIZOA J0836–43 (Cluver et al. 2010b), which is graced with copious amounts of molecular and atomic hydrogen to fuel the vigorous star formation stage it is now experiencing. Image credit: NASA/IPAC/Cluver.

Multiwavelength Image Construction: Creating color multiband graphics for astrophysical imaging is a powerful way to visualize complex data. The resulting images are not only pleasing to the eye (and to the public!) but convey information on the astrophysical mechanisms that produce the emission. If the imaging bands provide independent (i.e., uncorrelated) information, then a careful color combination will bring out the differences between the two or more bands.

There are graphics tools available where you can assign each image a mono-"color" transform—essentially one color with shades that trace brightness variations across the image. For example, using a "blue" transform results in the image appearing nominally blue but will brighten to a blue-white shade for the bright (luminous) emission in the image. Typically we do this color combination using the additive primaries: red, green, and blue (RGB), but it is also possible to use more hues of the rainbow if you have more than three bands. After combining the different color-transformed images, you get an additive result that is a new color (i.e., the combination of the primaries).

The example we provide here is the visually striking southern grand-design spiral galaxy M83 (NGC 5236). The imaging here comes from GALEX (ultraviolet bands of far- and near-UV) and WISE (the infrared bands of 3.4, 4.6, 12, and 22 μm). We apply the monochromatic color transform to each, respectively: magenta to FUV, blue to NUV, green-cyan to NIR, orange to the W3 12 μm, and red to the W4 22 μm. Hence, the progression in the "rainbow" is from blue: "hot stars," to green: "old, cooler stars," and to red: cold "ISM" revealing star formation).

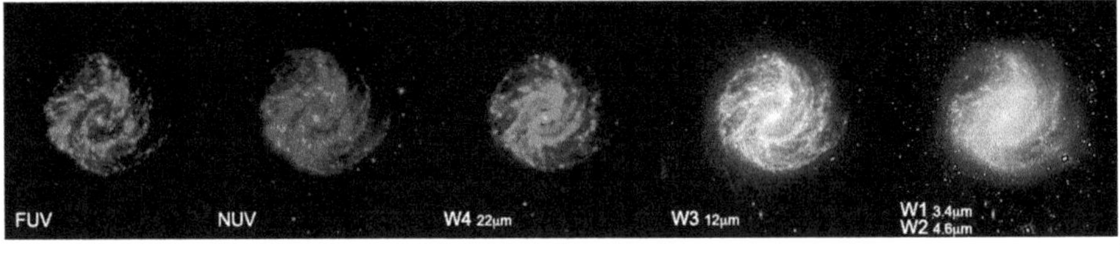

This combination of photometric bands is particularly powerful because of the astrophysics being traced—the UV is produced by hot young stars, the emission either escapes (so we see it in the images) or it is absorbed by the gas and dust of the star formation regions (i.e., the spiral arms). The absorbed UV emission is then reradiated in the mid-infrared, where the WISE 12 and 22 μm images show them clearly. Finally, the NIR bands of WISE (3.4 and 4.6 μm) are sensitive to the old and evolved stellar populations; that is, the past star formation. Hence, the resulting colorful images are a visual representation of the star formation history of M83 (see also Figure 3.15).

Source Characterization with Imaging: Unlike stars, galaxies are not simple objects to measure, and hence source characterization involves a number of measurements: position, shape and symmetry, size, photometry (integrated flux), surface brightness, compactness, radial profile, disk and bulge properties, etc.

Let us consider photometry and axisymmetric properties. We start by separating the target from the background—identify stars, background galaxies, and any other emission not associated with the target galaxy and remove as best from the image. The background or sky level (see **Background "Sky" Level Determination**) is then removed (or tracked). Consider the example NGC 4639 as imaged by WISE.

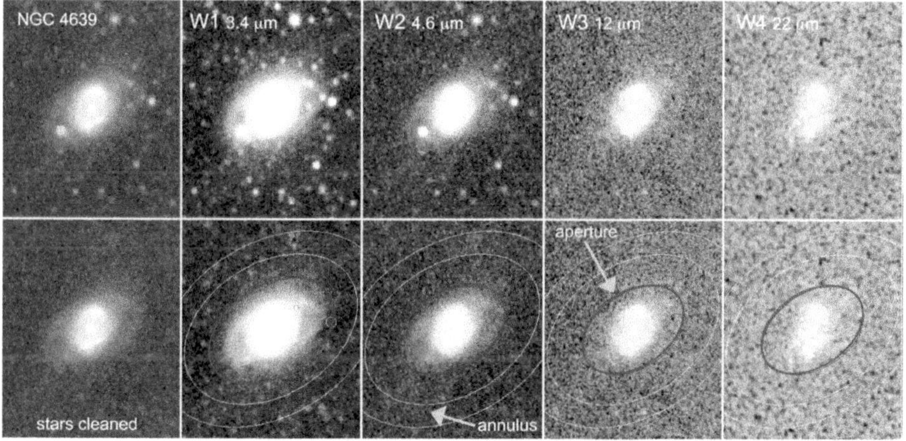

With a cleaned image and the background level and noise (rms of the background) determined, the coordinate position (often the centroid of the nucleus) and shape (inclination and position angle) are then measured. In this example, the elliptical isophote corresponds to the 3σ of the background (i.e., three times above the sky noise). Finally, we measure the integrated flux as close to the background noise as possible; in this case, we use the 1σ isophote (see image above, the blue aperture).

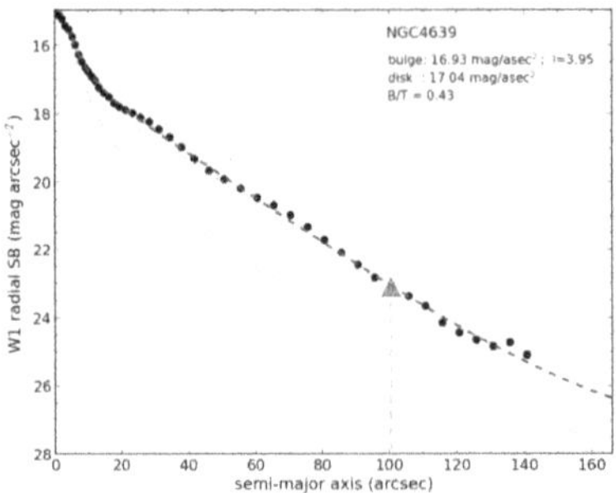

The isophotal flux represents a fraction of the "total" flux (in our WISE case, about 95%); to measure the total, either model the source or use a large asymptotic aperture. Each has advantages and downsides.

The axisymmetric elliptical radial profile can be used to assess the bulge and disk contributions, as well as help determine the total flux. Here we show the profile for NGC 4639 in W1 (3.4 μm) and fit a Sérsic function to the inner (bulge; blue dotted line) and outer (disk; orange dotted line) light, thus decomposing the two components. The red triangle (and dashed line) represents the 1σ isophote, i.e., where the Galaxy emission reaches the sky noise per pixel. We are down in the noise below this threshold, but because of axisymmetric averaging, we reach much deeper surface brightness, nearly 25 mag arcsec^{-2} (or 28 mag arcsec^{-2} in AB units).

Background "Sky" Level Determination: One of the most important steps toward source characterization with near- to mid-infrared imaging is to estimate the background level of your source. The "local background" is a good proxy for the actual background your source is located within. Background emission may be atmospheric (ground-based) or astrophysical (e.g., stars). For either mechanism, you must remove it from your target source in order to properly estimate the flux of your source.

Generally for discrete sources (stars and galaxies), you simply measure the emission around the source, using an annulus, either circular or elliptical, depending on the source shape. Complications arise from objects that may be in the annulus, such as stars and a nonconstant sky level. If the stars are not too bright, they should not affect the statistics of the background, unless computing the mean of the pixel-value distribution, which will be biased to stars in the annulus, thus biasing your photometry (i.e., your source will appear too faint), whereas using a simple median of the pixel-value distribution is often good enough. But if you want to improve the noise or uncertainty in this background estimation, more care must be taken.

Consider this example, NGC 4639, imaged by WISE in the W1 (3.4 μm) band. The source appears to be elliptical, bright in the center, and extended emission beyond. The background around the source has stars and some fuzzy objects (more distant galaxies). The key is to measure the "local" background, but not too close to the source as to be measuring the source itself—that is, the background should not be contaminated with the target source. Iteratively, we determine where the Galaxy light drops into the noise of the background, and then outside of this place is the annulus. The annulus should be large enough to get a robust and statistically significant measure of the background pixel-value distribution.

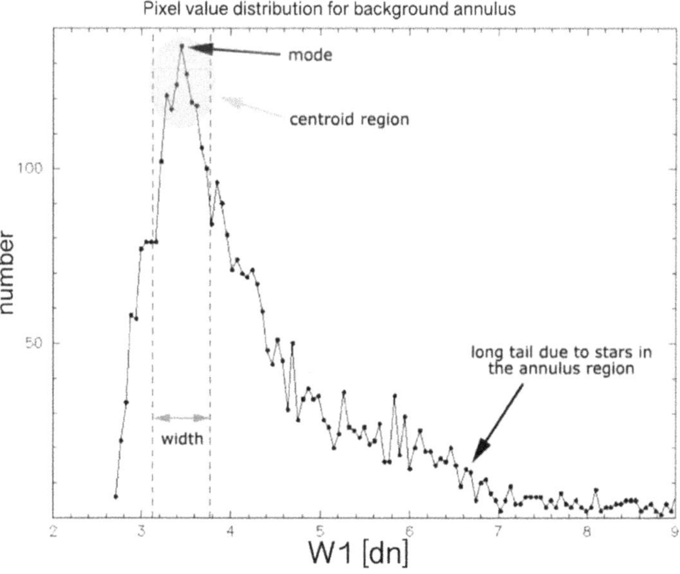

Because we want something better than a simple median, we first remove the bright stars in the annulus (and the source itself). Using a technique developed for the 2MASS and WISE surveys, we then (1) draw a histogram of the pixel-value distribution, (2) trim the extrema, (3) determine the mode, and (4) then compute the count-weighted first moment (centroid) about the mode. This value represents the sky level or background, which should be robust to bright stars and nearby sources in the local region. The uncertainty in this value, or the rms in the background level, may be computed from the 16% to 84% quartile width of the histogram, or more robustly, using the left side (16%) of the distribution from the centroid, because it is not contaminated with starlight. Both methods should give a value that is not too different from fitting a simple Gaussian and determining the sigma, or simply computing the standard deviation of the population about the centroid.

Magnitudes and Flux Density: Understanding the different magnitude systems, and how they relate to physical flux units, is an essential tool of any astronomer. Significant miscalculations may occur if care is not taken between differences (e.g., Vega and AB magnitudes).

Astronomers use different methods (sometimes for historical reasons) to define and calibrate the energy, luminosity, or brightness of a celestial object. For example, the X-ray window describes emission in terms of energy units (e.g., keV), the optical and NIR in terms of magnitudes, and the long wavelengths of the infrared to radio in terms of the radiation intensity, e.g., Janskys (Jy), or formally 10^{-26} W m^{-2} Hz^{-1}.

Generally, we measure an object in a "band," which is a frequency or wavelength range in which the instrument is designed to be sensitive to incoming photons or radiation (whether it is thought of as a particle or a wave). The monochromatic (where one frequency represents the band) magnitude is a logarithmic representation of the intensity measured in a band. Most importantly, it is calibrated against a celestial source that has well-characterized and understood properties, e.g., a commonly used magnitude system is based and referenced to the star Vega (Alpha Lyrae).

In the Vega system, each band (e.g., R) has a unique calibration based on Vega, and hence the "zero-point" magnitude for a standard photometric band has a known value that can be looked up and applied accordingly. The NIR uses the Vega system for some large surveys, including that of 2MASS. So for example, if you want to relate the 2MASS J-band magnitude to a physical intensity value, or the flux density (units of Jy), then you would look up the zero-point flux, F_0, which is 1592 Jy, and the flux of your source is then

$$f^{\nu}(Jy) = F_0(Jy) \times 10^{(-m(\mathrm{Vega})/2.5)}.$$

By relating magnitudes to flux density, the source brightness in a particular band can be compared to other (monochromatic) band measurements from the X-ray to radio. The combination of these is what is commonly referred to as a spectral energy distribution (SED). SEDs typically plot flux density (f_ν) versus wavelength (λ), but may also use "energy" units by scaling the flux density by the monochromatic frequency, i.e., $\nu\, f_\nu$. This is how spectral luminosities are computed, notably $L_\nu = (4\,\pi\, d^2)\ \nu f_\nu$, representing the monochromatic luminosity, often reported in solar units, $L_\odot$.

Finally, the most commonly used magnitude system since the 2000s is what is known as the AB system. It is very simply based on one zero-point flux density, 3631 Jy, used for all bands. So for example, the AB magnitude for a source of flux density f_ν (Jy) is

$$m(AB) = -2.5 \log\frac{f_\nu(J)}{3631}.$$

It is thus very straightforward to convert from magnitudes to flux density for all bands (unlike Vega magnitudes, which require a different zero-point flux conversion for each band). The SDSS used the AB magnitude system for their set of optical filters: *u*, *g*, *r*, *i*, *z* (0.35 to 1 μm). In the near and mid-IR, the difference between Vega and AB magnitudes is significant, and hence it is important to make this distinction when using magnitudes (better yet, convert to flux density and avoid confusion altogether). For example, in the WISE W1 (3.4 μm) band, the Vega magnitude has a value ~2.7 magnitudes less than the equivalent AB magnitude.

Spectral Energy Distributions: A compact description of the energy budget, the SED, graphically shows the intensity of the observed emission across the electromagnetic spectrum. It usually includes photometric measurements in standard bands, but may also include spectroscopic information. In either case, measurements all have to be consistent in what they are measuring. The photometry must be on the same system (e.g., Jansky units) and integrating the same area of the object. For galaxies, this is typically matched apertures or using the "total" integrated flux in a given band, calibrated in flux density units. In this way, we are able to compare across bands and across windows, such as the optical to infrared.

By comparing the observed SED with theoretical or semiempirical models (or templates), we are able to intuit the dominant physical mechanisms that are governing growth and evolution. For example, the SED may clearly show the signature of an old galaxy, whose energy budget is dominated by luminous, evolved stars. Or it may show that young stars are powering the current phase in its life, with gas and dust emission features indicative of energy redistribution.

A classic example is given below, the SED of the starburst galaxy M82. Imaging from the UV to the radio reveals M82 to be a highly dust-obscured galaxy, with a burst of hot massive stars producing copious amounts of UV photons. This is absorbed by the gas and dust, exciting the ISM and reemitting energy from the NIR to the submillimeter (as well as at radio wavelengths, not shown here). We observe a depressed UV window (due to extinction), the stellar scaffolding of old generations of stars (A–G stars in the optical, cool dwarfs and giants in the NIR), powerful PAHs in the mid-IR, deep silicate absorption at 10 μm (dust absorption), and a steeply rising continuum from warm (few hundred K) dust grains. Finally, a broad and luminous continuum from cool dust grains (5 to 30 K) represents the dense molecular cloud complexes where stars are born.

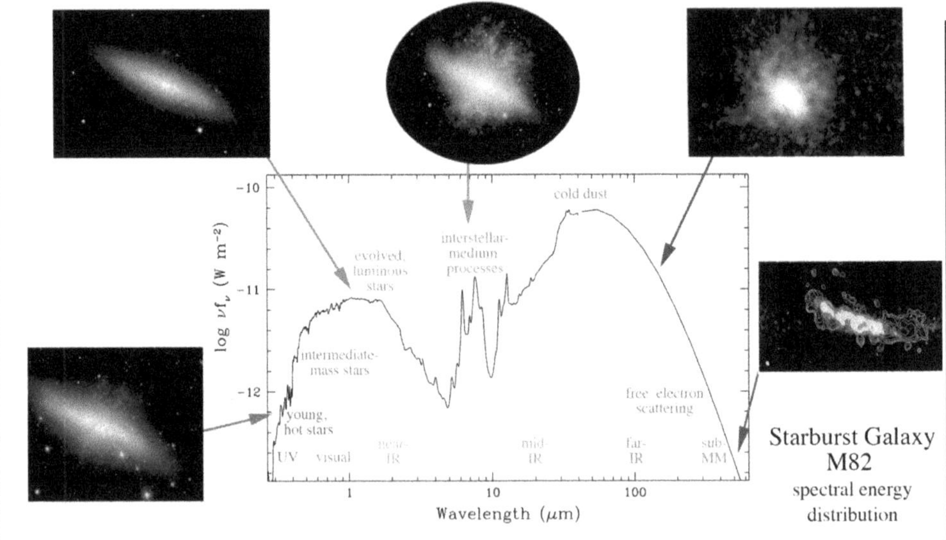

SED for the starburst galaxy M82. Wavelength spans from the UV to the submillimeter, demonstrating how the powerful (UV) energy from young and massive stars is reradiated at much lower energies in the infrared and submillimeter window.

References

Armus, L., Charmandaris, V., Bernard-Salas, J., et al. 2007, ApJ, 656, 148

Blain, A. W., Smail, I., Ivison, R. J., Kneib, J. P., & Frayer, D. T. 2002, PhR, 369, 111

Chiar, J. E., & Tielens, A. G. G. M. 2006, ApJ, 637, 774

Cluver, M. E., Appleton, P. N., Boulanger, F., et al. 2010a, ApJ, 710, 248

Cluver, M. E., Jarrett, T. H., Dale, D. A., et al. 2017, ApJ, 850, 68

Cluver, M. E., Jarrett, T. H., Hopkins, A. M., et al. 2014, ApJ, 782, 90

Cluver, M. E., Jarrett, T. H., & Kraan-Korteweg, R. C. 2010b, ApJ, 725, 1550

Cram, L., Hopkins, A., Mobasher, B., & Rowan-Robinson, M. 1998, ApJ, 507, 155

Dale, D. A., Smith, J. D. T., Armus, L., et al. 2006, ApJ, 646, 161

Fischer, J., Sturm, E., González-Alfonso, E., et al. 2010, A&A, 518, L41

Gautier, T. N. III, Fink, U., Treffers, R. R., & Larson, H. P. 1976, ApJ, 207, L129

Genzel, R., Lutz, D., Sturm, E., et al. 1998, ApJ, 498, 579

Ho, L. C., & Keto, E. 2007, ApJ, 658, 314

Jarrett, T. H. 2000, PASP, 112, 1008

Jarrett, T. H., Cluver, M. E., Brown, M. J. I., et al. 2019, ApJS, 245, 25

Jarrett, T. H., Masci, F., Tsai, C. W., et al. 2013, AJ, 145, 6

Kriek, M., van Dokkum, P. G., Labbé, I., et al. 2009, ApJ, 700, 221

Lacy, M., & Sajina, A. 2020, NatAs, 4, 352

Li, A. 2020, NatAs, 4, 339

Ogle, P., Antonucci, R., Appleton, P. N., & Whysong, D. 2007, ApJ, 668, 699

Peeters, E., Spoon, H. W. W., & Tielens, A. G. G. M. 2004, ApJ, 613, 986

Planck Collaboration,, Akrami, Y., Ashdown, M., et al. 2020, A&A, 641, A7

Rho, J., Jarrett, T. H., Cutri, R. M., & Reach, W. T. 2001, ApJ, 547, 885
Rieke, G. H. 2007, ARA&A, 45, 77
Rudolf, N., Günther, H. M., Schneider, P. C., & Schmitt, J. H. M. M. 2016, A&A, 585, A113
Sakamoto, K., Wang, J., Wiedner, M. C., et al. 2008, ApJ, 684, 957
Sanders, D. B., & Mirabel, I. F. 1996, ARA&A, 34, 749
Sawicki, M. 2002, AJ, 124, 3050
Soifer, B. T., Sanders, D. B., Madore, B. F., et al. 1987, ApJ, 320, 238

Chapter 4

Visual and Near-infrared Astronomy

Sebastian Gurovich, Jeffrey L Payne and Miroslav D Filipović

Electromagnetic radiation between 380 and 740 nm (the visual) and up to 2500 nm (near-infrared; nIR) stands out from the other wavelength domains covered in other chapters of this volume and Book 1 because of the continuous evolution of life on Earth and its tight dependence on the Sun, whose spectrum's peak energy/density distribution falls within the scope of this chapter! This is characterized by the Sun's color or, more simply and precisely, its spectral energy distribution, the peak of which falls seemingly, by the anthropic principle, in the visual part of the electromagnetic (EM) spectrum. The visible–nIR has and continues to mark humans and coached us toward scientific, technological, and cultural achievements. In this chapter, an attempt has been made to introduce some basic concepts of astronomy and in particular multimessenger astronomy related to in the visual part of the EM spectrum. Astronomers through the ages have made observations of the universe or modeled data to understand the physics underlying astronomical sources and their relations. So in this chapter, some of these aspects will be briefly covered, including some history of astronomy in the visual spectrum. On this short journey, we will attempt to connect naturally with the notion of multimessenger astronomy. A brief look at some historical (instrumental) milestones and their findings, as well as the important role of astronomical surveys for visual astronomy, is highlighted with reference to selected data sets and reduction and analysis procedures. Emphasis is made on ASTROPY, a community-driven resource of tools for astronomy, designed to facilitate data reduction and collaborative analysis. In this chapter, an attempt is made to tell a story about some wonders of the universe and the course of the inquiry, through multimessenger astronomy, of the sources' observational and physical parameter space, that also include some fundamental constants.

4.1 Context

Multimessenger astronomy (MMA) may be defined as "the coordinated measurement of astronomical sources through different particle messenger types." Visual astronomy has its origin in the collective and coordinated observations made by

doi:10.1088/2514-3433/ac2256ch4

humans, thousands of years ago, driven by the necessity of understanding the zodiac for agricultural, navigational, and spiritual motives. This resulted in the need for measurement of the relative position of the stars, planets, and their satellites in an attempt to predict relative motions and influence. The coming of seasons for example, among other events, could be predicted.

During the infancy of technology, humans were only able to see at "visual" wavelengths available through naked-eye observations. Observations were therefore biased by knowledge alone of the visual part of the EM spectrum. It is at these wavelengths, about 100 million years after the first cells started to specialize into organs, that the primitive eye evolved quickly in what has been termed an arm's race of evolution (Parker 2004). The eyes of many species, ourselves included, are optimized to perceive data between 380 and 740 nm, because the Sun's photosphere (with an effective temperature of $\approx$5770 K) demonstrates a blackbody distribution that peaks within this region—in fact, around 500 nm in the green portion of the spectrum, as given by Wien's law (see Book 1). This peak might have been shifted to longer wavelengths had our star been a K- or M-class object with a cooler surface temperature. Likewise, a shift to the ultraviolet may have occurred for species that evolved under an A- or F-class star.

However, the light from our Sun's blackbody spectrum has now been incorporated into the biological rhythms and psyche of most life on Earth, including, of course, humans. Figure 4.1 shows a representation, from a modern Aboriginal artist, of what appears to be an eye that sees constellations, a Moon, galaxies, and the Milky Way as an edge-on disk running through the origin. The coordinate system is marked on the land.

Around the mid-1800s, European astronomers set out to uncover the energy source of the Sun. Figure 4.2, taken from Hunt (2011), illustrates the Sun's continuous blackbody spectrum (top) matched to the absorption-line spectrum by Fraunhofer (bottom) from the outer atmosphere of the Sun. The line spectrum shows where energy was absorbed from the Sun's blackbody emission at specific atomic transitions, identified in this figure by letters.

Because observed starlight mostly conforms to a continuum spectrum without discrete physical cutoff frequencies, one could argue that this book should start here from a chronological evolutionary perspective. Nonetheless, this light has the same radiation physics from photons, as well as the shorter wavelength photons (low-energy UV), as discussed in Chapter 5. Thus, photons of different wavelengths may be produced, absorbed, or scattered by the same underlying physics, which do not need to be repeated here.

For humans and life in general, the Sun is a critical source of energy. Total solar eclipses and aurorae would have been among some of the first major astronomical MM observations by humans. These solar eclipses would have caused massive awe, wonder, and commotion, leading to precise observations and predictions, creating milestones in human scientific/cultural understanding and development across many millennia.[1]

[1] The reader is encouraged to read the work of Hipparchus, who used a solar eclipse to determine the Moon–Earth distance of ~429,000 km, only about 11% more than what scientists now accept as the average distance between the Moon and Earth!

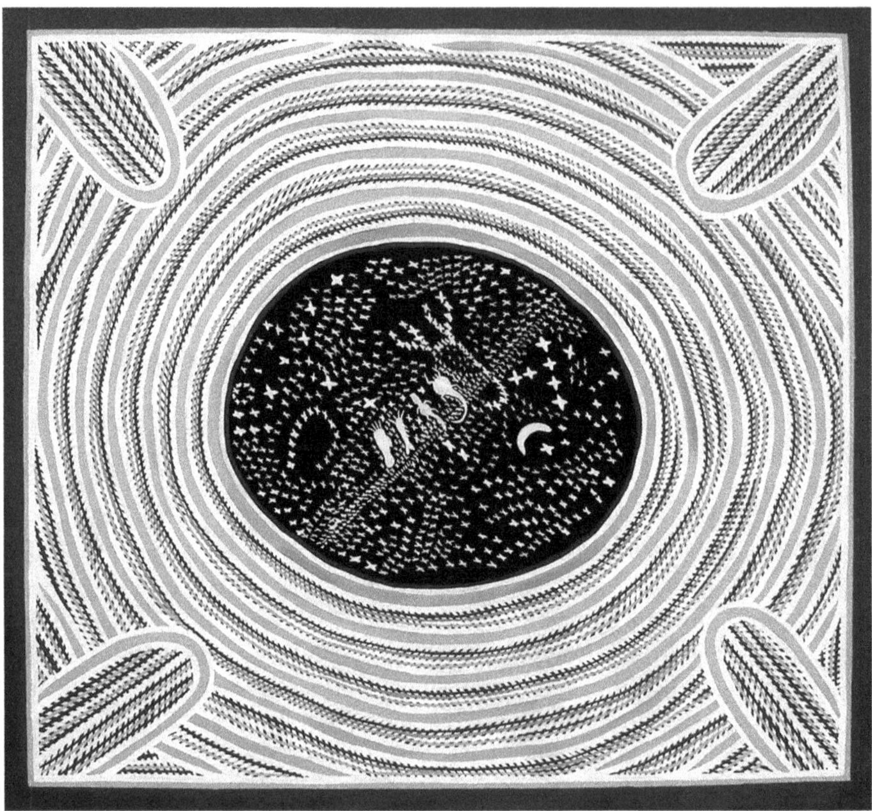

Figure 4.1. Aboriginal Milky Way star map, c. 2000, bush ochre and charcoal on canvas. Image credit: Bill Yidumduma Harney (Senior Wardaman Elder) (CCBY).

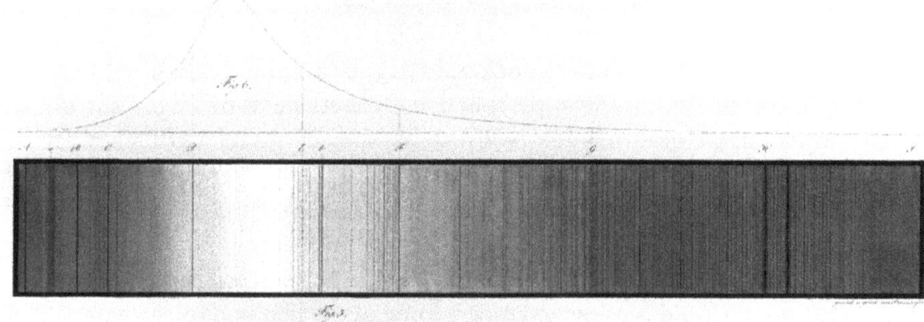

Figure 4.2. Illustration of the solar spectrum drawn and colored by Joseph von Fraunhofer with the dark lines named after him. Image credit: Fraunhofer.

In fact, even in our time, total solar eclipses cause comparable commotion, wonder, and awe. Astronomers and people (in general) still go to considerable lengths to observe eclipses to learn from them. During these events, the projected surface area of the Moon's planetary disk blocks the Sun's photosphere and its associated statistical noise, providing conditions that allow the otherwise hidden solar corona to be seen with the naked eye in all its splendid glory! The white pulsing light from the corona blooms out at totality, making our Sun appear more than five times larger than normal (Golub & Pasachoff 2009). This coronal light is very hot, around 1 million K—so hot, indeed, that all the Fraunhofer absorption lines in the solar atmosphere are Doppler-shifted outside of the visual spectrum, revealing a pure white continuum. That coronal light is a sight to see!

One of us had the opportunity to observe the total solar eclipse in 2017 August, where thousands upon thousands of people gathered to see it. As the eclipse approached, the daylight had a strange tint to it; street lights began to illuminate as if dusk was imminent with all the familiar animal and insect sounds. We suddenly noticed crescent shadows from leaves on the sidewalk, as seen in Figure 4.3. A relatively simple DSL digital camera available to anyone was able to capture the images shown in Figure 4.4. It truly was an unforgettable experience, and it is no wonder people spend thousands of dollars to travel the world to see them. It is also easy to see that the ability to predict these events in ancient times would have given great power to those who sought to rule over others.

Predating early European Surveys by tens of thousands of years are observational accounts by the First Nations people of stellar constellations. It is interesting to note that the Emu (in Australia) and Choike or Ñandu (in Argentina/Chile) are large flightless birds found "half a world away," yet to early observers represent(ed) an important constellation of stars within "The Galaxy." Could these be observations

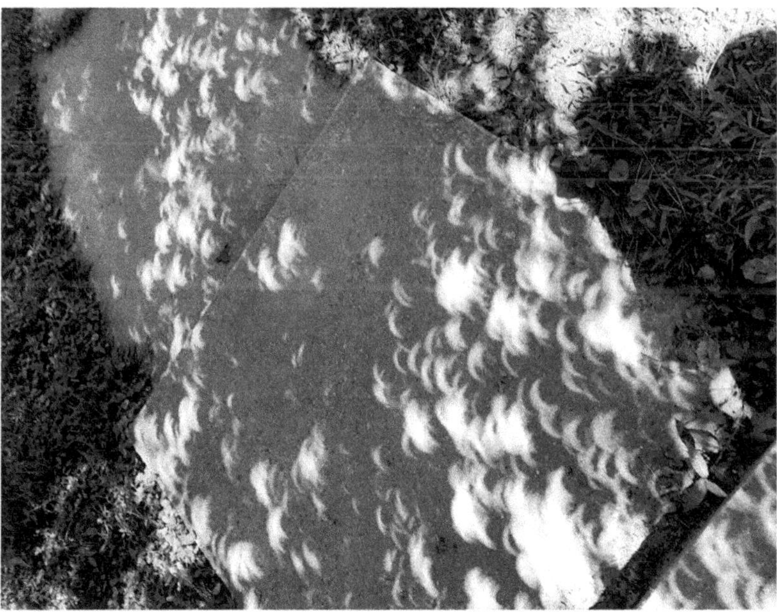

Figure 4.3. Shadow crescents from the 2017 August 21 total solar eclipse seen just before totality. Image obtained from iPhone 7 Plus, f/1.8, 1/289, 3.99 mm, ISO20. Image credit: Jeffrey L. Payne.

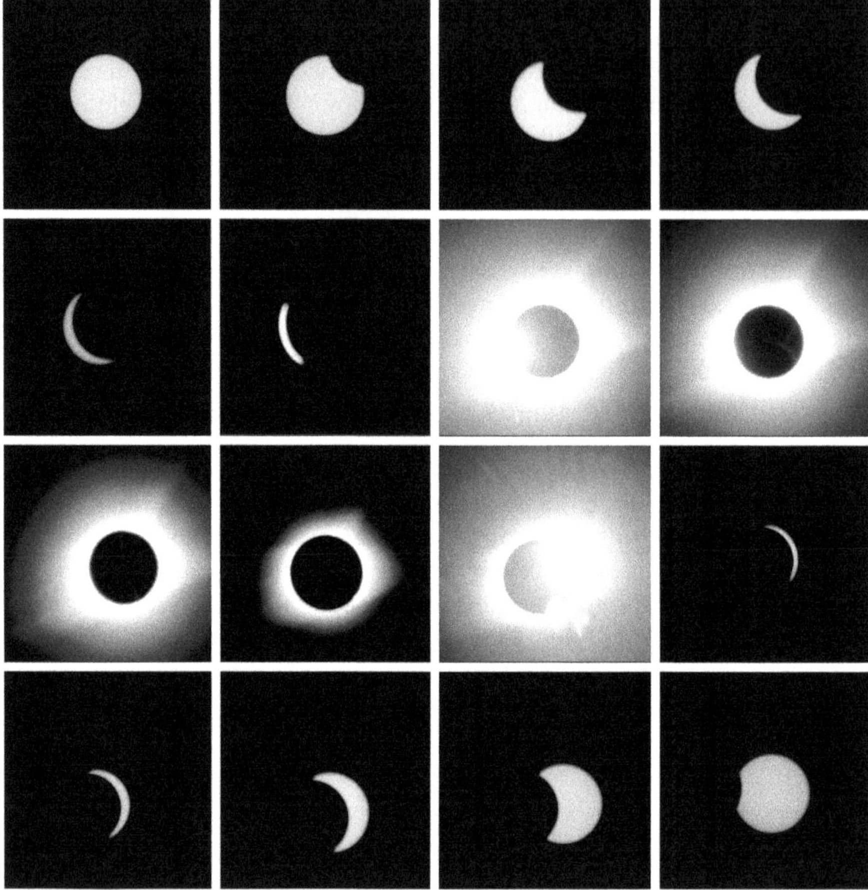

Figure 4.4. A composite image of the 2017 August 21 total solar eclipse as seen from Hopkinsville, Kentucky, USA. Individual images obtained from a Canon SLR EOS Rebel T3i, f/11, 1/80—2 seconds, 600 mm, Canon 70–300 mm, ISO100. Image credit: Jeffrey L. Payne.

for which agriculture, hunting, and cosmography all evolved from the consciousness of similar star dust?

Many thousands of years after the very early observations, astronomers continued surveying stars in the Galaxy in systematic studies of "the Galaxy." Large campaigns to map the stars and galaxies ensued. Examples of these early mappings, which had the underlying objective to reveal what powered the Sun, include the "Bonner Durchmusterung" (from Germany) and its southern extension "Cordoba Durchmusterung" (from Argentina) over two centuries ago (Figure 4.5; see references in https://en.wikipedia.org/wiki/ArgentineNational_Observatory).

Early telescopic stellar surveys made use of the then-new technology that allowed the fabrication of glass-cut astronomical filters that were crafted to preferentially sample emission from select frequency bands within the rest-wavelength visual spectrum. Because photons carry information of source radiation physics, it was

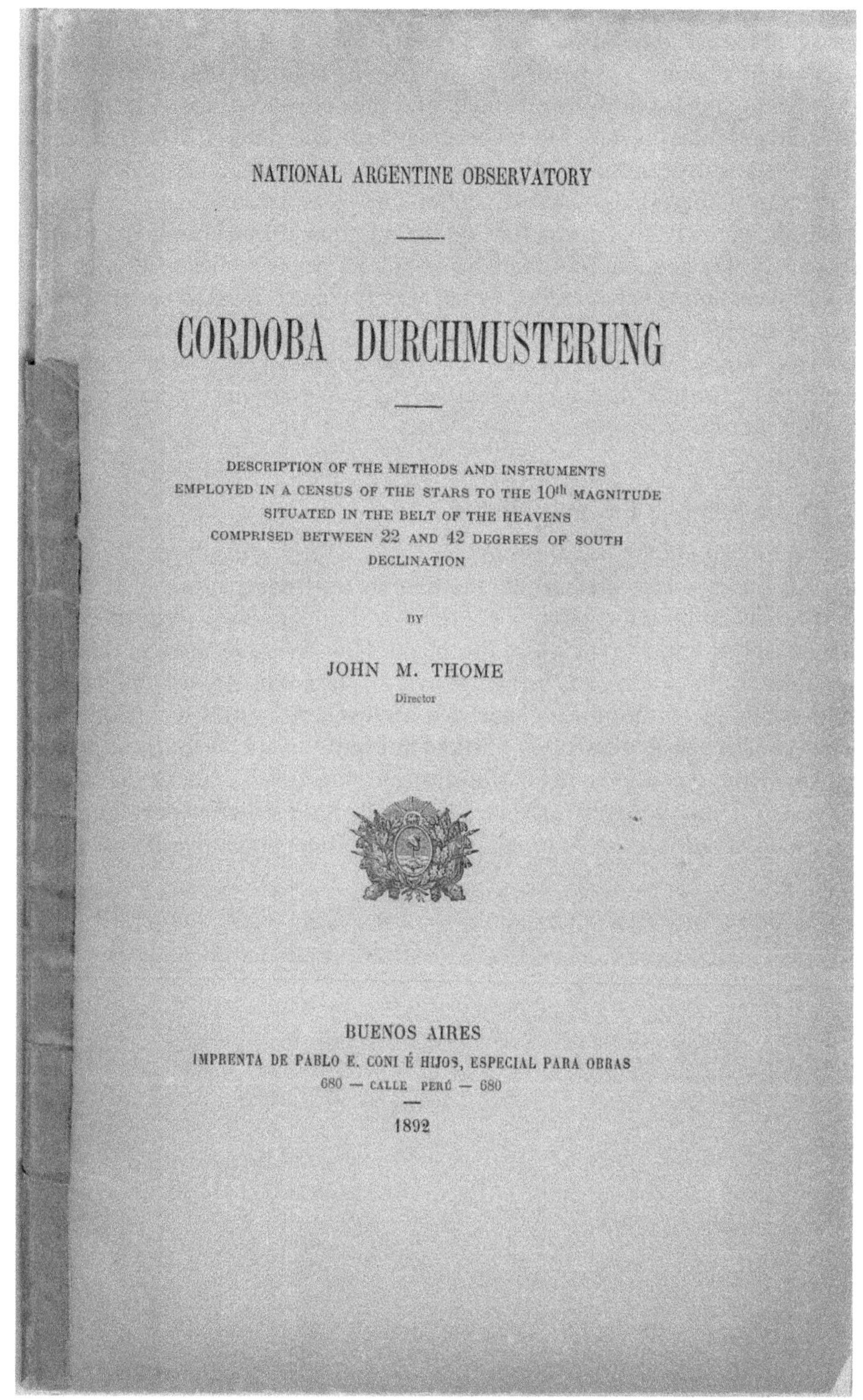

NATIONAL ARGENTINE OBSERVATORY

CORDOBA DURCHMUSTERUNG

DESCRIPTION OF THE METHODS AND INSTRUMENTS
EMPLOYED IN A CENSUS OF THE STARS TO THE 10th MAGNITUDE
SITUATED IN THE BELT OF THE HEAVENS
COMPRISED BETWEEN 22 AND 42 DEGREES OF SOUTH
DECLINATION

BY

JOHN M. THOME
Director

BUENOS AIRES
IMPRENTA DE PABLO E. CONI É HIJOS, ESPECIAL PARA OBRAS
680 — CALLE PERÚ — 680

1892

Figure 4.5. One of the earliest stellar surveys from the southern hemisphere (Argentina): The Cordoba Durchmusterung Survey. Image credit: Sofia Lacolla (OAC), formerly of the Argentine National Observatory.

realized that they held fossil evidence of the history of stellar populations in galaxies as well as Big Bang cosmology. Astronomers realized that combinations of magnitudes of sources in different filters or bands produced observational features in some observational planes, like color–magnitude diagrams (CMDs), which can be used to calculate distances. Filtered emission and spectral features gave testimony to the effect of foreground dust extinction, age, temperature, and metallicity or abundances, among other properties of the gas that formed some stars. In some early surveys, variable sources and their light-curve properties were also studied. Variability, via timing experiments, would later become critical to understanding the physics of the life cycle of stars. The presence of multiple star systems, as well as exoplanets, would also be revealed by light variability from star observations. Exploration of the time domain in observations would become an important part of MM observations.

4.1.1 Photon Emission Ensemble Statistics

4.1.1.1 Getting Light to the Detector

Figure 4.6 is a schematic diagram that represents the frequency/wavelength range of the EM spectrum on a logarithmic scale. Note that the visual spectrum near 1 μm represents only a limited portion of this plane. However, as shown in the following figure, it boasts the privilege of being relatively transparent, as with only a few other spectral windows of the atmosphere; the atmosphere is mostly opaque at other frequencies. The visual spectrum is marked in Figure 4.7 as one of these transparent windows where the atmospheric transmission function is above 50%. Relative transmission is increased at high elevation with little water vapor. This explains why most observatories can be found on mountain tops or plains where there is little atmosphere.

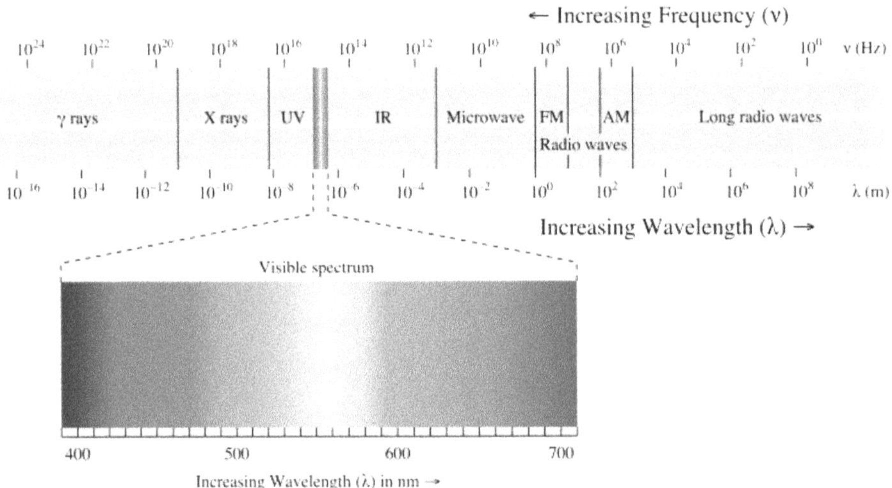

Figure 4.6. The visual part of the electromagnetic spectrum. Image credit: Wikipedia: Philip Ronan.

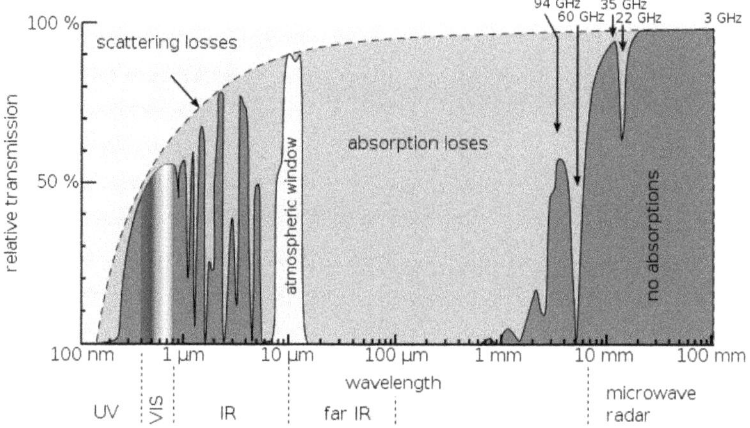

Figure 4.7. Atmospheric windows, depicting the atmospheric throughput function. Image credit: Wikipedia: Cephids.

Earth's atmosphere is relatively transparent at visual wavelengths, and as a result of this evolutionary fine-tuning of the early eye, most information received from beyond the atmosphere for much of the history of human life is from visual photons. Toward the nIR, the water vapor in the atmosphere becomes more intrusive to observations. Observation techniques are often adjusted to observe and remove the "sky signal" in any observing program; however, the associated noise of this signal cannot be removed. nIR sky brightness changes in timescales of 1–2 minutes. Toward the X-ray at shorter wavelengths, observations from the ground become very difficult because the atmosphere is also opaque. In Figure 4.7, observations at other frequencies, including parts of the near-UV, infrared, microwave, and radio are possible from the ground at different elevations; here, atmospheric extinction is not too costly for observations. As discussed in the previous and following chapters, observations from space have an extremely high cost (see Book 1, Chapter 1).

The color range sensitivity of the human eye not only coincides remarkably well with the visual spectrum, but the angular resolution of the eye is about 0.3 m in 1 km, or $1'$, depending on age. Therefore, given this resolution limit as governed by the Rayleigh criterion, telescopes and instruments are necessary tools to resolve much of the structure in the universe, including binary and multiple star systems, galaxies, and most galaxy clusters.

Instruments from the ground suffer from "seeing" effects or spatial distortions caused by time variations in the optical depth of Earth's atmosphere. Indeed, atmospheric seeing is a limiting factor to observations, typically $1.0''$ at most sites but may reach $0.8''$ at the best sites on excellent nights. Figure 4.8 shows that telescopes with apertures of 200 mm or greater have diffraction limits of the order of exceptional seeing conditions, so large-aperture telescopes mostly benefit from having an increased brightness sensitivity limit and not angular resolution. They are unable to reach the diffraction limit because of atmospheric seeing. However, some specialized telescopes have wave-front sensors that sample the effect of local

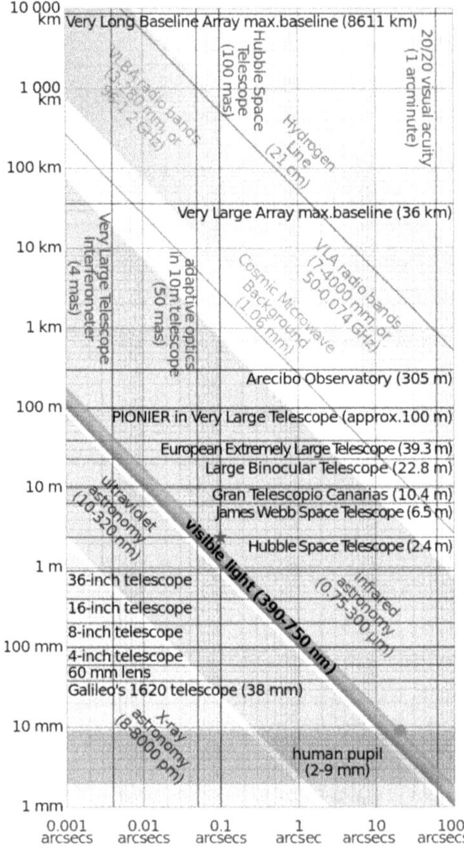

Figure 4.8. Diffraction limits for different astronomical instruments. Image credit: Wikipedia: Cmglee (CC BY-SA 3.0).

and atmospheric changes at 0.001 Hz–1 Hz (active) and from 1 Hz to up to 1000 Hz (adaptive) to correct for these distortions. Active optics operate over much longer timescales than adaptive optics but most use some form of deformable mirror system and fast computer-controlled actuators via feedback-controlled loops. A review article of adaptive optics techniques can be found in Davies & Kasper (2012).

4.1.1.2 Revealing Source Characteristics

Ensembles of photons from astronomical sources may be continuous or discrete in nature depending on the physical mechanism responsible for their creation, reemission/scattering, and/or (re)absorption. If the wavelength of photons is within the visual or optical part of the EM spectrum between, 380 and 740 nm, then provided the source is bright enough to be observed with the naked eye or perhaps via an optical telescope/instrument system, it can be detected and characterized. We must recall that the wavelengths of the observed photon and the emitted photon may be quite different, and one has to take into account many physical effects including cosmological effects like K-correction, some of which become more important when working with sources in other galaxies. See https://properphysics.wordpress.com/2015/04/25/k-correction/.

In order to understand the source emission physics and the type of feature or signal to be detected and studied, one must first identify the nature and purpose of the experiment. This might include a spectroscopic or continuum study over a broad or narrow band. Understanding must take into account the emission, absorption, and reemission of photons from the astronomical source, as well as "background" noise sources, sky noise, and instrumental noise.[2]

Source features may be from an ensemble of source data. For example, a stellar continuum feature in the spectra of an emission-line galaxy that is modeled for the age of the dominant stellar population or the stellar mass-to-light ratio for that galaxy may be extracted, or the features could be constrained to one or few objects.

Stellar coronographs and optical interferometers use the wave properties of light, including polarization and interference, to distinguish useful features from noise. Coronographs are used to block out light from the stellar photosphere and its noise to allow the detection of much fainter coronal light. Visual interferometers can be used to spatially resolve the disk of stars by creating a large effective aperture. An example is the Sydney University Stellar Interferometer (SUSI; Davis et al. 1999). This instrument has allowed the measurement of the diameter of stellar disks other than that of our Sun.

Image instruments from space offer the possibility of operation near the diffraction limit. Figure 4.8 shows instruments operating at visual wavelengths that follow the diagonal rainbow band. The human eye is marked by a red dot toward the lower-right corner of this band. It might be an interesting task to contemplate how this plane would extend for gravitational-wave telescope instruments (Arca Sedda et al. 2020)!

4.1.2 CCD and CMOS Detectors

Early photographic plate surveys were mechanically cumbersome to scale, used photochemical emulsifiers that had low quantum efficiencies, degraded relatively quickly, and required the manual operation of microdensitometers (Davis 2004).[3] Today, these are superseded by CCD/CMOS and nIR digital detector surveys. A review of these instruments and surveys can be found at Bessell (2005). These microelectronic detectors include ADU[4] circuitry that is cryogenically cooled and offers much higher quantum efficiencies and stable potential well charge linearity within a large dynamic range.

Effective data reduction procedures must be easily reproducible so data results can be quickly verified. This was a huge advantage of digitizing data with these new detectors. Calibration frames are usually (necessarily) obtained during each

[2] Further discussion of the signal-to-noise (S/N) ratio and how to calculate it can be found in Dereniak & Crowe (1984), and an online lecture may be found at http://www.ucolick.org/~bolte/AY257/.

[3] Also see Book 1, Chapter 9.

[4] Analog-to-digital units.

observing run in order to optimize the instrument throughput to beat down both the electrical and other background noise sources and for wave-length, flux and astrometric calibration. A high level of linearity is ideally maintained across the detectors using subtracted (Bias, Dark) and divided (FLAT) frames, the latter of which takes into account the pixel-to-pixel sensitivity variation across the image plane and nonuniform internal structure that can form as artifacts (e.g., vignetting). These detectors sample individual photons by filling microscopic light buckets counted as pixel intensities across the 2D detection area. An interesting lecture on this can be found at http://spiff.rit.edu/classes/ast613/lectures/ccds_kids/ccds_kids.html.

Newer devices offer higher quantum efficiencies (Q.E.), defined as the ratio of the number of counts collected by the cell to the number of photons at a given energy incident on it. Moreover, aside from reduction, digitization is a necessity for efficient data processing. This allows reproducible data reduction/cleaning and analysis methods that make the most of increasing computational power following Moore's law. The data-processing tasks are vast; without computers for humans, many individual tasks couldn't be completed manually in a lifetime!

Initially, the tendency was for observatories, universities, and both national and international research institutions to begin building telescopes and maintaining packages for the reduction and analysis of data in a more idiosyncratic fashion. Included were applications like STSCI, IRAF, FIGARO, etc. As time progressed, some standards quickly evolved, like the FITS standard (https://fits.gsfc.nasa.gov/standard10/fits_standard10.pdf) in the late 1980s and a movement started for efforts and resources to be pooled. Standard star CCD studies evolved for absolute photometric calibration, replacing wide-field mappings of photographic plate telescopes like the Astrograph. More advanced computing technology has now allowed the possibility of processing large data rates through pipelines.

Typical calibration frames include standard star frames, flats (used to normalize any pixel-to-pixel sensitivity variation), bias frames (pedestal level offsets), darks (electronic noise offsets), and science frames. Spectroscopic observations may also require wavelength calibration frames often made by standard arc lamps and spectroscopic flat-field frames. A full discussion of spectrographs, definitions, and associated resolution formulas can be found in Book 1 (Chapter 9).

Astrometric calibrations may also be critical to the success of an experiment, transforming the image plane on world coordinate systems by automatically calculating differential projections of the positions of sources. If proper motions of stars are to be measured, then very precise corrections must be made. Package standards are agreed upon by the astrometric community and "proprietary packages" with their algorithms have been incorporated via open-source initiatives into community-maintained package libraries known as "repos" or repositories. One example is heliocentric corrections or transformations that must be made when searching for periods of close-by variable stars (see https://docs.astropy.org/en/stable/coordinates/velocities.html, the ASTROPY online page for more).

4.1.2.1 ASTROPY coordinated packages

What more could astronomers want than for astronomical software to be developed from within the observational community? ASTROPY[5] has evolved to be the leading community-maintained library of astronomical modules. It is in Python and is perhaps the premiere open developer toolkit for astronomical observation research, reflected in metrics like the ASTROPY user and dev commits (see Tollerud 2020). It contains a wide range of useful packages that have evolved as a decentralized open-development resource and code repository from earlier community-maintained Python libraries like https://pythonhosted.org/Astropysics/.

The software development model for ASTROPY is based on the three-clause Berkeley Software Distribution (BSD) license, a permissive license that allows easy maintenance of the core package and the interoperability between astronomical data sets, most of which are compliant with the International Virtual Observatory Alliance (IVOA) standard. Outside the core package, ASTROPY also provides a platform of community development of more specialized affiliated packages so it could be considered a community development platform. ASTROPY is included as default in several Python stacks in Linux and macOS or can be easily installed on most systems. ASTROPY is highlighted here because a majority of observational astronomers use it, and its user base is continuously growing so students are encouraged to learn Python as a first programming language. Although knowledge of other programming languages is also advantageous for some problems, there are many excellent resources for learning Python online, including in the form of Jupyter notebooks (https://jupyter.org/), which are encouraged. To install ASTROPY, a possible course of action would be to install anaconda or miniconda to run some Jupyter notebook tutorials. Although Linux is recommended, ASTROPY can also be run on Windows or other scientific Python stack environments like Canopy, a free product. There are ASTROPY guides and tutorials from the ASTROPY link above. Also, by checking into observatory or institutional sites, other tutorials may be found for particular as well as guides for the instrumental setup and strategy for a typical observing program. Moreover, there are IVOA protocols like TAP and astroquery, which allow users to select and process data directly from within ASTROPY.

Core packages of ASTROPY include container classes representing gridded and tabular data as tables or multidimensional arrays, unit and physical quantity conversions, physical constants, celestial coordinates and time transformations, and world coordinate system support. Core operations also include file I/O support for FITS files, Virtual Observatory tables, common ASCII (American standard code for information interchange) table formats, and hierarchical data format (HDF5) files. There is also a framework for cosmological conversions and transformations. Statistical analyses packages are also included or can be used complementarily, like Scikit-learn, for machine-learning modules.

[5] A clever combination of "Astro" for astronomy and "py" for Python.

As mentioned earlier, ASTROPY uses the concept of "affiliated packages." These are not part of the core package but are provided more by individual efforts within the community that must also adhere to some standards of software development. They may include machine-learning and data-mining tools, or ported IDL (Interactive Data Language) astronomy routines, observing planning software, an astronomical plotting library (APLpy), ginga (an astronomical image viewer), and a montage-wrapper (mosaic engine). For more information about ASTROPY please refer to https://www.astropy.org.

The ASTROPY community is also always available for consultation. Options for help include multiple forums, including @astropy on Twitter, the Python Users in Astronomy Facebook Group, the Users Mailing List, the ASTROPY Slack Team (account required), the Developer Email List [astropy dev] and Discourse (see https://groups.google.com/g/astropy-dev/c/gxgtwU3Q1Kc). For more information and links, see the help page at https://www.astropy.org/help.html.

4.1.3 Timing is Everything

The classic timing experiment in the visible domain has been the observation of variable stars, which we assume the reader is quite familiar with and will not be extensively studied here. MMA can also require precise timing and coordination when high cadence observations and short or precise observation time-scales are required. Follow-up can be crucial in order to classify and understand some sources (see the discussion below). The historical evolution of the first photometric mappings more or less coincided with the first spectroscopic experiments. These experiments became a necessity after the detection of the element helium in the Sun by Pierre Jules César Janssen in 1868.

4.1.3.1 Photometric Sky Surveys

Modern multimessenger techniques in the last decade have matured substantially, accompanied by ever-precise photometric measurements across the time domain defining transient-type observations, made possible in part by advances in CCD/ detector technology and the internet. This has also been due to an exponential increase in visual data measurements from astronomical surveys.

A non-exhaustive list of some important or historic photometric telescopes/ surveys, with references, as a means of a brief introduction follows:

- MACHO (MAssive Compact Halo Objects; http://wwwmacho.anu.edu.au/)
- UKS (United Kingdom Schmidt; Globular Star Cluster study)
- DSS (Digitized Sky Survey 1 and 2)
- SDSS (Sloan Digital Sky Survey; several surveys)
- 2dFGRS (Two-Degree-Field Galaxy Redshift Survey)
- Pan-STARRS (Panoramic Survey Telescope and Rapid Response System)
- 2MASS (Two Micron All Sky Survey)
- Vista (Visible and Infrared Survey Telescope for Astronomy)
- TDSS (Time-Domain Spectroscopic Survey)
- Rubin Observatory/LSST (Legacy Survey of Space and Time)

The concluded MACHO survey began observations in the last decade of the last millennium and attempted to determine the nature of dark matter within the Galaxy. It was to use the concept of a gravitational lens, devised by Einstein, such that when the lens and the background object are point sources (stars) at some precise orientations or lines of sight, the system forms a lensed image where the image is displaced only microarcseconds from the source and, therefore, to most instruments the image is still unresolved but magnified. If a population of massive compact objects existed in the halo of galaxies, then MACHO, through microlensing, could hope to recover statistical properties of the distribution by analysis of the number statistics of the observed caustic (see https://microlensing-source.org/concept/caustics-overview/). So, microlensing events of stellar sources were measured both toward the SMC and LMC. This required quick follow-up observations as the study was looking for star brightening events associated with microlensing from the passage of massive objects. One explanation from the work of the collaboration is that MACHOs in the Milky Way halo comprise about 20% of the halo mass (Alcock et al. 2000) which is comparable to the stellar mass but significantly less than the dark matter required to explain the Milky Way rotation curve.

Another early survey, the Digitized Sky Survey (DSS) was an early all-sky visual survey made of scanned plate image data that was supplied to observatories via CD and used WEB1 and WEB2 infrastructure from the Association of Universities for Research in Astronomy, Inc. and included the scanning of photographic plates for older surveys (see https://archive.stsci.edu/cgi-bin/dss_form) that was also widely used by the community for studies of stars, galaxies, quasars, and large-scale structure among others. Other canonical surveys are described below.

The Sloan Digital Sky Survey (SDSS; https://www.sdss.org/) is a long ongoing imaging and spectroscopic (redshift) CCD survey using a dedicated 2.5 m wide-angle optical telescope at the Apache Point Observatory in New Mexico, United States. The project was named after Alfred P. Sloan and funded by the foundation named after him. This survey has revealed further details of the large-scale structure of the universe in the maps of galaxies and quasars and their dark matter environments determined by statistical parameters. Like the SDSS (e.g., Tegmark et al. 2004) the Two-Degree-Field (2dF) Galaxy Redshift Survey (http://www.2dfgrs.net/) has helped refine important cosmological parameters and was a survey from Australia (Percival et al. 2001).

The Time-Domain Spectroscopic Survey (TDSS) is a subproject of SDSS-IV, aimed at obtaining the identifiable spectra of ~220,000 optically variable objects, selected from SDSS/Pan-STARRS1 multiepoch imaging campaigns. Among the most interesting objects in this survey are quasars and M-dwarf population stars (Ruan et al. 2016).

The Panoramic Survey Telescope and Rapid Response System (Pan-STARRS1/2) makes for a very interesting base for case studies of visual multimessenger telescopes. The Pan-STARRS program includes a wide-field high-cadence telescope from Hawaii (US) surveying the sky for moving or variable transients and produces accurate astrometry and photometry of already known objects. Released in 2019 January, the second Pan-STARRS data release was the largest volume of astronomical data ever released at 1.6 PB. The Pan-STARRS main program relates to the detection of near-Earth objects (NEOs), yet the data have been used for a plethora of scientific objectives in galactic and extragalactic projects, too, including the detection of rare types of supernovae or gamma-ray bursts (GRBs).

Another interesting example of multimessenger timing is the EM verification of NS–NS (neutron star binary mergers) initiated by a LIGO gravitational-wave detection event (GW 170817). This was associated in time with a GRB (GRB 170817A) within the galaxy NGC 4993 as imaged by the Hubble Space Telescope (HST; see below for more and also Chapter 9).

Timing observations have also been important for the detection of transits in the discovery of exoplanets. The observation of a stellar occultation also led to the discovery of rings around the Kuiper-belt planet Chariklo, the largest confirmed centaur (small solar system body) of the outer solar system. Additional information about these surveys can be found both in this chapter and throughout the rest of this series.

The Rubin Observatory (LSST) is expected to start operations in October 2022 and includes an 8.4 m mirror, with a 3200 megapixel camera and a five-band filter wheel for the visual spectrum in the bands u, g, r, i, z, and y. Each image is the size of 40 full moons. More details of the operations, as well as technical information, can be found here: https://www.lsst.org/about/project-status. The survey has been designed to produce the deepest, widest image of the universe and is expected to image 17 billion stars and 20 billion galaxies and will last 10 years, via 11 data releases that will include up to 10 million source alerts per night, 1000 pairs of exposures, and produce 20 TB of data/night. The project will undergo a Legacy Survey, which is a 10 year timescale with science objectives that seek the resolution of the following questions:

- What is the nature of dark energy that is driving the acceleration of cosmic expansion?
- What is dark matter, how is it distributed, and how do its properties affect the formation of stars, galaxies, and larger structures?
- How did the Galaxy form, and how has its present configuration been modified by mergers with smaller bodies over cosmic time?
- What is the nature of the outer regions of the solar system? Is it possible to make a complete inventory of smaller bodies in the solar system, especially the potentially hazardous asteroids that could someday impact Earth?
- Are there new exotic and explosive phenomena in the universe that have not yet been discovered?

Important information with regard to the preprocessed output data quality including key numbers of the magnitude limits for each filter of the Legacy survey and of one image, the field of view, and MOK data can be found here: https://www.lsst.org/scientists/keynumbers. A link to the Legacy Survey Team paper is here: https://docushare.lsstcorp.org/docushare/dsweb/Get/Document-5462.

The Rubin Observatory/LSST is designing a broker service for the alerts of transient observations for MM follow-up observation. This includes streamed preprocessed data pipeline in a format compatible with IVOA protocols like TAP and astroquery. Those data will include catalog matching (astrometric) services as well as photometric services, making it easy for any astronomer to listen in and immediately use preprocessed data for MM follow-up observations. These broker services are crucial to big data surveys; because there is so much data, some will never be looked at or even have to be thrown out during the preprocess reduction process.

4.2 Imaging: Difference Imaging, Point Spread Function, or Aperture Photometry?

Instruments/telescope systems operating within the visual spectrum can be broadly characterized as either imagers or spectrographs. Spectrographs have much higher resolving power R in any given wavelength domain than do imagers. Spectrographs therefore permit measurement of "narrower" spectral features within source spectra and are covered below and further in Book 1.

Most multimessenger transient searches use imagers as instruments of detection as well as for studies of photometrically derived source parameters including the source's environment as broad aspects of the ensemble or the macroscopic physics of emission of a source system is contained over a relatively large wavelength domain. An example is the stellar continuum from the ensemble of starlight in galaxies.

Spectroscopic observations at high spectral resolution are much more expensive. We limit the discussion here to imager instruments that offer a much more limited spectral resolution and often use filter wheels, optimized for broad, white-light, or narrowband work, that are sufficient to obtain photometric line ratios to constrain the stellar population history of galaxies (Pforr et al. 2012) or separate quasar fields (Richards et al. 1997). Astrometric and photometric measurements of transient sources in imagers are obtained by mostly one of three methods: difference imaging, PSF, or aperture photometry.

Each of these three methods offers unique advantages and disadvantages. The choice for each method depends on the scientific requirements of the data as well as environmental factors. For example, in the VVV Survey,[6] both aperture and PSF photometric catalogs have been produced for "point" or unresolved sources. The PSF catalog offers more precise photometry, especially in highly crowded fields, but they do not go as deep as the aperture catalogs (Alonso-García et al. 2018). Therefore, the point spread function (PSF) catalog does not include as many sources. When working with extended sources, often it is useful for apertures to be fitted to the source brightness distribution as determined by isophotal analysis.[7]

Given the large data rates of modern optical and nIR photometric surveys and that source evolution for some sources can occur in seconds, real-time analysis is often crucial. One technique often used is called difference imaging analysis or difference

[6] A survey that mapped the bulge of the Galaxy and an adjacent section of the southern disk midplane in the nIR by ESO. This uses the advanced VISTA 4 m telescope that created a deep nIR atlas in five passbands and a catalog of more than 100 million sources.

[7] See ASTROPY photutils affiliated package for more information at https://photutils.readthedocs.io/en/stable/index.html.

imaging (DIA or DI). In DIA, a comparison is made between new images with similar statistical properties as the observed field imaged previously, during, or soon after the transient under near-identical conditions. Machine-learning and data-mining techniques (see Chapter 12) are also employed for the purpose of seeking out transient sources. Additionally, these techniques might aid in the determination of the precise location of a source. A binary inspiral for example could trigger a gravitational-wave detector and follow-up observations for localization at other observing facilities and then targeted spectroscopic observations to measure the source's evolution with different instrumental setups even at other wavelength domains.

Standard filter systems have been developed to sample different parts of the color–magnitude diagram. A typical example are Hα narrowband filters to target star-forming regions in galaxies (see Figure 5.1). Because no two bandpass sets of cut-glass filters are the same, before or during instrument commissioning runs, the photometric system is usually transformed to an original standard system filter set for more homogeneous comparison with earlier studies.

Perhaps the most promising system to do this is the LSST pipeline, which will use a template image from a difference image. Another technique employed is the creation of a master source list catalog using aperture or PSF photometry.

One example of the former technique is used in the TOROS project to search for optical gravitational-wave counterparts. An example of the latter may include VVV data, where the Cambridge Astronomical Survey Unit and the Vista Science Achieve both created aperture and PSF photometry catalogs of all tiled fields.

4.3 Spectroscopy

Astronomical spectra in which light intensity varies over different wavelengths or frequencies can be a continuum that has both emission and absorption features (see Book 1, Section 9.7). The resolving power of spectra is measured by resolution, which essentially is the difference ($\delta\lambda$) between two wavelengths that can just barely be "resolved" (Rayleigh's criterion). The resolving power is defined as $R \equiv \lambda/\delta\lambda$.

Resolving powers of ~10–100 are used for finding spectral energy distributions (SEDs), solid-state and molecular bands; $R \sim 1000$ (300 km s^{-1}) for atomic features, redshifts, and high-speed dynamics (e.g., stellar winds); $R \sim 10^4$ (30 km s^{-1}) for line profiles, radial velocities of stars, stellar abundances; and $R \sim 10^5$ (3 km s^{-1}) for interstellar medium (ISM) lines and molecular spectroscopy.

There are also many types of spectrographs used in astronomy. Object prism spectrographs are used to obtain low-resolution spectra of all stars in a field. Long-slit spectrographs, in which a slit is placed on the extended object that is to be observed, were, until recently, the workhorse of astronomical spectroscopy. A multiobject spectrograph is an instrument that allows the spectra of many objects to be simultaneously observed and may use fiber optics. Integral field spectrographs obtain near-continuous 2D spectra of a small field of view where each pixel in a data cube image represents a spectrum.

Dispersers can include prisms for low resolution, gratings for low to very high resolutions, grisms (prism and grating) for low to moderate resolutions, and

Fabry–Perot and Michelson interferometers and heterodyne detector methods for extracting information encoded as phase or frequency modulation.

The general course of data reduction of spectra includes identification of sky lines, defects, bad pixels, and ghosts along with background subtraction, flat calibration, and extraction. The spectroscopic data is often then calibrated in both the wavelength and flux domain done by taking exposures with comparison lamps and spectroscopic standard stars with known wavelength features.

After line or equivalent-width measurement, features within spectra are able to be calibrated to physical parameters by using models. Spectra tell astronomers much about the elemental composition, motion, density, temperature, and magnetic fields of astronomical objects. Understanding stellar spectra allowed Cecilia Payne-Gaposchkin (1900–1979) to deduce that stars were mostly hydrogen and helium, in 1925 (see Figure 4.9). She understood the transitions reflected in absorption spectra were about much more than just elemental composition, but also about other properties, especially temperature. Many, including Otto Struve (1897–1963), director of the National Radio Astronomy Observatories, have called her thesis one of the most brilliant ever written.

For observers planning their observations, instrumental sensitivity must be fine-tuned to the spectroscopic feature or features of interest. For example, if one wanted to investigate a model to explain the scatter of the total magnitude versus metal

Figure 4.9. Cecilia Payne-Gaposchkin was perhaps the first person to understand how to use spectra to understand stellar properties. Image credit: Wikipedia: Smithsonian Institution.

abundance—the [Fe/H] relation of galaxies—using photometric color indices, then defining the sample size and biases, for example, magnitude- versus volume-limited samples, and determining survey limits are important first tasks. As mentioned, most observatory facilities provide detailed information within their websites on how to plan for observations as well as basic instrumental setup instructions for their specific spectroscopic instruments. They often provide exposure time calculators, which are also true for image experiments as discussed above, allowing for the simulation of the expected S/N. The data obtained must be studied and the data reduction processes discussed with observatory technical staff including the community "channels" as described earlier in this chapter. Most instruments provide trial data and tutorials from which to begin.

4.4 Visible Follow-up to Multimessenger Events

4.4.1 GW 170817 and EM 170817A

Although GW 150914 marked the first confirmed detection of a gravitational wave, it was the GW 170817 event observed in the aLIGO 2017 O2 campaign two years later that was the first source with a confirmed visual GRB and kilonova (KN; also see Chapter 9). These were the "smoking gun" or first detected EM counterparts of a predicted NS–NS merger (see Figure 4.10), an event that was collectively observed by 204 different teams with a multitude of EM telescopes/instruments scattered across the world (Abbott et al. 2017).

For many of the aLIGO follow-up groups involved, target of opportunity (ToO) observational programs were executed in observation campaign schedules, as was the case of GW 170817 to first localize and quickly broadcast localization via semiautomatic/automatic messaging systems of the "astronomical cloud" for rapid follow-up by other observers using the GRB Coordinates Network (GCN). The necessity for sharing transient source information quickly in a standard unambiguous manner to other potential observers has meant that many MM programs have advanced in a decentralized, more organic manner.

4.4.1.1 GRB Follow-up

Visual multimessenger observations played a key role in unveiling the phenomena of GRBs. GRBs were first detected by satellites in 1967 in a nuclear atmospheric test monitoring program from space (Klebesadel et al. 1973). It was only in 1997 after associated GRB optical afterglows were detected and quickly confirmed via spectroscopic redshift measurements that refined details of the standard GRB model were given more credibility. It is now known that GRBs are among the most energetic events in the observable universe. GRB 080319B, for example, holds the record for one of the farthest objects ($z \approx 0.937$) in the universe and from a calculation of the light-travel time, was found to have occurred at about half the universe's current accepted age of 13.6 Gyr. Had one been looking at that line of sight in dark clear skies, GRB 080319B would have been easily observed with the naked eye for several seconds in an explosion that has been estimated to be about twice as energetic as the rest mass of the Sun. GRB 080319B paved the way to

GW170817

Binary neutron star merger

A LIGO / Virgo gravitational wave detection with associated electromagnetic events observed by over 70 observatories.

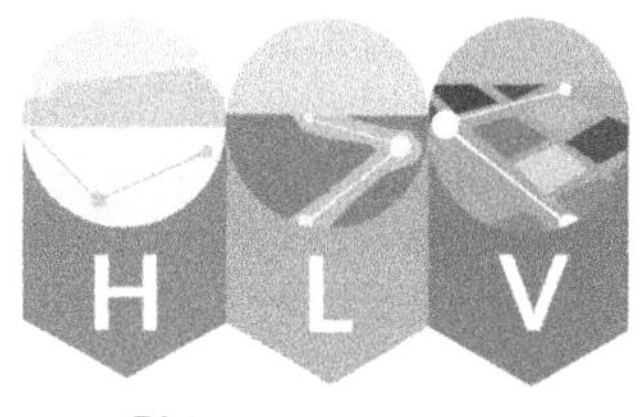

Distance
130 million light years

Discovered
17 August 2017

Type
Neutron star merger

12:41:04 UTC
A gravitational wave from a binary neutron star merger is detected.

gravitational wave signal
Two neutron stars, each the size of a city but with at least the mass of the sun, collided with each other.

gamma ray burst
A short gamma ray burst is an intense beam of gamma ray radiation which is produced just after the merger.

+ 2 seconds
A gamma ray burst is detected.

GW170817 allows us to measure the expansion rate of the universe directly using gravitational waves for the first time.

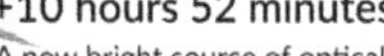

+10 hours 52 minutes
A new bright source of optical light is detected in a galaxy called NGC 4993, in the constellation of Hydra.

kilonova
Decaying neutron-rich material creates a glowing kilonova, producing heavy metals like gold and platinum.

Detecting gravitational waves from a neutron star merger allows us to find out more about the structure of these unusual objects.

+11 hours 36 minutes
Infrared emission observed.

+15 hours
Bright ultraviolet emission detected.

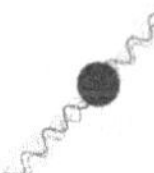

This multimessenger event provides confirmation that neutron star mergers can produce short gamma ray bursts.

+9 days
X-ray emission detected.

radio remnant
As material moves away from the merger it produces a shockwave in the interstellar medium - the tenuous material between stars. This produces emission which can last for years.

The observation of a kilonova allowed us to show that neutron star mergers could be responsible for the production of most of the heavy elements, like gold, in the universe.

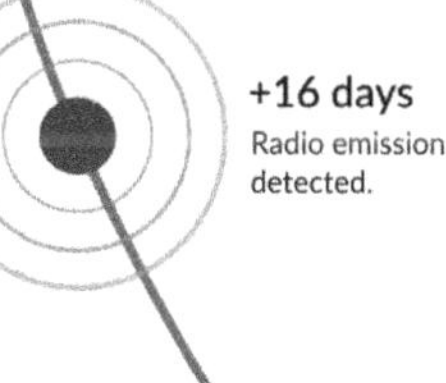

+16 days
Radio emission detected.

Observing both electromagnetic and gravitational waves from the event provides compelling evidence that gravitational waves travel at the same speed as light.

Figure 4.10. HLV LIGO/Virgo Network detection and follow-up of GW 170817 and GRB 170817A. Image credit: LSC/Daniel Williams.

classify features in GRB afterglow light curves, including spectral observations that related the high-energy physics involved with studies of its environment. Alerts required massive follow-up multimessenger coordination, and rapid response mechanisms were made possible because of the internet. Decades earlier, this would have been nearly impossible.

4.4.1.2 KN Follow-up

Multimessenger observations were combined intricately together to accurately constrain the light-curve evolution of the GW 170817 KN event, where each filter and spectrum was preferentially sensitive to different parts of the spectrum.

TOROS (Transient Optical Robotic Observatory of the South) was formed in 2014 to construct, deploy and manage a wide-field optical telescope on Corrdón Macón in the Atacama Plateau of northwestern Argentina. TOROS signed a memorandum of understanding with the LIGO/Virgo Collaboration and has participated in the LIGO observation runs prior to the TOROS deployment with the 1.5 m telescope at the Estación Astrofísica Bosque Alegre (EABA), the 0.83 m T80-South at Cerro Tololo Inter-American Observatory, and the 0.4 m telescope at the Dr. Cristina V. Torres Memorial Astronomical Observatory. TOROS is led by Diego Garcia Lambas, Lucas Macri, and Mario Diaz, who have designed and built the specialized telescopic system for TOROS for the detection of astronomical transients for rapid follow-up of gravitational-wave candidates. Torritos, in the initial phase of TOROS, was used in the aLIGO campaigns of 2012–2013, 2014–2017, and in the later campaign 2019.

When the EM 170817A KN was observed by the TOROS team, a bluer than expected initial light curve was found (Díaz & Macri 2017). Recent modeling had suggested the r-processes thought to occur in KN would give off more red emission because the heaviest r-process elements (greater than an atomic mass of 140 AMU) would absorb blue light, giving it a red color. The emission did eventually become redder, confirming that the heaviest r-process elements were created, thus also confirming the model. Data also suggested the merged binary NS–NS system did eventually collapse into a black hole (see Chapter 1 for more discussion).

This result was reported by several other follow-up teams that also observed quickly in the visual part of the EM spectrum. The most likely possibility, as suggested by newer KN models to reconcile older models with observations, resulted in the acceptance of the fact that a significant amount of elemental abundance of r-processed elements in the universe has been produced in compact mergers (Metzger et al. 2017).

This represents a great success story for MMA. Basically, NS–NS binary mergers result in the production of gravitational waves and a GRB. These compact objects release tremendous amounts of energy and create heavy elements at one time thought to only be the result of supernova explosions. The merged binary eventually collapses into a black holes. These types of insights into the physics of our universe are clearly a leap forward in understanding of the universe.

4.4.2 Size and Constraint Parameters from Stellar Occultations

Multimessenger observations have been and continue to be useful in setting constraints on planetary ringed systems of the extended solar system and beyond, including parameters of models of the distribution of mass in the Oort cloud. Early experiments were undertaken to measure the parameters of the rings of Neptune (French et al. 1987) and recently to constrain the parameters of the trans-Neptunian object 486958 Arrokoth[8] during the Horizon flyby mission to Pluto (Buie et al. 2020). Siraj & Loeb (2020) include an account of historic discoveries of stellar occultations and present a useful method for constraining asteroids and trans-Neptunians. The tracer particle here is the shadow cast onto observers. It is where the New Horizons spacecraft flew to, and the team, through these observations, provided important constraints for the flyby. Arrokoth is at the limit of the capability of HST and no ground-based facility has successfully detected Arrokoth.

4.4.3 Supernovae

One of the challenges for understanding supernova physics includes early-epoch data points in the supernova light curve as the rise times of the early light curve are significantly shorter than the fall time. For correct modeling of its evolution, one outstanding problem is type characterization and the determination of whether it is a single or double degenerate system. It is in the first few hours that supernova light curves rise most rapidly, so many important features occur and lie hidden in the early epochs, requiring rapid ToO coordinated observations. This is important for the refinement of spectrophotometric chemical modeling of SN.

For example, SN 1987A marked an important milestone for MMA. At 07:35 UT, the underground neutrino telescope Kamiokande II detected 12 antineutrinos in a burst lasting less than 13 s (Figure 1.5). This signal from Kamiokande II was localized in the direction of the LMC, well before any ground-based telescopes detected any optical signal from that same source (Arnett et al. 1989). Rapid sharing of positional and other source information was key—a central tenet of MMA was put to the test. because source evolution can occur in the span of days or even hours, coordinating campaigns that include optical telescopes with different instruments and aperture sizes across the world need to be done efficiently and in a standardized way.

4.5 Concluding Statements

Observational optical astronomy focuses on the unraveling of astrophysical information from the observation of source data via visual or optical telescope/instruments from the visual part of the EM spectrum. Although originally the first window available to astronomers, many other windows have recently opened up. Optical astronomy will no doubt continue to be critical for panchromatic studies. Multimessenger data usually follows three detection schemes: broadband imaging, narrowband imaging, and spectroscopic observations.

[8] Provisional name 2014 MU69. It is a contact binary 36 km long, composed of two planetesimals 21 km and 15 km across, which are joined along their major axes.

Before observations can be carried out, they should be planned. Typically, when planning observations, characteristic of the signal to require detection are defined, such as surface brightness limit or spectroscopic resolving power, etc. Telescope time at observatory facilities is awarded on a competitive basis via proposal submission or else archival data must be sought; most facilities have online observing information, tools, tutorials and technical staff to help in the planning.

The ever-increasing number of photometric and spectroscopic surveys in the visual and near-infrared domain have been useful and with LSST and other future surveys continue to identify and characterize a vast number of astronomical objects. Data reductions are greatly facilitated by computerized standard procedure reduction pipelines via IVOA compliant standards.

The internet allows multimessenger teams to be formed for the rapid dissemination of event information including time and coordinates. This allows additional observations in all domains to understand the physics of all the varieties of events and processes in the universe.

References

Abbott, B. P., Abbott, R., Abbott, T. D., et al. 2017, ApJ, 848, L12

Alcock, C., Allsman, R. A., Alves, D. R., et al. 2000, ApJ, 542, 281

Alonso-García, J., Saito, R. K., Hempel, M., et al. 2018, A&A, 619, A4

Arca Sedda, M., Berry, C. P. L., Jani, K., et al. 2020, CQGra, 37, 215011

Arnett, W. D., Bahcall, J. N., Kirshner, R. P., & Woosley, S. E. 1989, ARA&A, 27, 629

Bessell, M. S. 2005, ARA&A, 43, 293

Buie, M. W., Porter, S. B., Tamblyn, P., et al. 2020, AJ, 159, 130

Davies, R., & Kasper, M. 2012, ARA&A, 50, 305

Davis, A., Barkume, K., Springob, C., Tam, F., & Strelnitski, V. 2004, JAVSO, 32, 117

Davis, J., Tango, W. J., Booth, A. J., et al. 1999, MNRAS, 303, 773

Dereniak, E. L., & Crowe, D. G. 1984, Optical Radiation Detectors. Wiley Series in Pure and Applied Optics (Wiley: New York)

Díaz, M. C., Macri, L. M., Garcia Lambas, D., et al. 2017, ApJ, 848, L29

French, R. G., Jones, T. J., & Hyland, A. R. 1987, Icar, 69, 499

Golub, L., & Pasachoff, J. M. 2009, The Solar Corona (Cambridge: Cambridge Univ. Press)

Hunt, R. 2011, History of Spectroscopy—an essay, MSc (Astronomy) thesis, Melbourne

Kasen, D., Metzger, B., Barnes, J., Quataert, E., & Ramirez-Ruiz, E. 2017, Natur, 551, 80

Klebesadel, R. W., Strong, I. B., & Olson, R. A. 1973, ApJ, 182, L85

Parker, A. 2004, In The Blink Of An Eye (Cambridge, MA: Perseus)

Percival, W. J., Baugh, C. M., Bland-Hawthorn, J., et al. 2001, MNRAS, 327, 1297

Pforr, J., Maraston, C., & Tonini, C. 2012, MNRAS, 422, 3285

Richards, G. T., Yanny, B., Annis, J., et al. 1997, PASP, 109

Ruan, J. J., Andersonl, S. F., Green, P. J., et al. 2016, ApJ, 825, 137

Siraj, A., & Loeb, A. 2020, ApJ, 891, L3

Tegmark, M., Strauss, M. A., Blanton, M. R., et al. 2004, PhRvD, 69, 103501

Tollerud, E. J. 2020, in ASP Conf. Ser. 522, Astronomical Data Analysis Software and Systems XXVII, ed. P. Ballester, J. Ibsen, M. Solar, & K. Shortridge (San Francisco, CA: ASP), 491

Chapter 5

Ultraviolet Astronomy

Denis Leahy

Man's quest for knowledge naturally extends beyond the confines of Earth, hence the motivation for astronomy, the study of nature beyond Earth. As technology developed over the years, astronomy has gone beyond naked-eye observations and adopted new technologies to remain at the forefront of discovery. As the number of types of light that could be sensed increased, and as other messengers from afar besides light became detectable, the discipline of multimessenger astronomy emerged. This chapter on ultraviolet astronomy in this book was written to describe the nature of ultraviolet astronomy, the instruments that we utilize in ultraviolet astronomy, and the objects in the universe that are amenable to study by observing their ultraviolet emissions.

5.1 Introduction to UV Astronomy

The universe looks quite different at ultraviolet (UV) wavelengths from the majority of stars and galaxies seen in visible light. Most of the light in the universe is emitted by stars. Stellar surfaces for most stars are rather cool (for UV emission), emitting most of their energy in the visible and near-infrared parts of the spectrum. Only hot stars have surface temperatures high enough, $\gtrsim$10, 000 K, to emit UV radiation. Stars are normally sorted by their surface temperatures and luminosities using the Hertzsprung–Russell (H-R) diagram (see Figure 5.1). The main categories are main-sequence stars, giant and supergiant stars, and white dwarfs.

Main-sequence stars require masses greater than ~4 $M_{\odot}$ and high-enough surface temperatures. Because the birth mass distribution of stars declines with mass (the initial mass function), stars with mass >4 $M_{\odot}$ constitute only about 0.6% of all stars born. For the active life of a star, it undergoes fusion in its core and emits steady radiation from its surface. This lifetime is drastically shorter for higher-mass stars (see Figure 5.2). As a result, the number of hot stars compared to the number of cool stars is reduced by another factor of ~100. The rarity of hot stars is somewhat

doi:10.1088/2514-3433/ac2256ch5

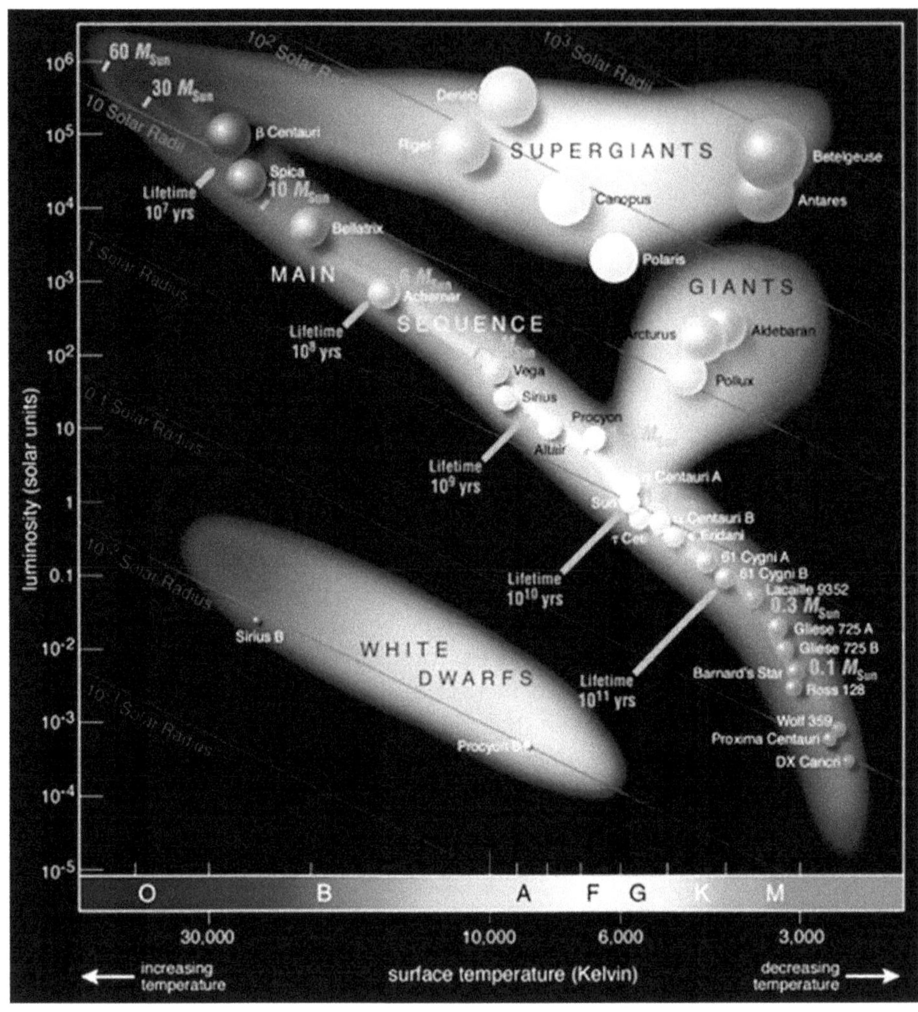

Figure 5.1. The Hertzsprung–Russell diagram for stars. Only stars hotter than 10,000 K emit any significant UV radiation. Image credit: Reproduced with permission © ESO.

Mass (solar masses)	Time (years)	Spectral type
60	3 million	O3
30	11 million	O7
10	32 million	B4
3	370 million	A5
1.5	3 billion	F5
1	10 billion	G2 (Sun)
0.1	1000s billions	M7

Figure 5.2. Representative stellar lifetimes for various masses. Image credit: Carlos A. Bertulani (CC BY-SA 3.0).

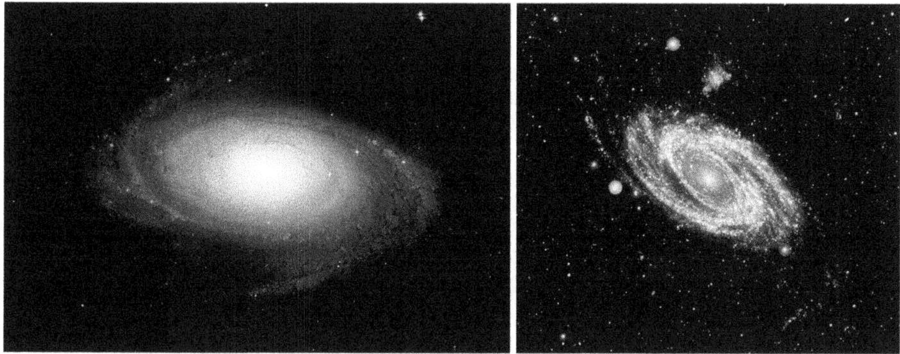

Figure 5.3. Left: M81 in visible light (HST) showing light from almost all stars. Right: M81 in UV light (GALEX) showing primarily the hot stars. The orientation is different in the two images: the visible light image is rotated about 20° counterclockwise with respect to the UV image. Image credits: STScI/AURA and NASA/JPL—Caltech.

compensated for by the fact that most hot stars have high luminosities: from 10^2 to 10^6 times the solar luminosity.

Figure 5.3 presents the image of the large spiral galaxy M81 in visible light (left panel) and in UV light (right panel). A comparison shows that the bulk of the stars, which are seen in visible light, are much more smoothly distributed than the rare hot, young, massive, and luminous stars, which emit strongly at UV wavelengths. The latter are concentrated in the spiral arms where the bulk of the dense gas and star formation are located.

UV astronomy cannot be carried out from Earth's surface, where optical, infrared, and radio telescopes are located. Earth's atmosphere absorbs radiation at various wavelengths primarily because of molecular and atomic transitions at the corresponding transition energies. The main constituents of dry air in the lower atmosphere are (molecules/atoms) N_2 (78.1%), O_2 (20.9%), argon (0.93%), CO_2 (0.041%), Ne (0.001 8%), He (0.000 52%), CH_4 (0.000 19%), and Kr (0.000 11%). In addition to dry air, normal air contains significant amounts of water vapor (0%–3%, average 0.25%). Many other substances of natural origin exist, such as mineral and organic dust, pollen, sea spray, and volcanic ash. Man-made substances are generally referred to as pollution. These other substances generally are only locally present and are highly time-variable.

The transmittance of the atmosphere to sea level as a function of wavelength between 3000 and 20,000 Å (roughly the range of the visible to mid-infrared) is determined by the absorption of molecules in the atmosphere. Examples of molecular absorption databases are MODTRAN4 (Berk et al. 1999; Vidal-Madjar et al. 1987) and HITRAN (https://lweb.cfa.harvard.edu/hitran/). Because the main molecules in the atmosphere, N_2 and O_2, are transparent in this range, the main absorbers are water vapor and CO_2. Both have their strongest absorption features in the infrared.

Most UV radiation is absorbed higher up in Earth's atmosphere, where it produces ozone from O_2. This ozone is the next most important absorber across

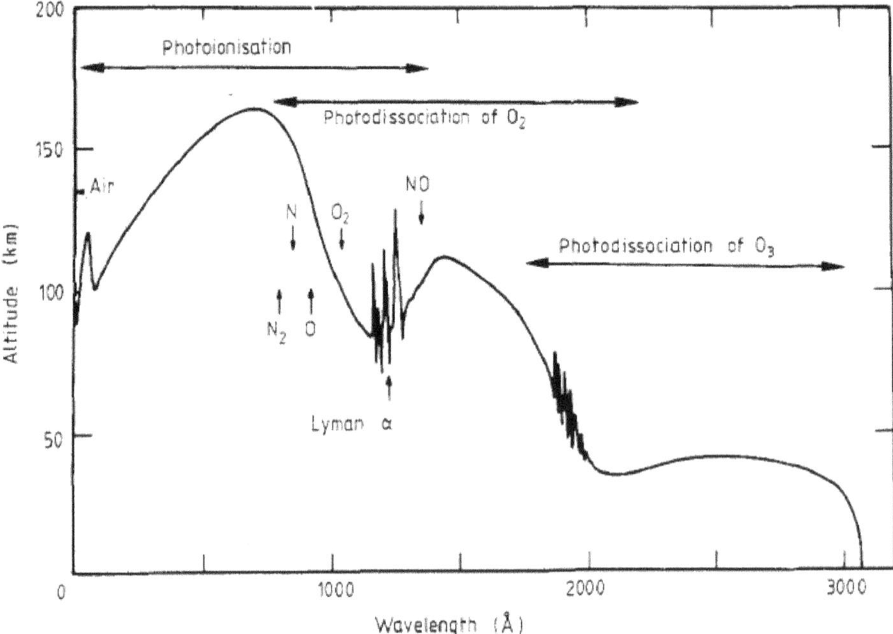

Figure 5.4. Penetration depth in the atmosphere of UV radiation as a function of wavelength below 3200 Å. Image credit: Vidal-Madjar et al. (1987).

the visible to mid-infrared band. Below 3000 Å, in the UV band, almost no radiation makes it to the surface. Figure 5.4 shows the penetration depth of UV radiation as a function of wavelength. At longer UV wavelengths, ozone is the major absorber. For shorter UV wavelengths, other species, including the abundant N_2 and O_2 and their dissociated atoms, are important.

Because of atmospheric absorption, UV astronomy has to be carried out from space or at high altitudes. Early experiments were done using high-altitude sounding rockets. Nearly all of the recent UV astronomy is done using instruments on spacecraft.

5.1.1 Line Transitions in the Ultraviolet

Line radiation is characteristic of energy transitions in atoms and ions, where the energy difference between two electronic energy levels in a given species is emitted as a photon. Line radiation is characterized by transition strength: permitted and forbidden lines, depending on the probability per unit time of a given transition and on excitation conditions. The general characteristics of line transitions are summarized in Book 1 and Kwok (2007).

The most important species (atoms, ions, and molecules) for line transitions come from the most common elements in the universe. Stellar surfaces, the winds from stars, and the interstellar medium (ISM) are composed of elements of roughly solar composition. The elements coming out of the Big Bang, resulting from early nucleosynthesis in the first few minutes of the universe were hydrogen, helium, and lithium. Since then, stellar nucleosynthesis has occurred in multiple generations of stars, which has resulted in the fusion buildup of heavier elements. The composition of the surface of our Sun is taken as representative of the composition of the elements in our Galaxy when the Sun was born, roughly 4.6 billion years ago.

The Sun's photospheric composition, in mass percentages, is approximately 73.5% hydrogen, 24.8% helium, 0.77% oxygen, 0.29% carbon, 0.16% iron, 0.12% neon, 0.09% nitrogen, 0.07% silicon, 0.05% magnesium, and 0.04% sulfur, with numerous other elements at lower concentrations. The Sun's composition is representative of that of other stars (with often small variations, and sometimes large deviations) and that of the ISM.

At stellar surface temperatures (~3000–50,000 K), many elements will exist in neutral form, or as negative or positive ions. Most of these neutrals or ions have electronic transitions in the multi-electron-volt range. The near- (NUV), far- (FUV), and EUV wavelength ranges are usually taken to be 2000–3000 Å (NUV), 1000–2000 Å (FUV), and 100–1000 Å (EUV), with wavelengths shorter than 100 Å taken as the X-ray region. One often-used database for atomic line transitions is the CHIANTI atomic database for emission lines (Dere et al. 1997; and its ongoing updates). Figure 5.5 shows the set of ions included in the atomic transition calculations in CHIANTI. Each ion has numerous electronic levels, and an ion's populations levels and emission-line intensities depend on the local conditions of element abundance, temperature, and electron density, as described in more detail in Dere et al. (1997).

5.1.2 Continuum Radiation in the Ultraviolet

Processes that produce continuum radiation include blackbody emission, synchrotron, free–free emission (also called bremsstrahlung), free–bound emission, and

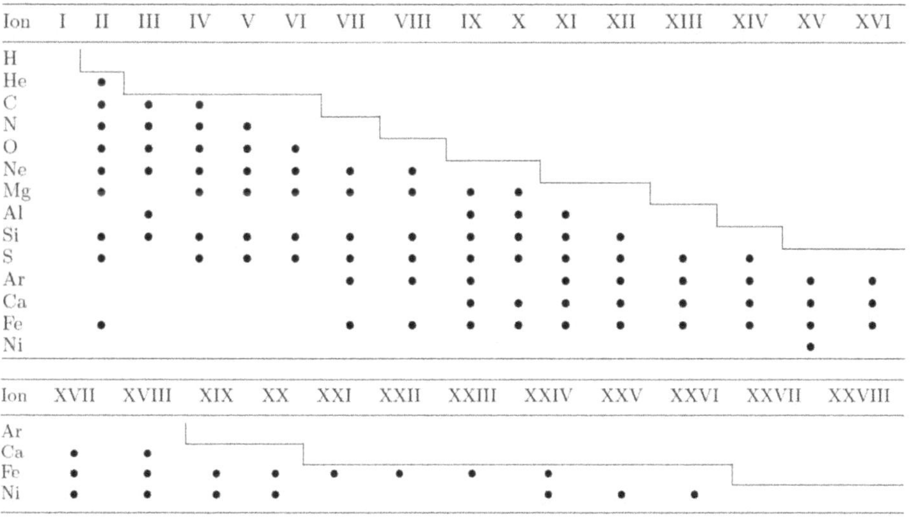

Figure 5.5. The set of ions included in the CHIANTI emission-line database. Image credit: Dere et al. (1997) Copyright EDP Sciences.

two-photon emission. For more details on the individual processes, many textbooks are suitable, e.g., Book 1 or Kwok (2007).

Ultraviolet continuum radiation originates in hot plasmas, which can be in stellar atmospheres, stellar winds, the ISM, or the atmospheres of accretion disks. The accretion disks are most commonly in interacting close binary stars but are also found around supermassive black holes at the centers of galaxies.

In the optically thick limit, plasma emission approaches that of a blackbody spectrum, which depends only on the temperature of the emitting body. For a blackbody spectrum, one can use Wien's displacement law, which gives the wavelength of peak emission, $\lambda_{\max} = (2.898 \times 10^{-3}$ mK)/T. This yields an estimate of the temperatures for which blackbody emission peaks in the NUV, FUV, and EUV bands. For NUV ($\lambda \simeq$ 2000–3000 Å), the temperature range is 9700 K (peak at 3000 Å) to 14,500 K (peak at 2000 Å). For FUV ($\lambda \simeq$ 1000–2000 Å), the temperature range is 14,500 K to 29,000 K (peak at 1000 Å). For EUV ($\lambda \simeq$ 100–1000 Å), the temperature range is 29,000 K to 290,000 K (peak at 100 Å). Clearly, for stars to emit NUV, FUV, or EUV, they require very high temperatures, which can only be achieved for stars above $\simeq 4\ M_{\odot}$ (see Figure 5.1).

5.1.3 Interstellar Extinction in the Ultraviolet

Interstellar extinction has long been realized as strongly affecting our view of the stars and their distribution in the Galaxy. The dimming of starlight is caused by interstellar dust particles and is also referred to as reddening, because shorter wavelengths are absorbed more strongly. This makes the observed light redder than the emitted light from a star. This is a similar manner to the effect of dust in Earth's atmosphere, which is commonly seen as the phenomenon of a red Sun at sunsets.

A typical amount of absorption of light by dust is a $\sim$1 magnitude reduction in the V band per kiloparsec of distance. The reduction in brightness for a particular star, in units of magnitudes as a function of inverse wavelength, is called the extinction curve. The extinction curve for different stars, i.e., in different directions in our Galaxy, is quite variable. This is strong evidence for the variability in composition and size of interstellar dust grains.

The basic characterization of the extinction curve uses a two-parameter description: $A(V)$, the extinction in magnitudes in the V band; and $R(V)$, the ratio of selective to total extinction, defined as $R(V) = A(V)/E(B - V)$. Here $E(B_V) = A(B) - A(V)$ is the excess extinction measured in the B band above that measured in the V band.

The extinction curve in general increases to shorter wavelengths with a broad feature of increased extinction around 2200 Å, referred to as the 2200 Å (or 2175 Å) bump. Fitzpatrick & Massa (2007) present a study of a large number (328) of Galactic extinction curves, measured toward main-sequence B and late O-type stars. These stars were chosen for their high temperature and moderate extinction, so that the extinction curves could be well measured. The data consisted of International Ultraviolet Explorer (IUE) (see below) spectrophotometry in the UV ($\lambda^{-1} > 3.3\ \mu\text{m}^{-1}$), Johnson optical UBV photometry, and 2-Micron All Sky Survey near-infrared *JHK* photometry. Figure 5.6 shows a sample of 10 observed extinction curves, of the 328, from Fitzpatrick & Massa (2007). The solid lines through each set of data points are the fits of their parameterized extinction curve in each data set. The 2175 Å ($\lambda^{-1} = 4.6\,\mu\text{m}^{-1}$) broad feature is clearly visible and is variable in strength, width, and peak position for different sight lines. The most variable part of the extinction curve is at the shortest wavelengths down to the shortest measured: $\lambda^{-1} \simeq 6 - 8.5\,\mu\text{m}^{-1}$ (1700 Å to 1200 Å). This makes it important to use an extinction curve appropriate for a given sight line, as use of an average Galactic extinction curve can introduce significant errors into the analysis of stellar spectra.

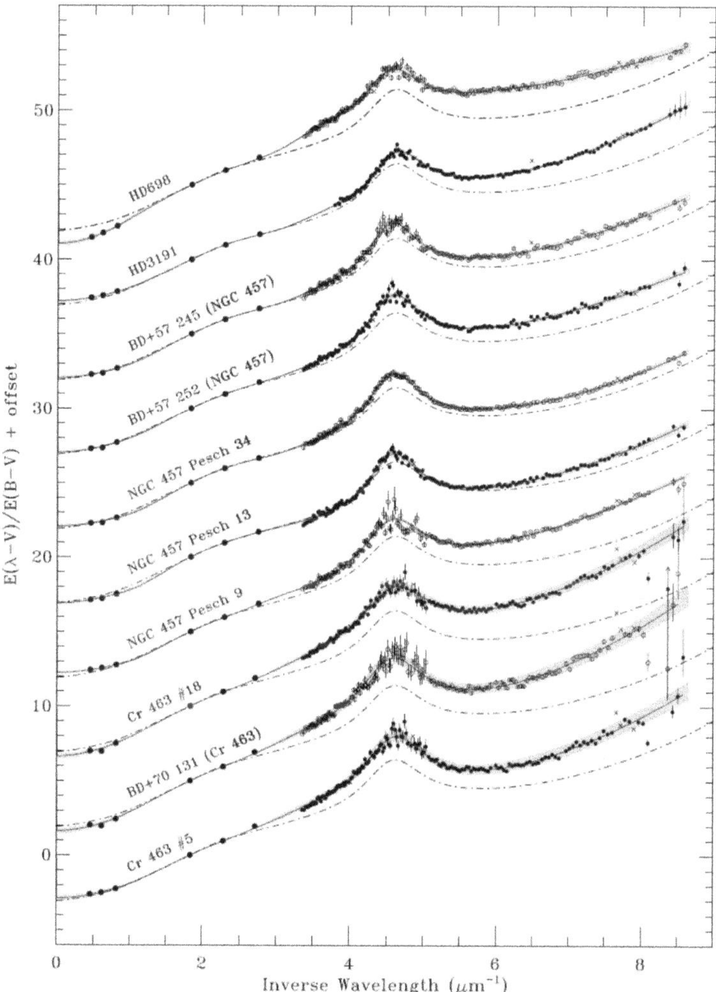

Figure 5.6. A sample of interstellar extinction curves for 10 sight lines to different Galactic O and B stars. The dotted–dashed line for each curve is an estimate of the average Galactic extinction using $R(V) = 3.1$. Image credit: Fitzpatrick & Massa (2007).

5.2 Ultraviolet Astronomy Observatories

The major facilities used for UV astronomy are described briefly here. The major science discoveries made using these facilities are described in Section 5.3.

5.2.1 Copernicus

The third Orbiting Astronomical Observatory (OAO-3), named Copernicus, was launched in 1972 and carried a telescope spectrometer for high-resolution studies of stellar spectra in the UV longward of about 1000 Å, with interstellar absorption lines a primary objective. The Copernicus spectrometer is described in Rogerson et al. (1973). The Copernicus telescope with its 32 inch diameter primary mirror and its Rowland circle grating spectrometer are shown schematically in Figure 5.7. Stellar spectra were scanned with a resolution of ~0.05 Å between 950 and 1450 Å, and twice this in first order between 1650 and 3000 Å. The pointing over several minutes is steady within 0.02″, and the measured photometric precision in the short-wavelength range is limited by the statistics of photon counts, with 14 s counts of about 10^3 on an unreddened Bl star, $m_V = 5.0$, at 1100 Å.

5.2.2 Voyager Ultraviolet Spectrometers

The two Voyager spacecraft each carried an Ultraviolet Spectrometer Experiment. The Voyager Ultraviolet Spectrometer (UVS) (Broadfoot et al. 1977) was an objective grating spectrometer covering the wavelength range of 500–1700 Å with 10 Å resolution. It was mounted on the right-side instrument boom (see Figure 5.8) along with several other primary instruments of the iconic Voyager spacecraft, with its prominent and large High-Gain Antenna. Its primary goal was the determination of the composition and structure of the atmospheres of Jupiter, Saturn, Uranus, and several of their satellites. UVS had two observational modes: airglow mode, which measured radiation due to resonant scattering of solar flux; and occultation mode,

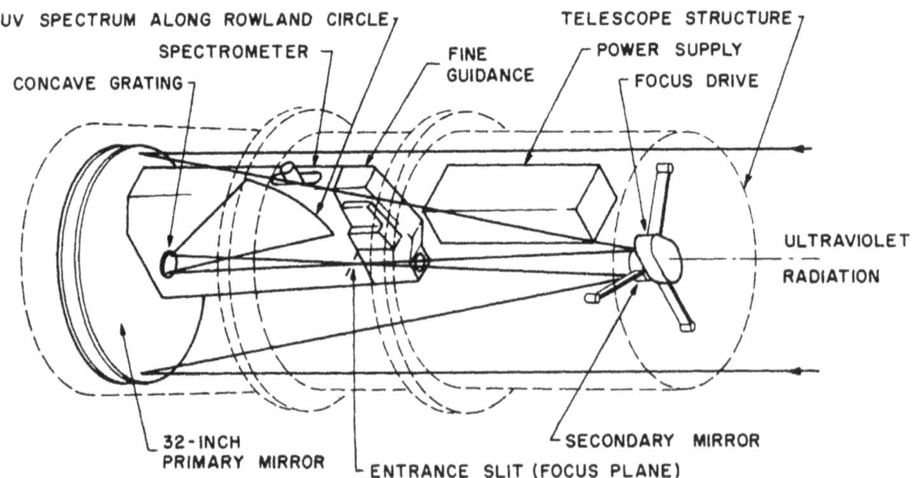

Figure 5.7. Schematic of the Copernicus telescope spectrometer. Image credit: Rogerson et al. (1973).

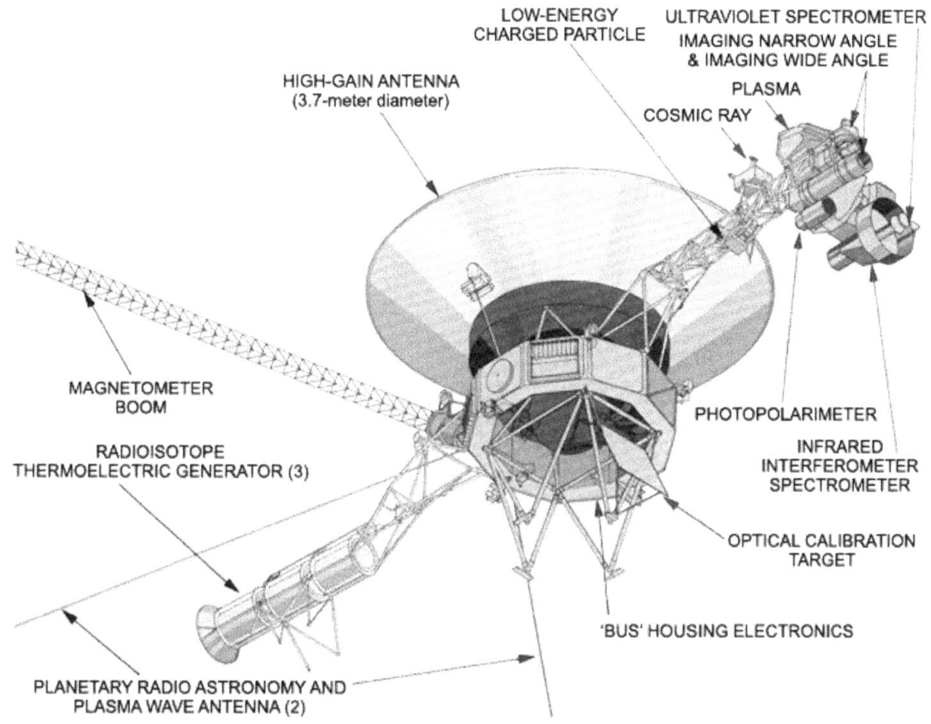

Figure 5.8. Schematic of the Voyager 1 and 2 spacecraft, showing the locations of the instruments. Image credit: NASA/JPL.

which measured atmospheric extinction of solar or stellar radiation. UVS made important solar system measurements and made valuable contributions to stellar astronomy at wavelengths below 1000 Å.

Because the period of operation of the UVS overlaps with many other UV missions, the calibration of the V1 and V2 UVS with other spectrometers is critical. Ben-Jaffel & Holberg (2016) revisited the UVS calibration. Using the Lyα airglow from Saturn, observed "in situ" by both Voyagers and remotely by the IUE, the Voyager values were matched to IUE, taking into account the shape of the Saturn Lyα line observed with the Goddard High Resolution Spectrograph on board the Hubble Space Telescope (HST). The resulting prescription is to keep the original calibration of the Voyager UVS with a maximum uncertainty of 30%, making both instruments some of the most stable EUV/FUV spectrographs in the history of space exploration.

5.2.3 International Ultraviolet Explorer

From 1978 to 1996, the IUE satellite has been acquiring a large collection of astronomical UV spectra. Figure 5.9 shows the IUE instrument schematic and light-path. The telescope is mounted on top and the gratings, cameras, and associated optical elements for the two spectrometers (the long-wavelength spectrograph, or

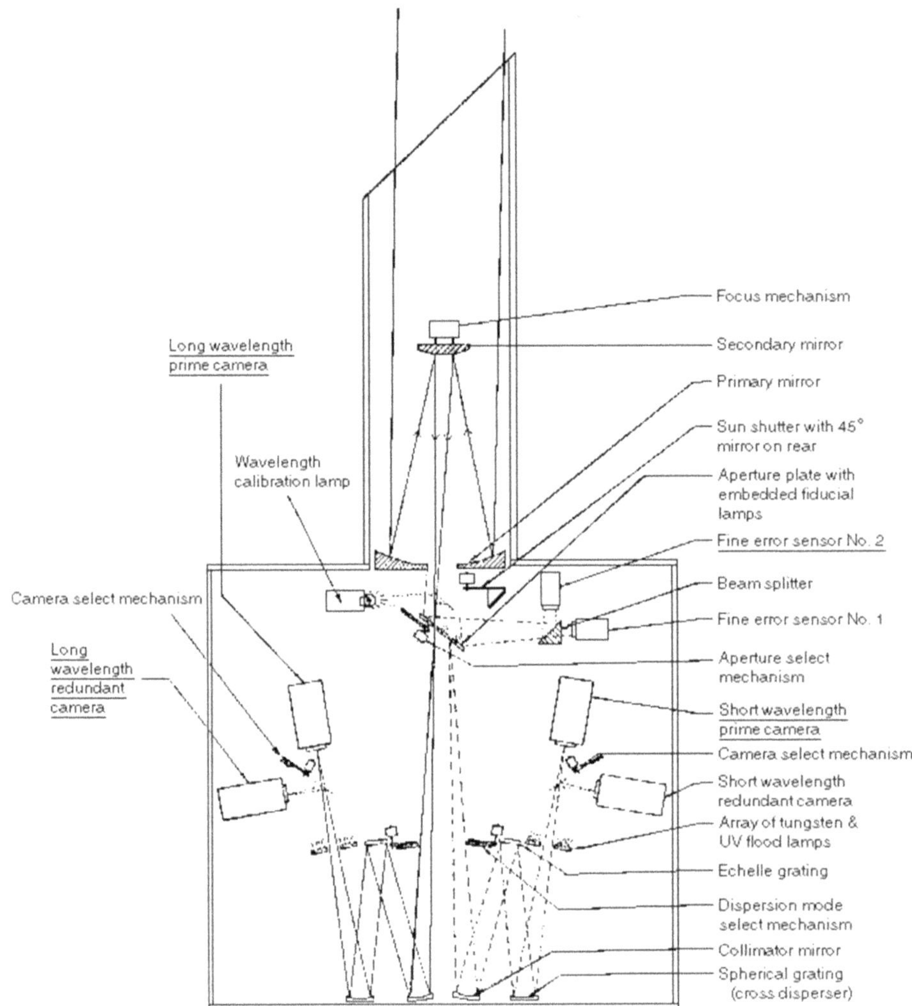

Figure 5.9. Instrument and light path for the International Ultraviolet Explorer. Image credit: Newmark et al. (1992)/ NASA.

LWS, and the short-wavelength spectrogram, or LWS) are mounted in the large box below the telescope.

The Final Archive of IUE is presented in Nichols & Linsky (1996). More than 100,000 spectral images, from which low- and high-dispersion spectra are extracted, have been observed. These include a broad range of sources from solar system objects, stars of all types, interstellar and Galactic halo gas, normal galaxies, and active galactic nuclei (AGNs). The Final Archive of IUE data contains the spectral images and extracted spectra, reprocessed with uniform software and calibrations that enhance the quality of the data products. The processing algorithms and calibration chronology of the IUE scientific instrument are presented in

Nichols & Linsky (1996). The Final Archive demonstrates an increased signal-to-noise ratio of 10%–50% for low-dispersion data and 100% or more for high-dispersion data compared to the old processing scheme; the usable spectral range was extended down to 1150 Å and up to 3400 Å.

5.2.4 Extreme Ultraviolet Explorer

The Extreme Ultraviolet Explorer (EUVE) mission, described in Bowyer & Malina (1991), Malina et al. (1994), and Bowyer et al. (1994) operated at extreme EUV wavelengths, in the range from 70 to 760 Å (Figure 5.10). To date, this is the only satellite mission to operate in the EUV, primarily because of the strong absorption of the ISM in the EUV. EUVE conducted an all-sky, all-band survey in the extreme

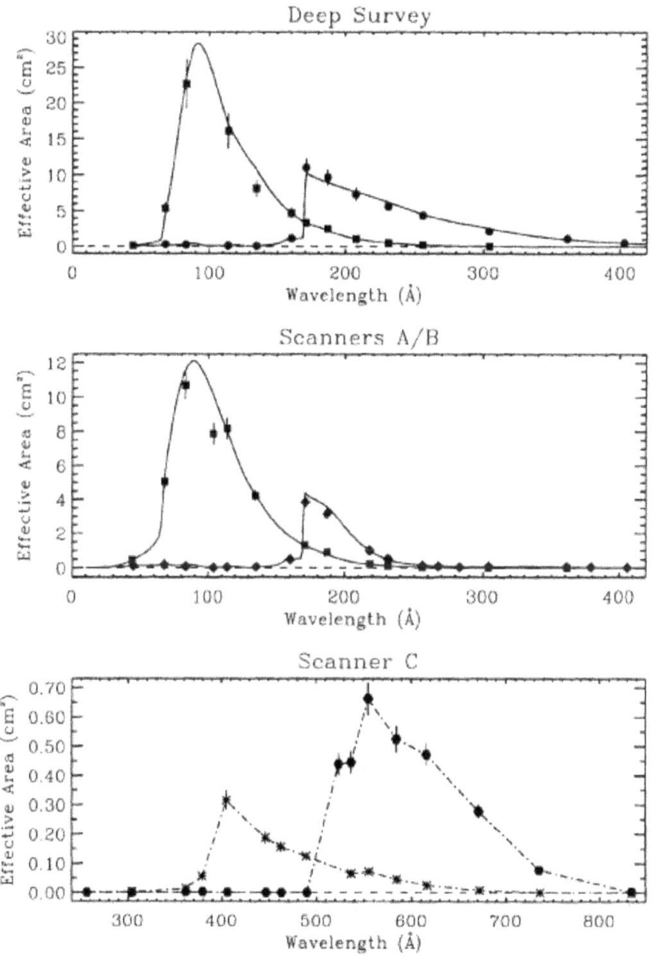

Figure 5.10. Effective area versus wavelength for the two EUV bandpasses of the Deep Survey Telescope (top panel) and for the four EUV bandpasses of the Scanner Telescopes (middle and bottom panels) on board the Extreme Ultraviolet Explorer. Image credit: Malina et al. (1994).

UV in four bands with an angular resolution of 6′ by 6′ with ~500 s average exposure between 1992 July and 1993 January.

The sky survey was conducted with three scanning telescopes, which were coaligned and pointed 90° away from the satellite spin axis during the survey. During that survey, 97% of the sky was surveyed at some level; about 80% of the sky received an exposure of 500 s or more. The Deep Survey telescope, pointed along the spin axis, operated during the survey to carry out a deeper exposure over a 2° by 180° strip of the sky along the ecliptic. After the survey and until mission end in January 2001, EUVE was operated primarily in spectroscopy mode, in which three spectrometers share the Deep Survey telescope. The Deep Survey telescope was pointed at individual targets selected by the EUVE Guest Observer Program. Additional survey observations were done to fill in gaps in the original sky survey.

The comprehensive catalog from the entire sky survey is the Second EUVE Source Catalog as presented by Bowyer et al. (1996). The data included:

1. All-sky survey detections from the initial 6 month scanner-survey phase,
2. Additional scanner detections made subsequently during specially programmed observations designed to fill in low-exposure sky areas of the initial survey,
3. Sources detected with deep-survey telescope observations along the ecliptic,
4. Objects detected by the scanner telescopes during targeted spectroscopy observations, and
5. Other observations.

Three lists of the EUV sources were given: the all-sky survey detections, the deep-survey detections, and sources detected during other phases of the mission. Each list gives positions and intensities in each wave band. The total number of objects listed is 734. For approximately 65% of these plausible optical, UV, radio, and/or X-ray identifications are given.

5.2.5 Far Ultraviolet Spectroscopic Explorer

The Far Ultraviolet Spectroscopic Explorer Mission is reviewed in Moos et al. (2000). FUSE observed in the FUV spectral region, 905–1187 Å, with a high spectral resolution. The instrument consisted of four coaligned prime-focus telescopes and Rowland spectrographs with microchannel plate detectors. The optical elements and light path for FUSE, with its large focal length, are shown in Figure 5.11.

FUSE was optimized for high sensitivity in the FUV. Two of the telescope channels used Al:LiF coatings for optimum reflectivity between approximately 1000 and 1187 Å, and the other two channels used SiC coatings for optimized throughput between 905 and 1105 Å. The gratings were holographically ruled to correct largely for astigmatism and to minimize scattered light. The microchannel plate detectors had KBr photocathodes and used photon counting to achieve good quantum efficiency with a low background signal.

The sensitivity was sufficient to examine reddened lines of sight within the Milky Way and also sufficient to use AGNs and QSOs for absorption line studies of both

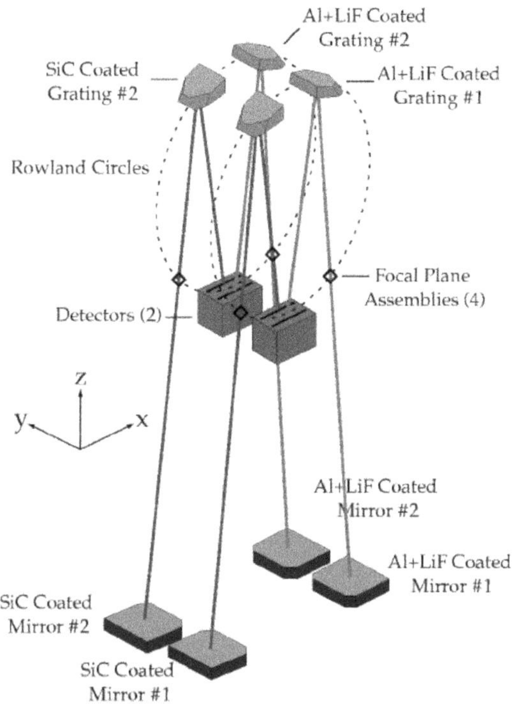

Figure 5.11. Far Ultraviolet Spectroscopic Explorer's light path; the telescope focal lengths are 2.245 m and the Rowland circle diameters are 1.652 m. Image credit: Moos et al. (2000).

Milky Way and extragalactic gas clouds. The FUV spectral region contains a number of key scientific diagnostics, including O vi, H i, D i, and the strong electronic transitions of H_2 and HD (hydrogen deuterium).

5.2.6 GALEX

The GALEX was a NASA Explorer Mission launched in 2003 (Martin et al. 2003). GALEX performed the first space UV sky survey, including imaging and grism surveys, in two bands (1350–1750 Å and 1750–2750 Å; Martin et al. 2005). It was designed with a large (1°) field of view in order to enable survey science to be carried out. The telescope assembly is shown in the left panel of Figure 5.12, and the light path, including the dichroic mirror, which splits the FUV and NUV channels, is shown in the right panel.

The primary goal of GALEX was to study star formation in galaxies and its evolution with time. The surveys included an all-sky imaging survey (limiting magnitude $m_{\rm AB} \simeq 20.5$), a medium imaging survey of 1000 deg^2 (limiting magnitude $m_{\rm AB} \simeq 23$), a deep imaging survey of 100 deg^2 (limiting magnitude $m_{\rm AB} \simeq 25$), and a nearby galaxy survey.

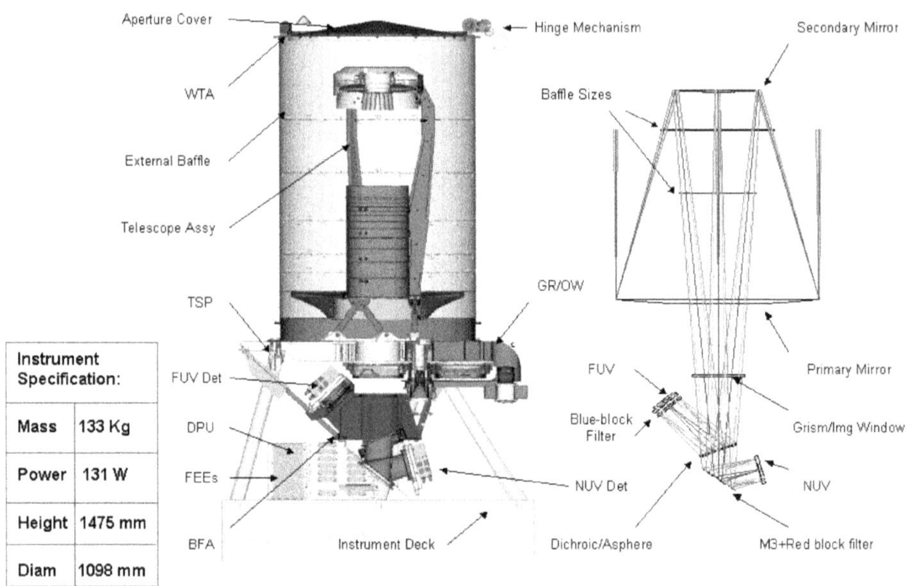

Figure 5.12. Galaxy Evolution Explorer instrument layout (left) and light path (right). Image credit: NASA.

5.2.7 HST's UV Instruments: Cosmic Origins Spectrograph, Space Telescope Imaging Spectrograph, and Wide Field Camera 3

The HST (often referred to as Hubble; Figure 5.15) was launched into low-Earth orbit in 1990 and remains in operation. HST has a 2.4 m mirror, and its instruments observe in the UV, visible, and near-infrared regions of the electromagnetic spectrum. HST accommodates five science instruments at a given time, plus the Fine Guidance Sensors, which are mainly used for aiming the telescope. When launched, the five science instruments on HST were the Wide Field and Planetary Camera (WF/PC), Goddard High Resolution Spectrograph (GHRS), High Speed Photometer (HSP), Faint Object Camera (FOC), and the Faint Object Spectrograph (FOS). Since the final servicing mission in 2009, the four active instruments have been the Advanced Camera for Surveys (ACS), the Cosmic Origins Spectrograph (COS), the Space Telescope Imaging Spectrograph (STIS), and the Wide Field Camera 3 (WFC3). Detailed information on the HST mission, its science, and its instruments can be found at the Space Telescope Science Institute web page (http://www.stsci.edu/). All of the HST instruments have fairly complex optics, compared to previous space observatories, because HST was designed to have highly flexible instruments, each with multiple configurations, to carry out a vast range of astrophysical studies.

COS (Hirschauer et al. 2021) was installed on HST in 2009 and was designed for UV (90–320 nm) spectroscopy of faint point sources with a resolving power of 1550–24,000. COS has two separate light paths and two separate camera systems, one for NUV and one for FUV (see Figure 5.13). STIS (Branton et al. 2021) has three 1024 by 1024 detector arrays. These are mounted at one end of the STIS instrument

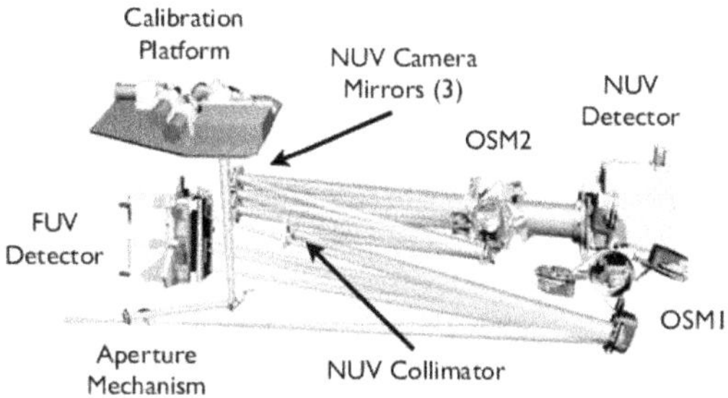

Figure 5.13. HST Cosmic Origins Spectrograph (COS) instrument layout. Image credit: Hirschauer et al. (2021).

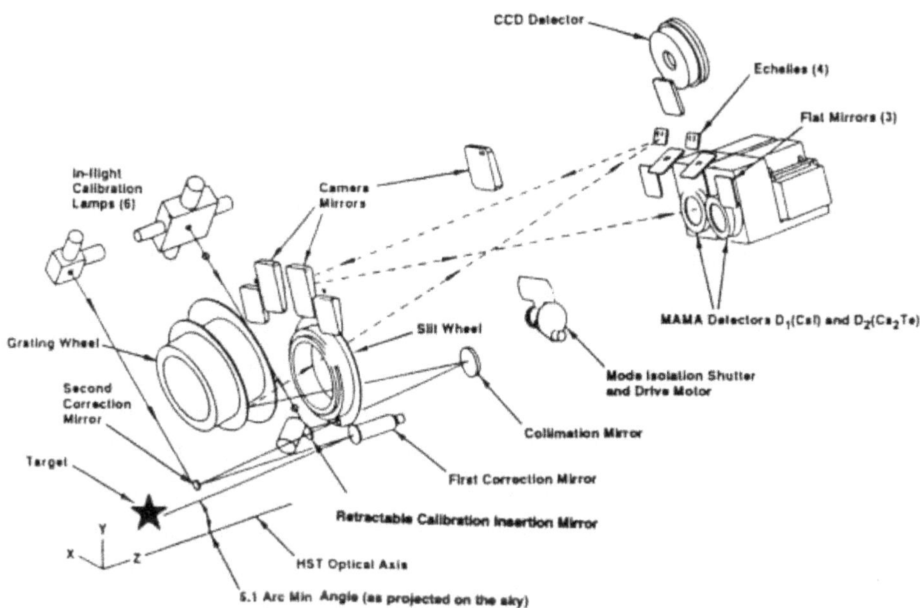

Figure 5.14. HST Space Telescope Imaging Spectrograph (STIS) instrument layout. Image credit: Branton et al. (2021).

package (the right side of Figure 5.14). The first is a charge-coupled device with a 52″ by 52″ field of view, covering the visible and near-infrared spectrum from 200 nm to 1030 nm. The other two detectors are optimized for detecting UV photons. They are Multi-Anode Multi-channel Arrays, each with a 25″ by 25″ field of view. One covers the NUV between 160 nm and 310 nm and the other covers the FUV between 115 nm and 170 nm. WFC3 (Dressel 2021) is HST's newest instrument and was installed in 2009. The WFC3 instrument has two independent light paths: an optical and NUV channel that uses a pair of charge-coupled devices (CCD) to record images from 200 to 1000 nm, and a near-infrared detector array that covers the wavelength range from 800 to 1700 nm. Both channels have a variety of broad- and narrowband filters, as well as prisms and grisms, which enable wide-field, very-low-resolution spectroscopy that is useful for surveys.

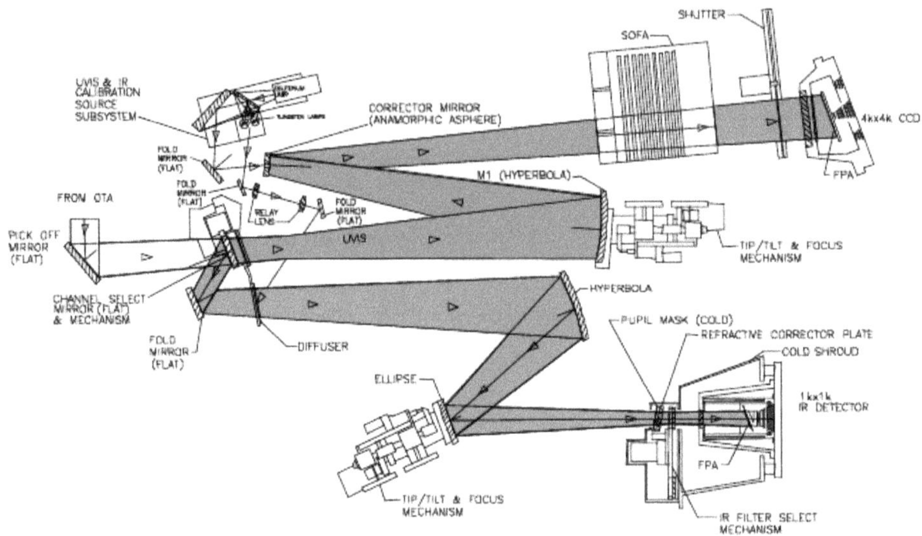

Figure 5.15. HST Wide Field Camera 3 (WFC3) instrument layout. Image credit: Dressel (2021).

5.2.8 Ultraviolet Optical Telescope on the Swift Gamma-Ray Burst Mission

The Ultraviolet Optical Telescope (UVOT) on board the Swift Gamma-Ray Burst Mission is sensitive to optical and NUV radiation. It was designed for studying nearby stellar populations and providing insight into the NUV properties of hot stars and the contribution of those stars to the integrated light of more distant stellar populations. UVOT has a wide field of view (17′ by 17′), three optical filters and three filters with NUV (~170–300 nm) sensitivity, 2.3″ spatial resolution, and is coaligned with Swift's X-ray Telescope. The passbands of the six UVOT filters are shown in Figure 5.16.

Siegel et al. (2014) describe a large survey of nearby stellar populations using UVOT. They reviewed the state of UV stellar photometry, outlined the survey, and discussed problems specific to wide and crowded-field UVOT photometry. Color–magnitude diagrams (CMDs) of the nearby open clusters M67, NGC 188, and NGC 2539, and the globular cluster M79 were presented. UVOT can discern the young- and intermediate-age main sequences, blue stragglers, and hot white dwarfs, producing results consistent with previous studies. They also found that UVOT characterized the blue horizontal branch of M79 and easily identified a known post-asymptotic giant-branch (post-AGB) star.

5.2.9 The Ultraviolet Imaging Telescope on the AstroSat Mission

The Ultraviolet Imaging Telescope (UVIT) instrument was launched in 2015 by the Indian Space Research Organization. UVIT has two 375 mm diameter telescopes: one for the FUV (1300–1800 Å), and the other for the NUV (2000–3000 Å) and the visible (VIS) (3200–5500 Å; Tandon et al. 2017). Figure 5.17 shows the telescope structure (left panel), with the optics entrance at the top (with the telescope doors

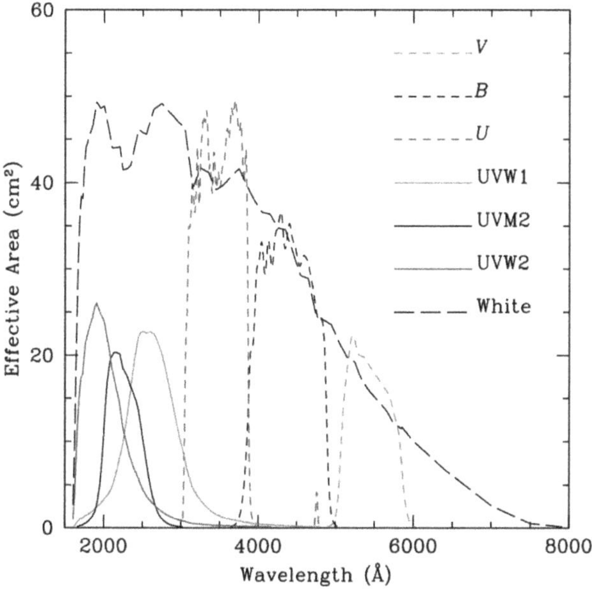

Figure 5.16. SWIFT Ultraviolet Optical Telescope (UVOT) filter bands. Image credit: NASA.

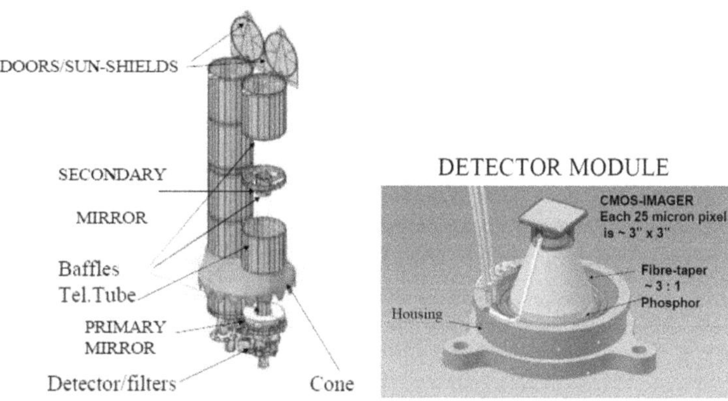

Figure 5.17. AstroSat Ultraviolet Imaging Telescope (UVIT): two-telescope assembly (left) and detector assembly (right). Image credit: AstroSat Handbook (SSPO/ISRO) http://www.iucaa.in/astrosat/AstroSat_handbook.pdf.

shown opened), the primary mirrors at the bottom of the telescope tubes, and the detectors and filter wheel assemblies below the primary mirrors. The right panel shows a picture of the detectors and electronics for the VIS and NUV cameras (left-hand telescope), where the light is split into VIS and NUV channels by a dichroic mirror. The right panel also shows the FUV camera with its dedicated telescope (the right-hand one).

UVIT is designed to make images, simultaneously in FUV, NUV, and VIS wavelengths, in a field of $\simeq 28'$ diameter, with a spatial resolution of $\simeq 1.5''$. The sensitivity in the FUV is $\simeq$ 20 mag in the AB magnitude scale for a 200 s exposure.

The VIS channel is used primarily for pointing the ASTROSAT spacecraft. The science targets observed during the first two years of operation included star clusters, galaxies, galaxy clusters, AGNs, Chandra deep fields, exoplanets, planetary nebulae (PNe), and supernova remnants (SNRs).

5.3 Ultraviolet Astronomy Science

5.3.1 UV Studies of the Heliosphere

Diffuse Lyα emission is now thought to primarily have a heliospheric origin. This was uncovered by the Voyager UVS team campaign of scanning the sky background over great circles that intersect the upwind and antisolar directions. An excess was found, showing a maximum in brightness close to the upwind direction, and interpreted as the scattering of Lyα photons from the hydrogen wall created as the heliosphere moves into the ISM. Figure 5.18 shows a diagram of the heliosphere. Illustrated are the termination shock where the fast solar wind is thermalized abruptly; the heliopause, which is the outer boundary of the shocked slow solar wind and is the contact discontinuity between the shocked solar wind and the shocked ISM; and the ISM shock located outside (right side of) the location of the Voyager spacecraft. The excess Lyα emission comes from the dense layer of shocked ISM just inside the ISM shock.

Ben-Jaffel & Holberg (2016) reassessed the excess Ly-α emission detected by Voyager UVS deep in the heliosphere and demonstrated its consistency with a heliospheric but not galactic origin. This confirms results obtained nearly two decades ago, namely, the UVS discovery of the distortion of the heliosphere and the corresponding obliquity of the local interstellar magnetic field (~40° from upwind)

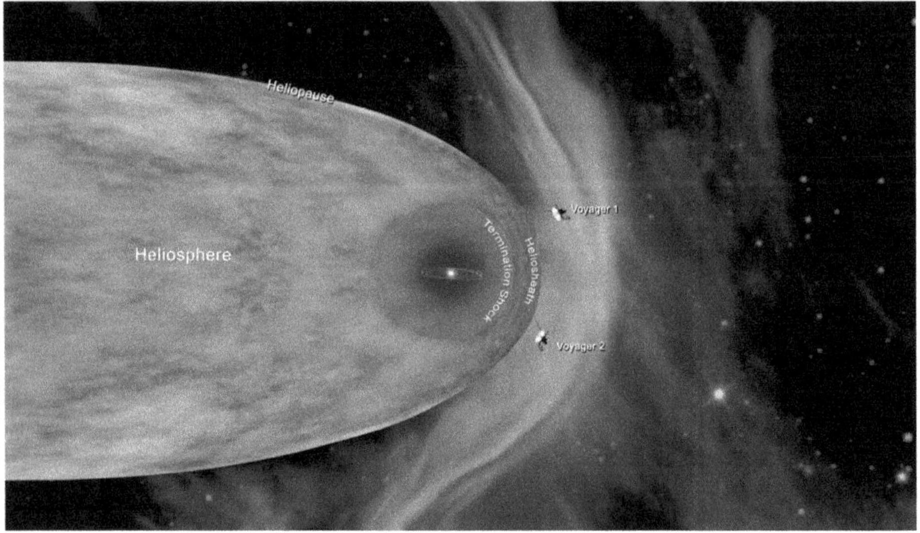

Figure 5.18. Schematic of the heliosphere: the interaction region between the solar wind and the ISM. Image credit: NASA/JPL–Caltech.

in the solar system neighborhood without requiring any revision of the Voyager UVS calibration.

5.3.2 UV Studies of the ISM

5.3.2.1 ISM Absorption Lines

Interstellar absorption lines of H I, C I, C II, N I, N II, O I, Mg I, Mg II, Si II, Si III, Si IV, P II, S I, S II, S III, S IV, Cl II, Ar I, Mn II, and Fe II were identified in Copernicus observations of UV spectra of the hot stars ξ Per, α Cam, λ Ori, ζ Oph, and γ Ara (Morton et al. 1973). A curve-of-growth analysis of abundances showed that the ratios of Mg, P, Cl, and Mn to hydrogen are less than in the Sun by factors of about 4 to 10. The depletions were consistent with the condensation of those elements into interstellar grains, although not in amounts in agreement with models of grain formation at that time.

Copernicus satellite observations were analyzed for the presence of molecular hydrogen in the ISM (Spitzer et al. 1973). Strong H_2 lines were measured in all 11 reddened stars ($E(B - V) > 0.10$) were observed. The fraction of hydrogen gas in molecular form exceeds 10^{-1}. The absorption spectrum of ξ Per, with $E(B - V) = 0.33$, is shown in Figure 5.19, showing electronic absorption levels for H_2 in different rotational and vibrational levels. The relatively large column densities in higher rotational levels, up to $J = 5$ or 6, correspond to H_2 excitation temperatures mostly between 150 K and 200 K. Measurement of two HD lines in nine stars yielded a ratio of HD to H_2 equal to $\sim 10^{-6}$. In a set of eight unreddened stars, with ($E(B - V) < 0.05$), no trace was found for H_2 absorption, with the fraction of molecular to atomic hydrogen less than 10^{-7}.

O VI (five times ionized O) is a good tracer of hot gas, $\sim 3 \times 10^5$ K, because of the high ionization potential of O V (113.9 eV). The Far Ultraviolet Spectroscopic Explorer (FUSE) surveyed O VI absorption from hot gas in and near the Galaxy (Wakker et al. 2003). The O VI 1031.926, 1037.617 absorption lines associated with hot gas in and near the Milky Way were detected in the spectra of a sample of 100 extragalactic targets and two distant halo stars. The sight lines cover most of the sky above Galactic latitude $|b| > 25°$. At lower latitudes, the UV extinction is generally too large at these short wavelengths to measure any absorption spectra. The velocity range -1200 to 1200 km s^{-1} was searched, and with a few exceptions, O VI was found only in the velocity range -400 to 400 km s^{-1}; the exceptions are likely intergalactic O VI.

The observed O VI absorption was separated into low-velocity components, $|v_{LSR}| <$ 100 km s^{-1}, associated with the Milky Way halo and high-velocity components, $|v_{LSR}| >$ 100 km s^{-1}. High-velocity O VI is very common, having equivalent width >65 mA in 50% of the sight lines and equivalent width >30 mA in 70% of the high-quality sight lines. The central velocities of high-velocity O VI components range from $|v_{LSR}| = 100$ to 330 km s^{-1}; they show no correlation between velocity and absorption strength. The extent of an O VI halo around M33 is limited to be <100 kpc at a 3σ detection limit of log N(O VI) $\simeq$ 14.0. The 50 km s^{-1} wide O VI channel maps show evidence for Galactic rotation. They also show two known HI high-velocity clouds:

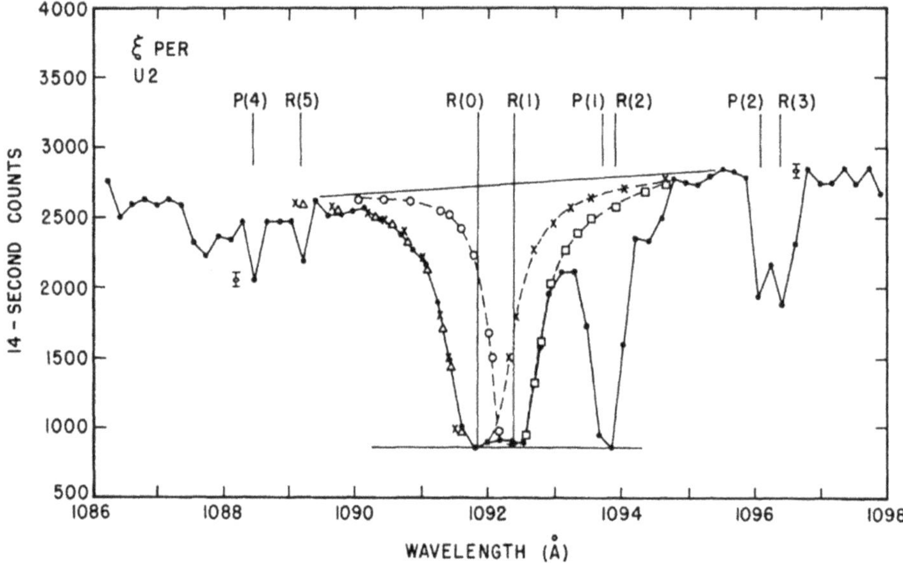

Figure 5.19. The Copernicus UV absorption spectrum of ξ Per. Image credit: Spitzer et al. (1973).

complex C and the Magellanic Stream. The channel maps show that O VI at velocities <200 km s^{-1} occurs along all sight lines in the region $l = 20° — 150°$, $b < -30°$, while O VI at velocities >200 km s^{-1} occurs along all sight lines in the region $l = 180° — 300°$, $b > 20°$.

Limits on interstellar deuterium abundance are determined by analysis of UV observations of absorption lines due to the Lyman series transitions in hydrogen and deuterium. Copernicus observed these UV lines in the spectrum of β Cen (Rogerson & York 1973). The derived ratio of deuterium to hydrogen, by number, was $1.4 \pm 0.2 \times 10^{-5}$. Ascribing this deuterium abundance to arise from Big Bang element synthesis, 1.5×10^{-31} g cm^{-3} was obtained for the present density of the universe.

FUSE observations of deuterium abundance along various lines of sight find a ratio of deuterium to hydrogen (D/H) that is constant inside the Local Bubble with the value of $\simeq$15 parts per million (ppm; Wood et al. 2004). The Local Bubble is a hot low-density region around the Sun extending out about 100 pc. However, at greater distances, the abundance is between 7 and 25 ppm, as shown in Figure 5.20. This shows that processes other than production in the Big Bang are important, including astration (destruction of deuterium by burning in stellar interiors) and condensation onto dust grains. The best determination of the cosmic deuterium abundance is indirect, from NASA's WMAP satellite, and is $\simeq$ 27 ppm.

5.3.2.2 ISM Extinction

Extinction curves for the interstellar material were measured from 7.5 to 10 μ^{-1} (1330–1000 Å) using Copernicus observations of the reddened stars ζ Oph, ζ Per, ξ Per, and α Cam (York et al. 1973). In all cases, the extinction from 7.5 to 9 μ^{-1}

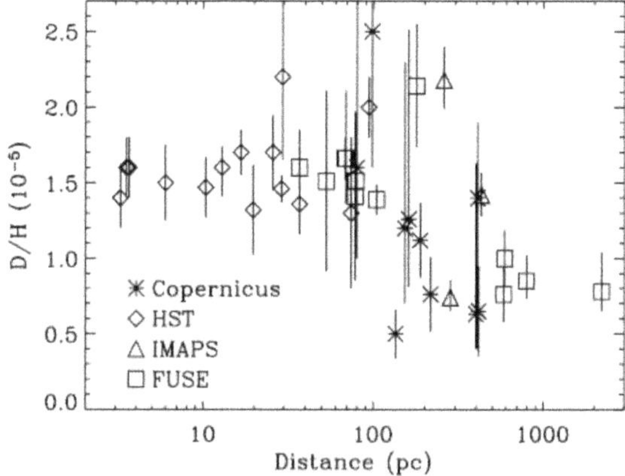

Figure 5.20. Measurements of the deuterium to hydrogen ratio as a function of distance from the Sun. Image credit: Wood et al. (2004).

agreed with previous work, and the extinction continues to increase to shorter wavelengths (see Figure 5.6 above). The new data between 9 and 10 μ^{-1} showed that very small grains (<350 Å for silicates, or <200 Å for iron particles) must be present in the general mix assumed for the interstellar material. The feature expected for graphite at 9.7 μ^{-1} (1030 Å) if graphite is the cause of the 4.6 μ^{-1} (2175 Å) is not seen. This limits how much graphite can contribute to the strong 2175 Å bump.

The UV interstellar extinction in the LMC was measured using IUE observations (Koornneef & Code 1981). Ten early-type supergiants in the LMC were observed by IUE. The LMC interstellar extinction law deviates significantly from the average Galactic law in two ways. The 2200 Å feature is deficient in strength. In the FUV ($\lambda <$ 2000 Å), the observed LMC extinction law is significantly above the Galactic curve. $E(\lambda - V)/E(B - V)$ is higher than the Galactic mean (see Figure 5.6) by a factor of ~1.5. This can be mainly explained by a lower dust-to-gas ratio in the LMC compared to the Galaxy (by factor ~3) and strong damping wings in the LMC from the interstellar Ly-α absorption line.

Sofia et al. (2005) presented FUSE measurements of extinction curves for nine Galactic sight lines, which include data from infrared up to FUV wavelengths (1050–1200 Å). The sight lines were chosen to sample a wide range of extinction properties. The large spread in their measured ratios of total to selective extinction, $R_V = 2.43$–3.81, provides evidence for the wide range of dust environments. All sight lines are well fit with an extrapolation of the Fitzpatrick & Massa extinction relationship (Fitzpatrick & Massa 1990). The Cardelli, Clayton, & Mathis extinction relationship (Cardelli et al. 1989) fails to fit the IR to UV and FUV extinction for two of the nine sight lines. For the other seven sight lines, it fits the IR to UV extinction but not FUV. Only for three sight lines did it fit the IR to UV and FUV extinction. Sofia et al. (2005) present a simple dust grain model with three types of grains (silicates, amorphous carbon, and graphite), each with their distribution of sizes. Using a maximum entropy method analysis they showed that seven of the nine sight lines in the study require a larger fraction of grain materials to be in dust when

FUV extinction is included in the models. Most of the added grain material was found to be in the form of small (radii $\leqslant$ 200 Å) grains.

5.3.2.3 ISM: SNRs

The first successful FUV spectroscopic observation of a filament in the Puppis A SNR was carried out using the Hopkins Ultraviolet Telescope by Blair et al. (1995). The filament was on the eastern side of the remnant near the brightest X-ray region. The spectrum covered the 820–1840 Å spectral range with 13 Å resolution and showed numerous lines of C, N, O, Ne, He, and possibly Si. Comparison of the spectrum with solar abundance shock models indicates a shock velocity of approximately 160–180 km s^{-1}. A slight overabundance of N at this position, consistent with a shocked ISM, conflicts with the much larger overabundance of N indicated by optical data from other regions of the outer shell in Puppis A. Comparison with optical CCD images obtained at Las Campanas Observatory and with Einstein X-ray HRI images indicates that the region observed corresponds to a very recent encounter between the blast wave and an interstellar cloud. The interaction of the shock wave with this cloud is responsible for the bright X-ray emission in this region.

5.3.3 UV Studies of Stars

The first UV studies of bright stars were carried out using the Copernicus satellite (see above). The improved instruments and sensitivity of IUE, operational from 1978 to 1996, led to great advances in stellar studies, with observations of the full range of stars hot enough to emit UV radiation. The IUE Final Archive (Nichols & Linsky 1996) is a valuable resource for studying UV emission from astronomical objects. Higher sensitivity and spectral resolution were enabled by FUSE, allowing more detailed spectra and the study of more distant objects. Given below is a small sample of the contribution of UV studies to the astrophysics of stars.

The EUV spectral range has been much less studied. An atlas of EUV spectra of stellar sources obtained with the EUVE spectrometers was presented by Craig et al. (1997). Ninety-five bright sources were observed in the short-wavelength (SW; 70–190 Å), medium-wavelength (MW; 140–380 Å), and long-wavelength (LW; 280–760 Å) bandpasses at ~0.5, 1, and 2 Å resolution, respectively. The sets of sources were mostly near the Sun and included stars, white dwarfs, and binaries. The stars were of the following types: 2 B, 1 A and 1 F star, 4 G dwarfs, 2 G giants, 1 pre-main-sequence star, a K giant, a K subgiant, 6 K dwarfs, and 10 M dwarfs. The white dwarfs (WDs) included 20 pure-hydrogen WDs, 4 helium-rich WDs, 7 metal-rich WDs, 5 intermediate-metallicity WDs, and 5 unusual WDs. The binaries included 11 RS CVn binaries, 3 dwarf novae in outburst, 10 AM Her–type cataclysmic variables (CVs), and 2 DQ Her–type CVs. A brief overview of the general nature and EUV spectral distribution of each group is presented along with spectra and a short discussion for each source.

5.3.3.1 Early-type Stars: Stellar Winds

The earliest FUV spectra of stars were obtained using UV spectrographs carried on sounding rockets. Morton (1967) obtained UV spectra of the bright stars δ, ϵ, ζ, η, ι, and κ Orionis with an $f/2$ objective spectrograph carried to 186 km by an Aerobee rocket in 1965. Emission lines of Si IV at 1402.8 Å and C IV at 1549.5 Å were identified in ϵ and ζ Orionis. Strong absorption lines of both components of the Si IV doublet and the unresolved C IV doublet were found in ϵ, ζ, and ι Orionis, shifted to shorter wavelengths by velocities of 1800, 1900, and 3800 km s^{-1}, respectively. The Si IV doublet was present in δ Orionis shifted in the same direction by 1400 km s^{-1}. These displaced resonance absorption lines were interpreted as forming in gas escaping from the stars at high velocities, indicating mass loss from these giants and supergiants.

Further UV observations of mass-loss effects in O and B main-sequence and giant stars were carried out using Copernicus (Snow & Morton 1976). The UV spectra were analyzed for P Cygni lines in 47 stars to determine the distribution of high-velocity mass loss in the H-R diagram. The UV spectra reveal an extensive sequence, showing the changing profiles of C III λ1176 and N V λ1240. With a few exceptions, stars brighter than $M_{\rm bol} \simeq -6.0$ are losing mass, while fainter stars are not, within the Copernicus detection limits. Maximum velocities up to five times the surface escape velocity were found, but no strong correlations with luminosity or effective temperature were seen. Some stars had a wide range of ionization states at all velocities, and some had narrow absorption components superposed on broader features. O VI absorption lines with shifts or asymmetries are present in several O and hot B stars, suggesting the existence of hot stellar coronae with electron temperatures $T_e \simeq 10^5$ K.

Copernicus UV spectra of OB supergiants with strong stellar winds were obtained with Copernicus (Hutchings 1976). Spectral scans at ~0.2 Å resolution in the FUV of eight stars showed P Cygni line profiles, which vary inversely as the mass-flow rate. In P Cygni, the C II λ1175 line showed no velocity shift or emission. This is evidence that higher mass-flow rates occur via a dense, slow-moving envelope in which collisional interactions are important.

UV spectra of the Wolf–Rayet stars γ^2 Vel, HD 50896, and HD 92740 were obtained by Copernicus (Johnson 1978). UV spectral features were classified as interstellar, photospheric absorption, UV-displaced P Cygni absorption, P Cygni emission, or simple emission. An inverse correlation was found between ejection velocities and excitation potentials in γ^2 Vel, the only star with enough data to study it. Intercomparisons were made between the Wolf–Rayets and between them and the OB stars that were previously known to have UV P Cygni profiles.

IUE spectroscopy was obtained for seven hot stars in each of the Magellanic Clouds (LMC and SMC; Hutchings 1982). The SMC UV extinction curve was found to be much steeper than in the Galaxy or the LMC. Stellar effective temperatures and luminosities for LMC and SMC stars were found to be similar to those in the Galaxy. Stellar wind phenomena were found to not be present and to be weak in the SMC, and stronger in the LMC, though still weak compared with Galactic stars.

FUV, UV, and optical spectra of late O-type and early B-type supergiants in the LMC were studied by Evans et al. (2004) using FUV FUSE, UV IUE/HST, and optical VLT-UVES spectroscopy. Temperatures, mass-loss rates, and CNO abundances were obtained using a non-LTE, spherical, line-blanketed model atmosphere code. The analysis supports other results that indicate lower temperatures for OB-type supergiants as a result of stellar winds and blanketing, which amounts to ~2000 K for spectral type B0 Ia.

5.3.3.2 Evolved Stages

The nearby globular cluster NGC 6752 was imaged at 1620 Å with the Ultraviolet Imaging Telescope (UIT) during the Astro-2 mission of the Space Shuttle Endeavour in 1995 March (Landsman et al. 1996). A UV-visible CMD was derived for 216 stars. The CMD provides a nearly complete census of the hot horizontal-branch (HB) population with good temperature and luminosity discrimination for comparison with theoretical tracks. The observed data show good agreement with the theoretical zero-age horizontal branch (ZAHB) for an assumed reddening of $E(B - V) = 0.05$ and a distance modulus of 13.05. The observed HB luminosity width is in agreement with theoretical models and supports the single-star scenario for the origin of extreme horizontal branch (EHB) stars. However, only four stars were identified as post-EHB stars, whereas almost three times this many were expected from the HB number counts. Some unmodeled effect is probably decreasing the post-EHB lifetime, resulting in the lower EHB star numbers.

Observational results on seven symbiotic systems, obtained by the ultraviolet spectrometer (UVS) on board of spacecraft Voyager 1 and Voyager 2, were presented by Li & Leahy (1997). The interacting winds model has been successfully used to explain many observational phenomena of symbiotic systems and planetary nebulae. The UVS observations provide strong evidence supporting the interacting winds model. Three of the systems observed by UVS show strong O VI λ1035 and other FUV lines, including N V λ1240. The one-dimensional photoionization program CLOUDY was used to calculate the luminosities of the FUV lines to compare with the observational results from UVS. The calculations showed that, based on a photoionization model, a thin dense shell is necessary to produce the observed lines. The density, radius, and thickness of the shell are consistent with those predicted by the interacting winds model. The calculations explain why some of the symbiotic systems have O VI λ1035 but others do not, based on the interacting winds model.

5.3.3.3 White Dwarfs and Neutron Stars

Hot WDs were studied using the EUVE All-Sky Survey by Vennes et al. (1997). Effective temperature and surface gravity determinations were presented for a sample of 90 EUV-emitting hot WDs. A larger sample of 110 EUV-selected DA WDs in the solar neighborhood was studied using optical Balmer line series spectroscopy. The optical data were obtained at the Michigan-Dartmouth-MIT Observatory, Mount Stromlo Observatory, Lick Observatory, and Cerro Tololo Inter-American Observatory. Using pure-hydrogen model atmospheres, the effective temperature and surface gravities were derived. The results constrain the space density as well as the population age and mass distribution for this DA WD sample. The mass spectrum is narrowly peaked around 0.56 $M_{\odot}$ indicative of a C–O core with a thin hydrogen layer, but there is a significant population of 10 ultramassive ($M \geqslant 1.1 M_{\odot}$ white dwarfs). All objects had effective temperatures between $\simeq$ 25,000 and $\simeq$ 75,000 K so that most are younger than 30 Myr. Using evolutionary models, the DA WD birth rate in the solar neighborhood is (0.7–1.0) $\times 10^{-12}$ pc^{-3} yr^{-1}. Most WDs are on normal C–O cooling tracks, but a few low-mass WDs and the population of ultramassive WDs may follow different paths with, respectively, He, and, likely, O–Ne–Mg cores.

Vennes et al. (1998) presented an analysis of optical, UV, and X-ray spectral properties of a sample of 13 hot binaries. This demonstrated a wide variety of companions and variations in WD properties. Each binary was selected to contain a hydrogen-rich (DA) WD, each paired with a bright unresolved companion (V magnitude $\leqslant$10.6), the spectral types ranged from K0 IV to G5 V to A1 III). Using low-dispersion IUE spectra, ROSAT photometry, and EUVE photometry and spectroscopy, the effective temperature, mass, and distance of the WDs were derived. The orbital properties for most binaries were established by means of high-dispersion optical spectroscopy obtained with the Hamilton echelle spectrograph at Lick Observatory. High-amplitude (>20 km s^{-1}) velocity variations were found in only two stars (HD 33959C and HR 8210), low-amplitude variations were found in four additional objects (HD 18131, HR 1608, θ Hya, and BD +27°1888), and no variations (<2 km s^{-1}) in the remainder. Ca H and K were detected in emission in four (BD+08°102, HD 18131, HR 1608, and EUVE J0702+129) of the six objects that were also detected in the 0.52 – 2.01 keV ROSAT PSPC band. The source of the hard X-ray emission in HD 33959C remains unknown. Properties of the WDs were also investigated. EUV spectroscopy showed the effect of a low heavy-element abundance in the atmosphere of the WD in HD 33959C and of a high heavy-element abundance in HD 223816. Measurements of all other objects are consistent with emission from pure-hydrogen atmospheres.

Most neutron stars are not significant EUV emitters. Korpela & Bowyer (1998) carried out a search for EUV emission from neutron stars with the EUVE Deep Survey and scanner telescopes. Twenty-one fields containing known neutron stars were observed in the Lexan/boron (40–190 Å) band. Of these, 11 fields were simultaneously observed in the aluminum/carbon (160–385 Å) band. The five neutron stars that were detected with EUVE had been reported previously. For those sources not detected, the observations were used to obtain upper limits on the spectral flux from the neutron stars in the EUV bands. The accreting neutron star binary, Hercules X-1 is an exception: it is a bright EUV source (Leahy et al. 2000). Observations of Hercules X-1 by the EUVE covering most of the 35 day cycle were reported by Leahy & Dupuis (2010). This was the only long EUV observation of Her X-1. Simultaneous X-ray observations with the Rossi X-ray Timing Explorer All-Sky Monitor (RXTE/ASM) X-ray showed that Her X-1 was in an X-ray anomalous low state. The first 4 days were also observed with the RXTE proportional counter array (PCA). The PCA data showed that the X-ray properties were nearly the same in anomalous low states as for normal low states in Her X-1. The flux is anomalous low is reduced by a factor of 2 compared to the normal low state. The EUV emission from Her X-1 was reduced by a factor of ~4 compared to normal low states. The twisted-tilted accretion disk, which is responsible for the normal 35 day X-ray cycle (Leahy 2002) was modified to explain this behavior. An increased disk twist reduces the X-ray illumination of HZ Her by a factor of ~2 and the illumination of the disk surface by a somewhat larger factor, leading to a larger reduction in EUV flux compared to X-ray flux.

5.3.4 UV Studies of Galaxies

5.3.4.1 Nearby Galaxies

M31 has been observed many times, including by GALEX in NUV and FUV and by HST in NUV (Williams et al. 2014). The central part of M31 has been observed with AstroSat/UVIT in multiple FUV and NUV bands (Leahy et al. 2018) as part of a survey program of M31. GALEX had a ~5″ spatial resolution compared to UVIT's ≃1.2″ resolution so that stellar photometry is much less subject to crowding in the UVIT data. The initial study of the UVIT data on the M31 bulge (Leahy et al. 2018) found 31 bright FUV point sources in M31's bulge, of which 26 were suitable for joint photometry with the HST PHAT survey of M31 (Williams et al. 2014). The multiband photometry allowed SED fitting and construction of an H-R diagram for the FUV-bright bulge sources. As shown in Figure 5.21, there are a number of hot stars detected consistent with main-sequence stars of mass between 5 and 20 $M_{\odot}$. This indicates recent star formation in M31's bulge, within the past 100 million years.

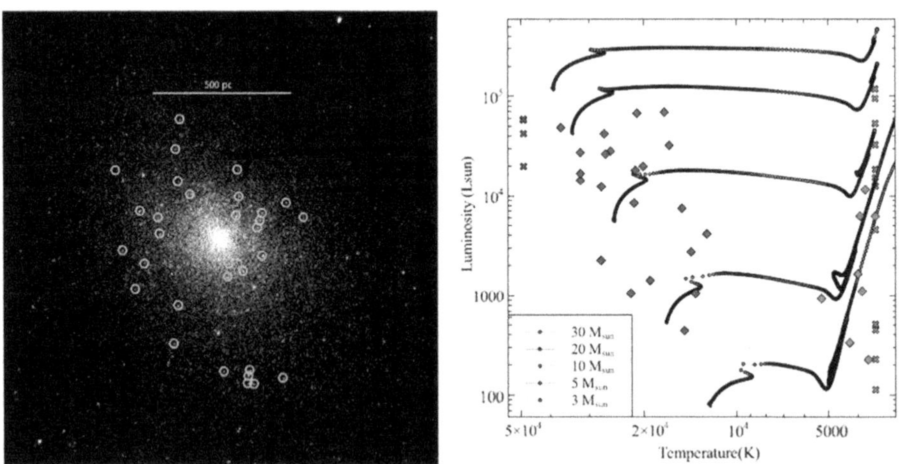

Figure 5.21. Left: The central part of the M31 bulge observed in the FUV CaF_2 filter (1300–1750 Å bandpass) showing bright FUV point sources. The area shown is the central 7 by 7 arcmin region of the bulge. Right: H-R diagram from fitting stellar atmosphere models to the combined UVIT-PHAT photometry. Image credit: Leahy et al. (2018).

Lee & Gil de Paz (2011) presented results from a GALEX UV survey of a complete sample of 390 galaxies within ~11 Mpc of the Milky Way. The UV data were a key component of the composite Local Volume Legacy, a UV-to-infrared imaging program designed to provide an inventory of dust and star formation in nearby spiral and irregular galaxies. The ensemble data set is an especially valuable resource for studying star formation in dwarf galaxies, which comprise over 80% of the sample.

The GALEX survey programs that obtained the data and the catalog of FUV (~1500 Å) and NUV (~2200 Å) integrated photometry were given by Lee & Gil de Paz (2011). Two measures of the global star formation efficiency were computed, the SFR per unit HI gas mass, and the SFR per unit stellar mass. That work illustrates the significant differences that can arise in our understanding of dwarf galaxies when the FUV is used to measure the SFR instead of Hα. Many late-type dwarf galaxies appeared to be devoid of star formation because they were not detected in previous Hα narrowband observations. These same galaxies were detected in the FUV and have FUV-derived SFRs that fall below the limit where the Hα flux is robust to Poisson fluctuations in the formation of massive stars. Otherwise, the UV colors and star formation efficiencies of Hα-undetected, UV-bright dwarf irregulars were found to be normal in comparison to those exhibited by the general population of star-forming galaxies. Thus, dwarf galaxies are not as inefficient in converting gas into stars as suggested by prior Hα studies.

The elliptical galaxy M87 was observed by the IUE to obtain a UV spectrum in the 1150–2000 Å spectral range (Bertola et al. 1980). This was combined with a spectrum in the 2000–3200 Å region and with observations obtained with the 5 m Hale telescope to create a broad spectral energy distribution from 1300 to 10,000 Å. M87 exhibits a flux density that drops from the optical rapidly into the UV, with a minimum at 2200 Å. It then increases sharply between 2200 and 1300 Å.

The 1300–2200 Å radiation comes from an extended source, which probably is made up of hot horizontal-branch stars. The ratio of hot stars to cool stars was shown to be different in M87, M31, and M32. UV emission lines are present in the spectrum of M87. The relative strengths of C IV, Hβ, and [O III] λ4959,5007 are consistent with those seen in planetary nebulae, but [O II] λ3727 is much too strong.

EUVE data of the central region of the Virgo Cluster was analyzed by Berghöfer et al. (2000). Diffuse EUV emission was found reaching a distance of $\simeq 13'$. The spatial distribution of this flux is incompatible with a thermal plasma origin. The EUV emission also cannot be produced by an extrapolation to lower energies of the observed synchrotron radio-emitting electrons. An additional component of low-energy relativistic electrons is required. The EUV emission could be due to inverse Compton scattering of relativistic electrons against the 3 K blackbody background. Separately, EUV emission originating from the jet in M87 was detected.

5.3.4.2 Low- (<0.3*) and Higher-redshift (*>0.3*) Galaxies*

A study of the most UV-luminous galaxies in the local universe is carried out by Hoopes et al. (2007). The sample of UV-luminous galaxies (UVLGs) is selected by matching the GALEX All-Sky Imaging and Medium Imaging Surveys with the

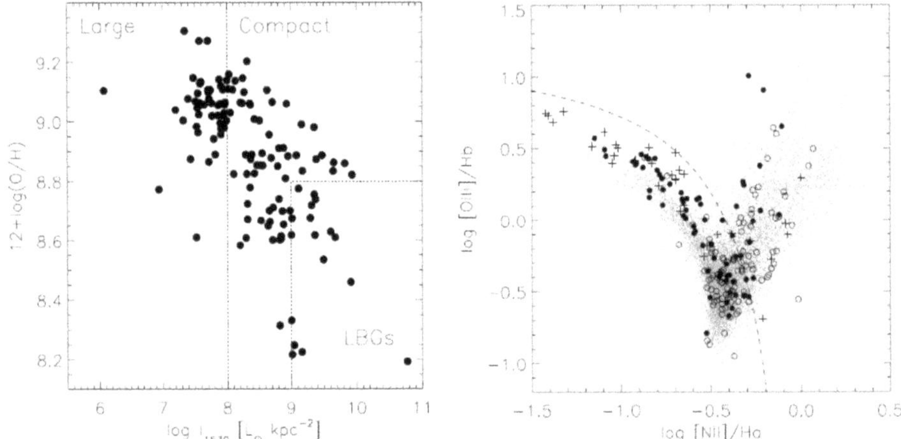

Figure 5.22. UV-luminous galaxies (UVLGs). Left: metallicity versus 1530 Å surface brightness showing the demarcation between large UVLGs and compact UVLGs. Luminous blue galaxies (LBGs) were selected using both a surface brightness and metallicity cutoff, as shown. Right: the BPT diagram (Baldwin et al. 1981) for the GALEX/SDSS galaxy sample with good line measurements. The dashed line marks the dividing line between starbursts and AGN from Kauffmann et al. (2003). The large UVLGs are shown as open circles, the compact UVLGs as filled circles, and the supercompact UVLGs as plus symbols. Image credit: Hoopes et al. (2007).

Sloan Digital Sky Survey third data release. The overlap between these two surveys is roughly 450 deg^2. Of 25,362 galaxies with SDSS spectroscopy in the range $0.0 < z < 0.3$ detected by GALEX, there are 215 galaxies with $L > 2 \times 10^{10} L_{\odot}$ at a 1530 Å observed wavelength.

The properties of this population are well correlated with UV surface brightness. The left panel of Figure 5.22 shows the UVLGs divided into three groups using the metallicity versus UV surface brightness plot. The galaxies with low UV surface brightness ($I_{\lambda 1530} < 10^8 L_{\odot}/kpc^2$) are primarily large spiral systems (large UVLGs) with a mixture of old and young stellar populations. The galaxies with high surface brightness ($I_{\lambda 1530} < 10^8 L_{\odot}/kpc^2$) consist primarily of compact starburst systems (compact UVLGs). The large galaxies appear to be the high-luminosity tail of the Galaxy star formation function and owe their large luminosity to their large surface area.

In terms of the behavior of surface brightness with luminosity, size with luminosity, the mass–metallicity relation, and other parameters, the compact UVLGs clearly depart from the trends established by the full sample of galaxies. The subset of compact UVLGs with the highest surface brightness ($I_{\lambda 1530} > 10^9 L_{\odot}/kpc^2$; called supercompact UVLGs) have characteristics that are remarkably similar to Lyman-break galaxies at higher redshift. They are much more luminous (and thus have much higher star formation rates) than typical local UV-bright starburst galaxies and blue compact dwarf galaxies. They have metallicities that are systematically lower than normal galaxies of the same stellar mass, indicating that they are less chemically evolved. In all these respects, they are the best local analogs for Lyman-break galaxies. The right panel of Figure 5.22 shows the BPT diagram (named after Baldwin, Phillips, & Terlevich) for the GALEX/SDSS sample. A significant fraction of the galaxies shows AGN-like ionization properties. For the full sample, ~50% are classified as AGNs, whereas for the UVLG subset ~30% are AGNs. This is likely because the UVLGs have high star formation rates and stronger line emission from star-forming regions so that a larger fraction of UVLGs have AGN emission hidden by bright star formation emission.

UV emission from galaxies with redshift $1 < z < 3$ has been studied with HST. Teplitz et al. (2013) give an overview of the 90 orbit HST treasury program to obtain NUV imaging of the Hubble Ultra Deep Field. The Wide Field Camera 3 UVIS detector was used, with the F225W, F275W, and F336W filters. That survey was designed to:

1. Investigate the episode of peak star formation activity in galaxies at $1 < z < 3$,
2. Probe the evolution of massive galaxies by resolving sub-galactic units (clumps),
3. Examine the escape fraction of ionizing radiation from galaxies at $z \sim 2 - 3$,
4. Greatly improve the reliability of photometric redshift estimates, and
5. Measure the star formation rate efficiency of neutral atomic-dominated hydrogen gas at $z \sim 1 - 3$.

Initial results are given by Teplitz et al. (2013), including a discussion of the data-reduction challenges and the necessity of specialized calibrations and the effects of charge transfer inefficiency. Results on number counts of UV-selected galaxies and morphology of galaxies at $z \sim 1$ are presented. The number density of UV dropouts at redshifts 1.7, 2.1, and 2.7 is measured, corresponding to the Lyman break falling in the F225W, F275W, and F336W filters, respectively. As an example, Figure 5.23 illustrates the selection of dropout galaxies for using F336W filter dropouts for $z \simeq 2.7$. The dropouts are consistent with the number predicted by published luminosity functions by Hathi et al. (2010) and Oesch et al. (2010). The WFC3

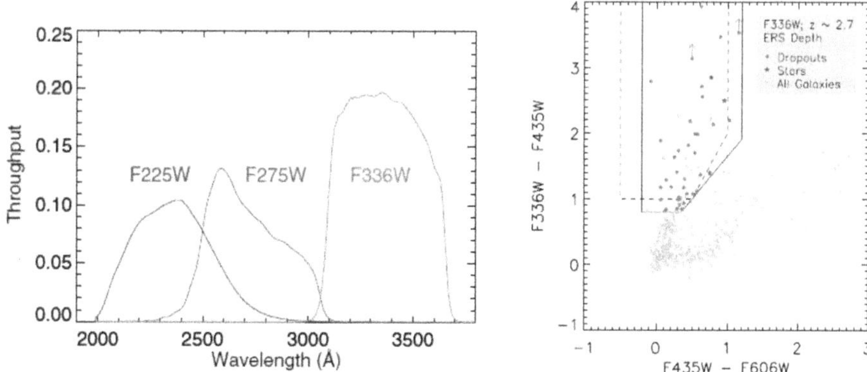

Figure 5.23. Left: bandpasses for the F225W, F275W, and F336W filters of WFC3. Right: color–color diagram showing the areas for selection of F336W dropouts. The solid line is the color selection region from Hathi et al. (2010); the dashed line is the region from Oesch et al. (2010). Red circles with arrows have F336W upper limits. Image credit: Teplitz et al. (2013).

image mosaics have sufficient sensitivity and resolution to enable analysis of the evolution of star-forming clumps, with 28th–29th magnitude depth at 5σ in a 0.2″ radius aperture.

HST data were used by Finkelstein et al. (2012) to study the evolution of galaxy rest-frame UV colors from $z = 8$ to $z = 4$. Near-infrared (NIR) data taken with the WFC3 on HST has allowed the discovery of large samples of galaxies at $z = 7 - 8$ (e.g., Bouwens et al. 2010), when the universe was only ~500 to 800 Myr old. At those redshifts, the NIR data are measuring rest-frame UV emission from these galaxies. Rest-frame UV color (spectral slope) can provide information on dust extinction in high-redshift galaxies, where spectroscopy is not yet feasible (e.g., see Reddy et al. 2012 and references therein). Data from Great Observatories Origins Deep Survey-South (GOODS-S) and the Hubble Ultra Deep Field (HUDF) 2009 data, and Early Release Science programs were analyzed.

A total of 2812 galaxies were selected with $z > 3.5$ and 113 galaxies with $7 < z < 8$. The UV spectral slope β was measured, with $f_\lambda \propto \lambda^\beta$. The evolution of β with z was modeled by dust extinction from small to negligible amounts at $z = 8$, increasing up to $A_v \simeq 0.5$ at $z = 4$. The onset of dust production at $z \sim 7$ is consistent with the evolution of $M \leqslant 3.5 M_\odot$ stars off the main sequence to the AGB phase, coupled with the fact that $M = 2 - 4 M_\odot$ AGB stars are the most efficient producers of dust. Thus, the observed increase of extinction from $z = 8$ to $z = 4$ is consistent with dust production from low-mass AGB stars. The most massive galaxies are redder at all redshifts. This is likely because more massive galaxies experience accelerated evolution, but also possibly they retain more of the high velocity dust from supernovae, which can be lost in low-mass galaxies.

5.3.5 UV Studies of Quasars and AGNs

AGNs come in many categories, depending on whether the classification is in radio, optical, or X-ray (see a recent review paper by Antonucci 2012 and references therein). Here a very brief, and incomplete, summary is given. The basic idea is that an AGN is a galaxy that has an actively accreting supermassive black hole at its center, which results in high-energy output and unusual spectral properties.

Radio-loud galaxies have $L_\nu = 10^{28} - 10^{36}$ erg s^{-1} Hz^{-1} and nearly perfectly correspond to elliptical galaxy hosts at low z. The resulting radio powers can exceed 1×10^{45} erg s^{-1} and the energy in the 100 kpc-scale radio lobes are as high as 10^{61} erg. The most luminous double radio sources have size >100 kpc and are called FR II. These show strong asymmetry on scales up to tens of kiloparsecs. The lower radio luminosity objects, also with size 100 kpc, are called FR I. These show weaker asymmetry, and only on small scales (<1 kpc). Compact steep spectrum (CSS): spectra peak at ~100 MHz, size <25 kpc (smaller than host galaxy). Gigahertz peaked spectrum (GPS): size <1 kpc, often self-absorbed sychrotron (up to GHz). Probably small because they are young ($1000 - 10^5$ yr). Only a very small fraction of small/short-lived sources grow to be huge bright long-lived sources. At low redshift, radio-quiet (but not silent) AGNs lie in spiral hosts and go by the name of Seyfert galaxies.

There are two types of observer-orientation effects at work for AGNs. The first is relativistic beaming: those viewed nearly along the beam direction are blazars, whereas those seen at higher inclination are radio-loud galaxies and quasars. At optical-UV wavelengths, AGNs exhibit a second orientation effect related to viewing angle with respect to optically opaque dusty structures referred to as the "AGN torus." The big blue bump is attributed to thermal radiation from optically thick accretion flows. The optical-UV broad emission lines (5000–10,000 km s^{-1} wide permitted emission lines) are attributed to line emission from the broad-line region (BLR). When both big blue bump and BLR are visible, the line of sight mostly bypasses the AGN torus. When they are not visible or seen very weakly as reprocessed light, usually polarized, the line of sight goes through a dense part of the AGN torus.

AGNs with a big blue bump are called thermal, and AGNs that show both a big blue bump and BLR are said to have a Type 1 (optical-UV) spectrum. A radio-loud AGN with no broad lines is called a narrow-line radio galaxy (NLRG) with a Type 2 spectrum. Radio-quiet and radio-loud objects with high ionization (in optical-UV lines) are virtually all visible (Type 1) or hidden (Type 2) Seyferts or quasars. At lower radio luminosities of all radio types, we find mostly LINERs (low ionization nuclear emission regions).

UV spectra of six quasars, ranging in redshift from 0.23 to 2.04, were obtained with IUE (Green et al. 1980). The average emission-line intensity ratio for Lyα to Hβ and the detection of higher-order Lyman emission lines were shown to be inconsistent with a simple optically thick recombination model. The Lyman and Balmer line velocity profiles were found to be identical. The emission lines in 3C 351 can be decomposed into sharp and broad components, with the sharp lines consistent with the recombination model. The emitted energy distributions of the three highest-redshift quasars show a sharp break in spectral slope around log $\nu_0 = 15.4$; the red side is fitted by $f_\nu \propto \nu^{-0.5}$, while the blue side goes as $\nu^{-2.3}$. This effect has been also seen in high-redshift quasars observed optically. The spectrum of PG 1247+268, at $z = 2.038$ contains an optically thick Lyα edge absorption system at $z = 1.218$. PG 1115+080 was detected at the 4.5σ level down to the violet limit of sensitivity, corresponding to rest λ432. No evidence for a He v resonance absorption trough or ionization edge was observed. Intergalactic helium must therefore be ionized, placing a lower limit on the temperature of the intergalactic gas of 3×10^5 K. Ionizing flux from quasars is a main contributor to the energy balance of the intergalactic medium.

Quasars are known to have the so-called big blue bump in the UV wave band. Shang et al. (2005) investigated the UV-to-optical spectral energy distributions of 17 AGNs using spectrophotometry spanning 900–9000 Å (rest frame). A sample of six of the spectra are shown in Figure 5.24. They show the same general line and continuum features as seen in the spectrum of PG 0953+415, shown in Figure 5.25. The UV coverage enabled a detailed study of the so-called big blue bump, the region in which the energy output peaks. Most objects exhibited a spectral break between 1000 Å and 1500 Å. There is strong evidence in the data that the FUV spectral region is below the extrapolation of the NUV-optical slope, indicating a spectral

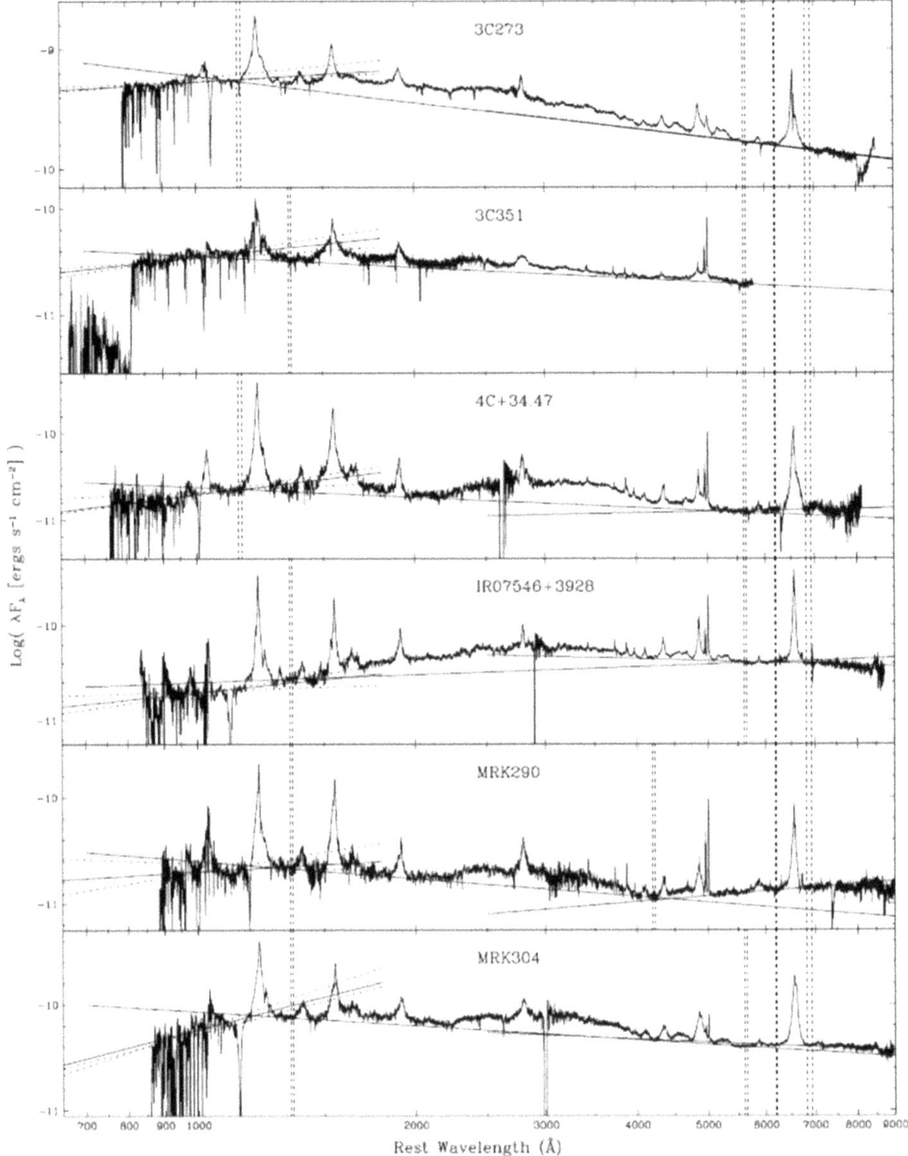

Figure 5.24. UV through optical spectra for six AGNs. The break in continuum slope in the region 1000 to 1500 Å is shown by the vertical dashed lines and is fit using two power laws. The spectra were obtained using FUSE, HST, and the 2.1 m telescope at Kitt Peak National Observatory. Image credit: Shang et al. (2005).

break around 1100 Å. The behavior of the sample data was compared to non-LTE thin-disk models covering a range in black hole mass, Eddington ratio, disk inclination, and other parameters. The distribution of UV-optical spectral indices redward of the break and FUV indices shortward of the break was shown to be in rough agreement with the models. There was no correlation between the FUV

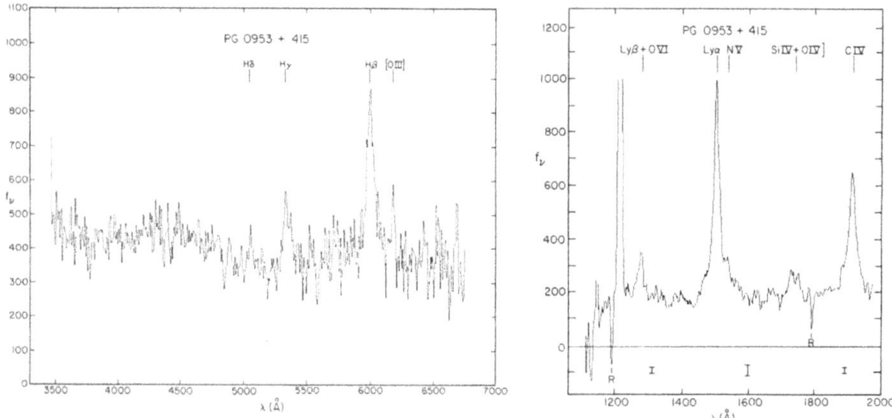

Figure 5.25. Left: Palomar spectrum of the quasar PG 0953+415 showing a strong Hβ line, a roughly flat optical continuum and higher-level Balmer lines. Right: IUE UV spectrum of the quasar PG 0953+415 showing strong Lyα and C IV lines, a flat UV continuum with a break at about 1200 Å, and other lines such as Lyβ, N V, Si IV, O VI, and O IV. The strong geocoronal Lyα line at 1216 Å is saturated; instrument features are marked with R. Image credit: Green et al. (1980).

spectral index and black hole mass, as predicted in some accretion disk models. Thus, the observed spectral break was argued to be intrinsic to AGNs, although intrinsic reddening, as well as Comptonization, can strongly affect the FUV spectral index.

AGNs have outflows that come in many observational classes. Blueshifted UV and X-ray absorption lines are hundreds of km s^{-1} wide for Seyfert galaxies. Quasar outflows show UV troughs that are tens of thousands of km s^{-1} wide, as in broad absorption line (BAL) quasars. A current major question for AGN studies is: What is the connection between the intrinsic luminosity of an AGN and the properties of the outflow, including terminal velocity and velocity width of the observed trough? Radiative acceleration is thought to be the principal driver of outflows. It predicts that terminal velocity should scale as $v \propto L^n$, with $0.25 < n < 0.5$. The observed fact that outflows in Seyfert galaxies terminate at 1000 km s^{-1}, while outflows in BALs extend out to 30,000 km s^{-1}, is roughly consistent with this. But a quantitative test of the relation is not well tested, given the lack of objects in between these two groups. In order to address this issue, Ganguly et al. (2007) carried out HST STIS observations of 14 low-redshift ($z < 0.5$) AGNs and found that their relatively small sample was consistent with the velocity–luminosity relationship.

References

Antonucci, R. 2012, A&AT, 27, 557
Baldwin, J. A., Phillips, M. M., & Terlevich, R. 1981, PASP, 93, 5
Ben-Jaffel, L., & Holberg, J. B. 2016, ApJ, 823, 161
Berghöfer, T. W., Bowyer, S., & Korpela, E. 2000, ApJ, 535, 615
Berk, A., Anderson, G. P., Bernstein, L. S., et al. 1999, Proc. SPIE, 3756, 348
Bertola, F., Capaccioli, M., Holm, A. V., & Oke, J. B. 1980, ApJ, 237, L65

Blair, W. P., Raymond, J. C., Long, K. S., & Kriss, G. A. 1995, ApJ, 454, L35
Bouwens, R. J., Illingworth, G. D., Oesch, P. A., et al. 2010, ApJ, 709, L133
Bowyer, S., Lampton, M., Lewis, J., et al. 1996, ApJS, 102, 129
Bowyer, S., Lieu, R., Lampton, M., et al. 1994, ApJS, 93, 569
Bowyer, S., & Malina, R. F. 1991, AdSpR, 11, 205
Branton, D., Riley, A., et al. 2021, STIS Instrument Handbook, v. 20.0 (Baltimore, MD: STScI)
Broadfoot, A. L., Sandel, B. R., Shemansky, D. E., et al. 1977, SSRv, 21, 183
Cardelli, J. A., Clayton, G. C., & Mathis, J. S. 1989, ApJ, 345, 245
Craig, N., Abbott, M., Finley, D., et al. 1997, ApJS, 113, 131
Dere, K. P., Landi, E., Mason, H. E., Monsignori Fossi, B. C., & Young, P. R. 1997, A&AS, 125, 149
Dressel, L. 2021, Wide Field Camera 3 Instrument Handbook, v. 13.0 (Baltimore, MD: STScI)
Evans, C. J., Crowther, P. A., Fullerton, A. W., & Hillier, D. J. 2004, ApJ, 610, 1021
Finkelstein, S. L., Papovich, C., Salmon, B., et al. 2012, ApJ, 756, 164
Fitzpatrick, E. L., & Massa, D. 2007, ApJ, 663, 320
Fitzpatrick, E. L., & Massa, D. 1990, ApJS, 72, 163
Ganguly, R., Brotherton, M. S., Arav, N., et al. 2007, AJ, 133, 479
Green, R. F., Pier, J. R., Schmidt, M., et al. 1980, ApJ, 239, 483
Hathi, N. P., Ryan, R. E. Jr, Cohen, S. H., et al. 2010, ApJ, 720, 1708
Hirschauer, A. S., et al. 2021, Cosmic Origins Spectrograph Instrument Handbook, v. 13.0 (Baltimore, MD: STScI)
Hoopes, C. G., Heckman, T. M., Salim, S., et al. 2007, ApJS, 173, 441
Hutchings, J. B. 1976, ApJ, 204, L99
Hutchings, J. B. 1982, ApJ, 255, 70
Johnson, H. M. 1978, ApJS, 36, 217
Kauffmann, G., Heckman, T. M., Tremonti, C., et al. 2003, MNRAS, 346, 1055
Koornneef, J., & Code, A. D. 1981, ApJ, 247, 860
Korpela, E. J., & Bowyer, S. 1998, AJ, 115, 2551
Kwok, S. 2007, Physics and Chemistry of the Interstellar Medium (University Science Books)
Landsman, W. B., Sweigart, A. V., Bohlin, R. C., et al. 1996, ApJ, 472, L93
Leahy, D. A. 2002, MNRAS, 334, 847
Leahy, D. A., & Dupuis, J. 2010, ApJ, 715, 897
Leahy, D. A., Marshall, H., & Scott, D. M. 2000, ApJ, 542, 446
Leahy, D. A., Bianchi, L., & Postma, J. E. 2018, AJ, 156, 269
Lee, J. C., Gil de Paz, A., & Kennicutt, R. C. Jr. 2011, ApJS, 192, 6
Li, P. S., & Leahy, D. A. 1997, ApJ, 484, 424
Malina, R. F., Marshall, H. L., Antia, B., et al. 1994, AJ, 107, 751
Martin, C., Barlow, T., Barnhart, W., et al. 2003, Proc. SPIE, 4854, 336
Martin, D. C., Fanson, J., Schiminovich, D., et al. 2005, ApJ, 619, L1
Moos, H. W., Cash, W. C., Cowie, L. L., et al. 2000, ApJ, 538, L1
Morton, D. C., Drake, J. F., Jenkins, E. B., et al. 1973, ApJ, 181, L103
Morton, D. C. 1967, ApJ, 147, 1017
Newmark, J. S., Holm, A. V., Imhoff, C. L., et al. 1992, IUENN, 47, 1
Nichols, J. S., & Linsky, J. L. 1996, AJ, 111, 517
Oesch, P. A., Bouwens, R. J., Carollo, C. M., et al. 2010, ApJ, 725, L150
Reddy, N., Dickinson, M., Elbaz, D., et al. 2012, ApJ, 744, 154

Rogerson, J. B., Spitzer, L., Drake, J. F., et al. 1973, ApJ, 181, L97
Rogerson, J. B., & York, D. G. 1973, ApJ, 186, L95
Shang, Z., Brotherton, M. S., Green, R. F., et al. 2005, ApJ, 619, 41
Siegel, M. H., Porterfield, B. L., Linevsky, J. S., et al. 2014, AJ, 148, 131
Snow, T. P. Jr., & Morton, D. C. 1976, ApJS, 32, 429
Sofia, U. J., Wolff, M. J., Rachford, B., et al. 2005, ApJ, 625, 167
Spitzer, L., Drake, J. F., Jenkins, E. B., et al. 1973, ApJ, 181, L116
Tandon, S. N., Subramaniam, A., Girish, V., et al. 2017, AJ, 154, 128
Teplitz, H. I., Rafelski, M., Kurczynski, P., et al. 2013, AJ, 146, 159
Vennes, S., Christian, D. J., & Thorstensen, J. R. 1998, ApJ, 502, 763
Vennes, S., Thejll, P. A., Génova Galvan, R., & Dupuis, J. 1997, ApJ, 480, 714
Vidal-Madjar, A., Encrenaz, T., Ferlet, R., et al. 1987, RPPh, 50, 65
Wakker, B. P., Savage, B. D., Sembach, K. R., et al. 2003, ApJS, 146, 1
Williams, B. F., Lang, D., Dalcanton, J. J., et al. 2014, ApJS, 215, 9
Wood, B. E., Linsky, J. L., Hébrard, G., et al. 2004, ApJ, 609, 838
York, D. G., Drake, J. F., Jenkins, E. B., et al. 1973, ApJ, 182, L1

Chapter 6

X-Ray Astronomy

Pierre Maggi

Since the 1960s, X-ray astronomy has used space-based instruments to avoid the absorption of our atmosphere. X-rays are a short-wavelength type of radiation and carry large amounts of energies. They are thus tracers of hot, energetic, and violent phenomena in the universe. This chapter presents a brief history of the field and the rapid development of detectors and optics used to study celestial X-rays. Then, the many types of X-ray-emitting objects found in our Galaxy and in the local universe are presented, including stellar sources of all sizes and ages and compact objects left at the end of stellar evolution. Finally, we review X-ray sources found at much larger distances, and the implications for our understanding of the evolution of the universe are briefly discussed.

6.1 Introduction—The Unveiling of the X-Ray Universe

While studying electrical discharge in gas in 1895 at Würzburg University, Wilhelm Conrad Röntgen (1845–1923) discovered a new form of radiation that he called "X-rays." In his honor, X-rays are still called "Röntgenstrahlung" (Röntgen radiation) in German. It is one of the most famous examples of a serendipitous discovery, that is, a discovery "by chance." No X-ray interference could be observed, though they were believed to be a wave phenomenon. Max von Laue (1879–1960), while at the Ludwig–Maximilians–Universität (LMU) of Munich, thought in 1912 that the reason why no interferences were seen was that slits in the gratings used were too large compared to the wavelengths of X-rays. He suggested that atoms in crystals could be used as gratings with much smaller spacings. The experiment by his assistants Walter Friedrich (1883–1968) and Paul Knipping (1883–1935) verified his predictions. Not only had they proved atoms were real, they also demonstrated that X-rays were indeed a form of electromagnetic radiation with very short wavelengths (between 10^{-8} m and 10^{-11} m).

This would turn out to be both the blessing and the curse of X-ray astronomy. Because of their short wavelength, X-ray photons carry large amounts of energies (of the order of keV). They are then naturally produced by hot and energetic

doi:10.1088/2514-3433/ac2256ch6

processes. On the other hand, X-rays interact easily with atoms: The many atoms in the atmosphere absorb efficiently X-ray photons. About half of the incident 1 keV radiation is stopped before reaching an altitude of 100 km. To observe celestial X-rays, it is therefore necessary to place telescopes and detectors high above the ground, using balloons, sounding rockets, or satellites. Consequently, it is not surprising that X-ray astronomy started only with the dawn of the space age (see Book 1, Chapter 1).

X-ray astronomy began in 1949 with the detection of X-rays from the hot corona of the Sun by the group of Herbert Friedman (Naval Research Laboratory) using detectors aboard V-2 rockets evacuated from Germany at the end of World War II (Friedman et al. 1951). The same group searched for nonsolar X-ray sources in the 1950s, though without success. The breakthrough came in 1962 June, when an Aerobee-150 rocket carried improved detectors above the atmosphere with the goal of detecting lunar X-rays (Figure 6.1). The experiment was led by Riccardo Giacconi of the American Science and Engineering (AS&E) company. Instead of detecting the Moon, a strong peak was recorded some 30° away, toward the Scorpius constellation. Upon studying several other explanations, they concluded that they had discovered the first cosmic X-ray source (Giacconi et al. 1962), later called Sco X-1 (meaning the first X-ray source discovered in Scorpius). It turned out that the sensitivity of Friedman's detectors, in the previous decade, was just above what would have been needed to make that discovery. After verification in subsequent rocket flights (Gursky et al. 1963) and the identification of a second source associated with the Crab Nebula (Bowyer et al. 1964), the field underwent very rapid growth (Hirsh 1983), embodied in 1970 by the launch of Uhuru, the first orbiting X-ray observatory, conducting in its 2.25 year lifespan the first all-sky X-ray survey and detecting 339 X-ray sources (Giacconi et al. 1972). The field reached a milestone in 1978 with Einstein, a satellite carrying the first X-ray telescope.

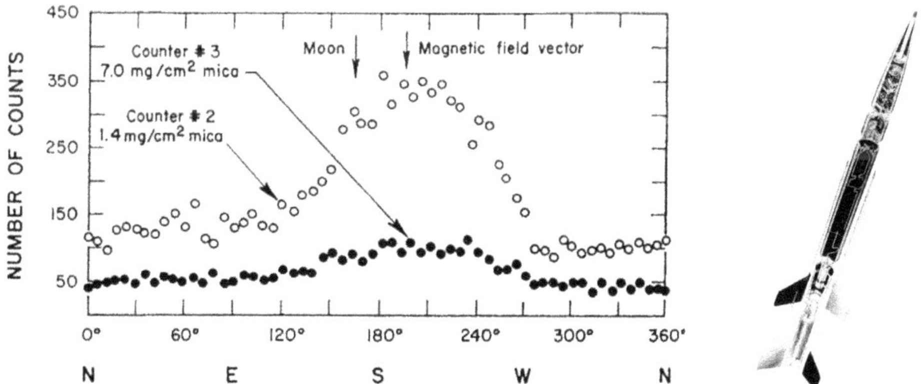

Figure 6.1. The birth of X-ray astronomy. The number of counts on two detectors is shown as they scanned the sky (Giacconi et al. 1962). The peak, clearly not at the position of the Moon, marks the discovery of the first extrasolar X-ray source, later called Sco X-1. On the right is an exploded view of the 9 m long, 38 cm wide Aerobee-150 rocket that carried these detectors above the atmosphere in 1962 June. Image credits: Smithsonian Institution/National Air and Space Museum.

With continuously improving optics and detectors, and long in-orbit lifetimes (Chandra and XMM-Newton operating for more than two decades), X-ray observatories have gathered a bonanza of discoveries both expected and unexpected. X-ray observations now constitute an integral part of multimessenger astronomy, providing an ideal window into the electromagnetic counterparts of gravitational-wave events, high-energy neutrino sources, and other transient sources. They provide key observations touching on all aspects of modern astrophysics.

In this chapter, we offer a panorama of X-ray astronomy, without the pretense of being fully comprehensive. Instead, we aim to present the main classes of X-ray sources, both Galactic and extragalactic sources (Sections 6.4 through 6.6). We highlight the many discoveries brought forth by X-ray observations, and, in keeping with the "in practice" of this book's title, we provide examples of key X-ray diagnostics applicable to each of these objects. But first, we begin in Sections 6.2 and 6.3 with the observational perspective, describing the tools, detectors, and telescopes, of X-ray astronomy.

6.2 Detecting and Characterizing X-Ray Photons

Because each individual X-ray photon carries a lot of energy, the typical count rate (number of photons per unit time) of sources in X-rays is much lower than in the infrared or optical domain. Thus, almost all X-ray detectors so far have been operated in a photon-counting mode, recording the properties of individual X-ray photons. This is in stark contrast with optical astronomy, where instead an average number, usually very large, of photons is accumulated at each position within each exposure on the sky.

6.2.1 What is an Ideal X-Ray Detector?

The first desirable quantity for a detector is high sensitivity, that is, the ability to detect at high signal-to-noise ratio (S/N) a source with flux (F; in erg s^{-1} cm^{-2})[1] with an exposure time (t_{exp}). This can be achieved by optimizing the combination of the following parameters:

- The quantum efficiency (QE) is the probability that an X-ray photon hitting the detector material will produce an electron (or electrons) that can then be recorded and/or amplified. The closer it is to 1, the higher the number of total signals that can be collected. In general, the QE is a function of the energy of the incident photon E.
- A large effective area A_{det} allows more photons to be collected in the same exposure time. We can then express the total number of photons collected from the source as

$$N_{src} = F \times \text{QE} \times A_{det} \times t_{exp}. \tag{6.1}$$

[1] In keeping with the common practice in astronomy of sacrificing consistency for the sake of tradition, most quantities in this chapter will use the cgs system, which was and remains the most commonly used of X-ray astronomy. For the conversion, see appendices in Book 1.

- A background as low as possible. In general, one can separate the contribution from the sky N_{sky} (astrophysical background) and an instrumental component N_{part} induced by particles hitting the spacecraft and detector:

$$N_{bg} = N_{sky} + N_{part} = j_{sky} \times \mathrm{QE} \times A_{det} \times \Omega \times t_{exp} + t_{exp} \times B_{part}, \tag{6.2}$$

with j_{sky} the sky photon intensity in photons per unit area per unit time and per unit solid angle, Ω the solid angle of the detector exposed to the sky, and B_{part} the count rate of the instrumental background. The fluctuations (noise) ΔN_{bg} around that value are given by $\sqrt{N_{bg}}$ because Poisson statistic applies. Therefore, the S/N of source is $N_{src}/\sqrt{N_{bg}}$ and the sensitivity is expressed as the minimal flux F_{min} that can be detected at a given S/N:

$$F_{min} = \frac{S/N}{\mathrm{QE} \times A_{det}} \sqrt{\frac{j_{sky} \times \mathrm{QE} \times A_{det} \times \Omega + B_{part}}{t_{exp}}}, \tag{6.3}$$

which is proportional to $\left(\mathrm{QE} \times A_{det} \times t_{exp}\right)^{-1/2}$ in the case where the astrophysical background dominates.

Second, a good detector should also be able to characterize the spectra of the detected sources. This is achieved first with a high energy resolution, i.e., the accuracy with which the energy of the incident radiation is recovered. In practice, even a monochromatic source at energy E is detected across several energy channels spanning ΔE, for reasons explained below. By convention, it is often the resolving power $R = E/\Delta E = \lambda/\Delta\lambda$, which is used to characterize the spectroscopic performance of a detector.

In addition to spectral resolution, the bandwidth, i.e., the energy band where the detector is sensitive, should be sufficiently large to distinguish between various emission processes possibly contributing to a source's flux.

Many of the common X-ray sources in the universe exhibit large variability or periodicity, often on very small timescales (e.g., 10^{-3} s). To be able to detect and measure said variability, detectors should have adequate time resolution. This encompasses both the accuracy with which the time of arrival of each X-ray photon can be measured, and also the time needed for the detector to "recover" and be able to record subsequent photons. If this so-called "dead time" is too long, variability patterns that are shorter cannot be probed, and the total number of photons collected from a source is significantly reduced.

Finally, some detectors may be able to measure the polarization angle of X-ray photons. This allows us to measure the orientation of magnetic fields in sources dominated by synchrotron emission, and the geometry of sources where scattering and reflection play a role. X-ray polarimetry is discussed briefly in Section 6.2.5.

Although there are several physical processes that limit each of these properties, as described below, there are also practical limitations that stem from the space-borne nature of such instrumentation. The detector needs to fit within both the size and weight limits of the spacecraft platform. It should not demand too much power, nor generate too much heat that cannot be evacuated easily. The telemetry rate is limited by the size of antennae on board and on the ground, and by the available downlink time. The amount of data produced by the detector cannot exceed that limit at the risk of losing data. Therefore, onboard computers and software may perform data processing to select useful data and reduce the total output. All of the complex electronics must be space-hardened and shielded to resist damage by the incessant bombardment of the spacecraft by energetic cosmic rays. Last but not least, a final limitation that applies to all space-borne instrumentation is the budget, as developing and building new instrumentation is often costly.

6.2.2 Gas Counters

Gas counters have been the workhorses of X-ray astronomy for four decades, from the first detection of a celestial X-ray source in 1962 to imaging the whole sky in the 1990s. They rely on the fundamental ability of X-rays to ionize matter. Their design is relatively simple: A container is filled with gas, inside of which an X-ray photon creates charges (e^--ion pairs). The electrons (often called photoelectrons) are then collected at electrodes inside the container. Depending on the voltage applied to electrodes, different modes can be used. X-ray detectors most often operate as proportional counters, where the photoelectrons are accelerated enough to further ionize other gas atoms, leading to a charge multiplication. The total number of electrons released, and thus the amplitude of the signal, is proportional (hence the name) to the energy of the incident X-ray photon. The average number of e^-–ion pairs generated by a photon with energy E is $N_0 = E/W$, with W the average energy for the creation of one pair. W depends on the gas used, typically a noble gas (Ar, Xe), a molecular gas (CO_2, CH_4), or a mixture thereof. For those gases, W is about 25–30 eV (Borges & Conde 1996; Pansky et al. 1997). Therefore, a 1 keV photon initially produces 35 to 40 photoelectrons.

The gas container must have an entrance window to allow X-rays to propagate through the gas. This window must be thin and large enough to let enough radiation through, but sturdy enough to hold the pressurized gas inside the counter and to resist the vibrations of a rocket launch. Many a detector have failed due to a ruptured window. Manipulating the thickness and composition of the entrance window affects the sensitivity to low-energy X-rays (0.2–2 keV).

The quantum efficiency of proportional counters is the product of two terms. The first term describes the transmission of X-rays through the entrance window as $T_w\, e^{-d_w \mu_w}$, where d_w is the window column density (in g cm^{-2}) and μ_w its mass attenuation coefficient. T_w is a constant, the geometric open fraction of the window and its support structure (usually, a metallic grid is used to reinforce the entrance window). The second term is the absorption of photons by the gas, written as $(1 - e^{-d_g \mu_g})$, with d_g the gas column density and μ_g its mass attenuation coefficient. Examples of window transmission and gas absorption are shown on Figure 6.2.

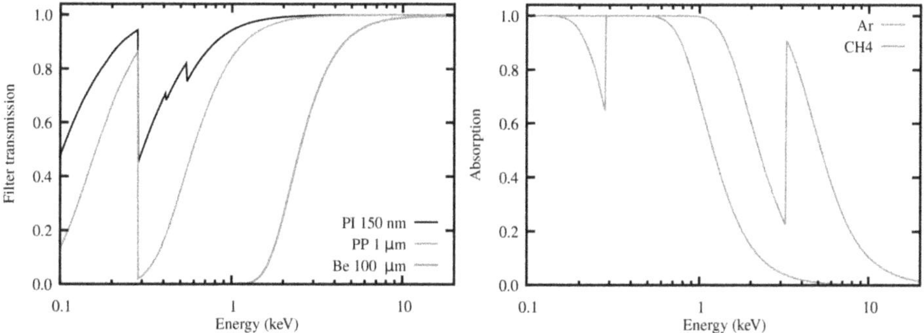

Figure 6.2. Left: transmission of filters made of 150 nm of polyimide (PI; black line), 1 μm of polypropylene (PP; orange), and 100 μm of beryllium (green). Note the strong carbon K absorption edge at 284 eV in the organic filter curves. Right: absorption coefficient as a function of energy for argon (orange) and methane (green). Data from Henke et al. (1993).

Energetic cosmic rays going through the detector are able to cause ionization in the gas, potentially producing a frequent noise. To reduce this background, anticoincidence techniques are used, surrounding as much as possible the X-ray detector with another detector opaque to X-rays (i.e., without a window). Cosmic rays trigger events in both detectors and can be thus be vetoed, while genuine X-ray photons produce events in the gas counter alone.

One of the advantages of proportional counters is that they can offer a large collecting area, either through large entrance windows made possible by using improved window material over time, or by stacking several units side by side in large arrays. Another strong suit of proportional counter is the good time resolution they can achieve: the downtime between events is mostly due to the charge's drift time, which is short. The time resolution can be improved further with an optimized geometry, for instance with multiple anode wires. The Proportional Counter Array (PCA) on the Rossi X-ray Timing Explorer (RXTE) reached a time resolution of 1 μs (Zhang et al. 1993).

The energy resolution of proportional counters is governed by the variance σ_e^2 of the total number of electrons N_e that is collected following the impact of a photon with energy E. For a constant energy, the initial interaction of the X-ray photon with the gas releases on average N_0 electrons with variance $\sigma_{N_0}^2$. In addition to the fluctuation of N_0, each of the photoelectrons is accelerated and causes an avalanche, multiplying the initial charge eN_0. These two processes being independent, the variance on N_e can be expressed as

$$\left(\frac{\sigma_{N_e}}{N_e}\right)^2 = \left(\frac{\sigma_{N_0}}{N_0}\right)^2 + \frac{1}{N_0}\left(\frac{\sigma_A}{\bar{A}}\right)^2, \tag{6.4}$$

where $\bar{A}$ is the charge multiplication factor of a single electron. For a quantity following the Poisson statistics (as does, for example, the count rate of incident X-ray photons from a source), $\sigma_{N_0}^2 = N_0$. But the observed fluctuations of N_0 are less than those of the Poisson distribution, because first the total energy absorbed must be equal to E and thus the multiple ionizations are not statistically independent, and second, some fraction of the energy can be used to excite the gas without causing an ionization. Following Fano (1947), $\sigma_{N_0}^2$ is given as $F \times N_0$, with F the Fano factor, which is the ratio of the observed variance of N_0 to the expected Poisson case ($\equiv N_0$). The Fano factor of a typical gas mixture is between 0.05 and 0.2 (Knoll 2000). For large values of the amplification factor $\bar{A}$, the statistical fluctuations $(\sigma_A/\bar{A})^2$ tend to a constant $b = 0.4$ to 0.5 for most gases used. The energy resolution then follows from Equation (6.4), as the measured energy is proportional to the number of collected electrons N_e by a factor W:

$$\frac{\sigma_E}{E} = \sqrt{\frac{F + b}{N_0}} = \sqrt{\frac{W(F + b)}{E}}. \tag{6.5}$$

Note that the energy resolution becomes better for higher-energy photons E and for smaller ionization threshold of the material W. The observed FWHM resolution is $\Delta E/E \approx 2.35(\sigma_E/E)$ and is shown in Figure 6.3.

Late-type proportional counters gained imaging capabilities by using multiwire anode grids. The charge avalanche is distributed on a limited number of wires. Reading the signal distribution on each wire allows us to reconstruct the two-dimensional position of the event, and therefore the celestial origin if the detector is placed at the focus of a telescope. The best examples are the Position Sensitive Proportional Counters (PSPCs), two identical detectors used by the German mission ROSAT, which performed the first imaging all-sky X-ray survey (Figure 6.4).

6.2.3 Solid-state Detectors

Starting with the Solid State Imaging Spectrometer (SIS) on the ASCA observatory (Tanaka et al. 1994), CCD detectors have replaced gas detectors as the main instruments to detect X-rays. The working principle is basically the same as gas detectors, with an incident X-ray producing photoelectrons whose total charge is

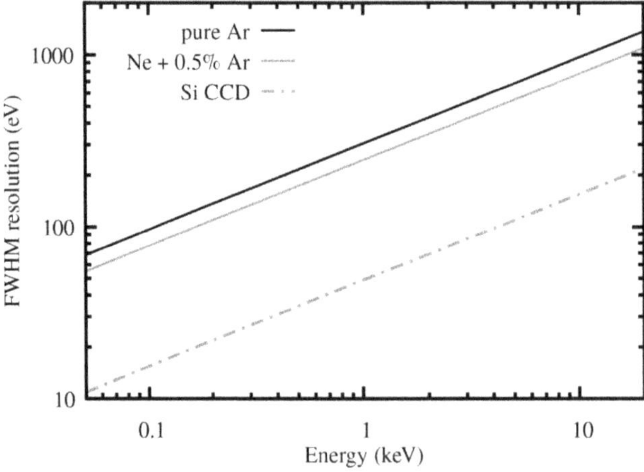

Figure 6.3. Limiting FWHM energy resolution of gas detectors (solid lines) using pure argon or a mixture Ne + 0.5 % Ar and of Si CCD detectors (dotted–dashed). Data from Knoll (2000).

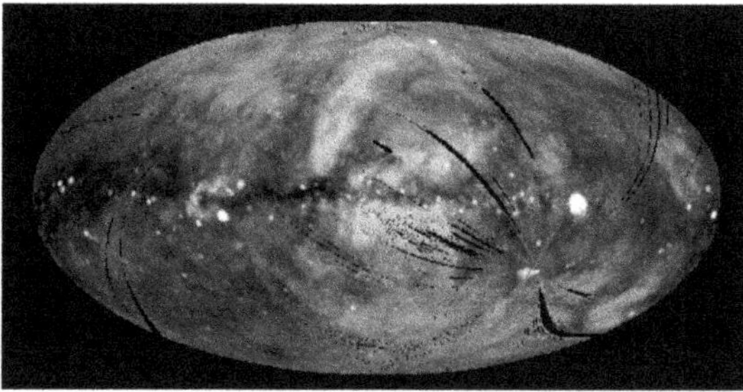

Figure 6.4. Left: one of the ROSAT PSPCs with the filter wheel and front end electronics removed. Right: X-ray color image from the ROSAT all-sky survey in Aitoff–Hammer equal-area projection in Galactic coordinates. Image credit: NASA, MPE (http://www.mpe.mpg.de/) and S. L. Snowden.

proportional to the photon energy. The detector material is this time a semiconductor solid, most of the time silicon. The detector is cooled down (80–120 K) to avoid thermal noise. Advances in manufacturing processes and miniaturization allowed large semiconductor detectors of high purity to be built, with electronic circuitry directly on the silicon wafer that creates an array of potential wells. This pixelated structure directly gives an imaging capability to CCD detectors. Charges are collected in the potential wells close to the interaction. By manipulating the potential over time, they are transferred row by row to be read out and amplified.

An immediate advantage of CCD detectors over gas counters is the lower W value of Si compared to most gases. While the Fano factor of Si is similar to most gas mixtures at about 0.12, it takes only 3.62 eV to create a charge pair, so there are many more charges for the same photon energy. This removes the need for charge amplification ($b = 0$ in Equation (6.5)). This leads to energy resolutions up to 10 times better than in gas counters (see Figure 6.3).

CCD detectors also achieve high quantum efficiencies in the X-ray-band energy, typically more than 60% and up to 95% for some detectors in the 1–8 keV band. The QE is controlled by the thickness of the sensitive layer, within which X-ray photons produce charges that can be collected, and by the X-ray attenuation length, the average distance it takes a photon to be absorbed, of the detector material. At low energies (below a few 100 eV), the attenuation length of Si is less than 0.1 μm and the X-rays might be absorbed too early, above the sensitive layer (e.g., within the on-chip circuitry). Some CCDs use backside illumination coupled with ultrathin entrance layers (0.02 μm; Hartmann et al. 1997) to achieve excellent QE in the 150–500 eV band, such as the EPIC-pn detector on board XMM-Newton. Above 10 keV, the X-ray attenuation length of Si exceeds 100 μm, but a sensitive thickness of up to 500 μm can be reached by manipulating the donor concentration in the semiconductor.

One limitation of CCD detectors is that they are subject to radiation damage. Intense particle radiation impacting the detector can create charge trappings over time. On the affected pixels, a fraction of the charge is trapped and not transferred to the readout electronics, increasing the so-called charge transfer inefficiency (CTI) of the detector. This results in a degraded energy resolution. Radiation damage can be mitigated with preventive measures such as by using a graded-Z shielding around the detectors. This shielding comprises an outer high-Z layer, designed to mostly stop particles (proton and electrons), while successive layers on the inner side absorb the resulting X- and γ-ray fluorescence (photons) to prevent them from hitting the detector. Such a design also reduces the particle-induced background (which we called $N_{\rm part}$ above). Nevertheless, for long-term operations, it is necessary to perform calibration observations to monitor the evolution of CTI and other key performances of CCD detectors.

Another limitation of CCD detectors is that they are slower than proportional counters, because of the relatively long time required for low-noise readout of the charges. This can be improved with complex electronics and the parallel treatment of columns, each with their own readout node. To reach better time resolution, CCD detectors can be operated in "windowed" modes, in which only a few pixel rows are used, allowing a faster readout time. Some cameras on board recent and future X-ray missions (e.g., eROSITA/SRG, SVOM) also use frame-store CCD detectors, which uses a two-zone chip. One zone is the traditional CCD chip, exposed to the sky, while the second one is shielded against X-rays. Following one frame exposure (when X-rays are collected on the first chip), the whole frame is transferred to the storage zone, a transfer that can be very fast ($\lesssim$0.1 ms) because charges are not read at that point. The frame store is then read out, as simultaneously the exposed chip accumulates the next frame, leading to a much improved duty cycle.

Athena, the next flagship X-ray observatory of the European Space Agency (ESA), will fly the Wide Field Imager (WFI), a new type of solid-state detector using active pixel sensors (Meidinger et al. 2015). Such sensors combine a sensitive silicon bulk, with which X-rays interact, and an internal gate beneath it, where charges can be collected and amplified when the pixel is switched on. Each sensor alternates between a passive phase, during which they collect the charge generated by incoming X-ray photons, and an active phase, when the signal charge is measured and amplified, removing the need for charge transfer as in CCD detectors. This improves the time resolution and prevents the effects of radiation damage that increase the CTI of CCD detectors over time.

6.2.4 High-resolution X-Ray Spectroscopy

Even in the absence of electron or thermal noise, the energy resolution of the detectors presented above is intrinsically limited by fluctuations in the number of charges created by photons of the same energy E (the Fano limit). To reach better energy resolution, grating spectrometers rely on diffraction or reflection gratings to disperse X-rays. The measurement of energy consists now in the measurement of a dispersion angle, instead of relating the strength of the signal to the energy of the incident photon. The dispersion angle θ is given as function of wavelength by the grating equation:

$$\sin\theta = \frac{m\,\lambda}{p}, \tag{6.6}$$

where m is the dispersion order ($m = 0, \pm1, \pm2, \ldots$, with the undispersed image in the zeroth order) and p the spatial period of the grating elements.

To be able to separate two closely spaced wavelengths, the dispersion angle θ must be large enough. Because X-ray wavelengths are so short (1–100 Å), the period p must be very small as well, and gratings with high line density (more than 1000 lines mm^{-1}) must be used. Dispersed X-rays are collected on a detector (for instance, a CCD camera) at the focus of the instrument. The spectral resolution is limited by the smallest dispersion angle that can be measured, which depends on the spatial size μ_{pix} of the detector pixels, the distance D between the grating and the detector, and the grating period p.

An example is the High Energy Transmission Grating Spectrometer (HETGS) on the Chandra observatory (Canizares et al. 2005). The High-Energy Grating (HEG) transmits X-rays with wavelengths between 15 Å and 1.2 Å (0.8 to 10 keV) and has a line density of 5000 lines mm^{-1} or a period $p = 0.2$ μm. The diffracted photons are measured on an array of CCD chip with linear pixel size $\mu_{pix} = 24$ μm. At $D = 8.63$ m behind the grating, the photons between 15 Å and 1.2 Å are spread (in the first dispersed order) over $d \approx D \times [\theta(15\text{Å}) - \theta(1.2\ \text{Å})] = 59.5$ mm, or 2480 pixels. Each pixel then covers about 5.6×10^{-3} Å, and the best resolution achievable is about twice that value, or 0.0112 Å. In practice, the finite point-spread function, background events, and counting statistics degrade the actual resolution. Another example is the Reflection Grating Spectrometer (RGS) of XMM-Newton (den Herder et al. 2001), which uses faceted gratings to reflect a fraction (about half) of the light directly behind two of the three X-ray telescopes of the satellite, while the rest is undispersed and focused on two CCD cameras.

The dispersion angle between two nearby wavelengths is independent of λ, as long as $\lambda \ll p$. Therefore, the resolving power $R = \lambda/\Delta\lambda$ is increasing for larger wavelengths (smaller E), the opposite behavior to that in gas or solid-state spectrometer based on charge multiplication, where the resolving power increases with E (toward shorter wavelengths; see Equation (6.5)). Grating spectrometers thus remain the best instruments to perform high-resolution spectroscopy in the soft X-ray band.

However, the main limitation of such spectrometers stems from their operating principle: They work by spreading on the detectors an already limited photon rate. Furthermore, the throughput of the whole instrument is also reduced either by the covered fraction and interferences of blocking elements in transmission gratings, or the limited reflectivity of reflection gratings. Only the brightest sources can produce high-enough count rates per pixel (or wavelength) element to provide meaningful measurements. Extended sources like supernova remnants (SNRs) or galaxy clusters cannot benefit from grating observations because this second source of dispersion on the detectors (due to the sky extent) severely degrades the spectral resolution.

To reach even better spectral resolution across a wider energy band and for extended sources as well, a new type of detector is needed. Those are microcalorimeters, which measure the heat deposited by individual X-ray photons in microscopic devices cooled down to extremely low temperatures (<0.1 K). Such detectors can reach an energy resolution of a few eV or less, rivaling that of grating spectrometers and even exceeding them at $E \gtrsim 5\,\mathrm{keV}$. Their small size enables arrays of microcalorimeters to be built and placed at the focus of X-ray telescopes, effectively turning the instrument into an integral field unit (IFU), where each pixel can deliver a high-resolution spectrum. A string of rocket, spacecraft, and cryogenic failures have so far thwarted all attempts at stable operations of such instruments (Triendl 2000; Kelley et al. 2007; Takahashi et al. 2018; Adams et al. 2020).

The second main instrument of Athena will be the X-IFU (Pajot et al. 2018), an array of 3840 microcalorimeters with a planned energy resolution $\Delta E = 2.5\mathrm{eV}$ up to 7 keV, in the early 2030s. Its core technology relies on the Transition Edge Sensor (TES; Irwin & Hilton 2005), where the sensor is cooled down just below the superconducting limit where the electrical resistance is null. Following the absorption and thermalization of an X-ray photon, the sensor is heated to above the transition (nonzero resistance). As the sensor is biased, an electrical current is thus measured until the sensor is cooled down back below the transition where the current vanishes. The shape of the pulse allows the increase of the temperature of the sensor, and thus the total energy of the incident photons, to be measured.

6.2.5 X-Ray Polarimetry

Besides the 2D position, time of arrival, and energy of the X-ray photon, its polarization is the fifth property that a detector with polarimetric capabilities could measure. This would reveal the polarization degree of sources, that is, the fraction of emitted radiation that is polarized, and the polarization angle, the orientation of the polarization on the sky.

A relatively high degree of polarization is expected for sources dominated by synchrotron emission, because the emission is polarized in the plane perpendicular to the magnetic field lines, or by scattering and reflection processes (with polarization perpendicular to the scattering plane). Several classes of astrophysical sources, such as pulsar wind nebulae (PWNe) and SNRs (Sections 6.5.1 and 6.5.2), or accreting neutron stars (NSs), black holes (BHs), and active galactic nuclei (AGNs) (Sections 6.5.3 and 6.6.1) fall in this category. Extending the polarimetric measurements available from radio to optical wavelengths to the X-ray domain can reveal the magnetic field structure in these sources, as well as their geometry.

The eighth Orbiting Solar Observatory, OSO-8, carried in its payload the first X-ray polarimeter to observe cosmic X-ray sources. It used a crystal Bragg polarimeter, rotating with the spacecraft about the line of sight to a polarized source. The maximal reflection amplitude is obtained when the polarization vector of a photon satisfying the Bragg condition (Equation (6.9)) is perpendicular to the scattering plane defined by the incident and reflected flux. Therefore, as the satellite rotates with a period Ω, the signal from a polarized source is modulated at a period 2Ω. The degree and angle of polarization are determined by the amplitude and phase of the signal modulation, respectively.

The polarimeter of OSO-8 led to a single measurement of polarization of a cosmic X-ray source so far, in the Crab Nebula (Weisskopf et al. 1978, 1976). The relatively high polarization degree (integrated over the nebula) of 19% at 2.6 and 5.2 keV firmly established the synchrotron origin of the nebular emission. Only upper limits to the polarization degree were obtained for a few other bright sources (Hughes et al. 1984), and no other X-ray polarimeter has been flown since OSO-8. One critical issue with crystal Bragg polarimeters is that they are by design very photon hungry. Bragg conditions only allow a tiny energy band to be used, and enough counts must be collected over the spacecraft rotation to allow the signal modulation to be accurately measured.

Several missions, such as the Imaging X-ray Polarimeter Explorer (IXPE; Weisskopf et al. 2016; launched in December 2021) and the enhanced X-ray Timing and Polarimetry mission (eXTP; Zhang et al. 2019; planned for launch in 2027), will reopen the X-ray polarimetric window that was shut for more than 40 years. Their main innovation is to use gas pixel detectors (Costa et al. 2001), an evolution of the gas proportional counter where both the origin and direction of the photoelectron track are measured (Figure 6.5). As the photoelectron is emitted preferentially in the direction of the polarization, the reconstruction of the track leads to a measurement of the polarization of each individual photon. Meanwhile the origin of the track relates to the celestial origin, and imaging polarimetry can be performed.

6.3 Tracking or Focusing X-Rays

Measuring the position of the origin of X-rays has been a key problem of the field since its inception, because distinguishing any potential X-ray sources from the Sun, Moon, or Earth was required to show that such celestial X-ray sources did in fact exist. Several techniques exist to localize sources. In addition, X-ray telescopes have

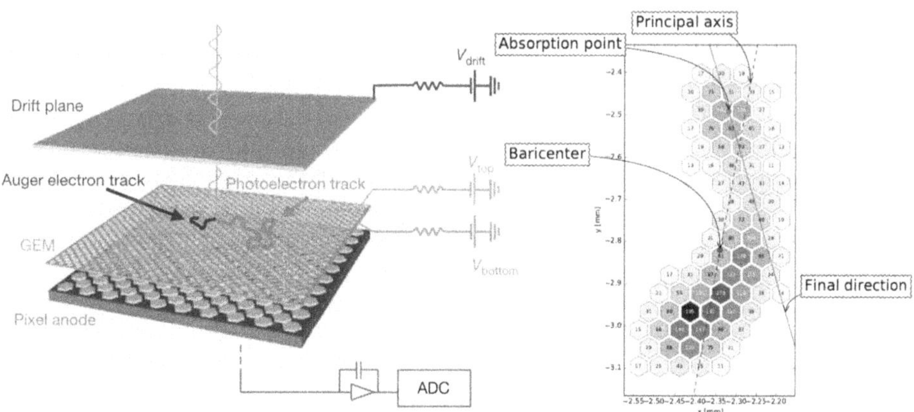

Figure 6.5. Left: working principle of a gas pixel detector for X-ray polarimetry. The incident photon ionizes the gas, and the direction of the photoelectron track is recorded on the pixelated anode at the bottom, after charge multiplication in the gas electron multiplier (GEM). Reproduced from Costa et al. (2001), with permission of Springer. Right: ionization track of a 5.9 keV photon recorded on the anode. The number and color of each hexagonal pixel denote the collected charge after going through the GEM. The reconstructed impact point and polarization direction are shown by the green dot and line, respectively. Image credit: IXPE team.

been developed. They concentrate and focus celestial X-rays, thus greatly increasing the S/N and allow sky positions to be measured with higher accuracy. However, the refracting lenses and reflecting mirrors used in longer-wavelength astronomy cannot be used to focus X-rays, which simply pass through the materials used in traditional telescope designs. We present here those various observing techniques.

6.3.1 Nonfocusing Techniques

The spatial resolution of the first detectors simply stemmed from their design. Gas counters have entrance windows that naturally limit the field of view of the detector assembly. The crude measurement of the origin of incoming X-rays can be inferred by knowing the pointing direction of the detector as it scans across the sky. This technique can be improved by collimators, usually a set of plates or grids that limits the incident angle of X-rays that can reach the detector. With a scan in a second spatial direction, the celestial position of an X-ray source can then be measured with greater accuracy.

An alternative is to employ a rotation modulation collimator, a set of grids that modulate the signal from a source in one direction on the detector. The instrument rotates around the pointing direction, producing a unique modulation curve as a function of time. The sky image is then reconstructed by the cross-correlation technique (in the Fourier domain). This method is also called temporal modulation as it relies on changing the detector footprint on the sky over time.

In aperture modulation, it is the opening (or aperture) of the instrument, which we manipulate. In most cases a coded mask is used, which has a pattern of holes (a "code") in an X-ray-absorbing mask, placed in front of a position-sensitive detector. Sources in the field of view project shadows on the detector behind the coded mask, and these shadows are used to reconstruct the sky distribution. They remain heavily in use for hard X-rays (for instance above 100 keV) when even grazing-incidence reflectivity becomes negligible (see below). As such, they are more relevant to gamma-ray astronomy, which is discussed in Chapter 7.

Finally, lunar occultations, that is, observing an X-ray source as it disappears behind the limbs of the Moon, have also been used to measure X-ray positions with accuracy greatly exceeding the angular resolution of the detectors used. The flux distribution of an extended source (or even the very fact that it is extended) can also be inferred by recording the count rate of the source as the lunar disk moves in front of it, as done most famously for the Crab Nebula on two occasions (Bowyer et al. 1964; Palmieri et al. 1975). Use of this method remains limited because of practical reasons: rarity of occurrences, limited sky coverage, and strong scheduling constraints.

6.3.2 X-Ray Telescopes

6.3.2.1 Origin and Principle

X-rays can be reflected off a smooth surface in the case of total external reflection, when the incident angle is below a critical angle θ_{crit}, which is typically less than a few degrees for keV photons. Using this principle, Hans Wolter (1911–1978) (Wolter

1952) noted in 1952 that a segment of a paraboloid of revolution could be used to focus X-rays, though no image can be formed because of severe aberrations. He further demonstrated that, using a system with an even number of mirrors (limited in practice to two), images could be formed. He proposed three configurations using combinations of paraboloid, hyperboloid, and ellipsoid mirrors.

In the pioneering years of X-ray astronomy, Riccardo Giacconi (1931–2018) and Bruno Rossi (1905–1993) (Giacconi & Rossi 1960) realized that the idea of Wolter, originally developed for the purpose of X-ray microscopy, was exactly what was needed to build X-ray telescopes. Most importantly, they suggested "nesting" several mirrors of decreasing diameter to increase the collecting area. The design usually adopted is the so-called Wolter-I configuration (the only one of the three designs of Wolter that allows nesting). X-rays are first reflected off a paraboloid before undergoing a second reflection off a hyperboloid. In practice, a conical approximation of both shapes is often used. Such telescopes were first tested for solar observations in 1963 and 1965, aboard sounding rocket flights. Further developments culminated with the launch of Einstein in 1978, the first satellite carrying an X-ray telescope (effective area of 100 cm^2 at 1 keV) for extrasolar observations. This opened a new page in X-ray astronomy. Since then, most orbiting X-ray observatories use Wolter telescopes, which have been vastly improved in terms of collecting area and angular resolution.

6.3.2.2 Properties and Performances

For very small grazing angles, most photons are reflected off smooth surfaces. The reflectivity decreases with both increasing grazing angle and photon energy. At each energy E, the reflectivity falls abruptly above a critical angle $\theta_{crit}(E)$ (Figure 6.6, left).

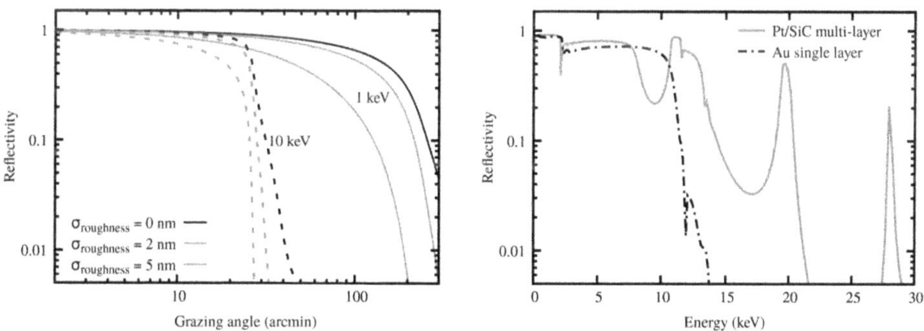

Figure 6.6. Left: reflectivity of a single 50 nm-thick gold layer at 1 keV and 10 keV (solid and dashed lines, respectively), as a function of the grazing angle in arcminutes, measured with respect to the surface. The effect of a surface roughness with a standard deviation between 0 (perfectly smooth) and 5 nm is shown by the three curves of various colors. Right: comparison of the reflectivity at a grazing angle of 0.4° of a single Au layer (same as in left panel) with that of a Pt/SiC multilayer, as a function of the photon energy. The multilayer is a stack of 100 Pt/SiC bilayers with 10 nm thickness. Data from Henke et al. (1993).

This angle is inversely proportional to the photon energy and proportional to the square root of the surface mass density ρ:

$$\theta_{\rm crit}(E) = {\rm cst.} \times \sqrt{\rho}\, E^{-1}. \qquad (6.7)$$

To increase the critical angle, and thus the total amount of photons focused by the telescope, mirror surfaces are coated with thin layers of heavy materials, such as gold or iridium.

X-rays can be scattered by the high spatial frequency fluctuations of the mirror surface (at scale lengths below a micron), called microroughness. This scattering and interference effect lower the reflectivity at any given incidence angle, as shown in Figure 6.6 (left) for microroughnesses of various amplitudes. Through polishing of the substrate and a careful coating procedure, the surface roughness can now be kept below a few nanometers.

Wolter-I X-ray telescopes focus paraxial photons (parallel to the optical axis) to a focal point at a distance f behind the mirrors. Following the double reflection, the ratio r/f between the radius r of a mirror shell and the focal length f is given by

$$r/f = \sin(4\alpha), \qquad (6.8)$$

with α the slope angle of the first mirror element with respect to the optical axis. For paraxial rays, we can equate α to $\theta_{\rm crit}(E_{\rm max})$ to estimate the highest energy $E_{\rm max}$ that can be focused by a given telescope size. Combining Equations (6.7) and (6.8), we find that $E_{\rm max} \propto \sqrt{\rho} \times f/r$ (by design r/f is not very large, so we can approximate $\arcsin(r/f)$ as r/f). For a gold-coated mirror shell of 1.2 m radius and 12 m focal length (as planned for the Athena X-ray Observatory), this translates to a maximal energy of about 3.5 keV.

There are several strategies to reach higher energies. First, most Wolter-I telescopes use nested shells. The inner shells of smaller radii have larger $E_{\rm max}$. Following the previous Athena example, the innermost shell of radius 0.26 m allows photons up to about 15 keV to be reflected. Obviously, the inner rows of shells will contribute only a small fraction of the overall collecting area, so the total efficiency above 10 keV remains limited. Second, the focal length, usually limited by the weight and size of the spacecraft that it is possible to launch, can be increased, either by using telescopic structures that deploy once in orbit (as used on NuSTAR; Harrison et al. 2013) or by using two spacecraft flying in formation. Both solutions introduce technical challenges, especially the latter.

A third strategy consists of manipulating the coating of the mirrors. A stack of alternating layers of high-Z material (for instance gold, platinum, and tungsten) with lower density material (carbon, silicon, or other light alloys) acts as an interference reflector: The thickness of the layers is chosen so that the path length difference between reflections on the successive layer interfaces is equal to one or a multiple of the wavelength, resulting in constructive interferences. Thus, the interfaces in the multilayer stack act like the atoms in a crystal lattice, and the same equation, Bragg's law, can be used to relate the layer thickness d, the grazing angle θ, and the wavelength λ:

$$n\lambda = 2d \sin \theta, \tag{6.9}$$

with n an integer, the order of diffraction. We can use the small-angle approximation to estimate the wavelength (or photon energy $E_{\text{enhanced}} = hc/\lambda$) that is enhanced by constructive interference on a layer of thickness d as

$$E_{\text{enhanced}} \approx 36 \left(\frac{d}{1 \text{ nm}}\right)^{-1} \left(\frac{\theta}{1 \text{ deg}}\right)^{-1} \text{ keV}. \tag{6.10}$$

Stacks with varying layer thicknesses can be built to improve the reflectivity over the spectral range of interest. A prime example is the multilayer coating of the segmented mirror shells of NuSTAR telescopes (Christensen et al. 2011), increasing the collecting area between 10 and 78 keV (the energy of the K-shell absorption edge of platinum). This results in the highest effective area of a focusing telescope for hard X-rays (Figure 6.7).

The effective area of a single mirror shell of aperture $2r$ is roughly the surface seen from the incident photon, i.e., a thin annulus with a projected area

$$a_{\text{eff}} \approx 2\pi r \times L \sin \alpha, \tag{6.11}$$

with L the length of the mirror shell. Because the slope angle α must be small to reflect X-ray photons, a_{eff} is much smaller than the total polished mirror area. Given Equation (6.8), a_{eff} is roughly $(\pi r^2/2) \times (L/f)$. Even summing the contributions by all nested shells, the total effective area of X-ray telescope is much less than the one that can be achieved by optical or near-infrared telescopes (using normal-incidence

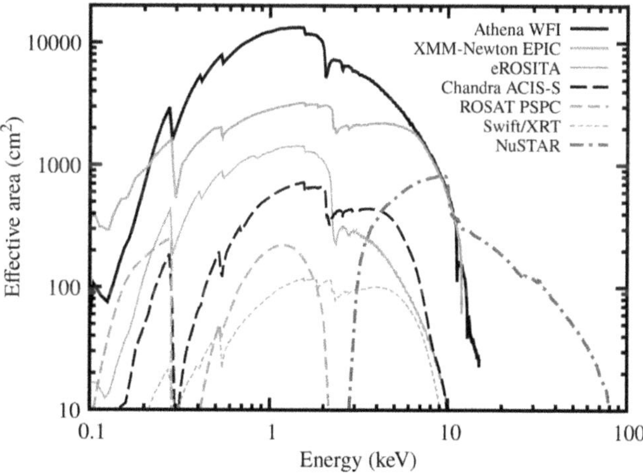

Figure 6.7. Effective area curves as a function of energy of the main past, current, and future X-ray missions. The effective area shown includes the telescope response, optical blocking filters, and detector quantum efficiency. For eROSITA and NuSTAR, the combined response of all seven and two telescopes are shown, respectively. For XMM-Newton, the three telescopes and EPIC detectors are combined. The low-energy response of Chandra ACIS has decreased significantly over time,and the curve shown corresponds to that of 2008. Data from PIMMS (Mukai 1993), CXC (https://cxc.harvard.edu/proposer/), and SIXTE (Dauser et al. 2019).

reflection) for the same total telescope weight, which is often the limiting factor for space-borne mission instruments.

As in all optical systems, the effective area decreases for off-axis sources (non-paraxial rays). This effect, known as vignetting, is further compounded in X-rays by the strong dependence with energy of the reflectivity. The loss of effective area off axis is thus stronger for higher-energy photons.

The image quality, or spatial resolution, of an X-ray telescope, is often measured by the size of the point-spread function (PSF), the characteristic instrumental broadening of a point source. The PSF of an X-ray telescope is contributed by both the figure error of the mirror shape and the degrading effect of the micro-roughness of the mirror surface. The imperfect alignment of the nested shells or segments of shells introduce further degradation of the image quality. Past and current X-ray missions have achieved a wide range of PSF sizes, from several arcminutes down to just under an arcsecond for the telescope on board Chandra.[2] The PSF size increases significantly with off-axis angle. This is due to the conical approximation of the mirror surface, which results in a hyperbolically curved focal surface. As most detector surfaces so far have been flat, only on-axis rays are optimally focused, and the best image quality is only achieved in the central part of the field of view.

6.3.2.3 Manufacturing Techniques and Future Developments

The different fabrication techniques of Wolter telescopes result in various strengths and weaknesses in term of final performances. The earlier telescopes were prepared from thick glass blanks that could be thoroughly ground and polished to obtain the desired shape and low microroughness, before coating. Such a design may lead to excellent angular resolution (just under 1″ on axis for Chandra). However, the total area is seriously limited by the thickness of the mirror shells, in terms of both weight and low nesting level.

A second technique that was developed to improve on that point was that of replicated optics. A mandrel (the opposite of a mold) is fabricated with the required figure for a Wolter mirror shell, and a metal, usually nickel, is deposited on the mandrel in an electrolytic bath (electroforming). The resulting thin shell is relatively lightweight and can be coated with an X-ray-reflective material. Many more shells can be nested than in the first approach, leading to high collecting area. Furthermore, a mandrel can be used to produce many shells, making this technique well suited for producing several telescopes. This has been the approach used for the three and seven identical telescopes of XMM-Newton and eROSITA, respectively (see Book 1, Figure 1.16). The angular resolution is not as good as before, but FWHM PSFs of 15″ to 30″ were achieved in the cases cited above.

Another way to gain a high nesting level and effective area is to use segmented thin foil telescopes. The mirror shells are broken into several segments, each made of a thin foil. Likewise, the double curvature of Wolter's design is reproduced by a set

[2] For reference, the PSFs achieved in the visible band from a space-based telescope (like the HST) or a ground-based telescope with adaptive optics, have size of the order of a few tenths of an arcsecond.

of two foils. The foils are made of glass or light metal (e.g., aluminum), so the total weight remains low. Several missions such as Japanese's Suzaku and NASA's NuSTAR used this approach for their mirrors, with high throughput at moderate cost and weight. The PSF is however worse than with other techniques, often between half and a few arcminutes. Segmented/thin foils (as in Suzaku and NuSTAR) are light, with high effective areas but mediocre PSFs (alignment and figure).

To build the largest X-ray telescope that Athena will carry, a new method was chosen, called Silicon Pore Optics (SPO). The approach is similar to the segmented thin foil one, but uses industry-grade silicon wafers as a substrate for the foils. The raw silicon plates are cut, bent, and coated before being stacked to form one mirror module. Industrial production methods are necessary to produce, treat, and assemble the roughly 100,000 mirror plates needed to populate the 1.4 m aperture of Athena's telescope. Stringent constraints for the production and alignment of the plates should allow the on-axis PSF to reach 5″.

New types of non-Wolter optics are also being developed or studied. One approach is the lobster-eye telescope, inspired by crustacean's eyesight and conceptualized for X-ray astronomy in the late 1970s (Angel 1979). The telescope is composed of multiple glass plates, each with millions of microscopic square pores covering the aperture of the telescope. X-rays can be focused following two reflections on two orthogonal sides of the pores. As such, they are a subset of the Kirkpatrick–Baez design (Kirkpatrick & Baez 1948). Lobster-eye optics have the advantage of allowing a large field of view at a very low weight: the Microchannel X-ray Telescope (MXT; Wei et al. 2016) on the Chinese–French mission SVOM (Götz et al. 2016) has a field of view of almost a square degree for an optics weight of only 10 kg. The inconvenience of this design is the nonstandard, cruciform PSF, with only half the source's flux concentrated in the PSF core (after two reflections), and the rest in a diffuse area (no reflection) and orthogonal cross-arms (single reflection).

Other designs for harder X-rays ($\gtrsim$10 keV) use refractive prisms arranged in stacks of concentric disks with a varying number of prisms as a function of radius, producing the desired focusing effect. Those are called stacked prism lenses (SPLs; Mi et al. 2019). Finally, X-rays at even higher energies may also be deviated using Bragg diffraction in transmission configuration, the so-called Laue geometry, using crystals or multilayer stacks arranged in the appropriate geometry (Frontera & von Ballmoos 2010; Bajt et al. 2017).

6.4 Stellar X-Ray Sources

In this section, we present stellar X-ray sources and the vast knowledge gained with the instruments described above (Sections 6.2 and 6.3). X-rays have been detected from many types of stellar objects, spanning widely different spectral types and ages, from young stars still in their infancy (although not covered here) to late stages such as white dwarfs.

6.4.1 The Sun and Other Main-sequence Stars

It is only natural that our Sun, being the closest star, was the first target for the first rocket-borne X-ray detectors. The Sun was first detected in X-rays in 1949 (Friedman et al. 1951) using Geiger counters (a type of gas counter that operates at saturation, recording only the count rate of all photons that can enter the detector). It was rather surprising that the Sun emitted a detectable amount of X-ray photons, as a blackbody at an effective temperature of 5800 K like the solar photosphere would not. Instead, the observed emission implied temperatures of about 10^6 K.

Such high temperatures were also needed to produce the high-ionization species that emit the so-called "coronal lines." The Sun's corona is the "halo" extending beyond the solar disk as seen during a total solar eclipse, from which several unidentified spectral lines (in the optical range) have been detected since the 19th century, most notably a green 5303 Å line and ascribed to a putative new element called coronium. Later, Walter Robert Wilhelm Grotrian (1890–1954) and Bengt Edlén (1906–1993) (Grotrian 1939; Edlén 1943) showed it came from Fe XIV, i.e., iron stripped of 13 of its electrons. Similarly, the X-ray spectra of the Sun was shown to be dominated by spectral lines of high-ionization species (Blake et al. 1965).

How to produce and sustain a million-Kelvin plasma around the Sun has been a long-standing challenge of solar physics. The heating mechanism is most likely the turbulent magnetic field loops at the surface of the Sun, which evolve and reconnect, producing electric currents that accelerate and heat the plasma. Given the temperature and density conditions, the X-ray emission is thermal in nature, and its spectrum is calculated in the aptly named coronal approximation. This type of emission is often found in the X-ray universe and is therefore discussed in more detail below (Section 6.4.2).

6.4.1.1 X-Ray-emitting Single Stars

Regardless of its origin, the Sun's X-ray emission represents only 10^{-7}–10^{-6} of its total bolometric luminosity $L_{\rm bol}$ (with significant enhancements during solar flares). Even the nearest solar-like stars could only be detected after X-ray telescopes were used, starting with the Einstein mission. Subsequent surveys showed that essentially all late-type stars (F to M types) exhibit X-ray emission, down to a visual magnitude $M_V = 20$ (in a volume-limited sample; e.g., Schmitt & Liefke 2004).

The magnetic field in those stars is generated by a dynamo produced by differential rotation in the stellar interior, at the interface of diffusive and convection zones. Consequently, their X-ray luminosities L_X are well correlated to their rotation velocities, v (Pallavicini et al. 1981). Specifically, $L_X \propto (v \sin i)^2$, with i the inclination of the rotation axis with respect to the line of sight, or, when the rotation period $P_{\rm rot}$ itself (independent of inclination) is known, then $L_X \propto P_{\rm rot}^{-2}$ (Figure 6.8, left). Below a certain $P_{\rm rot}^{\rm sat}$ (i.e., for the fastest rotators), this quadratic X-ray–rotation period relationship breaks down, and L_X reaches saturation around 10^{-3} of the bolometric luminosity. The saturation period $P_{\rm rot}^{\rm sat}$ increases with decreasing stellar mass (Pizzolato et al. 2003). This can be interpreted as the threshold when all of the stellar

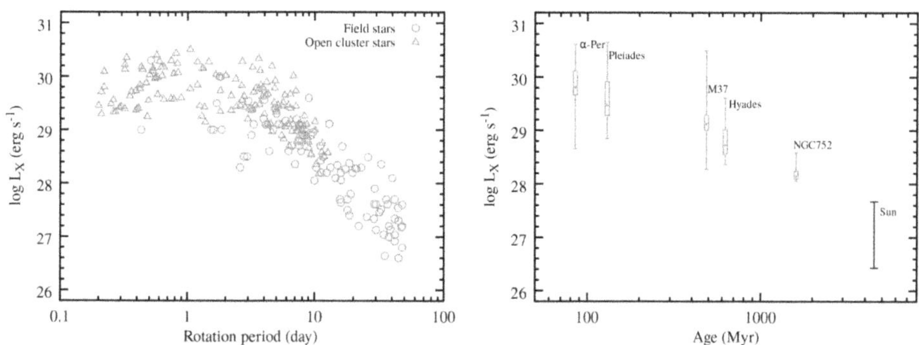

Figure 6.8. Left: X-ray luminosity of late-type main-sequence stars in the field and open clusters as a function of their rotation period (data from Pizzolato et al. 2003). Right: evolution of the average X-ray luminosity of solar-like main-sequence stars in open clusters of increasing age (data from Núñez et al. 2015). The box lower and upper boundaries mark the first and third quartiles, respectively, and the whiskers show the entire range. The median value for each cluster is shown by a dot inside the box. At 4.5 Gyr, the error bar indicates the range of L_X covered by the Sun between solar minimum and maximum.

surface is filled by active regions that are at the footprints of the magnetic coronal loops providing the coronal heating (Vilhu 1984). Another effect that can limit the coronal X-ray emission is the distortion of the corona by centrifugal forces, which can become dominant in the magnetic loop of fast rotators (Jardine & Unruh 1999).

Over time, the rotation of stars slows down via magnetic braking, leading to a decrease of their X-ray emission, as illustrated by the decline of the average L_X from stars in open clusters of increasing age (Figure 6.8, right). By the same token, giant stars, which are slow rotators with cooler surfaces, also have less X-ray activity.

In early-type stars (i.e., more massive stars) the situation is quite different. A-type stars are thought to lack an outer convection zone and thus the dynamo activity needed to power coronal X-ray emission. Indeed, little to no X-ray emission is found from these stars, which are among the brightest in the optical. Stellar X-ray emission is detected again from earlier (O and early B) type stars, something that was unexpected (Harnden et al. 1979; Seward et al. 1979). This time L_X is proportional to the bolometric luminosity (Sciortino et al. 1990; Nazé 2009). Their emission is different in origin: It is produced by the strong, fast stellar winds driven by UV absorption lines. Such winds are unstable, and shocks develop that heat the plasma to million-degree temperatures (Lucy & White 1980; shock heating is discussed in Section 6.5.1). The spatial origin of the emission, be it at the base of the wind or in the outer regions, is a matter of debate. However, the former is disfavored because the strong absorption of soft X-rays (below 1 keV) by the optically thick portion of the inner wind is absent in most cases (see, e.g., Feldmeier et al. 1997 for a discussion of various models).

6.4.1.2 Binary Systems

The discussion above applied to single stars or those with a companion distant enough as to have no measurable impact on their X-ray properties. Here, we present the impact of binarity on the X-ray emission of some classes of late- and early-type stars.

As seen above, rotation is one of the key ingredients to explain the X-ray coronal emission of stars. In binaries close enough, the two components become tidally locked, with rotation periods equal to the orbital period, as is the case for our own Moon. Consequently, stars in some binary systems retain a relatively fast rotation, possibly faster than in single stars of the same type, for a much longer time. The best examples are stars of the RS CVn–type, a class of binary stars comprising a main-sequence, usually an F- or G-type, star and a K0 IV subgiant. Just off the main sequence, the latter starts to expand and develop a stronger and deeper convective envelope. Combined with the fast rotation, magnetic activity is much enhanced, leading to X-ray coronal emission four orders of magnitude above that of our Sun. The coronal origin of their X-ray emission is confirmed by the detection of starspots, which modulate their optical light curves (Hall 1981), with an activity cycle of about 10 years, similar to the solar cycle. As for single stars, there is a limit of about one per thousand for $L_X/L_{\rm bol}$, indicating again the onset of saturation (Dempsey et al. 1993). Other types of binaries are also relatively X-ray bright for the same reason (fast rotation due to close binarity). Like the RS CVn–type, they are named after their prototypical systems: BY Dra–type binaries are composed of two main-sequence K or M stars, and W UMa types also have two late-type stars, this time in contact in an even closer orbit with a common envelope.

Binarity is even more prone to be found for massive, early-type stars (Moe & Di Stefano 2017). The stellar winds from both components will collide and produce a dense shock at the interface of the two winds, the shape of which is roughly conical and set by the balance of the wind momenta $\dot{M}_i v_{\infty,\,i}^2$, with $\dot{M}_i$ and $v_{\infty,\,i}^2$ the mass-loss rate and terminal velocity of the wind from stars $i = 1, 2$, respectively. This occurs in a system with O stars, as well as binaries including one or two Wolf–Rayet stars, which are evolved O stars having lost their hydrogen-rich envelopes and producing the strongest and fastest stellar winds. The X-ray emission from such colliding wind binaries was expected and calculated even before that of single early-type stars was detected (Prilutskii & Usov 1976; Cherepashchuk 1976). Because the X-ray emission of the shocks roughly scales with the density squared, the dense shocks in the colliding winds make early-type binaries brighter than single stars of the same type (Pollock 1987; Chlebowski & Garmany 1991). Over an orbital period, the X-ray emission is modulated by absorption when the colliding wind shock is obscured by the dense, optically thick inner stellar winds as well as by the dense shock itself (Pollock et al. 2005). Such an X-ray variability is also useful to recognize binarity in those systems, where the variability of optical/UV spectral lines of a companion may be masked by the bright and broad lines of a more massive primary Wolf–Rayet star (Pittard & Parkin 2010).

6.4.2 X-Ray Emission from a Coronal Plasma and High-resolution Spectral Diagnostics

Plasmas with temperatures exceeding a million degrees are ubiquitous in the universe. They are found in stellar coronae (see above), in the remnants of supernovae (Section 6.5.1), and at larger scales, they fill the space between galaxies in clusters (Section 6.6.3). If certain conditions forming the coronal approximation are met, the plasma spectrum can be calculated and compared to the observed emission to derive the physical conditions (temperature, density, and elemental abundances) of the plasma. An essential assumption is that the plasma is optically thin (low or negligible optical depth). Therefore, any emitted X-ray escapes the system and can be observed, without affecting the ionization or heating balance of the plasma nor the population of electrons in the ionic levels. It also implies that the source of plasma heating must be nonradiative in nature (for instance, it is magnetic turbulence in the solar corona). Another important criterion is that the plasma density is low, such that electron–ion collisions are rare. Therefore, electrons may excite ions (from their ground state), but they will spontaneously deexcite radiatively (emitting X-ray photons) before they can deexcite from another collision. The electrons and ions are relaxed to a Maxwell–Boltzmann distribution with a common electron temperature T_e, i.e., they follow a velocity distribution:

$$\propto (k_B T_e)^{-3/2} v^2 \exp(-m_e v^2/2k_B T_e). \tag{6.12}$$

Finally, the plasma is in a steady state, with balanced ionizations and recombinations. That balance is set by the electron density n_e, which controls how many electrons can recombine with ions or produce further ionization, and the temperature T_e, which affects the ionization and recombination rates of various ions.

The X-ray emission from such a coronal plasma has line and continuum components. The lines are from the radiative deexcitation of ions following a collision with a free electron, or after a recombination to an excited state. The energy difference between the two levels of the transition sets the energy of the emitted photon. Therefore, the line energy allows the identification of the ionic species present in the emitting plasma. In turn, the ionization state and the flux ratio between lines are controlled by the electronic temperature and elemental abundances of the plasma, which can therefore be inferred from observations.

A further diagnostic is offered if several lines of the same ion are detected. The best example is the triplet of helium-like ions (ions with only two electrons), which falls in the soft X-ray band for all abundant elements from carbon to iron and is worth discussing in more detail. The "triplet" contains three components linking the $n = 1$ (ground) and $n = 2$ shells: a resonance line denoted w; an intercombination line, actually a blend of two lines very close in energy, denoted $x + y$; and a forbidden line denoted z (Figure 6.9, top left). Only high-resolution instruments such as gratings or microcalorimeters (Section 6.2.4) are able to resolve the triplet. For instance, the resolving power R needed to distinguish the forbidden line from the intercombination line blend ranges from 56 for C v (at 0.3 keV) to 215 for Fe xxv (at 6.8 keV). These lines have strengths that depend on the electron density, temperature, and ionization processes occurring in the plasma (Porquet et al. 2010).

First, the forbidden to intercombination line ratio $R = z/(x + y)$ is sensitive to electron density in the 10^8 to 10^{12} cm^{-3} range, applicable to stellar corona, while being only moderately affected by the temperature. In the low-density regime, $n = 2$ levels are populated only by collisional excitation from the ground $n = 1$ level (either

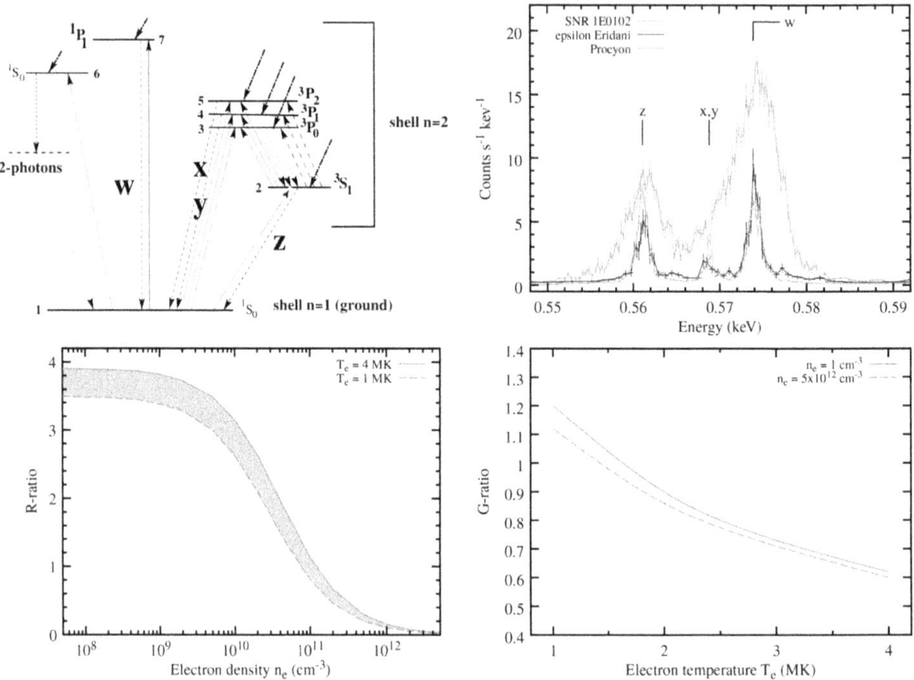

Figure 6.9. Top left: energy level diagram for He-like ions. Solid and dashed arrows represent collisional excitations and radiative transitions, respectively. The resonance (w), intercombination ($x + y$), and forbidden (z) lines are marked. The thick dotted–dashed lines indicate recombination processes (figure from Porquet & Dubau 2000, reproduced with permission © ESO). Top right: The O vii He-like triplet observed with XMM-Newton RGS (see Section 6.2.4) in stars ϵ Eridani and Procyon, and supernova remnant SNR 1E0 102.2–7219. The latter is an extended source, which broadens the lines. Bottom left: the O vii R ratio as a function of electron density in the 1 to 4 MK temperature range. Bottom right: the O vii G ratio of a collision-dominated plasma as a function of electron temperature at high (solid line) and low density (dashed line). Data for the ratios from Porquet et al. (2001).

directly or by subsequent radiative cascade from higher levels) and by recombination from H-like ions. As electron density increases, electrons in $n = 2$ levels may be collisionally excited to nearby states before they radiatively decay. In particular, this will depopulate the 1s2s $^3S_1$ level to the 1s2p $^3P_{0,\,1,\,2}$ level, which will suppress the forbidden z line in favor of the intercombination x and y lines, and therefore lead to a reduction of the R ratio as n_e increases (Figure 6.9; bottom left). In intense UV fields, photoexcitation from $^3S_1$ to $^3P_{0,\,1,\,2}$ has the same effect of lowering the R ratio, which can mimic a high-density plasma. This effect can occur for instance around massive stars if the emitting plasma is close to the very hot photosphere (e.g., Kahn et al. 2001).

Second, the forbidden+intercombination over resonance ratio $G = (z + x + y)/w$ is a good diagnostic both for the temperature and the source of ionization of the plasma. In collision-dominated plasma, the 1s2p $^1P_1$ level is preferentially populated by collisional excitation from the ground due to a high value of the effective collision strength for that transition. This results in an intense resonance line and a G ratio $\sim$1.

Furthermore, the G ratio changes as a function of T_e (Figure 6.9; bottom right), because the collisional excitation rates for the forbidden and intercombination lines have a different dependence on temperature than that for the resonance line. Conversely, in the case of a photoionized plasma, where ionization is mostly by interaction with high-energy photons of an external source, it is radiative recombination from H-like ions that populates the upper $n = 2$ levels.[3] Recombination to the ^3S or ^3P triplet levels is favored compared to the $^1P_1$ level owing to the higher statistical weights of the triplet levels. Therefore, the resonance line is lower than in the collision-dominated case. The G ratio is then in excess of 4. In between these two regimes (the so-called hybrid plasma), both recombination and collisional processes occur.

The continuum emission is often dominated by the thermal bremsstrahlung (or free–free emission; see Book 1, Chapter 6), that is, the emission of free electrons accelerated in the potential of ions. The emissivity of bremsstrahlung for a Maxwellian population of electrons is proportional to $T_e^{-1/2}\exp(-h\nu/k_B T_e)$, with ν the frequency of the emitted photon. The exponential cutoff offers an alternative way to measure the electronic temperature, particularly when the temperature $k_B T_e$ exceeds 10 keV, at which point all abundant elements except iron are fully ionized and do not contribute line emission. Other forms of continuum emissions are the two-photon and radiative recombination processes. The former occurs when an electron is in a metastable level and the direct radiative transition to the ground state is forbidden (for instance 2^1S_0 to the ground 1^1S_0 of helium-like ions), but the emission of two photons sharing the energy difference between the levels is possible. In the latter, an electron recombines with an ion and emits a photon in the process, which has an energy equal to the sum of the kinetic energy (not quantized) of the free electron and the ionization potential of the level the electron recombined to. This produces the radiative recombination continuum that can be the prevalent form of continuum emission, for instance in the case of cooling gas, where most ions are recombining.

6.4.3 White Dwarfs

WDs represent the endpoint of stellar evolution for the vast majority of stars, below a zero-age main-sequence mass $M_{\rm ZAMS}$ (i.e., "initial mass") $\lesssim 8\ M_\odot$. Following various nuclear-burning stages, their envelopes are expelled, leaving the core exposed. That object is supported by electron degeneracy pressure, which leads to sizes of the order $R_{\rm WD} \sim 10^4$ km (i.e., similar to the Earth's radius), but a mass ranging from $\sim 0.2\ M_\odot$ to about $1.4\ M_\odot$ (see Section 6.5.1). The observed X-ray properties of WDs vary greatly whether the WD is isolated or part of a binary system, and further refinement is needed for binaries where the WDs have high magnetic fields. These objects are discussed below, and for now, we focus on isolated WDs.

[3] A photoionized plasma is not a coronal plasma and is only discussed here for the sake of completeness.

WDs are inactive in terms of nuclear burning. They mostly comprise heavy elements synthesized during prior stages of stellar evolution, in particular carbon and oxygen, which sinks to the inner part of the star through gravitational settling. The remaining (small) fraction of He and H forms a thin atmosphere in the outer layer. Thanks to efficient thermal conduction, WDs are roughly constant in temperature inside and have initial temperature exceeding 10^5 K. The surface radiates as a blackbody (see Book 1, Section 2.2.6); the small emitting surface (compared to main-sequence stars) leads to very long cooling times through this process alone, taking several gigayears to bring the WD to about 10^3 K.

The first thermally emitting WD detected in X-rays was Sirius B (Figure 6.10)—the WD companion to α Canis Majoris (Sirius A, the brightest star in the sky) known since 1862, and also the WD nearest to us at only 2.5 pc. Emission from the Sirius A hot corona or accretion onto Sirius B was ruled out (Mewe et al. 1975) and an origin in thermal emission from the WD atmosphere was proposed instead (Shipman 1976).

Indeed, if the surface is above several 10^4 K to a few 10^5 K, the blue end to the peak of the blackbody radiation is falling in the extreme UV/soft X-ray band (Section 5.3.3). With the thin hydrogen layer fully ionized and thus transparent to X-rays, the luminosities expected are thus

$$L = 4\pi R_{\rm WD}^2 \sigma T^4 \approx 7 \times 10^{34} \left(\frac{R_{\rm WD}}{10^4 \text{ km}} \right)^2 \left(\frac{T}{10^5 \text{ K}} \right)^4 \text{erg s}^{-1}, \tag{6.13}$$

which are only accessible with state-of-the-art instrumentation in a limited Galactic volume, further reduced by interstellar extinction to which very soft X-rays are extremely sensitive.

Large-area surveys, even when sensitive to very soft X-rays down to 0.1 keV (120 Å), have found a dearth of thermally emitting isolated WDs compared to the known local space density of such object (Fleming et al. 1996). The likely explanation is that a small quantity of He and heavier elements, which are not fully ionized, are lifted by radiation pressure in the outer layers where they increase the X-ray opacity. The relative lack of X-ray-emitting WDs is more pronounced for stars with effective temperature above 5×10^4 K, in agreement with such a scenario (Barstow et al. 1993) as radiative lifting becomes more prominent at higher temperatures. High-resolution grating spectra have been obtained to detect directly absorption by heavy elements (Vennes & Dupuis 2002) in some WD atmospheres, but such measurements remain rare and observed metal abundances are difficult to reconcile with model predictions (Adamczak et al. 2012).

6.4.3.1 Cataclysmic Variables and Novae

Stellar evolution predicts that some binaries will at some point contain a WD in close orbit (period of a few hours) with a nuclear-burning companion. This configuration may enable mass transfer to the WD, which powers X-ray emission in a variety of ways. Such binary is known as a cataclysmic variable (CV). Further classification arises depending on the magnetic field strength of the WD. In the

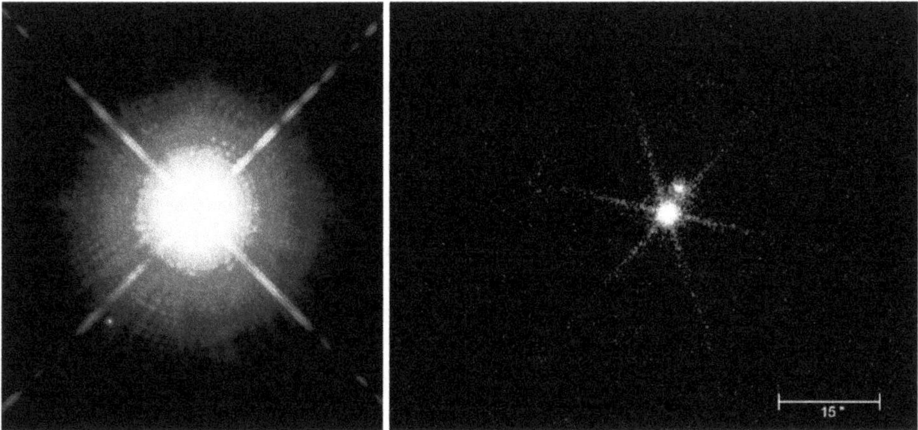

Figure 6.10. A comparison of the Sirius binary in optical and X-ray light. The left image shows the optical emission dominated by Sirius A, the brightest star in the sky (V magnitude of −1.46) outshining its WD companion Sirius B ($m_V = 8.41$, i.e., about 10^4 times fainter), seen as a tiny dot below and to the left, at about 5 arcsec. The right-hand image, obtained with Chandra, shows the reverse situation as Sirius B is far brighter in the X-ray band. Spikes in both images are image artifacts due to the extreme brightness of the components in their respective bands. Image credits: (left) NASA/ESA/H. Bond (STScI) and M. Barstow (University of Leicester), (right) NASA/SAO/CXC.

low-magnetic case, matter provided by the companion settles in an accretion disk[4] that reaches down to the WD. The temperature of the disk increases inward from a few 10^3 K to typically 4×10^4 K, radiating copiously in the UV band. The innermost accretion disk, in roughly Keplerian orbit, connects with the WD in a "boundary layer," and the deceleration provides additional heating. Depending on the flow rate of matter in the disk, the boundary layer is either optically thin or thick (Patterson & Raymond 1985a, 1985b). In the former case, at low-mass-flow rate (CV in "quiescence"), the temperature above the boundary layer is hot, $kT > 10^8$ keV, and the resulting thermal bremsstrahlung (Book 1, Section 6.1.1) extends to hard X-rays. If the mass-flow rate increases dramatically (CV in "outburst"), the boundary layer becomes optically thick and radiates as a blackbody at a temperature of a few 10^5 K, emitting very soft X-rays.

Accreted hydrogen-rich matter piles up on the WD. The temperature and density of that envelope grow over time, eventually becoming high enough to ignite hydrogen burning in the CNO cycle at the base of the envelope. If this occurs under electron-degenerate conditions, the matter does not expand in reaction to the increased heat from revived nuclear burning (as the equation of state becomes independent of temperature). The temperature thus keeps increasing, leading to an explosive, thermonuclear runaway in the accreted envelope. The mass of the envelope at ignition $M_{\rm env}$ ranges from 10^{-6} to 10^{-3} $M_\odot$, depending on the temperature and gravity of the WD (Yaron et al. 2005). It is inversely correlated with the WD mass (Nomoto 1982). The total energy available is thus $0.007 M_{\rm env}\, c^2 \approx 10^{47}(M_{\rm env}/10^{-5}\ M_\odot)$ erg.

[4] Accretion processes are presented in more detail in Section 6.5.3 in the context of X-ray binaries.

The nuclear flash ejects a fraction of the envelope at high velocity, in an optically thick hot photosphere that dramatically enhances the optical appearance of the binary, by up to 12 magnitudes. Such bright events known as novae (as in "new" stars) have been recorded for at least 2000 years by, e.g., Chinese astronomers (Stephenson 1976), but undoubtedly noticed by other civilizations even before.

As the photosphere recedes inward due to expansion of the envelope, deeper layers become visible and the source fades out in the optical, until the WD reappears. A fraction of the hot, unejected envelope may proceed with steady H burning, heating the surface to above 10^5 K. Thus, a postnova WD may appear as a very soft X-ray source (often called a supersoft X-ray source, or SSS) until H burning stops.

Pure blackbody emission models perform poorly to reproduce the observed spectra (Krautter et al. 1996), instead requiring detailed atmosphere models, as absorption lines have been directly observed on top of the soft thermal continuum in high-resolution spectra (Ness et al. 2003). Broad emission lines (from carbon to silicon) have been detected as well (Ness et al. 2005), even after the disappearance of the soft thermal continuum following the switching off of shell H burning. This indicates an origin in the collisionally excited (see Section 6.4.2) expanding shell ejected in the nova outburst.

After an outburst, accretion can continue, starting another cycle, until eventually the next outburst. The recurrence time may exceed 10^6 years or be as short as a few years; the smallest recurrence time observed so far is about one year, for a system in our neighbor M31 (a.k.a Andromeda) galaxy (Henze et al. 2014, 2018).

Perhaps counterintuitively, WDs accreting at a higher rate may be steadily burning hydrogen without degenerate flashes. They appear as a constant SSS. This only occurs on massive WDs, in a relatively narrow range of mass accretion rate and companion mass (van den Heuvel 1992). The lifetime of SSSs is also relatively short, a few million years, and they are thus rare objects. They, nevertheless, gain interests as one of the potential progenitors of Type Ia supernovae (Section 6.5.1), albeit with outstanding observational challenges (Woods & Gilfanov 2013; Kuuttila et al. 2019). Novae are also a potential type of progenitors, but considering the uncertain amount of ejected mass in the nova outburst, it remains unclear whether the WD will gain net mass over time (Yaron et al. 2005) and thus approach the Chandrasekhar limit (1.4 $M_{\odot}$ or 2.765 $\times 10^{30}$ kg).

If the WD in a CV is strongly magnetized, the magnetic field can take control of the accretion flow at a radius that depends on field strength and mass accretion rate (see details in Equations (6.39) and (6.40)). If that radius exceeds the binary separation, no disk is formed at all, and matter follows the magnetic field line from the companion to the pole(s) of the WD, and such systems are called "polars" (Krzeminski & Serkowski 1977). If that radius is between the WD and the companion, a small accretion disk truncated at that radius is formed, hence their name "intermediate polar." The high freefall velocity attained by accreted material and the shock produced as it collides on the WD lead to high temperatures and thus hard X-ray spectra. Heating of the surface by hard X-rays provide additional emission in soft X-rays. Both components can be modulated at the orbital period (if the disk of an intermediate polar eclipses the magnetic poles) or at the rotation period of the WD (both periods are locked and equal in the high magnetic field case).

6.5 X-Ray Sources from the Stellar Graveyard

In this section, we will discover and discuss the large variety of X-ray sources produced at the endpoint of stellar evolution, primarily that of massive stars. Such stars die in gigantic explosions known as SNe, leaving only a compact remnant behind in the form of an NS or a BH. The imprints of SNe on the interstellar medium are called SNRs (Section 6.5.1; also see Sections 2.7.2, 5.3.2 and 7.5.1), and remain visible in X-rays long after the explosion. Then, the NSs left are visible through the thermal emission of their hot surfaces, or by the nebula of energetic electrons accelerated in their extremely high magnetic fields (Section 6.5.2; also see Section 7.5.2). Finally, the exchange of matter between the NS or BH and a stellar binary companion results in bright accretion-powered X-ray binaries (Section 6.5.3; also see Sections 2.7.2 and 7.5.4).

6.5.1 Supernova Remnants

6.5.1.1 Supernovae: The Swan Song of Dying Stars

While the metallicity Z, rotation, and presence of a binary companion also play important roles, at first order, the evolution and final fate of a star depend most critically on its initial mass M_{ZAMS}. This sets the central temperature T and density ρ and thus the nuclear reactions that are able to occur and power the star. All stars above 0.08 $M_{\odot}$ spend most of their life (the main sequence) in the core H-burning phase, and those above 0.5 $M_{\odot}$ then go through a core He-burning phase. Stars of initial masses between 0.5 and 7 $M_{\odot}$ end up in white dwarfs composed mostly of carbon and oxygen, and supported by electron degeneracy pressure. In such objects, the mass–radius relationship implies that more massive WDs are more compact and thus have higher densities. The equation of state is polytropic, with $P \propto \rho^{\eta}$. The index $\eta = 5/3$ for nonrelativistic electrons and tends to 4/3 for ultrarelativistic electrons, which is the case as the density increases. Consequently, there is a maximum mass that can be supported by electron degeneracy pressure, known as the Chandrasekhar mass ($M_{\mathrm{Ch}} \approx 1.4\ M_{\odot}$). If the density increases suddenly, for instance in the merger of two WDs, or gradually, in the accretion of matter on the WD from a stellar companion, the ignition of carbon burning may occur. In those degenerate conditions, the heat generated by the nuclear reaction is not well dispersed by the expansion of the star, as the pressure does not depend on temperature as in a normal star, but instead increases the temperature locally. This leads to a sudden rise of the C-burning reaction rate, as it scales sharply with temperature (up to T^{40}), and thus to a further increase of temperature. The runaway process results in the complete nuclear incineration of the star, burning its material all the way to ^{56}Ni, which later radioactively decays to iron. The whole carbon burning of a Chandrasekhar mass releases about 2×10^{51}

erg, 10, 000 times more than the typical energy of nova outbursts (Section 6.4.3). That event leads to the complete disruption of the star and is known as a thermonuclear SN, or, observationally, a Type Ia SN. These explosions are the main source of iron (Fe) in the universe.

The most massive and thus much rarer stars ($M_{ZAMS} \gtrsim 8\ M_{\odot}$) reach the central conditions, enabling the steady burning of carbon ($T > 7 \times 10^8$ K) and subsequently more massive nuclei (O, Ne, Si, etc.). The successive burning stages last shorter and shorter. For instance, after a H-burning phase of a few million years, the C-burning phase lasts only a few hundred years while the Si-burning phase only lasts a day. After this process, the core of the massive stars ends up with an "onion-layer" structure of heavier elements, from H and He in the outer layer, through C, O, Ne, and finally Si and Fe in the center. Because ^{56}Fe has the highest binding energy per nucleon, iron fusion would take energy away rather than produce it. Thus, a core of iron builds up in the center of the star.

As the central temperature continues to rises, thermal photons now have sufficient energy to disintegrate iron nuclei into helium nuclei and then into protons and neutrons. This consumes almost 0.5 GeV per iron nucleus or about 10^{52} erg for the whole iron core. Furthermore, at such high densities the electron-capture reaction $e^{-}+p \rightarrow n + \nu_e$ is more common than the inverse reaction, leading to the neutronization of the matter. This takes away more energy from the star, as most neutrinos escape and deplete the core of the electrons, which were providing most of the core's gravitational support via degeneracy pressure. The neutronization and iron photodisintegration result in the loss of thermal pressure and thus the rapid gravitational collapse of the core. The matter of the collapsing core is converted mostly to neutrons in electron-capture reactions, creating a (proto-)neutron star that is now supported by the neutron degeneracy pressure. The energy of that gravitational collapse is $G\, M_{NS}^2/R_{NS}$, with M_{NS} and R_{NS} the mass and radius of the NS, respectively, amounting to a few 10^{53} erg. Although most of that energy is carried away by neutrinos, a small fraction is harnessed by the star and exceeds its gravitational binding energy ($\sim 10^{51}$ erg). That again disrupts the star, producing a so-called core-collapse supernova (CCSN), which marks the end of the life of massive stars, and the formation of an NS or even a BH if the mass of the proto-NS and fallback material exceed the maximum mass of NS (about 3 $M_{\odot}$). In the explosion, a shock wave propagate outwards and nuclear burning occurs in the very hot region behind the shock. In this explosive nucleosynthesis, large quantities of elements between a mass number $A = 26$ (i.e., aluminum) and the iron-group elements (Fe, Co, Ni, and Zn) are produced. Even heavier nuclei are produced by the rapid capture of neutrons.

6.5.1.2 Life after Death: Evolutionary Phases of SNRs

In both kinds of SNe, the explosion energy E_{SN} is tapped from the gravitational binding energy of the star and is first converted into the kinetic energy of the stellar ejecta of mass (M_{ej}). An estimate of the ejecta velocity V_{ej} can be made by equating E_{SN} to $(M_{ej}V_{ej}^2)/2$:

$$V_{\rm ej} \approx 10^4 \left(\frac{E_{\rm SN}}{10^{51}\ {\rm erg}}\right)^{1/2} \left(\frac{M_{\rm ej}}{M_\odot}\right)^{-1/2} \ {\rm km\ s^{-1}}. \tag{6.14}$$

Such high velocities are indeed measured in the optical spectra of the early phases of supernovae.

As the ejecta move at velocities highly in excess of the sound speed in the ISM (about 10 km s^{-1}), a shock wave develops and propagates ahead of the ejecta, often called the blast wave or forward shock (FS), which sweeps up, compresses, and heats the ambient medium (see Figure 7.11). The shocked ambient medium pushes back on the ejecta, driving a reverse shock (RS) backwards into the ejecta (in the observer's frame, however, the reverse shock radius is still increasing). The shocked ejecta are separated from the shocked ambient medium by a contact discontinuity. From the center (close to the SN) outwards, there is therefore a stratification of unshocked stellar ejecta, shocked ejecta, shocked ISM, and unshocked ISM. This whole structure forms an SNR, an object visible long after (up to 150,000 years) the supernova event, which allows both the explosion itself and the ISM of the host galaxy to be studied.

At a shock front, the interstellar gas is highly perturbed, and physical values (density, velocity, and pressure) have physical discontinuities at the interface between the shocked and unshocked gas. The properties of the gas just behind the shock can be obtained by the fluid equations applied in the shock frame (traveling at velocity v_s in the ISM frame). We note the density, velocity, and pressure ahead of the shock ("upstream") as ρ_1, v_1, p_1, respectively, and those behind the shock ("downstream") as ρ_2, v_2, p_2. We further assume that the pressure is only the thermal pressure of the gas. We can then write the jump conditions:

$$\rho_1 v_1 - \rho_2 v_2 = 0 \tag{6.15}$$

$$\left(\rho_1 v_1^2 + P_1\right) - \left(\rho_2 v_2^2 + P_2\right) = 0 \tag{6.16}$$

$$\left[v_1\left(\frac{1}{2}\rho_1 v_1^2 + \frac{\gamma}{\gamma - 1}P_1\right)\right] - \left[v_2\left(\frac{1}{2}\rho_2 v_2^2 + \frac{\gamma}{\gamma - 1}P_2\right)\right] = 0, \tag{6.17}$$

where γ is the ratio of specific heats ($\gamma = 5/3$ in an ideal monoatomic gas or nonrelativistic fully ionized plasma). When the Mach number[5] is sufficiently high ($v_s \gg c_{\rm sound}$), one can use the strong shock approximation. The preshock pressure is ignored ($P_1 = 0$). The upstream velocity is simply $\mathbf{v}_1 = -\mathbf{v}_s$ and one replaces ρ_1 with the ambient mass density ρ_0. Equations (6.15) to (6.17) are recasted to form the Rankine–Hugoniot relations[6]:

[5] The Mach number (M or Ma) is a dimensionless quantity in fluid dynamics representing the ratio of flow velocity past a boundary to the local speed of sound.

[6] The Rankine–Hugoniot relations describe the relationship between the states on both sides of a shock wave or a combustion wave (deflagration or detonation) in a one-dimensional flow in fluids or a one-dimensional deformation in solids. They are named in recognition of the work carried out by Scottish engineer and physicist William John Macquorn Rankine (1820–1872) and French engineer Pierre Henri Hugoniot (1851–1887).

$$v_2 = -\frac{\gamma - 1}{\gamma + 1} v_s \overset{\gamma=5/3}{=} -0.25\, v_s \tag{6.18}$$

$$\rho_2 = \frac{\gamma + 1}{\gamma - 1} \rho_0 \overset{\gamma=5/3}{=} 4\, \rho_0 \tag{6.19}$$

$$P_2 = -\frac{2}{\gamma + 1} \rho_0 v_s^2 \overset{\gamma=5/3}{=} 0.75\, \rho_0 v_s^2. \tag{6.20}$$

The temperature in the shocked region is obtained from the ideal gas law $P = (\rho/\mu)kT$ with μ the mean particle mass. In a fully ionized plasma with He/H = 0.1 (by number) there will be one proton (mass m_p), 0.1 helium nucleus (mass $4 \times m_p$), and $1 + 0.1 \times 2$ electrons (negligible mass). Therefore, μ is $(1.4/2.3)m_p \approx 0.61\, m_p$. Combining Equations (6.19) and (6.20) yields

$$kT_s = \frac{2(\gamma - 1)}{(\gamma+1)^2} \mu\, v_s^2 \overset{\gamma=5/3}{=} \frac{3}{16} \mu\, v_s^2. \tag{6.21}$$

Initially, the mass $M_{\rm ISM}$ of the ISM swept up by the FS is much smaller than $M_{\rm ej}$. This early stage is called the ejecta-dominated phase. To estimate the duration of this phase, one can compute the time $t_{\rm ED}$ when the swept-up mass is equal to the ejecta mass, assuming (at the first order) a constant velocity $V_{\rm ej}$ for the ejecta. This gives

$$t_{\rm ED} = V_{\rm ej}^{-1} \left(\frac{3\, M_{\rm ej}}{4\pi\rho_0} \right)^{1/3}, \tag{6.22}$$

which for most parameters ($V_{\rm ej} \sim 10,\,000$–$15{,}000$ km s^{-1}, $M_{\rm ej} \sim 1$–$30\, M_\odot$, $\rho_0 \sim 0.1$–1 g cm^{-3}) is less than 1000 years, i.e., only the youngest SNRs are in the ejecta-dominated phase.

Once the swept-up mass becomes larger than the mass of the ejecta, the outer shock decelerates. Still, the radiative losses are negligible, so that the expansion remains adiabatic. The behavior of the remnant can be described by the Sedov (or Sedov–Taylor) self-similar solution. This is the Sedov phase: The total energy of the supernova $E_{\rm SN}$ is shared (and constant) between the kinetic energy and the internal energy ($P_2\, V_2$) of the shocked ISM:

$$E_{\rm SN} = E_{\rm kin} + E_{\rm internal} = \frac{1}{2} M_{\rm ISM}\, v_2^2 + P_2\, V_2. \tag{6.23}$$

Using the Rankine–Hugoniot relations, one can show that the shock radius R_s will only depends on $E_{\rm SN}$ and ρ_0 as

$$R_s \propto E_{\rm SN}^{1/5}\, \rho_0^{-1/5}\, t^{2/5} \tag{6.24}$$

with the time t since explosion. An equivalent solution is to write the shock velocity $v_s = dR_s/dt$ as

$$v_s = \frac{2}{5}\frac{R_s}{t}. \tag{6.25}$$

This velocity can be calculated given the postshock temperature (Equation (6.21)), itself measured from the X-ray spectrum. One then obtains the dynamical age t_{Sedov} of an SNR with size R_s:

$$t_{\mathrm{Sedov}} = \frac{2}{5}\sqrt{\frac{3\,\mu}{16\,kT_s}}\,R_s \approx 400\left(\frac{R_s}{1\ \mathrm{pc}}\right)\left(\frac{kT_s}{1\ \mathrm{keV}}\right)^{-1/2}\ \mathrm{yr}. \tag{6.26}$$

As more and more ambient medium is engulfed by the expanding blast wave, the shock velocity, and therefore postshock temperature, decreases. This cooler material loses more energy, because the recombining material is able to efficiently radiate away energy in the UV and optical lines (Cox & Daltabuit 1971): Between 10^5 K and 10^7 K, the ISM cooling function scales as $n^2T^{-1/2}$. This last phase is called the radiative phase and starts when shocks have velocities slower than about 250 km s^{-1} ($T_s \lesssim 10^6$ K), at which point they become too cold to emit X-rays.

6.5.1.3 X-Ray Emission of SNRs

SNRs appear as extended objects with large differences in morphology and spectral features (Figure 6.11). They may exhibit both thermal and nonthermal X-ray emission. The former is produced by collisionally excited plasma heated to X-ray-emitting temperatures by the forward and reverse shocks. This emission is dominated by high-excitation ion lines plus some forms of continua, as presented in Section 6.4.2. The main differences compared to stellar corona are the lowest densities involved and, in some cases, the elemental abundances of the plasma. The typical ISM density within which SNRs expand is 1 to 100 cm^{-3}, compared to the 10^8–10^{12} cm^{-3} of the stellar corona, and can be down to 10^{-3}–10^{-2} cm^{-3} in cavities blown by the winds of the supernova progenitor or other neighboring massive stars. As both ionization and recombination processes are proportional to n_e, collisions with electrons are rare, and it takes a long time for the plasma to reach ionization equilibrium at its given electronic temperature. Consequently, young SNRs are often underionized with respect to T_e. This skews the spectrum toward lower energies as less-ionized species are more abundant than at collisional–ionization equilibrium (Smith & Hughes 2010).

When the shocked SN ejecta dominate the X-ray emission, the emission-line spectrum can be used to measure the elemental abundance of material produced and released by SNe. Thus, the type of explosion (thermonuclear or core collapse) or the mass of the progenitor (in the latter case) may sometimes be inferred from X-ray observations long after the explosion. However, this ejecta-dominated phase is relatively short (see Equation (6.22)). Thermal emission from older SNRs entering the Sedov phase is increasingly dominated by shocked ISM. This can be used to probe the gas-phase abundances of nearby galaxies (Maggi et al. 2016, 2019).

In some cases nonthermal emission is also present, mostly from the synchrotron radiation of relativistic electrons gyrating in magnetic fields. At an energy E, an

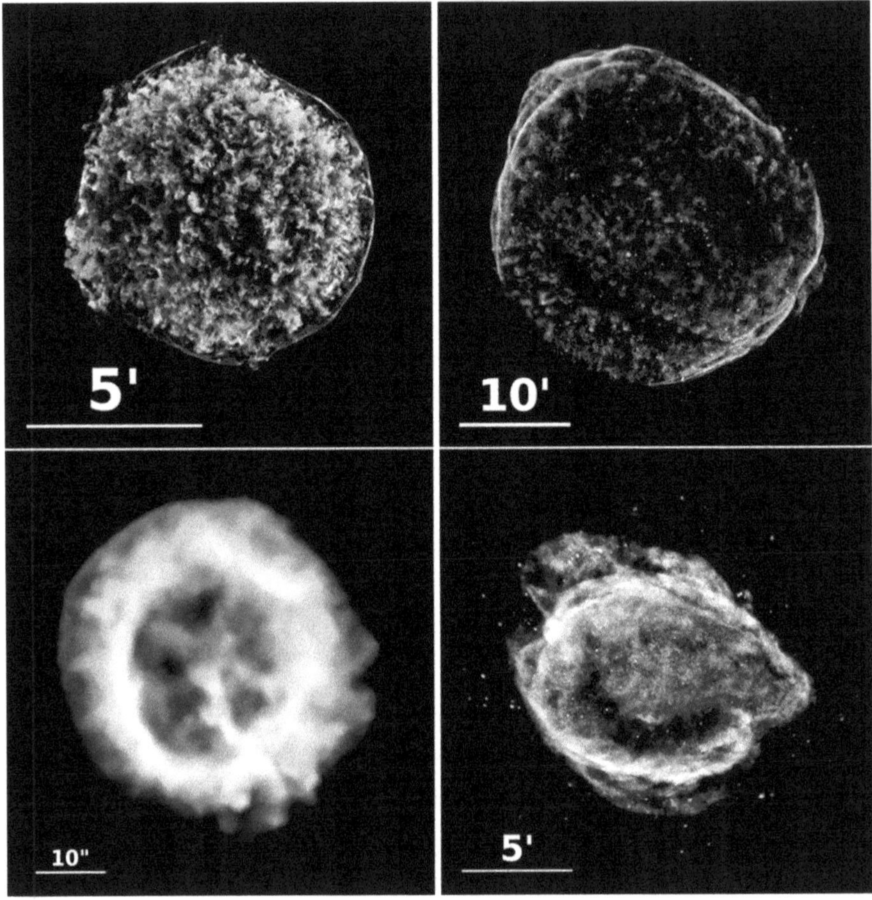

Figure 6.11. X-ray images of supernova remnants obtained with the Chandra X-ray observatory. Clockwise from the top-left corner are Tycho's SNR, SN 1006, SNR G299.2–2.9, and 1E 0102–72.3. Low-, medium-, and higher-energy X-rays are colored red, green, and blue respectively. Synchrotron emission is evident as thin bluish rims from Tycho's SNR and SN 1006. The physical scale varies across pictures, as indicated by the white bars. Image credit: Chandra X-ray Observatory https://www.chandra.si.edu/.

electron radiates photons at a characteristic frequency $\nu_{\rm ch}$ with energy (Ginzburg & Syrovatskii 1965):

$$h\nu_{\rm ch} = 0.19\left(\frac{B_\perp}{100\ \mu{\rm G}}\right)\left(\frac{E}{10\ {\rm TeV}}\right)^2\ {\rm keV}. \tag{6.27}$$

Most SNRs are observed at radio wavelengths to be synchrotron sources (Section 2.7.2), indicating the presence of GeV-energy electrons. When X-ray instruments with sufficient spatial and spectral resolution became available, some SNR shells have revealed synchrotron spectra extending in the X-ray band (Koyama et al. 1995), requiring electrons with energies upwards of 100 TeV. These electrons have a power-law distribution of energy owing to the acceleration process they underwent

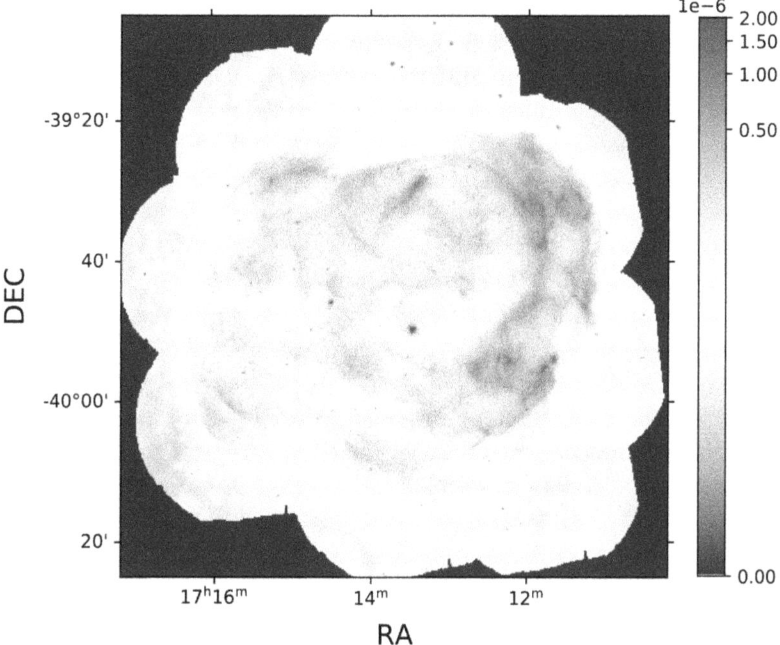

Figure 6.12. X-ray image of synchrotron-dominated SNR RX J1713.7−3946 obtained with a mosaic of XMM-Newton observations. The color bar indicates the intensity in units of photon cm^{-2} s^{-1} $pixel^{-1}$ (each pixel being 4″ across) in the 1–10 keV band. The NS produced by the supernova explosion is visible as the bright point source near the center. Image courtesy of Dr. Fabio Acero.

at the SNR shock wave, and therefore the emitted photon spectrum is a power law as well. Those electrons are accelerated at the fast shocks where the magnetic field is compressed, and possibly amplified by streaming instabilities (Caprioli & Spitkovsky 2014).

Some SNRs are completely dominated by nonthermal synchrotron emission over their whole extent (e.g., RX J1713.7−3946 in Figures 6.12 and 7.9, Acero et al. 2009, 2017; Sano et al. 2020; or RX J0 852.0−4622, Slane et al. 2001; Maxted & Filipović 2018). SNR G1.9+0.3 is also dominated by X-ray synchrotron (Reynolds et al. 2008), although faint thermal emission is seen as well (Borkowski et al. 2010.) It is still expanding at a tremendous velocity of about 14,000 km s^{-1} and is the youngest SNR in the Milky Way: It is believed to be the remnant of a Type Ia supernova that exploded around 1900 CE (Carlton et al. 2011), at a time when many telescopic observations were routinely made. However, it occurred close to the Galactic Center and was likely masked by the large interstellar extinction of that region. After a century long-wavelength (radio) and short-wavelength (X-rays, γ-rays) observations are able to detect the aftermath of the supernova (Luken et al. 2020; Brose et al. 2019).

In other SNRs, X-ray synchrotron is confined into very thin rims (1 arcsec to 3 arcsec) tracing the outer shocks (e.g., Tycho's SNR and SN 1006, Figure 6.11). In those cases, the width of the synchrotron rims can be used as a proxy to measure the magnetic field strength: At first order, electrons move with the shocked plasma (they are advected) from the shock front at a relative velocity v_s/r, with r the shock compression, taken to be 4 in this discussion and in Equation (6.18). Therefore, they do not gain energy anymore and only lose it through synchrotron radiation losses,[7] after a typical timescale (Helder et al. 2012)

$$\tau_{\rm loss} = 125\left(\frac{B_\perp}{100\ \mu{\rm G}}\right)^{-2}\left(\frac{E}{10\ {\rm TeV}}\right)^{-1}\ {\rm yr}, \tag{6.28}$$

giving the corresponding advection length

$$l_{\rm adv} = v_s r^{-1}\tau_{\rm loss} = 4.95\times 10^{17}\frac{v_s}{5000\ {\rm kms}^{-1}}\left(\frac{B_\perp}{100\ \mu{\rm G}}\right)^{-2}\left(\frac{E}{10\ {\rm TeV}}\right)^{-1}\ {\rm cm}. \tag{6.29}$$

Equating the observed width of the filament (corrected for projection effect and instrumental broadening) at a given characteristic energy (Equation (6.27)) to the advection length, the estimated magnetic field is thus

$$B = 365\left(\frac{v_s}{5000\ {\rm kms}^{-1}}\right)^{2/3}\left(\frac{h\nu_{\rm ch}}{1\ \ {\rm keV}}\right)^{-1/3}\left(\frac{l_{\rm adv}}{0.01\ {\rm pc}}\right)^{-2/3}\ \mu{\rm G}. \tag{6.30}$$

A more complex (and accurate) description should use the diffusion length $l_{\rm diff}$ of the accelerated electrons, but in the physical conditions found in those young SNRs the two methods give equivalent results (Ballet 2006).

6.5.2 Neutron Stars and Pulsar Wind Nebulae

The NS produced in the CCSN of a massive star is a stable compact remnant that survives the supernova event and the SNR phase. Barring interaction with a binary companion (Section 6.5.3), which may even lead to a dramatic merger (Chapter 9), this is the end of the road for the star as far as stellar evolution is concerned. However, an NS still produces significant emission, even at high energy. The thermal cooling of the hot surface (in the first thousand years, the surface temperature is above 10^6 K) provides a small contribution, but the main power sources of high-energy emission are the intense magnetic field and the fast rotation of the star. Both parameters reach extreme values in NSs because of the conservation of magnetic flux BR^2 and of angular momentum $MR^2\Omega$ during core collapse, with B and Ω the surface magnetic field and rotation rate, respectively. As NSs have tiny radii (of the order 10 km) thanks to the dense packing of neutrons, a rotation period smaller than 1 s can be obtained, and the magnetic field can easily exceed 10^{12} G. For comparison, the strongest continuous magnetic field produced in laboratory is $<10^7$ G.

Observationally, NSs were discovered first in radio (see Chapter 1 and Section 2.7.2) in the late 1960s as fast periodic signals, hence their name "pulsar" for "pulsating stars." This brilliantly confirmed the existence of such object, first

[7] Formally, they also lose part of their energy via inverse-Compton (iC) scattering off low-energy photons, mainly from the cosmic microwave background (CMB), in proportion to the ratio of photon energy density to magnetic energy density.

predicted to form in supernovae by Fritz Zwicky and Walter Baade three decades earlier (Baade & Zwicky 1934; Zwicky 1939). Their stellar masses have been measured in a few orbiting systems and most accurately in double pulsar binaries (Burgay et al. 2003), and indeed $M_{\rm NS}$ clusters around 1.4 $M_\odot$ (Zhang et al. 2011). Likewise, the strong magnetization of NSs was first confirmed indirectly by the slowing down of their rotation, attributed to magnetic dipole radiation (Gunn & Ostriker 1969). Later, X-ray detection of electron cyclotron resonant scattering features (CRSFs; Truemper et al. 1978) allowed the first direct measurement of magnetic field strength. CRSFs are produced because electrons gyrating in strong magnetic fields have discrete energy levels (called Landau levels). This modifies the cross section for electron–photon scattering, with only photons having energies close to the energy difference E_c between Landau levels able to be scattered, leading to absorption lines around that energy. Expressed as function of the critical field $B_{\rm crit} = \frac{m_e^2 c^3}{e\hbar} = 44 \times 10^{12}$ G, where cyclotron energy equals the electron rest-mass energy, E_c is

$$E_c = B\frac{e\hbar}{m_e c} = 11.6\,\frac{B}{10^{12}\ \mathrm{G}}\ \mathrm{keV}. \tag{6.31}$$

Because CRSFs are produced in the strong gravitational field close to the NS, the lines are observed at a lower energy $E_c^{\rm obs} = E_c/(1+z)$, with $z = 1/\sqrt{2GM/rc^2} - 1$ the gravitational redshift. At or just above the NS surface, z is between 0.1 and 0.3. CRSFs have been observed between 9 keV and 80 keV in a few dozen NSs (Revnivtsev & Mereghetti 2015; Maitra 2017), requiring magnetic fields of the order $10^{12} - 10^{13}$ G.

High-energy emission, in particular X-rays, are produced in isolated neutron stars[8] via three mechanisms. First, the rotating magnetosphere induces strong currents that lift electrons and ions off the NS surface. Thus, the NS surroundings is filled with a conducting plasma. Charges of opposite sign may be separated in regions of the magnetosphere called “gaps” (Cheng 2009), leading to a net electric field that accelerates particles (mostly electrons and positrons), which radiate from optical to γ-rays through several processes, e.g., curvature radiation and iC scattering. This magnetospheric emission is often pulsed, modulated at the rotation frequency of the NS. Second, the NS is born hot and cools down, first slowly by neutrino processes, before the temperature T falls below $\sim 10^6$ K, where the much more rapid photon cooling starts to dominate. Thus, for about 10^6 yr, an NS surface can radiate as a blackbody, with the peak of its spectrum in the very soft X-rays (0.1 keV to 0.5 keV). Very soft X-ray sources are readily absorbed by neutral gas and thus cannot be detected beyond a few hundred parsecs. Consequently, only a small sample of isolated cooling NSs is known (Haberl 2007). They are however critically important: If the distance is measured, for instance by the parallax method (Walter 2001), then the radius of the NS is deduced from the Stefan–Boltzmann law. Smooth X-ray pulsations are detected in several of these sources, ascribed to an inhomogeneous surface temperature distribution that betrays a third kind of X-ray emission: Some

[8] See Section 6.5.3 for the case of accreting NSs.

accelerated particles stream back and hit the NS surface, preferentially at the magnetic poles. It provides an additional source of heating that create hot spots on the NS.

In addition to their point-like emission, many NSs are associated with extended sources around them. Pulsars drive relativistic winds that drain their rotational energy. As the wind expands, it is slowed down and confined by the ambient pressure in a wind termination shock, which reaccelerates leptons to ultrarelativistic energies. These electrons and positrons, which are in the 10^{-5} G to 10^{-3} G magnetic field far from the NS, radiate synchrotron emission from radio to X-rays with a typical power-law spectrum, i.e., with a number of photon $N(E) \propto E^{-\alpha}$.

Most rapidly rotating magnetized NSs are then surrounded by a so-called PWN. The most famous example is the Crab Nebula (a.k.a. M 1), powered by a 33 ms pulsar, itself the product of a well-documented supernova from 1054 CE, particularly in the Chinese records (Stephenson & Green 2003; Payne et al. 2019). Only the most energetic electrons, of energy $E \gtrsim 10$ TeV, emit synchrotron in the X-ray range (Equation (6.27)). But because they radiate X-ray photons, each carrying a large amount of energy, they suffer radiative losses on timescales that scale with $E^{-1}B^{-2}$, which can become shorter than the time it takes for electrons to diffuse outwards. Therefore, electron populations further from the pulsar become more and more depleted of their most energetic members, and the observed X-ray spectrum steepens (α increases) outward. The overall morphology of PWNe is thus more compact in X-rays than at radio frequencies, which are emitted by lower-energy electrons of much longer lifetimes. The small-scale appearance in X-rays reveal compact filamentary and knotty structures, which are regions where tangled magnetic field in the nebula is compressed and amplified by the expanding nebula, enhancing the synchrotron emissivity. Spatial inhomogeneities are further enhanced by the Doppler boosting of the emission of relativistic electrons moving toward the observer. Spatially resolved polarimetric X-ray observations have the potential to reveal the magnetic field structures in PWNe and thus probe the interplay between the pulsar wind, field, and ambient medium. The instrumental challenge of that approach is mentioned in Section 6.2.5.

The violent origin of an NS in a supernova is also expected to impart a large "kick" velocity (of several hundreds of km s^{-1}; Hobbs et al. 2005) on it via two possible phenomena. In one case, hydrodynamical instabilities cause asymmetric ejection of stellar ejecta (Wongwathanarat et al. 2013; Janka 2017), launching the NS in the opposite direction. In the other case, anisotropic neutrino emission during proto-NS formation carries away a large fraction of its gravitational binding energy on one side (Socrates et al. 2005; Fryer & Kusenko 2006), while both the stellar ejecta and the NS are accelerated in the other direction. Both scenarios might explain the several hundred kilometer per second recoil velocities that have been measured from several NS by proper motion measurement (e.g., Holland-Ashford et al. 2017; Mayer et al. 2020). Indeed, the angular displacement on the sky $\delta\theta$ of an NS at a distance D over a period t is

$$\delta\theta = 0.11\left(\frac{v_{\perp}}{500\ \mathrm{km\ s^{-1}}}\right)\left(\frac{t}{1\ \mathrm{yr}}\right)\left(\frac{D}{1\ \mathrm{kpc}}\right)^{-1}\ \mathrm{arcsecond} \tag{6.32}$$

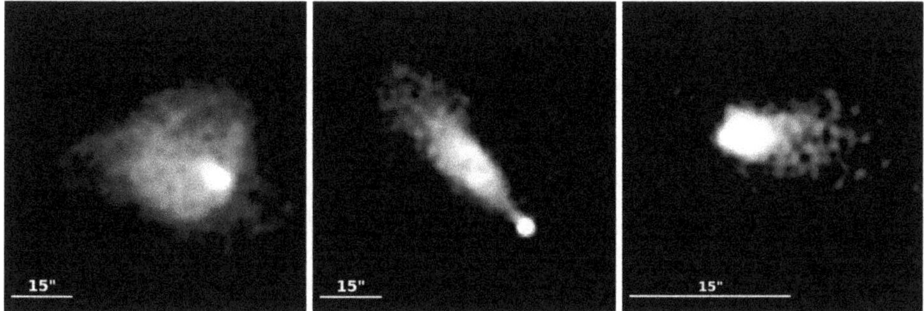

Figure 6.13. X-ray images obtained with Chandra of a supersonic PWN. From left to right: PSR B1951+32 inside the supernova remnant CTB 80, PSR J1101−6101 (a.k.a. the Lighthouse PWN; Pavan et al. 2014; Pavan & Pühlhofer 2016), and PSR J1747−2958 (a.k.a. the Mouse PWN). A scale of 15″ is shown by the white bars. Images have been smoothed with a Gaussian kernel size of 3 to 5 pixels (1.5″ to 2″) and are displayed with a square-root stretch.

with $v_\perp$ the velocity of the NS projected on the sky. These minute proper motions can nevertheless be detected thanks to the high angular resolutions of some observatories, like Chandra, which has operated for more than 20 years, offering a long baseline. Those X-ray measurements are especially valuable for NSs not detected in radio, as they offer the only way of directly measuring (projected) kick velocities.

As the expansion of SNRs slows down over time (Section 6.5.1), NSs launched at high velocities eventually plow through an unshocked ISM, a medium where the sound speed is much less (about 10 km s^{-1}). The NS motion then becomes highly supersonic. If the NS is still energetic enough to be surrounded by a relativistic wind, the ram pressure $P_{\rm ram} = \rho v^2$ (with ρ the density of the ambient medium and v the NS velocity) will confine the PWN in the opposite direction of motion and create bow shocks. Supersonic PWNe then exhibit an elongated, cometary morphology, sometimes akin to the propagation of a bullet. Striking examples are displayed in Figure 6.13. A wide variety of morphology is observed from these sources, depending on the orientation of the pulsars' propagation, rotation, and magnetic field axes (Barkov et al. 2019). Alsaberi et al. (2019) presented the first extragalactic case of the runaway pulsar (and PWN) DEMS 5 in the SMC.

6.5.3 X-Ray Binaries

It was realized after the discovery of Sco X-1 and the first X-ray sources that a process much different than that taking place in the solar corona should be invoked to power their emission. Breakthrough evidence came in the form of the detection of rapid regular pulsations (period of a few seconds) that were modulated at a longer period (a few days). The short coherent pulsations could be interpreted as a rotation modulation of the emission around an extremely compact object such as a spinning NS. Even a compact star like a WD would be torn by centrifugal force at rotation periods so low. The longer period was evidence of the orbital motion in a binary system. The detection of regular X-ray eclipses, when the X-ray source was obscured by a then-unseen stellar companion, finally confirmed this interpretation.

Aptly name XRBs, these sources, comprising a compact object (NS or BH) and a stellar counterpart, constitute one of the main types of point sources seen in the X-ray sky. Hundreds are known in our own Milky Way (Liu et al. 2006, 2007; Coleiro & Chaty 2013), and many more from nearby galaxies (Haberl & Sturm 2016; Garofali et al. 2018; Lazzarini et al. 2021). Collectively, they are the main contribution to the unresolved emission of galaxies seen up to cosmological distances (Section 6.6.2). In this section, we describe their nature, their power source, and focus on the specific properties of XRBs hosting either an NS or a BH.

6.5.3.1 Accretion: The Ultimate Energy Source

Accretion, which is the transfer of mass onto a massive object, can be one of the most efficient energy-generating processes. A test mass m falling from a large distance r onto a massive object of mass M will convert half its gravitational potential energy $E_{gr} = -GMm/r$ into kinetic energy,[9] the other half being available as thermal energy E_{th} and ultimately radiation. The total gain is thus

$$\Delta E_{\rm th} = \frac{1}{2}\frac{GMm}{R}, \tag{6.33}$$

with R the final radius, in our case the size of the accreting object. If a stable mass accretion rate $\dot{m} = dm/dt$ can be transferred from a binary companion, the total available accretion luminosity is

$$L_{\rm acc} = \frac{d\Delta E_{\rm th}}{dt} = \frac{1}{2}\frac{GM\dot{m}}{R} = \eta\dot{m}c^2 = 5.7\times10^{38}\,\eta\left(\frac{\dot{m}}{10^{-8}\,M_\odot\,{\rm yr}^{-1}}\right)\,{\rm erg\ s^{-1}}. \tag{6.34}$$

In the equation above, $\eta = GM/2Rc^2$ represents the accretion radiative efficiency, i.e., the fraction of total available rest-mass energy $\dot{m}c^2$ that is converted into radiation. Evidently, the more massive and more compact an object, the higher the accretion efficiency. While WDs can only reach $\eta_{\rm WD} \sim 10^{-4}$, a canonical NS with $M_{\rm NS} = 1.4\,{\rm M}_\odot$ and $R_{\rm NS} = 10$ km has a much higher $\eta_{\rm NS} = 0.10$ if the material reach down to the NS surface. For a BH, there is no hard surface for the matter to fall unto, but there exists a last stable orbit at $3r_s$ (for a nonrotating BH), with $r_s = 2GM_{\rm BH}/c^2$ the Schwarzschild radius (or event horizon). Therefore, $\eta_{\rm BH}$ becomes independent of the BH mass and is 1/12. For rapidly rotating BHs, the last stable orbit moves inward and the efficiency is correspondingly increased. Accretion onto an NS or BH thus provides an efficient energy source, far exceeding the rest-mass energy conversion efficiency of nuclear burning, $\eta = 0.007$ (for hydrogen burning, the dominating reaction powering most stars).

Mass particles provided by the companion possess angular momentum and will not travel directly to the compact object. Instead, they will exchange energy with one another and settle in an accretion disk, where they follow coplanar circular orbits at

[9] A consequence of the virial theorem.

the local Keplerian velocity $V_K = \sqrt{GM/r}$. Mechanisms such as turbulence or magnetorotational instabilities (Shakura & Sunyaev 1973; Balbus & Hawley 1991) increase the viscosity between adjacent layers, allowing angular momentum to be transported outwards and mass particles to move to inner orbits. The viscous stress heats the disk and this heat is radiated away. Rewriting Equation (6.33) for the thermal gain of a particle m between two adjacent Keplerian orbits separated by dr, one finds

$$L_{\rm disk}(r) = \frac{d\Delta E_{\rm th}}{dt} = \frac{GM\dot{m}}{2}\frac{dr}{r^2} \tag{6.35}$$

as a function of radius r in the disk, measured from the compact object. If the disk is geometrically thin, and therefore dense, one can assume that it is optically thick and the heat is homogenized. The thermal energy is then radiated as a blackbody, with a well-known relation of radiation power to disk temperature $L_{\rm disk} = A\sigma T^4$. For a differential annulus at radius r with width dr, the area A is $2 \times 2\pi r dr$ (counting radiation from both the over- and underside), and therefore

$$T(r) = \left(\frac{GM\dot{m}}{8\pi\sigma}\right)^{1/4} r^{-3/4}. \tag{6.36}$$

In a steady state, the mass flow $\dot{m}$ is constant through the disk and the radial temperature profile is purely proportional to $r^{-3/4}$. The emitted spectrum is the sum of blackbody emission from each annulus at radius r with the temperature profile $T(r)$ of Equation (6.36), often called a multi-color disk model. The maximum temperature of the disk, and therefore the wavelength at which the peak power is emitted, is that of its innermost portion $T_{\rm in} = T(r_{\rm in})$. For a NS the disk may proceed all the way to the NS surface ($r_{\rm in} \sim R_{\rm NS}$), so:

$$T_{\rm in}^{\rm NS} = 0.24\left(\frac{M_{\rm NS}}{M_\odot}\right)^{1/4}\left(\frac{\dot{m}}{10^{-8}\ M_\odot\ {\rm yr}^{-1}}\right)^{1/4}\left(\frac{r_{\rm in}}{10\ {\rm km}}\right)^{-3/4}\ {\rm keV}, \tag{6.37}$$

which is in the soft X-ray band. For a BH we again take $r_{\rm in} \sim 3r_s$, and therefore

$$T_{\rm in}^{\rm BH} = 0.26\left(\frac{\dot{m}}{10^{-8}\ M_\odot\ {\rm yr}^{-1}}\right)^{1/4}\left(\frac{M_{\rm BH}}{M_\odot}\right)^{-1/2}\ {\rm keV}. \tag{6.38}$$

For stellar-mass BHs, ($M_{\rm BH} = 3 - 10\ M_\odot$), this is still in the soft X-ray band, but the peak temperature scales with the inverse square root of the BH mass (due to the linear relation of r_s with $M_{\rm BH}$). For the supermassive black hole (SMBH) at the heart of an AGN, the accretion disk peaks in the optical/UV.

6.5.3.2 Stellar Counterparts and Types of XRBs

Stellar formation and evolution predict many possibilities for binary systems: A massive star capable of leaving an NS or BH may be accompanied by another massive star or a much lighter, solar-like star. Thus, XRBs are broadly classified as low-mass X-ray binaries (LMXBs) and high-mass X-ray binaries (HMXBs), with the "mass" qualifier applying to the donor companion.

In LMXBs, the companion is a late-type star, mostly $\lesssim 1\ M_{\odot}$, in close orbit with the compact object. If the companion fills its Roche lobe, the surface of constant potential energy in the rotating frame of the binary system that connects on the line separating the two stars, at a point known as the inner Lagrange point L_1, matter crossing that point can be transferred to the more massive compact object. This provides the mass supply to fuel the accretion disk with properties given above. Late-type stars are optically faint and are often hard to characterize in detail,[10] so the LMXB's output is completely dominated by the X-ray emission. This class of XRB is associated with old stellar populations. Finally, due to the small size of the companion tiny orbits are possible. Some LMXBs have orbital periods shorter than an hour!

HMXBs evolve from massive stellar binaries where both components are early-type stars. The secondary star outlives—for a short period of time compared to the companions of LMXBs—the more massive primary, which ends up in an NS or BH. Mass transfer can occur in two ways: First, if the companion is a massive O- or B-type supergiant star ($M \gtrsim 15\ M_{\odot}$), its strong UV emission[11] launches powerful stellar winds, with mass-loss rates up to $\dot{m}_w \sim 10^{-5}\ M_{\odot}\ \mathrm{yr}^{-1}$. Only a small fraction of that wind can be captured by and accreted onto the compact object, enough to produce a bright X-ray source (Equation (6.34)).

Second, the NS or BH may orbit around Be stars, a class of early-type stars (B to late O-type stars, $M \gtrsim 5\ M_{\odot}$) that exhibits some variable emission (hence the suffix "e") and an infrared excess. These properties are now understood to originate in an equatorial disk around the star, in which matter is ejected by the star rotating close to its maximum velocity. In Be/X-ray binaries (BeXRBs), the compact object, often in an eccentric orbit around the Be star, will periodically have close encounters with the equatorial disk, which provides a reservoir of material for accretion. Thus, BeXRBs are often transient X-ray sources with quasi-regular outbursts separated by about one orbital period (type I outbursts). Sometimes, conditions in the Be star disk may change (Okazaki et al. 2013; Martin et al. 2014) and dramatically enhance the accretion rate, powering brighter, so-called, type II outbursts. In BeXRBs, the compact object is almost always an NS. It is likely that the higher mass of BHs and the smaller orbit of Be/BH binaries truncate the Be star disks and inhibits accretion (Zhang et al. 2004). Only one Be/BH binary has been confirmed as of yet (Casares et al. 2014), and its X-ray emission is indeed very faint (Ribó et al. 2017). The X-ray

---

[10] For instance, the Sun at the distance of the Galactic Center ($\approx$8.5 kpc) would be fainter than the visual magnitude $m_V > 19$, on top of which there would be a large extinction (Book 1, Section 4.4).

[11] The emission of an O star with surface temperature $T \gtrsim 30,000$ K peaks in the far-UV, at $\lambda_{\max} \lesssim 100$ nm.

temporal behavior of BeXRBs has large variations owing to different orbital periods, eccentricities, properties of the Be star disk, etc.

As both components of the binary are short lived, HMXBs are found only within areas where recent star formation occurs, typically within the last 60 Myr (Antoniou et al. 2010; Antoniou & Zezas 2016). In our Milky Way, this is occurring in the spiral arms, and thus from our vantage points all HMXBs are close to the Galactic Plane (Coleiro & Chaty 2013).

6.5.3.3 Accretion-powered Neutron Stars

In addition to variations depending on the type of companion, the nature of the companion also plays an important role. We present here unique properties found in XRBs hosting an NS. The key difference is of course the presence of the magnetic field of the NS.

In old XRBs, i.e., LMXBs, the magnetic field had time to decay and the binary may thus contain a weakly magnetized NS ($B < 10^{10}$ G). In HMXBs, however, NSs are younger and still possess strong magnetic fields ($B > 10^{11}$ G). This prevents matter in the accretion disk from proceeding freely down to the NS. Instead, when a magnetic NS is accreting, matter follows the magnetic field lines and is channeled toward the magnetic poles, in a so-called accretion column. Much of the X-ray emission is produced close to the NS, when matter is decelerated from a supersonic velocity to essentially a hard stop on the NS surface (Figure 6.14). The strong shock ensuing in the column thermalizes the kinetic energy of the flow, and electrons radiate that energy as hard X-rays. Such hard X-ray radiation may then heat up the NS surface at the base of the accretion column, producing a "hot spot" around the pole, radiating as a hotter blackbody than the rest of the NS. Electrons escaping the column also transfer part of their energy to the soft blackbody photons from the NS surface (via iC scattering, see Book 1, Chapter 7), resulting in a harder power-law spectrum. These two components

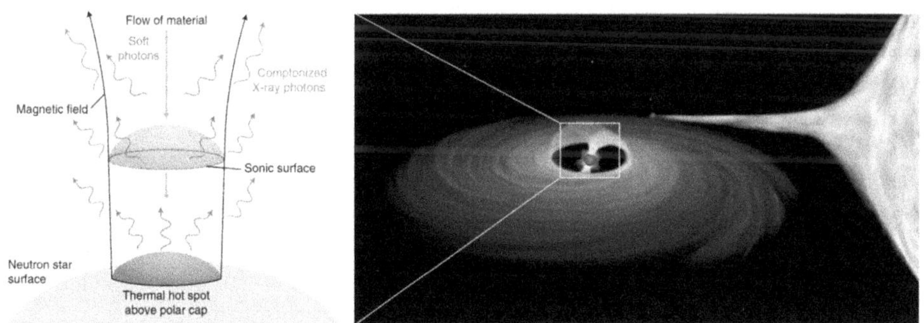

Figure 6.14. Artist impression (right) of an NS capturing matter from a companion (in blue), which settles in an accretion disk (orange) before being funneled by the magnetic field line close to the NS. On the left is a sketch of an accretion column above a magnetic pole, where the flow undergoes a transition from supersonic velocity to subsonic at the sonic surface. Energetic electrons scattering on soft X-ray photons from the hot column or heated hot spot at the base of the column (red wiggles) produce harder "Comptonized" X-ray photons (blue wiggles). Image from Baring (2018), with permission of Springer, and adapted from Becker & Wolff (2007).

of X-ray emission are produced in a small region around the NS poles. As the magnetic and rotation axes are often misaligned, those regions are shifting in and out of the field of view over the NS rotation, leading to the detection of X-ray pulsations.

Radiation and particles are subject to the strong magnetic field, which further modify (and complicate) the outgoing radiation, with the addition, for instance, of cyclotron resonance features as described in Section 6.5.2.

Observations in the X-ray band are thus essential to measure not only the spin period of the NS, but also the distribution of magnetic field strengths and orientations, and to gain insight into the small-scale processes occurring in and around the accretion column, which all affect the observed X-ray spectra and light curves of accreting NSs in HMXBs.

The conditions for a strongly magnetized NS to accrete, and thus be observed in X-rays, are set by its magnetic field strength B, rotation period P, and mass accretion rate $\dot{m}$.[12] With a dipolar field $B = \mu/r^3$ (where μ is the dipole moment), the magnetic pressure increases sharply inwards, as $P_{\rm mag} = \mu r^{-6}/8\pi$. The kinetic energy density of the flow, $E_{\rm kin} = \rho v^2/2$ is estimated assuming a freefall velocity $v_{\rm ff} = \sqrt{2GM/r}$ and a density $\rho = \dot{m}/4\pi r^2 v_{\rm ff}$. Thus, $E_{\rm kin} \propto r^{-5/2}$ increases inwards but more slowly than the magnetic pressure. The Alfvén radius $R_{\rm A}$ is defined (Pringle & Rees 1972) as the radius where both balance out:

$$R_{\rm A} = \left(\frac{\mu^4}{2GM\dot{m}^2}\right)^{1/7}. \tag{6.39}$$

Expressed as a function of B_0, the dipolar field at the surface of the NS, we find

$$R_{\rm A} \sim 1800\left(\frac{B_0}{10^{12}\ {\rm G}}\right)^{4/7}\left(\frac{R_{\rm NS}}{10\ {\rm km}}\right)^{12/7}\left(\frac{M_{\rm NS}}{1.4\ M_\odot}\right)^{-1/7}\left(\frac{\dot{m}}{10^{-8}\ M_\odot\ {\rm yr}^{-1}}\right)^{-2/7}\ {\rm km}. \tag{6.40}$$

Note that the Alfvén radius is a function of the mass accretion rate. Depending on theoretical prescriptions to describe the disk–magnetosphere boundary, the size of the magnetosphere $R_{\rm M}$ is obtained through a slight modification of the equation above, giving $R_{\rm M} = \xi R_{\rm A}$, with a unitless factor $\xi \sim 0.1$–1 (Campana et al. 2018).

At $R_{\rm M}$, matter leaves the disk and follows the magnetic field lines to the NS. For this "magnetic accretion" to take place, however, the magnetospheric radius (and thus inner edge of the disk) must be inside the corotation radius $R_{\rm co}$, where the Keplerian velocity of the disk equals the rotational velocity of the magnetic field $v_{\rm mag} = 2\pi r/P$, with P the spin period of the NS. If $\xi R_{\rm A} > R_{\rm co}$ material in the disk is prevented from accreting by this centrifugal barrier, in the so-called "propeller" state (Illarionov & Sunyaev 1975). Interestingly, the corotation radius

$$R_{\rm co} = P^{2/3}\left(\frac{G\ M_{\rm NS}}{4\pi^2}\right)^{1/3} \sim 1700\left(\frac{P}{1\ {\rm s}}\right)^{2/3}\left(\frac{M_{\rm NS}}{1.4\ M_\odot}\right)^{1/3}\ {\rm km} \tag{6.41}$$

[12] The NS mass and radius formally play a role as well; however, the mass of an NS shows much less variations due to common formation mechanisms, being limited to roughly 1 $M_\odot$ to 3 $M_\odot$, while the radii are also in a narrow range depending on the exact equation of state of matter in such conditions.

is only a function of the NS spin period, not of the accretion rate. Therefore, for a given NS rotating with period P, the accreting condition $\xi R_A < R_{co}$ is met above a critical mass accretion rate $\dot{m}_{crit}$, obtained by equating Equations (6.39) and (6.41):

$$\dot{m}_{crit} = \sqrt{2}(2\pi)^{7/3}(GM)^{-5/3}\ \xi^{7/2}\ P^{-7/3}\ B_0^2\ R_{NS}^6. \tag{6.42}$$

Because $\dot{m}$ is related to the radiated luminosity, observable in X-rays (Equation (6.34)), we conclude that there exists a lower limit to the luminosity that a given accreting NS can produce:

$$L_{crit} \approx 4.5 \times 10^{36}\ \xi^{7/2}\left(\frac{B_0}{10^{12}\ \mathrm{G}}\right)^2\left(\frac{P}{1\ \mathrm{s}}\right)^{-7/3}\left(\frac{M_{NS}}{1.4\ M_\odot}\right)^{-2/3}\left(\frac{R_{NS}}{10\ \mathrm{km}}\right)^5\ \mathrm{erg\ s^{-1}}. \tag{6.43}$$

X-ray observations of several objects with variable accretion rate, and thus luminosity, have established that this basic interpretation holds truth. An example is shown in Figure 6.15 (left, data from Lutovinov et al. 2017): the light curve of the HMXB SMC X-2 in the SMC suddenly dropped below the detection limit at a luminosity consistent with Equation (6.43) for the spin period of the NS and its magnetic field known independently from the detection of a CRSF (Equation (6.31), Jaisawal & Naik 2016). Observations of a minimum X-ray luminosity for several classes of accreting magnetized objects, from strongly magnetized NSs in XRBs to white dwarfs in CV (Section 6.4.3) and even young stars still in formation, support the interpretation of the onset of the "propeller" state over more than 10 orders of magnitude in B, independently of the nature of the accretor or companion (Campana et al. 2018).

In the old NSs of low-mass X-ray binaries, the magnetic field has decayed significantly. If the NS has not been spun up significantly, as is the case for recycled millisecond pulsars (Wijnands & van der Klis 1998; van der Klis 2000), accretion can proceed all the way down to the surface in the disk. The accreted material, mostly

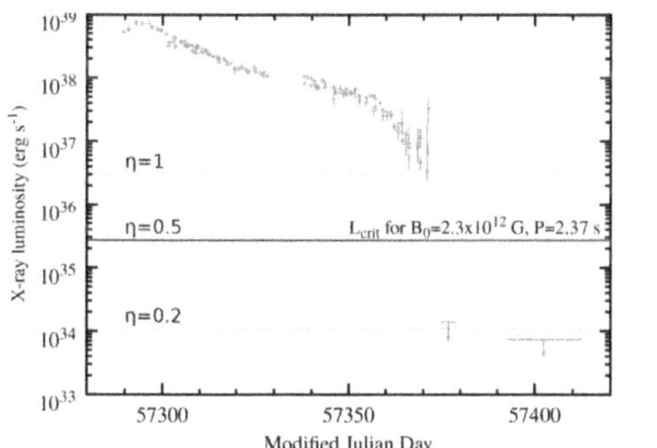

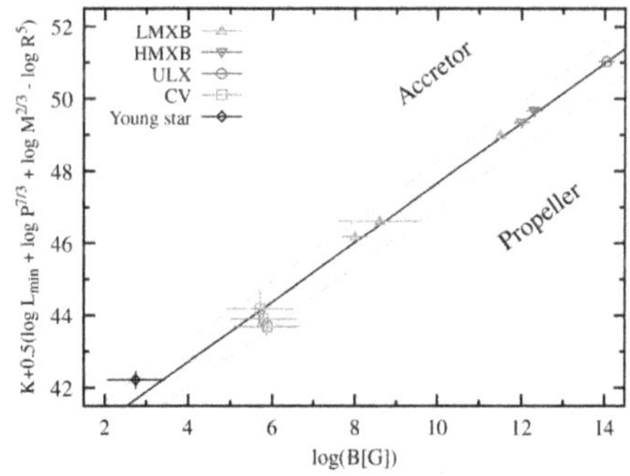

Figure 6.15. Observations of a critical X-ray luminosity for accreting objects. Left: X-ray light curve of the X-ray binary SMC X-2 during its 2015 outburst. The source declined until a sharp drop below the detection limit, consistent with the onset of the "propeller" state for the magnetic field and spin period of the NS, at a critical luminosity given by the black lines for various η parameters (see text) between 0.2 and 1. Data from Lutovinov et al. (2017). Right: correlation between magnetic field strength and a combination of observed minimum X-ray luminosities and stellar parameters set by Equation (6.43) for various classes of magnetic accretors (XRBs, CVs, and young stars). Objects above the line are in the accreting state, and those below are in the "propeller" state. Data from Campana et al. (2018) except for the ultraluminous X-ray source (ULX) M82 X-2 Tsygankov et al. (2016).

hydrogen, forms a layer around the compact star. Temperature and density conditions enable hydrogen nuclear burning under nondegenerate and therefore steady conditions, as opposed to the explosive fusion in novae (Section 6.4.3). A layer of helium-rich material develops. When sufficient helium has been synthesized and mounting temperature and density reach a critical value, helium burning starts, and this process is unstable. It results in a thermonuclear flash, incinerating the rest of the accumulated layer in less than a few seconds. The nuclear energy from the helium flash is released as thermal radiation, increasing the X-ray emission 10-fold compared to the persistent (accretion-driven) level. The source quickly cools down and come back to the preflash level within a few tens of seconds. A new cycle of accretion then starts until the next flash.

Both the blackbody area and ratio of nuclear energy versus accretion energy in bursts and persistent periods, respectively, support that the compact objects in such sources are indeed NSs, and they are known as X-ray bursters. The typical time between bursts is a few hours. The recurrence time is shorter during periods of higher persistent flux (Galloway et al. 2004). This can easily be understood as a higher persistent flux implies a higher accretion rate, and thus a faster replenishment of the fuel for the next burst. The luminosity distribution of bursts extends up to about the Eddington limit (Equation (6.46)). For the brightest cases, the observed blackbody area is increased during the bursts, consistent with an expansion brought by the radiation pressure during the peak of the burst.

6.5.3.4 Black Hole Binaries

Through optical spectroscopy of the companions and radial velocity modulated by orbital motion, the mass of the components in XRBs can be measured. One of the most striking results of the early days of X-ray astronomy was that some sources, for instance Cygnus X-1, must host a compact object more massive than the heaviest of NSs (Webster & Murdin 1972). This was the first observational clue to the existence of astrophysical BHs. There are now more than 20 XRBs that host a confirmed BH in the Milky Way (Remillard & McClintock 2006; Casares & Jonker 2014) with well-known binary properties (Figure 6.16), plus several in nearby galaxies (e.g., Orosz et al. 2007) and many more bona fide candidates (Corral-Santana et al. 2016). Without a hard surface for the material to fall onto and no magnetic field to shear the disk, one could expect that the X-ray emission of BH XRBs is markedly different than that from their NS counterparts. Indeed, except for Cygnus X-1 and a few others, most of these sources are not persistent sources but usually spend their lives in quiescent states, with very low X-ray luminosity. Occasionally, they undergo bright outbursts that last for about a year, during which most of these sources are discovered.

BH binaries, persistent or transients, have variable emission properties and evolve between several spectral states. Historically the main states were known as the high/soft state, when the X-ray emission is bright and peaks below a few keV, and the low/hard state, when the source is fainter and peaks in hard X-rays ($\sim$100 keV). The soft emission is interpreted as the thermal emission from the accretion disk (Equation (6.38)). In the hard state, the disk is truncated farther outward, and the inner flow is hot and optically thin. The spectrum is dominated by a hard tail, with a power-law spectrum of varying slope, that originates from either thermal emission from very hot

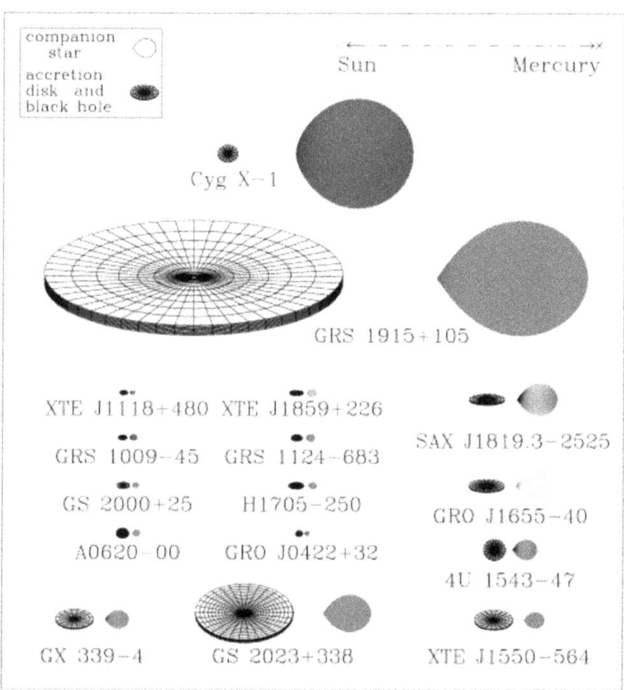

Figure 6.16. A collection of Galactic BH binaries. The companions are in color, distorted by the more massive BHs. The accretion disk is shown as a mesh around the BH, with the inclination of the system with respect to the plane of the sky indicated by the tilt of the disk. The orbital separation and sizes of the companions and disks are shown to scale, given by the Sun–Mercury separation on the top right. Image credit: J. Orosz Remillard & McClintock (2006).

electrons in the inner part of the accretion flow, or by soft disk photons iC-scattered by those hot electrons. A compact jet is observed in radio, launched from the inner region in the direction perpendicular to the disk. The complex interactions between the components in the region close to the BH lead to the changing look of BH binaries on timescales of days to months, with attempts at providing a unified model in Fender et al. (2004) or displayed in Figure 6.17 (Done et al. 2007).

Hard X-ray emission from the inner accretion flow (often called the corona) is also illuminating the underlying disk. That emission is processed in the disk through absorption, scattering, and fluorescence,[13] and radiated back toward us in a reflection component (Basko et al. 1974; Done et al. 2001). This adds to the already complex spectral modeling of BH binaries, but it also provides us with a nice probing tool. The properties of the reflection component are indeed related to the illuminating source, the physical conditions such as temperature and ionization of

[13] Fluorescence is the process following ionization of an inner-shell electron in an atom or ion, leaving it in an excited state. A bound electron from the outer shell then transits to the inner-shell gap, emitting a photon with energy equal to the energy difference between the initial and final levels.

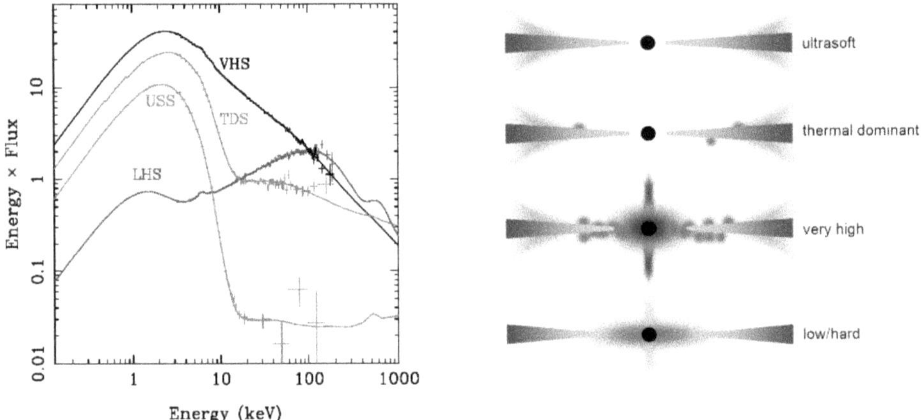

Figure 6.17. Spectral states observed in the BH binary GRO J1655–40 (left panel) and proposed changes in accretion flow to explain these spectra (right panel). In the ultrasoft state, the spectra are completely dominated by the disk (orange). From the thermal dominant to the very high state, contributions from active regions above the disk and hot inner flow (small and large blue dots, respectively) start to dominate the spectra above a few keV. In the hard state the disk is truncated and the spectra are dominated by the optically thin inner flow. In the latter two states, jets are launched from the inner flow. Image credit: Done et al. (2007) with permission of Springer.

the reflecting material (i.e., the disk), and geometry. The fluorescence spectrum, dominated by the Fe K-shell fluorescence line at 6.4 keV, is used to probe strong gravity fields, as it is broadened by general relativistic effects if it is coming from a region very close to the BH event horizon (Fabian et al. 1989).

6.5.3.5 Ultraluminous X-Ray Sources

Because photons provide radiation pressure (Book 1, Section 2.2.3), compact sources may not be arbitrarily bright. This is particularly relevant to XRBs: Outgoing radiation pressure from a source too luminous may prevent the accretion of matter, the very source of power in such sources. To establish this limiting luminosity, one can consider as the main process the Thompson scattering (Book 1, Section 4.3.4) of outgoing photons on the electrons of the infalling matter. At a distance r from a compact source, an electron undergoes $n_{\mathrm{ph}} c \sigma_T$ scattering per unit time, with the $n_{\mathrm{ph}} = L_\nu / 4\pi r^2 c h\nu$ the density of photons with energy $E = h\nu$, and σ_T the electron Thompson cross section. Each scattering transfers the momentum E/c of the photon, so that the total radiation force is calculated by integrating over all frequencies:

$$F_{\mathrm{rad}} = \int_0^\infty \frac{n_{\mathrm{ph}} c \sigma_T h\nu}{c} d\nu = \int_0^\infty \frac{L_\nu \sigma_T}{4\pi r^2 c} d\nu = \frac{L \sigma_T}{4\pi r^2 c}. \tag{6.44}$$

Protons and ions are dragged with electrons through electrostatic interaction. They bear most of the gravitational force, which for a proton of mass m_p is

$$F_{\mathrm{g}} = \frac{GMm_p}{r^2}. \tag{6.45}$$

There is a limiting luminosity where both forces balance out, now called the Eddington luminosity L_{Edd}, after Sir Arthur Eddington[14] who first devised it, albeit in the context of stars:

$$L_{\mathrm{Edd}} = \frac{4\pi G M m_p c}{\sigma_T} = 1.3 \times 10^{38} \frac{M}{M_\odot} \text{ erg s}^{-1}. \tag{6.46}$$

One type of source that has puzzled X-ray astronomers is that exceeding the Eddington luminosity, with luminosity of 10^{39} to 10^{42} erg s^{-1}. These so-called ULXs are rare point sources, one or a few per galaxy (but more for "starburst"; Liu 2005), and are found in correlation with regions of recent star formation. They are likely to be an extreme type of X-ray binary. However, such high luminosities should not be allowed by simple accretion theory: The mass of BHs in XRBs measured dynamically clusters at 10 $M_\odot$ (Kovetz et al. 2017; Casares & Jonker 2014), which places $L_{\mathrm{Edd}} \lesssim 2 \times 10^{39}$ erg s^{-1}. Suggestions to explain ULXs include supercritical accretion (Begelman 2002), where accretion and emission are from and to different directions, or beaming effects, where outgoing radiation is enhanced in the viewing direction (Okazaki et al. 2013), leading to a significant overestimate of the luminosity if isotropic emission is assumed. However, one of the more alluring options was that ULXs were powered by intermediate-mass BHs with $M = 10^2 - 10^4$ $M_\odot$ and not, in fact, above their L_{Edd}. This would bridge the gap between the "lightweight" BHs observed in XRBs and the supermassive kind ($M > 10^6$ $M_\odot$) powering active galactic nuclei.

Recently, pulsations have been detected from several ULXs (Bachetti et al. 2014; King & Lasota 2019), which proved that some ULXs are in fact powered by super-Eddington accretion onto magnetized NSs. Depending on models, a very strong magnetic field ($B > 10^{13}$ G) may be needed to reduce Thompson cross section and sustain the high luminosity (Mushtukov et al. 2015) or standard fields ($10^{11} - 10^{13}$ G) may be sufficient, but the emission is collimated by outflows launched from the inner accretion disk (but outside the magnetospheric radius R_{M}; King et al. 2017; King & Lasota 2019).

The intermediate-mass BH origin may however remain valid for the most extreme ULXs, for instance ESO 243–49 HLX-1 (for hyperluminous X-ray source 1), which has been observed above 10^{42} erg s^{-1} and is thought to host a BH more massive than 500 $M_\odot$ (Farrell et al. 2009).

6.6 X-Rays at Cosmological Scales

6.6.1 Active Galactic Nuclei in X-Rays

6.6.1.1 Accretion on Supermassive Black Holes

As evidenced by their presence in every chapter of this book, AGNs are truly ubiquitous in astronomy. In X-rays, AGNs are often seen as bright point sources, variables on a timescale down to only a few minutes. This is clear evidence that they are bright compact sources and not a collection of many fainter objects like stars. To

[14] The same who first tested general relativity as presented in Book 1 (Section 8.3).

explain their large X-ray luminosity (10^{40}–10^{47} erg s^{-1}) from a small spatial region, accretion again provides a handy energy source. Even if the luminosity is Eddington limited (Equation (6.46)), accreting SMBHs with masses of the order 10^6–10^{10} $M_\odot$ are sufficient to power AGNs. Evidence for the existence of such astronomical monsters has steadily increased over the last decades to the point of becoming overwhelming, from the study of bulges' stellar dispersion, broadening in infrared and optical emission lines from AGNs, to the direct imaging of the 6.5×10^9 $M_\odot$ BH of M87 (Event Horizon Telescope Collaboration et al. 2019) and weighing of our own Milky Way's 4.3×10^6 $M_\odot$ SMBH (Gillessen et al. 2009; Nobel Prize for Physics 2020; see also Section 7.5.5 and Figures 2.20 and 7.17).

Much of the various properties of AGNs at all photon wavelengths (from radio to gamma rays) are explained in a single unified framework (Antonucci 1993; Urry & Padovani 1995) where the appearance of an AGN depends on the viewing direction (Section 1.2.3 and Figure 1.7): A large dusty torus surrounds the central engine, the SMBH, and its accretion disk, impeding our view of the inner regions for a certain range of directions.

Accretion onto SMBHs is not much different than on stellar-mass BHs (Gilfanov & Merloni 2014), differing only in scale and in feeding ground. We know from Equation (6.34) that a 10^8 $M_\odot$ BH, to emit at its Eddington limit (1.3×10^{46} erg s^{-1}), needs to "swallow" between 0.7 $M_\odot$ and 2.5 $M_\odot$ of mass per year (depending on its spin). Obviously, no single stellar companion may be sufficient. The central regions of galaxies, however, have a much higher density of stars, and those approaching too close to the BH are disrupted entirely by tidal forces (Section 1.2.9). Stellar gas released in the disruption orbits about the BH and can fuel its accretion disk.

We know from Equation (6.38) that the innermost and hottest temperature of the disk around a SMBH is in the optical/UV rather than in X-rays. However, the inner accretion flow is hotter and a source of energetic electrons that Comptonize the lower-energy photons from the disk. Thus, the X-ray spectra of AGNs comprise the hard X-ray emission of the corona and the reprocessed emission (reflection and fluorescence), much like for stellar-mass BH XRBs (Section 6.5.3). Ejection processes works similarly around SMBH, albeit at much larger scale: Jets are launched close to the speed of light, reaching thousands of parsecs (e.g., Figure 1.6) A correlation between X-ray and radio and luminosity is observed for accreting BHs, spanning many orders of magnitude (Merloni et al. 2003; Gallo et al. 2006).

6.6.1.2 Cosmic X-Ray Background, AGN Surveys, and Cosmology

Because almost all massive galaxies host a central SMBH, and because of the large number of galaxies in the universe, it should be no surprise that AGNs are the most common point-like sources in the X-ray sky. They are so numerous that the vast majority of X-ray observations are affected by the collective effect of unresolved AGNs, appearing as one uniform, diffuse source filling the field of view. This component is the so-called cosmic X-ray background (CXB), which had been detected already in the first successful experiment by Giacconi and his team in 1962 (Giacconi et al. 1962). The origin of this component was puzzling, but given its isotropy, was most likely extragalactic in nature. The bright X-ray emission from

AGNs was proposed as the source of the CXB early on (Setti & Woltjer 1973). The observed X-ray spectrum of the CXB peaks at about 30 keV and is well explained by the emission from AGNs (specifically the Compton hump of their reflected component, see above), convolved with a broad distribution of absorbing column densities (Gilli et al. 2007). Thanks to the increasing spatial resolution and sensitivity of X-ray instruments, the CXB has been resolved almost entirely into individual point sources, the vast majority being AGNs, in soft and hard X-rays (Worsley et al. 2005; Hickox & Markevitch 2006; Harrison et al. 2016).

Very distant AGNs may only provide a few counts on a detector during a deep exposure. If they cannot be characterized in exquisite detail as their bright, relatively nearby counterparts, their sheer number still provides invaluable information via the analysis of large statistical samples. Cosmic surveys are observing programs designed to obtain large, clean, and homogeneous samples of types of sources, for instance AGNs, given the observing instrument and constraints. They might be narrow with deep exposures, focusing on a single region observed up to the limit allowed by the observatory, or shallow but covering a large area on the sky, using either a mosaic of individual pointed observations or a scanning technique, or a combination of both. Two examples at both extreme ends are shown in Figure 6.18.

The classification of detected point sources as AGNs is done either by association with their multiwavelength counterparts (e.g., X-ray over optical or infrared flux ratios) or through X-ray properties only (e.g., spectral shape, temporal variability). At any rate, complementary data from other wavelengths are almost always needed, most importantly to obtain the redshift of the host galaxy, and therefore the third dimension in the survey: cosmological distance, or equivalently, lookback time. AGN X-ray emission traces the history of accretion of matter onto SMBHs and allows galaxies to be tracked over cosmological distances. Complemented with

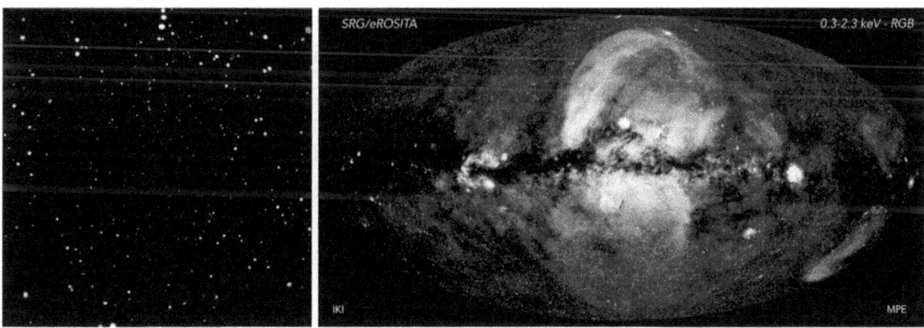

Figure 6.18. From the deepest to the widest: On the left, the central 8′ ×8′ Chandra Deep-Field South is the region with the deepest X-ray exposure, totaling 7 Ms or 81 days (!) of combined exposure time accumulated over 15 years (Luo et al. 2017). Hundreds of point sources, mostly AGNs, are detected, with various colors depending on their intrinsic spectra. The larger size of sources on the edges is due to the degradation of spatial resolution with off-axis angle (Section 6.3.2). Right: the whole sky in X-rays, displayed in Galactic Aitoff projection, as covered by the eROSITA X-ray telescope on board SRG (Book 1, Figure 1.16) during a 6 month scanning (Predehl et al. 2021). The sky-averaged effective exposure time (accounting for vignetting) is 170 s. Image credits: NASA/CXC/Penn State/B.Luo et al.; Jeremy Sanders, Hermann Brunner and the eSASS team (MPE); Eugene Churazov, Marat Gilfanov (on behalf of IKI).

accurate redshifts, X-ray surveys of AGNs provide a wealth of crucial information on the nature of large-scale structures and thus the energy content of the universe, as well as on the history of accretion and galaxy formation.

6.6.2 X-Ray Emission of Normal Galaxies

The X-ray emission of normal galaxies, understood here as those with a nondominant contribution by a central AGN, comes from all types of sources that have been encountered in this chapter, including nuclear-burning stars (late- and early-types), isolated WDs, SNRs and PWNe, and accreting X-ray binaries. The total light is however dominated by the latter sources. Stars and isolated WDs are out of reach of most reasonable surveys with even the best instruments available, beyond a horizon of a few tens and hundreds kiloparsecs, respectively. Consequently they are not even resolved in galaxies outside our own Milky Way or immediate neighbors, the MCs. But unresolved stellar X-ray sources collectively contribute only to a small fraction of the total X-ray luminosity of non-AGN galaxies: Even a billion Sun-like stars give out less X-ray light than a single XRB. SNRs are typically brighter ($L_X \gtrsim 10^{35}$ erg s^{-1}) and a large population of SNRs can be detected as extended sources within 5 Mpc (Long et al. 2010; Sasaki et al. 2012; Long et al. 2014), while the brightest (a few 10^{38} erg s^{-1}; Binder et al. 2015; Yew & Filipović 2018) may be seen up to 40 Mpc.

With a luminosity of 10^{36}–10^{39} erg s^{-1}, or up to 10^{41} erg s^{-1} for ULXs, only large XRBs populations may be studied in detail in the local universe (e.g., Haberl & Sturm 2016; Garofali et al. 2018), with the brightest objects detected farther out to within ~100 Mpc, beyond which the host galaxies themselves become hard to spatially resolve. Because the lifetime of high-mass XRBs is short, their presence requires star formation to have occurred recently (<100 Myr). The integrated luminosity of all high-mass XRBs in a galaxy can thus be used to estimate the star formation rate of the Galaxy. Above a certain threshold, the HMXB luminosity scales linearly with the SFR (Grimm et al. 2003; Mineo et al. 2012a). Low-mass XRBs on the other hand have a contribution that scales with the stellar mass of the Galaxy and starts to dominate at lower SFR/M (Lehmer et al. 2010).

Another component not yet discussed is the diffuse emission from the hot interstellar medium. Among the several phases of the ISM, it is large by volume (≈30%), but negligible in terms of total mass as it is so tenuous (density $n \lesssim 0.1$ cm^{-3}). Heating of the gas to multimillion-degree temperatures is provided by the winds of massive stars and subsequent SNR shocks. The resulting X-ray emission is that from a coronal gas (Section 6.4.2), enabling the measurement of gas-phase abundances of elements having emission lines, most notably oxygen, neon, magnesium, silicon, and iron. As the hot ISM heating is provided by massive star populations, the luminosity of that component also scales with recent SFR (Mineo et al. 2012b). In high-SFR cases, i.e., massive disk galaxies or starburst galaxies, the hot gas is expelled in large winds into the circumgalactic medium (Li & Wang 2013) enriching it with heavy elements newly formed in supernova explosions (Crain et al. 2013).

Beyond a hundred megaparsecs, galaxies become hard or even impossible to resolve in individual point sources. Instead, the integrated X-ray light can be seen.

Complementary data at other wavelengths become necessary to identify and exclude AGNs and to measure the redshift and other properties (SFR, mass) of the galaxies. The evolution of X-ray emission of normal galaxies can thus be followed up to cosmological distances, using deep surveys (for instance, the Chandra Deep Fields, Figure 6.18). Beyond a redshift of ~ 1, multiple galaxies have to be stacked to obtain meaningful constraints. This has suggested that the scaling of low-mass and high-mass XRBs with galactic mass and SFR, respectively, increase with redshift (Basu-Zych et al. 2013; Lehmer et al. 2016). In other words, similar galaxies in the past formed more or brighter XRBs than their present-day analogs (in terms of SFR/M). This effect might be due to the younger age of stellar populations (Zhang et al. 2012) or the lower metallicity of high-redshift galaxies (Douna et al. 2015). This has an important consequence for the rate of compact object binary mergers detected by gravitational-wave observatories (Section 9.7) and as a source of additional heating contributing to the reionization of the universe in its early days (Mirabel et al. 2011; Fragos et al. 2013).

6.6.3 Galaxy Clusters

Galaxies are not located randomly in the universe. Because of primordial inhomogeneities and gravitational interactions, they aggregate into galaxy clusters at the knots of the Cosmic Web (Figure 3.9). Clusters may contain only a few member galaxies (usually better called galaxy groups) up to hundreds of members of various mass, spanning typically 1 Mpc. They evolve in their common gravitational potential well and have a velocity dispersion that allows an order of magnitude to be estimated for their total mass of $10^{14}\ M_{\odot}$. Using the independent method of gravitational lensing (Book 1, Section 8.3) of galaxies behind the clusters, similar masses are deduced. Galaxy clusters are thus the most massive bound structure in the universe.

Optical and infrared observations show us gravitational lensing and the galaxies inside clusters (mostly the light of stars in those galaxies). X-ray observations have long revealed clusters in a different light (Meekins et al. 1971; Forman et al. 1972) and paint quite a different picture. Galaxy clusters are extended X-ray sources, with X-rays emitted by the ICM. This is the gas filling the space between galaxies that is also bound in the gravitational potential well of the cluster. Although the ICM densities are very low (10^{-3}–10^{-2} cm^{-3}), measurements show that the total ICM gas typically weighs 10 times more than the galaxies (i.e., stars). Both ICM and stars are themselves outweighed by the dark matter (Chapter 10), which accounts for about 90% of the cluster mass.

The infall speed of the ICM gas during cluster formation is supersonic, and shocks in the gas and compression heat it to multimillion-degree temperatures ($kT \sim$ 1–10 keV). The total mass of the cluster can be obtained from the observed X-ray temperature and emissivity profile. Assuming that the gas is in hydrostatic equilibrium, its pressure gradient should match the potential gradient, or, in the spherically symmetric one-dimensional case:

$$\frac{dP}{dr} = -\rho \frac{GM(r)}{r^2} \tag{6.47}$$

with $M(r)$ is the mass (mostly dark matter) enclosed within radius r and ρ the density of the ICM gas. Using again the ideal gas law as equation of state, we find

$$M(r) = -r\frac{kT(r)}{G\mu}\left(\frac{d\ln\rho}{d\ln r} + \frac{d\ln T}{d\ln r}\right). \tag{6.48}$$

The temperature is straightforward to obtain from X-ray spectroscopy, as the gas produces an optically thin coronal emission, which we encountered before (Sections 6.4.2 and 6.5.1). The luminosity of that emission is proportional to the density squared, so the density profile is inferred from the X-ray image. However, even if the gas follows a 1D spherical symmetry, we observe only its projected emission on the sky (and thus we see a combination of temperature along the line of sight). Special care must be taken to account for that.

Scaling relations can be used to derive the mass of the cluster if spatially resolved modeling is not possible due to low-S/N data, poor spatial resolution, etc. Still, with the assumption of hydrostatic equilibrium and ideal gas, the temperature scales as $T \sim M/R$, and the emitting volume is $R^3 \sim M$. This combines into $M \propto T^{3/2}$. A surprisingly good agreement with that naive estimate has been found (Arnaud et al. 2005; Vikhlinin et al. 2006) if the central regions, sensitive to strong cooling (Fabian et al. 1994) are excluded. Next, the X-ray luminosity–temperature relation predicts $L_X \sim T^2$ for pure bremsstrahlung (appropriate if T exceeds several 10^7 K). Combined with the $M-T$ relation, one expects $M \sim L_X^{3/4}$. Observations have established a slightly steeper $L_x - T$ relation (Allen et al. 2001; Vikhlinin et al. 2009) and therefore a shallower $M \sim L_X^{0.63}$ scaling relation. Thus, the important parameter that is the total mass of the cluster (which is mostly the dark matter mass) can be obtained to a fair degree of accuracy from simple X-ray observables. As such, galaxy clusters are particularly appealing to constrain some cosmological parameters.

The main observable for cluster-based cosmology is the cluster mass function which, combined with the expansion history of the universe, predicts the number of clusters as a function of mass and redshift. Observationally, this is measured as the number of cluster per unit solid angle in various mass and redshift bins. Because clusters are the collapsed remnants of primordial overdensities, they are sensitive to the amount of said overdensities and to the amount of matter (dark and baryonic) that binds the clusters together. This is parameterized in cosmology as the present rms matter fluctuation averaged over a comoving distance of 8 Mpc, or σ_8 for short, and the matter density parameter Ω_m, respectively. Furthermore, the dark energy content and equation of state parameter w intervene as well: An accelerated expansion due to dark energy will result in a smaller cluster mass function at higher redshift compared to a dark-energy-free model, simply because of the added "dilution" of the universe. Many studies have used X-ray samples of galaxy clusters to constrain those parameters (e.g., Allen et al. 2011 and references therein). They are particularly powerful when combined with other tracers (Type Ia SNe, CMB, baryonic acoustic oscillations) to provide joint constraints. An example is shown in Figure 6.19. So far, cluster-based constraints on dark energy agree with a parameter $w = -1$, i.e., dark energy as a cosmological constant.

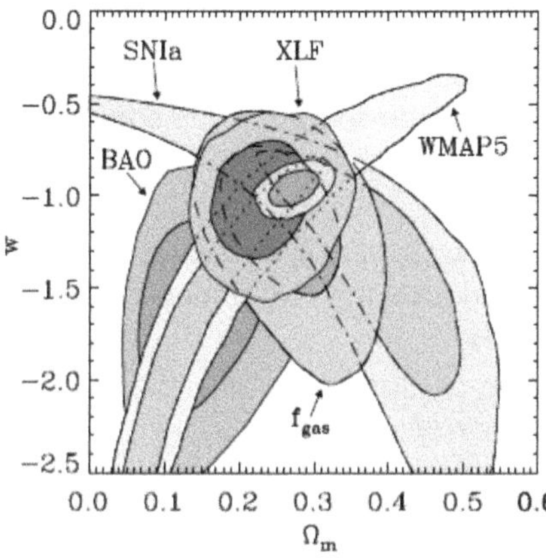

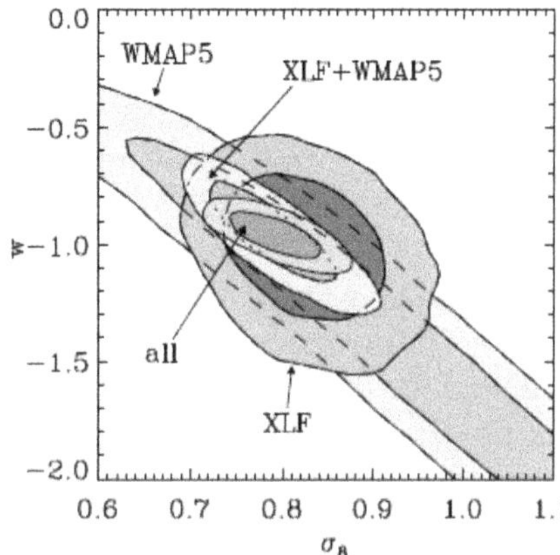

Figure 6.19. Constraints on cosmological parameters w, Ω_m, and σ_8 (see text), obtained from X-ray-detected galaxy clusters in combination with other tracers. The 1σ and 2σ confidence regions are shown by the light and dark shades. Image credit: Mantz et al. (2010), © 2010 The Authors. Journal compilation © 2010 RAS, and Allen et al. (2011).

Much like AGN surveys (Section 6.6.1), cluster studies benefit from both ultradeep narrow surveys to find the first groups and clusters of galaxies at high redshift and large area to all-sky surveys to maximize the total number of detected clusters. While Athena (Section 6.2.3) will do the former, detecting and characterizing clusters at redshift $z \gtrsim 2$ (Pointecouteau et al. 2013) in the 2030s, eROSITA will detect a total of approximately 10^5 clusters in the extragalactic sky over a four-year survey (Clerc & Ramos-Ceja 2018), providing improved cosmological constraints (Pillepich et al. 2018) with great impact.

6.7 Concluding Statement: X-Rays in the Multimessenger Context

Six decades after the discovery of Sco X-1, it is clear that X-ray astronomy is a mature discipline, having discovered several types of celestial objects and bringing renewed interest for many more. We have covered in this chapter such discoveries, from classical sources such as stars, to more exotic ones like NSs and BHs. The development of detectors and optics geared to X-ray detection have driven many of these discoveries. Improved or breakthrough technology, most notably X-ray polarimetry and microcalorimeters, will bring forth even more in the near future.

However, it is perhaps fair to say that, like other once "niche" branches, X-rays astronomy will become less and less done in isolation. While complementary multiwavelength observations have always been needed, it is anticipated that X-rays will in turn provide a fantastic windows of opportunity to support multi-messenger observing campaigns.

For the binary NS mergers that are seen as gravitational-wave sources (Book 1, Section 1.3.6), the electromagnetic counterparts are short gamma-ray bursts that have afterglows starting in the X-ray domain. Thus, X-ray observations are the first to offer a means of searching for the electromagnetic counterparts and localizing them to relatively good accuracy (a few arcminutes versus tens to potentially thousands of square degrees), allowing to pin point the host galaxy of the event.

Galactic accelerators like PWNe and SNRs are quintessential X-ray sources from which parameters most relevant to the production of high-energy particles (e.g., magnetic field strength and configuration, shock speed, etc.) can be inferred. Similarly, if AGNs are the sources of ultra-high-energy charged particles (observed after decay in γ-rays) or neutrinos, X-rays hold the key to understanding the conditions in the immediate vicinity of the SMBH.

Finally, the dark matter content of the universe is also affecting the X-ray sky. Some models of dark matter annihilation predicts an electromagnetic signal at X-ray energies, although none have been conclusively observed yet. The largest reservoir of dark matter is galaxy clusters, which are bright X-ray sources. Their distribution directly probes the amount and fluctuation of dark matter and the properties of dark energy, as we briefly touched upon.

References

Acero, F., Aloisio, R., Amans, J., et al. 2017, ApJ, 840, 74
Acero, F., Ballet, J., Decourchelle, A., et al. 2009, A&A, 505, 157
Adamczak, J., Werner, K., Rauch, T., et al. 2012, A&A, 546, A1
Adams, J. S., Baker, R., Bandler, S. R., et al. 2020, JLTP, 199, 1062
Allen, S. W., Schmidt, R. W., & Fabian, A. C. 2001, MNRAS, 328, L37
Allen, S. W., Evrard, A. E., & Mantz, A. B. 2011, ARA&A, 49, 409
Alsaberi, R. Z. E., Maitra, C., Filipović, M. D., et al. 2019, MNRAS, 486, 2507
Angel, J. R. P. 1979, ApJ, 233, 364
Antoniou, V., & Zezas, A. 2016, MNRAS, 459, 528
Antoniou, V., Zezas, A., Hatzidimitriou, D., & Kalogera, V. 2010, ApJ, 716, L140
Antonucci, R. 1993, ARA&A, 31, 473
Arnaud, M., Pointecouteau, E., & Pratt, G. W. 2005, A&A, 441, 893
Baade, W., & Zwicky, F. 1934, PNAS, 20, 259
Bachetti, M., Harrison, F. A., Walton, D. J., et al. 2014, Natur, 514, 202
Bajt, S., Prasciolu, M., Fleckenstein, H., et al. 2017, LSA, 7, 17162
Balbus, S. A., & Hawley, J. F. 1991, ApJ, 376, 214
Ballet, J. 2006, AdSpR, 37, 1902
Baring, M. G. 2018, NatAs, 2, 282
Barkov, M. V., Lyutikov, M., & Khangulyan, D. 2019, MNRAS, 484, 4760
Barstow, M. A., Fleming, T. A., Diamond, C. J., et al. 1993, MNRAS, 264, 16
Basko, M. M., Sunyaev, R. A., & Titarchuk, L. G. 1974, A&A, 31, 249
Basu-Zych, A. R., Lehmer, B. D., Hornschemeier, A. E., et al. 2013, ApJ, 762, 45
Becker, P. A., & Wolff, M. T. 2007, ApJ, 654, 435
Begelman, M. C. 2002, ApJ, 568, L97
Binder, B., Williams, B. F., Eracleous, M., et al. 2015, AJ, 150, 94
Blake, R. L., Chubb, T. A., Friedman, H., & Unzicker, A. E. 1965, ApJ, 142, 1
Borges, F. I. G. M., & Conde, C. A. N. 1996, NIMPA, 381, 91
Borkowski, K. J., Reynolds, S. P., Green, D. A., et al. 2010, ApJ, 724, L161
Bowyer, S., Byram, E. T., Chubb, T. A., & Friedman, H. 1964, Sci., 146, 912

Brose, R., Sushch, I., Pohl, M., et al. 2019, A&A, 627, A166
Burgay, M., D'Amico, N., & Possenti, A. 2003, Natur, 426, 531
Campana, S., Stella, L., Mereghetti, S., & de Martino, D. 2018, A&A, 610, A46
Canizares, C. R., Davis, J. E., Dewey, D., et al. 2005, PASP, 117, 1144
Caprioli, D., & Spitkovsky, A. 2014, ApJ, 794, 46
Carlton, A. K., Borkowski, K. J., Reynolds, S. P., et al. 2011, ApJ, 737, L22
Casares, J., & Jonker, P. G. 2014, SSRv, 183, 223
Casares, J., Negueruela, I., & Ribó, M. 2014, Natur, 505, 378
Cheng, K. S. 2009, in Astrophysics and Space Science Library, Vol. 357, High Energy Emission from Pulsars and Pulsar Wind Nebulae, 481
Cherepashchuk, A. M. 1976, SvAL, 2, 138
Chlebowski, T., & Garmany, C. D. 1991, ApJ, 368, 241
Christensen, F. E., Jakobsen, A. C., Brejnholt, N. F., et al. 2011, Proc. SPIE, 8147, 81470U
Clerc, N., Ramos-Ceja, M. E., Ridl, J., et al. 2018, A&A, 617, A92
Coleiro, A., & Chaty, S. 2013, ApJ, 764, 185
Corral-Santana, J. M., Casares, J., Muñoz-Darias, T., et al. 2016, A&A, 587, A61
Costa, E., Soffitta, P., Bellazzini, R., et al. 2001, Natur, 411, 662
Cox, D. P., & Daltabuit, E. 1971, ApJ, 167, 113
Crain, R. A., McCarthy, I. G., Schaye, J., Theuns, T., & Frenk, C. S. 2013, MNRAS, 432, 3005
Dauser, T., Falkner, S., Lorenz, M., et al. 2019, A&A, 630, A66
Dempsey, R. C., Linsky, J. L., Fleming, T. A., & Schmitt, J. H. M. M. 1993, ApJS, 86, 599
den Herder, J. W., Brinkman, A. C., Kahn, S. M., et al. 2001, A&A, 365, L7
Done, C., & Nayakshin, S. 2001, ApJ, 546, 419
Done, C., Gierliński, M., Kubota, A., et al. 2007, A&ARv, 15, 1
Douna, V. M., Pellizza, L. J., Mirabel, I. F., & Pedrosa, S. E. 2015, A&A, 579, A44
Edlén, B. 1943, ZAp, 22, 30
Event Horizon Telescope Collaboration, Akiyama, K., Alberdi, A., et al. 2019, ApJ, 875, L1
Fabian, A. C., Crawford, C. S., Edge, A. C., & Mushotzky, R. F. 1994, MNRAS, 267, 779
Fabian, A. C., Rees, M. J., Stella, L., & White, N. E. 1989, MNRAS, 238, 729
Fano, U. 1947, PhRv, 72, 26
Farrell, S. A., Webb, N. A., Barret, D., Godet, O., & Rodrigues, J. M. 2009, Natur, 460, 73
Feldmeier, A., Puls, J., & Pauldrach, A. W. A. 1997, A&A, 322, 878
Fender, R. P., Belloni, T. M., & Gallo, E. 2004, MNRAS, 355, 1105
Fleming, T. A., Snowden, S. L., Pfeffermann, E., Briel, U., & Greiner, J. 1996, A&A, 316, 147
Forman, W., Kellogg, E., Gursky, H., Tananbaum, H., & Giacconi, R. 1972, ApJ, 178, 309
Fragos, T., Lehmer, B. D., Naoz, S., Zezas, A., & Basu-Zych, A. 2013, ApJ, 776, L31
Friedman, H., Lichtman, S. W., & Byram, E. T. 1951, PhRv, 83, 1025
Frontera, F., & von Ballmoos, P. 2010, XROI, 2010, 215375
Fryer, C. L., & Kusenko, A. 2006, ApJS, 163, 335
Gallo, E., Fender, R. P., & Miller-Jones, J. C. A. 2006, MNRAS, 370, 1351
Galloway, D. K., Cumming, A., Kuulkers, E., et al. 2004, ApJ, 601, 466
Garofali, K., Williams, B. F., Hillis, T., et al. 2018, MNRAS, 479, 3526
Giacconi, R., Gursky, H., Paolini, F. R., & Rossi, B. B. 1962, PhRvL, 9, 439
Giacconi, R., Murray, S., Gursky, H., et al. 1972, ApJ, 178, 281
Giacconi, R., & Rossi, B. 1960, JGR, 65, 773
Gilfanov, M., & Merloni, A. 2014, SSRv, 183, 121

Gillessen, S., Eisenhauer, F., Trippe, S., et al. 2009, ApJ, 692, 1075
Gilli, R., Comastri, A., & Hasinger, G. 2007, A&A, 463, 79
Ginzburg, V. L., & Syrovatskii, S. I. 1965, ARA&A, 3, 297
Götz, D., Meuris, A., Pinsard, F., et al. 2016, Proc. SPIE, 9905, 99054L
Grimm, H.-J., Gilfanov, M., & Sunyaev, R. 2003, MNRAS, 339, 793
Grotrian, W. 1939, NW, 27, 214
Gunn, J. E., & Ostriker, J. P. 1969, Natur, 221, 454
Gursky, H., Giacconi, R., Paolini, F. R., & Rossi, B. B. 1963, PhRvL, 11, 530
Haberl, F., & Sturm, R. 2016, A&A, 586, A81
Haberl, F. 2007, Ap&SS, 308, 181
Hall, D. S. 1981, in NATO ASIC Proc. 68: Solar Phenomena in Stars and Stellar Systems, ed. R. M. Bonnet, & A. K. Dupree (Dordrecht: Reidel), 431
Harnden, F. R. Jr+, Branduardi, G., Elvis, M., et al. 1979, ApJ, 234, L51
Harrison, F. A., Aird, J., Civano, F., et al. 2016, ApJ, 831, 185
Harrison, F. A., Craig, W. W., Christensen, F. E., et al. 2013, ApJ, 770, 103
Hartmann, R., Strüder, L., & Kemmer, J. 1997, NIMPA, 387, 250
Helder, E. A., Vink, J., Bykov, A. M., et al. 2012, SSRv, 173, 369
Henke, B. L., Gullikson, E. M., & Davis, J. C. 1993, ADNDT, 54, 181
Henze, M., Darnley, M. J., Williams, S. C., et al. 2018, ApJ, 857, 68
Henze, M., Ness, J. U., Darnley, M. J., et al. 2014, A&A, 563, L8
Hickox, R. C., & Markevitch, M. 2006, ApJ, 645, 95
Hirsh, R. F. 1983, Glimpsing an Invisible Universe. The Emergence of X-ray Astronomy (Cambridge: Cambridge Univ. Press)
Hobbs, G., Lorimer, D. R., Lyne, A. G., & Kramer, M. 2005, MNRAS, 360, 974
Holland-Ashford, T., Lopez, L. A., Auchettl, K., Temim, T., & Ramirez-Ruiz, E. 2017, ApJ, 844, 84
Hughes, J. P., Long, K. S., & Novick, R. 1984, ApJ, 280, 255
Illarionov, A. F., & Sunyaev, R. A. 1975, A&A, 39, 185
Irwin, K. D., & Hilton, G. C. 2005, in Transition-Edge Sensors, Vol. 99 (Berlin: Springer-Verlag), 63
Jaisawal, G. K., & Naik, S. 2016, MNRAS, 461, L97
Janka, H.-T. 2017, ApJ, 837, 84
Jardine, M., & Unruh, Y. C. 1999, A&A, 346, 883
Kahn, S. M., Leutenegger, M. A., Cottam, J., et al. 2001, A&A, 365, L312
Kelley, R. L., Mitsuda, K., Allen, C. A., et al. 2007, PASJ, 59, 77
King, A., & Lasota, J.-P. 2019, MNRAS, 485, 3588
King, A., Lasota, J.-P., & Kluźniak, W. 2017, MNRAS, 468, L59
Kirkpatrick, P., & Baez, A. V. 1948, JOSA, 38, 766
Knoll, G. F. 2000, Radiation Detection and Measurement (New York: Wiley)
Kovetz, E. D., Cholis, I., Breysse, P. C., & Kamionkowski, M. 2017, PhRvD, 95, 103010
Koyama, K., Petre, R., Gotthelf, E. V., et al. 1995, Natur, 378, 255
Krautter, J., Oegelman, H., Starrfield, S., Wichmann, R., & Pfeffermann, E. 1996, ApJ, 456, 788
Krzeminski, W., & Serkowski, K. 1977, ApJ, 216, L45
Kuuttila, J., Gilfanov, M., Seitenzahl, I. R., Woods, T. E., & Vogt, F. P. A. 2019, MNRAS, 484, 1317
Lazzarini, M., Williams, B. F., Durbin, M., et al. 2021, ApJ, 906, 120

Lehmer, B. D., Alexander, D. M., Bauer, F. E., et al. 2010, ApJ, 724, 559
Lehmer, B. D., Basu-Zych, A. R., Mineo, S., et al. 2016, ApJ, 825, 7
Li, J.-T., & Wang, Q. D. 2013, MNRAS, 428, 2085
Liu, Q. Z., & Mirabel, I. F. 2005, A&A, 429, 1125
Liu, Q. Z., van Paradijs, J., & van den Heuvel, E. P. J. 2006, A&A, 455, 1165
Liu, Q. Z., van Paradijs, J., & van den Heuvel, E. P. J. 2007, A&A, 469, 807
Long, K. S., Blair, W. P., Winkler, P. F., et al. 2010, ApJS, 187, 495
Long, K. S., Kuntz, K. D., Blair, W. P., et al. 2014, ApJS, 212, 21
Lucy, L. B., & White, R. L. 1980, ApJ, 241, 300
Luken, K. J., Filipović, M. D., Maxted, N. I., et al. 2020, MNRAS, 492, 2606
Luo, B., Brandt, W. N., Xue, Y. Q., et al. 2017, ApJS, 228, 2
Lutovinov, A. A., Tsygankov, S. S., Krivonos, R. A., Molkov, S. V., & Poutanen, J. 2017, ApJ, 834, 209
Maggi, P., Haberl, F., Kavanagh, P. J., et al. 2016, A&A, 585, A162
Maggi, P., Filipović, M. D., Vukotić, B., et al. 2019, A&A, 631, A127
Maitra, C. 2017, JApA, 38, 50
Mantz, A., Allen, S. W., Rapetti, D., & Ebeling, H. 2010, MNRAS, 406, 1759
Martin, R. G., Nixon, C., Armitage, P. J., Lubow, S. H., & Price, D. J. 2014, ApJ, 790, L34
Maxted, N. I., Filipović, M. D., & Sano, H. 2018, ApJ, 866, 76
Mayer, M., Becker, W., Patnaude, D., Winkler, P. F., & Kraft, R. 2020, ApJ, 899, 138
Meekins, J. F., Fritz, G., Chubb, T. A., & Friedman, H. 1971, Natur, 231, 107
Meidinger, N., Nandra, K., Plattner, M., et al. 2015, JATIS, 1, 014006
Merloni, A., Heinz, S., & di Matteo, T. 2003, MNRAS, 345, 1057
Mewe, R., Heise, J., Gronenschild, E. H. B. M., et al. 1975, Natur, 256, 711
Mi, W., Nillius, P., Pearce, M., & Danielsson, M. 2019, NatAs, 3, 867
Mineo, S., Gilfanov, M., & Sunyaev, R. 2012a, MNRAS, 419, 2095
Mineo, S., Gilfanov, M., & Sunyaev, R. 2012b, MNRAS, 426, 1870
Mirabel, I. F., Dijkstra, M., Laurent, P., Loeb, A., & Pritchard, J. R. 2011, A&A, 528, A149
Moe, M., & Di Stefano, R. 2017, ApJS, 230, 15
Mukai, K. 1993, Legacy, 3, 21
Mushtukov, A. A., Suleimanov, V. F., Tsygankov, S. S., & Poutanen, J. 2015, MNRAS, 454, 2539
Nazé, Y. 2009, A&A, 506, 1055
Ness, J. U., Starrfield, S., Burwitz, V., et al. 2003, ApJ, 594, L127
Ness, J. U., Starrfield, S., Jordan, C., Krautter, J., & Schmitt, J. H. M. M. 2005, MNRAS, 364, 1015
Nomoto, K. 1982, ApJ, 253, 798
Núñez, A., Agüeros, M. A., Covey, K. R., et al. 2015, ApJ, 809, 161
Okazaki, A. T., Hayasaki, K., & Moritani, Y. 2013, PASJ, 65, 41
Orosz, J. A., McClintock, J. E., Narayan, R., et al. 2007, Natur, 449, 872
Pajot, F., Barret, D., Lam-Trong, T., et al. 2018, JLTP, 193, 901
Pallavicini, R., Golub, L., Rosner, R., et al. 1981, ApJ, 248, 279
Palmieri, T. M., Seward, F. D., Toor, A., & van Fland ern, T. C. 1975, ApJ, 202, 494
Pansky, A., Breskin, A., & Chechik, R. 1997, JAP, 82, 871
Patterson, J., & Raymond, J. C. 1985a, ApJ, 292, 535
Patterson, J., & Raymond, J. C. 1985b, ApJ, 292, 550

Pavan, L., Bordas, P., Pühlhofer, G., et al. 2014, A&A, 562, A122
Pavan, L., Pühlhofer, G., & Bordas, P. 2016, A&A, 591, A91
Payne, J. L., Filipovic, M. D., Longo, G., et al. 2019, AAS Meeting 234, 131
Pillepich, A., Reiprich, T. H., Porciani, C., Borm, K., & Merloni, A. 2018, MNRAS, 481, 613
Pittard, J. M., & Parkin, E. R. 2010, MNRAS, 403, 1657
Pizzolato, N., Maggio, A., Micela, G., Sciortino, S., & Ventura, P. 2003, A&A, 397, 147
Pointecouteau, E., Reiprich, T. H., Adami, C., et al. 2013, arXiv:1306.2319
Pollock, A. M. T. 1987, ApJ, 320, 283
Pollock, A. M. T., Corcoran, M. F., Stevens, I. R., & Williams, P. M. 2005, ApJ, 629, 482
Porquet, D., & Dubau, J. 2000, A&AS, 143, 495
Porquet, D., Dubau, J., & Grosso, N. 2010, SSRv, 157, 103
Porquet, D., Mewe, R., Dubau, J., Raassen, A. J. J., & Kaastra, J. S. 2001, A&A, 376, 1113
Predehl, P., Andritschke, R., Arefiev, V., et al. 2021, A&A, 647, A1
Prilutskii, O. F., & Usov, V. V. 1976, AZh, 53, 6
Pringle, J. E., & Rees, M. J. 1972, A&A, 21, 1
Remillard, R. A., & McClintock, J. E. 2006, ARA&A, 44, 49
Revnivtsev, M., & Mereghetti, S. 2015, SSRv, 191, 293
Reynolds, S. P., Borkowski, K. J., Green, D. A., et al. 2008, ApJ, 680, L41
Ribó, M., Munar-Adrover, P., Paredes, J. M., et al. 2017, ApJ, 835, L33
Sano, H., Inoue, T., Tokuda, K., et al. 2020, ApJ, 904, L24
Sasaki, M., Pietsch, W., Haberl, F., et al. 2012, A&A, 544, A144
Schmitt, J. H. M. M., & Liefke, C. 2004, A&A, 417, 651
Sciortino, S., Vaiana, G. S., Harnden, F. R. Jr+, et al. 1990, ApJ, 361, 621
Setti, G., & Woltjer, L. 1973, in X- and Gamma-Ray Astronomy, Vol. 55, ed. H. Bradt, & R. Giacconi (Dordrecht: Reidel), 208
Seward, F. D., Forman, W. R., Giacconi, R., et al. 1979, ApJ, 234, L55
Shakura, N. I., & Sunyaev, R. A. 1973, A&A, 500, 33
Shipman, H. L. 1976, ApJ, 206, L67
Slane, P., Hughes, J. P., Edgar, R. J., et al. 2001, ApJ, 548, 814
Smith, R. K., & Hughes, J. P. 2010, ApJ, 718, 583
Socrates, A., Blaes, O., Hungerford, A., & Fryer, C. L. 2005, ApJ, 632, 531
Stephenson, F. R. 1976, QJRAS, 17, 121
Stephenson, F. R., & Green, D. A. 2003, JAHH, 6, 46
Takahashi, T., Kokubun, M., Mitsuda, K., et al. 2018, JATIS, 4, 021402
Tanaka, Y., Inoue, H., & Holt, S. S. 1994, PASJ, 46, L37
Triendl, R. 2000, Natur, 403, 693
Truemper, J., Pietsch, W., Reppin, C., et al. 1978, ApJ, 219, L105
Tsygankov, S. S., Mushtukov, A. A., Suleimanov, V. F., & Poutanen, J. 2016, MNRAS, 457, 1101
Urry, C. M., & Padovani, P. 1995, PASP, 107, 803
van den Heuvel, E. P. J., Bhattacharya, D., Nomoto, K., & Rappaport, S. A. 1992, A&A, 262, 97
van der Klis, M. 2000, ARA&A, 38, 717
Vennes, S., & Dupuis, J. 2002, in ASP Conf. Ser. 262, in The High Energy Universe at Sharp Focus: Chandra Science, ed. E. M. Schlegel, & S. D. Vrtilek (San Francisco, CA: ASP), 57
Vikhlinin, A., Burenin, R. A., Ebeling, H., et al. 2009, ApJ, 692, 1033
Vikhlinin, A., Kravtsov, A., Forman, W., et al. 2006, ApJ, 640, 691

Vilhu, O. 1984, A&A, 133, 117
Walter, F. M. 2001, ApJ, 549, 433
Webster, B. L., & Murdin, P. 1972, Natur, 235, 37
Wei, J., Cordier, B., Antier, S., et al. 2016, arXiv:1610.06892
Weisskopf, M. C., Cohen, G. G., Kestenbaum, H. L., et al. 1976, ApJ, 208, L125
Weisskopf, M. C., Silver, E. H., Kestenbaum, H. L., Long, K. S., & Novick, R. 1978, ApJ, 220, L117
Weisskopf, M. C., Ramsey, B., & O'Dell, S. 2016, Proc. SPIE, 9905, 990517
Wijnands, R., & van der Klis, M. 1998, Natur, 394, 344
Wolter, H. 1952, AnP, 445, 94
Wongwathanarat, A., Th, H., Janka,, & Müller, E. 2013, A&A, 552, A126
Woods, T. E., & Gilfanov, M. 2013, MNRAS, 432, 1640
Worsley, M. A., Fabian, A. C., Bauer, F. E., et al. 2005, MNRAS, 357, 1281
Yaron, O., Prialnik, D., Shara, M. M., & Kovetz, A. 2005, ApJ, 623, 398
Yew, M., Filipović, M. D., & Roper, Q. 2018, PASA, 35, e015
Zhang, C. M., Wang, J., Zhao, Y. H., et al. 2011, A&A, 527, A83
Zhang, F., Li, X. D., & Wang, Z. R. 2004, ApJ, 603, 663
Zhang, S. N., Santangelo, A., Feroci, M., et al. 2019, SCPMA, 62, 29502
Zhang, W., Giles, A. B., Jahoda, K., et al. 1993, Proc. SPIE, 2006, 324
Zhang, Z., Gilfanov, M., & Bogdán, Á. 2012, A&A, 546, A36
Zwicky, F. 1939, PhRv, 55, 726

Chapter 7

Gamma-Ray Astronomy

Gavin Rowell

Gamma-ray astronomy has advanced tremendously over the past couple of decades with high-impact results in a variety of scientific areas such as high-energy astrophysics, particle cosmology, Standard Model, and beyond-Standard-Model particle physics. This chapter provides an overview of the field by looking at the key results over the past few years and the connections with other areas of astronomy and identifies the key breakthroughs made. It concludes with a summary of the next generation of new gamma-ray facilities that promise to expand the field further into a mainstream astronomical activity.

7.1 Gamma Rays as Tracers of High-energy Particles and New Physics

Gamma rays are nature's highest-energy photons. They can trace elements, fundamental particle physics processes, and accelerated particles, and they can be used to indirectly probe for dark matter and its properties. For example, the fundamental particle physics of gamma-ray production from particles links them to processes producing radio to X-ray photons and neutrinos. Thus, gamma-ray studies of astrophysical particle accelerators like supernova remnants (SNRs), pulsars and their wind nebulae (PWN), black hole environments and jets (both active galactic nuclei (AGNs) and microquasars), and massive star clusters require a multiwavelength and multimessenger approach.

As the highest-energy photons, gamma rays can constrain important parameters of particle acceleration such as shock speed, acceleration timescale, magnetic fields, and the maximum particle energies attained. Gamma rays often represent a significant fraction of nonthermal energy in extreme objects and are extremely important in understanding transient and variable sources. The absorption of gamma rays from distant objects probes the extragalactic background light and allows the study of star formation and galaxy populations.

doi:10.1088/2514-3433/ac2256ch7

Gamma-ray astronomy is the most recent photon window into the universe to be exploited. Gamma rays are the highest-energy photons and theoretically extend to energies beyond $\sim 10^{20}$ eV (matching the highest-energy cosmic rays detected), although to date the highest-energy gamma rays actually measured have energies of about 10^{14} eV. Figure 7.1 illustrates the multimessenger spectrum and highlights the various underlying physics processes.

We will concentrate our discussion here on gamma-ray astronomy at energies above about 100 MeV, which has seen great advances in the past two decades. We will, however, include a historical overview of MeV to multi-TeV gamma-ray astronomy. The electron rest mass at 0.51 MeV is generally considered to be the starting energy of gamma rays although the term "soft" gamma rays is often applied to facilities working down to 0.1 MeV or so. Gamma rays in the energy range $\sim$0.1 to 10s of MeV (MeV gamma-ray astronomy) are associated with nuclear spectral lines from supernova and massive stars, the annihilation of electron–positron pairs, and broadband gamma rays from low-energy cosmic rays and electrons (via the same processes that produce >GeV gamma rays, which we will discuss later). Some further discussion of MeV gamma-ray astronomy results and perspectives may be found in Pinkau (2009), Roland (2016), and De Angelis et al. (2017).

7.1.1 A Brief History of MeV to Multi-TeV Gamma-Ray Astronomy

It could be argued that the field attracted serious attention in the mid-1960s with the observations of bursts of gamma rays by the Vela satellites launched for the US military (Klebesadel et al. 1973). This followed the first detection of gamma rays in 1962 by the Explorer XI satellite (Kraushaar & Clark 1962). The mid-1960s also saw many efforts with high-altitude balloons that proved unsuccessful (see review by Weekes 1988 as well as Book 1, Chapter 1.3.6). However, astronomy was only possible, at least in part, once precise locations of the gamma-ray emission could be pinpointed, leading to the situation today where many gamma-ray sources are now spatially resolved down to the arcminute level.

The OSO-3 space mission (Kraushaar et al. 1972) provided the first glimpse of the gamma-ray sky in MeV to GeV gamma rays, revealing large-scale emission along

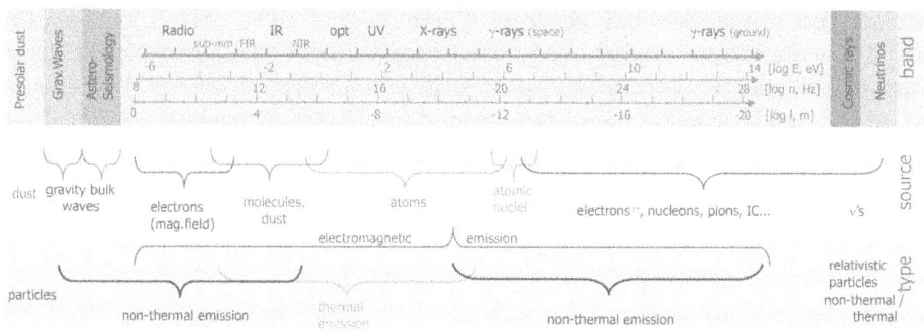

Figure 7.1. Multimessenger spectrum with underlying physics processes responsible for the photons and particles. Image credit: Roland (2016).

the plane of the Milky Way, clearly indicating its origin in our own galaxy. This was followed by the SAS-2 (Fichtel et al. 1975) and COS-B (Bignami & Hermsen 1983) missions. The latter established about 25 discrete gamma-ray sources and permitted the first detailed counterpart studies. These groundbreaking detectors utilized the conversion of gamma rays into a cascade of electrons, positrons, and lower-energy gamma rays via a spark chamber. Using this approach with better technology, the late 1990s saw the launch of the EGRET mission on board the Compton Gamma-Ray Observatory (CGRO).

EGRET greatly expanded the number of MeV to GeV gamma-ray sources to over 200 (Hartman et al. 1999) and enabled the first detailed views of the diffuse emission along the Galactic Plane and made significant contributions to the understanding of pulsars. Although EGRET's capabilities allowed the discrete gamma-ray sources to be pinpointed to better than 1°, the first real hints of the challenges that lay ahead came with the fact that a large fraction of the Galactic Plane sources were unidentified in that they had no obvious counterparts, a situation that continues today for GeV gamma-ray astronomy (see, e.g., Thompson 2015).

The Fermi-LAT (Large Area Telescope; Atwood et al. 2009) was launched in 2008 and replaced EGRET, following de-orbiting of the CGRO in 2000. Still operating today, Fermi-LAT utilizes the pair-creation tracking technique to reconstruct the arrival direction of gamma rays. In this process, incoming gamma rays create electron–positron pairs, which then go on to create ion tracks in the silicon strip tracking detectors. The silicon strip detectors measure the direction of these pairs, which enables reconstruction of the incoming gamma-ray to a precision approaching 0.1° at energies >10 GeV. Fermi-LAT has been a great success in revolutionizing GeV gamma-ray astronomy with the detection of over 5000 sources in the 0.1 to 100 GeV-energy band (Abdollahi et al. 2020).

Moving to higher energies, gamma-ray astronomy from multi-GeV to beyond 100 TeV energies relies on detection with ground-based telescopes. This is due to extraterrestrial gamma-ray photon fluxes following a steeply falling power law in their energy distribution and hence space-based telescopes do not have sufficient collection area to gather useful photon statistics.

Gamma rays of energy above a few GeV will, upon entering Earth's atmosphere, initiate an extensive air shower (EAS) of secondary particles (predominantly electron and positrons) that generate Čerenkov radiation. The most sensitive ground-based telescopes today image this Čerenkov light to reconstruct the arrival direction and energy of the gamma-ray. For gamma rays of several TeV energies and above, and depending on the ground-level altitude above sea level, the secondary EAS particles can be detected at ground level, and several telescopes have exploited this avenue effectively.

The huge size of the Čerenkov light pool on the ground (approaching 1 km^2) and large spread of the EAS particles nicely compensate for the low photon fluxes in the TeV energy range, enabling the ground-based telescopes to carry out highly sensitive studies of steady and time-varying sources. For some recent detailed reviews of the history and various techniques, see Hillas (2013), Mirzoyan (2014), de Naurois & Mazin (2015), and Fegan (2019).

The early days of ground-based gamma-ray astronomy were devoted to establishing the observational existence of Čerenkov light from EAS and in 1953, the first such detection was made using a photomultiplier (PMT) coupled to a searchlight mirror (Galbraith & Jelley 1953). Nesterova & Čudakov (1955) were to later repeat this in 1955. The late 1950s and early 1960s saw the first multipixel arrays (about 10 to 20) of PMTs being used to revealed the somewhat extended nature of the Čerenkov EAS image (Brennan et al. 1958; Boley & Macoy 1961; Millar et al. 1962). However, the use of an image intensifier and phosphor storage system by Hill & Porter (1961), who photographed details of the Čerenkov image elongation, clearly demonstrated that the EAS arrival direction could be determined to within about 0.5°.

Following Hillas' (1985) proposal to use a detailed parameterization of the Čerenkov image shape (the angular distribution), and related ideas put forward by Pliasheshnikov & Bignami (1985), it was not until 1989 that the first convincing detection of TeV gamma rays came (9σ statistical significance above the cosmic ray, CR, background; Weekes et al. 1989). The source in question was the Crab Nebula (a PWN), recognized as the standard candle for TeV astronomy. This was achieved using the 37 PMT pixel camera installed on the Whipple telescope in Arizona, USA. Interestingly in the 1970s, a related method to detect Čerenkov light using widely separated telescopes in Narrabri, Australia (the "double-beam" technique) provided a tantalizing hint for a signal from our nearest AGN, Centaurus A (Grindlay et al. 1975), but this method was never developed further.

By the late 1990s, the power of stereoscopic Čerenkov imaging with more than one telescope was proven by the HEGRA IACT System (Daum et al. 1997), as well as the benefits of a high-resolution camera pixelation on a single telescope by CAT (Barrau et al. 1998). Besides the imaging Čerenkov technique, several other methods have proved themselves, with the most successful being the direct detection of EAS particles at ground level via an array of large water tanks in the MILAGRO experiment (Abdo et al. 2007b), and to a certain extent, Čerenkov detection via its lateral distribution with the use of thermal solar towers (e.g., Gingrich et al. 2005; Paré et al. 2002).

Today, ground-based gamma-ray observatories such as HESS (High energy stereoscopic system), https://www.mpi-hd.mpg.de/hfm/HESS/; VERITAS (Very energetic radiation imaging telescope array system), https://veritas.sao.arizona.edu/; and MAGIC (Major atmospheric gamma imaging cherenkov), https://magic.mpp.mpg.de/ (stereoscopic air Čerenkov arrays); ARGO-YGJ (Aielli et al. 2006) (Resistive Plate Counter array); HAWC, High-altitude water cherenkov, www.hawc–observatory.orghttps://www.hawc-observatory.org/ (water Čerenkov array following on from MILAGRO; Smith 2005) lead the way at TeV energies, while the Fermi-LAT space mission at GeV energies still operates.

7.2 The Status of MeV to TeV Gamma-Ray Astronomy

In this section we will review the present-day situation in gamma-ray astronomy, starting with an overview of the catalogs of sources across the gamma-ray bands.

Figure 7.2 presents an all-sky view of some all-sky views based on selected source catalogs spanning the sub-MeV to the multi-TeV gamma-ray bands.

The catalogs shown here (from Swift-BAT, Fermi-LAT, and TeVCat (combining results from all TeV gamma-ray facilities TevCat: catalog of TeV gamma-ray sources, http://tevcat2.uchicago.edu/) focus primarily on discrete sources, either

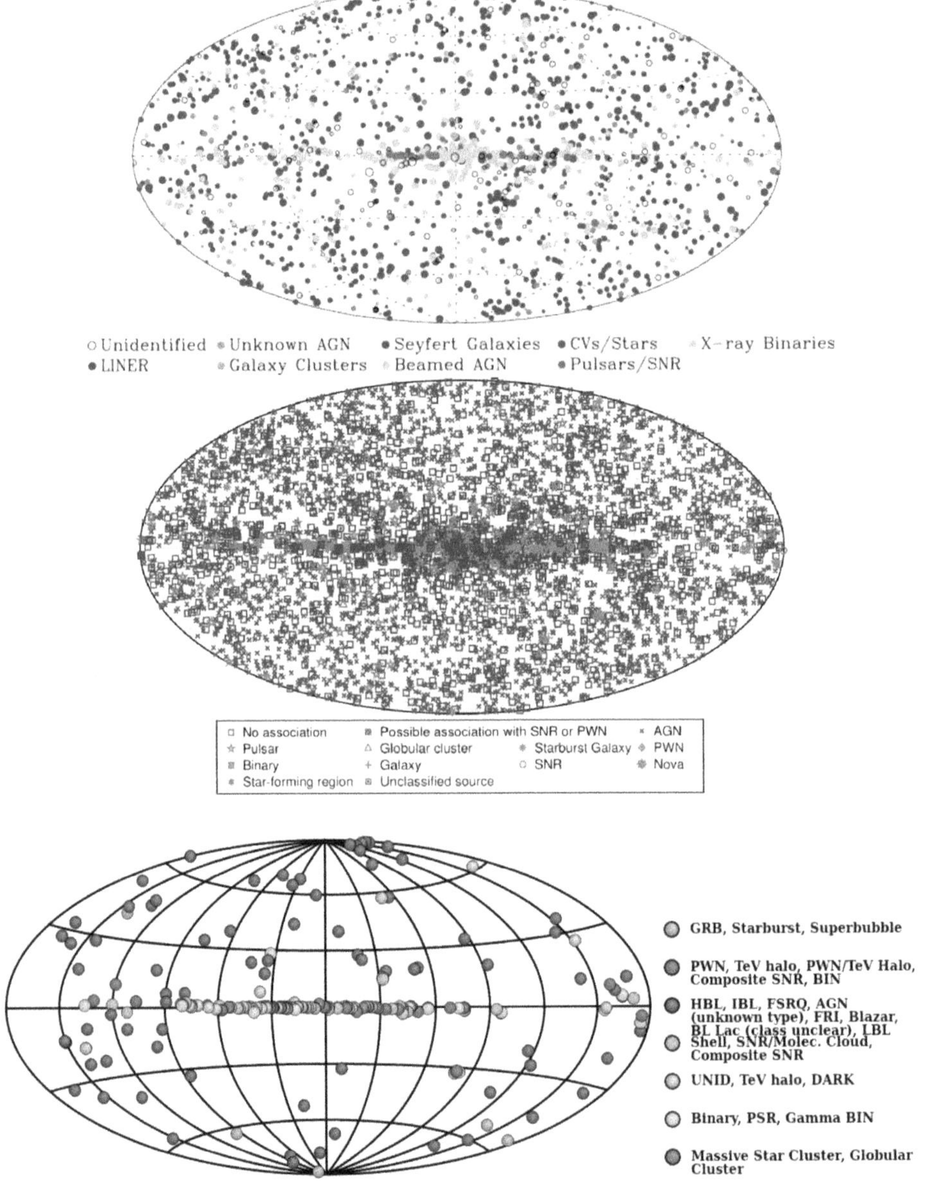

Figure 7.2. All-sky gamma-ray source catalogs in various bands: sub-MeV (top: 14 keV to 0.2 MeV; Oh et al. 2018), GeV (middle: 0.1 to 100 GeV; Abdollahi et al. 2020) and TeV (bottom: >0.1 TeV; TeVCat: catalog of TeV gamma-ray sources, http://tevcat2.uchicago.edu/).

steady or variable in flux. Another important catalog is that from COMPTEL (0.7–30 MeV; Schönfelder et al. 2000), which provided the first detailed look at the MeV gamma-ray sky and paved the way for the Swift-BAT and INTEGRAL missions, which have cataloged many hundreds of sources up to a few MeV (Oh et al. 2018; Bird et al. 2007; Bouchet et al. 2008). The AGILE mission overlaps Fermi-LAT in the GeV band and has provided its catalog (Bulgarelli et al. 2019).

We note a number of these facilities (e.g., Fermi-LAT, Ajello et al. 2019; Fermi-GBM, von Kienlin et al. 2020; and Swift-BAT, Lien et al. 2016) have separate catalogs devoted to one-off transient sources such as gamma-ray bursts (GRBs) in the hard X-ray to soft gamma-ray band up to 1 MeV.

Additionally, there are several degree-scale features not tied to individual sources, such as the ^{26}Al line at 1.809 MeV from COMPTEL and INTEGRAL (Diehl et al. 1995; Bouchet et al. 2015) tracing nuclear reactions in massive star and core-collapse supernovae, the e^{+}/e^{-} annihilation line at 0.511 MeV (Knödlseder et al. 2005) tracing electron production, and the Galactic Plane diffuse emission tracing the broad population of Galactic cosmic rays and electrons (Ackermann et al. 2012).

What is abundantly clear from these catalogs is the wide variety of source types that emit gamma rays. In the sub-MeV and MeV bands, accreting sources such as X-ray bursts (XRBs; high mass and low mass) and AGNs tend to dominate the Galactic and extragalactic populations, respectively. In the GeV and TeV bands, AGNs (with various subtypes) dominate the extragalactic population, while the Galactic population is more varied with PWNe, SNRs, pulsars, gamma-ray binaries, globular clusters, star-forming regions being represented.

Notably, the fraction of unidentified gamma-ray sources has become significant in the GeV and TeV bands, particularly in the Galactic population. These sources have no obvious counterparts in other wavelengths, and this fact has motivated news efforts to probe deeper with renewed observations across the radio to X-ray domains. We will leave further discussion to later sections devoted specifically to the Galactic and extragalactic source populations.

It is, however, worth noting the growth in source numbers over time. Figure 7.3 charts the growth of X-ray and gamma-ray source numbers up to 2019 in what is known as the "Kifune" plot, a logarithmic format first shown by Kifune (1996).

What is obvious is that the growth rate of source numbers across the X-ray and gamma-ray domains is rather similar, reflecting the development of new facilities. To date there are now over 5000 and 200 sources in the GeV and TeV gamma-ray domains, respectively.

7.3 GeV to TeV Gamma-Ray Production

In this section we will briefly review the physics behind GeV to TeV gamma-ray production and see how such gamma-ray emission is deeply connected to relativistic particles and to the production of other photons spanning radio waves to X-rays.

The mechanism most often invoked to explain the existence of relativistic particles is shock acceleration (in some extreme environments, particle acceleration by electric fields and by reconnection in magnetic fields is also discussed). More specifically, the so-called diffusive shock acceleration process (e.g., Drury 1983) appears to operate in a variety of astrophysical sources, as we will highlight later. An overview of shock acceleration is presented in Book 1 (Section 7.3). We will here provide just a summary of the spectral properties of the accelerated particles to underpin our discussion of gamma-ray production. Diffusive shock acceleration leads to a nonthermal energy distribution of relativistic particles, described mathematically as a power law. A simple form for the resulting differential particle flux number versus energy dN/dE or particle spectrum, is given by

$$dN/dE = KE^{-\Gamma} \exp(-E/E_c), \tag{7.1}$$

where Γ is the power-law spectral index, E_c is a cutoff energy determining the strength of the exponential term, and K a normalization. The spectral index can be expressed in the form $\Gamma \approx (r + 1)/(r - 1)$, where the shock compression factor $r = 4$ for strong shocks, thus giving $\Gamma = 2$ (variation to this are predicted according to shock properties and nonlinear effects). The exponential term encapsulates the upper-energy limit of the accelerator, and such limits are imposed by the shock environment (e.g., shock speed, size, and magnetic field—see Hillas 1985) and/or

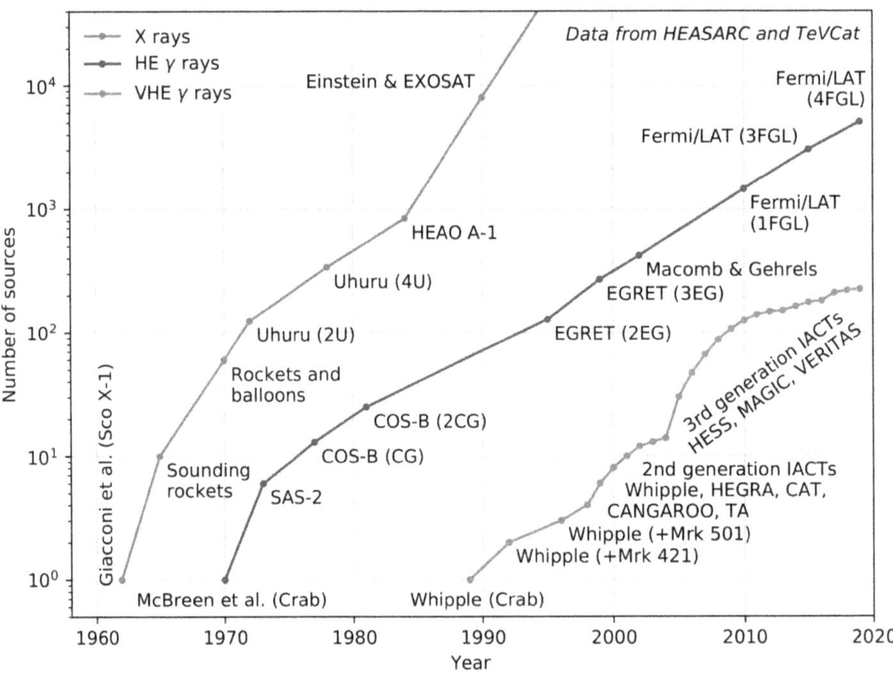

Figure 7.3. The latest "Kifune" plot charting the growth of X-ray and gamma-ray source numbers over time to 2019, available from https://github.com/sfegan/kifune-plot, Copyright 2017, Stephen Fegan. The data for the X-ray and high-energy (HE gamma-ray) source counts are from "How Many Known X-Ray (and Other) Sources Are There?", https://heasarc.gsfc.nasa.gov/docs/heasarc/headates/how_many_xray.html and the very high-energy (VHE) or TeV gamma-ray source counts are from TeVCat: catalog of TeV gamma-ray sources, http://tevcat2.uchicago.edu/.

radiative losses such as synchrotron emission usually for electrons (see Zirakashvili & Aharonian 2007). We refer the reader to Aharonian (2004) for an in-depth look at these topics.

GeV to TeV gamma rays are produced by relativistic cosmic rays (protons, nuclei) and electrons in hadronic and leptonic radiative processes respectively. The spectral shape of the resulting gamma rays is heavily influenced by the particle spectrum properties (Equation (7.1)).

The hadronic process involves the inelastic collision of cosmic rays with ambient protons or nuclei and is often labeled proton–proton (or pp) collisions. The ambient protons/nuclei (N) are often in the form of interstellar gas either, neutral (predominantly), or ionized and proceed according to the following reactions:

Hadronic–Cosmic rays p of energy $E_{\rm p}$ colliding with matter nuclei N:

$$p + N \rightarrow N' + \pi^{\circ}, \pi^{+}, \pi^{-} \ (\pi \approx \text{equal numbers}) \tag{7.2}$$

$$\pi^{\circ} \rightarrow \gamma\,\gamma \ \left(E_{\gamma} \sim 0.17 E_{\rm p}\right) \tag{7.3}$$

$$\pi^{+} \rightarrow \mu^{+} + \nu_{\mu} \rightarrow e^{+} + \nu_{e} + \overline{\nu_{\mu}} \tag{7.4}$$

$$\pi^{-} \rightarrow \mu^{-} + \overline{\nu_{\mu}} \rightarrow e^{-} + \overline{\nu_{e}} + \nu_{\mu}. \tag{7.5}$$

Thus, the GeV to TeV gamma rays with energy E_{γ} result from π° decay and the decay of charged $\pi^{\pm}$ leads to muons, electrons, and their associated neutrinos. The neutrinos are detected by telescopes such as IceCube and KM3Net (see Chapter 8). Interestingly, the electrons generated via the decay of muons can then provide secondary radio to X-ray synchrotron emission.

The energy-loss rate for cosmic-ray protons interacting with a ambient medium of number density n (typical units cm^{-3}) is given by $dE_p/dt = (n\sigma_{\rm pp} fc)E_{\rm p}$, where $f \sim 0.5$ (Gaisser et al. 2016) is the inelasticity for single interactions, $\sigma_{\rm pp}$ is the total cross section for pp collisions, and c is the speed of light. Figure 7.4 presents $\sigma_{\rm pp}$ versus proton kinetic energy $T_{\rm P} = E_{\rm P} - m_{\rm p}c^2$.

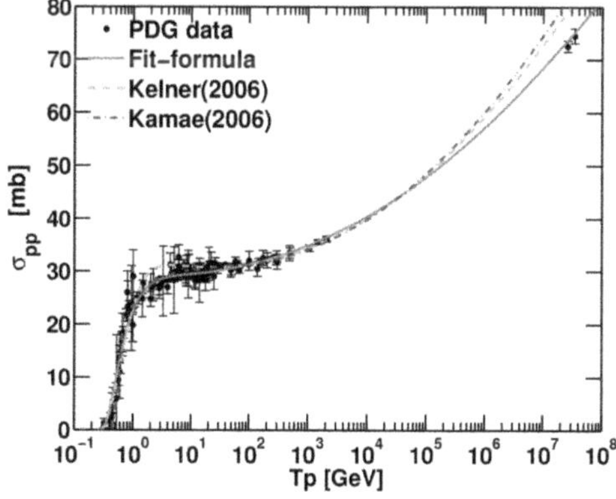

Figure 7.4. Total pp cross section $\sigma_{\rm pp}$ versus proton kinetic energy $T_{\rm P}$ with functional parameterizations given by Kafexhiu et al. (2014). Image adapted with permission from Kafexhiu et al. (2014), Copyright 2014 by the American Physical Society.

From this we can see that *pp* collisions become important for energies above a few GeV. Below this energy, ionization losses are dominant (see Padovani et al. 2009).

For electrons, several processes are relevant, namely inverse-Compton (IC) scattering, synchrotron emission, and bremsstrahlung.

Leptonic–Inverse-Compton Scattering:

TeV electrons will upscatter soft photons such as the cosmic microwave background CMB and local IR, optical, and UV components to TeV energies via the IC process[1]:

$$e^{-}+\gamma_{\rm soft} \rightarrow e^{'-}+\gamma_{\rm TeV}. \tag{7.6}$$

Assuming a monoenergetic soft photon energy ω with number density $n_{\rm ph}$, the IC energy-loss rate for electrons with Lorentz factor $\gamma = E_{\rm e}/(m_{\rm e}c^2)$ is $dE_{\rm e}/dt = \frac{4}{3}\sigma_{\rm T}c\omega n_{\rm ph}\gamma^2$. Here, σ_T is the Thompson cross section, and the energy-loss rate expression given above is valid for the Thompson regime where $b = 4\omega\gamma < <1$ (typical for the CMB, and IR soft photon targets and TeV electrons). Following Aharonian (2004), the Klein–Nishina (KN) limit occurs for $b > >1$ (often on optical, UV fields, or >100 TeV electrons on the CMB) in which case the energy-loss rate follows $dE_{\rm e}/dt = \frac{3}{8}(\sigma_{\rm T}cn_{\rm ph}(\ln b - 11/6))/\omega$. An extreme version of IC is synchrotron self-Compton (SSC) emission (often invoked in AGN and GRB jets), whereby electrons upscatter their own X-ray synchrotron photons to TeV energies.

Leptonic–Synchrotron Radio to X-Rays (electron acceleration in magnetic fields):

Electrons will be accelerated in a magnetic field via the Lorentz force (see Book 1 for a discussion). Here, we just note the electron's energy-loss rate given by the expression $dE_{\rm e}/dt = \frac{4}{3}\sigma_T c U_{\rm B}\gamma^2$, where $U_B = B^2/8\pi$ is the magnetic field energy density (cgs units) and B is the magnetic field flux density (Gauss). As a rule of thumb in a magnetic field typically found in our Galaxy (few microgauss), GeV and TeV electrons will give rise to radio and X-ray synchrotron emission respectively. For extreme magnetic fields such as that around neutron stars, a related emission process known as curvature radiation becomes important as relativistic electrons follow their highly curved magnetic field lines.

Leptonic–Bremsstrahlung:

Electrons can also convert their energy to TeV photons upon scattering in the field of a nucleus (within ambient matter of density $n_{\rm p}$). The energy-loss rate is $dE_{\rm e}/dt = cm_{\rm p}n_{\rm p}(E_{\rm e}/X_o)$ for the radiation length $X_o = \frac{7}{9}(\rho/(n_{\rm p}\sigma_o)$ (gm cm^{-2}). Here, σ_o is the pair production cross section, m_p is the proton mass, and $\rho = m_{\rm p}n_{\rm p}$ is the matter mass density. Bremsstrahlung is usually subdominant unless the ambient matter density is very large, $n > 10^4$ cm^{-3}, and the B-field low (<few microgauss), which may be found in some environments, such as large quiescent clouds. The IC and synchrotron process will therefore generally dominate and thus compete for the electron's energy.

[1] Also see Section 2.1.3 and Book 1, Section 7.3.1.

An extremely useful way to characterize the importance of these process is via their so-called cooling time $t = E/\dot{E}$, which is the time taken for a population of particles to radiate all of their energy away at a loss rate $\dot{E} = dE/dt$. A short cooling time implies fast conversion of particle energy to photon energy, and hence, such a process will dominate over other competing processes with a longer cooling time. This issue is particularly important for electrons where synchrotron and IC radiation losses usually compete. The cooling times for all four processes discussed above in units of years are

$$t_{\rm pp} = \left(n\sigma_{\rm pp}fc\right)^{-1} \approx 5.3 \times 10^7 (n/{\rm cm}^3)^{-1} \ {\rm yr}, \tag{7.7}$$

$$t_{\rm IC} \approx 3 \times 10^5 (U_{\rm rad}/{\rm eV/cm}^3)^{-1} (E_{\rm e}/{\rm TeV})^{-1}) f_{\rm KN}^{-1} \ {\rm yr}, \tag{7.8}$$

$$t_{\rm Sync} \approx 12 \times 10^6 (B/\mu{\rm G})^{-2} (E_{\rm e}/{\rm TeV})^{-1} \ {\rm yr}, \tag{7.9}$$

$$t_{\rm Br} \approx 4 \times 10^7 (n/{\rm cm}^3)^{-1} \ {\rm yr}, \tag{7.10}$$

where $E_{\rm e}$ is the electron energy in TeV units. The KN suppression factor $f_{\rm KN} \approx (1 + b)^{-1.5}$ is included in the IC cooling time.

7.4 Gamma Rays and Multimessenger Links

From the previous section we can now appreciate the intimate links between gamma-ray emission and nonthermal emission at lower energies from radio up to X-rays. Here, we will discuss some relevant examples.

The close connection between the TeV gamma-ray IC $F_{\rm IC}$ and X-ray synchrotron $F_{\rm sync}$ fluxes was pointed out by Aharonian et al. (1997). By taking the ratio of their cooling times $t_{\rm IC}/t_{\rm Sync}$, a simple expression for the flux ratios can be found such that $F_{\rm IC}/F_{\rm sync} = 10(B/10\mu{\rm G})$. Here, a δ-function approximation for the synchrotron and IC cross sections, and IC scattering of the CMB in the Thompson regime are assumed (Book 1, Chapter 4).

Thus, a comparison of the TeV gamma-ray and X-ray synchrotron fluxes can provide an independent estimate of the magnetic field B in an extreme environment. Because the synchrotron process also extends down to the radio band, we can see the close connections across the radio to TeV gamma-ray bands in the context of leptonic processes. Furthermore, the secondary synchrotron component as a by-product of the hadronic process provides additional links to the radio to X-ray bands.

Multiwavelength photon spectra from the hadronic and leptonic processes can be computed by integrating over the underlying particle spectrum (Equation (7.1)) multiplied by the relevant cross sections and particle to photon conversion efficiencies. For leptonic processes the relevant cooling times can be smaller than the age of the source (e.g., often for synchrotron emission versus a few kiloyear age for an SNR). This will lead to a significant alteration or evolution of the electron spectrum, and this should be tracked to ensure the correct "evolved" electron spectrum and subsequent synchrotron, IC, and bremsstrahlung photon components are calculated. Another influence will be the time-dependent nature of the particle accelerator, and these are broadly categorized into "impulsive" (e.g., SNRs operating for a few thousand years) or "continuous" (e.g., particle injection from a pulsar in a PWN) types. The reader is referred to Aharonian & Atoyan (1996), Kelner et al. (2006), Manolakou et al. (2007), and Hinton & Hofmann (2009) for further discussion and useful expressions.

As an illustrative example, Figure 7.5 presents some conceptual energy spectra arising from a particle spectrum of the form given by Equation (7.1).

From here, we can see that observations in the radio, X-ray, and gamma-ray bands are crucial in constraining the properties of the underlying relativistic particles. An overriding scientific goal is to discriminate the leptonic and hadronic processes in order to progress our understanding of accelerator properties. Because, as is apparent in Figure 7.5, there are a variety of spectral shapes characteristic of the leptonic and hadronic processes, detailed spectral studies across the radio, X-ray, and gamma-ray bands can help discriminate the two processes.

However, spectral discrimination is not always feasible due to limited statistics, narrow energy coverage within a band, limited gamma-ray angular resolution, competing thermal sources of radio and X-ray emission, and complex particle propagation properties that can create a variety of spectral shapes (e.g., see Aharonian & Atoyan 1996). The next-generation gamma-ray facilities such as the Cherenkov Telescope Array (CTA Consortium et al. 2019) aims to address many of these issues in the TeV gamma-ray band with its arcminute angular resolution and wide energy coverage (spanning about three decades). Further discussion of CTA and other next-generation facilities is left for a later section.

Another crucially important way to discriminate hadronic from leptonic processes is via their distinct spatial signatures expected in relation to the ambient interstellar gas. As shown in Equations (7.2) and (7.7), gamma-ray emission is directly coupled to the ambient gas properties such as its density n. More specifically,

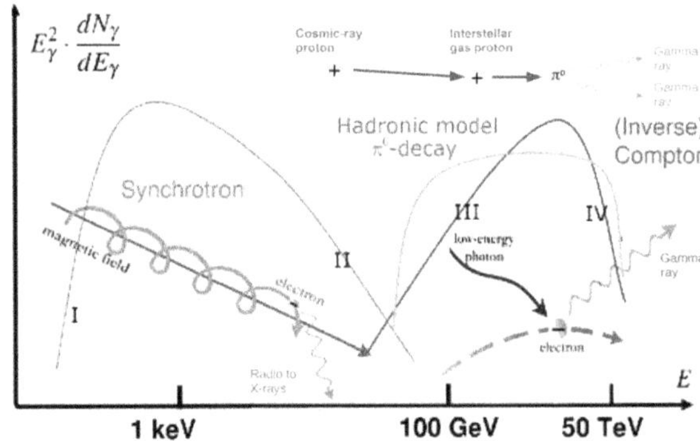

Figure 7.5. Conceptual spectral energy distributions of photons from leptonic (synchrotron, inverse Compton) and hadronic (π^0-decay) processes. Image adapted from Spurio (2015), with permission of Springer. Leptonic bremsstrahlung (not shown) is often subdominant compared to the inverse-Compton component.

the gamma-ray flux from this process $F_\gamma \propto n_{\rm CR}(E_{\rm p})n$, where $n_{\rm CR}$ is the energy-dependent density of cosmic rays interacting with the ambient gas. Thus, the gamma-ray emission to first order is expected to spatially correlate with the ambient gas density (or its mass), assuming $n_{\rm CR}(E_{\rm p})$ does not vary too much with the gas (we will later discuss how this can occur). Because the ambient gas can be traced by a variety of spectral lines, usually in the radio band, this relationship outlines a further important linkage between gamma-ray and radio astronomy.

Already there are several examples of gamma-ray sources (discussed later) spatially matching associated gas clouds traced by spectral lines such as the carbon monoxide (CO) and carbon monosulfide (CS) molecules (tracing molecular hydrogen H_2), as well as the H I spin–flip transition tracing atomic hydrogen. A number of interstellar gas surveys (Dame et al. 2001; Mizuno & Fukui 2004; Jackson et al. 2006; Walsh et al. 2011; Jones et al. 2012; Taylor et al. 2003; McClure-Griffiths et al. 2005; Stil et al. 2006) have been exploited so far, as well as recent surveys (e.g., Braiding et al. 2018; Umemoto et al. 2017; Wang et al. 2020; Dickey et al. 2013) reaching higher resolution ($0.5'$; not all yet completed). In such sources the hadronic process is implied for the gamma-ray emission, signaling the presence of cosmic-ray accelerators. We note that the leptonic bremsstrahlung process (Equation (7.10)) follows a similar relationship with the ambient gas density, but as discussed earlier, this is very rarely dominant due to stronger electron energy losses via the synchrotron and IC processes.

The hadronic process via Equations (7.4) and (7.5) also signals an obvious linkage to neutrino astronomy (another process via cosmic-ray and photon interactions that also produces neutrinos—see Chapter 8). The neutrino emission will also spatially correlate with the ambient gas density under the same assumptions given above for the gamma-ray emission. Additionally, the spectral signature of neutrinos will be somewhat similar to that of hadronic gamma rays (see, e.g., Kelner et al. 2006). As neutrinos are an unambiguous signature of hadronic interactions, they are seen as the smoking gun to confirm the presence of cosmic rays, and this issue has been a major driver behind facilities like IceCube and KM3NeT, as discussed in Chapter 8.

7.5 Galactic TeV Gamma-Ray Sources

In this section we will review the Galactic TeV gamma-ray source population and highlight some key results that have emerged. Where necessary, we will make reference to the GeV gamma-ray source population as revealed by the Fermi-LAT mission which now lists over 5000 sources (Abdollahi et al. 2020). Figure 7.6 presents Galactic surveys maps at TeV gamma-ray energies from the HESS and HAWC facilities (HESS Collaboration et al. 2018a; Albert et al. 2020d) covering the southern and northern skies, respectively. These are to date the most comprehensive Galactic Plane surveys at these energies. Smaller-scale surveys have been published by HESS (Galactic Center region, HESS Collaboration et al. 2018e), VERITAS (Cygnus region, Abeysekara et al. 2018a; Galactic Center

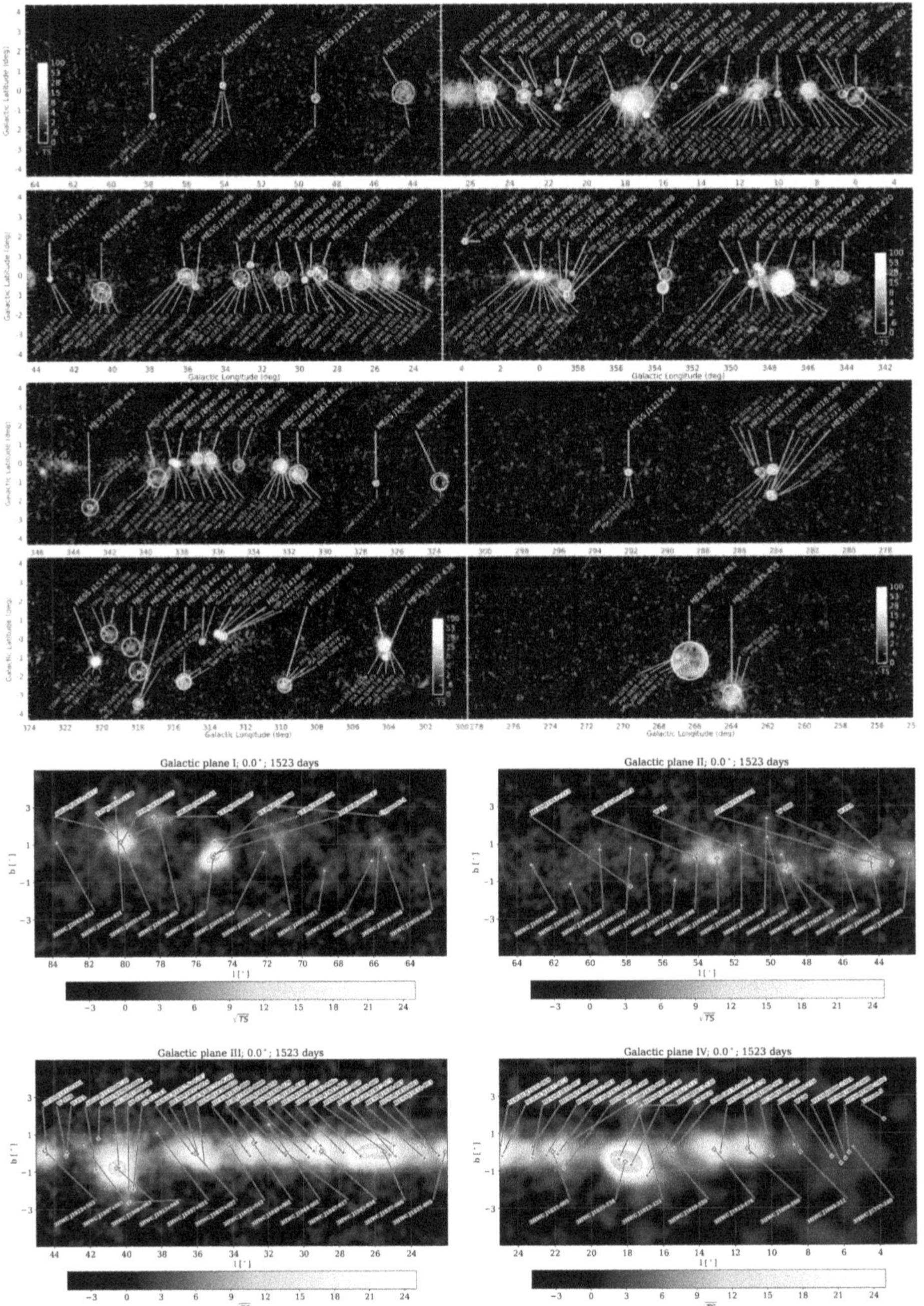

Figure 7.6. TeV gamma-ray survey sky maps from HESS (top four rows) (HESS Collaboration et al. 2018a, reproduced with permission © ESO) and HAWC (bottom four rows) (Albert et al. 2020d) covering the southern and northern Galactic Plane, respectively (note: not all coverage is shown here). Counterpart names are also listed. The sky-map color scales are in units of $\sqrt{\mathrm{TS}}$, representing the statistical significance of gamma-ray features above background fluctuations due to cosmic-ray events.

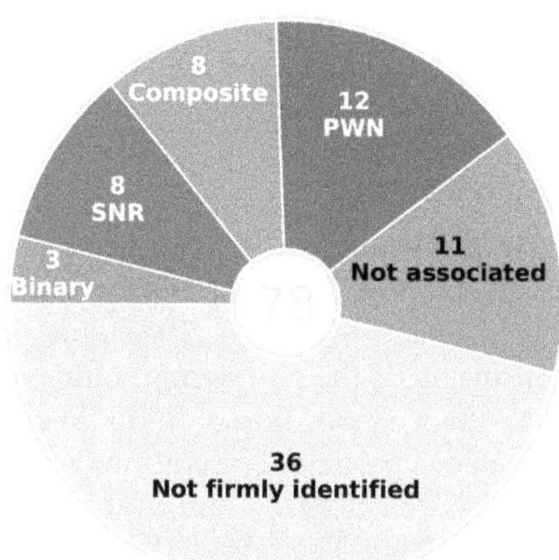

Figure 7.7. TeV gamma-ray source identifications from the HGPS (HESS Collaboration et al. 2018a, reproduced with permission © ESO).

region, Archer et al. 2016), and MAGIC (Galactic Center region, MAGIC Collaboration et al. 2020b).

The HESS Galactic Plane Survey (HGPS; HESS Collaboration et al. 2018a), focusing on the southern hemisphere, has revealed 78 Galactic TeV gamma-ray sources, the bulk of the Galactic population so far. A total of 31 of these sources are associated with multiwavelength counterparts of the following types: SNRs (shell-type), PWNe, composite (composite SNR with interior PWN), binary (binary systems containing an interacting compact object and massive star). These associations follow from morphological, positional, and variability evidence that link the TeV sources to their counterparts at lower energies, usually in the radio and X-ray bands. The good angular resolution of HESS, in the 3′ to 5′ range, has enabled many of these identifications.

A striking aspect of the survey is the large fraction of TeV gamma-ray sources that remain unidentified, with 36 sources listed as not firmly associated with any counterpart although there may be one nearby (for example, an adjacent SNR or a powerful pulsar potentially driving a PWN), and 11 sources with no hints for any association at all. The second HAWC catalog (Abeysekara et al. 2017b) focusing on the northern hemisphere lists 39 TeV sources. Due to HAWC's large (~1 sr) field of view, it also covered the extragalactic sky. To date, 65 sources have been cataloged, with 20 new sources and the rest associated with known TeV sources (from HESS, VERITAS, or MAGIC). Although its angular resolution at 12′ to 60′ is not as good as that for HESS, HAWC's large field of view and 24 hr operation has enabled it to reveal a number of low-brightness, degree-scale gamma-ray sources, as well as some source reaching energies beyond 50 TeV. Indeed, HAWC's highest-energy catalog has revealed nine sources with emission >56 TeV and three with emission >100 TeV (Abeysekara et al. 2020). All of these sources are spatially close to sources previously identified by HAWC, HESS, VERITAS, and/or MAGIC, and thus, they likely represent their upper-energy extremes. We will discuss some of these extreme sources in the context of PeVatrons later in Section 7.5.6.

7.5.1 Shell-type Supernova Remnants

Shell-type SNRs have been discussed as the main accelerators of Galactic cosmic rays for over 80 years. It was Baade & Zwicky (1934) who first raised this possibility, and then was then followed up by Morrison (1958) and Ginzburg & Syrovatskii (1964), who considered the origin of the nonthermal radio emission in SNRs. Later, diffusive shock acceleration theory operating in SNR shock fronts was put forward as the ideal mechanism to achieve particle acceleration (Krymskii 1977; Axford et al. 1977; Bell 1978; Blandford & Ostriker 1978), and over time, this has been developed further (e.g., Drury 1983; Bell 2004). The reviews by Hillas (2005), Letessier-Selvon & Stanev (2011), and Aharonian (2013) are suggested reading for further details on this topic.

Some key fundamental arguments favoring SNRs are

1. The luminosity of Galactic cosmic rays ($\sim 10^{41}$ erg s^{-1}) can be met by the supposed Galactic SNR population (roughly three per century) requiring about 10% to 30% of an SNR's kinetic energy of 10^{51} erg being converted to cosmic rays, and
2. Diffusive shock acceleration operating in SNRs can provide a cosmic-ray power-law index of $\Gamma \approx 2.0$, in agreement with the observed cosmic-ray spectral index after accounting for propagation.

The first predictions of hadronic TeV gamma-ray emission (heavily dependent on the factor n/d^2, for gas density n and SNR distance d) from SNRs from Drury et al. (1994), Naito & Takahara (1994), and Aharonian et al. (1994) provided impetus for studies by the first Čerenkov imaging facilities. Following a number of unconvincing claims and upper limits, the first solid detection of the very young SNR Cas A was made by the stereoscopic HEGRA IACT System (Aharonian et al. 2001).

Today, over a dozen SNRs are identified as TeV gamma-ray sources (and over 20 as GeV gamma-ray sources), either via their shell-like morphology, positional overlap with the radio or X-ray shell, or overlap with adjacent interstellar gas. The gamma-ray shell-like morphology, seen in the very young (<500 yr) and young SNRs ($<10^4$ yr) is interpreted via leptonic and/or hadronic processes. The leptonic process is invoked due to the presence of shell-like nonthermal synchrotron X-ray emission (identified via featureless broadband X-ray spectra), signaling the presence of multi-TeV electrons. Evidence for the hadronic process in the very young and young SNRs has come from consideration of the clumpy gas inside and adjacent to the SNR shocks. The mature SNRs ($>10^4$ yr) have gamma-ray emission that tend to overlap large adjacent interstellar gas clouds, and this has been interpreted as evidence for the hadronic process in which cosmic rays are escaping the SNR shock and propagating into the surrounding gas. These SNRs have likely evolved beyond the so-called Sedov–Taylor phase in which the bulk of the cosmic rays are accelerated. The hadronic interpretation is further backed up by the general lack of nonthermal X-ray emission associated with the mature SNRs because the X-ray synchrotron cooling time is typically less than 10^4 yr (that is, the nonthermal X-ray emission only lasts for a few thousand years). The gamma-ray spectral shapes of the SNRs have provided additional clues. Figure 7.8 summarizes a number of GeV to TeV gamma-ray spectra for SNRs spanning a range of ages.

Here, we can see some clear differences in spectral shapes. The mature SNRs ($>10^4$ years) tend to exhibit GeV-peaked spectra with rather soft spectral indices ($\Gamma \approx 2.6$) in the TeV band. Such a spectral shape is consistent with the energy-dependent escape and diffusion of cosmic rays and their interaction with adjacent gas clouds (e.g., Aharonian & Atoyan 1996; Gabici 2011). The presence of the "pion bump" or sharp rolloff below 135 MeV (the rest mass–energy of π°; Ackermann et al. 2013) for some mature SNRs is also seen as strong evidence for the hadronic process.

For the young SNRs with strong peaks in the TeV band, both hadronic and leptonic processes have supporting evidence. The hard GeV spectral shape ($\Gamma < 2$) is particularly suggestive of the IC process. However, the energy-dependent interaction of cosmic rays with the clumpy gas is believed to suppress the GeV emission, thereby allowing a hadronic interpretation.

The very young SNRs (two examples so far: Cas A and Tycho) show broad GeV to TeV emission with a peak at a few GeV. Their GeV to TeV spectral broadness is consistent with the hadronic process (in contrast to the leptonic process producing a narrower IC peak). A tentative hint of increasing GeV emission from SN 1987A, now believed to have entered the SNR phase, was recently reported by Malyshev et al. (2019). We should also point out that gamma-ray emission from supernova events (SNe) linked to the initial cataclysmic collapse and powered by the SN ejecta interacting with the progenitor circumstellar medium has attracted attention (Waxman & Loeb 2001; Murase et al. 2011). Indeed, a recent hint for fading transient GeV emission has been reported from SN2004dj (d = 3.5 Mpc) (Xi et al.

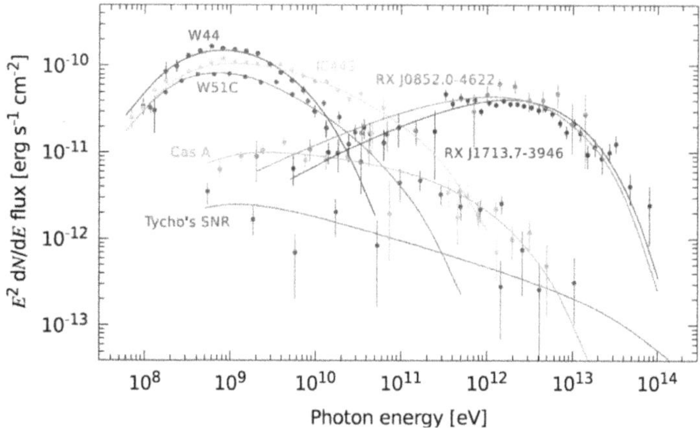

Figure 7.8. GeV to TeV gamma-ray spectra from a number of shell-type SNRs. Very young (<500 yr, Cas A, Tycho) SNRs are shown in green, young SNRs (<10^4 yr, RXJ1 713.7–3946, RXJ0 852.0–4622) in red and mature SNRs (>10^4 yr, W44, W51C, IC 443) shown in blue/cyan. Image credit: Becker Tjus & Merten (2020), with permission from Elsevier.

2020a), which is a Type IIP SNe where such interactions might occur. We will now discuss some SNR examples to highlight the progress made so far.

The young, 1600 yr old TeV-bright SNR RXJ1713.7−3946 is probably the most studied of all gamma-ray SNRs to date. It exhibits a clear shell-like morphology in the radio, X-rays, GeV, and GeV to TeV gamma-ray bands. Formally, both pure leptonic and hadronic scenarios have been proposed using feasible parameters for the total particle energy budgets ($\sim 10^{47-48}$ erg in electrons and $\sim 10^{49-50}$ erg in cosmic-ray protons), although somewhat low magnetic fields and/or intense optical photon fields are required in the leptonic scenario (HESS Collaboration et al. 2018c). A mixed scenario (e.g., Zirakashvili & Aharonian 2010; Gabici & Aharonian 2014) in which the hadronic process arises from dense gas clumps inside the SNR has gained acceptance. Significant effort has gone into studying the interstellar gas toward RXJ1713.7−3946 in an effort to further examine the viability of any hadronic interpretation and disentangle the hadronic and leptonic components (e.g., Moriguchi et al. 2005; Fukui et al. 2012; Maxted et al. 2012; Sano et al. 2013, 2020). Figure 7.9 presents the latest HESS >2 TeV and earlier XMM-Newton X-ray images from HESS Collaboration et al. (2018c) and Acero et al. (2009b). The GeV gamma-ray and radio continuum images can be found in Lazendic et al. (2004) and Federici et al. (2015).

The study by Fukui et al. (2012) revealed a general correspondence between the interstellar gas and the TeV gamma-ray emission in an annular region focusing on the SNR rim, providing the first suggestion of a global hadronic component. In recent years, attention has focused on the energy-dependent penetration of cosmic rays into dense gas clumps (first pointed out by Gabici et al. 2007). In the limited age of the SNR, the GeV cosmic rays do not penetrate far into the dense clumps due to magnetic turbulence from the SNR shock disturbance (Zirakashvili & Aharonian 2010; Inoue et al. 2012; Celli et al. 2019), thereby providing a natural explanation for the suppressed GeV emission. The most recent work by Sano et al. (2020) mapped the interstellar molecular gas on arcsecond scales and in fact confirmed the very clumpy nature of the gas downstream from the SNR shock.

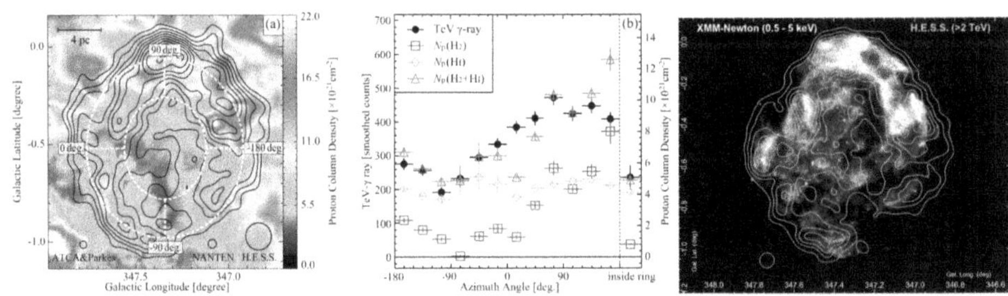

Figure 7.9. Left and middle panels: interstellar gas column density (N_p) image with HESS TeV contours (from Aharonian et al. 2007b) and their resulting azimuthal comparison from Fukui et al. (2012). Right panel: HESS >2 TeV (cyan contours; 10 to 40 counts/pixel) and XMM-Newton (0.5 to 5 keV X-ray; 0 to 1.27×10^{-4} photons cm^{-2} s^{-1} $arcmin^{-2}$) images using data from HESS Collaboration et al. (2018c) and Acero et al. (2009b). The HESS point-spread function (68% containment radius) is shown as the cyan circle at the bottom left. The XMM-Newton resolution at 6″ is too small to show.

The mature SNR W28 is well known for the extensive adjacent molecular gas surrounding its radio shell (and center filled with thermal X-ray emission). A cluster of 1720 MHz OH masers toward the northeast cloud is evidence of an SNR shock and interstellar gas interaction (Frail et al. 1994; Claussen et al. 1997). Since the discovery of TeV emission overlapping the molecular gas (Aharonian et al. 2008c), W28 has become a prime example to explore the escape of cosmic rays from SNRs and diffusive propagation to the ambient gas as the OH maser emission confirms the close proximity of the gas. Figure 7.10 illustrates the multiwavelength picture toward W28, including the radio SNR shell, TeV gamma-ray emission, and surrounding interstellar gas from various tracers (Maxted et al. 2017).

The TeV gamma-ray emission around W28 remains the best example of a TeV/gas overlap and along with the lack of any nonthermal X-ray emission, the clearest example of where the hadronic process is at work. Interpretation of this and the GeV emission (also from other SNRs with adjacent gas) has looked at the details of cosmic rays escaping from SNRs, the effect of diffusion properties, and the role of accelerated ambient cosmic rays (e.g., Fujita et al. 2009; Yan et al. 2012; Nava & Gabici 2013; Gabici 2014; Cui et al. 2018; Tang 2019).

Another potential example of a mature SNR is HESS J1832−085, which was until recently thought to be an unidentified TeV gamma-ray source. Using the low-frequency radio GLEAM survey by MWA (Hurley-Walker et al. 2017) and VLA survey by Anderson et al. (2017), a radio SNR counterpart of G23.11+0.18 was identified by Maxted et al. (2019). The same authors also found the SNR overlaps with interstellar gas surrounding a bubble in the 4–5 kpc distance range and no shell-like nonthermal X-ray emission. These observations generally support a mature SNR origin and highlight the close relationship between the hadronic gamma-ray

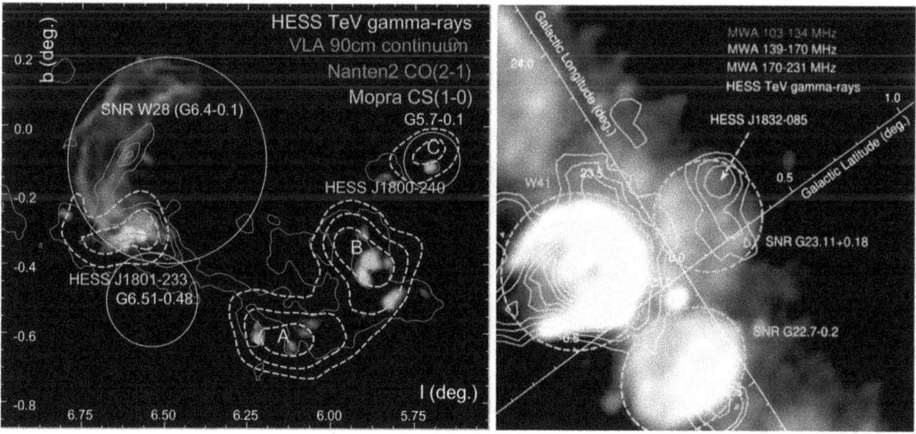

Figure 7.10. Left panel: multiwavelength view of the W28 SNR (G6.4−0.1). The HESS TeV gamma-ray contour (dashed cyan) are overlaid. Other information includes the VLA radio continuum and molecular gas traced by the Nanten2 and Mopra radio telescopes. Other SNRs are indicated with yellow circles. Image credit: Maxted et al. (2017), with the permission of AIP Publishing. Right panel: MWA radio image of the SNR G23.11+0.18 (dashed cyan) with HESS TeV contours (solid yellow) are overlaid. Image credit: Maxted et al. (2019).

and radio synchrotron emission, which have similarly long cooling times ($>10^4$ yr) as per Equations (7.7) and (7.9).

Finally, we want to highlight three additional shell-like TeV gamma-ray sources (HESS Collaboration et al. 2018d), which are suspected to be mature SNRs. At the time of its discovery, HESS 1534−571 was identified as a likely counterpart to the candidate radio SNR G232.7−1.0 (Green et al. 2014; Maxted et al. 2018), making this the first SNR to be discovered in gamma rays. HESS J1912+101 is linked to a TeV "halo" HAWC source (2HWC J1912+099), a GeV source (Zhang et al. 2020), and may also have a polarized radio counterpart (Reich & Sun 2019). On the other hand, HESS J1614−518 still has no obvious counterpart(s).

7.5.2 Pulsar Wind Nebulae (PWNe) and Gamma-Ray Halos

Pulsar-powered PWNe represent the most numerous type of identified Galactic TeV gamma-ray source (Figure 7.7) and a small fraction (< few percent) of the identified GeV Galactic sources (HESS Collaboration et al. 2018a; Abdollahi et al. 2020). The gamma-ray emission is attributed to the IC process from multi-TeV electrons accelerated at a termination shock well away from pulsar's magnetosphere. A termination shock is set up by a pressure balance between the pulsar's electron wind and that of the surrounding SNR ejecta and ambient gas and electrons accelerated at this shock create the PWN. The Crab Nebula (a.k.a. M1 or SN 1054) is the archetype PWN and the first established TeV gamma-ray source (Weekes et al. 1989). Much of the theory behind PWNe and the relativistic flow of electrons and positrons driving them was developed to explain the Crab Nebula's steady emission from the radio to the X-ray bands (e.g., Pacini & Salvati 1973; Rees & Gunn 1974; Kennel & Coroniti 1984; Reynolds & Chevalier 1984 and review by Gaensler & Slane 2006). The Crab Nebula emission is well explained by the SSC model due to the high magnetic field, ~0.3 mG, required, whereas external IC emission (on soft fields such as the CMB) is invoked for other PWNe.

Since the recent discovery of two degree-scale low-surface-brightness TeV sources from nearby pulsars by HAWC (Abeysekara et al. 2017a), the concept of "gamma-ray halos" (or specifically, "TeV halos"; Linden et al. 2017) has become accepted as a probable late stage of PWNe. Figure 7.11 provides a sketch (Giacinti et al. 2020) of the three main stages of PWNe evolution discussed today (see also Blondin et al. 2001).

During stage 1 ($\lesssim$ 10 kyr), the SNR forward shock (FS) propagates outward and a reverse shock (RS) forms as it slows down. The contact discontinuity (CD) separates the SNR ejecta and the forward-shocked interstellar medium (ISM). Beyond the pulsar wind termination shock (TS), the electrons, driven by the pulsar wind, form the PWN which is still inside the SNR. This stage is often when we would see "composite" SNRs (in radio to gamma rays), where the SNR shock is visible as shell-like radio emission and the interior harbors the PWN, which is bright in X-rays and gamma rays (via synchrotron and IC emission, respectively). In stage 2 (10 to 100 kyr), the SNR has evolved beyond the Sedov–Taylor phase and has started to dissipate into the ISM. The PWN morphology is influenced by the incoming RS and

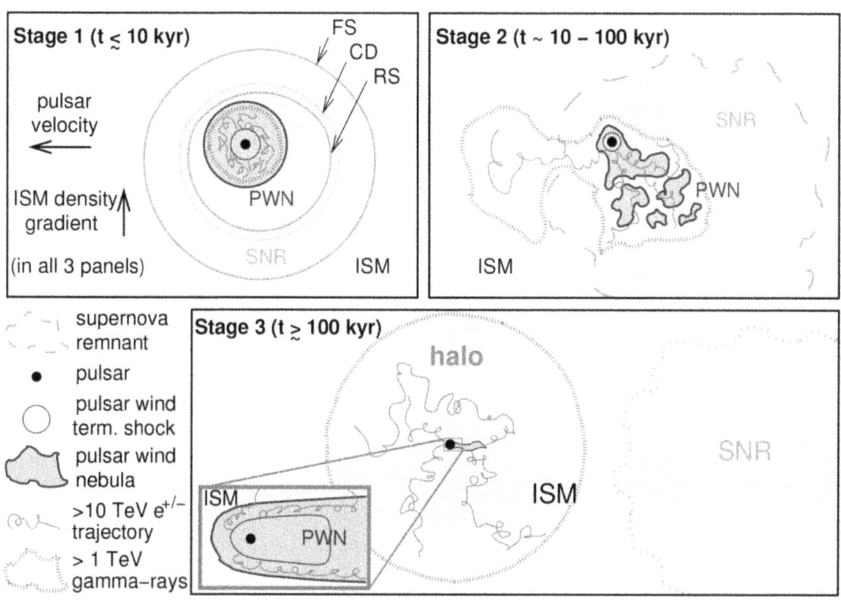

Figure 7.11. Three stages in the evolution of a PWN. Image credit: Giacinti et al. (2020), reproduced with permission © ESO.

can become irregular. The PWN follows the kick velocity of the pulsar and has likely migrated to the edge of the SNR. The PWN electrons begin to escape the PWN boundary and propagate away. In stage 3 (>100 kyr), the PWN has migrated well away from the SNR, most of the PWN electrons have escaped and now propagate diffusively in the undisturbed interstellar gas, and little or no sign of the SNR remains. The "TeV halo" signature of a PWN can be a feature from stage 2 onwards.

The evolution described above provides the morphological clues to identify PWNe, with the most obvious ingredient being a nearby pulsar. The X-ray morphologies of PWNe are extremely varied, reflecting the unique conditions under which they were formed and evolve in (e.g., interstellar gas, supernova properties, pulsar kick velocity; see Reynolds et al. 2017). The associated TeV gamma-ray emission is usually much more spatially extended than that of the X-rays, reflecting the short synchrotron cooling times near the pulsar (high B-field well above 10 μG and decreasing with distance from the pulsar) versus the longer IC cooling time, which is independent of the magnetic field.

From the HGPS, the identified and candidate PWNe are so far mostly linked to relatively young pulsars with spin-down power $\dot{E} > 10^{35}$ erg s^{-1}, and their TeV gamma-ray efficiency improves with pulsar characteristic age τ_c and physical offset from the TeV PWN emission peak (HESS Collaboration et al. 2018b; Figure 7.12). This behavior follows an accepted model of PWN evolution (see HESS Collaboration et al. 2018b) consistent with stage 1 and/or stage 2 as explained above. The trends can also be explained in simple terms via the PWN's decreasing magnetic field with age giving rise to an increased fraction of electron energy going

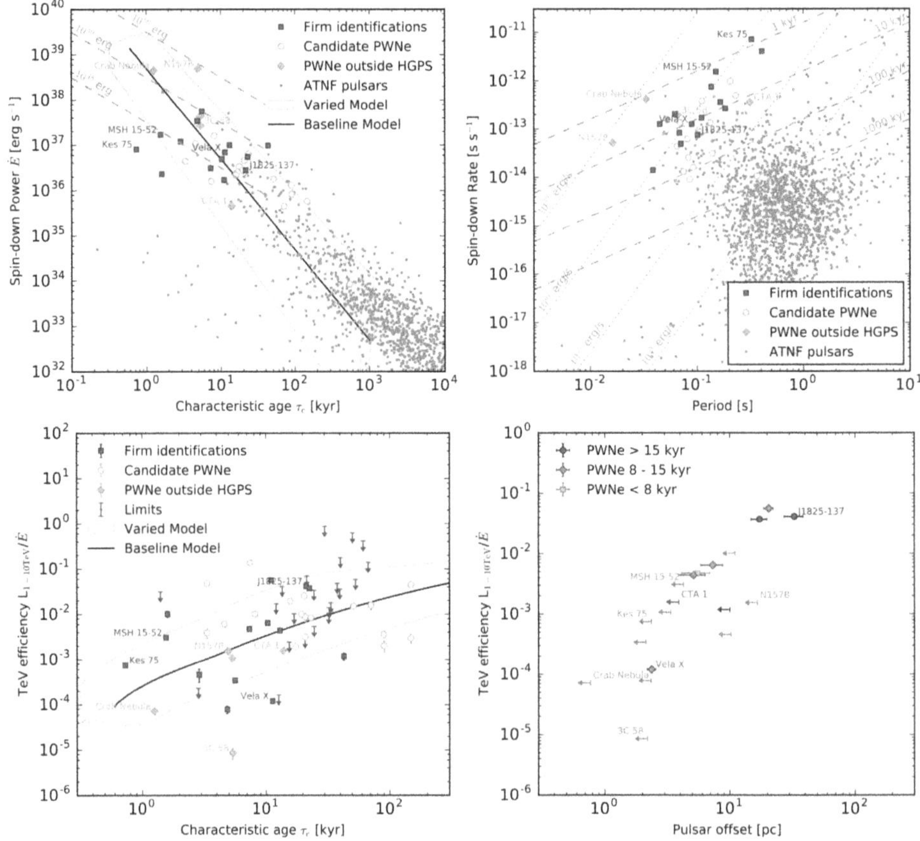

Figure 7.12. Summary of pulsar properties according to their TeV PWN status. Top left to bottom right: pulsar spin-down power $\dot{E}$ versus characteristic age τ_c; pulsar spin-down rate versus period; TeV luminosity efficiency ($L_{1-10\,\mathrm{TeV}}/\dot{E}$) versus τ_c; and TeV luminosity efficiency versus pulsar physical offset from the TeV PWN center. Images credit: HESS Collaboration et al. (2018b), reproduced with permission © ESO.

into the IC process at the expense of X-ray synchrotron emission (see discussion in Section 7.4).

The offset aspect is explained of course by a pulsar's migration distance increasing with age. The TeV halos extending up to 5° in diameter detected by HAWC are found toward two rather old ($\tau_c > 10^5$ yr) nearby pulsars, Geminga and PSR B0656 +14 (associated with the Monogem SNR), less than 300 pc distant with moderate spin-down power (few $\times 10^{34}$ erg s^{-1}). These two examples would be clear outliers against the trends determined by HESS Collaboration et al. (2018b) in Figure 7.12 as expected, given they represent the much older stage 3. We note the recent study by Di Mauro et al. (2021) suggests the presence of GeV halos around some of the bright HAWC TeV sources, indicating that lower-energy GeV electrons may be escaping the PWN region during stage 3. Additionally, Sudoh et al. (2019) has suggested many of the unidentified TeV sources could be related to TeV halos.

Identifying PWNe among the large number of unidentified TeV sources is a key challenge. An obvious morphological signature, at least for stage 1 and stage 2, is a gamma-ray spectral hardening (higher fraction of high-energy gamma rays) toward the underlying pulsar. The most prominent example here is HESS 1825−137 (Aharonian et al. 2006d), the brightest and one of the largest TeV PWN, where GeV spectral evolution has also recently been seen (Principe et al. 2020). Similar evidence has been seen in HESS J1303−631 and the Vela PWN (HESS Collaboration et al. 2012; Reynolds et al. 2017). Figure 7.13 shows the compelling spectral evolution in the TeV band for HESS J1825−137.

The spectral evolution can be used in a relatively simple way to infer the leptonic nature of the TeV gamma-ray emission, and we can use HESS J1825−137 as an example. This PWN is powered by the pulsar PSR J1826−1334 with period $P = 101$ ms, spin-down power $\dot{E} = 3 \times 10^{36}$ erg s^{-1}, characteristic age $\tau_c \sim 20$ kyr, and dispersion-measure distance $d \sim 4$ kpc. The spectral steepening with distance from the pulsar requires emission cooling times to be less than or roughly equal to the source age (20 kyr). Looking at the cooling times for IC $t_{\rm IC}$ on the CMB photon field and synchrotron $t_{\rm Sync}$ emission (Equations (7.8) and (7.9)) for electrons of energy 60 TeV (producing 1 TeV gamma rays and few keV X-rays) and a magnetic field of $B = 10\ \mu$G (certainly expected near the pulsar), we can see that $t_{\rm Sync} \sim 1$ kyr <<

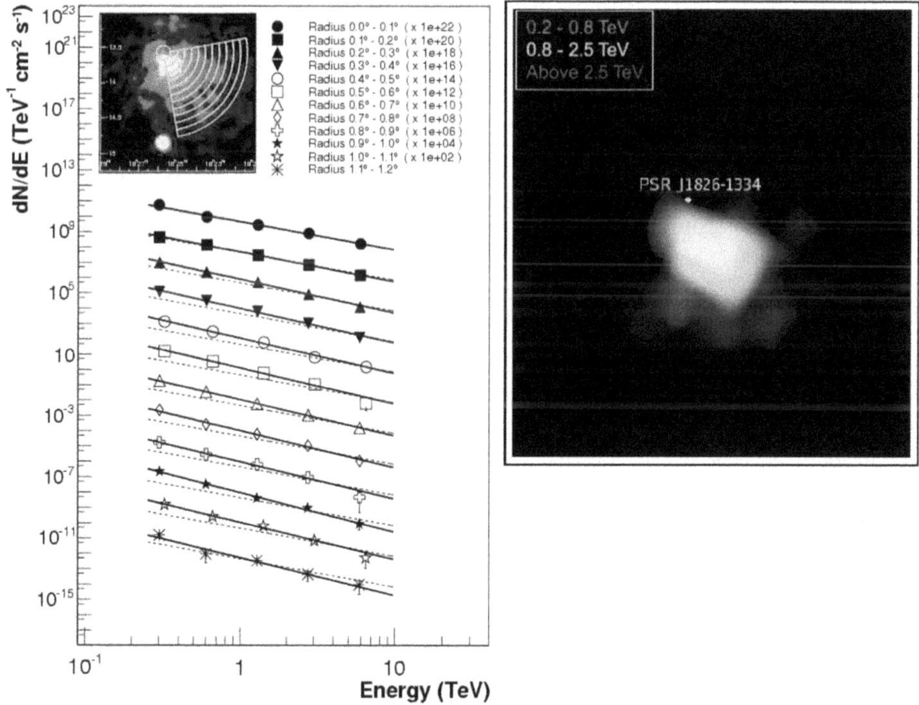

Figure 7.13. Left: spectral steepening of HESS J1825−137 versus distance from the pulsar PSR J1826−1334 (Image credit: Aharonian et al. 2006d, reproduced with permission © ESO). Above: multi-energy-band image of HESS J1825−137 graphically showing the spectral steeping. Image credit: Funk et al. (2008).

$t_{iC} \sim 20$ kyr. Thus, both IC and synchrotron cooling will occur, with the latter being dominant near the pulsar. The PWN magnetic field will decrease with distance (e.g., $B \sim r^{-0.7}$ for radius r) from Van Etten (2011), and so we expect the IC process to gradually become more dominant farther from the pulsar. This gives rise to a compact X-ray synchrotron nebula close to the pulsar with the TeV gamma-ray emission extending much farther out. This is indeed what is seen toward HESS J1825 −137 and many other PWNe. In a hadronic scenario, we note that $t_{pp} >> 20$ kyr by some orders of magnitude as the gas density should be very low ($n < 0.1$ cm^{-3}) as argued below. Thus, the hadronic scenario is implausible unless extreme turbulence can diffusively trap cosmic rays for a very long time in the PWN, an unlikely situation, and at odds with the large PWN extension. In fact, additional transport (advection) beyond diffusion may be needed to explain the 100 pc diameter of the PWN (HESS Collaboration et al. 2019).

Another way to help identify PWNe is by studying the interstellar gas toward them. Because PWNe are expanding into a core-collapse SNR bubble, the interstellar gas density there should be quite low ($n < 0.1$ cm^{-3} or so). Thus, a void or bubble in the interstellar gas is a telltale sign. Additionally, the general asymmetric morphology of PWNe is largely influenced by the inhomogenous interstellar gas surrounding the SNR (because the SNR reverse shock is influenced by this gas) and thus moderate to large gas clouds can be expected to be found adjacent to PWNe. The expected distance to the gas or a void in the gas (measured by a radio spectral line with a distance-dependent Doppler shift) can further be constrained according to the underlying pulsar's dispersion measure. Indeed, a large ($>10^4$ M$_\odot$) gas cloud has been found to the north of HESS J1825−137 (Lemiere et al. 2005; Voisin et al. 2016), which helps explain the development of this PWN in a southerly direction. Additional studies of the gas toward other PWN and PWN candidates from the HGPS have recently been carried out (Voisin et al. 2019), revealing several potential gas cloud "partners" to these sources.

During the 2007 to 2010 period, the AGILE and Fermi-LAT missions detected several large flares from the Crab Nebula (Tavani et al. 2011; Abdo et al. 2011). Further flares have been detected since then up to late 2014 (Arakawa et al. 2020). These flares came as a complete surprise, and immediately changed the notion of PWNe as steady, reliable sources. Indeed, up to that point, the Crab Nebula was long considered the standard candle in gamma-ray astronomy (as it is in many other photon bands). The rapid nature of the flares suggests a powerful acceleration mechanism requiring a very strong magnetic field ~1 mG (slightly above the field invoked in the SSC model to explain the steady PWN emission as discussed earlier). Even more revealing was the maximum energy of the flares, which reached 500 MeV, challenging the theoretical maximum energy that electrons could attain due to synchrotron losses under induced electric field conditions in magnetospheric shocks. Acceleration via magnetic reconnection within current sheets was a way to avoid this problem, and this mechanism (well studied as an acceleration mechanism in solar flares) is now often invoked.

We conclude this section with a brief comment on the emerging class of magnetar wind nebulae (MWNe; reviewed by Reynolds et al. 2017). Magnetars are extremely magnetic neutron stars (fields ~1000× stronger than regular neutron stars) powered largely by magnetic field energy, rather than spin down due to rotational losses. As a result, their outflows and nebulae are quite temporal in nature and often associated with giant flaring and other transient events. Small-scale X-ray nebulae have been detected around several magnetars (although some may be interpreted as dust-scattered halos) and thus present opportunities for the potential to generate gamma-ray emission. We discuss one example magnetar (SGR 1806–20) in Section 7.5.5 due to the presence of overlapping extended TeV emission.

7.5.3 Pulsars

The pulsed emission from neutron stars is driven by their rotation and originates from within the neutron star's magnetosphere. The GeV Fermi-LAT mission has dramatically expanded the number of GeV gamma-ray pulsars over the past decade (Caraveo 2014; Grenier & Harding 2015; Abdollahi et al. 2020), and the current list[2] extends to 253 examples as of April 2020. Among these, 118 are millisecond pulsars, making them a significant fraction of the total, with the rest labeled as "young" pulsars. About 30% of the young pulsars are radio quiet in that their pulsations are detected in the gamma-ray band, not the radio. The MeV to GeV gamma-ray emission is generally believed to originate farther out (in the so-called outer gap and/or slot gap regions) from the pulsar compared to radio emission region thought to be associated close to the neutron star surface in the polar cap region. In recent years, however, gamma-ray emission models involving reconnection acceleration in a current sheet just beyond the light cylinder (where the tangential speed of magnetic field lines reach the vacuum speed of light) have gained some favor. The beam-opening angle of the gamma-ray emission is therefore somewhat wider than that of the radio beam.

The MeV to GeV gamma-ray emission mechanism is usually discussed as curvature radiation, although synchrotron emission has been discussed in the context of acceleration within current sheets. Inevitably, the extreme magnetic field experienced by the gamma rays in this region will mean that they can be readily converted into electron/positron pairs, which can then generate their own synchrotron emission and create secondary lower-energy gamma rays in a cascade process (e.g., Daugherty & Harding 1982, 1983). The end result is that the gamma-ray spectrum exhibits a strong exponential cutoff in the multi-GeV-energy regime and can often be parameterized as

$$dN/dE = KE^{-\Gamma}\exp(-E/E_c)^b, \tag{7.11}$$

where b is the "sharpness" of the cutoff, and the other terms are defined as per Equation (7.1). From the second Fermi-LAT pulsar catalog (Abdo et al. 2013), E_c is typically a few GeV, Γ is in the range 1 to 1.8, and b is unity for all pulsars except for

[2] Public list of LAT pulsars: https://confluence.slac.stanford.edu/display/GLAMCOG/Public+List+of+LAT-Detected+gamma-ray+Pulsars.

a few where $b < 1$, suggesting a subexponential cutoff (see also Abdollahi et al. 2020 for a somewhat different functional form).

Due to the GeV exponential cutoff, detecting pulsed emission at TeV gamma-ray energies has been a much greater challenge owing to the >20 GeV threshold of ground-based gamma-ray telescopes. However, the past decade has seen rapid improvement in the <50 GeV performance of telescopes such that four pulsars are now detected—Crab Pulsar (Aliu et al. 2008; Ansoldi et al. 2016), Vela Pulsar (HESS Collaboration et al. 2018g), PSRB 1706−44 (Spir-Jacob & Djannati-Ata 2019), and Geminga (MAGIC Collaboration et al. 2020a). Not surprisingly, these four are the brightest pulsars in the Fermi-LAT pulsar catalog. Figure 7.14 shows the pulsed energy spectra for three of these pulsars.

The most remarkable aspect is the clear evidence for an additional pulsed spectral component extending beyond the exponential cutoff form determined by Fermi-LAT for the Crab and Geminga pulsars. The extra component is now measured up to 1 TeV in the case of the Crab. The curvature radiation (and synchrotron) model used to explain the pulsed GeV emission is somewhat challenged by these results, as it seems to require rather extreme conditions to reach TeV energies (e.g., Viganò & Torres 2015). Instead, the IC upscattering of X-ray photons by electrons accelerated in the outer gap region at

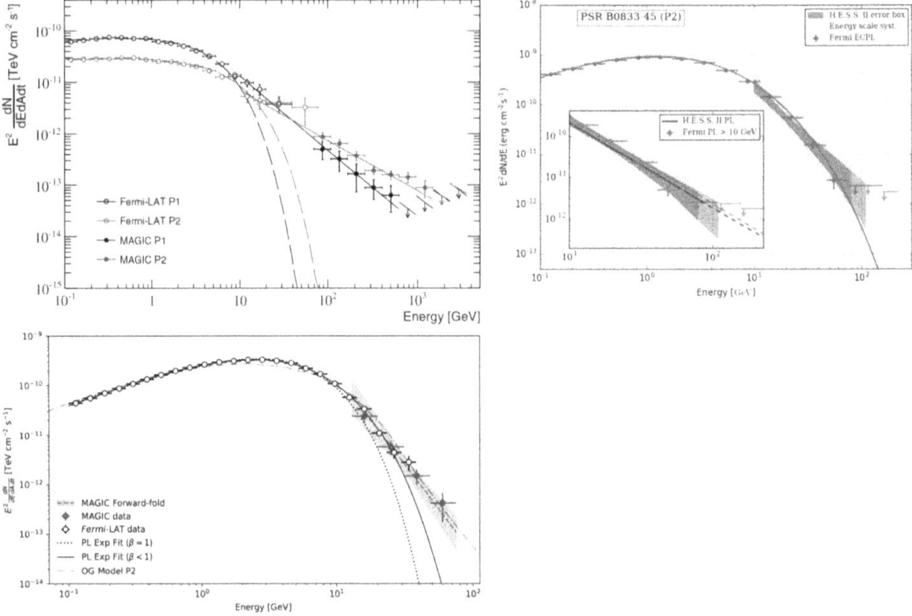

Figure 7.14. Top left: pulsed gamma-ray spectrum (peaks P1 and P2) from the Crab Pulsar with exponential cutoff fits as dashed lines. Image credit: Ansoldi et al. (2016), reproduced with permission © ESO. Bottom left: pulsed gamma-ray spectrum (peak P2) from Geminga with exponential cutoff fits (replace $\beta = b$ in Equation (7.11)) and an outer gap model (OG model). Image credit: MAGIC Collaboration et al. (2020a), reproduced with permission © ESO. Top right: pulsed gamma-ray spectrum from the Vela Pulsar P2 peak with an exponential cutoff fit as the green line. Image credit: HESS Collaboration et al. (2018g), reproduced with permission © ESO.

the edge of the magnetosphere (e.g., Hirotani 2011, 2013; Harding & Kalapotharakos 2015), or from ultrarelativistic "cold" wind electrons extending out to the wind termination shock at the PWN boundary (e.g., Bogovalov & Aharonian 2000; Aharonian et al. 2012; Mochol & Petri 2015) has met with some success.

7.5.4 Gamma-Ray Binary Systems and Related Microquasars

Gamma-ray binaries are often considered a subset of the compact binary class in which one of the binary members is a compact object (white dwarf, neutron star, or black hole) and the other member is a "regular" star powered by fusion (low to high mass). In recent years, the discovery of gamma rays from a massive stellar binary stellar system (η-Car) has extended this definition further.

There are a number of ways in which gamma-ray emission may be produced. Accretion of stellar material toward the compact object, or the interaction of winds from the stellar and neutron star, can set up particle acceleration conditions leading to gamma rays and nonthermal radio to X-ray emission. Strong thermal X-ray emission is also generated by the same processes, giving rise to the terms high-mass X-ray binary (HMXB) and low-mass X-ray binary (LMXB) that have been used for decades.[3] Particle acceleration via rotation of the neutron star may also take place in a similar fashion to that found in pulsars. Additionally, the accretion and rotation processes can lead to the formation of collimated outflows or jets (sometimes reaching relativistic speeds) that can accelerate particles. The presence of jets in such systems has led to the term "microquasar" and "microblazar" in an analogous situation to their much larger extragalactic cousins (discussed in Section 7.6.1). The often highly eccentric orbits in such binary systems lead to particularly active particle acceleration around the periastron phase when the binary members' separation is minimized. Similarly, the rather random nature of accretion also leads to a highly variable and/or transient output, such as in the case of classical novae where gamma rays result from the thermonuclear detonation of accreting material onto a white dwarf.

Here, we will provide a broad overview of the GeV to TeV observations and the various source classes while the reader is referred to the following reviews for more details: Mirabel & Rodríguez (1999), Dubus (2013), Paredes & Bordas (2019), and Chernyakova & Malyshev (2020) (also see Section 6.5.3). At present, there are a total of 13 gamma-ray binaries detected in TeV gamma rays (TeVCat: catalog of tev gamma-ray sources, http://tevcat2.uchicago.edu/) and a similar number detected at GeV energies (Abdollahi et al. 2020).

We will first look at the gamma-ray binaries involving a suspected neutron star compact object and a massive star (OB or Be types), which represent the most numerous subclass. These are often termed high-mass gamma-ray binaries (HMGBs). A key point is that the gamma-ray emission can be modulated by the orbital period (typically months to decades) of the binary system due to orbital-dependent absorption and production. Gamma rays of energy >0.1 TeV (γ)

[3] Indeed, the first extrasolar X-ray source discovered, Sco X-1, is a low-mass X-ray binary.

produced near the neutron star can readily interact with lower-energy optical/UV photons ($\gamma_{\rm Opt-UV}$) from the massive stellar companion and create electron/positron pairs ($\gamma\ \gamma_{\rm Opt-UV} \rightarrow e^{-}e^{+}$). This interaction is angle dependent, and the pairs then create lower-energy secondary GeV gamma rays in a cascade process. Additionally, IC scattering by electrons to produce the gamma rays (IC by electrons is usually the proposed mechanism given the presence of nonthermal X-ray and radio emission) is highly anisotropic, giving rise to an orbital dependence in the gamma-ray production. Finally, the underlying particle acceleration and hence gamma-ray emission production are strongly linked to the separation of the binary companion and usually maximizes around the periastron phase of closest separation. Putting all of these effects together leads to a quite complex orbital modulation of the GeV to TeV gamma-ray emission. On the other hand, this modulation has enabled the discovery of new HMGBs using gamma rays (e.g., Corbet et al. 2019) and unique investigations into the properties of the binary companions. Here, we will look at some examples.

The first HMGB detected was the LS 5039 system, which has a 3.9 day orbital period (Aharonian et al. 2005b, 2006e). A clear spectral difference was seen for the orbital phase at inferior conjunction (O star behind the neutron star) versus superior conjunction (O star in front of the neutron star). The situation is illustrated in Figure 7.15. The superior conjunction phase creates the maximum pair absorption in the 0.1 to 1 TeV gamma-ray band, and the periastron phase is slightly offset from superior conjunction. Another example is the recently discovered HMGB linked to PSR J2032+4127 (Abeysekara et al. 2018b; ~50 year period), and its X-ray and TeV gamma-ray light curves (versus orbital phase) are shown in Figure 7.15.

Interestingly, PSR J2032+4127 is located at the edge of the unidentified TeV source TeV J2032+4130, which exhibits weak, extended emission and is thought to be associated with the massive stellar cluster Cyg OB2 (Aharonian et al. 2002, 2005a). The peak emission from PSR J2032+4127 during the previous periastron phase in 2017 was about 10 times larger than the extended emission from TeV J2032 +4130. PSR J2032+4127 and PSR B1259–63/LS 2883 are only HMGBs where the compact object is confirmed as a neutron star due to their pulsar nature. We note though that the compact object in LS 5039 was recently suggested to be a magnetar (Yoneda et al. 2020).

Another binary class detected is the enigmatic stellar binary system η Car (Humphreys & Martin 2012), comprising two massive companions in a 5.52 year orbit. One of the companions is a luminous blue variable (LBV) of ~100 $M_\odot$ exhibiting extreme mass loss via its winds. The MeV to GeV gamma-ray emission detected by Fermi-LAT has been detected around the full orbit but the 10 to 300 GeV emission most noticeably peaked around the previous periastron phase (early 2015; Farnier et al. 2011; Reitberger et al. 2015). Just recently, HESS detected TeV gamma-ray emission of up to 0.5 TeV (HESS Collaboration et al. 2020c) in observations concentrated around the previous periastron. The gamma-ray emission is generally believed to come from particles shock-accelerated at the wind-collision region. The hadronic process has been somewhat favored (HESS Collaboration

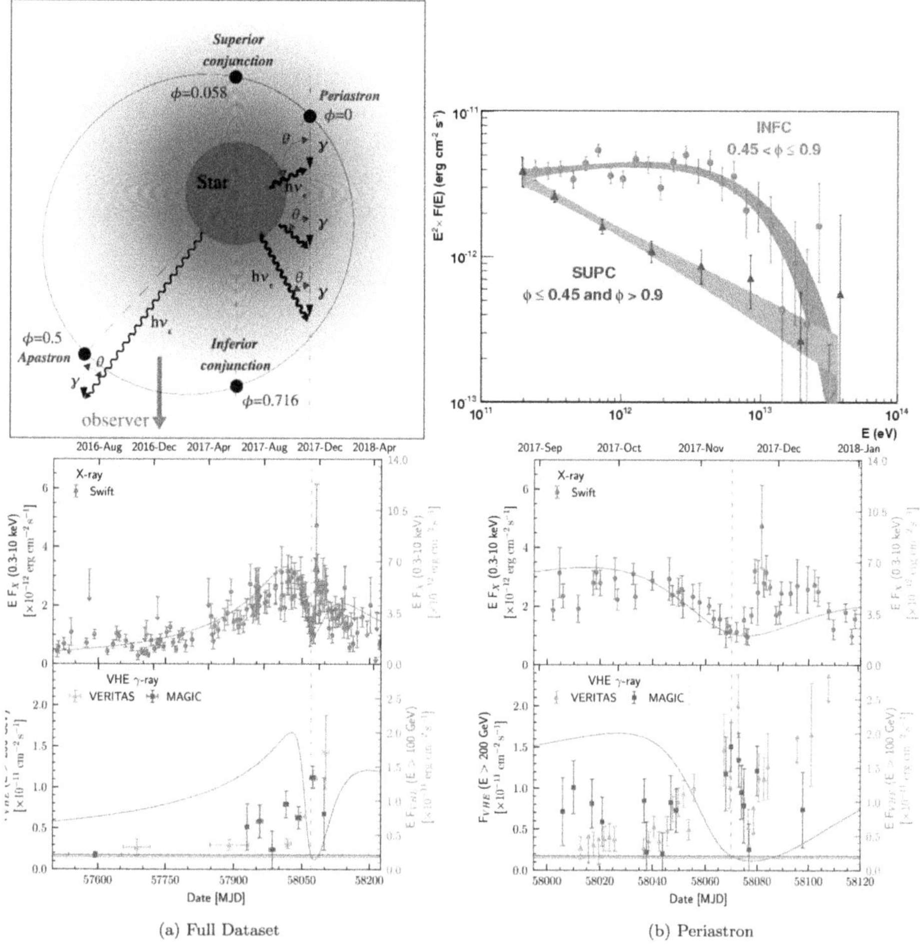

(a) Full Dataset

(b) Periastron

Figure 7.15. Top panels: LS 5039 binary-viewing situation versus orbital phase ϕ (neutron star is the black dot) and gamma-ray spectra for the inferior conjunction (INFC) and superior conjunction (SUPC) phases (Images from Aharonian et al. 2006e, reproduced with permission © ESO). Bottom: X-ray and TeV gamma-ray light curves versus orbital phase for PSR J2032+4127 with periastron as the vertical dashed line. The solid line is a model expectation. Image credit: Abeysekara et al. (2018b).

et al. 2020c) for the multi-GeV to TeV emission due to the intense synchrotron and IC losses electrons might suffer in the wind-collision zone that would prevent them from reaching sufficiently high energies.

The microquasar SS 433 has for decades attracted attention as one of the most powerful objects in our Galaxy (Margon 1984). Its bilobal jets reaching speeds of about $0.3c$ extending for about a degree on either side of the central object (likely a Wolf–Rayet (WR) star accreting onto a black hole in a 13 day orbit) require up to 10^{41} erg s^{-1} to sustain. The jets exhibit several lobes or knots seen mostly in X-rays (Safi-Harb & Ögelman 1997), and also, they are the only jets with evidence for a baryonic flow (Migliari et al. 2002), making them an ideal laboratory to study extreme cosmic-ray "beam dumps." SS 433 is embedded within the biconical-shaped SNR W50 (Dubner et al. 1998). Extended gamma-ray emission at about 1 GeV was first detected by Bordas et al. (2015), which broadly centered on SS 433 and W50, but the weak signal and resolution of Fermi-LAT prevented firm conclusions as to its origin. The HAWC telescope then discovered two multi-TeV gamma-ray sources at about 25 TeV energy overlapping the inner X-ray knots (Abeysekara et al. 2018c) as shown in Figure 7.16.

A leptonic scenario was favored by Abeysekara et al. (2018c) to explain the TeV emission and suggested electrons are locally accelerated at the knots where the TeV emission is seen. Following the TeV detection by HAWC, Sun et al. (2019) provided an improved Fermi-LAT image of SS 433 in GeV gamma rays, indicating an extension similar to the W50 SNR but centered on the central SS 433 source. The GeV emission is possibly related to the W50 SNR rather than to SS 433.

Additional GeV sources were recently reported by Li et al. (2020a), with one of them reported to show modulation at the ~160 day precessional period of the SS 433 system. However, this has been questioned by Bordas (2020), who notes that modulation would be very unlikely to be maintained over the >30 pc distance to this source from the central system. Overall, the HAWC and Fermi-LAT detections

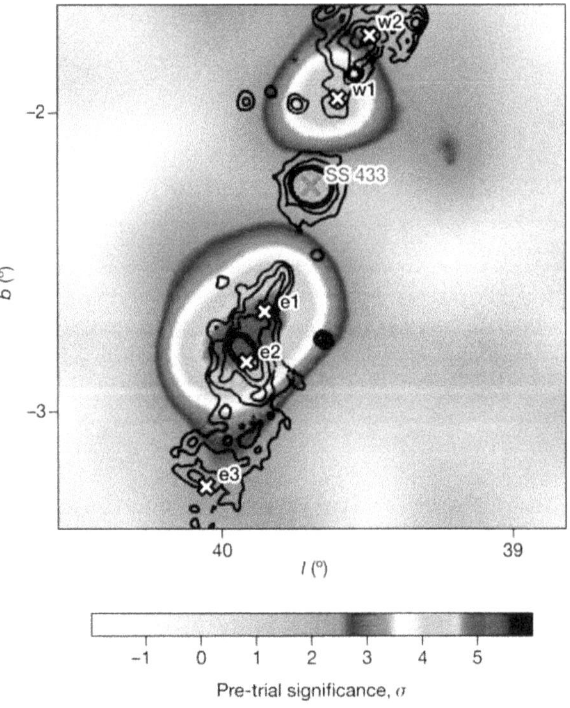

Figure 7.16. HAWC TeV gamma-ray image of SS 433 with X-ray contours (black solid) and knots from Safi-Harb & Ögelman (1997) labeled. Image credit: Abeysekara et al. (2018c), with permission of Springer.

have sparked renewed theoretical activity devoted to extreme particle acceleration in microquasar jets leading to both leptonic and hadronic gamma-ray emission. We will discuss some of these in the context of PeVatrons (Section 7.5.6), but we recommend the recent review devoted to particle acceleration in astrophysical jets (Matthews et al. 2020), which is also relevant to our discussion on AGN (Section 7.6.1).

GeV gamma-ray emission associated with microquasar jets has been detected in three other objects, Cyg X-3, Cyg X-1, and V404 Cygni (Fermi-LAT Collaboration et al. 2009; Tavani et al. 2009; Sabatini et al. 2010; Bodaghee et al. 2013; Malyshev et al. 2013; Zdziarski et al. 2017; Xing 2006). The first three are HMXBs while notably, V404 Cygni is an LMXB (1 $M_\odot$ companion). To date, several flares from Cyg X-1 and Cyg X-3 have been detected by Fermi-LAT and/or AGILE. Hints of TeV gamma-ray emission coinciding with an X-ray flare from Cyg X-1 was also reported by MAGIC (Albert et al. 2007) in 2006, but not coinciding with any of the GeV flares and so far not repeated.

We will conclude our discussion of gamma-ray binaries with a brief look at novae that Fermi-LAT has now established as a class of gamma-ray sources. To date, 12 novae are seen in GeV gamma rays (e.g., Xing 2006; Ackermann et al. 2014; Cheung et al. 2016; Li et al. 2020b), all of which were identified by their optical flaring. The gamma-ray flaring episodes can last up ~20 days (usually just after the initial optical flare) with luminosities of 10^{34-36} erg s^{-1} reaching up to ~10 GeV in energy. So-called "classical" novae are powered accretion of material from a donor star onto a white dwarf (WD) compact binary companion. Blobs of accreting material can detonate in a thermonuclear explosion. The gamma-ray emission suggests that this process can create the conditions for particle acceleration to near-TeV energies, and both leptonic and hadronic processes can explain the emission with the latter generally expected given the high density environment within the nova ejecta. Most of the gamma-ray novae are of the classical type, whereas V404 Cyg as a symbiotic system and the gamma-ray emission are thought to be due to interactions of the nova ejecta with the red giant donor star wind. A recent review of this developing source class can be found in Chomiuk et al. (2020).

7.5.5 Other Galactic Sources

In this section we will review the other types of Galactic sources that are established gamma-ray emitters.

Massive Stars and Their Stellar Clusters: The enormous energy liberated by massive stars (up to 10^{39} erg s^{-1}) in O-type wind kinetic energy and subsequently in their clusters has for many years made them ideal places to accelerate particles (Casse & Paul 1980; Voelk & Forman 1982; Cesarsky & Montmerle 1983). Additionally, the longevity of massive stars (10 $^{6-7}$ yr) could mean that their cluster may make significant contributions to the Galactic cosmic-ray flux (e.g., Aharonian et al. 2019). Several massive stellar clusters detected in GeV to TeV gamma rays—NGC 3603, Westerlund 1, Westerlund 2, W43 (HESS J1848−018), and Cyg OB2 (HESS Collaboration et al. 2011; Abramowski et al. 2012; HESS Collaboration et al. 2018a; Yang & Aharonian 2017; Yang et al. 2018; Aharonian et al. 2002, 2005a)—are the most confident associations so far based on the extended nature of the gamma-ray emission, its spectrum, and in some cases, a lack of any obvious alternative accelerators such as SNRs or sufficiently powerful pulsars. Recently, another cluster, RSGC1, was linked to GeV gamma-ray emission (Sun et al. 2020).

These clusters can be considered somewhat extreme in certain aspects of their nature, such as a very large number of O and B stars (NGC 3603, Cyg OB2); over 20 WR stars (Westerlund1); a combination of many O, B, and WR stars; a large number of red supergiant stars (RSGC1); and a "mini" starburst of massive star birth (W43). Another interesting example is Cl1806−20, where marginally extended TeV emission has been detected (HESS Collaboration et al. 2018f). This cluster harbors the magnetar SGR 1806−20 and the LBV star LBV 1806−20, which is one of the most-luminous stars in our Galaxy. The TeV emission extension may follow a similar morphology to the radio nebula powered by LBV 1806−20, thus hinting that this single star may play a prominent role in the multi-TeV particle acceleration. Cyg OB2 is also rather special for its tight concentration of 50 to 100 O-type stars, several WR stars, and the LBV Cyg OB2#12, making it one of the most powerful sources of stellar winds known. Just recently, the gamma-ray binary PSR J2032+4127 at the edge of Cyg OB2 was detected as a point-like source Abeysekara et al. (2018b), as discussed earlier. It is however resolved in TeV gamma rays from the extended gamma-ray emission overlapping Cyg OB2, and thus, this cluster remains a viable counterpart.

Superbubbles: Following on from the previous discussion on stellar clusters, superbubbles are blown out by the collective winds of massive stellar clusters (the more mature and sparse versions usually known as OB associations) and SNRs over a typical 10^{5-7} year timeframe. Their outer boundaries contain large-scale shocks of over 100 pc in diameter, giving rise to some the largest shock structures in our Galaxy. As a result, they have long-been proposed as particle accelerators (Montmerle 1979; Cesarsky & Montmerle 1983; Bykov & Toptygin 2001; Parizot et al. 2004) to extreme energies. To date there have been two gamma-ray detections that are associated with superbubbles: the Cygnus superbubble (Ackermann et al. 2011; Bartoli et al. 2014) and 30Doc C in the LMC (HESS Collaboration et al. 2015). There is also speculation that the very extended TeV emission (2° diameter) around the extreme cluster Westerlund1 (Abramowski et al. 2012) could also be related to a superbubble process.

The GeV gamma-ray emission in the Cygnus superbubble appears to overlap somewhat and thread in between the massive stellar clusters Cyg OB2, NGC 6910, and the other star-forming regions in a kind of cocoon-like emission. This is attributed to multi-TeV cosmic rays accelerated by the clusters, which then fill in the cavities the clusters have carved out. Additional acceleration may also come from the local SNR γ Cygni. Extended TeV gamma-ray emission seen by the ARGO-YBJ telescope (Bartoli et al. 2014) has a similar morphology to that of the GeV emission and is considered to be the TeV counterpart. Including the MILAGRO source MGRO J2031+41 (Abdo et al. 2007a), which is thought to be related, the energy spectrum extends to >20 TeV.

The 30 Dor C is a superbubble in the LMC and is one of most luminous massive star formation regions known with nonthermal X-ray emission along parts of its rim. This X-ray rim emission is considerably more luminous and larger in size than that typically found in Galactic SNRs. The TeV emission (HESS Collaboration et al. 2015) found by HESS is positioned toward a massive star cluster toward the northwest region of the bubble and overlaps molecular gas (Sano et al. 2017), although the HESS angular resolution prevents a detailed morphological study. A hadronic process for the TeV emission may be at work, suggesting cosmic rays above 100 TeV in energy although a leptonic process is also not ruled out. Recent X-ray observations confirm the presence of nonthermal X-ray emission to >20 keV in energy, and hence electron acceleration to over 100 TeV energies around parts of the superbubble rim (Lopez et al. 2020).

Galactic Center Region and Diffuse Emission: Following the first hints for TeV gamma rays from the Galactic Center region by Kosack et al. (2004), the gamma-ray emission over the inner 500 pc of our Galaxy has now been separated into diffuse and local components (made possible by the 5′ to 6′ resolution at TeV energies). The local components are attributed to the central source Sgr A* (Acero et al. 2010), the composite SNR G0.9+0.1 (Aharonian et al. 2005b), and the PWN G0.13–0.11 (Archer et al. 2016; HESS Collaboration et al. 2018e). The diffuse components comprise the inner 70 pc surrounding Sgr A*, which requires a central PeVatron cosmic-ray accelerator (discussed in Section 7.5.6), and the larger-scale component extending for a few hundred parsecs along the Galactic Plane (Aharonian et al. 2006a; Archer et al. 2016; Ahnen et al. 2017; HESS Collaboration et al. 2018e; MAGIC Collaboration et al. 2020b).

The TeV emission toward Sgr A* appears to be related to our Galaxy's central black hole region although the nearby PWN G359.95–0.04 cannot yet be ruled out as a contribution (Acero et al. 2010). So far, the Sgr A* TeV emission has shown no indication for variability (including during a 2017 pericenter passage of a large molecular cloud), in contrast to the situation in X-rays where a number of flares have been seen over the decades (see Ahnen et al. 2017 and references therein). Immediately upon its discovery by HESS (Aharonian et al. 2006a), the diffuse TeV component overlapping the Galactic "Ridge" showed clear indications for an overlap with the dense interstellar gas (see Figure 7.17), providing the first evidence for multi-TeV cosmic rays within the central few hundred parsecs of the Galactic Center region.

This region hosts a large fraction of our Galaxy's interstellar gas (with over 10^6 $M_\odot$ in mass present) and a number of potential cosmic-ray accelerators nearby besides the central black hole. In 2018, HESS showed that the inner 70 pc of the TeV emission surrounding Sgr A* does not have any indication for a high-energy cutoff in its spectrum, suggesting the presence of cosmic rays reaching PeV energies (HESS Collaboration et al. 2016). When comparing the spatial profile of this TeV emission with that of the dense gas, the inferred cosmic-ray density follows a 1/radius dependence, indicating a quasi-continuous accelerator operating for at least 1000 years. This seems to point toward the central black-hole-powered region as the

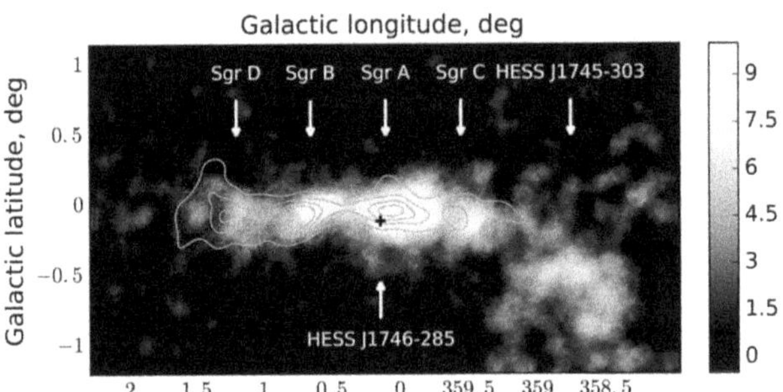

Figure 7.17. TeV gamma-ray emission from the Galactic Center region as detected by HESS The cyan contours depict the dense molecular gas traced by the CS(1–0) transition (HESS Collaboration et al. 2018e) and various other features (identified in radio to X-rays) are labeled. Image from HESS Collaboration et al. (2018f), reproduced with permission © ESO.

accelerator, and moreover, the energy available in the central accelerator (if operating sporadically over the past 10^6 years) could account for a good fraction of the total PeV cosmic rays in our Galaxy. Alternative accelerator scenarios for this PeVatron region and wide diffuse emission have been discussed: millisecond pulsars (Guépin et al. 2018), a combination of millisecond pulsars and dark-matter (Lacroix et al. 2016), and cosmic rays escaping SNRs (Jouvin et al. 2020).

Looking out to wider scales, the diffuse MeV to GeV gamma-ray emission along the Galactic Plane is now a ubiquitous feature (Ackermann et al. 2012) since being revealed in the early days of space-based gamma-ray astronomy (Fichtel et al. 1975). This diffuse emission arises from the long-term resident or "sea" of cosmic rays and electrons pervading the Galaxy. These have long since moved away (via diffusion or otherwise) from their acceleration sites. The broad distribution of the diffuse gamma-ray emission matches that of the Galactic latitude, and the longitude distribution of SNRs, pulsars, massive star clusters, and the interstellar molecular gas provides some clues as to where these particles come from (the IC component from electrons is expected to be somewhat broader, reflecting the distribution of soft infrared photon targets). The only direct measurements of this sea of particles are via satellites in Earth orbit and out to the Voyager probes at the heliopause and just beyond where they have sampled MeV interstellar cosmic rays without solar wind influences for the first time (Stone et al. 2013, 2019).

Measuring the in situ cosmic-ray and electron sea across the Galaxy requires an understanding of secondary by-products such as gamma rays, and detailed models and specific observation strategies have been developed over the years for this purpose. The GALPROP model (Strong & Moskalenko 1998) has been used a basis to estimate the diffuse MeV to GeV gamma-ray background in order to identify MeV to GeV sources in Fermi-LAT data (Abdollahi et al. 2020). Recent updates to GALPROP (Porter et al. 2017; Jóhannesson et al. 2018) set up a 3D distribution of accelerators (e.g., SNRs and pulsars) and follow the trajectories of cosmic rays and electrons via a transport equation (e.g., see Strong & Moskalenko 1998) interacting with the interstellar gas 3D gas distribution (CO and H I for molecular and atomic gas) and radiation fields (infrared and optical) to produce gamma rays via the three main processes (pp collisions, bremsstrahlung, and IC; synchrotron emission is also calculated). Other models (Kissmann et al. 2015; Evoli et al. 2017) with 3D geometry in various aspects are also advancing (see Becker Tjus & Merten 2020 for a detailed review of various codes). A key driver of these models is the improving sensitivity of TeV gamma-ray facilities (e.g., CTA), which will reach arcminute resolution, demanding more precise 3D grids over which to solve particle transport and interaction.

At TeV energies, the rapid cooling times of electrons in particular mean that the diffuse gamma-ray emission could be quite sporadic and vary considerably for 10^{3-4} years after local acceleration events like SNRs. The most recent enhancements to GALPROP (Porter et al. 2019) have been motivated by this issue. Some indirect observations of the GeV to TeV cosmic-ray sea have come from looking at the Fermi-LAT MeV to GeV gamma-ray emission from passive individual giant molecular clouds that are believed to be not too close to recent accelerators like SNRs and young pulsars. Both Aharonian et al. (2020) and Baghmanyan et al. (2020) recently exploited this idea, which was proposed some time ago (Aharonian 2001; Casanova et al. 2010), and they found cosmic-ray densities up to twice the Earth-like value for clouds in the 4 to 6 kpc galactocentric distance range and for some very nearby clouds (<500 pc from Earth).

The TeV gamma-ray diffuse emission has so far proven to be more difficult to measure. However, evidence for the TeV Galactic Plane diffuse emission has been reported by Abdo et al. (2008), Abramowski et al. (2014a) and Bartoli et al. (2015). The fraction of unresolved sources has been a key uncertainty as to whether this large-scale TeV emission detected so far is from truly diffuse particles or from those accelerated not far from their accelerators as unresolved sources. Some recent estimates (Steppa & Egberts 2020; Cataldo et al. 2020) suggest this fraction could be up to 60%. As further evidence of this diffuse or unresolved source emission, in the analysis of its Galactic Plane data (HESS Collaboration et al. 2018a), HESS had to implement an ad hoc large-scale diffuse component in order to successfully model the complete TeV Galactic Plane emission observed by HESS.

No discussion of diffuse gamma-ray emission would be complete without mentioning the Fermi Bubbles, the multidegree GeV gamma-ray features extending above and below the Galactic Plane (discovered by Su et al. 2010). Their discovery had an immediate impact on the view of our Galaxy's history. There are currently two schools of thought hotly debated as to their origin, based also on prior bubble-like or wisp features seen from radio to X-rays: (1) prior AGN activity driven by our Galaxy's supermassive black hole (e.g., Zubovas et al. 2011; Guo & Mathews 2012; Mou et al. 2014; Bland-Hawthorn et al. 2019; Zhang et al. 2020) or (2) starburst activity driven by the winds and death of massive stars in the inner few hundred parsecs of our Galaxy (e.g., Crocker & Aharonian 2011; Lacki 2014; Crocker et al. 2015). In addition, a fundamental question still surrounds the nature of the underlying particles (cosmic rays and/or electrons) generating the Fermi Bubble emission, with both types of models being put forward (e.g., Crocker & Aharonian 2011; Herold & Malyshev 2019). In any case, the bubbles require a considerable

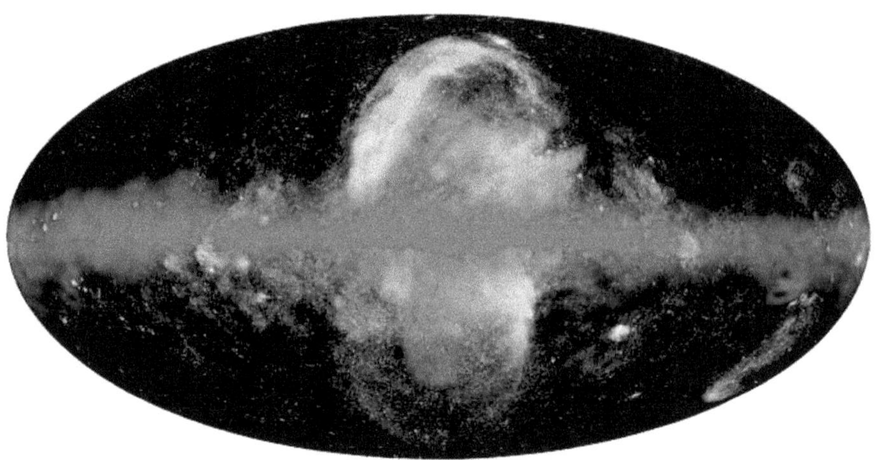

Figure 7.18. eROSITA 0.6–1 keV X-ray emission (cyan) versus Fermi-LAT GeV gamma-ray emission (red) revealing new X-ray bubbles wrapping around the Fermi GeV bubbles. Image credit: Predehl et al. (2020), with permission of Springer.

amount of energy to be injected ($\sim 10^{55}$ erg). The most recent new result in relation to the bubbles was the discovery of soft X-ray emission by eROSITA (Predehl et al. 2020) appearing to wrap around the northern and southern gamma-ray bubbles as shown in Figure 7.18 (this also clearly shows the diffuse gamma-ray emission along the Galactic Plane).

The X-ray emission shows a clear outer boundary, and its total energy is about 10× larger than that in the GeV gamma rays. Current speculation (Predehl et al. 2020) is that the X-rays are revealing a forward shock (through the halo gas of the Milky Way) driven by the Fermi Bubbles, which themselves are driven by energy from the Galactic Center region.

Globular Clusters: Globular clusters are a rapidly growing population of GeV gamma-ray sources with 30 examples now identified (Abdollahi et al. 2020), following on from earlier discoveries (Kong et al. 2010; Abdo et al. 2010e). Only one globular cluster, Terzan5, is detected at TeV gamma-ray energies (HESS Collaboration et al. 2011), and interestingly, the TeV emission appears slightly displaced from the optical center of the cluster. Although dark matter scenarios have been discussed, the gamma-ray emission is most likely attributed to the large population of millisecond pulsars present in globular clusters (and possibly related to XRB emission).

Unidentified Sources: As discussed earlier, the unidentified gamma-ray sources represent typically 30% of the GeV and TeV source population and have been a major area of interest since the first catalogs were released (e.g., Reimer 2001). Today, a majority of the unidentified extragalactic GeV sources detected by Fermi-LAT are thought to be AGN blazars of an unknown class based on their point-like nature, variability, application of machine-learning techniques, and newly found radio to optical counterparts (e.g., Acero et al. 2013; Doert & Errando 2014; Chiaro et al. 2016; Einecke 2016; Schinzel et al. 2017; Mandarakas et al. 2019). We will leave further discussion of these sources to Section 7.6.1.

The unidentified Galactic gamma-ray sources have, however, presented a much greater challenge. This is due to their extended nature and hence potential to be confused with many potential counterparts along the line of sight plus the ability of single particle accelerators to generate rather specific extended gamma-ray features due to particle transport and interact with the unique interstellar gas situation surrounding them. Additionally, a growing number of unidentified Galactic gamma-ray sources from the COS-B and EGRET missions (Bignami & Hermsen 1983; Hartman et al. 1999); Gehrels et al. 2000) at MeV to GeV energies came with rather large positional uncertainties (degrees or so), complicating counterpart searches. Despite this, the obvious potential connections to SNRs (plus massive star clusters and pulsars) and their surrounding interstellar gas have been noticed for quite some time (Montmerle 1979; Esposito et al. 1996; Romero et al. 1999; Torres et al. 2003).

With Fermi-LAT's angular resolution improving with energy, for example $\sim 0.1^{\circ}$ above 10 GeV, Galactic sources detected at much higher energies (Ajello et al. 2017) have much better localization when compared to results using lower-energy analyses. Moreover, many of them can be cross-referenced against the growing number of TeV gamma-ray sources cataloged at similar or better resolution (HESS Collaboration et al. 2018a). The number of unidentified TeV gamma-ray sources has grown steadily (Aharonian et al. 2008a, 2006b; HESS Collaboration et al. 2018a) since the first one, TeV J2032+4130, was discovered in 2002 (Aharonian et al. 2002, 2005a). This source has proved particularly enigmatic and has been linked to the extreme stellar cluster Cyg OB2 although the leptonic and/or hadronic nature of the gamma-ray emission is yet to be firmly established (as for the majority of the Galactic TeV sources). As discussed earlier, this source has an additional component associated with a high-powered pulsar in a binary system (Abeysekara et al. 2018b).

Following the first HESS surveys, many of the unidentified TeV sources appeared not far away from mature SNRs or moderate-power pulsars. A related interpretation was built around these ideas (e.g., de Jager et al. 2009; Yamazaki et al. 2006; Chang et al. 2008; Aharonian et al. 2008b), especially as it was becoming clearer that SNRs with escaping (or runaway) cosmic rays interacting with adjacent or more distant interstellar gas could lead to detectable emission. In more recent times, links to old-pulsar-powered TeV halos have been proposed (Sudoh et al. 2019), as the gamma-ray halo concept has become more widely understood. As a result, the interstellar gas, deeper X-ray observations, and new radio continuum surveys have hence gathered more attention as a way to help identify these TeV sources (e.g., Horns et al. 2007; Matsumoto et al. 2008; Eger et al. 2011; Voisin et al. 2016; Lau et al. 2017b; de Wilt et al. 2017; Lau et al. 2017a, 2019; Maxted et al. 2019; MAGIC Collaboration et al. 2020c; Feijen et al. 2020).

7.5.6 PeVatrons—Extreme Accelerators in the Milky Way

A special mention in this section of Galactic gamma-ray sources is devoted to the concept of accelerators known as PeVatrons, where particle acceleration is able to reach PeV energies or greater. This notion has held great mystery since the "knee" at PeV energies in the local cosmic-ray spectrum was first revealed over 50 years ago (Kulikov & Khristiansen 1958). The knee is suspected to be a signpost of the maximum limits of particle accelerators in our Galaxy (Peters 1961). The knee is a slight steepening of the cosmic-ray spectral index Γ (see Equation (7.1)), and its position in cosmic-ray energy increases proportionally to its charge. Thus, heavier elemental cosmic rays like iron reach energies a factor of 20 higher than proton cosmic rays (Figure 7.19), although the precise spectral limits of the higher-mass cosmic rays are not well known at these energies (Alves Batista et al. 2019). Well above the knee at about 10^{18} eV, the cosmic-ray spectrum flattens slightly in a feature usually called the "ankle." The ankle supposedly represents where the extragalactic cosmic-ray component starts to take over.

In a seminal paper, Hillas (1984) calculated the theoretical maximum energy $E_{\max}$ a charged particle can be diffusively shock-accelerated to by noting that the size of the accelerating shock should be at least twice the particle's Larmor radius (otherwise the particle will escape and no longer be accelerated):

$$E_{\max} \approx 10^{18} Z\beta RB \text{ eV} \tag{7.12}$$

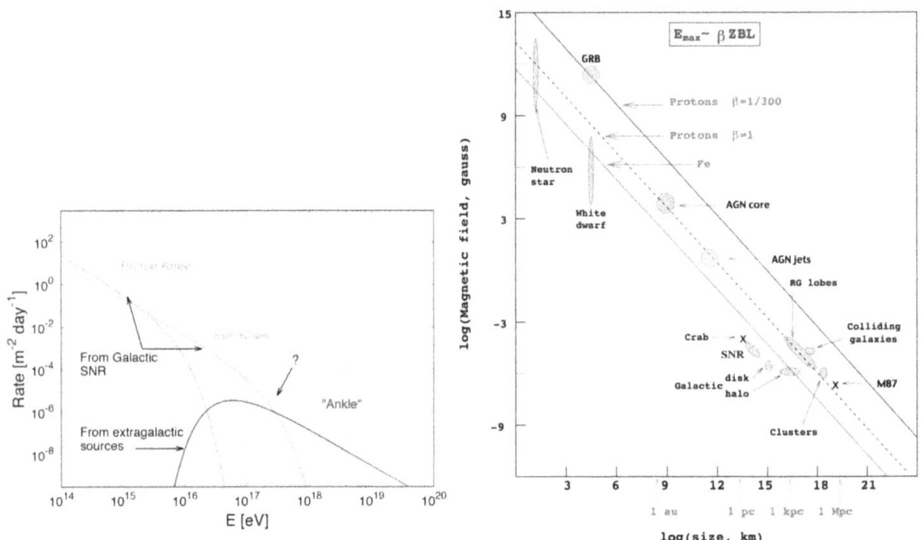

Figure 7.19. Left: cosmic-ray spectrum schematic showing the "knee" and "ankle" features, and the Galactic components from proton, iron, and extragalactic cosmic rays. The latter two are not observationally confirmed, hence the "?" above. Image credit: Alves Batista et al. (2019) with permission of Frontiers. Right: Hillas plot from Letessier-Selvon & Stanev (2011), Copyright 2011 by the American Physical Society, showing the maximum energy attainable by an accelerator with shock speed β, size (L in this figure), and magnetic field B.

for the particle charge Z, shock radius R in kiloparsec, magnetic field B in microgauss, and $\beta = v/c$ for the shock velocity v. With this, Hillas created his famous "Hillas plot," which has been reproduced many times (Figure 7.19 shows a version from Letessier-Selvon & Stanev 2011). It plots the locus of $E_{\max}$ points versus R and B for an assumed β and provided an immediate indication as to what types of objects could accelerate cosmic rays to various energies.

From Figure 7.19 and Equation (7.12), we can see that SNRs with size 1 pc and shock speed 10,000 km s^{-1} early in their evolution, when they are expected to provide their strongest acceleration, can in principle accelerate cosmic rays to 1 PeV provided their postshock magnetic field $B > 100\ \mu$G. Such a high magnetic field was a problem under the diffusive shock acceleration picture for quite some time (Lagage & Cesarsky 1983), until Bell (2004) showed that additional amplification is possible due to the effects of the cosmic rays themselves. Thus, at present, SNRs are considered potential PeVatrons. There are a number of other Galactic source types theoretically able to reach such energies. An incomplete list includes superbubbles (>100 pc scale shocks associated with massive stellar winds and many SNRs), massive stellar clusters, neutron star electric fields, magnetar flares, Galactic Center region, and microquasars.

An ideal way to search for PeVatrons is via the interstellar clouds around SNRs and through other accelerators (Gabici & Aharonian 2007). Here, the expected energy-dependent diffusion properties of cosmic rays toward the surrounding clouds can yield very specific gamma-ray spectra depending on the distance to the cloud and the age of the accelerator (i.e., how long ago did the cosmic-ray escape and then start to diffuse toward the cloud). An expectation is that the more distant clouds will pick up the highest-energy cosmic rays first and thus yield a very hard gamma-ray spectrum. A related scenario exploits the energy-dependent diffusion into dense interstellar gas clouds (Gabici et al. 2007). Here, due to the turbulence inside many dense clouds (especially those disrupted by star formation processes), the diffusion coefficient for low-energy particles will be suppressed. Thus, the penetration depth inside clouds could be shorter for low-energy cosmic rays (Inoue et al. 2012), reducing the gamma-ray emission at low energies, say <1 TeV, compared to that at higher energies. The dense clouds can then act a "reservoirs" of higher-energy cosmic-ray interactions.

Observationally, there are a number of hints for PeVatrons. First, the GeV synchrotron gamma rays detected from the Crab Nebula PWN is an unambiguous indicator of PeV electrons. The large scale of massive stellar clusters (and their long lifetimes as discussed earlier) may help them to accelerate particles to PeV energies, making them potential PeVatrons. The recent work by Aharonian et al. (2019) looked at the GeV emission toward Cyg OB2 and Westerlund 1 and found that a continuous injection of cosmic rays to PeV energies is a feasible interpretation.

The other more convincing cases for PeVatrons are the Galactic Center region surrounding the central black hole (HESS Collaboration et al. 2016), the TeV sources detected by HAWC with emission beyond 56 TeV and 100 TeV (Abeysekara et al. 2020), and diffuse emission along the Galactic Plane detected by the TibetASγ array (Amenomori et al. 2021). The central black hole source has been discussed earlier but we would note that the intrinsic cosmic-ray spectral cutoff energy of about 0.4 PeV as determined by the HESS Collaboration et al. (2016) could in fact be up to 2× higher due to the noticeable absorption of >10 TeV gamma rays on the interstellar infrared fields (Porter et al. 2018).

There are a number of TeV sources seen by HESS, MAGIC, and/or VERITAS exhibiting hard gamma-ray spectra or with little to no constraint on any exponential cutoff in their spectral shape (e.g., Abramowski et al. 2014b; HESS Collaboration et al. 2020b; and see HESS Collaboration et al. 2018a). Many of these have HAWC counterparts >56 TeV. Already, these have obviously been linked to PeVatrons with potential counterparts such as SNRs, e.g., HAWC J2227+610 and its association with SNR G106.3+2.7 (Albert et al. 2020c), and very recently, HAWC J1825−134 (Figure 7.20; Albert et al. 2021), which reaches gamma-ray energies of 200 TeV and is

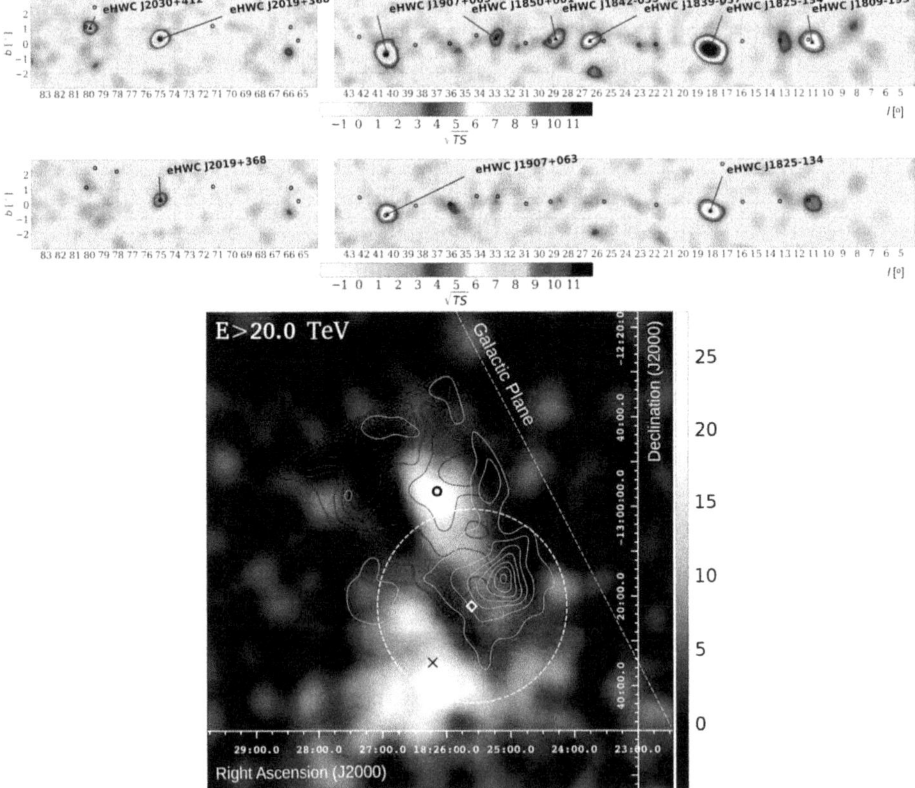

Figure 7.20. Top and middle: HAWC sky maps of statistical significance ($=\sqrt{TS}$) for energies >56 TeV (top) and >100 TeV (middle). Images from Abeysekara et al. (2020), Copyright 2020 by the American Physical Society. Bottom: HESS sky map toward HESS J185−137 and HESS J1826−130 for energies >20 TeV. Other objects are labeled as PSR J1826−1256 (black circle) and PSR J1826−1334 (black cross), eHWC J1825−134 >100 TeV (extension and location as white dashed circle and diamond from Abeysekara et al. 2020). The cyan and blue contours represent the total column density from the Nobeyama ^{12}CO(1–0) and SGPS H I surveys in the velocity ranges 45 to 60 km s^{-1} (4 kpc) and 60 to 80 km s^{-1} (4.6 kpc), respectively (see HESS Collaboration et al. 2020b for references). Image credit: HESS Collaboration et al. (2020b), reproduced with permission © ESO.

found in between the bright PWN HESS J1825−137 and the HESS PeVatron candidate source HESS J1826−130 (Figure 7.20). Interestingly, this highest-energy emission is very close to a dense molecular region associated with a young stellar cluster and star formation region in Voisin et al. (2016), suggesting a hadronic process is at work. If this is the case, it would certainly imply cosmic rays exceeding PeV energies.

We also note a recent discussion on the potential for microquasars to accelerate PeV cosmic rays (e.g., Cooper et al. 2020; Sudoh et al. 2020; Kimura et al. 2020) given the multi-TeV emission seen toward the jet lobes of SS 433 (Abeysekara et al. 2018c), and methods to identify PeVatrons via the spectral properties of unidentified TeV gamma-ray sources (Spengler 2020) and combinations of known SNRs and cataloged molecular clouds as potential targets for cosmic-ray collisions (Mitchell et al. 2021).

The IceCube neutrino detector, in combination with the Ice-Top array, is also sensitive to PeV gamma rays, and their most recent detections (Aartsen et al. 2020) have placed the tightest constraints so far on gamma rays in the 0.6 to 100PeV energy range. At the most extreme energies, $>10^{18}$ eV, gamma-ray flux constraints have been set by the Pierre Auger Observatory (Aab et al. 2017) and the Telescope Array (Abbasi et al. 2020).

Finally, we note the exciting discovery by the LHAASO facility (Cao et al. 2021) of a population of Galactic γ-rays with energies reaching at least 1 PeV. This result certainly opens the door to new insights into our galaxy's most extreme particle accelerators.

7.6 Extragalactic Gamma-Ray Sources

The extragalactic gamma-ray sources represent roughly half of the total population in both the GeV and TeV bands. Here, we will review the catalog of these sources and highlight some of the key results.

7.6.1 AGNs

The dominant extragalactic source class is the AGN of the “blazar” type, where an accretion-powered jet is directed toward Earth. The accretion is powered by a super-massive black hole of mass $>10^7$ $M_\odot$, which is believed to then form a perpendicular jet outflow that usually reaches relativistic speeds. Because the jet direction is toward Earth, strong Doppler effects that significantly boost the observed gamma-ray flux come into play. The observed flux usually attributed to a blob of material in the jet is boosted by a factor δ^4, where the Doppler factor δ is given by

$$\delta = [\Gamma(1 - \beta \cos \theta)]^{-1} \tag{7.13}$$

for the jet observation angle θ, Lorentz factor $\Gamma = (\sqrt(1 - \beta^2))^{-1}$, and $\beta = v/c$ for the jet speed v. Note that for head-on viewing ($\theta = 0$) we have $\delta \approx 2\Gamma$. The boosting factor δ^4 comes from three aspects (combining multiplicatively) resulting from special relativity, (1) the energy of each photon is boosted by a factor δ, (2) the effective contraction of time between successive photons means the photon rate is boosted by a factor δ, and (3) the jet angle is tightened such that the photon/solid-angle emitted is boosted by a factor δ^2. Over the past few decades, an AGN unification picture has been developed (Antonucci 1993; Urry & Padovani 1995) to explain various observational properties according to the viewing orientation. Figure 1.7 shows a version (Reynolds et al. 2014) developed to include radio-quiet AGNs viewed from the side with a weak jet disrupted by the accretion disk, or, with no jet at all (see, e.g., Madejski & Sikora 2016; Padovani et al. 2017; Panessa et al. 2019).

Based on this, AGN classes such as Seyfert, radio-loud/-quiet quasars, blazars, radio galaxies (broad and narrow line) can be defined, although many of the details and class overlaps are still debated. An even broader label is the Fanaroff–Riley (FR) class (Fanaroff & Riley 1974) to distinguish radio-loud AGNs more centrally peaked (FRI class) from radio-loud AGNs with strong emission in jet lobes (FR II class), in cases where the jet and central region are separately visible.

The blazar AGN class itself has a number of subclasses originally based on their radio luminosities, in what is known as the blazar sequence (Fossati et al. 1998). A more recent version (Ghisellini 2016; Ghisellini et al. 2017) uses the third Fermi-LAT AGN catalog (Ackermann et al. 2015b) and predefined spectral components to group blazars according to their GeV gamma-ray luminosity, given the large number of examples detected (>1500) (the recently released fourth Fermi-LAT AGN has over 2800 examples). Figure 7.21 shows this sequence, illustrating the double-peaked nature of the blazar spectra spanning the radio, X-ray, and gamma-ray domains.

The first broad peak represents the radio to X-ray synchrotron emission, while the second peak covering the hard X-ray to gamma-ray band usually represents the IC process, so that the entire spectrum results from accelerated electrons. However, a bit later in this section, we will note that a hadronic process for at least parts of the gamma-ray emission has been put forward. A further broad classification separates blazars into BL Lacs (after the archetype BL Lacertae) and flat spectrum radio

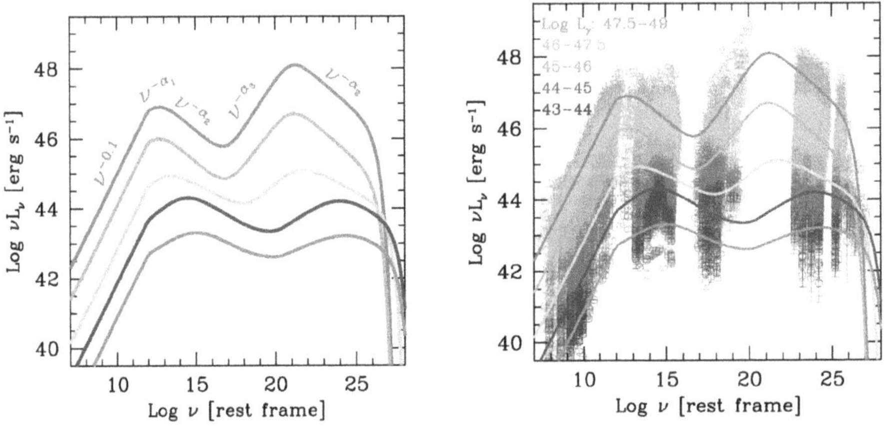

Figure 7.21. Updated version of the AGN blazar sequence based on their GeV gamma-ray luminosities (erg/s) shown as different color bands. The left panel shows the spectral components assumed and the right panel shows the average spectral shapes determined along with observed data. The top two curves represent FSRQs, and the bottom three curves represent BL Lacs (approximately LBL, IBL, HBL/EBL subtypes; cyan, blue, purple). Image credit: Ghisellini (2016), with permission of MDPI.

quasars (FSRQs) based on optical, radio, and variable properties and general spectral shapes. From the previous blazar sequence (Fossati et al. 1998), the labels low-frequency BL Lac, intermediate-frequency BL Lac, high-frequency BL Lac, and extreme-frequency BL Lac (LBL, IBL, HBL, and EBL) were coined according to the radio frequencies where their synchrotron component maximizes. The FSRQs display a relatively constant spectral shape but varying luminosities across whole spectra. In terms of the new blazar sequence by Ghisellini et al. (2017), the FSRQs are represented by the top two curves in Figure 7.21 (red and dark green), whereas the BL Lacs are represented by the bottom three curves (cyan—LBL, blue—IBL, purple—HBL/EBL). We note that the alternative names low-synchrotron-peaked blazar (LSP), intermediate-synchrotron-peaked blazar (ISP), and high-synchrotron-peaked blazar (HSP) have also been used (Ajello et al. 2020a).

At GeV energies, the latest Fermi-LAT catalog (Ajello et al. 2020a) reports 2863 AGN sources with 655 FSRQs (dominantly LBLs), 1067 BL Lacs (evenly split across LBL, IBL, and HBL), 1077 blazars of unknown type, and 64 nonblazar AGNs comprising 38 radio galaxies. At TeV energies (TeVCat: catalog of TeV gamma-ray sources, http://tevcat2.uchicago.edu/) 80 AGN sources are cataloged comprising 74 blazars, 4 radio galaxies, and 2 unknown types. Within the blazars, 63 are BL Lacs dominated by the HBL class (52 examples) and 8 are FSRQs. As expected, the HBLs are a prominent BL Lac class at TeV energies due to the higher-energy photons (and hence particles) being sampled.

The strong Doppler-boosting of the gamma-ray emission enables the detection of AGNs out to cosmological distances with redshift reaching $z = 4$ at GeV energies. Figure 7.22 illustrates the Fermi-LAT AGN blazar redshift distributions for AGNs in the FSRQ and BL Lac classes (Ajello et al. 2020a).

At TeV energies, AGN blazars are presently seen out to a redshift limit of $z \sim 1$ (TeVCat: catalog of TeV gamma-ray sources, http://tevcat2.uchicago.edu/). This limit is mostly due to the absorption of gamma rays on the extragalactic infrared photons, which can significantly reduce the gamma-ray flux for energies >1 TeV (a topic we will cover below). FSRQs dominate the TeV gamma-ray AGNs detected above redshifts $z = 0.5$. Nonblazar AGNs (often termed misaligned AGNs) where the jet does not point toward Earth, and hence the Doppler-boosting of the emission is weak or nonexistent, tend to be found much closer. At GeV energies they are mostly found within redshift $z < 0.2$ reaching up to $z \sim 1$, while at TeV energies the most distant example is at $z \sim 0.02$ (3C264; Archer et al. 2020). Within the AGN unification scheme outlined earlier, the FR I and FR II radio galaxy types are often considered the nonblazar versions of BL Lac and FSRQs blazars, respectively (e.g., Torresi 2020).

Among the nonblazars seen in gamma rays (see review by Sahakyan et al. 2018), the most prominent examples are Centaurus A (Cen A), M87 (Virgo A), and Fornax A. Importantly, the proximity and side-on view of these objects permit a probe of quite different particle acceleration/interaction regions and the potential to identify the host galaxy core from the jet. Indeed, extended gamma-ray emission detected from Cen A (GeV and TeV energies; Abdo et al. 2010b; HESS Collaboration et al. 2020a) and Fornax A (GeV energies only; Ackermann et al.

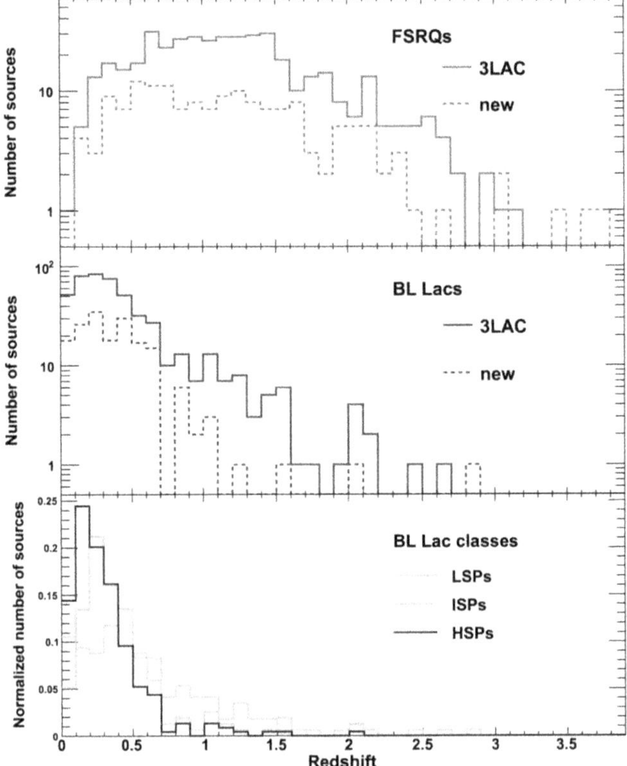

Figure 7.22. Redshift distribution of Fermi-LAT AGN blazars at GeV energies, separated into FSRQs and BL Lac types. Image credit: Ajello et al. (2020a).

2016) has permitted investigation into such issues. The extended emission tracks along the bilobal jets, and the most recent result for Cen A at TeV energies (HESS Collaboration et al. 2020a) from HESS is shown in Figure 7.23.

Here, we can see that the intrinsic source shape is highly elliptical aligned to the orientation of the jets (blue ellipse in Figure 7.23). Explaining this extension (along with similar features in X-rays and radio) appears to require electron acceleration along the jet up to at least 100 TeV in energy. Here, the electrons are responsible for the radio to X-ray synchrotron emission, and the IC emission at TeV gamma-ray energies (with the dominant upscattered photon field from infrared emission). Based on its spectral shape, the GeV gamma-ray emission detected by Fermi-LAT is likely dominated by the core of Cen A. For Fornax A, the GeV gamma-ray emission overlapping the jet lobes might be a combination of leptonic (IC) and hadronic (cosmic-ray and gas collisions) processes.

AGN Variability and Flaring: The nonsteady emission from AGNs is one of their most defining characteristics. The AGN blazars dominate the population of variable sources at gamma-ray energies although the FR II GeV gamma-ray fluxes are usually detected during high states in other wavebands (Torresi 2020). The variability and flaring episodes reflect the inherently stochastic nature of the

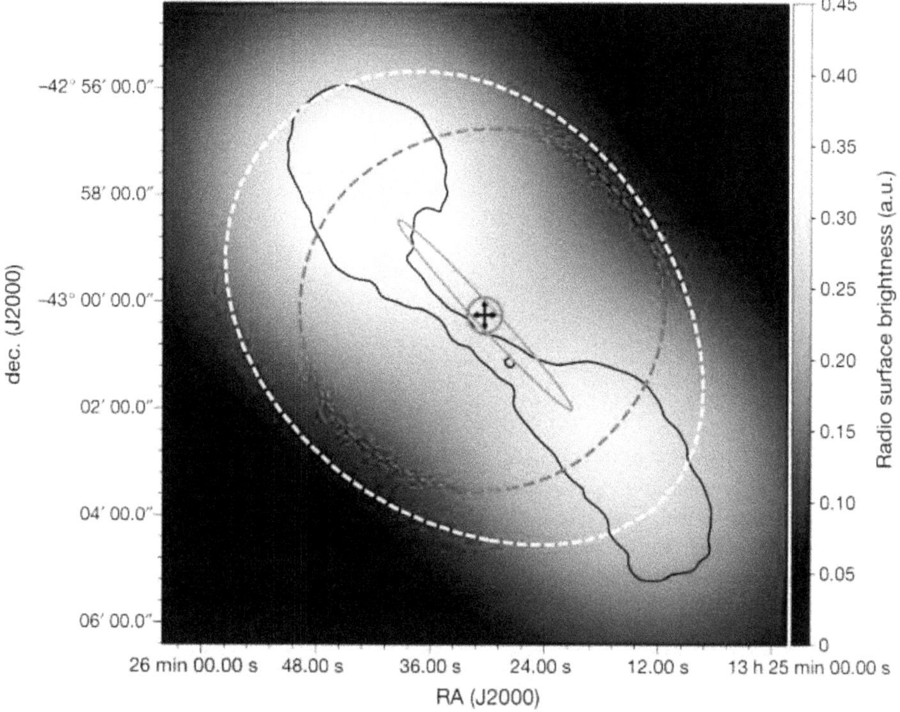

Figure 7.23. VLA 21 cm image of Centaurus A with contours in black (0.5 and 4 Jy beam^{-1}). The white dashed line indicates the TeV gamma-ray morphology convolved with the HESS point-spread function (green dashed circle). The blue-gray ellipse is the intrinsic shape (1σ level) of the TeV gamma-ray emission. The black arrows represent 1σ uncertainties on the TeV gamma-ray position and the red circle the systematic pointing uncertainly of HESS Image credit: HESS Collaboration et al. (2020a), with permission of Springer.

conditions leading to particle acceleration, and especially for blazars, the variability timescales are shortened by Doppler effects as discussed earlier. Variability down to minute timescales has been seen at TeV energies, and from causality arguments, such timescales Δt_{obs} can be used to constrain the physical size of the comoving (i.e., in the flow of the AGN jet presumably) source region R'_{src} emitting the photons such that

$$R'_{\text{src}} \leqslant c\ \Delta t_{\text{obs}}\ \delta = 10^{15}(\delta/10)(\Delta t_{\text{obs}}/1\ \text{h})\ \text{cm}. \tag{7.14}$$

Figure 7.24 illustrates the radio to TeV gamma-ray SEDs for several AGN blazars in the HBL/EBL BL Lac class that show a range of variability across these energy bands.

Here, we can clearly see significant variability in the X-ray and gamma-ray bands for Mkn 501 and 1ES 1426+428 in contrast for the generally stable gamma-ray emission from 1ES 0229+200 (with mild changes in X-ray emission). The latter is part of the emerging class of extreme TeV blazars. The wide variety of variability for AGNs is further illustrated in Figure 7.25, which looks at a detailed

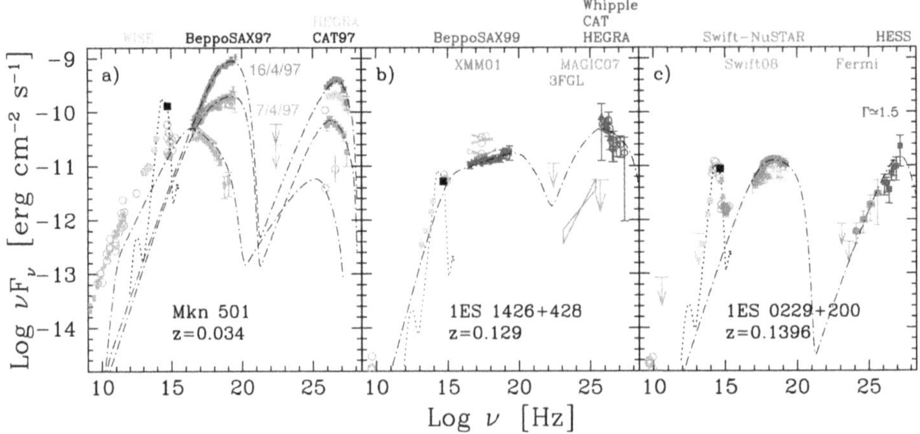

Figure 7.24. SEDs for several AGN blazars of the HBL/EBL BL Lac type. Contemporaneous data (1σ statistical errors; upper limits at 95% confidence level) are labeled with the same color, with gray indicating archival data. Dashed–dotted lines represent synchrotron self-Compton models, and dotted lines thermal emission from host galaxies. The gamma-ray data are corrected for extragalactic background light absorption. Image credit: Biteau et al. (2020), with permission of Springer.

multiwavelength study of two TeV flares from the IBL BL Lac S50716+714 (z = 0.31) (MAGIC Collaboration et al. 2018) during 2015.

From this figure we can see that the first TeV gamma-ray flare (phase A) is accompanied by activity in many other bands. However, a particularly noticeable feature is the dramatic change in the optical polarization position angle (formally, the electric vector position angle EVPA) accompanying the phase A TeV flare. This behavior may be attributed to a knot of material entering a shock region in the jet (with the phase B flare occurring later when the knot leaves the shock region). Such EVPA changes have been seen overlapping gamma-ray flares from other AGNs, thus optical polarimetry has gained in popularity as an important tool to help understand the complex high-energy physics occurring in AGN flares, along with regular monitoring and deep follow-up in the radio (e.g., TANAMI project Ojha et al. 2010; Müller et al. 2018) and in the X-ray bands (e.g., Wierzcholska & Wagner 2016).

Another significant flaring event was exhibited by TXS 0506+056 (z = 0.336 5), a blazar of unknown type. The most notable aspect of this event was that the GeV and TeV gamma-ray flares seen by Fermi-LAT and MAGIC correlated in direction and time with an energetic neutrino detected by IceCube (IceCube Collaboration et al. 2018a, 2018b). Further analysis of the IceCube data revealed a burst of neutrinos from the same direction, hinting at an energetic event a few years prior, although this was not accompanied by any gamma-ray activity. The interpretation of this situation is led to much debate and further discussion can be found in Section 8.4.7.

Some of the strongest source constraints come from the two extraordinary flaring episodes in 2006 of the HBL BL Lac PKS 2155−304 (z = 0.116; Aharonian et al. 2007a, 2009). The first flare, on 2006 July 28, had a peak flux reaching about 15 Crab units, with luminosity $L_\gamma \sim 10^{12} L_{\rm Crab}$ in the TeV band. The most rapid variation

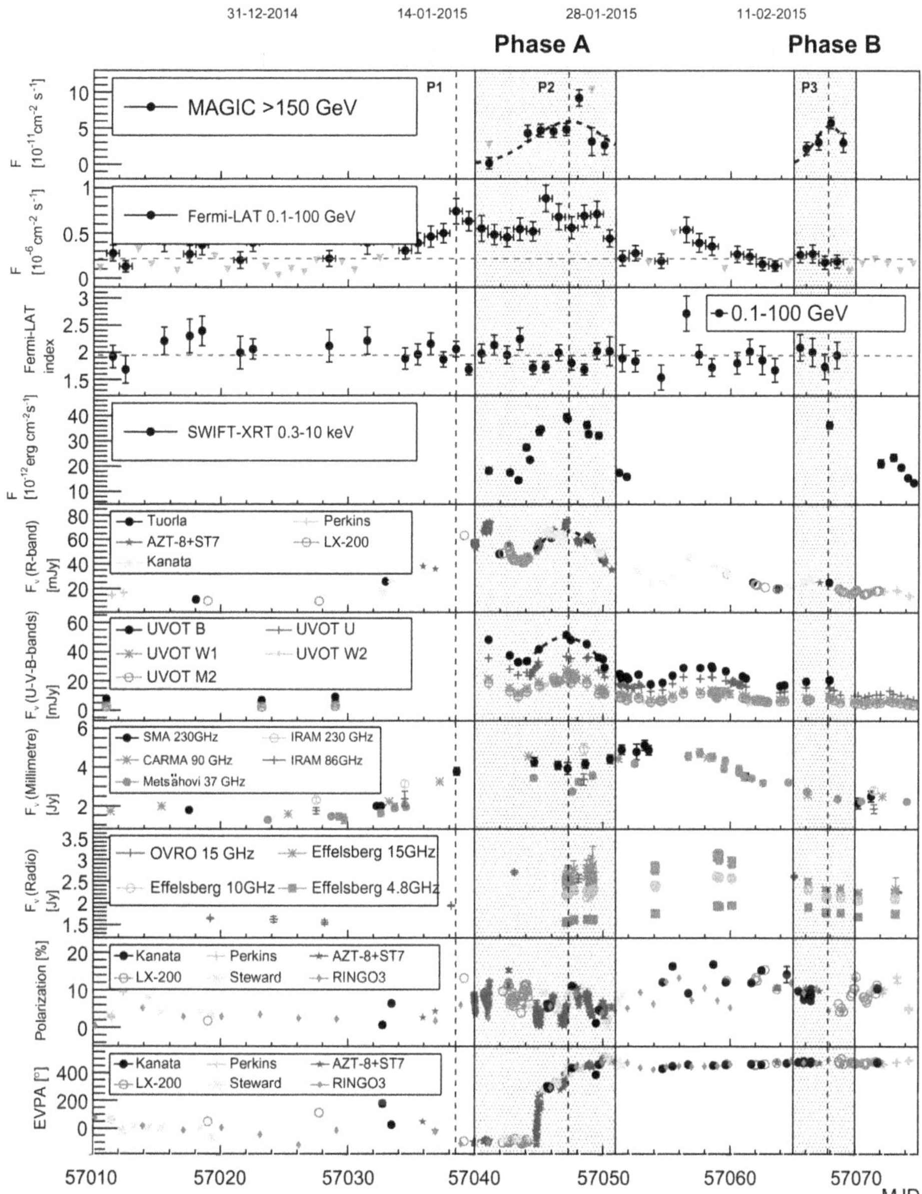

Figure 7.25. Multiwavelength light curve overlapping two TeV gamma-ray flares (top panel) from S50716 +714. The polarization is measured in the optical and near-infrared bands, and its EVPA is shown in the bottom panel. Image credit: MAGIC Collaboration et al. (2018), reproduced with permission © ESO.

timescale detected was $\Delta t_{\rm obs} \sim 2 - 3$ minutes. A second strong flare the next day was accompanied by dedicated X-ray coverage. Figure 7.26 illustrates the first flare light curve and the correlation between the TeV gamma-ray and X-ray emission during the second flare decrease.

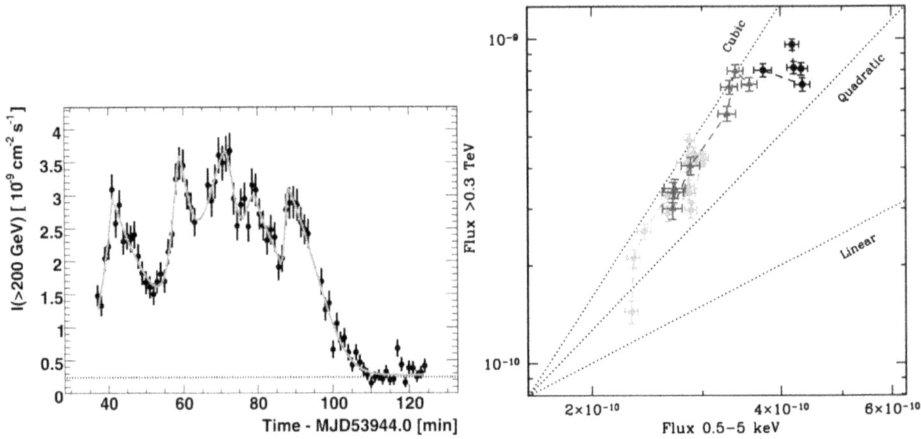

Figure 7.26. Left: TeV gamma-ray light curve of the 2006 July 28, flare from PKS 2155−304 in 1 minute intervals. The solid curve is a fit to the five subpeaks within the flare. Right: TeV gamma-ray and X-ray emission correlation from the second flare, July 29, decay. The colors (black—early; blue—mid; green—late) represent different time intervals within the flare, and the dotted lines represent different slopes for the relation $F_\gamma \propto F_X^\beta$. Image credit: Aharonian et al. (2007a, 2009), reproduced with permission © ESO.

A second flare revealed a roughly cubic relationship between the gamma-ray and X-ray fluxes F_γ F_X^β with $\beta = 3$ during the flare decay phase. This challenges models based on a single population of electrons responsible for the flare (the so-called one-zone models) and suggests a complex source region, perhaps containing several electron populations and/or a dynamically changing source region (e.g., expanding blob with a changing magnetic field).

Using Equation (7.14) and $\Delta t_{\rm obs}\tilde{}2 - 3$ minutes, the comoving source size can be constrained to $R'_{\rm src} \leqslant 0.3\delta$ au. If we now consider the scale of the underlying central engine in the rest frame of the AGN host galaxy, which is effectively the supermassive black hole's Schwarzschild radius R_S, we can infer the black hole's mass, avoiding the Doppler and Lorentz factor terms This is achieved by noting the rest-frame observed timescale $\Delta t_{\rm obs} = \Delta t'(1 + z)/\delta$, for $\Delta t' = R'_{\rm src}/c$, $R_{\rm src} = R'_{\rm src}/\Gamma$ via length contraction and a head-on view such that $\delta \sim 2\Gamma$. We then note that $R_{\rm src} \sim R_s$ and find that $\Delta t_{\rm obs} \geqslant (1 + z)R_s/(2c)$ (see Rieger 2019). Thus, the variability timescale constrains the black hole radius and hence its mass. With $\Delta t_{\rm obs}\tilde{}2 - 3$ minutes, we find $M \leqslant 4 \times 10^7\ M_\odot$, about a factor of 100 below the mass estimated from the host galaxy luminosity, $\sim 10^9\ M_\odot$. However, the situation is likely to be more complex as summarized by Rieger (2019), and we will look at some acceleration scenarios below.

Besides the random/stochastic flaring on short timescales, in recent years a number of AGNs have displayed evidence of quasi-periodic oscillations (QPOs) with periods of 1 to 2 years (see Rieger 2019) although for many only a few cycles have been observed and thus statistical caution is warranted. A prominent example is the HBL PG 1553+113 ($z = 0.5$), which shows some evidence for a ~2.2 yr cycle in

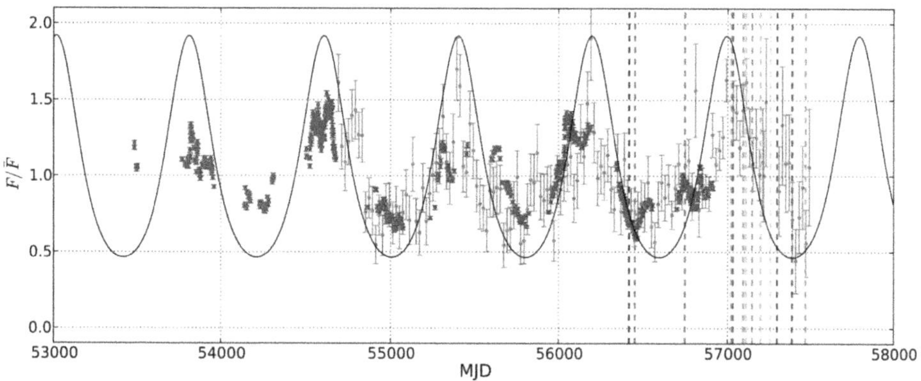

Figure 7.27. Long-term GeV gamma-ray (green points; 20 day bins) and optical (blue points) activity of PS 1553+113. The black line is a periodicity model, and the vertical dashed lines are epochs of spectral calculations. Image credit: Sobacchi et al. (2016) © 2016 The Authors, Published by Oxford University Press on behalf of the Royal Astronomical Society.

its GeV gamma-ray and optical activity (Ackermann et al. 2015a; Sobacchi et al. 2016) as shown in Figure 7.27.

Interpretation of this year-scale QPO behavior has led to the suggestion of binary supermassive black holes in orbit around each other (which would result from historical galaxy mergers) that may lead to peculiar patterns in the circumbinary accretion disk surrounding both black holes. The year-scale periodicity however would demand very close separations and hence significant gravitational-wave emission that would limit the lifetime of the binary system (a merger might be expected on a 10^4 yr timescale as a result). Thus, alternative explanations have been put forward for QPO behavior such as regular motions inside a rotating jet giving rise to periodic changes in the Doppler factor or modulation of the accretion flow onto the black hole. Further unbiased sampling of the gamma-ray and other waveband light curves for many AGN blazars will be needed to advance this topic.

Particle Acceleration in AGNs: The energy for the jet formation and subsequent particle acceleration is ultimately driven by the accretion process as the supermassive black hole converts the gravitational potential energy of surrounding matter to kinetic energy. It is one of the most powerful processes in physics that can generate radiation via heating of the accreting material. For matter falling onto a radius R toward a black hole with Schwarzschild radius R_S at an accretion rate $\dot{M}$, the accretion radiation luminosity can be written as

$$L_{\mathrm{acc}} = 0.5\dot{M}c^2(R_S/R) = \eta\dot{M}c^2 \tag{7.15}$$

where $\eta = 0.5(R_S/R)$ is the accreting efficiency to covert the matter rest mass–energy into radiation. Thus, accretion can be remarkably efficient. For AGNs with supermassive black holes we typically find $L_{\mathrm{acc}} \sim 10^{46}$ erg/s, which requires an accretion rate of $\dot{M} \sim 1\ M_{\odot}$/yr. This is to be compared to stellar-sized black hole accretion in

XRBs and microquasars (as discussed earlier) with $L_{\text{acc}} \sim 10^{30-40}$ erg/s and $\dot{M} \sim 10^{-7} M_{\odot}$/yr. The formation of the jet associated with the accretion disk is still the topic of much debate. A general principle is that the magnetic field attached to the rotating accretion disk is wound up into a helical pattern, acting as a guide for particles to be accelerated at the jet base or along the jet, perpendicular to the accretion flow. From the Hillas plot shown earlier (Figure 7.19), we can easily see that the size and magnetic field strength of the accretion region and jet are ideal places for particle acceleration to extreme energies, hence the long-held notion that AGNs are the sources of the highest-energy cosmic rays beyond 10^{20} eV. For a more in-depth discussion, the reader is referred to the recent review by Blandford et al. (2019).

Following the review by Rieger (2019), for illustration of some possibilities, we present in Figure 7.28 several scenarios that have been put forward to explain the time-varying particle acceleration and subsequent variable photon emission.

The three shown here (see Rieger 2019 and references therein) all involve small-scale features within the jet or at the jet base close to the black hole horizon. They include electron/positrons accelerated in an electric field near the black hole horizon, jet-in-jets involving mini-jet features that locally accelerate particles via reconnection, and a matter interaction model involving cosmic-ray interactions with photon fields from stars embedded in the jet. The latter involves the hadronic process while the former two involve the leptonic (IC gamma-ray and synchrotron radio to X-ray) process. For the leptonic process, the gamma-ray emission is often attributed to the SSC process but external IC (upscattering of external photon fields such as the CMB and infrared photons farther out in the jet or in a luminous broad line region around the accretion disk) has been considered. For hadronic models, the scenario as shown in Figure 7.28 is one such geometric setup. Generally, the acceleration of protons to over 10^{19} eV in extremely intense magnetic fields of 20 G or more (compared to the milligauss fields required in leptonic SSC models) is required. In this case the gamma-ray emission results from synchrotron emission directly from the protons or from the π° decay generated when these protons interact with low-energy photons ($p\gamma$ process as outlined in Chapter 8 as this is also a major production channel for

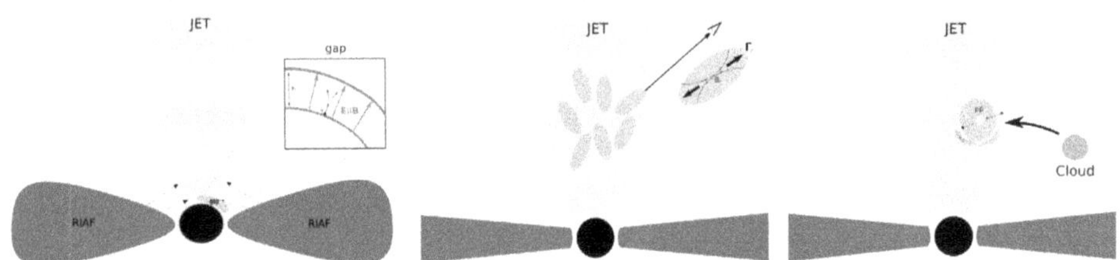

Figure 7.28. Several scenarios to explain TeV gamma-ray variability from AGNs. Left: electron/positrons generated in electric fields near the black hole horizon lead to inverse-Compton gamma rays. Middle: "Jet-in-jet" where many mini-jet plasmoids can accelerate particles via reconnection Right: clouds or stars within the jet act as a dense source of cosmic-ray protons. Image credit: Rieger (2019) with permission of MDPI.

neutrinos). The fast synchrotron cooling time might be a more natural explanation (versus $p\gamma$ processes) to explain the rapid (minute-wise scale) flaring episodes discussed above (e.g., Barkov et al. 2012).

Extragalactic Background Light Constraints: An important issue affecting the observed high-energy flux from AGN blazars is the absorption of gamma rays on the low-energy (or soft) photon fields permeating the universe known as the extragalactic background light (EBL). The level of EBL photons reflects the star formation history of the universe and is thus a fundamental signature of its evolution.

Gamma rays will annihilate to electron/positron pairs upon interaction with soft photon fields such that the observed flux at Earth $F_{\rm obs}(E)$ will be attenuated according to the following relation:

$$F_{\rm obs}(E) = F_{\rm int}(E)\exp\left(-\tau\left(E_{\gamma},\, z\right)\right), \tag{7.16}$$

where $F_{\rm int}(E)$ is the intrinsic flux at the source, and $\tau(E,\, z)$ is the absorption optical depth versus gamma-ray energy E_{γ} and distance/redshift z. The optical depth is dependent on the distribution of EBL photons and the pair production cross section. The pair production cross section peaks for the interaction of an isotropic soft photon field of blackbody temperature T (eV units, or peak wavelength $\lambda_{\mu{\rm m}}$) with gamma rays (TeV units) of energy $E_{\gamma} = 0.9E_{\rm T,eV} \approx 0.7\lambda_{\mu{\rm m}}$. Thus, optical to infrared photons are relevant to GeV gamma-ray absorption, while mid-infrared photons are relevant to TeV gamma-ray absorption (see, e.g., Aharonian et al. 2006c, 2008d).

The EBL across the optical to infrared bands is quite difficult to measure directly owing to foreground contamination. However, in recent years the EBL has been quantified by fitting Equation (7.16) to the observed flux from an ensemble of blazar spectra spanning a range of redshifts and a variety of EBL models in a statistical approach. Earlier methods adopted the approach that the intrinsic gamma-ray spectral index would be $\Gamma \geqslant 1.5$ based on fundamental particle acceleration and IC scattering theory. Figure 7.29 illustrates the most recent compilations of results from Fermi-LAT at GeV energies (Abdollahi et al. 2018), and those from TeV gamma-ray telescopes such as HESS, MAGIC, and VERITAS (Biteau 2015; HESS Collaboration et al. 2017; Acciari et al. 2019; Abeysekara et al. 2019; Desai et al. 2019).

Here we can see that the EBL constraints from the gamma-ray measurements are generally at the bottom end of those from the direct measurements. The complementary roles of the GeV and TeV bands are also evident with the former constraining the EBL up to a few micrometers and the latter taking over above those wavelengths up to about 100 μm. Ever since the first gamma-ray constraints on the EBL were published (Aharonian et al. 2006c), the GeV to TeV gamma-ray results have suggested that the universe is somewhat more transparent to gamma-ray photons than initially thought, giving promise to future TeV facilities such as the CTA to probe many sources beyond redshift $z = 2$ (as we will highlight briefly at the end of this chapter).

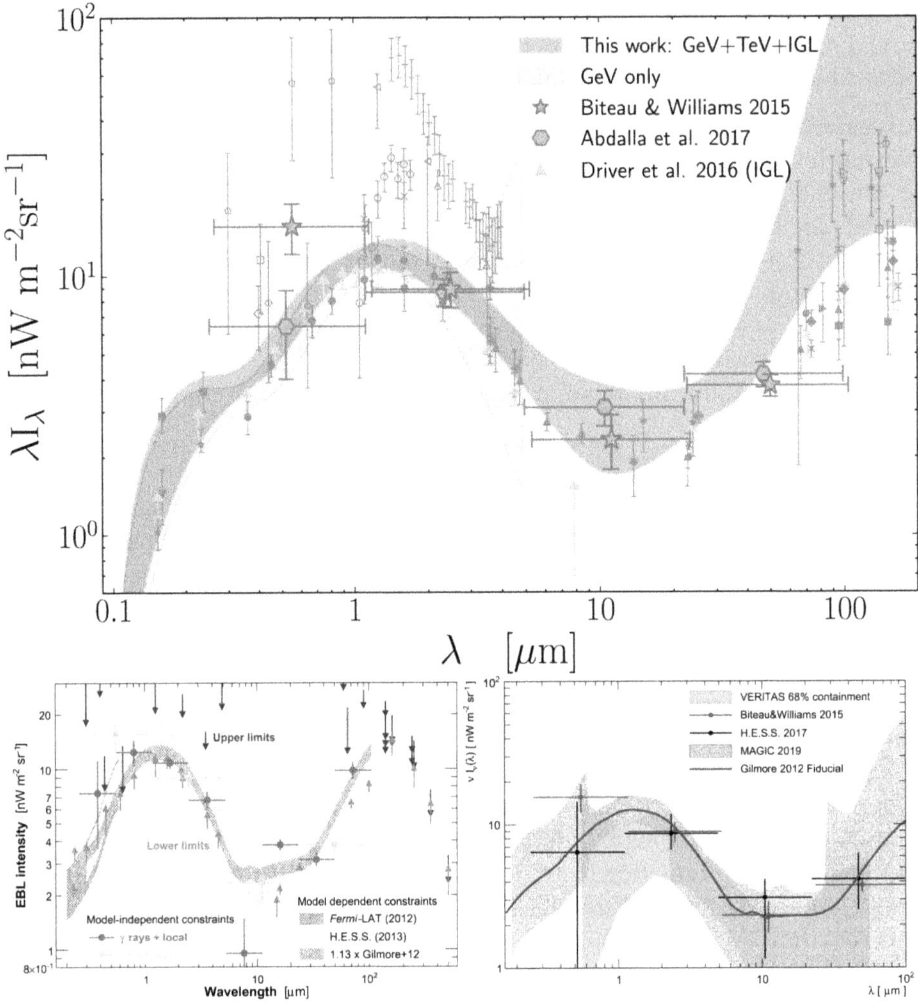

Figure 7.29. Compilation of recent EBL constraints from gamma-ray blazar fluxes and other direct measurements. Top: from Fermi-LAT, separated according to blazar redshift (Desai et al. 2019). Bottom left: from the combination of HESS, MAGIC, and VERITAS up to 2015 at $z = 0$ (Biteau 2015). Bottom right: most recent comparison from VERITAS (Abeysekara et al. 2019) including the latest results from the HESS Collaboration et al. (2017) and MAGIC (Acciari et al. 2019), at $z = 0$.

7.6.2 Non-AGN Galaxies

Among the non-AGN galaxies detected in gamma rays, starburst galaxies (with high rates of massive star formation) represent a prominent type. About a dozen starburst galaxies are seen in the GeV band (Abdollahi et al. 2020; Ajello et al. 2020c) with two of them, NGC 253 and M82, seen at TeV energies (Acero et al. 2009a; VERITAS Collaboration et al. 2009; HESS Collaboration et al. 2018h).

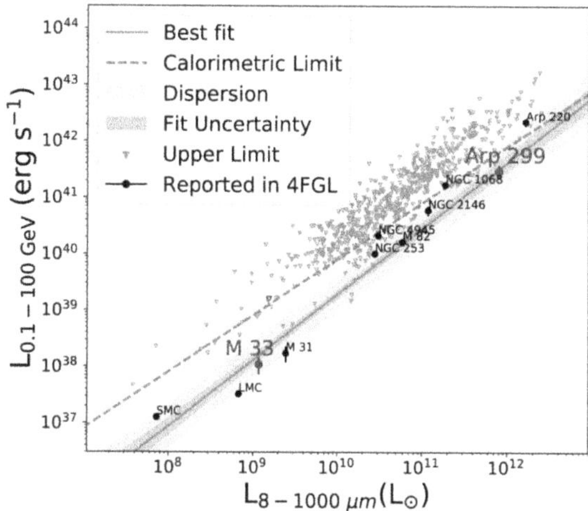

Figure 7.30. GeV gamma-ray and infrared luminosity correlation for starburst galaxies (and normal galaxies with some level of starburst activity). Image credit: Xi et al. (2020b).

At the lower end of starburst activity, several "normal" galaxies have been seen in the GeV band (besides our own Milky Way with disk-dominated GeV emission) such as the Andromeda Galaxy (M31; Abdo et al. 2010a; Ackermann et al. 2017b); the LMC (Abdo et al. 2010c; which was first detected in GeV gamma rays by EGRET Sreekumar et al. 1992); the SMC (Abdo et al. 2010d), which are relatively active in star formation for their small sizes; and M33, a satellite of the Andromeda galaxy. Most recently, GeV gamma-ray emission from M33 and Arp 229 (Xi et al. 2020b) has also been reported. The GeV to TeV gamma-ray emission is thought to be tied to the level of massive star formation (leading to cosmic-ray acceleration and their interaction with the interstellar gas). Evidence of this comes from the general correlation between the GeV and infrared emission, with the latter being a long-recognized tracer of massive star formation (note that the 1.4 GHz radio continuum emission is also a good tracer of star formation).

The GeV emission from two starburst galaxies also appears to be variable (Peng et al. 2019), suggesting the potential for an AGN-like component powered by a central supermassive black hole. Given the expected presence of a high level of cosmic-ray acceleration in starburst galaxies and their recent links to arrival directions of the highest-energy cosmic rays (Aab et al. 2018), they may contribute to the underlying TeV to PeV cosmic rays responsible for the astrophysical neutrinos detected by IceCube (Aartsen et al. 2013) and the extragalactic gamma-ray background (see, e.g., Ajello et al. 2020c and references therein). A suggestion for a large-scale (several kiloparsecs) GeV gamma-ray halo around M31 (Karwin et al. 2019) may indeed have further implications for the connection to the neutrino events seen by IceCube (Recchia et al. 2021).

7.6.3 Gamma-Ray Bursts

GRBs are quite likely to have been the first gamma-ray sources to be detected in the 1960s by US military satellites (e.g., Klebesadel et al. 1973, and as discussed in Section 7.1.1). Today, they are probably the gamma-ray source type that has attracted the most attention. The groundbreaking BATSE detector on board the CGRO, which cataloged over 1600 GRBs (Paciesas et al. 1999), showed that GRBs can be separated into two main types based on the duration of their MeV gamma-ray flares. While all GRBs exhibit a rapid falloff in flux over time, their falloff times are generally characterized as T_{90}, the time taken for their integral photon counts to increase from 5% to 95% of the total counts. The so-called short GRBs (SGRBs) and long GRBs (LGRBs) have $T_{90} \leqslant 2$s and $T_{90} \geqslant 2$ s, respectively, which define their so-called prompt emission. Weaker afterglow emission has been observed from many GRBs up to timescales of months, and often over a wider range of wavebands compared to the prompt emission (usually dominated by low-energy gamma rays).

More than 2000 GRBs are now cataloged by the Swift/Gehrels-BAT and Fermi-GBM telescopes (Lien et al. 2016; von Kienlin et al. 2020). Above 0.1 GeV, Fermi-LAT has detected of order 170 GRBs (Ajello et al. 2019), with a few of them emitting photons above 50 GeV. At TeV energies, GRBs were finally unambiguously detected in 2019 after several decades of searching. An early hint for a GRB detection was reported by MILAGRITO (Atkins et al. 2000), a precursor to the MILAGRO water Čerenkov telescope. The compelling TeV detections by MAGIC and HESS of GRB 190114C, GRB 180729B, and GRB 201216C (MAGIC Collaboration et al. 2019; Abdalla et al. 2019; de Naurois et al. 2019; Blanch et al. 2020) have so far been of LGRBs (up to $z \sim 0.8$), with HESS detecting GRB 180729B up to 10 hr after the prompt emission. However, just recently, MAGIC reported a marginal detection of the nearby ($z = 0.162$) SGRB GRB 160821B (Acciari et al. 2021) starting from 24 s after the burst lasting for 4 hr.

Today, it is accepted that GRBs are generated by relativistic jet outflows generated by two distinct types of objects. The LGRBs are established as the result of the core collapse of a massive star (Woosley & Bloom 2006; often called the "collapsar" model). SGRBs were for many years theoretically linked (e.g., Mészáros 2002) to the merger of compact objects, usually a pair of neutron stars (Berger 2014). It was not until the gravitational-wave signal from the binary neutron star (BNS) merger GW 170817 was linked to the SGRB GRB 170817A (Abbott et al. 2017) and a radio jet (Mooley et al. 2018) that this theory was confirmed. It is recognized that this event was an SGRB viewed slightly off axis from its jet direction (~20°), in what is termed a kilonova, an r-process supernova powered by a BNS or neutron star and black hole merger, with generally isotropic emission that includes an SGRB jet component. The isotropic emission is generated by high-mass element nucleosynthesis, leading to an optical/infrared counterpart. Several other SGRBs are also associated with kilonovae (e.g., GRB 160821B was tentatively detected in the TeV band by MAGIC, as discussed above).

In the GeV band, the LGRBs dominate with less than 20 of the ~170 Fermi-LAT GRBs being SGRBs, suggesting they are somewhat less able, compared to LGRBs, to accelerate particles above multi-GeV energies. The general picture of particle acceleration in all GRBs involves a jet process, not unlike that found in AGN blazars, although there are some differences. The jet formation is induced by the core collapse or compact object merger process but many aspects remain unclear. In any case, the gamma-ray energy released in GRBs reaching 10^{55} erg (isotropically equivalent) in a few seconds makes them the most extreme explosive events since the Big Bang, thus requiring extraordinary conditions to accelerate particles.

The particle acceleration in a GRB is likely related to the formation of shocks at various locations and times within the jet. What causes these shocks is still the subject of much debate, such as from photon radiative pressure (the fireball model; Meszaros & Rees 1993) or magnetic pressure (see review by Pe'er 2019). Figure 1.4 presents a general picture illustrating shock in a GRB jet (from an LGRB or SGRB progenitor), as well as the prompt and afterglow emission regions. The prompt emission is generated in internal shocks set up when shells of material in the jet (moving at slightly different speeds) interact. The afterglow emission is related to a shock front at the leading edge of the expanding GRB ejecta. Additionally, a "reverse" shock may be generated when the external shock encounters the interstellar medium or the circumstellar medium from the progenitor (e.g., Piran 1999). The reverse shock is believed to be responsible for radio and optical flashes or the brightening detected in some GRBs. Additional components unique to SGRBs are time-extended X-ray emission possibly related to the still-active postmerger object (e.g., a magnetar) and the kilonova emission discussed earlier.

The prompt and afterglow emission components up to GeV energies are generally attributed to the synchrotron process from relativistic electrons. The so-called "Band" function (Band et al. 1993), a broken power law with an exponential cutoff, has been successful in empirically describing the spectral shape of this emission up to about 0.1 GeV. The function parameters (spectral slopes, exponential cutoff, and break energy) can follow the evolution of the GRB spectrum as it evolves from the prompt to the afterglow phase. However, additional high-energy components have been required since the detections of multi-GeV photons by EGRET and Fermi-LAT, and even more so since the recent TeV detections.

We will now take a closer look at the recent TeV gamma-ray results, as they have provided some further insight into GRB physics. The recent TeV detections of the mostly afterglow emission have provided the first observational evidence for an additional IC component (MAGIC Collaboration et al. 2019), and detailed time evolution of the afterglow emission across a wide range of energies. Figure 7.31 compares the light curve and energy spectra evolution for GRB 190114C. We can see that the TeV light curve follows a generally similarly smooth time evolution decay law (flux $\propto t^{-\alpha}$ with α in the range -1.5 to -1.3) seen in the X-ray and GeV bands. The range of α values reflects the different times postburst the measurements concentrate on and the expected evolution of spectral features (e.g., breaks) appearing in the underlying electron spectrum (see Ajello et al. 2020b for a more in-depth discussion).

The appearance of a double peak in the spectrum across the X-ray to gamma-ray bands is a signature of synchrotron and an additional IC process to account for the

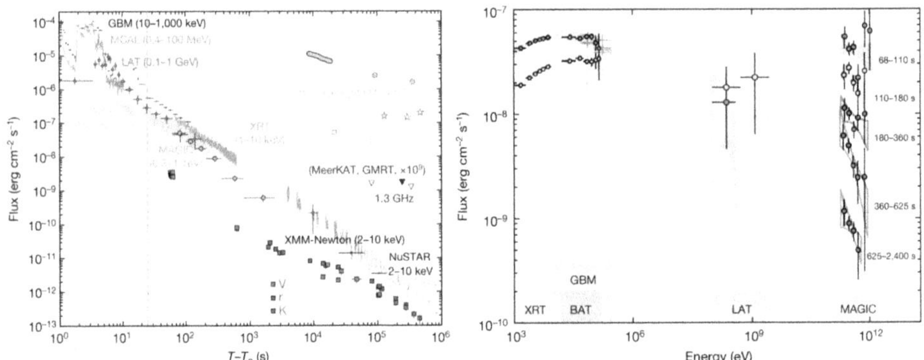

Figure 7.31. Left: light curve (energy flux versus time reference against the initial burst time T_o) for GRB 190114C. The MAGIC TeV fluxes (green) are compared with fluxes from the radio to GeV gamma-ray bands. Right: energy spectra for the interval 68 s to 2400 s after the initial burst, in the X-ray, GeV, and TeV gamma-ray bands. Images from MAGIC Collaboration et al. (2019), with permission of Springer.

TeV emission. Specifically, the TeV emission is thought to arise from the SSC process (upscattering X-ray synchrotron photons via the IC process as discussed earlier for AGN and some other gamma-ray source types). Alternative processes are synchrotron emission directly from electrons or protons. In both cases, particles of extreme energies are required (PeV or higher energies), which would not be necessarily unexpected from GRB jets. However, the total energy budget for proton synchrotron is uncomfortably large (due to the large proton mass) and for electron synchrotron up to TeV energies, the initial jet Lorentz factor would need to exceed values of $\sim$1000 (linked to the maximum energy electrons can attain due to synchrotron losses), also uncomfortably large in the context of how such jets are expected to form. The SSC TeV emission keeps the maximum synchrotron emission to sub-GeV values, thereby constraining the initial jet Lorentz factor to a range 300–700, more in line with the expectation for GRB jets.

7.6.4 Other Flaring Gamma-Ray Sources

We conclude our look at extragalactic sources with a quick look at some additional topics.

The second Fermi-LAT flare catalog (Abdollahi et al. 2017) from 2017 lists over 4000 GeV flare events associated with over 500 sources. The vast majority of these sources are AGNs (BL Lac and FSRQs) as expected, and just over 10 are associated with Galactic sources (XRBs, novae, and pulsars). Interestingly, 77 of the flares have no association and the majority of these are extragalactic with just two found within 5° of the Galactic Plane. Furthermore, the photon spectral index of these nonassociated sources tends to be rather hard ($\Gamma \leqslant 1.5$,) which might suggest a new class of gamma-ray transients that could be studied in more detail by future TeV facilities.

Finally, Fermi-LAT has very recently revealed exciting evidence for GeV emission from a magnetar giant flare (Fermi-LAT Collaboration et al. 2021). The burst location is consistent with several galaxies, IC 1576, IC 1578, and IC 1582, and the starburst galaxy NGC 253 (the Sculptor Galaxy) at 3.5 Mpc distance. It is linked to GRB 200415a, which was originally classified as a short GRB based on its MeV emission as seen by the Fermi-GBM. The unusually long delay (19 s) to the GeV emission and QPO seen in the MeV signal (see references in Fermi-LAT Collaboration et al. 2021) point to a giant magnetar flare. The delayed GeV emission is currently suggested to arise from the ultrarelativistic outflow from the giant flare interacting with dense material surrounding the magnetar. This result may suggest exciting new prospects for GeV to TeV gamma-ray emission to be detected from other magnetar giant flares from within the Milky Way, and other nearby galaxies (as has been long suggested, e.g., Ioka et al. 2005). Giant flares are thought to be related to a significant rupture of neutron star crust or deformation of the neutron star magnetosphere and given they are so far seen only once from different magnetars, they may represent a major step in their evolution.

7.7 Additional Physics Topics with Gamma Rays

So far our discussion has centered on astrophysical situations where particle acceleration is taking place, giving rise to relevant gamma-ray emission. In this section, we will briefly review the range of additional physics topics impacted by gamma-ray studies beyond this discussion so far. These additional topics are within the emerging fields of astroparticle physics and particle cosmology.

Since the Big Bang, the universe has left behind traces of its evolution via many so-called cosmic relics. The elusive dark matter is the best known of these relics, accounting for over 75% of the universe's matter we see today. Its detection via direct and indirect methods is discussed broadly in Chapter 10. Here, we will just emphasize the role that gamma rays play in indirect detection methods.

Weakly interacting massive particles (WIMPs) have been the most popular idea for dark matter. The self-annihilation of WIMPs (such as supersymmetric neutralino particles) can give rise to gamma rays in the form of line emission (with energy matching that of the WIMP) or broadband emission from secondary hadrons. Gamma rays in the GeV to TeV band are considered the most fruitful avenue to probe given the theoretical mass of WIMPs mostly appears in the same energy range. Observationally, gamma-ray emission is ideally searched for in regions that are astrophysically boring. As a result, dwarf galaxies, which are not expected to host strong astrophysical particle accelerators, are considered ideal sources. On the other hand, the intense gravitational attraction at the center of our galaxy makes it a compelling place to look, despite the confusion with astrophysical sources. To date, no convincing GeV to TeV gamma-ray evidence for dark matter has been seen from dwarf galaxies or from our Galactic Center looking for broadband and line emission (e.g., Abdallah et al. 2018; Abdalla et al. 2018; Abdallah et al. 2020; Abdollahi et al. 2020; Hoof et al. 2020).

A well-debated topic has been an excess of emission around GeV energies above the astrophysical background expectation seen by Fermi-LAT and its relation to dark matter (see Ackermann et al. 2017a, and references therein). However, several alternative astrophysical explanations appear feasible, such as an unresolved population of millisecond pulsars, star formation processes, or the central supermassive black hole (e.g., Bartels et al. 2016; Macias et al. 2018). One particular aspect disfavoring the dark matter interpretation is the generally nonspherical and potentially clumpy morphology of this GeV excess (e.g., Yang & Aharonian 2016). The systematic uncertainties in the astrophysical background models that can lead to large-scale GeV gamma-ray features are also being investigated in more detail, shedding light on the need for further understanding of potential particle sources and their time-dependent nature (e.g., Porter et al. 2019).

Besides WIMPs, there are a number of alternative ideas for dark matter across the range of mass scales (see e.g., Bertone & Tait 2018; Doro 2017). Some of them, such as axions (or light axion-like particles, ALPs) and primordial black holes (PBHs), could have distinct signatures in the gamma-ray band. Under certain conditions, gamma-ray photons may oscillate into ALP states (Hooper & Serpico 2007; de Angelis et al. 2011), which can avoid absorption on the EBL over cosmological distances. Thus, a lack of EBL absorption on the GeV to TeV spectra of distant AGN blazars is a way to search for evidence of such ALPs. However, for AGN blazar spectra so far studied in detail at GeV and TeV energies, there is at present no indication for such an effect (e.g., Sanchez et al. 2013; Domínguez & Ajello 2015). The gamma-ALP oscillation may also impart energy-dependent oscillations (dependent on the mass of the ALP and fluctuations in the intervening intergalactic magnetic field) in the gamma-ray flux potentially within the energy resolution of telescopes. So far, rather tight constraints on the ALP mass in the range of 4 to 100 neV have been set using Fermi-LAT and HESS observations of NGC 1275 (radio galaxy) and PKS 2155 −304 (AGN blazar), respectively (Ajello et al. 2016; Abramowski et al. 2013; Zhang et al. 2018).

PBHs across a huge mass range (Planck mass up to $\sim 10^5\ M_\odot$) are a predicted consequence of density fluctuations in the radiation-dominated era of the early universe from inflation up to about 50 kyr after the Big Bang and are a potential source of dark matter (see, e.g., Carr & Kühnel 2020; Ballesteros et al. 2020). The gravitational-wave signals from mergers of black holes with mass $>10\ M_\odot$ (Abbott et al. 2016) and the very recent suggestion for a stochastic gravitational-wave background (Arzoumanian et al. 2020) have sparked renewed interest in PBH studies and their potential as dark matter candidates. PBHs lose energy or evaporate via Hawking radiation according to the PBH's thermal temperature, which follows a $1/M_{\rm BH}(t)$ dependence, where $M_{\rm BH}(t)$ is the PBH's mass varying with time. As the PBH nears its endpoint with ever decreasing mass, its thermal temperature can rise to GeV energies, producing a burst of fundamental particles (quarks, gluons, electrons, neutrinos, photons) and gamma-ray photons. At the current age of the universe, it's expected that PBHs of mass $\sim 5 \times 10^{14}$ g could emit seconds-duration pulses of GeV to TeV gamma-ray emission at their final stages. To date, no evidence for such gamma-ray signals have been seen, and the most recent searches have been conducted using data from the wide-field Fermi-LAT and HAWC (Ackermann et al. 2018; Albert et al. 2020b) telescopes.

Gamma rays can be used to probe for Lorentz invariance violations (LIVs), a consequence of quantum gravity and string theory models (e.g., Alfaro 2005; Kostelecký & Samuel 1989; Amelino-Camelia 2001). The LIV signatures are suggested to appear via a variety of avenues such as a speed-of-light dependency on photon energy, decay of photons, production of Čerenkov light in the vacuum, and changes in the threshold for pair production leading to reduced EBL absorption (e.g., Amelino-Camelia et al. 1998; Stecker & Glashow 2001; Stecker 2003; Kifune 1999). The LIV effect can be characterized by the energy scale at which it is invoked $E_{\rm LIV}^{(n)}$, and this is of order the Planck energy ($\sim 1.22 \times 10^{28}$ eV) or higher. A popular LIV effect to search for has been the temporal delay in photon arrival times from pulsed or flaring sources such as the Crab Pulsar, and various AGNs or GRBs across the GeV and TeV bands (e.g., MAGIC Collaboration et al. 2017; Abdalla et al. 2019; Vasileiou et al. 2013). This technique depends on assuming there are no intrinsic temporal delays in the photon production at the source, something that is not always viable.

The strongest constraints on the LIV mass scale of $>2.2 \times 10^{31}$ eV (1800 times the Planck energy) for a so-called LIV effect of order $n = 1$ have so far been set by HAWC using gamma-ray photons of energy $\geqslant$100 TeV (Albert et al. 2020a) looking for the effects of superluminal photon decay giving rise to unusually strong cutoffs in gamma-ray spectra. A more complete review of the various potential LIV gamma-ray photon signatures, LIV theory, and comparison of experimental constraints is given by Martínez-Huerta et al. (2020).

Magnetic monopoles are a fundamental consequence of magnetic quantization (suggested by Dirac in 1931). They are thought to be produced during the very early universe, resulting from Grand Unified Theories of all the fundamental forces, and also specifically during the electroweak phase transition, which saw the separation of the electroweak forces at about 10^{-32} s after the Big Bang (see reviews by Patrizii & Spurio 2015; Patrizii et al. 2019; Mavromatos & Mitsou 2020). Monopole masses ranging from a few TeV up to $\sim 10^{13}$ GeV are suggested and because they are charged (magnetically charged), they can be accelerated to relativistic speeds in cosmic magnetic fields. The makeup of monopoles (a combination or "soup" of fundamental quarks, gluons, W and Z particles, their antiparticles, and photons—e.g., Patrizii et al. 2019) means that they can also lose energy via well-known processes such as ionization, and at extreme energies (with monopole Lorentz factors $>10^7$) via pair production and photonuclear reactions. A fundamental upper limit on the monopole flux, known as the Parker bound, is set according to the maximum flux of monopoles allowed to prevent quenching or short-circuiting of Galactic magnetic fields at a rate faster than such fields are created in galaxies (a modified Parker bound comes from scaling this limit with the monopole mass).

For sufficiently high energy (velocity greater than the local speed of light in a medium such as air or water), monopoles can also create Čerenkov photons at a rate a few 1000× greater than that for a single electric charge (Tompkins 1965). As a result, ground-based gamma-ray telescopes detecting Čerenkov light can probe for ultrarelativistic monopoles (with velocity $v/c \sim 1$) via their unique linear appearance (Doro 2017), and the first upper limits have been set (Spengler & Schwanke 2011) a factor of ~100 higher than the Parker bound. Much tighter limits (below the Parker bound) are available to neutrino telescopes detecting Čerenkov light in water or ice, as the lower local speed of light permits Čerenkov light to be produced at much lower energies down to $v/c > 0.5$. Additionally, the particle air showers generated by ultrarelativsitic monopoles could be detected by cosmic-ray telescopes like the Pierre Auger Observatory (Aab et al. 2016; Patrizii et al. 2019).

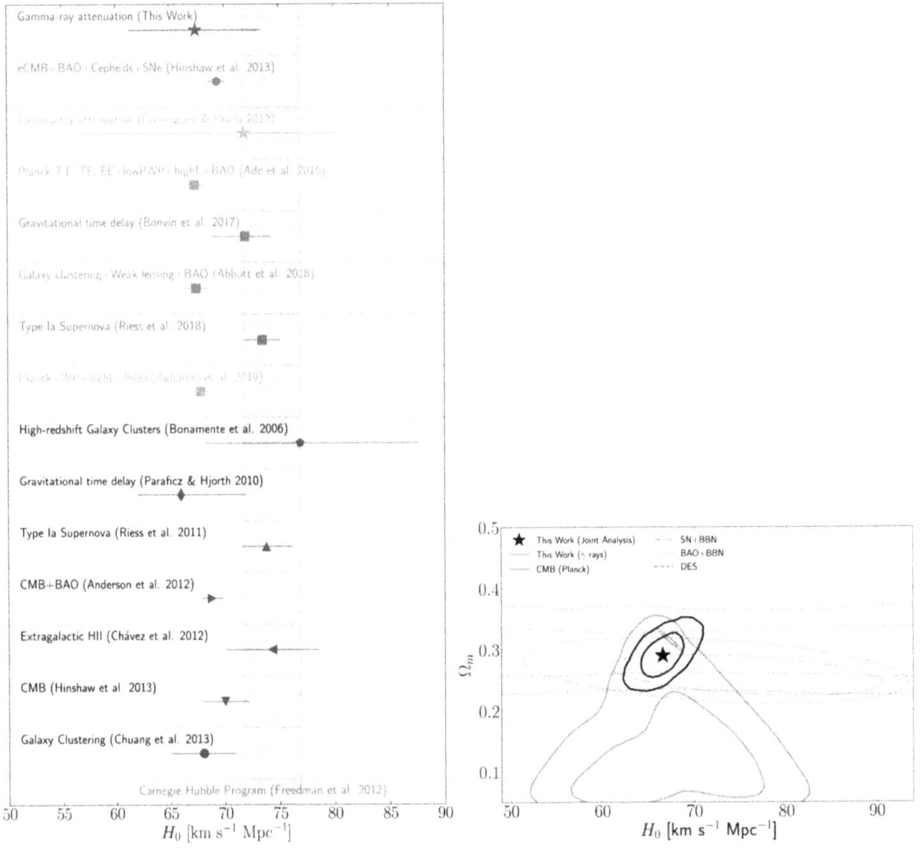

Figure 7.32. Left: Hubble constant H_o constraints from gamma-ray observations (top measurement) compared to results from other techniques (see Domínguez et al. 2019 and references therein). Right: H_o and matter density Ω_m constraints from gamma-ray measurements (green contours) and a joint fit from gamma rays and other techniques (supernova plus Big Bang nucleosynthesis SN+BBN; weak lensing from the Dark Energy Survey, DES; cosmic microwave background, CMB; and baryon acoustic oscillations, BAOs). Images from Domínguez et al. (2019).

Finally, we will examine the determination of cosmological parameters such as the Hubble constant, H_o and the matter content in the universe Ω_m. The level of EBL (discussed earlier) in the universe is a fundamental consequence of galaxy evolution as a function of redshift. As a result, constraints on the EBL optical depth can then lead to constraints on H_o and Ω_m when interpreted in terms of a model evolution for the universe (e.g., the standard ΛCDM model). The recent results from Domínguez et al. (2019), shown in Figure 7.32, used an ensemble of GeV spectra from 739 AGN blazars (and one GRB) measured by Fermi-LAT (Abdollahi et al. 2018) and TeV spectra from 38 AGN blazars by various ground-based gamma-ray facilities (Desai et al. 2019).

We can see that the gamma-ray measurements of the EBL can offer another measurement of H_o, in this case with a value of about 65 to 67 km s^{-1} Mpc^{-1} with a 5% to 10% error. This appears consistent with those from the CMB fluctuations and from baryon acoustic oscillations and potentially further supports the tension in the H_o measurement arising from supernova and Cepheid measurements. The matter density parameter Ω_m is found to be 0.29 within a percent when jointly combined with other measurements.

7.8 Next-generation Gamma-Ray Observatories and Future Perspectives

In this final section we will briefly look at the next generation of gamma-ray facilities under consideration, construction, or recently commenced and include an outlook on what might come next. Figure 7.33 illustrates the differential flux sensitivity comparisons for various past, current, and proposed gamma-ray facilities.

The MeV band has been undersampled since the end of the COMPTEL telescope in 2000 (it was on board the CGRO satellite along with the EGRET and BATSE telescopes). The MeV telescope proposals eASTROGAM (de Angelis et al. 2018) and AMEGO (McEnery et al. 2019) are gathering momentum (the AMEGO

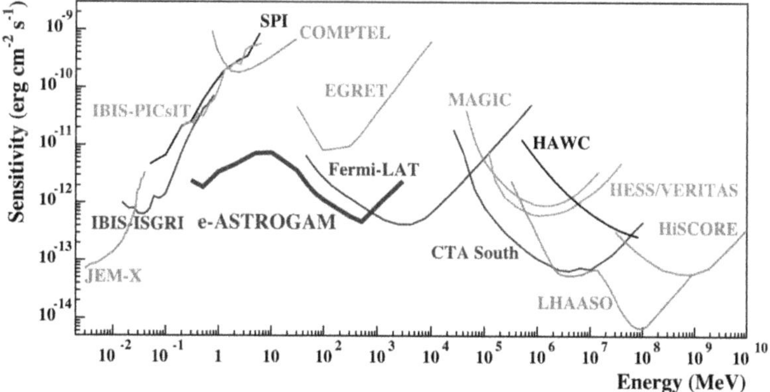

Figure 7.33. Differential flux sensitivities for various MeV to PeV gamma-ray facilities (past, current, and proposed). Tracking facilities like HESS assume a 50 hr observation time while wide-field facilities like HAWC and LHAASO assume a 1 yr observation time. SPI, IBIS, and JEM-X are detectors on board the INTEGRAL satellite. Image credit: de Angelis et al. (2018), Copyright 2018, with permission from Elsevier.

sensitivity and energy coverage are somewhat similar to that of eASTROGAM as shown in Figure refsens). They both aim to significantly improve the angular resolution in the MeV band down to $\sim 1°$ or better, and energy resolution to 20% or better. A new aspect will be their ability to measure the polarization of gamma rays. This will open up new ways to discriminate the leptonic and hadronic origins of the gamma-ray emission and the ability to track magnetic field structures in relativistic jets (see review by Ilie 2019). Additionally, the THESEUS mission (Amati et al. 2018) has been discussed as the next major facility (following on from Swift-Gehrels-BAT) for transient sources such as GRBs sampled across the X-ray to MeV band with the smaller-scale SVOM mission (Claret 2018) being a potential pathfinder.

In the GeV band, the Fermi-LAT remains the main instrument for the foreseeable future (and can in principle operate for another decade due to its lack of expendables on board). The GAMMA-400 mission (Topchiev et al. 2016) has been proposed as a potential successor to Fermi-LAT, and the AMS100 mission (Schael et al. 2019), designed for cosmic-ray studies, will also have excellent gamma-ray detection performance in the GeV band. The DAMPE gamma-ray telescope (see Li et al. 2021) has a focus on high GeV-energy resolution studies for dark matter searches and the detection of some transients and is expected to operate well into the current decade.

In the multi-GeV band and above, the next-generation ground-based CTA, www.cta–observatory.org is the key facility. CTA will initially comprise 64 telescopes situated across northern and southern sites (Canary Islands, Spain in the north, and Paranal, Chile in the south; the "Alpha" configuration). About 80% of the telescopes will be located at the southern site. The comparatively greater focus on Galactic sources at the southern site requires a wide energy coverage detecting gamma rays beyond 100 TeV in order to probe for extreme Galactic particle accelerators or PeVatrons as discussed earlier. The CTA sensitivity (CTA-North and CTA-South for 50 hr observation; sensitivity versus time and energy) is shown in Figures 7.33 and 7.34. Here we can clearly see CTA is expected to be at about 10 times more sensitive than current instruments in the "workhorse" 1 to 10 TeV energy range and extends the energy coverage from about 20 GeV to beyond 100 TeV. CTA's angular resolution, reaching arcminute scales for gamma rays >10 TeV or so, will also provide the most detailed morphological studies in gamma rays so far. Importantly, CTA will close the energy coverage gap with Fermi-LAT and thus will open up a new window in the study of gamma-ray transients and variable sources (e.g., AGN blazars, GRBs, XRBs, novae, etc.). Figure 7.34 (right panel) shows CTA's time-limited sensitivity improvement over that of Fermi-LAT. This is due to CTA's huge effective collection area ($>10^4$ m^2) in the multi-GeV-energy domain, which may enable CTA to detect AGN blazar flares to redshift $z \sim 2$ and bright GRBs out to redshift $z \sim 4$.

A large fraction of CTA's initial observing time will be devoted to large key science projects (KSPs) focusing on a wide range of topics including systematic surveys of the Galactic Plane, Galactic Center region, extragalactic sky, and LMC, and targeted projects to study transients, AGNs, star formation regions, particle accelerators and PeVatrons, dark matter, and beyond-Standard-Model physics.

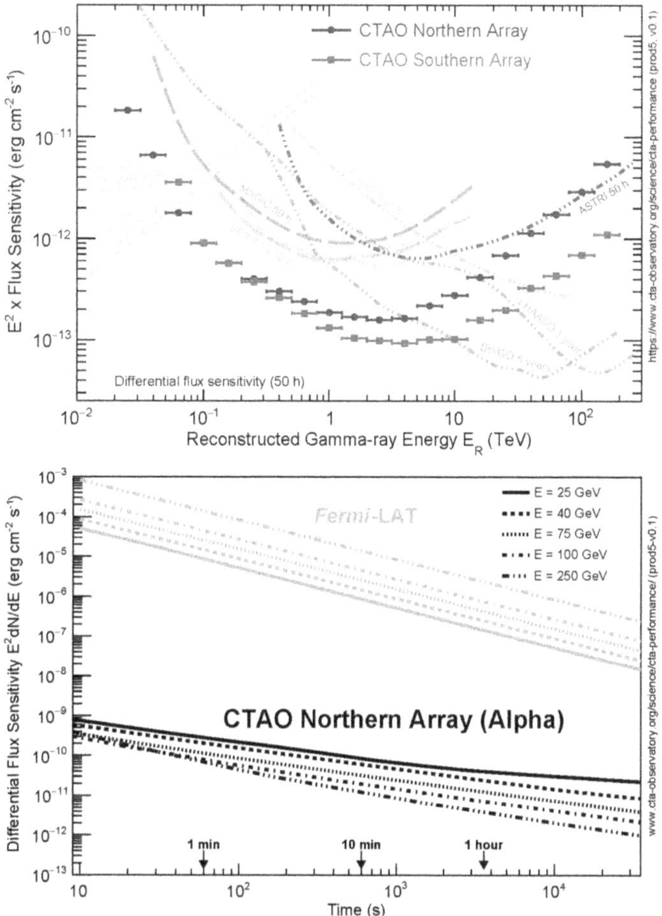

Figure 7.34. Differential flux sensitivities vs. energy and time for CTA ('Alpha' configuration), Fermi-LAT, and several other gamma-ray facilities. Images credit: Cherenkov Telescope Array Observatory (CTAO) Northern/Southern Array, and Cherenkov Telescope Array Consortium (CTAC), CTA performance, https://www.flickr.com/photos/cta_observatory/26079109147/in/album-72157673853528865/ and https://www.cta-observatory.org/wp-content/uploads/2019/04/CTA-Performanceprod3b-v2-South-20deg-ShortTerm.png.

Non-gamma-ray science such as optical intensity interferometry and optical photometry are also being pursued because CTA will represent a large array of optical telescopes in the world. Further details of these science programs can be found in CTA, www.cta-observatory.org. More recently, CTA's potential to probe fundamental physics (such as the EBL absorption, related cosmology, Lorentz invariance in quantum gravity, and dark matter at the Galactic Center) was updated in detail (Abdalla et al. 2021; Acharyya et al. 2021).

Regarding new wide-field ground-based gamma-ray telescopes, the Large High Altitude Air Shower Observatory (LHAASO; http://english.ihep.cas.cn/lhaaso/; Bai et al. 2019) facility has recently commenced operation. LHAASO operates from a high (>4400 m) plateau in China and comprises an array of solid particle detectors, water-Čerenkov detectors, and wide-field telescopes detecting Čerenkov and fluorescence light. The large collection area LHASSO achieves (more than 10 times larger than that of HAWC) will enable LHASSO to push very deep into the >1PeV gamma-ray band and thus further advance our understanding of PeVatrons. Its large collection also permits a highly sensitive search for transients. In the southern hemisphere, the Southern Wide-field Gamma-ray Observatory (SWGO; Albert et al. 2019) has been proposed and is based primarily on the water-Čerenkov technique. SWGO is designed to operate with a collection area at least five times larger than that of HAWC. Being in the southern hemisphere (high-altitude sites in Chile and Peru are under consideration), SWGO will access the Galactic Plane and thus the rich variety of Galactic gamma-ray sources. As for LHAASO, SWGO is expected to further improve our understanding of PeVatrons and transient sources. Similarly, the HiScore concept (Tluczykont et al. 2014) based on an array of air-Čerenkov detectors could offer excellent angular resolution (0.1°) for PeV gamma rays over a wide field of view. It is currently under development in Russia.

Although not a wide-field array, the TenTen concept (Rowell et al. 2008) considers a "sparse" array of at least 50 imaging Čerenkov telescopes over a wide ground area to achieve a 10km^2 collection area for energies >10 TeV up to at least 500 TeV in energy. The imaging Čerenkov telescopes can achieve an excellent angular resolution (<0.1°) and so are required for detailed morphological studies and comparisons with radio to X-ray images. With PeVatrons becoming a recognized class of gamma-ray sources and CTA's performance expected to have impact in this area, such large sparse array concepts will no doubt attract further attention over the next decade.

Following on further from the opportunities that the next generation of facilities will offer, serious consideration is now going into the development of TeV gamma-ray facilities distributed across Earth to provide 24 hr coverage of transients and variable sources. One such idea is the Čerenkov Ring (Ruhe et al. 2019) concept, which calls for Čerenkov imaging telescopes located across longitude bands in both the northern and southern hemispheres to enable 24 hr coverage. The telescopes would ideally be robotic for ease of operation, have a rapid response, and carry out automated searches or surveys for sources. These telescopes or small arrays of a few telescopes should be sufficiently sensitive to detect and characterize (i.e., determine energy spectra) sources in the 0.1 to 10s of TeV energy band. A threshold energy in the range of 0.1 TeV to 1 TeV is highly desirable to ensure the study of extragalactic sources suffering absorption on the EBL (e.g., AGN blazars and GRBs). Small arrays of two to five imaging Čerenkov telescopes have already clearly proved their worth (e.g., the HEGRA IACT System, Daum et al. 1997; HESS, https://www.mpi-hd.mpg.de/hfm/HESS/; VERITAS, https://veritas.sao.arizona.edu/; MAGIC, https://magic.mpp.mpg.de/) and can achieve arcminute-scale resolution under certain conditions (HESS Collaboration et al. 2018c; HESS Collaboration 2020), which is ideal for point-like transient sources. In the >10 TeV range, excellent performance is also predicted for similarly small arrays (e.g., Stamatescu et al. 2011). Further studies are warranted to explore the potential of small arrays utilizing Čerenkov cameras with large fields of view (similar to those of the CTA) and optimized for high resolution. For full 24 hr coverage in the southern hemisphere, sites in Australia and Southern Africa are essential to complement CTA-South in Chile.

References

CTA: Cherenkov telescope array, https://www.cta-observatory.org/
CTA performance, https://www.flickr.com/photos/cta_observatory/26079109147/in/album-72157673853528865/
Cta performance, https://www.cta-observatory.org/wp-content/uploads/2019/04/CTA-Performance-prod3b-v2-South-20deg-ShortTerm.png
CTA Consortium 2019, Science with the Cherenkov Telescope Array (Singapore: World Scientific)
HAWC: High-altitude water cherenkov, https://www.hawc-observatory.org/
HESS: High energy stereoscopic system, https://www.mpi-hd.mpg.de/hfm/HESS/
How many known X-ray (and other) sources are there?, https://heasarc.gsfc.nasa.gov/docs/heasarc/headates/how_many_xray.html
LHAASO: Large high-altitude air shower observatory, http://english.ihep.cas.cn/lhaaso/
MAGIC: Major atmospheric gamma imaging cherenkov, https://magic.mpp.mpg.de/
TeVCat: Catalogue of tev gamma-ray sources, http://tevcat2.uchicago.edu/
VERITAS: Very energetic radiation imaging telescope array system, https://veritas.sao.arizona.edu/
Aab, A., Abreu, P., Aglietta, M., et al. 2016, PhRvD, 94, 082002
Aab, A., Abreu, P., Aglietta, M., et al. 2018, ApJ, 853, L29
Aab, A., Abreu, P., Aglietta, M., et al. 2017, JCAP, 2017, 009
Aartsen, M. G., Abbasi, R., Abdou, Y., et al. 2013, PhRvL, 111, 021103
Aartsen, M. G., Ackermann, M., Adams, J., et al. 2020, ApJ, 891, 9
Abbasi, R. U., Abe, M., Abu-Zayyad, T., et al. 2020, MNRAS, 492, 3984
Abbott, B. P., Abbott, R., Abbott, T. D., et al. 2016, PhRvL, 116, 061102
Abbott, B. P., Abbott, R., Abbott, T. D., et al. 2017, ApJ, 848, L12
Abdalla, H., Abe, H., Acero, F., et al. 2021, JCAP, 2021, 048
Abdalla, H., Adam, R., Aharonian, F., et al. 2019, Natur, 575, 464
Abdalla, H., Aharonian, F., Ait Benkhali, F., et al. 2018, JCAP, 2018, 037
Abdalla, H., Aharonian, F., Ait Benkhali, F., et al. 2019, ApJ, 870, 93
Abdallah, H., Abramowski, A., Aharonian, F., et al. 2018, PhRvL, 120, 201101
Abdallah, H., Adam, R., Aharonian, F., et al. 2020, PhRvD, 102, 062001
Abdo, A. A., Ackermann, M., Ajello, M., et al. 2010a, A&A, 523, L2
Abdo, A. A., Ackermann, M., Ajello, M., et al. 2011, Sci, 331, 739
Abdo, A. A., Ackermann, M., Ajello, M., et al. 2010b, Sci, 328, 725
Abdo, A. A., Ackermann, M., Ajello, M., et al. 2010c, A&A, 512, A7
Abdo, A. A., Ackermann, M., Ajello, M., et al. 2010d, A&A, 523, A46
Abdo, A. A., Ajello, M., Allafort, A., et al. 2013, ApJS, 208, 17
Abdo, A. A., Allen, B., Aune, T., et al. 2008, ApJ, 688, 1078
Abdo, A. A., Allen, B., Berley, D., et al. 2007a, ApJ, 658, L33
Abdo, A. A., Allen, B., Berley, D., et al. 2007b, ApJ, 664, L91
Abdo, A. A., Ackermann, M., Ajello, M., et al. 2010e, A&A, 524, A75
Abdollahi, S., Acero, F., Ackermann, M., et al. 2020, ApJS, 247, 33

Abdollahi, S., Ackermann, M., Ajello, M., et al. 2017, ApJ, 846, 34
Abdollahi, S., Ackermann, M., Ajello, M., et al. 2018, Sci, 362, 1031
Abeysekara, A. U., Albert, A., Alfaro, R., et al. 2017a, Sci, 358, 911
Abeysekara, A. U., Albert, A., Alfaro, R., et al. 2017b, ApJ, 843, 40
Abeysekara, A. U., Albert, A., Alfaro, R., et al. 2020, PhRvL, 124, 021102
Abeysekara, A. U., Archer, A., Aune, T., et al. 2018a, ApJ, 861, 134
Abeysekara, A. U., Archer, A., Benbow, W., et al. 2019, ApJ, 885, 150
Abeysekara, A. U., Benbow, W., Bird, R., et al. 2018b, ApJ, 867, L19
Abeysekara, A. U., Albert, A., Alfaro, R., et al. 2018c, Natur, 562, 82
Abramowski, A., Acero, F., Aharonian, F., et al. 2013, PhRvD, 88, 102003
Abramowski, A., Aharonian, F., Ait Benkhali, F., et al. 2014a, PhRvD, 90, 122007
Abramowski, A., Aharonian, F., Ait Benkhali, F., et al. 2014b, ApJ, 794, L1
Abramowski, A., Acero, F., Aharonian, F., et al. 2012, A&A, 537, A114
Acciari, V. A., Ansoldi, S., Antonelli, L. A., et al. 2019, MNRAS, 486, 4233
Acciari, V. A., Ansoldi, S., Antonelli, L. A., et al. 2021, ApJ, 908, 90
Acero, F., Aharonian, F., Akhperjanian, A. G., et al. 2010, MNRAS, 402, 1877
Acero, F., Aharonian, F., Akhperjanian, A. G., et al. 2009a, Sci, 326, 1080
Acero, F., Ballet, J., Decourchelle, A., et al. 2009b, A&A, 505, 157
Acero, F., Donato, D., Ojha, R., et al. 2013, ApJ, 779, 133
Acharyya, A., Adam, R., Adams, C., et al. 2021, JCAP, 2021, 057
Ackermann, M., Ajello, M., Albert, A., et al. 2015a, ApJ, 813, L41
Ackermann, M., Ajello, M., Albert, A., et al. 2017a, ApJ, 840, 43
Ackermann, M., Ajello, M., Albert, A., et al. 2014, Sci, 345, 554
Ackermann, M., Ajello, M., Albert, A., et al. 2017b, ApJ, 836, 208
Ackermann, M., Ajello, M., Allafort, A., et al. 2013, Sci, 339, 807
Ackermann, M., Ajello, M., Allafort, A., et al. 2011, Sci, 334, 1103
Ackermann, M., Ajello, M., Atwood, W. B., et al. 2015b, ApJ, 810, 14
Ackermann, M., Ajello, M., Atwood, W. B., et al. 2012, ApJ, 750, 3
Ackermann, M., Ajello, M., Baldini, L., et al. 2016, ApJ, 826, 1
Ackermann, M., Atwood, W. B., Baldini, L., et al. 2018, ApJ, 857, 49
Aharonian, F., Akhperjanian, A., Barrio, J., et al. 2001, A&A, 370, 112
Aharonian, F., Akhperjanian, A., Beilicke, M., et al. 2002, A&A, 393, L37
Aharonian, F., Akhperjanian, A., Beilicke, M., et al. 2005a, A&A, 431, 197
Aharonian, F., Akhperjanian, A. G., Anton, G., et al. 2009, A&A, 502, 749
Aharonian, F., Akhperjanian, A. G., Aye, K. M., et al. 2005b, Sci, 309, 746
Aharonian, F., Akhperjanian, A. G., Barres de Almeida, U., et al. 2008a, A&A, 477, 353
Aharonian, F., Akhperjanian, A. G., Barres de Almeida, U., et al. 2008b, A&A, 483, 509
Aharonian, F., Akhperjanian, A. G., Bazer-Bachi, A. R., et al. 2007a, ApJ, 664, L71
Aharonian, F., Akhperjanian, A. G., Bazer-Bachi, A. R., et al. 2008c, A&A, 481, 401
Aharonian, F., Akhperjanian, A. G., Bazer-Bachi, A. R., et al. 2006a, Natur, 439, 695
Aharonian, F., Akhperjanian, A. G., Bazer-Bachi, A. R., et al. 2006b, ApJ, 636, 777
Aharonian, F., Akhperjanian, A. G., Bazer-Bachi, A. R., et al. 2006c, Natur, 440, 1018
Aharonian, F., Akhperjanian, A. G., Bazer-Bachi, A. R., et al. 2007b, A&A, 464, 235
Aharonian, F., Akhperjanian, A. G., Bazer-Bachi, A. R., et al. 2006d, A&A, 460, 365
Aharonian, F., Akhperjanian, A. G., Bazer-Bachi, A. R., et al. 2006e, A&A, 460, 743
Aharonian, F. A. 2001, SSRv, 99, 187

Aharonian, F. A., & Atoyan, A. M. 1996, A&A, 309, 917
Aharonian, F. A., Atoyan, A. M., & Kifune, T. 1997, MNRAS, 291, 162
Aharonian, F. A., Bogovalov, S. V., & Khangulyan, D. 2012, Natur, 482, 507
Aharonian, F. A., Drury, L. O. C., Voelk, H. J., et al. 1994, A&A, 285, 645
Aharonian, F. A. 2013, Gamma-Ray Emission of Supernova Remnants and the Origin of Galactic Cosmic Rays (Dordrecht: Springer)
Aharonian, F., Peron, G., Yang, R., Casanova, S., & Zanin, R. 2020, PhRvD, 101, 083018
Aharonian, F., Yang, R., & de Oña Wilhelmi, E. 2019, NatAs, 3, 561
Aharonian, F. A. 2004, Very High Energy Cosmic Gamma Radiation: A Crucial Window on the Extreme Universe (Singapore: World Scientific)
Aharonian, F. A., Khangulyan, D., & Costamante, L. 2008d, MNRAS, 387, 1206
Aharonian, F., Akhperjanian, A. G., Aye, K.-M., et al. 2005b, A&A, 432, L25
Ahnen, M. L., Ansoldi, S., Antonelli, L. A., et al. 2017, A&A, 601, A33
Aielli, G., Assiro, R., Bacci, C., et al. 2006, NIMPA, 562, 92
Ajello, M., Albert, A., Anderson, B., et al. 2016, PhRvL, 116, 161101
Ajello, M., Angioni, R., Axelsson, M., et al. 2020a, ApJ, 892, 105
Ajello, M., Arimoto, M., Axelsson, M., et al. 2020b, ApJ, 890, 9
Ajello, M., Arimoto, M., Axelsson, M., et al. 2019, ApJ, 878, 52
Ajello, M., Atwood, W. B., Baldini, L., et al. 2017, ApJS, 232, 18
Ajello, M., Di Mauro, M., Paliya, V. S., & Garrappa, S. 2020c, ApJ, 894, 88
Albert, A., Alfaro, R., Alvarez, C., et al. 2020a, PhRvL, 124, 131101
Albert, A., Alfaro, R., Alvarez, C., et al. 2020b, JCAP, 2020, 026
Albert, A., Alfaro, R., Alvarez, C., et al. 2020c, ApJ, 896, L29
Albert, A., Alfaro, R., Alvarez, C., et al. 2020d, ApJ, 905, 76
Albert, A., Alfaro, R., Alvarez, C., et al. 2021, ApJL, 907, L30
Albert, A., Alfaro, R., Ashkar, H., et al. 2019, arXiv:1902.08429
Albert, J., Aliu, E., Anderhub, H., et al. 2007, ApJ, 665, L51
Alfaro, J. 2005, PhRvL, 94, 221302
Aliu, E., Anderhub, H., Antonelli, L. A., et al. 2008, Sci, 322, 1221
Alves Batista, R., Biteau, J., Bustamante, M., et al. 2019, FrASS, 6, 23
Amati, L., O'Brien, P., Götz, D., et al. 2018, AdSpR, 62, 191
Amelino-Camelia, G., Ellis, J., Mavromatos, N. E., Nanopoulos, D. V., & Sarkar, S. 1998, Natur, 393, 763
Amelino-Camelia, G. 2001, Natur, 410, 1065
Amenomori, M., Bao, Y. W., Bi, X. J., et al. 2021, PhRvL, 126, 141101
Anderson, L. D., Wang, Y., Bihr, S., et al. 2017, A&A, 605, A58
Ansoldi, S., Antonelli, L. A., Antoranz, P., et al. 2016, A&A, 585, A133
Antonucci, R. 1993, ARA&A, 31, 473
Arakawa, M., Hayashida, M., Khangulyan, D., & Uchiyama, Y. 2020, ApJ, 897, 33
Archer, A., Benbow, W., Bird, R., et al. 2020, ApJ, 896, 41
Archer, A., Benbow, W., Bird, R., et al. 2016, ApJ, 821, 129
Arzoumanian, Z., Baker, P. T., Blumer, H., et al. 2020, ApJ, 905, L34
Atkins, R., Benbow, W., Berley, D., et al. 2000, ApJ, 533, L119
Atwood, W. B., Abdo, A. A., Ackermann, M., et al. 2009, ApJ, 697, 1071
Axford, W. I., Leer, E., & Skadron, G. 1977, ICRC, 11, 132
Baade, W., & Zwicky, F. 1934, PhRv, 46, 76

Baghmanyan, V., Peron, G., Casanova, S., Aharonian, F., & Zanin, R. 2020, ApJ, 901, L4
Bai, X., Bi, B. Y., Bi, X. J., et al. 2019, arXiv:1905.02773
Ballesteros, G., Coronado-Blázquez, J., & Gaggero, D. 2020, PhLB, 808, 135624
Band, D., Matteson, J., Ford, L., et al. 1993, ApJ, 413, 281
Barkov, M. V., Aharonian, F. A., Bogovalov, S. V., Kelner, S. R., & Khangulyan, D. 2012, ApJ, 749, 119
Barrau, A., Bazer-Bachi, R., Beyer, E., et al. 1998, PhRA, 416, 278
Bartels, R., Krishnamurthy, S., & Weniger, C. 2016, PhRvL, 116, 051102
Bartoli, B., Bernardini, P., Bi, X. J., et al. 2014, ApJ, 790, 152
Bartoli, B., Bernardini, P., Bi, X. J., et al. 2015, ApJ, 806, 20
Becker Tjus, J., & Merten, L. 2020, PhR, 872, 1
Bell, A. R. 1978, MNRAS, 182, 147
Bell, A. R. 2004, MNRAS, 353, 550
Berger, E. 2014, ARA&A, 52, 43
Bertone, G., & Tait, T. M. P. 2018, Natur, 562, 51
Bignami, G. F., & Hermsen, W. 1983, ARA&A, 21, 67
Bird, A. J., Malizia, A., Bazzano, A., et al. 2007, ApJS, 170, 175
Biteau, J., Prandini, E., Costamante, L., et al. 2020, NatAs, 4, 124
Biteau, J., & Williams, D. A. 2015, ApJ, 812, 60
Blanch, O., et al. 2020, ATel, 14275
Bland-Hawthorn, J., Maloney, P. R., Sutherland, R., et al. 2019, ApJ, 886, 45
Blandford, R. D., & Ostriker, J. P. 1978, ApJ, 221, L29
Blandford, R., Meier, D., & Readhead, A. 2019, ARA&A, 57, 467
Blondin, J. M., Chevalier, R. A., & Frierson, D. M. 2001, ApJ, 563, 806
Bodaghee, A., Tomsick, J. A., Pottschmidt, K., et al. 2013, ApJ, 775, 98
Bogovalov, S. V., & Aharonian, F. A. 2000, MNRAS, 313, 504
Boley, F. I., & Macoy, N. H. 1961, RScI, 32, 1359
Bordas, P., Yang, R., Kafexhiu, E., & Aharonian, F. 2015, ApJ, 807, L8
Bordas, P. 2020, NatAs, 4, 1132
Bouchet, L., Jourdain, E., Roques, J. P., et al. 2008, ApJ, 679, 1315
Bouchet, L., Jourdain, E., & Roques, J.-P. 2015, ApJ, 801, 142
Braiding, C., Wong, G. F., Maxted, N. I., et al. 2018, PASA, 35, e029
Brennan, M. H., Malos, J., Millar, D. D., & Wallace, C. S. 1958, Natur, 182, 973
Bulgarelli, A., Fioretti, V., Parmiggiani, N., et al. 2019, A&A, 627, A13
Bykov, A. M., & Toptygin, I. N. 2001, AstL, 27, 625
Cao, Z., Aharonian, F. A., An, Q., et al. 2021, Natur, 594, 33
Caraveo, P. A. 2014, ARA&A, 52, 211
Carr, B., & Kühnel, F. 2020, ARNPS, 70, 355
Casanova, S., Aharonian, F. A., Fukui, Y., et al. 2010, PASJ, 62, 769
Casse, M., & Paul, J. A. 1980, ApJ, 237, 236
Cataldo, M., Pagliaroli, G., Vecchiotti, V., & Villante, F. L. 2020, ApJ, 904, 85
Celli, S., Morlino, G., Gabici, S., & Aharonian, F. A. 2019, MNRAS, 487, 3199
Cesarsky, C. J., & Montmerle, T. 1983, SSRv, 36, 173
Chang, C., Konopelko, A., & Cui, W. 2008, ApJ, 682, 1177
Cherenkov Telescope Array Consortium, Acharya, B. S., Agudo, I., et al. 2019 Science with the Cherenkov Telescope Array (Singapore: World Scientific)

Chernyakova, M., & Malyshev, D. 2020, in Multifrequency Behaviour of High Energy Cosmic Sources—XIII, 45
Cheung, C. C., Jean, P., Shore, S. N., et al. 2016, ApJ, 826, 142
Chiaro, G., Salvetti, D., Mura, G. L., et al. 2016, MNRAS, 462, 3180
Chomiuk, L., Metzger, B. D., & Shen, K. J. 2020, arXiv:2011.08751
Claret, A. 2018, in 42nd COSPAR Scientific Assembly, Vol. 42, E1.15-15-18
Claussen, M. J., Frail, D. A., Goss, W. M., & Gaume, R. A. 1997, ApJ, 489, 143
Cooper, A. J., Gaggero, D., Markoff, S., & Zhang, S. 2020, MNRAS, 493, 3212
Corbet, R. H. D., Chomiuk, L., Coe, M. J., et al. 2019, ApJ, 884, 93
Crocker, R. M., & Aharonian, F. 2011, PhRvL, 106, 101102
Crocker, R. M., Bicknell, G. V., Taylor, A. M., & Carretti, E. 2015, ApJ, 808, 107
Cui, Y., Yeung, P. K. H., Tam, P. H. T., & Pühlhofer, G. 2018, ApJ, 860, 69
Dame, T. M., Hartmann, D., & Thaddeus, P. 2001, ApJ, 547, 792
Daugherty, J. K., & Harding, A. K. 1982, ApJ, 252, 337
Daugherty, J. K., & Harding, A. K. 1983, ApJ, 273, 761
Daum, A., Hermann, G., & Heß, M. 1997, APh, 8, 1
de Angelis, A., Tatischeff, V., Grenier, I. A., et al. 2018, JHEAp, 19, 1
De Angelis, A., Tatischeff, V., Tavani, M., et al. 2017, ExA, 44, 25
de Angelis, A., Galanti, G., & Roncadelli, M. 2011, PhRvD, 84, 105030
de Jager, O. C., Ferreira, S. E. S., & Djannati-Ataï, A. 2009, arXiv:0906.2644
de Naurois, M., & Mazin, D. 2015, CRPhy, 16, 610
de Naurois, M., et al. 2019, ATel, 13052
de Wilt, P., Rowell, G., Walsh, A. J., et al. 2017, MNRAS, 468, 2093
Desai, A., Helgason, K., Ajello, M., et al. 2019, ApJ, 874, L7
Di Mauro, M., Manconi, S., Negro, M., & Donato, F. 2021, PhRvD, 104, 103002
Dickey, J. M., McClure-Griffiths, N., Gibson, S. J., et al. 2013, PASA, 30, e003
Diehl, R., Dupraz, C., Bennett, K., et al. 1995, A&A, 298, 445
Doert, M., & Errando, M. 2014, ApJ, 782, 41
Domínguez, A., Wojtak, R., Finke, J., et al. 2019, ApJ, 885, 137
Domínguez, A., & Ajello, M. 2015, ApJ, 813, L34
Doro, M. 2017, EPJWC, 136, 01003
Drury, L. O. 1983, RPPh, 46, 973
Drury, L. O., Aharonian, F. A., & Voelk, H. J. 1994, A&A, 287, 959
Dubner, G. M., Holdaway, M., Goss, W. M., & Mirabel, I. F. 1998, AJ, 116, 1842
Dubus, G. 2013, A&ARv, 21, 64
Eger, P., Rowell, G., Kawamura, A., et al. 2011, A&A, 526, A82
Einecke, S. 2016, Galax, 4, 14
Esposito, J. A., Hunter, S. D., Kanbach, G., & Sreekumar, P. 1996, ApJ, 461, 820
Evoli, C., Gaggero, D., Vittino, A., et al. 2017, JCAP, 2017, 015
Fanaroff, B. L., & Riley, J. M. 1974, MNRAS, 167, 31P
Farnier, C., Walter, R., & Leyder, J. C. 2011, A&A, 526, A57
Federici, S., Pohl, M., Telezhinsky, I., Wilhelm, A., & Dwarkadas, V. V. 2015, A&A, 577, A12
Fegan, D. 2019, Cherenkov Reflections. Gamma-Ray Imaging and the Evolution of TeV Astronomy (World Scientific: Singapore)
Feijen, K., Rowell, G., Einecke, S., et al. 2020, PASA, 37, e056
Fermi-LAT Collaboration, Abdo, A. A., Ackermann, M., et al. 2009, Sci, 326, 1512

Fermi-LAT Collaboration, Ajello, M., Atwood, W. B., et al. 2021, NatAs, 5, 385
Fichtel, C. E., Hartman, R. C., Kniffen, D. A., et al. 1975, ApJ, 198, 163
Fossati, G., Maraschi, L., Celotti, A., Comastri, A., & Ghisellini, G. 1998, MNRAS, 299, 433
Frail, D. A., Goss, W. M., & Slysh, V. I. 1994, ApJ, 424, L111
Fujita, Y., Ohira, Y., Tanaka, S. J., & Takahara, F. 2009, ApJ, 707, L179
Fukui, Y., Sano, H., Sato, J., et al. 2012, ApJ, 746, 82
Funk, S., Hinton, J. A., & De Jager, O. C. 2008, ICRC, 2, 605
Gabici, S., & Aharonian, F. A. 2014, MNRAS, 445, L70
Gabici, S. 2011, MmSAI, 82, 760
Gabici, S. 2014, in IAU Symp. 296, Supernova Environmental Impacts, ed. A. Ray, & R. A. McCray (Cambridge: Cambridge Univ. Press), 320
Gabici, S., & Aharonian, F. A. 2007, ApJ, 665, L131
Gabici, S., Aharonian, F. A., & Blasi, P. 2007, Ap&SS, 309, 365
Gaensler, B. M., & Slane, P. O. 2006, ARA&A, 44, 17
Gaisser, T., Engel, R., & Resconi, E. 2016, Cosmic Rays and Particle Physics (2nd ed.; Cambridge: Cambridge Univ. Press)
Galbraith, W., & Jelley, J. V. 1953, Natur, 171, 349
Gehrels, N., Macomb, D. J., Bertsch, D. L., Thompson, D. J., & Hartman, R. C. 2000, Natur, 404, 363
Ghisellini, G., Righi, C., Costamante, L., & Tavecchio, F. 2017, MNRAS, 469, 255
Ghisellini, G. 2016, Galax, 4, 36
Giacinti, G., Mitchell, A. M. W., López-Coto, R., et al. 2020, A&A, 636, A113
Gingrich, D. M., Boone, L. M., Bramel, D., et al. 2005, ITNS, 52, 2977
Ginzburg, V. L., & Syrovatskii, S. I. 1964, The Origin of Cosmic Rays (Amsterdam: Elsevier)
Green, A. J., Reeves, S. N., & Murphy, T. 2014, PASA, 31, e042
Grenier, I. A., & Harding, A. K. 2015, CRPhy, 16, 641
Grindlay, J. E., Helmken, H. F., Brown, R. H., Davis, J., & Allen, L. R. 1975, ApJ, 197, L9
Guépin, C., Rinchiuso, L., Kotera, K., et al. 2018, JCAP, 2018, 042
Guo, F., & Mathews, W. G. 2012, ApJ, 756, 181
Kulikov, G. V., & Khristiansen, G. B. 1958, JETP, 35, 635
HESS Collaboration 2020, NatAs, 4, 167
HESS Collaboration, Abdalla, H., Abramowski, A., et al. 2018e, A&A, 612, A9
HESS Collaboration, Abdalla, H., Abramowski, A., et al. 2018d, A&A, 612, A8
HESS Collaboration, Abdalla, H., Abramowski, A., et al. 2017, A&A, 606, A59
HESS Collaboration, Abdalla, H., Abramowski, A., et al. 2018b, A&A, 612, A2
HESS Collaboration, Abdalla, H., Abramowski, A., et al. 2018c, A&A, 612, A6
HESS Collaboration, Abdalla, H., Abramowski, A., et al. 2018f, A&A, 612, A11
HESS Collaboration, Abdalla, H., Abramowski, A., et al. 2018a, A&A, 612, A1
HESS Collaboration, Abdalla, H., Adam, R., et al. 2020c, A&A, 635, A167
HESS Collaboration, Abdalla, H., Adam, R., et al. 2020a, Natur, 582, 356
HESS Collaboration, Abdalla, H., Adam, R., et al. 2020b, A&A, 644, A112
HESS Collaboration, Abramowski, A., Aharonian, F., et al. 2016, Natur, 531, 476
HESS Collaboration, Abramowski, A., & Acero, F. 2011, A&A, 531, L18
HESS Collaboration, Abdalla, H., Aharonian, F., et al. 2018g, A&A, 620, A66
HESS Collaboration, Abdalla, H., Aharonian, F., et al. 2018h, A&A, 617, A73
HESS Collaboration, Abdalla, H., Aharonian, F., et al. 2019, A&A, 621, A116

HESS Collaboration, Abramowski, A., Acero, F., et al. 2012, A&A, 548, A46
HESS Collaboration, Abramowski, A., Acero, F., et al. 2011, A&A, 525, A46
HESS Collaboration, Abramowski, A., Aharonian, F., et al. 2015, Sci, 347, 406
Harding, A. K., & Kalapotharakos, C. 2015, ApJ, 811, 63
Hartman, R. C., Bertsch, D. L., Bloom, S. D., et al. 1999, ApJS, 123, 79
Herold, L., & Malyshev, D. 2019, A&A, 625, A110
Hill, D. A., & Porter, N. 1961, Natur, 191, 690
Hillas, A. M. 1984, ARA&A, 22, 425
Hillas, A. M. 1985, ICRC, 3, 445
Hillas, A. M. 2005, JPhG, 31, R95
Hillas, A. M. 2013, APh, 43, 19
Hinton, J. A., & Hofmann, W. 2009, ARA&A, 47, 523
Hirotani, K. 2011, ApJ, 733, L49
Hirotani, K. 2013, ApJ, 766, 98
Hoof, S., Geringer-Sameth, A., & Trotta, R. 2020, JCAP, 2020, 012
Hooper, D., & Serpico, P. D. 2007, PhRvL, 99, 231102
Horns, D., Hoffmann, A. I. D., Santangelo, A., Aharonian, F. A., & Rowell, G. P. 2007, A&A, 469, L17
Humphreys, R. M., & Martin, J. C. 2012, in Eta Carinae and the Supernova Impostors (Berlin: Springer), 1
Hurley-Walker, N., Callingham, J. R., Hancock, P. J., et al. 2017, MNRAS, 464, 1146
IceCube Collaboration, M. G., Aartsen, M., & Ackermann, M. 2018a, Sci, 361, 147
IceCube Collaboration, M. G., Aartsen, M., & Ackermann, M. 2018b, Sci, 361, eaat1378
Ilie, C. 2019, PASP, 131, 111001
Inoue, T., Yamazaki, R., Inutsuka, S. i., & Fukui, Y. 2012, ApJ, 744, 71
Ioka, K., Razzaque, S., Kobayashi, S., & Mészáros, P. 2005, ApJ, 633, 1013
Jackson, J. M., Rathborne, J. M., Shah, R. Y., et al. 2006, ApJS, 163, 145
Jóhannesson, G., Porter, T. A., & Moskalenko, I. V. 2018, ApJ, 856, 45
Jones, P. A., Burton, M. G., Cunningham, M. R., et al. 2012, MNRAS, 419, 2961
Jouvin, L., Lemière, A., & Terrier, R. 2020, A&A, 644, A113
Kafexhiu, E., Aharonian, F., Taylor, A. M., & Vila, G. S. 2014, PhRvD, 90, 123014
Karwin, C. M., Murgia, S., Campbell, S., & Moskalenko, I. V. 2019, ApJ, 880, 95
Kelner, S. R., Aharonian, F. A., & Bugayov, V. V. 2006, PhRvD, 74, 034018
Kennel, C. F., & Coroniti, F. V. 1984, ApJ, 283, 710
Kifune, T. 1996, NCimC, 19C, 953
Kifune, T. 1999, ApJ, 518, L21
Kimura, S. S., Murase, K., & Mészáros, P. 2020, ApJ, 904, 188
Kissmann, R., Werner, M., Reimer, O., & Strong, A. W. 2015, APh, 70, 39
Klebesadel, R. W., Strong, I. B., & Olson, R. A. 1973, ApJ, 182, L85
Knödlseder, J., Jean, P., Lonjou, V., et al. 2005, A&A, 441, 513
Kong, A. K. H., Hui, C. Y., & Cheng, K. S. 2010, ApJ, 712, L36
Kosack, K., Badran, H. M., Bond, I. H., et al. 2004, ApJ, 608, L97
Kostelecký, V. A., & Samuel, S. 1989, PhRvD, 39, 683
Kraushaar, W. L., & Clark, G. W. 1962, PhRvL, 8, 106
Kraushaar, W. L., Clark, G. W., Garmire, G. P., et al. 1972, ApJ, 177, 341
Krymskii, G. F. 1977, DoSSR, 234, 1306

Lacki, B. C. 2014, MNRAS, 444, L39
Lacroix, T., Silk, J., Moulin, E., & BÅ‘hm, C. 2016, PhRvD, 94, 123008
Lagage, P. O., & Cesarsky, C. J. 1983, A&A, 125, 249
Lau, J. C., Rowell, G., Burton, M. G., et al. 2017a, MNRAS, 464, 3757
Lau, J. C., Rowell, G., Voisin, F., et al. 2019, MNRAS, 483, 3659
Lau, J. C., Rowell, G., Voisin, F., et al. 2017b, PASA, 34, e064
Lazendic, J. S., Slane, P. O., Gaensler, B. M., et al. 2004, ApJ, 602, 271
Lemiere, A., Terrier, R., & Djannati-Ataï, A. 2005, ICRC, 4, 105
Letessier-Selvon, A., & Stanev, T. 2011, RvMP, 83, 907
Li, J., Torres, D. F., Liu, R.-Y., et al. 2020a, NatAs, 4, 1177
Li, K.-L., Hambsch, F.-J., Munari, U., et al. 2020b, ApJ, 905, 114
Li, X., Duan, K., Jiang, W., Shen, Z., & Salinas, M. M. 2021, PoS, 358, 576
Lien, A., Sakamoto, T., Barthelmy, S. D., et al. 2016, ApJ, 829, 7
Linden, T., Auchettl, K., Bramante, J., et al. 2017, PhRvD, 96, 103016
Lopez, L. A., Grefenstette, B. W., Auchettl, K., Madsen, K. K., & Castro, D. 2020, ApJ, 893, 144
Macias, O., Gordon, C., Crocker, R. M., et al. 2018, NatAs, 2, 387
Madejski, G., & Sikora, M. 2016, ARA&A, 54, 725
MAGIC Collaboration, Ahnen, M. L., Ansoldi, S., et al. 2017, ApJS, 232, 9
MAGIC Collaboration, Acciari, V. A., Ansoldi, S., et al. 2020a, A&A, 643, L14
MAGIC Collaboration, Acciari, V. A., Ansoldi, S., et al. 2019a, Natur, 575, 455
MAGIC Collaboration, Acciari, V. A., Ansoldi, S., et al. 2020b, A&A, 642, A190
MAGIC Collaboration, Acciari, V. A., Ansoldi, S., et al. 2020c, MNRAS, 497, 3734
MAGIC Collaboration, Acciari, S., Ansoldi, S., et al. 2019b, Natur, 575, 459
MAGIC Collaboration, Ahnen, M. L., & Ansoldi, S. 2018, A&A, 619, A45
Malyshev, D., Pühlhofer, G., Santangelo, A., & Vink, J. 2019, arXiv:1903.03045
Malyshev, D., Zdziarski, A. A., & Chernyakova, M. 2013, MNRAS, 434, 2380
Mandarakas, N., Blinov, D., Liodakis, I., et al. 2019, A&A, 623, A61
Manolakou, K., Horns, D., & Kirk, J. G. 2007, A&A, 474, 689
Margon, B. 1984, ARA&A, 22, 507
Martínez-Huerta, H., Lang, R. G., & de Souza, V. 2020, Symm, 12, 1232
Matsumoto, H., Uchiyama, H., Sawada, M., et al. 2008, PASJ, 60, S163
Matthews, J. H., Bell, A. R., & Blundell, K. M. 2020, NewAR, 89, 101543
Mavromatos, N. E., & Mitsou, V. A. 2020, IJMPA, 35, 2030012–191
Maxted, N., Rowell, G., de Wilt, P., et al. 2017, in AIP Conf. Ser. 1792, High Energy Gamma-Ray Astronomy (Melville, NY: AIP), 040034
Maxted, N. I., Braiding, C., Wong, G. F., et al. 2018, MNRAS, 480, 134
Maxted, N. I., Filipović, M. D., Hurley-Walker, N., et al. 2019, ApJ, 885, 129
Maxted, N. I., Rowell, G. P., Dawson, B. R., et al. 2012, MNRAS, 422, 2230
McClure-Griffiths, N. M., Dickey, J. M., Gaensler, B. M., et al. 2005, ApJS, 158, 178
McEnery, J., van der Horst, A., Dominguez, A., et al. 2019, BAAS, 51, 245
Mészáros, P. 2002, ARA&A, 40, 137
Meszaros, P., & Rees, J. 1993, ApJ, 405, 278
Migliari, S., Fender, R., & Méndez, M. 2002, Sci, 297, 1673
Millar, D. D., Malos, J., Wallace, C. S., & McCusker, C. B. A. 1962, JPSJS, 17, 114
Mirabel, I. F., & Rodríguez, L. F. 1999, ARA&A, 37, 409
Mirzoyan, R. 2014, APh, 53, 91

Mitchell, A. M. W., Rowell, G. P., Celli, S., & Einecke, S. 2021, MNRAS, 503, 3522
Mizuno, A., & Fukui, Y. 2004, in ASP Conf. Ser. 317, Milky Way Surveys: The Structure and Evolution of our Galaxy, ed. D. Clemens, R. Shah, & T. Brainerd (San Francisco, CA: ASP), 59
Mochol, I., & Petri, J. 2015, MNRAS, 449, L51
Montmerle, T. 1979, ApJ, 231, 95
Mooley, K. P., Deller, A. T., Gottlieb, O., et al. 2018, Natur, 561, 355
Moriguchi, Y., Tamura, K., Tawara, Y., et al. 2005, ApJ, 631, 947
Morrison, P. 1958, NCim, 7, 858
Mou, G., Yuan, F., Bu, D., Sun, M., & Su, M. 2014, ApJ, 790, 109
Müller, C., Kadler, M., Ojha, R., et al. 2018, A&A, 610, A1
Murase, K., Thompson, T. A., Lacki, B. C., & Beacom, J. F. 2011, PhRvD, 84, 043003
Naito, T., & Takahara, F. 1994, JPhG, 20, 477
Nava, L., & Gabici, S. 2013, MNRAS, 429, 1643
Nesterova, N. M., & Čudakov, A. E. 1955, JETP, 28, 384
Oh, K., Koss, M., Markwardt, C. B., et al. 2018, ApJS, 235, 4
Ojha, R., Kadler, M., Böck, M., et al. 2010, A&A, 519, A45
Paciesas, W. S., Meegan, C. A., Pendleton, G. N., et al. 1999, ApJS, 122, 465
Pacini, F., & Salvati, M. 1973, ApJ, 186, 249
Padovani, M., Galli, D., & Glassgold, A. E. 2009, A&A, 501, 619
Padovani, P., Alexander, D. M., Assef, R. J., et al. 2017, A&ARv, 25, 2
Panessa, F., Baldi, R. D., Laor, A., Padovani, P., Behar, E., & McHardy, I. 2019, NatAs, 3, 387
Paré, E., Balauge, B., Bazer-Bachi, R., et al. 2002, NIMPA, 490, 71
Paredes, J. M., & Bordas, P. 2019, in Frontier Research in Astrophysics—III, 44
Parizot, E., Marcowith, A., van der Swaluw, E., Bykov, A. M., & Tatischeff, V. 2004, A&A, 424, 747
Patrizii, L., & Spurio, M. 2015, ARNPS, 65, 279
Patrizii, L., Sahnoun, Z., & Togo, V. 2019, RSPTA, 377, 20180328
Pe'er, A. 2019, Galax, 7, 33
Peng, F.-K., Zhang, H.-M., Wang, X.-Y., Wang, J.-F., & Zhi, Q.-J. 2019, ApJ, 884, 91
Peters, B. 1961, NCim, 22, 800
Pinkau, K. 2009, ExA, 25, 157
Piran, T. 1999, PhR, 314, 575
Pliasheshnikov, A. V., & Bignami, G. F. 1985, NCimC, 8, 39
Porter, T. A., Jóhannesson, G., & Moskalenko, I. V. 2017, ApJ, 846, 67
Porter, T. A., Jóhannesson, G., & Moskalenko, I. V. 2019, ApJ, 887, 250
Porter, T. A., Rowell, G. P., Jóhannesson, G., & Moskalenko, I. V. 2018, PhRvD, 98, 041302
Predehl, P., Sunyaev, R. A., Becker, W., et al. 2020, Natur, 588, 227
Principe, G., Mitchell, A. M. W., Caroff, S., et al. 2020, A&A, 640, A76
Recchia, S., Gabici, S., Aharonian, F. A., & Niro, V. 2021, ApJ, 914, 135
Rees, M. J., & Gunn, J. E. 1974, MNRAS, 167, 1
Reich, W., & Sun, X.-H. 2019, RAA, 19, 045
Reimer, O. 2001, Unidentified Gamma-Ray Sources (Berlin: Springer)
Reitberger, K., Reimer, A., Reimer, O., & Takahashi, H. 2015, A&A, 577, A100
Reynolds, C., Ueda, Y., Awaki, H., et al. 2014, arXiv:1412.1177
Reynolds, S. P., & Chevalier, R. A. 1984, ApJ, 278, 630
Reynolds, S. P., Pavlov, G. G., Kargaltsev, O., et al. 2017, SSRv, 207, 175

Rieger, F. 2019, Galax, 7, 28
Roland, D. 2016, JPhCS, 703, 012001
Romero, G. E., Benaglia, P., & Torres, D. F. 1999, A&A, 348, 868
Rowell, G. P., Stamatescu, V., Clay, R. W., et al. 2008, NIMPA, 588, 48
Ruhe, T., Elsässer, D., Rhode, W., Nöthe, M., & Brügge, K. 2019, EPJWC, 207, 03002
Sabatini, S., Tavani, M., Striani, E., et al. 2010, ApJ, 712, L10
Safi-Harb, S., & Ögelman, H. 1997, ApJ, 483, 868
Sahakyan, N., Baghmanyan, V., & Zargaryan, D. 2018, A&A, 614, A6
Sanchez, D. A., Fegan, S., & Giebels, B. 2013, A&A, 554, A75
Sano, H., Inoue, T., Tokuda, K., et al. 2020, ApJ, 904, L24
Sano, H., Tanaka, T., Torii, K., et al. 2013, ApJ, 778, 59
Sano, H., Yamane, Y., Voisin, F., et al. 2017, ApJ, 843, 61
Schael, S., Atanasyan, A., Berdugo, J., et al. 2019, NIMPA, 944, 162561
Schinzel, F. K., Petrov, L., Taylor, G. B., & Edwards, P. G. 2017, ApJ, 838, 139
Schönfelder, V., Bennett, K., Blom, J. J., et al. 2000, A&AS, 143, 145
Smith, A. J. 2005, ICRC, 10, 227
Sobacchi, E., Sormani, M. C., & Stamerra, A. 2016, MNRAS, 465, 161
Spengler, G., & Schwanke, U. 2011, ICRC, 5, 105
Spengler, G. 2020, A&A, 633, A138
Spir-Jacob, M., Djannati-Ataï, A., & Mohrmann, L. 2019, arXiv:1908.06464
Spurio, M. 2015, The Sky Seen in Gamma-rays (Cham: Springer)
Sreekumar, P., Bertsch, D. L., Dingus, B. L., et al. 1992, ApJ, 400, L67
Stamatescu, V., Rowell, G. P., Denman, J., et al. 2011, APh, 34, 886
Stecker, F. W. 2003, APh, 20, 85
Stecker, F. W., & Glashow, S. L. 2001, APh, 16, 97
Steppa, C., & Egberts, K. 2020, A&A, 643, A137
Stil, J. M., Taylor, A. R., Dickey, J. M., et al. 2006, AJ, 132, 1158
Stone, E. C., Cummings, A. C., McDonald, F. B., et al. 2013, Sci, 341, 150
Stone, E., Cummings, A., Heikkila, B., & Lal, N. 2019, NatAs, 3, 11
Strong, A. W., & Moskalenko, I. V. 1998, ApJ, 509, 212
Su, M., Slatyer, T. R., & Finkbeiner, D. P. 2010, ApJ, 724, 1044
Sudoh, T., Inoue, Y., & Khangulyan, D. 2020, ApJ, 889, 146
Sudoh, T., Linden, T., & Beacom, J. F. 2019, PhRvD, 100, 043016
Sun, X.-N., Yang, R.-Z., Liu, B., Xi, S.-Q., & Wang, X.-Y. 2019, A&A, 626, A113
Sun, X.-N., Yang, R.-Z., & Wang, X.-Y. 2020, MNRAS, 494, 3405
Tang, X. 2019, MNRAS, 482, 3843
Tavani, M., Bulgarelli, A., Piano, G., et al. 2009, Natur, 462, 620
Tavani, M., Bulgarelli, A., Vittorini, V., et al. 2011, Sci, 331, 736
Taylor, A. R., Gibson, S. J., Peracaula, M., et al. 2003, AJ, 125, 3145
Thompson, D. J. 2015, CRPhy, 16, 600
Tluczykont, M., Hampf, D., Horns, D., et al. 2014, APh, 56, 42
Tompkins, D. R. 1965, PhRv, 138, 248
Topchiev, N. P., Galper, A. M., Bonvicini, V., et al. 2016, JPhCS, 675, 032010
Torres, D. F., Romero, G. E., Dame, T. M., Combi, J. A., & Butt, Y. M. 2003, PhR, 382, 303
Torresi, E. 2020, in IAU Proc. 342, Perseus in Sicily: From Black Hole to Cluster Outskirts (Cambridge: Cambridge Univ. Press), 158

Umemoto, T., Minamidani, T., Kuno, N., et al. 2017, PASJ, 69, 78
Urry, C. M., & Padovani, P. 1995, PASP, 107, 803
Van Etten, A., & Romani, R. W. 2011, ApJ, 742, 62
Vasileiou, V., Jacholkowska, A., Piron, F., et al. 2013, PhRvD, 87, 122001
VERITAS Collaboration, Acciari, V. A., Aliu, E., et al. 2009, Natur, 462, 770
VERITAS Collaboration, Aliu, E., Arlen, T., et al. 2011, Sci, 334, 69
Viganò, D., & Torres, D. 2015, MNRAS, 449, 3755
Voelk, H. J., & Forman, M. 1982, ApJ, 253, 188
Voisin, F., Rowell, G., Burton, M. G., et al. 2016, MNRAS, 458, 2813
Voisin, F. J., Rowell, G. P., Burton, M. G., et al. 2019, PASA, 36, e014
von Kienlin, A., Meegan, C. A., Paciesas, W. S., et al. 2020, ApJ, 893, 46
Walsh, A. J., Breen, S. L., Britton, T., et al. 2011, MNRAS, 416, 1764
Wang, Y., Beuther, H., Rugel, M. R., et al. 2020, A&A, 634, A83
Waxman, E., & Loeb, A. 2001, PhRvL, 87, 071101
Weekes, T. C. 1988, PhR, 160, 1
Weekes, T. C., Cawley, M. F., Fegan, D. J., et al. 1989, ApJ, 342, 379
Wierzcholska, A., & Wagner, S. J. 2016, MNRAS, 458, 56
Woosley, S. E., & Bloom, J. S. 2006, ARA&A, 44, 507
Xi, S.-Q., Liu, R.-Y., Wang, X.-Y., et al. 2020a, ApJ, 896, L33
Xi, S.-Q., Zhang, H.-M., Liu, R.-Y., & Wang, X.-Y. 2020b, ApJ, 901, 158
Xing, Y., & Wang, Z. 2006, arXiv:2006.15790
Yamazaki, R., Kohri, K., Bamba, A., et al. 2006, MNRAS, 371, 1975
Yan, H., Lazarian, A., & Schlickeiser, R. 2012, ApJ, 745, 140
Yang, R.-z., & Aharonian, F. 2016, A&A, 589, A117
Yang, R.-z., & Aharonian, F. 2017, A&A, 600, A107
Yang, R.-z., de Oña Wilhelmi, E., & Aharonian, F. 2018, A&A, 611, A77
Yoneda, H., Makishima, K., Enoto, T., et al. 2020, PhRvL, 125, 111103
Zdziarski, A. A., Malyshev, D., Chernyakova, M., & Pooley, G. G. 2017, MNRAS, 471, 3657
Zhang, C., Liang, Y.-F., Li, S., et al. 2018, PhRvD, 97, 063009
Zhang, H.-M., Xi, S.-Q., Liu, R.-Y., et al. 2020, ApJ, 889, 12
Zirakashvili, V. N., & Aharonian, F. 2007, A&A, 465, 695
Zirakashvili, V. N., & Aharonian, F. A. 2010, ApJ, 708, 965
Zubovas, K., King, A. R., & Nayakshin, S. 2011, MNRAS, 415, L21

Chapter 8

Neutrino Astronomy

Clancy W James

This chapter considers multimessenger astrophysics with neutrinos, the most elusive of the standard model particles. It begins with an introduction to their basic physical properties, which are key for understanding how they are detected. It then ascends the neutrino energy scale, reviewing known and prospective astrophysical neutrino sources, and the methods by which they are detected. This covers 10^{20} orders of magnitude in energy, from the cosmic neutrino background to ultra-high-energy neutrinos. The greatest part of the chapter is given to those neutrino fluxes that have already been identified: neutrinos from the stellar life cycle, which have been observed from our Sun and supernova 1987A; and "high-energy astrophysical neutrinos," which have only been identified in the last decade.

8.1 Introduction—Neutrinos

Neutrinos are fundamental particles of nature. They are leptons, having half-integer spin, are neutral, and are almost massless. They come in three flavors, or generations: electron neutrinos, ν_e; muon neutrinos, ν_μ; and tau neutrinos, ν_τ, together with their antiparticles.[1] They were first proposed in 1930 by Wolfgang Pauli (1900–1958) to explain the missing energy and momentum observed in nuclear β decays. Electron neutrinos were not detected until 1956, however (Cowan 1956), followed by muon neutrinos in 1962 (Danby et al. 1962), and the tau neutrino in 2000 (DONUT Collaboration et al. 2001). They now form part of the standard model of particle physics, shown in Figure 8.1.

Of all the standard model particles, neutrinos are the most mysterious: their masses are still unknown, but must be tiny, at most about 1 eV each; they rarely interact, with most being able to pass through the entire Earth; and they change flavor, or "oscillate," as they propagate. This makes neutrinos simultaneously hard

[1] Unless otherwise stated, throughout this chapter, "neutrino" will be used to refer to both neutrinos and antineutrinos.

doi:10.1088/2514-3433/ac2256ch8

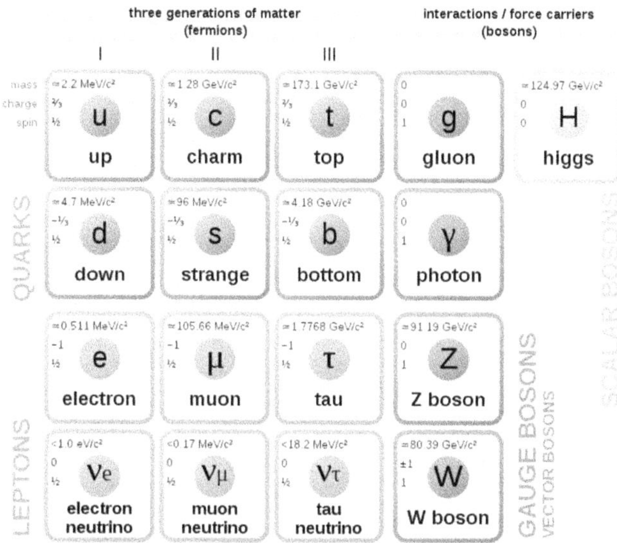

Figure 8.1. The standard model of particle physics, giving (from top to bottom for each particle) their masses, charge (units of elementary charge e), and spin (units of the reduced Planck constant $\hbar$). Image credit: Wikipedia/MissMJ.

to detect and very useful astronomical messengers. Unlike photons, neutrinos will pass unattenuated through intervening matter and radiation fields, providing unfiltered information on the high-energy astrophysical processes that create them.

The sources of neutrinos in the universe vary greatly, as shown in Figure 8.2. This should not be unexpected—imagine writing a book such as this, with Chapters 2–7 dedicated to different astrophysical neutrino sources, and Chapter 8 titled "photons"! Neutrino astronomy is still in its infancy, however: While the energy range of observed neutrinos spans 10^{10} orders of magnitude, only three astrophysical sources have been confirmed: our Sun, supernova 1987A, and a blazar, TXS 0506+056.

Before diving into neutrino astrophysics, we will first don our particle physics hats, and consider the fundamental properties of neutrinos in Section 8.2. From there, we will work our way up the neutrino energy scale, beginning in Section 8.3 at MeV energies with neutrinos from stellar nucleosynthesis and supernovae. We then focus on what is currently the most relevant aspect of multimessenger neutrino astrophysics: "high-energy astrophysical neutrinos," neutrinos in the TeV–PeV (10^{12}–10^{15} eV) range thought to be generated from the collisions of cosmic rays. Section 8.4 presents the detection methods, recent discoveries, and outstanding problems. Finally, Section 8.5 looks at so-called ultra-high energy (UHE) neutrinos from cosmic-ray interactions with the cosmic microwave background (CMB).

8.2 The Physics of Neutrinos

8.2.1 Interactions

Neutrinos are famous for almost never interacting—right now, you have about 10^{14} neutrinos from fusion reactions in the Sun passing through your body every second.

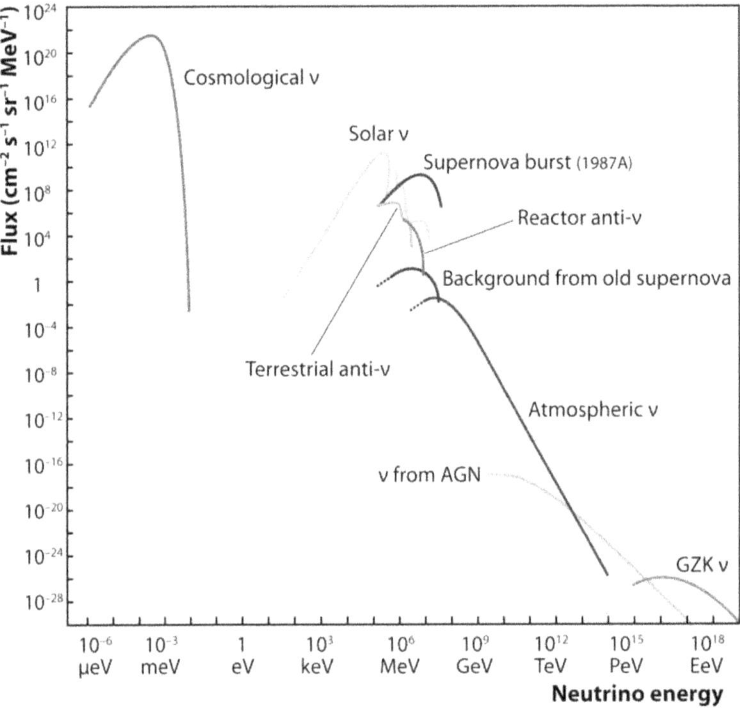

Figure 8.2. Neutrino fluxes from different sources, as a function of energy. Of these, cosmological, supernova background, and GZK fluxes have not yet been observed, while the AGN flux has been identified only for a single blazar. Image credit: Mimouni (2015).

This is because they only interact via the weak force, i.e., the exchange of a W or Z boson. These are heavy particles, meaning that it is difficult for neutrinos to interact at low energies, resulting in a low cross section. While the interaction probability rises with energy, even the ultra-high-energy neutrinos discussed in Section 8.5 will penetrate hundreds of kilometers of rock.

Neutrinos interact via the exchange of the weak-force mediators, W and Z bosons. Interactions exchanging a W are called "charged-current" (CC) interactions, due to the charge on the W, while Z boson exchanges are, unsurprisingly, termed "neutral-current" (NC) interactions. In CC interactions, a neutrino is transmuted into its corresponding lepton—an e, μ, or τ—while in NC interactions, the neutrino continues, albeit with reduced energy. These are shown in the first two panels of Figure 8.3.

A CC interaction with a nuclear quark—either the constituent (or "valence") u and d quarks of protons and neutrons, or the virtual ("sea") quarks arising from the strong force interactions between them—involves the neutrino ν_ℓ turning into its corresponding lepton ℓ, and the quark becoming a new quark $q^{'}$. An example might be $\nu_e + d \rightarrow e^{-}+u$. At the MeV scale, this can turn neutrons into protons (and vice versa at tens of MeV); at high energies, the transmuted quark carries so much energy that a "jet" of new particles is produced. CC interactions of ν_e with atomic electrons,

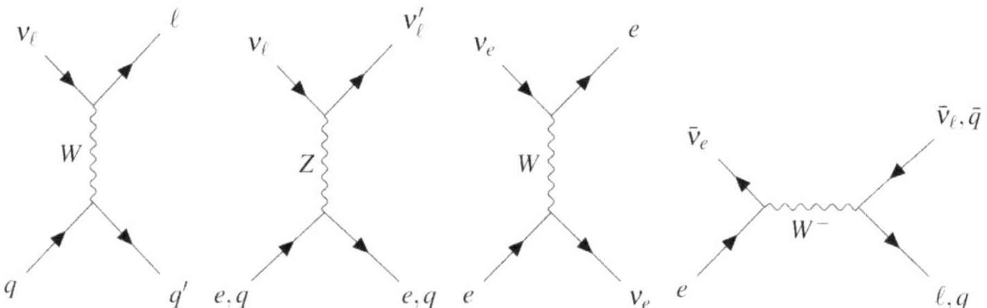

Figure 8.3. Leading-order Feynman diagrams of key neutrino interactions, with time running from left to right. From left to right: a charged-current (CC) interaction, a neutral-current (NC) interaction, neutrino–electron scattering, and a W–resonance interaction.

shown in the third panel of Figure 8.3, are also possible, although because a ν_e and e exist before and after the interaction, it behaves more like an NC interaction.

An NC interaction involves a neutrino exchanging a (neutral) Z with either an electron or quark, as shown in panel 2 of Figure 8.3. In either case, the particles remain the same, albeit with different energies. How this interaction manifests also depends on the energies involved—at MeV energies, an electron will be energetic enough to emit light and be detected, while at TeV–PeV energies, a large cascade of secondary particles will be initiated.

The final interaction channel occurs only for $\bar{\nu}_e$ interactions with electrons and results in the formation of a W^-. Depending on its energy, it can then decay to a quark (e.g., $\bar{u}d$) or lepton–neutrino (e.g., $\mu\bar{\nu}_\mu$) pair. When the center-of-mass energy of the system is equal to the mass of the W (about 80 GeV), there is a steep increase in the cross section for this interaction, known as a "resonance"—in this case, the Glashow resonance, which occurs for $\bar{\nu}_e$ at 6.3 PeV. In other circumstances, the interactions of neutrinos at GeV energies and above with atomic electrons are subdominant to those with nucleons, as shown in Figure 8.4.

8.2.1.1 Cross Section and Interaction Length

The probability for a neutrino to interact is given by its cross section, σ. It has units of area, which can be understood in the following way. Consider a beam of N neutrinos spread over an area A that includes a single interaction target. If the target had a physical area of σ and blocked all neutrinos passing through it, we would expect $M = N\sigma/A$ neutrinos to hit it. Quantum mechanically, we would say that each neutrino had a probability $p = M/N = \sigma/A$ to interact.

Because particles are small, cross sections are measured in "barns": 1 b $=10^{-24}$ cm^2, which is a typical scale for nuclear fission reactions. For 1 MeV solar neutrinos, however, the cross section per atomic electron is about 10^{-17} mb, or 10^{-44} cm^2.

The characteristic distance that a neutrino can travel in matter is given by its interaction length ℓ, which can be calculated simply from the cross section σ and the target density ρ:

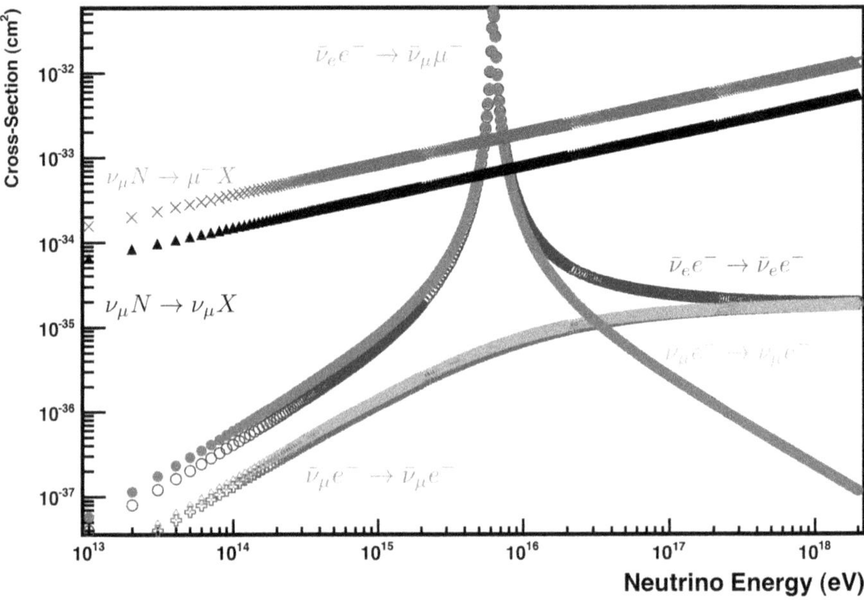

Figure 8.4. Neutrino cross section (cm^2) as a function of energy, eV, for several kinds of neutrino interactions. Other than $\bar{\nu}_e$, the cross sections for neutrinos, and their antiparticles, are almost identical for all three flavors. The feature near 6×10^{16} eV is due to the Glashow resonance. Image credit: Formaggio & Zeller (2012) with permission, Copyright 2012 by the American Physical Society.

$$\ell \equiv \frac{\sigma}{\rho}. \tag{8.1}$$

For interactions with nucleons, the target density (cm^{-3}) is closely related to the matter density (g cm^{-3}) by the Avogadro constant, $N_A = 6.022\ 140\ 86 \times 10^{23}$ mol^{-1}, which is an excellent approximation for nucleons per gram. For interactions with electrons, the ratio between atomic number Z and atomic mass A is also required. Including these constants leads to a formula:

$$\frac{\ell_{\rm int}}{1\ {\rm km}} \approx 166 f \left(\frac{\rho}{1\ {\rm g\ cm^{-3}}}\right)^{-1} \left(\frac{\sigma}{10^{-31}\ {\rm cm}}\right)^{-1}, \tag{8.2}$$

where $f = 1$ for nucleon targets and $f = A/Z$ for electron targets. Because water has a very useful density of $\rho \approx 1$ g cm^{-3}, it is common to write interaction lengths in terms of water equivalent, or W.E. Using Equation (8.2), we find our 1 MeV solar neutrino has $\ell_{\rm int} = 1.66 \times 10^{15}$ km, or 300 lt-yr W.E. At 10 TeV, a $\bar{\nu}_\mu$ has a total (nuclear) cross section of about 2×10^{-34} cm^2, i.e., an interaction length of 84,000 km W.E.

8.2.2 Neutrino Oscillations

Neutrinos are produced in weak interactions in definite flavor states, as ν_e, ν_μ, or ν_τ. Unlike other standard model particles, however, neutrino flavor eigenstates are not the same as their mass eigenstates, m_1, m_2, and m_3. Rather, the flavor eigenstates "mix" via the Pontecorvo–Maki–Nakagawa–Sakata, or "PMNS" matrix, $\mathbf{U}$:

$$\begin{bmatrix} \nu_e \\ \nu_\mu \\ \nu_\tau \end{bmatrix} = \mathbf{U} \begin{bmatrix} \nu_1 \\ \nu_2 \\ \nu_3 \end{bmatrix}. \tag{8.3}$$

$\mathbf{U}$ is usually written as the product of three matrices, representing mixing between all combinations of two flavors:

$$\mathbf{U} = \begin{bmatrix} 1 & 0 & 0 \\ 0 & c_{23} & s_{23} \\ 0 & -s_{23} & c_{23} \end{bmatrix} \begin{bmatrix} c_{13} & 0 & s_{13}e^{-i\delta_{\mathrm{CP}}} \\ 0 & 1 & 0 \\ -s_{13}e^{i\delta_{\mathrm{CP}}} & 0 & c_{13} \end{bmatrix} \begin{bmatrix} c_{12} & s_{12} & 0 \\ -s_{12} & c_{12} & 0 \\ 0 & 0 & 1 \end{bmatrix}. \tag{8.4}$$

Here, $s_{ij} = \sin\theta_{ij}$ and $c_{ij} = \cos\theta_{ij}$. The angles θ_{12}, θ_{23}, θ_{13} are known as "mixing angles" and describe the degree to which the flavor states mix into the mass states. When all $\theta_{ij} = 0$, then mass and flavor eigenstates have a one-to-one correspondence, while when $\theta_{ij} = 45°$, mixing is maximal between states. δ_{CP} is a phase term that results in $\mathbf{U}$ being asymmetric under CP, i.e., the simultaneous operations of charge conjugation C (switching all particles for antiparticles and vice versa) and parity P (reversing all coordinates). All electromagnetic and strong interactions conserve CP, while it is not conserved in some weak interactions. Only very recently has it been conclusively shown that $\delta_{\mathrm{CP}} \neq 0$ (Abe et al. 2020). The values of parameters of the PMNS matrix are not theoretically predicted in the standard model. Best estimates based on several experimental measures are given in Table 8.1.

As neutrinos propagate, the different mass eigenstates oscillate at different frequencies. In a simplified two-neutrino case in vacuum, a neutrino with energy E initially created in flavor state α, after propagating some distance L, will be found in state β with probability:

$$P\left(\nu_\alpha \rightarrow \nu_\beta\right) = \sin^2\left(2\theta_{i,j}\right)\sin^2\left(X_{i,j}\right) \tag{8.5}$$

$$X_{i,j} = 1.267\left(\frac{\Delta m_{ij}^2}{\mathrm{eV}^2}\right)\left(\frac{L/E}{\mathrm{km/GeV}}\right). \tag{8.6}$$

Table 8.1. Best-fit Neutrino Oscillation Parameters Assuming a Normal Mass Hierarchy (Tanabashi et al. 2018)

θ_{12}	$33.6°^{+0.8}_{-0.7}$
θ_{23}	$45.7°^{+1.1}_{-1.2}$
θ_{13}	$8.49°^{+0.14}_{-0.14}$
Δm_{21}	7.53×10^{-5} eV^2
Δm_{32}	$2.444^{+0.034}_{-0.034} \times 10^{-3}$ eV^2
δ_{CP}	$78.5°^{+10.3}_{-9.2}$

Notes. Several experiments are currently aiming to further constrain the parameters of this table, which is likely to become rapidly out of date.

Necessarily, $P(\nu_\alpha \to \nu_\alpha) = 1 - P(\nu_\alpha \to \nu_\beta)$. When $X_{i,j} = \pi$, the probability has oscillated through one full cycle. It is often useful therefore to define the "oscillation length" between two flavor states i, j as

$$\frac{L_{i,j}^{\text{osc}}}{\text{km}} \equiv 2.48 \frac{E}{\text{GeV}} \frac{\text{eV}^2}{\left|\Delta m_{i,j}^2\right|}. \tag{8.7}$$

An example of neutrino oscillations is given in Figure 8.5, where effects due to both large and small mass differences can be observed. The two-neutrino picture is still relevant however due to the relative sizes of the mass differences ($\Delta m_{32}^2 \gg \Delta m_{21}^2$) as discussed below.

8.2.2.1 Atmospheric Neutrino Oscillations

The decay of $\pi^+ \to \mu^+ + \nu_\mu$ and $\mu^+ \to e^+ + \nu_e + \bar{\nu}_\mu$ (and similarly for their antiparticles) from cosmic-ray interactions in Earth's atmosphere produces a flux of neutrinos in the ratio ν_μ: ν_e = 2: 1. At energies near the 1 GeV range, the large mass difference Δm_{32}^2 becomes significant ($L_{32}^{\text{osc}} \approx 1000$ km), while the small mass difference has little effect. Because ν_e mixes very weakly into ν_3, this oscillation exhibits itself predominantly via a disappearance of ν_μ and appearance of ν_τ. This phenomenon is known as "atmospheric neutrino oscillations" and allows the measurement of the mass difference between ν_3 and ν_1/ν_2. See Kajita et al. (2016) for a review of the confirmation of atmospheric neutrino oscillations by the Super-Kamiokande detector, for which the Nobel Prize in 2015 was awarded to Takaaki Kajita (1959–).

8.2.2.2 Solar Neutrino Oscillations and the MSW Effect

Neutrinos produced by the Sun are exclusively ν_e. The dominant contribution from the p–p chain has sub-MeV energies, while other channels produce fluxes in the 1–10 MeV range. At these energies, the oscillation lengths are much smaller than

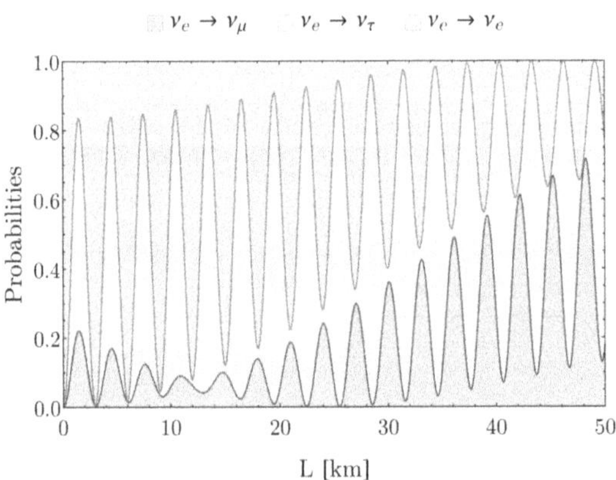

Figure 8.5. Oscillation probability of a 3 MeV ν_e. The probability initially oscillates between ν_e (red) and an admixture of ν_μ (blue) and ν_τ (green), due to the large Δm_{32}^2 mass difference ($L^{\text{osc}} = 3$ km at 3 MeV). The small mass difference, Δm_{21}^2, becomes important only on longer length scales ($L^{\text{osc}} = 100$ km). Image credit: Fantini et al. (2018) with permission of World Scientific.

1 au, and the initial ν_e flux should be fully mixed. However, we have only considered vacuum mixing. In matter, the potential seen by neutrinos is not that of a vacuum, due to the possibility of interacting with atomic electrons. This modifies oscillation probabilities (Wolfenstein 1979). In particular, a slow decrease in the density through which a neutrino is traveling can resonantly enhance mixing (Mikheyev & Smirnov 1985). Known as the Mikheyev, Smirnov, and Wolfenstein (or "'MSW") effect, this results in solar neutrinos above 2 MeV being mostly in the ν_2 state upon exiting the Sun, which mixes almost equally into ν_e, ν_μ, and ν_τ states ($P(e \rightarrow e) \sim 30\%$), resulting in a lower ν_e flux than predicted from vacuum oscillations ($P(e \rightarrow e) \sim 55\%$) (Tanabashi et al. 2018). For further details of matter effects, and neutrino oscillations in general including the latest experimental results, see the articles in Ohlsson (2016).

8.2.2.3 Neutrino Masses

The masses of neutrinos are unknown. Neutrino oscillation measurements can constrain the mass differences, Δm_{32}^2 and Δm_{21}^2 ($\Delta m_{i,j} = m_i - m_j$), through Equation (8.7). Furthermore, the resonance effect for solar neutrinos identifies the sign of Δm_{21} as $m_2 > m_1$, while their mean mass is constrained as $\frac{1}{3}(m_1 + m_2 + m_3) < 1.1$ eV through tritium decay (Aker et al. 2019). This leaves a twofold ambiguity in the mass ordering, allowing both a "normal" hierarchy, $m_3 > m_2 > m_1$, and an "inverted" ordering, $m_2 > m_1 > m_3$. The absolute mass scale (mass of the lightest neutrino) also remains unresolved. Figure 8.6 illustrates both the neutrino mass ordering and flavor mixing for each hierarchy.

8.2.2.4 A Final Word: Neutrinos and Physics beyond the Standard Model

Neutrinos are perhaps the most enigmatic of the standard model particles. While this chapter focused on their use in high-energy astrophysics, there are many fields of research using neutrinos to probe for answers to some of the most fundamental questions in particle physics. Within standard model physics, the neutrino mass hierarchy, and the charge-parity ("CP") violating phase, remain poorly constrained (Tanabashi et al. 2018). Beyond the standard model, "sterile" neutrinos, postulated to have opposite chirality to standard model neutrinos, would interact only via gravity and are predicted from some grand unified theories (GUTs) invoking the Seesaw mechanism (Miranda & Valle 2016). Supersymmetry (SUSY), invoked to explain hierarchy problems in particle physics, tends to predict the lightest supersymmetric particle to be a supersymmetric neutrino, or "neutralino," which would also make a good dark matter candidate. See Nagashima (2014) for an in-depth, but relatively accessible, review of some of these issues.

8.3 Low-energy Neutrinos from the Stellar Life Cycle

Stellar nucleosynthesis is the process by which stars fuse elements to produce energy. Because, on average, this involves turning protons into neutrons (heavier elements have a high neutron fraction), e^+/ν_e pairs must be produced. The typical energy range of these neutrinos (0.3–30 MeV) requires detection methods that are quite distinct from those used for detecting much higher-energy neutrinos in the GeV–PeV range.

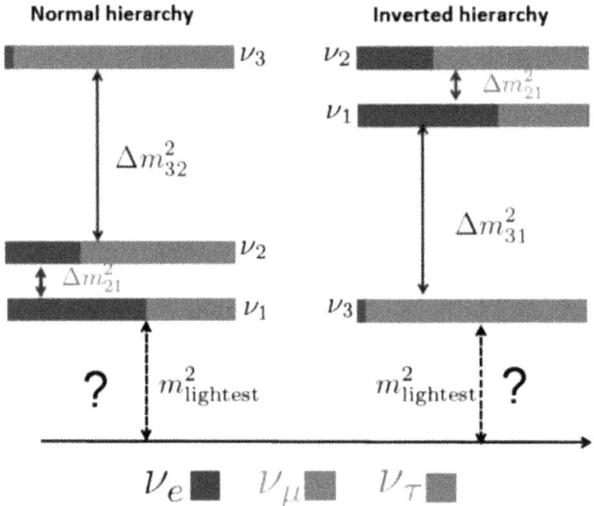

Figure 8.6. Illustration of neutrino masses and flavor mixing. Shown are the mass eigenstates ν_1, ν_2, ν_3 and their two possible orderings: the normal hierarchy ($\nu_3 > \nu_2 > \nu_1$) and the inverted hierarchy ($\nu_2 > \nu_1 > \nu_3$). The colors reflect their mixing into neutrino flavor states ν_e, ν_μ, ν_τ, while the mass differences as measured from atmospheric ($\Delta m^2_{32} \approx \Delta m^2_{31}$) and solar ($\Delta m^2_{21}$) neutrino oscillations. The mass of the lightest neutrino is not constrained by oscillation measurements. Image credit: KM3NeT collaboration (Adrián-Martínez et al. 2016).

8.3.1 Neutrino Production in Stars

8.3.1.1 The Solar Neutrino Flux

Our star, the Sun, is currently fusing hydrogen into helium. Three main reaction chains contribute to a net output of $4p \rightarrow {}^4He + 2\nu_e + 26.73$ MeV, as shown in Figure 8.7.

Most of the energy produced takes the form of gamma rays and electrons and is quickly converted to thermal energy to provide thermodynamic support against gravitational collapse. On average, only about 2% is carried by neutrinos. The solar neutrino spectrum is shown in Figure 8.8. About 90% of all neutrinos produced come from the ppI chain (Altmann et al. 2001). However, these neutrinos have energies less than 1 MeV and are very difficult to detect. The much rarer flux from the β-decay of boron to an excited state of beryllium (${}^8B \rightarrow {}^8Be^* + e^+ + \nu_e$), which produces only 0.02% of solar neutrinos, is much easier to detect.

Observations of solar neutrinos have been used to confirm the standard solar model (SSM), detect neutrino oscillations, and constrain neutrino oscillation parameters. Precision studies of solar neutrinos are ongoing (Borexino Collaboration et al. 2018).

8.3.1.2 Neutrino Fluxes from Massive Stars

Main-sequence stars with masses only a little greater than the Sun's fuse hydrogen into helium predominantly via the CNO cycle, with 7% of the total luminosity carried by neutrinos. Helium fusion proceeds via the triple-alpha process, in which helium fuses into ^{8}Be and then ^{12}C, without the emission of a neutrino. In stars

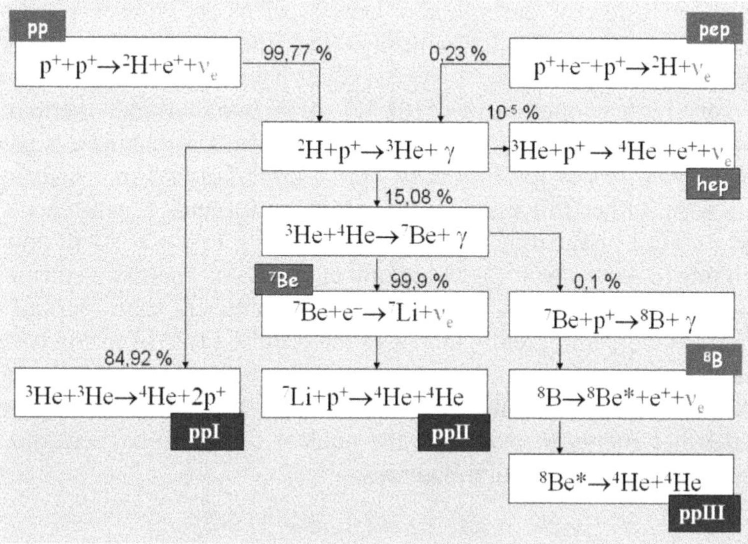

Figure 8.7. Proton fusion in the Sun, showing the reactions and branching fractions for the three reaction chains, ppI, ppII, and ppIII. Image credit: Wikipedia/Dorottya Szam (CC BY-Sa 2.5).

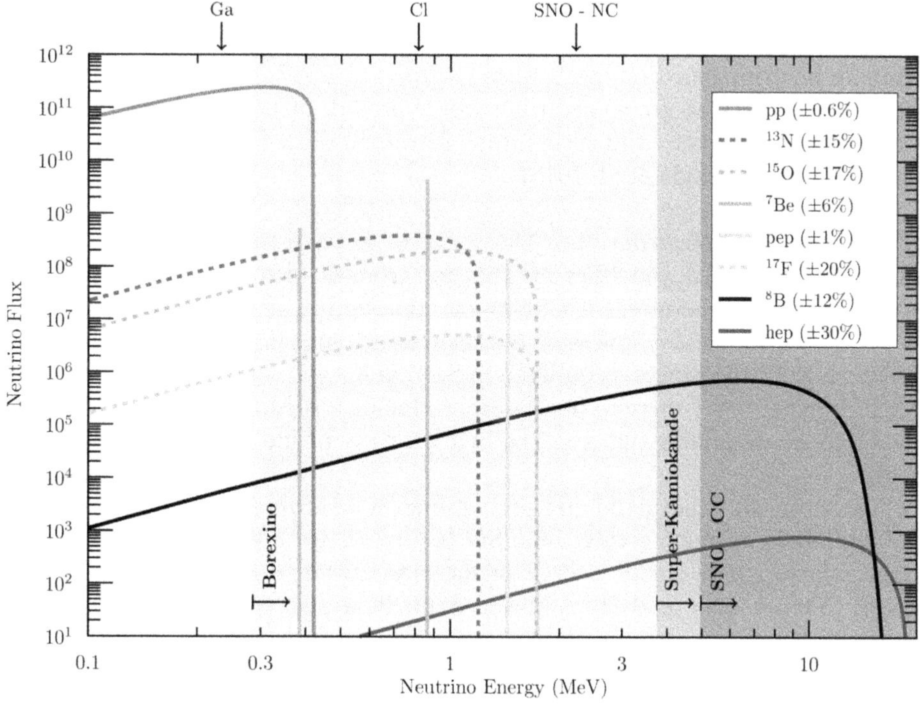

Figure 8.8. Solar neutrino spectrum as calculated from the standard solar model (Grevesse & Sauval 1998). The sensitivity of different experimental methods is shown in gray shading. Image credit: Vissani (2019), copyright 2019 World Scientific.

undergoing the fusion of carbon and heavier elements, however, the neutrino luminosity exceeds the optical luminosity—i.e., these are neutrino stars with a secondary optical component.

At the very high temperatures ($\gtrsim 10^9$ K) of carbon burning, thermal neutrino production dominates nuclear processes. The dominant mechanism is usually pair production ($\gamma \rightarrow e^+ + e^-$), followed by annihilation to neutrino pairs ($e^+ + e^- \rightarrow \nu + \bar{\nu}$). Other processes include plasmon (quanta of plasma oscillations) decays ($\gamma^* \rightarrow \nu + \bar{\nu}$), neutrino photoemission ($\gamma + e^- \rightarrow e^- + \nu + \bar{\nu}$), and thermal bremsstrahlung ($e^- \rightarrow e^- + \nu + \bar{\nu}$) (Munakata et al. 1985). Thermal neutrino emission provides a cooling mechanism that is much more efficient than thermal radiation from the photosphere and greatly increases the required rate of fusion necessary for thermal equilibrium.

At the onset of silicon burning, the core becomes hot enough that electrons have sufficient thermal energy to overcome the nuclear Coulomb barrier, and initiate a large number of electron-capture reactions:

$$\begin{array}{c} e^- + (A, Z) \rightarrow (A, Z-1) + \nu_e \\ \uparrow \quad \downarrow \\ e^- + \bar{\nu}_e + (A, Z) \leftarrow (A, Z-1) \end{array} \tag{8.8}$$

(and similarly for e^+; Odrzywolek & Heger 2010). The cycle shown in Equation (8.8) converts thermal energy into neutrinos, for no net change in nuclear composition. From this stage until collapse—a period of perhaps only one hour—this is the dominant neutrino emission mechanism (Kato et al. 2017), as shown in Figure 8.9. Detecting such pre-supernova neutrinos would provide an early-warning mechanism of a nearby supernova, with near-future detectors being sensitive out to approximately 200 pc (Yoshida et al. 2016).

8.3.1.3 Supernova Neutrinos

Both classes of supernovae—Type Ia (SN Ia), from the detonation of white dwarfs, and core-collapse supernovae (CCSNe)—produce a flux of neutrinos. In the case of SN Ia, the neutrino energy release is expected to be only $\sim$1% of the total luminosity of $\sim 10^{51}$ erg. Of the processes outlined in Section 8.3.1.2, weak nuclear interactions should dominate over thermal processes, producing a predominantly ν_e flux. Such a signal may be detectable out to 1 kpc by existing neutrino detectors (Wright et al. 2016).

CCSNe, on the other hand, are predominantly neutrino explosions. While their total luminosity is much greater than SN Ia, at 3×10^{53} erg, only $\sim$1% of the total energy release is the kinetic energy of the ejecta, and $\sim$0.01% is electromagnetic (Smartt 2009).[2] This leads to a lower optical output, despite the greater total luminosity.

[2] With the exception of pair-instability supernovae, which occur in stars with extremely massive carbon-oxygen cores. Here, neutrino emission is expected to be similar to that in SN Ia, albeit on a much larger scale, and dominated by pair production and annihilation (Wright et al. 2017).

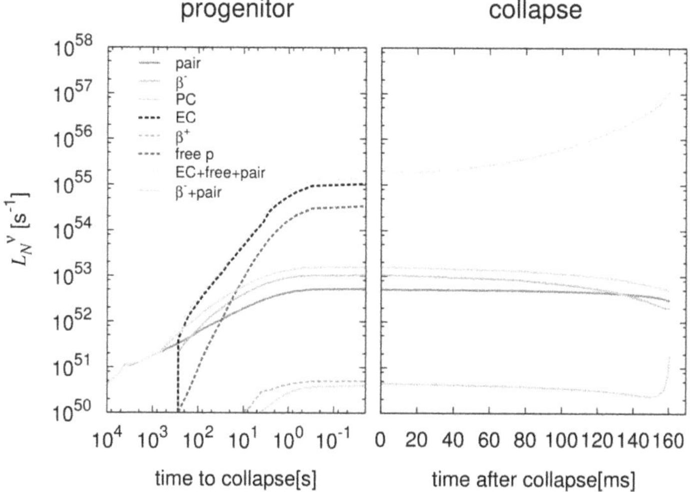

Figure 8.9. Emission of neutrinos by a 12 $M_{\odot}$ star immediately prior to (left), and post (right), collapse at $t = 0$ (Kato et al. 2017). Shown are contributions from "pair" production, β^{-} decay, electron/positron capture (EC/PC), β^{+} decay, electron capture on free protons ("free p"), and combinations thereof. Image credit: Kato et al. (2017).

The dominant source of neutrino emission is the newly formed neutron star or "proto-neutron star." Initially at a temperature of 10^{11} K, and an ill-defined radius of ~100 km, thermally produced neutrinos escape and cool the proto-neutron star on a timescale of 10 s (Janka 2017). This flux of neutrinos is so intense that it is instrumental in powering the supernova ejecta (Müller 2016).

8.3.1.4 Supernova 1987A

The standard description of CCSNe was famously proven with the optical observation of SN 1987A on February 24 in the Large Magellanic Cloud at a distance of 51.4 kpc. A posteriori analysis revealed the supernova to have occurred on February 23. It also showed that three neutrino detectors—Kamioka, Irvine–Michigan–Brookhaven (IMB), and Baksan—had observed a large excess of neutrinos two or three hours prior to the arrival of optical light (Hirata et al. 1987; Bionta et al. 1987; Alexeyev et al. 1988). This early arrival of the neutrinos was due to the time taken for the optical signal to break out of the stellar envelope. The neutrino signal was consistent with the formation of a neutron star and placed limits on the mass of the electron neutrino and the total number of neutrino flavors. An optical image, and the neutrino detections, is shown in Figure 8.10.

8.3.2 Detection Methods

Detection methods for neutrinos from the stellar life cycle are dominated by their low energies (0.3–30 MeV), which limits their available interactions. The μ and τ mass (105.66 MeV and 1.777 GeV) is too high for these particles to be created in CC interactions, so only NC and the ν_e CC interactions are available. The detection

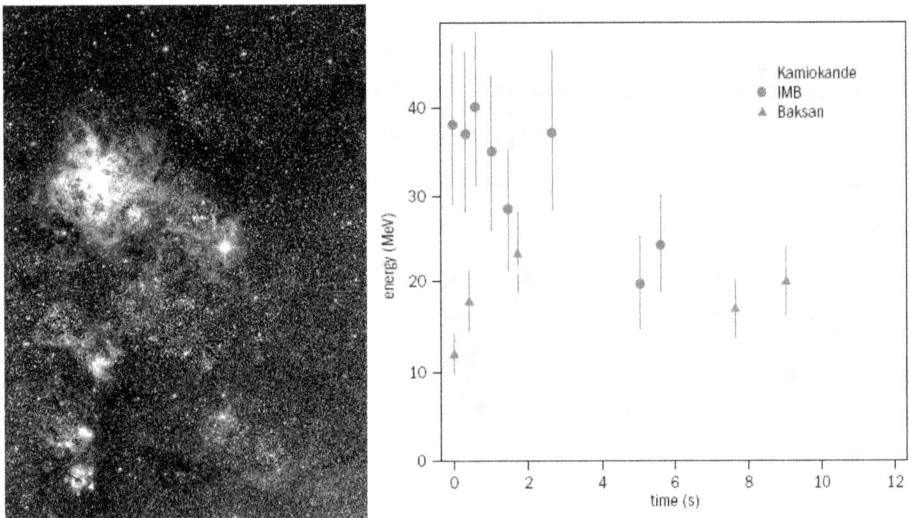

Figure 8.10. Left: optical image of SN 1987A (just right of center) and the nearby Tarantula Nebula, taken with the Schmidt Telescope. Right: neutrino detection times (relative) and energies (MeV), according to the detecting instrument. Image credits: (left) Wikipedia/ESO. (right) CERN Courier, 2007 January 30, SN1987A heralds the start of neutrino astronomy; https://cerncourier.com/a/sn1987a-heralds-the-start-of-neutrino-astronomy/.

method determines the minimum accessible energy and also the neutrino flavors to which it is sensitive. As more-sensitive methods are usually more expensive and difficult to implement, this must be traded off against the total detector volume.

8.3.2.1 Chlorine Detectors

The first experiment to observe solar neutrinos was the Homestake experiment in South Dakota, USA (Davis et al. 1968). It used 615 tons of C_2Cl_4 and the reaction $^{37}Cl + \nu_e \rightarrow {}^{37}Ar + e^-$. Measuring the rate of argon production (which was performed once every few months) measured the reaction rate and, hence, the neutrino flux. While the reaction has a threshold of 814 keV, making it sensitive to all but *pp* neutrinos, the event rate is dominated by ν_e from ^8B (77%), with the next contribution from ^7Be (15%).

The rate detected by Homestake was very close to one-third of that expected from theoretical calculations (Cleveland et al. 1998; Bahcall et al. 1968). The resolution of this discrepancy was the discovery of neutrino oscillations, which cause the exclusively ν_e solar flux to transmute into ν_μ and ν_τ. Raymond Davis, Jr. (for Homestake) and Masatoshi Koshiba (for the Kamioka detector) shared half of the 2002 Nobel Prize in Physics for this discovery.

8.3.2.2 Gallium Detectors

The only feasible method for detecting neutrinos from the *pp* cycle is the reaction $^{71}Ga + \nu_e \rightarrow {}^{71}Ge + e^-$, which has an energy threshold of 233 keV (Kuz'min 1966).

The e^- produced is not sufficiently energetic to be detected and hence this method relies on chemically extracting and counting the produced ^{71}Ge.

Two detectors have been built using this method: SAGE (Soviet–American Gallium Experiment; 1990–present) in Baksan, Russia, consists of 50–57 tons of metallic Ga (Abdurashitov et al. 1999), and GALLEX (Gallium Experiment; 1991–1997) at Gran Sasso, Italy, with 30 ton Ga in 101 ton $GaCl_3$; it was subsequently renamed GNO (Gallium Neutrino Observatory) and ran from 1998 to 2003 (Anselmann et al. 1992).

This method is sensitive only to ν_e, and the low rates of interactions observed by SAGE and GALLEX were some of the first experimental hints of neutrino oscillations.

8.3.2.3 Liquid Scintillator

A scintillation detector uses a material that emits light (i.e., it scintillates) in the presence of an energetic charged particle. This allows the detection of low-energy electrons from the NC scattering reaction $\nu + e^- \rightarrow \nu + e^-$, making such a detector sensitive to all flavors of neutrinos with a low detection threshold. The Borexino experiment, located at the site of GALLEX at the Gran Sasso underground laboratory, Italy, uses about 320 m^3 of liquid scintillator viewed by 2212 PMTs, allowing it to make spectral measurements of all solar neutrino fluxes (Alimonti et al. 2009). Borexino has measured a spectrum of solar neutrinos consistent with expectations from the *pp* process (Borexino Collaboration et al. 2014).

8.3.2.4 Water Čerenkov Detectors

Water Čerenkov detectors have unique importance, in that they are not only used for the detection of MeV solar and supernova neutrinos but also for high-energy astrophysical neutrinos, which are the subject of Sections 8.4 and 8.4.5. Thus, their operation will be described in slightly more detail than the methods given above.

Water Čerenkov detectors identify relativistically charged particles. For MeV neutrinos, this is primarily from the scattering of ν_e off of atomic electrons: $\nu_e + e \rightarrow \nu_e + e$ (Figure 8.3; two central panels). At higher energies, neutrinos of all flavors can be detected, as described in Section 8.4.2. If the recoil electron exceeds the velocity of light in the medium (about 2.3×10^8 m s^{-1} in water, corresponding to an electron energy of 0.8 MeV), an electromagnetic shock will be produced. This shock is analogous to a sonic boom from an aircraft traveling faster than the local sound velocity. The shock emits radiation, known as Čerenkov radiation, predominantly as blue and ultraviolet light. The radiation is emitted at the Čerenkov angle θ_C: $\cos\theta_C = (n\beta)^{-1}$, where n is the medium's refractive index, and $\beta = v/c$ is the particle velocity v relative to the speed of light in vacuum, c. The number N of Čerenkov photons emitted per distance traveled L per wavelength range λ is given by the Frank–Tamm formula:

$$\frac{d^2N}{dLd\lambda} = 2\pi\alpha q^2\lambda^{-2}\sin^2\theta_C, \tag{8.9}$$

where $\alpha \approx 1/137$ is the fine-structure constant and q is the charge in units of the elementary charge, e. The formula is valid for $\beta n > 1$. Detecting these photons allows the energy, position, and direction of the emitting particle to be reconstructed.

Photomultiplier tubes (PMTs) are the standard device used for detecting emitted Čerenkov photons (Figure 8.11). When an incident photon strikes the photocathode on the front surface of the PMT, an electron is knocked out with a probability depending on the photon wavelength. This is known as the PMT "quantum efficiency," as it typically peaks at 30%–40%. This electron is then accelerated by a high voltage (usually about 1 kV) to impact a series of metal surfaces or dynodes. At each dynode stage, the incident electrons knock out several more, so that at the anode sufficiently many electrons are produced to be detectable with electronic circuitry.

A charged particle in water ($n \sim 1.35$) emits about 390 Čerenkov photons per centimeter in the 300–700 nm range in which PMTs are typically sensitive. As many photons must be detected to accurately reconstruct an event, a very high density of PMTs is required, and the detection threshold is significantly higher than the theoretical threshold for Čerenkov light of 0.8 MeV. The advantages of this method are that the electron direction will be correlated with the initial neutrino direction; detection is instantaneous and does not require periodic chemical analysis, and it is relatively cheap to build large volumes of water.

The largest current experiment employing this method to search for solar and supernova neutrinos is Super-Kamiokande, located 1 km under Mount Ikeno near Hida, Japan (Fukuda et al. 2003)—see Figure 8.12. SuperK, as it is known, uses 13,014 PMTs observing 5×10^4 m^3 of pure water in a cylinder 40 m in diameter and 40 m high. Čerenkov light from individual particles appears as "rings" when projected onto the detector surface (see Figure 7.17, Book 1). With a threshold electron energy of 4 MeV, SuperK detects about 20 solar neutrinos per day, almost entirely from ^{8}B (Abe et al. 2016). SuperK also detects neutrinos from cosmic-ray interactions in the atmosphere, and observations of this flux were used to discover

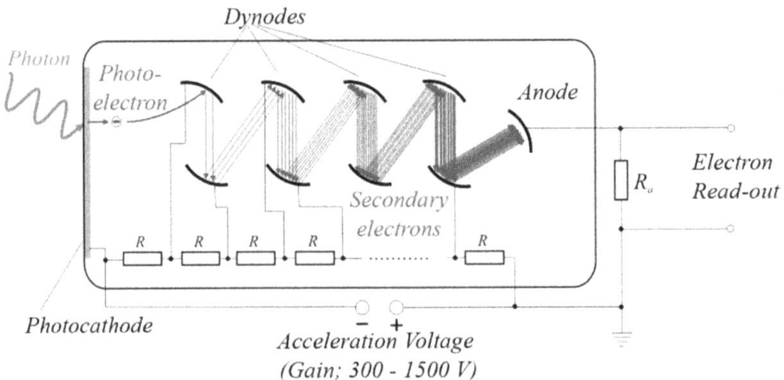

Figure 8.11. Diagram of a photomultiplier tube (PMT). Image credit: de Benutzer, Jkrieger, translation by Wikipedia/Dietzel65.

Figure 8.12. Experiments measuring the solar neutrino flux. Left: photograph of the SuperK detector, taken during detector construction, looking upwards. Middle: photograph of the now-decommissioned Sudbury Neutrino Observatory, SNO. Right: photograph of the Borexino experiment prior to being filled. Image credit: (left) The Super-Kamiokande Collaboration, ICRR (Institute for Cosmic Ray Research), The University of Tokyo, (middle) Ernest Orlando Lawrence Berkeley National Laboratory, (right) Borexino Collaboration.

evidence for muon neutrino oscillations (Fukuda et al. 1998). SuperK, which itself is an upgrade of the original Kamioka detector, will shortly undergo an upgrade to HyperK, with a total volume of 10^6 m^3 (Hyper-Kamiokande Proto-Collaboration et al. 2018).

Historically important water Čerenkov detectors include the Sudbury Neutrino Observatory, SNO, in Ontario, Canada (Boger et al. 2000). Decommissioned in 2006, SNO used 1000 tons of heavy water, allowing the reaction $D_2O + \nu_\ell \rightarrow$ [NC]DHO $+ \nu_\ell + n$. Subsequently, the emitted neutron can be captured by deuterium (D), emitting a 6 MeV gamma-ray. The gamma-ray will produce an e^+, e^- pair, and producing detectable Čerenkov light. This made SNO sensitive to neutrinos of all flavors, allowing it to observe the ν_μ and ν_τ flux from the initially pure ν_e solar neutrino flux, resolving the solar neutrino problem (Ahmad et al. 2002). Another such detector was the IMB detector, which used about 9450 m^3 of pure water, and helped to detect the neutrino flux from SN 1987A (Haines et al. 1986).

For the discoveries of neutrino oscillations, Arthur B. McDonald (SNO) and Takaaki Kajita (SuperK) received the 2015 Nobel Prize in Physics.

8.3.2.5 Relic Neutrinos from the Big Bang

There is one further flux of low-energy neutrinos. When the universe was very young, hot, and dense, neutrinos were produced in nuclear interactions. These neutrinos were in thermal equilibrium with electrons until about 0.2 s after the Big Bang (technically, a redshift of about 10^{10}), at which point the universe cooled and expanded so that these neutrino decoupled from matter. Subsequently, primordial e^+/e^- annihilation heated the universe by 40%, so that these neutrinos are now at a temperature slightly lower than the CMB: 1.95 K, compared to 2.725 K. The cosmic neutrino background, or CνB, has a current number density estimated at 330 ν cm^{-3}. These are sometimes termed "relic neutrinos," and while they have never been observed, their existence is predicted by Big Bang cosmology.

Being of such low energy ($\approx 1.7 \times 10^{-4}$ eV), there are no prospects of directly detecting these neutrinos. However, their gravitational influence in the early universe leaves a detectable signature in the CMB (Follin et al. 2015; Baumann et al. 2019) and determines the primordial elemental abundances (Mathews et al. 2017). Future galaxy surveys should also be able to detect their effects on the structure on galaxy clusters (Marulli et al. 2011) and voids (Kreisch et al. 2019).

8.4 High-energy Astrophysical Neutrinos: TeV Regime

The phrase "high-energy astrophysical neutrinos" refers primarily to neutrinos in the TeV–PeV (10^9–10^{12} eV) range. The motivation for searching for neutrinos in this range comes from observations of cosmic rays. The spectrum of these particles—mostly protons and atomic nuclei—is shown in Figure 8.13 and extends to 10^{20} eV. Evidence suggests that cosmic rays up to at least 10^{16} eV are produced by sources such as supernova remnants (SNRs) within our Galaxy (Apel et al. 2012), while the flux above 10^{19} eV is almost certainly extragalactic in origin (Aab et al. 2018a). However, no source of cosmic rays has ever been definitively identified, and the origin of these highest-energy particles in nature remains a mystery.

8.4.1 The Cosmic-Ray/Neutrino/Gamma-Ray Connection

Cosmic rays are expected to be accelerated through Fermi first-order acceleration, also known as "diffusive shock acceleration," or DSA (Fermi 1949; Bell 1978). In its simplest form, DSA yields a power-law spectrum:

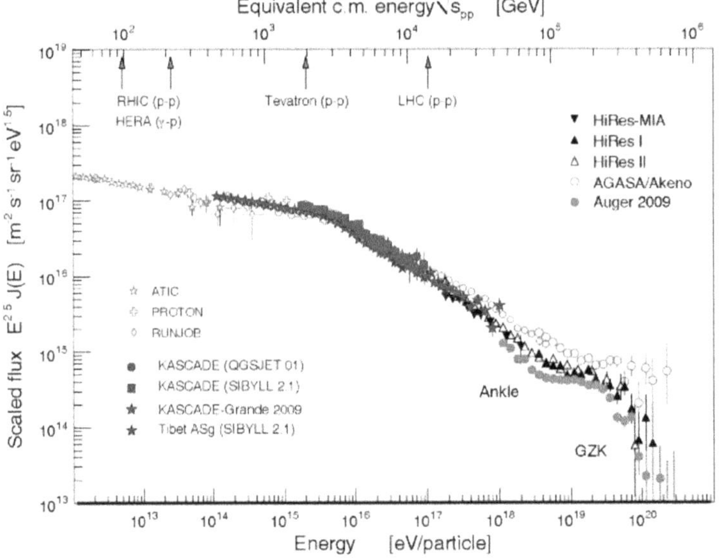

Figure 8.13. The spectrum of high-energy cosmic rays, showing data from several experiments (Matthiae 2010). The spectrum is weighted by $E^{2.5}$ to illustrate features—thus, an apparently "constant" spectrum over a factor of 10 in cosmic-ray energy reduces the flux by a factor of 316. Also shown is the center-of-mass (c.m.) energy equivalent assuming proton–proton collision, $\sqrt{s_{\rm pp}}$, and the maximum energies reached by several particle colliders. Image credit: Matthiae (2010).

$$\frac{dN_{\rm CR}}{dE_{\rm CR}} \propto E_{\rm CR}^{-\alpha}, \tag{8.10}$$

with $\alpha = 2$. Extensions of the theory to include modified shocks tend to predict steeper power-law spectra ($\alpha > 2$) (Urošević et al. 2019).

Cosmic-ray protons, $p_{\rm CR}$, of all energies will interact with gas in the interstellar medium (ISM) to produce (typically several) pions, in "pp" interactions:

$$p_{\rm CR} + p_{\rm ISM} \rightarrow n/p'_{\rm CR} + \pi^{+}, \pi^{0}, \pi^{-}..... \tag{8.11}$$

The decay of pions—and any cosmic-ray neutrons—will produce, among other things, neutrinos:

$$n \rightarrow p + e^{-} + \overline{\nu}_{e} \tag{8.12}$$

$$\pi^{+} \rightarrow \mu^{+} + \nu_{\mu} \tag{8.13}$$

$$\mu^{+} \rightarrow e^{+}\overline{\nu}_{\mu} + \nu_{e} \tag{8.14}$$

$$\pi^{0} \rightarrow 2\gamma. \tag{8.15}$$

The physics of these interactions intimately links the fluxes of high-energy cosmic rays, gamma rays, and neutrinos. The secondary $e^{\pm}$ will also be visible via their synchrotron radiation. The flavor ratio expected from pion decay is $\{\nu_e\colon \nu_\mu\colon \nu_\tau\} = \{1\colon 2\colon 0\}$, although an almost-uniform flavor ratio is expected at Earth due to neutrino oscillations (see Section 8.4.6.1).

The interaction cross section for cosmic rays with protons changes slowly with energy, and the number of pions produced rises only with the logarithm of the cosmic-ray energy. Therefore, neutrinos produced by pp interactions are expected to also follow a power-law spectrum, with an index only slightly flatter than the cosmic-ray index α (Kamae et al. 2006).

Photons can also interact with cosmic rays to produce typically a single pion, in $p\gamma$ interactions:

$$p_{\rm CR} + \gamma \rightarrow n/p'_{\rm CR} + \pi^{+}/\pi^{0}. \tag{8.16}$$

The threshold energy $E_{\rm th}$ for this interaction is

$$E_{\rm th} = \frac{m_p m_\pi \left(1 + \frac{m_\pi}{2m_p}\right)}{2E_\gamma} \tag{8.17}$$

$$\approx \frac{7 \times 10^{16}\ {\rm eV}^2}{E_\gamma}, \tag{8.18}$$

for photons of energy E_γ. The $p\gamma$ cross section is strongly peaked near the production threshold. For infrared photons of the CMB, with $E_\gamma = 10^{-3}$ eV,

$E_{th} = 7 \times 10^{19}$ eV, and hence $E_\nu \sim 3.5 \times 10^{18}$ eV (Greisen 1966; Zatsepin & Kuz'min 1966).

However, if cosmic rays are being accelerated in photon-rich environments—such as active galactic nuclei (AGNs) and gamma-ray burst (GRB) jets—then optical or UV photons with E_γ = 1–10 eV can interact with cosmic rays of $E_{CR} \approx 10^{16-17}$ eV, producing neutrinos with 100 TeV to 1 PeV of energy, with a spectrum closely following the target photon spectrum.

Observations of high-energy astrophysical neutrinos, therefore, probe these environments.

8.4.1.1 Atmospheric Neutrinos

The same neutrino-producing interactions that occur when a cosmic ray hits interstellar gas also happen when a cosmic ray hits the top of Earth's atmosphere. This presents by far the largest flux of neutrinos at Earth between 1 GeV and 100 TeV (see Figure 8.2). The key difference between this flux and the astrophysical flux from *pp* interactions is that secondary $\pi^\pm$ will tend to interact and lose energy before they decay, while $\mu^\pm$ lose almost all their energy prior to decaying. Because the relativistic gamma factor, and associated time dilation, scales with energy, the decay probability for $\pi^\pm$ scales as E^{-1}. Hence, instead of the neutrino spectrum being similar to the cosmic-ray spectrum at Earth (with a power-law index of −2.7 below 10^{15} eV), it is steepened to a power-law of index −3.7. This allows the flux of neutrinos from astrophysical sources to dominate above about 100 TeV.

8.4.2 Detection Methods

All detectors studying high-energy astrophysical neutrinos are water (or ice) Čerenkov detectors, described in Section 8.3.2.4. Neutrinos are identified via the production of secondary particles generated in their interactions (Section 8.2.1), and the identification of the Čerenkov light produced by their passage through the detector.

Due to the low rate of astrophysical neutrinos, neutrino telescopes (Figure 8.15) require very large volumes (of order 1 km^3) of optically transparent natural material to search for the light signature of neutrino interactions. It must also be extremely dark, because the Čerenkov flash is not very bright, and deep, to reduce the radioactive background from atmospheric cosmic-ray interactions. Photomultiplier tubes (PMTs) are used to count individual photons of Čerenkov light and their time of arrival. So far, deep water and glacial ice sheets have proven to be suitable material.

Almost all of the secondary particles produced in CC and NC interactions will either lose energy or decay rapidly, producing a particle cascade, or "shower," a few meters in length. The main exception to this is in ν_μ CC interactions, where the outgoing $\mu^\pm$ can travel several kilometers before losing all its energy, becoming mildly relativistic, and decaying (see Figure 8.14). This leads to two primary neutrino detection channels.

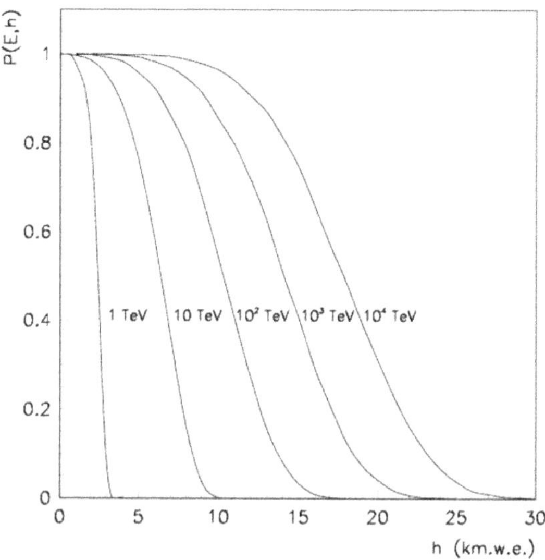

Figure 8.14. Survival probability of μ in standard rock as a function of their energy, as calculated by the MUSIC code (Antonioli et al. 1997), copyright 1997, with permission from Elsevier.

8.4.2.1 Track Events

The "track" channel consists of searching for throughgoing μ generated by ν_μ CC interactions in the medium surrounding the detector. The rest-frame lifetime of a μ is 2.2×10^{-6} s, which is boosted in an observer's frame by relativistic time dilation. Therefore, a μ will not decay until its energy has fallen below 1 GeV, and it can travel large distances. This extends a detector's effective volume to be much larger than the detector itself. Furthermore, the direction of the track generated by the μ is easy to reconstruct—existing detectors have a resolution of 0.3–1°(see Table 8.2). It is thus very good for identifying the source of detected neutrinos. However, the original neutrino energy is obscured by the stochastic processes setting the initial muon energy, the energy lost by the muon on its passage to the detector, and the energy lost by the muon on its passage through the detector. Due to the high rate of downgoing μ from cosmic-ray interactions in the atmosphere above the detector, the track channel is mostly only sensitive to upcoming events generated by neutrinos that have traversed most of Earth and interacted underneath the detector. In this sense, neutrino detectors are telescopes that look downwards. An example of a track event is given in Figure 8.16 (left).

8.4.2.2 Shower Events

The "shower" or "cascade" channel searches for NC interactions, and the CC interactions of ν_e and ν_τ, inside the detector volume itself. In these interactions, the recoil quark receiving a fraction of the neutrino's energy and momentum will produce a "jet" of new particles, mostly consisting of mesons (quark anti-quark pairs, such as $\pi^\pm$ and π^0). These particles initiate a hadronic cascade through their subsequent interactions and decays. This cascade loses its energy over the course of a

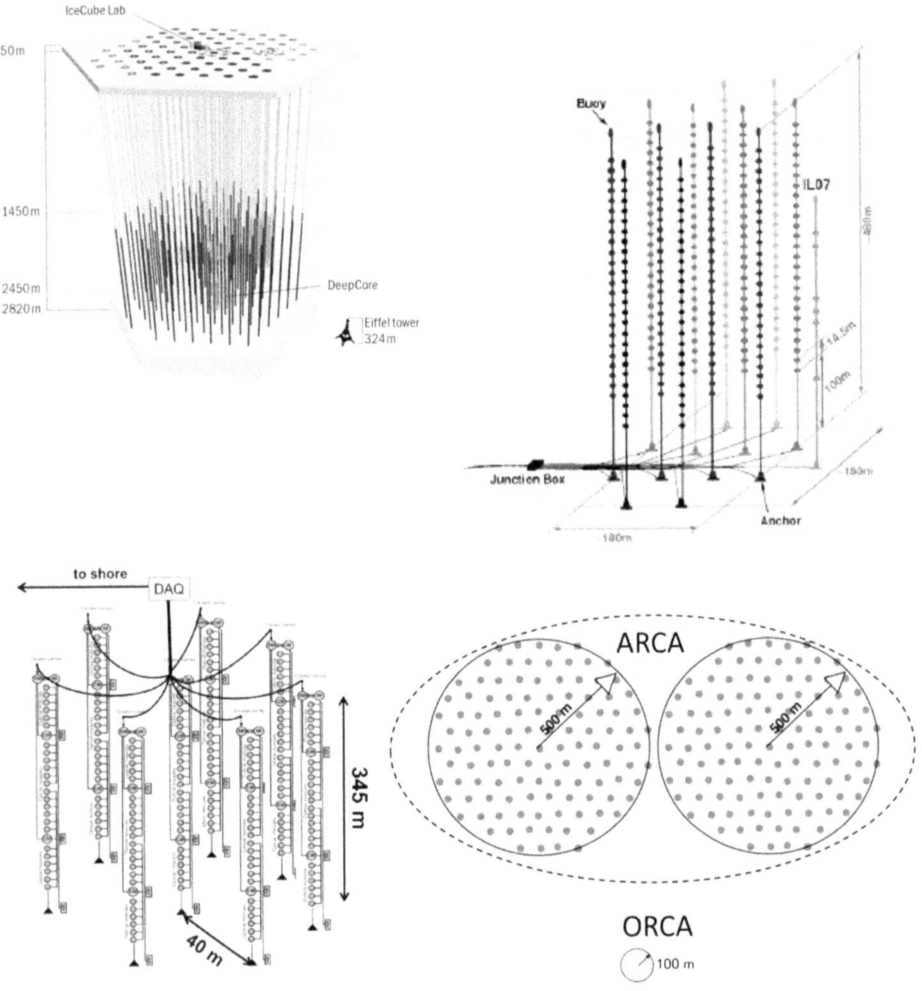

Figure 8.15. Illustrations of neutrino telescopes. Top left: IceCube; top right: ANTARES; bottom left: Dubna, the first cluster of GVD; bottom right: seafloor layout of the two KM3NeT detectors, ARCA and ORCA. Image credits: IceCube Collaboration, Schnabel (2015) with permission from Elsevier, Avrorin et al. (2015), and http://www.km3net.org (KM3NeT Collaboration).

few meters, resulting in a large, localized burst of Čerenkov light from the many charged particles produced. In ν_e CC interactions, the outgoing $e^{\pm}$ will initiate an electromagnetic cascade of γ and $e^{\pm}$, and the entire neutrino energy will be deposited at once.

While the angular resolution to such events is not as good as with track events, energy resolution is improved, because a greater fraction of the initial neutrino energy is visible. Furthermore, neutrino events can be identified arriving from any direction, because only neutrinos can cause the sudden appearance of a particle cascade inside the detector without also generating an incoming light signature.

Table 8.2. Characteristic Angular Resolutions to Track and Cascade Events from the ANTARES and IceCube Detectors and the Estimated Resolution from KM3NeT Once It is Complete

Detector	Tracks	Cascades	Reference
ANTARES	0.3°	3°	Albert et al. (2017a)
IceCUbe	0.7°	10–20°	Aartsen et al. (2019b); IceCube Collaboration (2013)
KM3NeT	0.1°	1.5°	Adrián-Martínez et al. (2016)

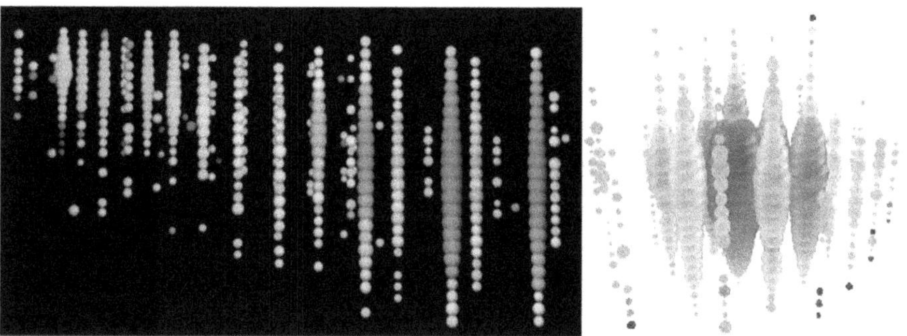

Figure 8.16. Track (left) and cascade (right) events observed by IceCube. Both are famous events. The cascade event, dubbed "Bert" after the Sesame Street character, was one of the first two high-energy astrophysical neutrinos ever identified (IceCube Collaboration et al. 2018a). The track event, titled IceCube-170922A, identified the first-ever point source of HE astrophysical neutrinos, the blazar TXS 0506+056 (Aartsen et al. 2013). Image credit: Left: IceCube Collaboration. Right: Aartsen et al. (2013) with permission, Copyright 2013 by the American Physical Society.

These characteristics make the shower channel relatively better for identifying diffuse neutrino fluxes. An example of a shower event is given in Figure 8.16 (right).

8.4.2.3 Starting Track Events

Track and cascade events are by far the most common event topologies to be observed. A subclass of these occurs for a ν_μ CC interaction, or a ν_τ CC interaction with the τ decaying to a μ (see below), inside the detector volume. This is known as a "starting track" event, where both the cascade at the interaction point, and an outgoing μ, will be visible. Such an event provides both a good energy and direction resolution.

8.4.2.4 Tau Neutrino Topologies

The τ produced in a ν_τ CC interaction decays with a lifetime of 2.9×10^{-13} s. The apparent decay time for a relativistic τ is extended due to time-dilation according to its γ factor. Given its rest mass of 1.776 GeV, a 1 PeV τ ($\gamma = 5.6 \times 10^5$) will decay in 162 ns in the lab frame, i.e., in a distance of about 50 m. The primary τ decay modes are

$$\tau \rightarrow \mu + \nu_\tau + \bar{\nu}_\mu \qquad [17.4\%] \tag{8.19}$$

$$e + \nu_\tau + \bar{\nu}_e \qquad [17.8\%] \qquad (8.20)$$

$$\{\pi^\pm, \pi^0, K^\pm, K^0 \ldots\} + \nu_\tau \qquad [64.8\%], \qquad (8.21)$$

where the third decay mode represents many different hadronic decays involving one or more mesons. Because the τ is very heavy, it is unlikely to lose much energy before its decay. First, if the τ interacts near to the detector, it can produce either a track event (if the τ decays to a μ) or a cascade event. Second, if the τ has very high energies, it can decay at distances far enough from the initial interaction point to initiate a second cascade in the detector from the e or hadronic decay products. Such an event is known as a "double-bang" event, although one has never been observed. Third, a ν_τ CC interaction always produces an outgoing, lower-energy ν_τ from the τ decay. This means that ν_τ are never absorbed in Earth but "regenerate" through multiple NC and CC interactions until the ν_τ is of sufficiently low energy that Earth becomes transparent (a similar effect occurs for ν_μ, but they regenerate to sub-GeV energies due to the long lifetime of the μ). A flux of ν_τ also therefore produces a small accompanying flux of ν_e and ν_μ from these respective decay modes.

8.4.2.5 Glashow Resonance Events

At 6.3 PeV, $\bar{\nu}_e$ interactions with atomic electrons have a center-of-mass energy equal to the rest mass of a W^- (80 GeV). This leads to a "resonance"—a greatly increased cross section, as shown in Figure 8.4. The interaction itself is shown in Figure 8.3 (right), and is known as the Glashow resonance. The W^- decays into either an $\ell\bar{\nu}_\ell$ pair with probability slightly less than 1/9th for each flavor (for a total of 1/3rd), or a $q\bar{q}$ pair with a total probability of 68% (2/3$^{\rm rds}$). This ratio is interesting in and of itself, as it is due to the existence of color charge in the strong interaction.

The $e\bar{\nu}_e$, and all $q\bar{q}$ decays, leads to a cascade at the interaction point; the $\mu\bar{\nu}_\mu$ leads to a track event with no initial cascade, and the $\tau\bar{\nu}_\tau$ lead to its own interesting mix of event topologies, as described above. Because the cross section at the Glashow resonance is so high, only downgoing events are expected in a neutrino detector, with the exception of regenerated ν_τ from initial $W^- \rightarrow \tau\bar{\nu}_\tau$ decays.

8.4.3 Event Identification

Astrophysical neutrinos must be identified against three kinds of background. The first is the natural optical background light, typically from radioactive decays, although underwater detectors also face occasional bursts of bioluminescence from underwater organisms. Filtering out this background requires searching for "events," with several PMTs detecting light within a short time interval, consistent with relativistic particles generating Čerenkov light in the detector. These events are recorded and processed to identify the direction and energy of the particles, and the interaction type (track or shower). An example is given in Figure 8.17.

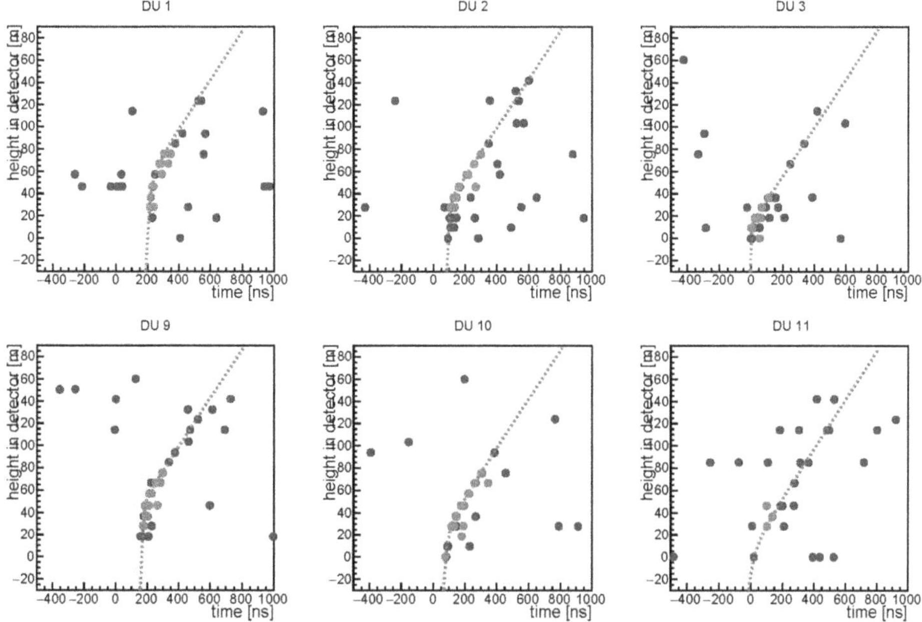

Figure 8.17. Example of an event identification. A candidate neutrino event in the first six detection units of KM3NeT/ORCA (Adrián-Martínez et al. 2016). Blue dots indicate photon detections as a function of time and detection height. Red dots indicate those photons identified in real time as being due to a Čerenkov cone, shown by magenta dotted lines. The event is due to an upgoing μ produced by a ν_μ interacting in the seabed. Image credit: http://www.km3net.org.

8.4.3.1 Downgoing Muons

Most detected events—approximately 99.999%—are due to downgoing atmospheric muons. These are produced predominantly via $\pi^{\pm}$ decay from cosmic-ray interactions in Earth's atmosphere. These can be easily removed by rejecting downgoing track-like events. Because misreconstructing even 1 event in 10,000 leads to these muons dominating data samples, very strict cuts on well-reconstructed events must be implemented to ensure that only neutrino events survive. This is usually accomplished by assessing the goodness of fit of the light profile of the event. Cascade events inside the detector volume provide a cleaner, but smaller, data sample. They can also be "impersonated" by a throughgoing muon that is undetected by PMTs at a detector's outer edge and loses the majority of its energy in a single, point-like interaction.

8.4.4 Source Identification

After removing downgoing muon events, the resulting sample of neutrino events will still largely be dominated by the atmospheric neutrino flux described in Section 8.4.1.1, with a small contamination from misreconstructed atmospheric muons.

Identifying astrophysical neutrinos from this sample requires statistical techniques, utilizing the following information:

- **Direction:** An excess of neutrinos coming from a particular direction on the sky is a clear indication of an astrophysical origin. Sources might be point like (e.g., a blazar), extended, such as the Milky Way's expected dark matter halo, or very broad, such as the Galactic Plane. This criterion works best with track-like events due to their better angular resolution.
- **Energy:** The atmospheric neutrino spectrum is very steep, while astrophysical neutrino sources are expected to produce a flatter spectrum. Selecting for high-energy events therefore allows the identification of even a completely diffuse (all-sky) neutrino flux. This criterion works best with cascade events, due to their better energy resolution.
- **Time:** The atmospheric neutrino background is almost constant. However, multimessenger observations can identify the times when neutrinos might be expected, such as during blazar flares (Section 8.4.7), or coincident with short transients such as GRBs (Section 8.4.9.3) or gravitational-wave mergers. This allows the rejection of background events lying outside a narrow time range.
- **Flavor:** Atmospheric neutrinos in the TeV range do not oscillate over Earth baselines and have a flavor ratio of ν_e: ν_μ: $\nu_\tau \approx 0.1$: 1: 0 (Honda et al. 2007). The astrophysical flux however should be almost flavor-uniform (Section 8.4.6.1) and will therefore have a relative excess of cascade events. Any ν_τ identified through a "double-bang" event (Section 8.4.2.4) would be guaranteed to be of cosmic origin.
- **Veto:** A cosmic-ray interaction above the detector will produce many muons. Detecting this muon cascade—either with the buried detector itself, or using a surface array—in coincidence with a neutrino will exclude that event as being of astrophysical origin. This tends to only work for downgoing events at high zenith angles, where accompanying muons have a chance of being detected.

These techniques, to a greater or less degree, have all been applied to produce our current knowledge of high-energy astrophysical neutrinos, discussed in the following section.

8.4.5 Experimental History

The first attempts to deploy an astrophysical neutrino telescope—indeed, the very notion of an astrophysical neutrino telescope—began in the 1970s (Spiering 2019). Early discussions culminated with the Deep Underwater Muon And Neutrino Detector (DUMAND) project in Hawaii (Roberts 1992), which ran until 1995. While no significantly sized detector was ever deployed, measurements made by test arrays of PMTs were valuable and proved the feasibility of future experiments.

The first completed detector was AMANDA, the Antarctic muon and neutrino detector array (Andres et al. 2000). It was deployed at depths of 1.5–1.9 km in deep ice at the South Pole. Consisting of 677 optical modules, each housing a 20 cm PMT, it monitored an approximately cylindrical volume 200 m in diameter and 400 m high. AMANDA detected about 30 downgoing cosmic-ray muons per second, and a few atmospheric ν per day, but could not identify any astrophysical sources of neutrinos.

Its successor experiment, IceCube, is also located at the South Pole (Aartsen et al. 2017b). IceCube is the largest current astrophysical neutrino telescope, consisting of 5160 optical modules housing 10" PMTs, and encompassing 1 km^3 of ice in a hexagonal cylinder at depths between 1.45 and 2.45 km. IceCube began deployment in 2005, with full operation in 2010. It is famous for its discovery of high-energy astrophysical neutrinos in 2012 (see Section 8.4.6). IceCube also uses a surface array, IceTop, consisting of (frozen) water tanks with PMTs to detect cosmic-ray cascades and veto these events for neutrino detection. IceCube has been upgraded with an infill, DeepCore (Abbasi et al. 2012), to study neutrino oscillation physics in the 10–100 GeV range (see Section 8.2.2). Further upgrades to allow IceCube to study neutrino oscillations in the 1–10 GeV range (IceCube-Gen2 Collaboration et al. 2019), and extend the detector to greater volumes, are also planned.

ANTARES—Astronomy with a Neutrino Telescope and Abyss environmental RESearch—is an underwater neutrino telescope at a depth of 2.475 km in the Mediterranean, off the coast of Marseilles, France (Ageron et al. 2011). The longest-running such instrument, it was completed in 2008 and is due to be decommissioned in the near future. Of similar size to AMANDA, with 885 10" PMTs, it observes a volume 350 m in height and about 0.03 km^2 in cross section, for a total enclosed volume of 0.01 km^3.

Two new detectors are currently under construction. Due for completion around 2026–2027, KM3NeT is a network of underwater detectors located at two sites in the Mediterranean (Adrián-Martínez et al. 2016). KM3NeT/ORCA, off the southern coast of France near ANTARES, will target 1–100 GeV atmospheric neutrinos. KM3NeT/ARCA, which is of most relevance for astrophysics, will be deployed 100 km off the coast of Sicily. At a depth of 3.5 km, ARCA (Astrophysics Research with Cosmics in the Abyss) will consist of two "blocks" of detectors, of total volume of 1.2 km^3. While only a slightly larger size than IceCube, it will have improved angular resolution, and being in the northern hemisphere, will have complementary sky coverage. GVD, the Gigaton Volume Detector, is being constructed in Lake Baikal in Russia (Avrorin et al. 2014). It will also have a final volume of $\sim$1 km^3 and will similarly study astrophysical sources of high-energy neutrinos. The first full cluster of detectors was completed in 2016, with 288 optical modules deployed at depths between 750 m and 1275 m (Avrorin et al. 2015).

8.4.6 The High-energy Diffuse Flux

High-energy astrophysical neutrinos were first identified in 2013 by the IceCube Collaboration (2013). The search looked for high-energy cascade events inside the detector, one of which is shown in Figure 8.16 (right). This allowed atmospheric muons to be excluded with high precision and took advantage of the steeply decreasing atmospheric neutrino spectrum to identify an astrophysical component "poking out" above 100 TeV in energy, as shown in Figure 8.18. These high-energy

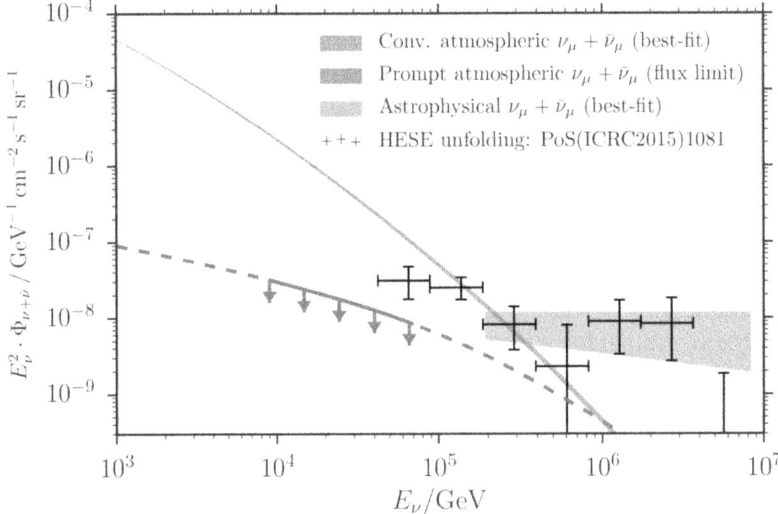

Figure 8.18. The high-energy astrophysical neutrino flux measured by IceCube in 2015 (IceCube Collaboration et al. 2015; orange), compared to the atmospheric neutrino flux (green), and the expected "'prompt" contribution from charm meson decay (dotted). Image credit: Reimann (2019), with permission of MDPI, and the IceCube Collaboration.

events are rare: only 54 in four years of data, with 7–17 of them estimated as atmospheric muon or neutrino events.

This flux has subsequently been examined in detail, using a range of different data samples, as shown in Figure 8.19. The spectrum appears to be consistent with a power law of the form:

$$\frac{dN}{dE} = \phi\left(\frac{E}{100\ \text{TeV}}\right)^{-\gamma}, \tag{8.22}$$

with best-fit parameters $\gamma = 2.53 \pm 0.07$ and $\phi = 4.98^{+0.75}_{-0.81} \times 10^{-18}\,\text{Gev}^{-1}\,\text{cm}^{-2}\,\text{s}^{-1}\,\text{sr}^{-1}$ (Aartsen et al. 2020b).

The origin of the flux is currently unknown. It is usually termed the "diffuse" flux because it is consistent with a uniform spectrum over the entire sky. The inconsistency between the spectrum when measured in (mostly upgoing) μ and (more likely downgoing) cascades may be due to different spectra between the northern and southern hemispheres, in particular due to the Milky Way being mostly in the south. A different argument is that each of the analyses shown in Figure 8.19 probes a different energy range, so that a uniform flux with a spectral break would result in the discrepancies observed in Figure 8.19.

Both IceCube and ANTARES have searched for the source of this flux, with no firm indications as to its origin (Albert et al. 2020). While one blazar (see Section 8.4.7 below) has been detected as a neutrino emitter, blazars are unlikely to be responsible for the majority of the observed neutrino emission (Aartsen et al. 2017a).

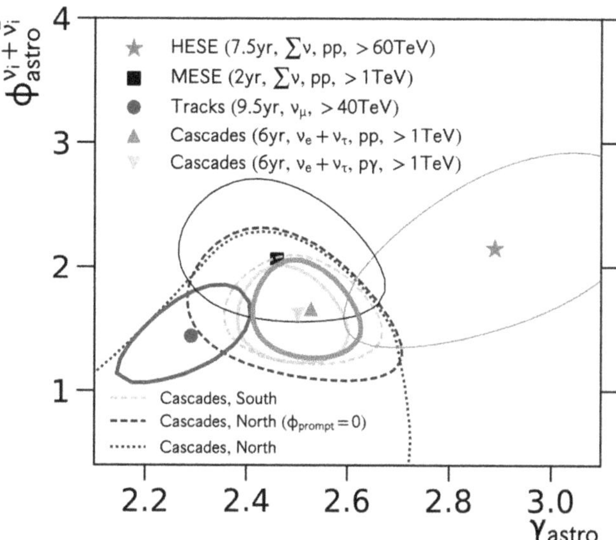

Figure 8.19. Best fit to the normalization Φ_{astro} and power-law index γ_{astro} of the diffuse astrophysical neutrino flux defined Equation (8.22). The event samples are "HESE" (high-energy starting events), "MESE" (medium-energy starting events), cascade events assuming pp and $p\gamma$ production mechanisms, and predominantly throughgoing muons from ν_μ. See Aartsen et al. (2020b) and references therein for further definitions of these samples. Here, $\phi_{astro} = 3 \times 10^{-18}\phi$ from Equation (8.22). Image credit: Aartsen et al. (2020b) with permission, copyright 2020 by the American Physical Society.

Resolving the origin and nature of the diffuse flux observed by IceCube will be a major challenge for neutrino detectors in the next decade.

8.4.6.1 Flavor Analysis

One clue as to the origin of the flux comes from flavor analysis. The neutrino flux from an astrophysical source will arrive fully mixed because over astrophysical distances, small variations in initial energies will lead to random oscillation phases at Earth. However, the best-fit neutrino oscillation parameters do not allow one initial neutrino flavor to equally mix into all three flavors. This preserves some of the initial neutrino flavor flux, and hence the production physics, from the source to Earth.

Lipari et al. (2007) consider three initial production scenarios: a standard pion-decay scenario from Equation (8.11), a neutron-decay scenario whereby only neutral neutrons escape the source, and a muon-damped scenario, i.e., where μ loses effectively all their energy prior to decay. The observable neutrino flavor ratios at Earth predicted from each of these scenarios are shown in Table 8.3.

Analysis of the energy and topology of events allows a fit to the neutrino flavor ratio. Primarily, this involves fitting the fraction of track events from ν_μ CC and ν_τ CC, where the subsequent τ decays to μ, but it also includes searching for double-bang events and fitting the energy and inelasticity distributions, giving some power to differentiate between ν_e and ν_τ fluxes. The current best fit from IceCube is shown in Figure 8.20. The fit is consistent with all expectations from astrophysical sources, and it is unlikely that significant improvements will be achieved in the near future.

Table 8.3. Predicted Flavor Ratios (ν_e: ν_μ: ν_τ) at Earth as a Function of Initial Flavor Ratios at Source, for Each of Three Scenarios, as Calculated by Aartsen et al. (2015) Using the Oscillation Parameters of Gonzalez-Garcia et al. (2014) in the Case of an Inverted Hierarchy

Scenario	Initial Ratio	Ratio at Earth
Standard	1:2:0	0.93:1.05:1.02
Neutron decay	1:0:0	0.19:0.43:0.38
Muon-damped	0:1:0	0.55:0.19:0.26

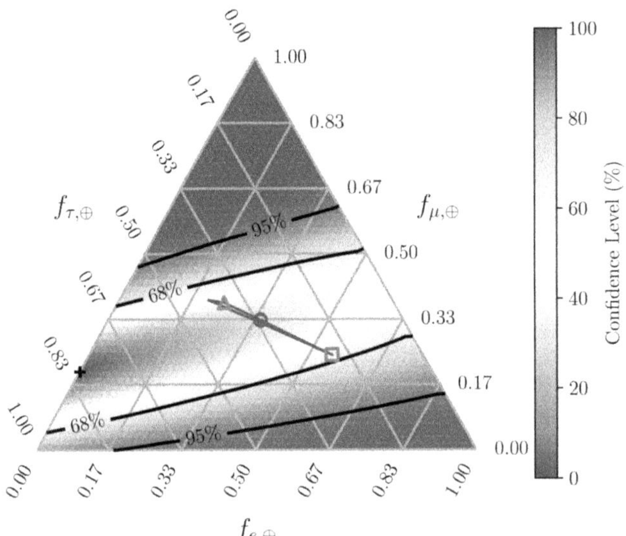

Figure 8.20. Limits on the flavor ratio of the diffuse astrophysical neutrino flux, from the IceCube experiment (Aartsen et al. 2019a). The shading indicates the confidence level at which any given composition can be excluded; red, blue, and green markers indicate expectations from astrophysical fluxes; and solid black lines indicate exclusion levels. Silver lines show constant compositions for each flavor. Current constraints cannot distinguish between different neutrino production scenarios. Image credit: Aartsen et al. (2019a) with permission, copyright 2019 by the American Physical Society.

8.4.7 Blazar TXS 0506+056

If multimessenger astronomy was born on 1987 February 23, with the discovery of neutrinos from SN 1987A in the Large Magellanic Cloud (see Section 8.3.1.4), it was reborn 30 years later, on 2017 August 17 with multimessenger observations of the binary neutron star merger GW 170817. Yet barely 40 days after this discovery, on 2017 September 22, an equally significant event occurred.

The IceCube Collaboration observed the upgoing neutrino shown in Figure 8.16 (left): IC 170922A. The Fermi gamma-ray satellite then reported that the neutrino's arrival direction pointed back to a blazar undergoing an energetic gamma-ray flare, TXS 0506+056, as shown in Figure 8.21 (IceCube Collaboration et al. 2018a). This triggered a swathe of follow-up observations, producing the blazar spectral energy distribution shown in Figure 8.22.

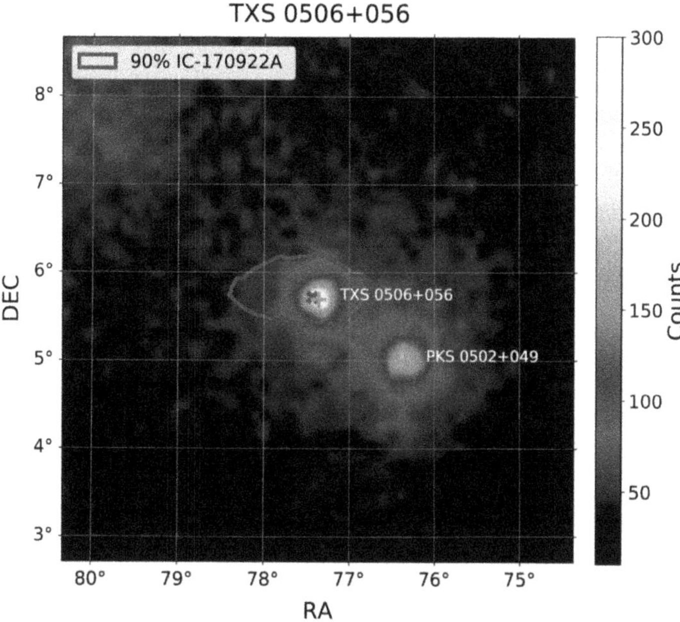

Figure 8.21. Localization of IC 170922A to the blazar TXS 0506+056, showing the likely reconstructed direction (green contour) compared to the gamma-ray flux measured by Fermi. A second blazar, PKS 0502 +049, is shown in the bottom right. Image credit: Garrappa et al. (2019).

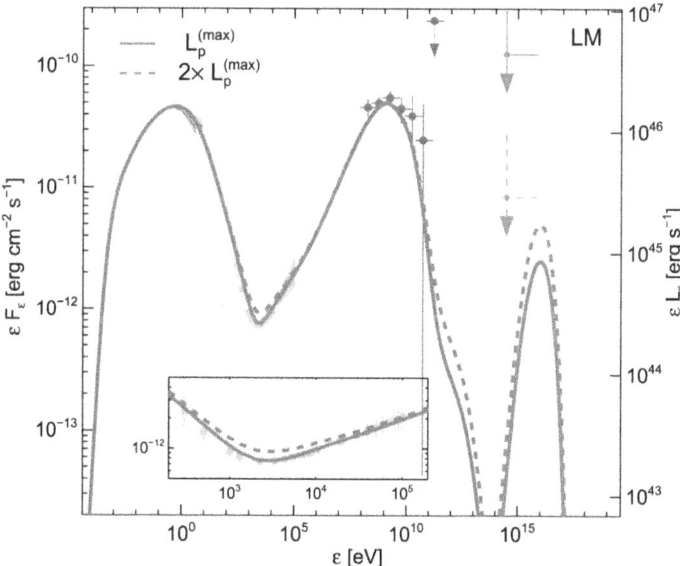

Figure 8.22. SED of blazar TXS 0506+056, using data from UV (green/blue), X-rays (orange), and gamma rays (purple), as well as a fit using only leptonic processes ('L_P'; gray). A subdominant hadronic component produces a predicted neutrino flux (red)—doubling this component produces the dashed fluxes, leading to more neutrinos, but overpredicting the observed X-ray flux. Image credit: Keivani et al. (2018).

On its own, this detection had a chance probability of 0.3%, after accounting for 10 other high-energy events that were not in coincidence with blazars (IceCube Collaboration et al. 2018a). However, a lookback analysis by IceCube revealed that TXS 0506+056 had undergone a "neutrino flare" of several months' duration, centered on 2014 December 14 (IceCube Collaboration et al. 2018b). Shown in Figure 8.23, the flare consisted of an excess of 13 ± 5 events above atmospheric background expectations. Combined, the evidence strongly points toward this blazar being the third identified astrophysical neutrino source, after the Sun and SN 1987A, and the first to produce high-energy neutrinos.

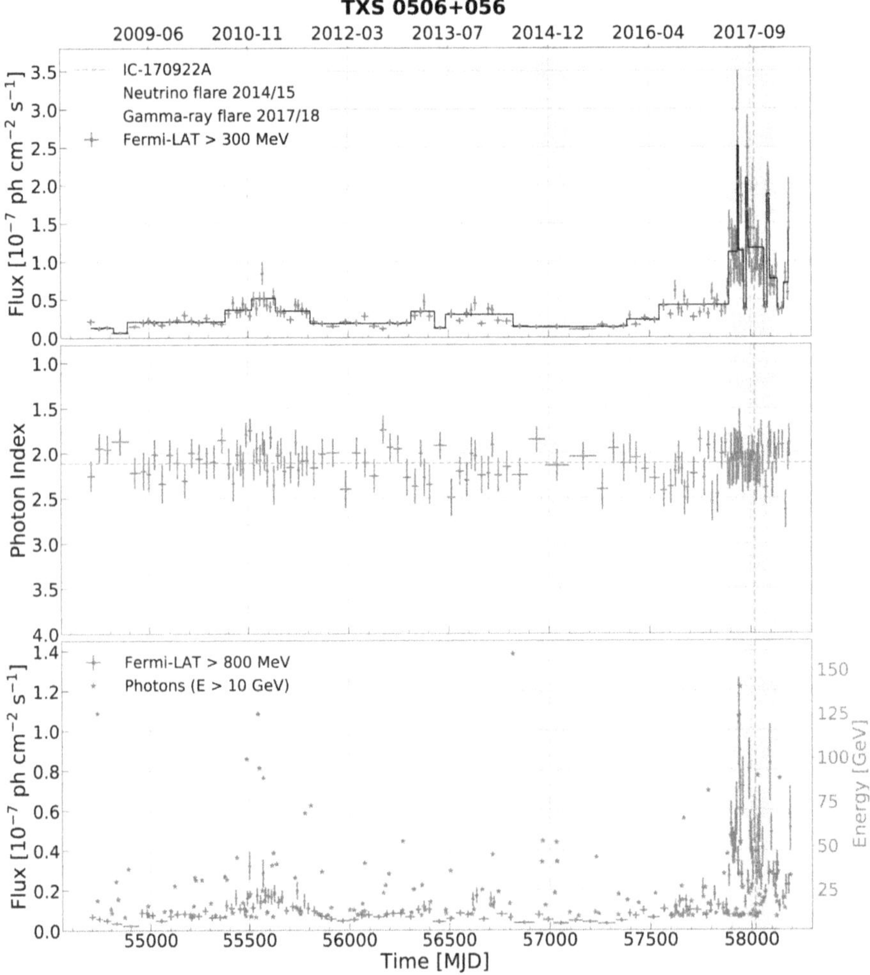

Figure 8.23. Measured gamma-ray flux (top and bottom panels) and photon index (middle) from TXS 0506+056 over several years. The time of observation of IC 170922A is shown by a red line, with the corresponding gamma-ray flare highlighted in green. The time of the neutrino flare is highlighted in orange. Image credit: Garrappa et al. (2019).

Unlike multimessenger observations of GW 170817, this observation raised more questions than it answered. For instance: why did this particular blazar—only the 45th brightest gamma-ray-emitting blazar (Ackermann et al. 2015)—produce the first neutrino association and not another? Extensive searches for associations with other blazars have so far produced at best tentative results (Garrappa et al. 2019; Kadler et al. 2016; Albert et al. 2017a). And how did this blazar produce both the high-energy neutrino IC 170922A and the neutrino flare? Various attempts at modeling the emission cannot simultaneously produce a high-energy neutrino and the observed SED (see, e.g., Keivani et al. 2018; Petropoulou et al. 2020), and have greater difficulty reproducing the flare (e.g., Reimer et al. 2019; Xue et al. 2019). The main issue is that sufficient $\pi^{\pm}$ production to produce the observed neutrinos implies π^0 production. The electromagnetic cascades resulting from $\pi^0 \rightarrow \gamma\gamma$ decays typically overshoot the observed X-ray spectrum of the blazar (Murase et al. 2018).

Answering these questions is currently a subject of intense investigation.

8.4.8 Prospective Galactic Sources

The observation of Galactic cosmic rays up to at least PeV energies motivates the search for their source, dubbed "PeVatrons." Neutrinos carry typically 5% of the interacting cosmic-ray energy (Kelner et al. 2006), so the spectrum continues to within 5% of the peak cosmic-ray energy. For cosmic-ray accelerators in our Galaxy, with energies capable of producing PeV cosmic rays (such as the supernova remnant RX J1713; see Figure 6.12 and Section 7.5.1), the neutrino spectrum is expected to extend to 10–100 TeV. Figure 8.24 shows some models for neutrino spectra from Galactic sources, which are best identified by the emission of TeV gamma rays.

Northern hemisphere neutrino telescopes, such as ANTARES, KM3NeT, and GVD, are best placed for observing the majority of Galactic neutrino sources,

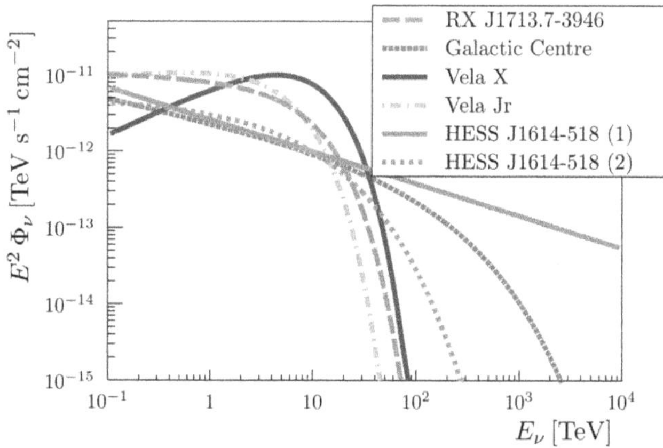

Figure 8.24. Predictions of neutrino fluxes from potential cosmic-ray accelerators within our Galaxy (Aiello et al. 2019). These models typically attribute most of the observed high-energy gamma-ray flux to π^0 decay, implying neutrinos from $\pi^{\pm}$. Image credit: Aiello et al. (2019), Copyright 2019, with permission from Elsevier.

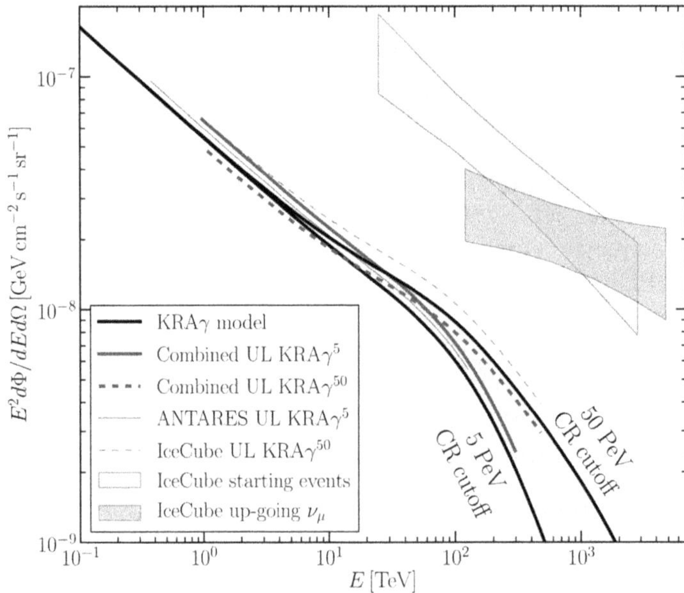

Figure 8.25. Predictions of, and limits on, the diffuse Galactic neutrino flux (Albert et al. 2018). Also shown are the diffuse neutrino fluxes measured by IceCube. Image credit: Albert et al. (2018).

(which lie in the southern hemisphere. This is because ν_μ will be visible in the more-sensitive upcoming track channel, which also provides a better angular resolution. Even in optimistic neutrino production scenarios, however, it will take the future neutrino telescope KM3NeT 5–10 years to identify individual sources (Aiello et al. 2019).

While the current rate of neutrino production from Galactic sources is highly uncertain, something has produced Galactic cosmic rays. This leads to a more robust prediction for a diffuse neutrino flux from the Galactic Plane. Expectations of, and current limits on, this flux are shown in Figure 8.25. Current detector sensitivities are very close to predictions.

8.4.9 Extragalactic Sources

The Milky Way is not special, and any source of neutrinos in our Galaxy will have counterparts in the billions of other galaxies in the universe. However, the proximity of the Milky Way means that its neutrino flux should dominate over the background from other, similar galaxies. There are several classes of sources that do not occur in the Milky Way or are not currently occurring. Many of these have been suggested as the origin of ultra-high-energy cosmic rays. Detector sensitivity to these source classes is independent of detector location because the extragalactic sky is similar from both hemispheres. However, the sensitivity to any given source will depend on its declination and the detector location, as shown in Figure 8.26. The status of searches for neutrinos from extragalactic sources is given in this section.

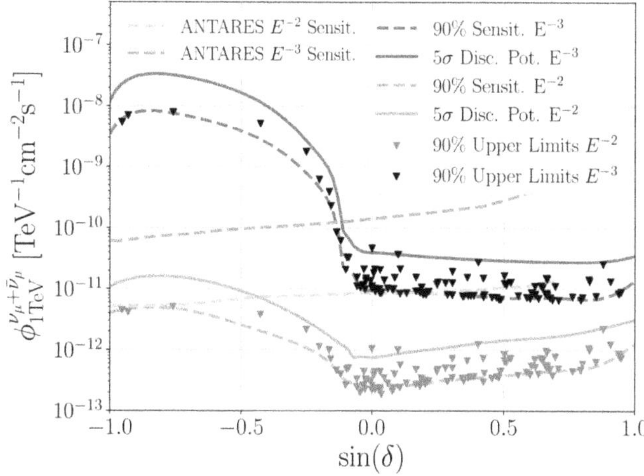

Figure 8.26. Sensitivity of neutrino searches with the IceCube and ANTARES neutrino telescopes to point-like sources of neutrinos as a function of declination δ, together with individual source limits (points). Image credit: Aartsen et al. (2020a) with permission, copyright 2020 by the American Physical Society.

8.4.9.1 AGNs

AGNs have long been thought to be sites of high-energy particle acceleration. The high photon densities near the base of AGN jets are expected to generate many $p\gamma$ interactions, producing both high-energy gamma rays and neutrinos. These neutrinos will be beamed in the same manner as photons so that blazars—AGNs with jets pointed toward Earth—have long been held as good neutrino source candidates.

As well as the detection of TXS 0506+056, there have been many tentative associations of neutrino events with blazars (Kadler et al. 2016; Garrappa et al. 2019; ANTARES Collaboration et al. 2015; Krauß et al. 2014; Adrián-Martínez et al. 2012). To date, however, only TXS 0506+056 remains a firm candidate.

8.4.9.2 Star-forming Galaxies

The processes leading to a flux of neutrinos from cosmic-ray interactions in the Milky Way will also occur in other galaxies. In particular, galaxies with a high rate of star formation (star-forming galaxies, or SFGs) and hence particle acceleration in associated shocks (see the sources detected by TeV gamma-ray instruments in Chapter 7) will produce a higher-than-average neutrino flux. There is also some hint of ultra-high-energy cosmic rays being associated with SFGs (Aab et al. 2018b).

The maximum neutrino flux from such galaxies is limited by the γ-ray flux through the π^0–$\pi^\pm$ mechanism. Observations of nearby SFGs, such as NGC 253, reveal gamma-ray emission in both GeV and TeV bands (HESS Collaboration et al. 2018). However, the corresponding neutrino flux is small compared to the diffuse flux observed by IceCube (Bechtol et al. 2017), and no association with SFGs has been found in neutrino searches (Aartsen et al. 2020a; Albert et al. 2018).

8.4.9.3 GRBs

The canonical explanation of long-duration GRBs is the "collapsar" model (MacFadyen & Woosley 1999), whereby the direct collapse of a massive star to a black hole emits a highly relativistic jet. Shocks in the jet accelerate electrons and lead to the observed burst of gamma rays (Rees & Meszaros 1992). If protons are also present in the jet, this may explain the source of ultra-high-energy cosmic rays (see Bustamante et al. 2013, and references contained therein), while $p\gamma$ interactions in the intense photon environment would also lead to an associated flux of neutrinos (Waxman 1995).

Predictions of the neutrino flux from GRBs vary greatly. Key uncertainties are the hadronic loading of the jet—i.e., how many cosmic-ray protons are present—and the bulk Lorentz factor Γ (Hümmer et al. 2012). Furthermore, the radius of the emission zone also strongly influences the predicted neutrino spectrum (Bustamante et al. 2015). This leads to varying expectations for the neutrino flux—see Figure 8.27.

No neutrino emission has yet been associated with a GRB (Aartsen et al. 2017c; Albert et al. 2017b). Only time—and perhaps luck—will tell if this is the next breakthrough in neutrino astrophysics.

8.4.9.4 Dark Matter

In the standard ΛCDM (Lambda cold dark matter) cosmology, the universe is composed of about 26% dark matter (Planck Collaboration et al. 2016). While the nature of dark matter (DM) particles X_{DM} is so far unknown, one method of searching for it is through its potential annihilation to standard model particles. Channels that would produce neutrinos, directly or via subsequent decay products, are

$$X_{\mathrm{DM}} + X_{\mathrm{DM}} \rightarrow b\bar{b},\ W^{+}W^{-},\ \tau^{+}\tau^{-},\ \mu + \mu^{-},\ \nu\bar{\nu}. \tag{8.23}$$

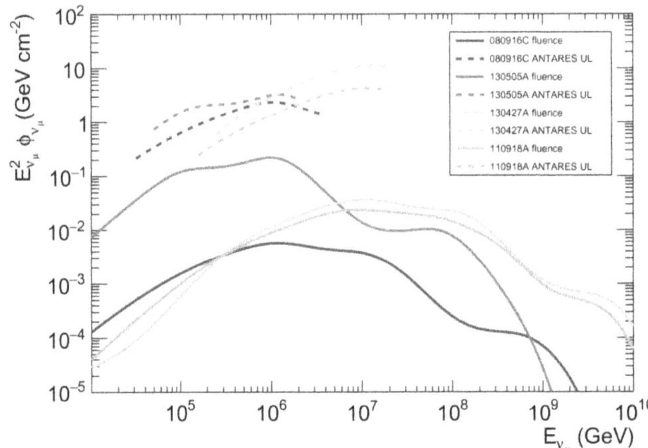

Figure 8.27. Expected spectra from, and limits on, some particularly bright GRBs from the ANTARES neutrino telescope. Image credit: Albert et al. (2017b) © 2017 The Authors, Published by Oxford University Press on behalf of the Royal Astronomical Society.

DM is expected to congregate in the centers of galaxies or could be trapped by our Sun. If they then decay or annihilate, this would produce an excess of neutrinos in those directions, allowing neutrino telescopes to search for them. While limits placed on the dark matter annihilation cross section by neutrino telescopes tend to be weaker than those from γ-ray instruments (see Figure 8.28), they are sensitive to different annihilation products and hence probe a different part of the DM parameter space.

8.5 Ultra-high-energy (UHE) Neutrinos

Ultra-high energy (UHE) neutrinos generally refer to those created through the interactions of cosmic-ray protons with energies at or above 5×10^{19} eV with background light in the universe, particularly the CMB (Greisen 1966; Zatsepin & Kuz'min 1966). From an experimental point of view, UHE neutrinos are those at sufficiently high energies, and correspondingly low fluxes, that km^3 detectors like IceCube will be unlikely to detect them. This calls for new detection methods, which take over from Čerenkov detectors at energies typically above 100 PeV.

While these neutrinos hold great astrophysical interest, they are unlikely to generate the kind of multimessenger detection as lower-energy neutrinos. This is for the simple reason that it is very difficult to accelerate cosmic rays to energies above 100 PeV. If the $p\gamma$ or pp interactions necessary for neutrino production occur at the source, then the energy-loss rate will be too great, and cosmic rays will not attain ultrahigh energies. Conversely, the energy-loss rate from pion photoproduction with the CMB is so low that cosmic rays must traverse distances of order 100 MPc in order to lose a large fraction of their energy, meaning that they will have escaped their source region. Nonetheless, neutrinos continue to be a source of new and

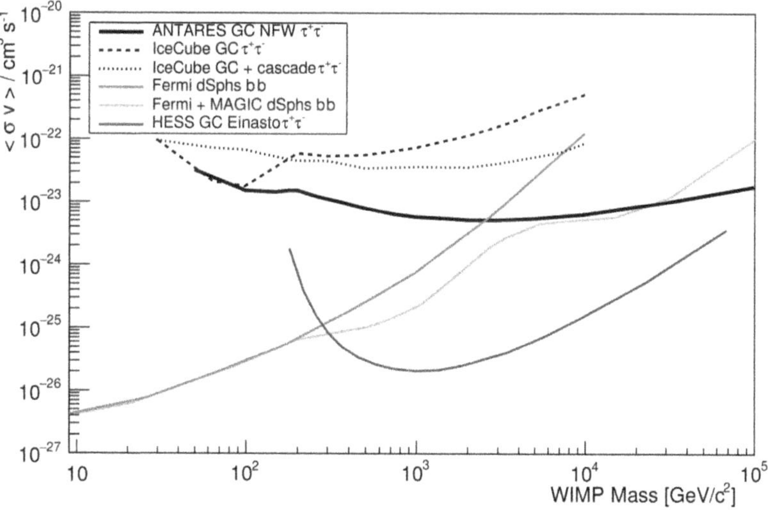

Figure 8.28. Limits on dark matter annihilation rate–cross section from observations of the Galactic Center as a function of DM particle mass. Image credit: Albert et al. (2017c), Copyright 2017, with permission from Elsevier.

unexpected discoveries, and so it is worthwhile briefly reviewing the experimental methods at ultrahigh energies.

8.5.1 Extensive Air Showers

Earth's atmosphere presents only 10 m of W.E. interaction target. However, it is very easy to monitor, and the 3000 km^2 Pierre Auger (cosmic ray) Observatory already searches for the extensive air showers caused by cosmic-ray interactions in the atmosphere above it. Using a combination of surface particle detectors measuring the ground pattern of secondary particles and optical telescopes to search for nitrogen fluorescence induced by the passage of cascade particles through the atmosphere, Auger can identify neutrino cascades as being those having penetrated too far to be due to cosmic rays or gamma rays hitting the top of the atmosphere (Aab et al. 2019). This channel is most sensitive to ν_e CC interactions, which deposit the greatest energy upon interaction. This technique provides a greater effective area for neutrino detection than IceCube at energies above about 10^{17} eV, as shown in Figure 8.29.

8.5.1.1 Tau Neutrinos

UHE neutrinos cannot penetrate more than about 1000 km of rock, preventing them from arriving at a detector from below. They can, however, "skim" Earth's edge and arrive at a detector traveling almost horizontally, or slanted slightly upward. They can also penetrate large mountain ranges to interact on the outgoing side. Such an

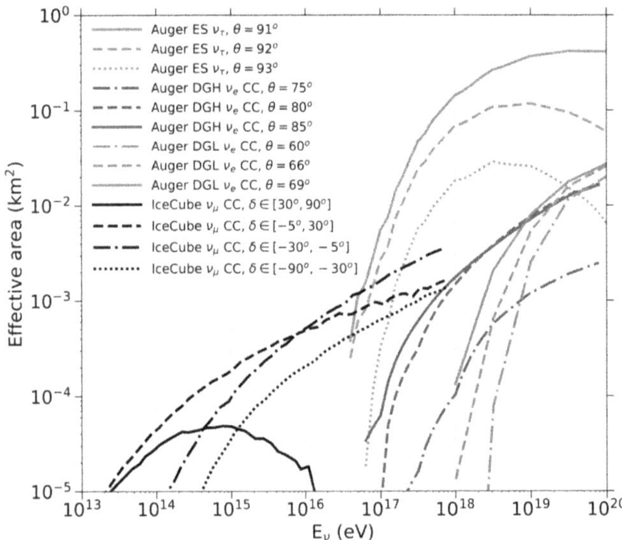

Figure 8.29. Comparison of effective areas of IceCube in different declination ranges δ, and the Pierre Auger Observatory (PAO) at different zenith angles θ, to UHE neutrinos. The PAO results are divided into Earth-skimming (ES), and downgoing events at high (DGH) and low (DGL) zenith angles (Aab et al. 2019). Image credit: Aab et al. (2019).

interaction will be unobservable because any emitted radiation will be absorbed in the rock.

At UHE however, the τ produced from a ν_τ CC event will decay tens of km from the interaction point (about 50 km at 1 EeV). Approximately 82% of these decays—i.e., those not producing a μ—will lead to a particle cascade at the decay point. This allows a much larger volume of material than Earth's atmosphere to be used as an interaction target. While the particles in the cascade will be directed upwards, fluorescence light is emitted isotropically and will be detectable by ground-based telescopes. This provides a much higher detection probability, but only for the small range of angles where a neutrino can successfully skim the Earth (see Figure 8.29).

8.5.2 Radio Detection

Above 10^{17} eV, the radio emission from particle cascades becomes easier to detect than the optical emission. When a high-energy particle cascade occurs in a medium, low-energy electrons are knocked into the cascade, while some positrons annihilate out. This leads to an excess negative charge propagating with the shower front, leaving a positively ionized medium in its wake. The excess negative charge will radiate coherently at wavelengths greater than the dimensions of the particle cascade (Askar'yan 1962, 1965). Known as the Askaryan effect, this leads to emission peaked at 100 MHz for particle cascades in air and 1 GHz for cascades in dense media (Schröder 2017). Furthermore, in air, oppositely charged particles (e^- and e^+) will be deflected in opposite directions via the Lorentz force from Earth's magnetic field. The resulting "geomagnetic" emission is coherent and dominates the Askaryan mechanism for atmospheric cascades (Huege 2016). In both cases, the emission takes the form of a very short pulse of inverse-bandwidth duration (typically 1–100 ns). In order to detect it, a large volume of radio-transparent material is required to have sufficient neutrino target mass and allow the radio emission to escape and be detected. The target media of current experimental interest are Earth's atmosphere, the Arctic and Antarctic ice sheets, and the Moon.

8.5.2.1 Atmospheric Radio Detection

Radio emission in air is directed forwards so that emission from the decay of Earth-skimming ν_τ would not normally be detectable. Placing detectors on a mountain range, however, allows the observation of upcoming radiation from interactions in the ground nearby, or cascades emerging from interactions in the mountains themselves. The Giant Radio Array for Neutrino Detection (GRAND) is a planned radio detector using 200,000 radio antennas deployed over 200,000 km^2 (Álvarez-Muñiz et al. 2020). It will be deployed over several mountainous sites of suitable radio quietness in China. The first subarray of 10,000 antennas has a target deployment date of 2025, with a 300 antenna prototype currently under construction.

8.5.2.2 Detection in Ice

While water absorbs and reflects radio waves, ice is remarkably radio transparent. This allows large natural volumes of ice—specifically, the Greenland and Antarctic ice sheets—to be used as suitable detection media for UHE neutrino interactions. Radio signals generated in the ice can travel for several kilometers and to be detected either from within the ice itself, or via detectors located above it.

The first experiment to use this technique was RICE, the Radio Ice Čerenkov Detector (Kravchenko et al. 2006). It consisted of 20 radio receivers buried at depths of a few hundred meters in the South Pole ice immediately above the much deeper AMANDA detector (now the current location of IceCube). This array, with intradetector spacings of 10–100 m, provided some ability to reconstruct the arrival direction of the in-ice radio wavefront and distinguish between events arriving from within the ice and from the surface.

Current experiments using in-ice detection are the Antarctic Ross Ice-Shelf antenna neutrino array, ARIANNA (Anker et al. 2019), the Askaryan Radio Array, ARA, at the South Pole in the vicinity of IceCube (Ara Collaboration et al. 2012), and a detector that may be built in Greenland (Wissel et al. 2015). These detectors are all in their initial stages, with plans for significantly larger upgrades. A total detection volume of at least 10 km^3 will be needed to detect the very rare UHE neutrinos.

Instead of placing detectors in ice, a detector can be located above it. The Fast On-orbit Rapid Recording of Transient Events (FORTE) satellite passed over the Greenland ice sheet and was able to search for transient radio bursts from neutrino interactions in it. The one candidate event that survived all cuts was assumed to be background, and this limited the UHE neutrino flux (Lehtinen et al. 2004).

ANITA is a high-altitude balloon-borne experiment that takes advantage of the South Pole (wind) vortex during southern hemisphere summers to orbit the South Pole for weeks at a time (Gorham et al. 2010). ANITA consists of three levels of 200 MHz–1 GHz radio antennas covering 360° in azimuthal angle—see Figure 8.30. Flying at an altitude of about 37 km, ANITA views vast volumes of ice, albeit with less sensitivity than a detector placed within the ice itself. Each flight of ANITA is destructive to the experiment, with four flights having been completed between 2006 and 2017.

Radio emission in ice is peaked at the Čerenkov angle, at about 56° to the direction of the incident neutrino. However, transmission at the ice–air interface bends the signal toward the horizontal, in the direction of the original particle. Emission from downgoing neutrinos will be totally internally reflected at the air–ice interface, while UHE neutrinos cannot penetrate Earth. Therefore, ANITA is expected to be most sensitive to Earth-skimming neutrinos arriving almost horizontally.

ANITA has, however, detected radio emission from many cosmic-ray events, either directly, or when reflected off the ice. A reflected event can be distinguished from a direct event due to a parity flip. Once during its first flight, and once during its third (see Figure 8.30), ANITA has detected upcoming events that show no signs of a parity flip (Gorham et al. 2016, 2018). These events resemble the signatures of a τ decaying in the atmosphere after a ν_τ CC interaction in or beneath the ice. However, such a ν_τ would have had to penetrate most of Earth to produce a steep upgoing arrival direction, which is completely precluded by standard model physics. The origin of these events is a current topic of debate.

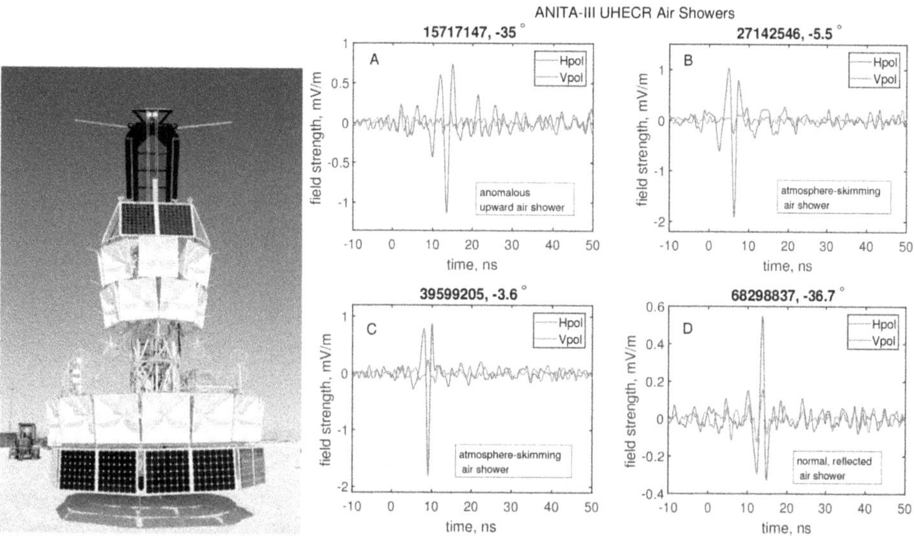

Figure 8.30. Left: the ANITA I experiment, suspended from cables, just prior to launch in 2006. The radio antennas are three rings of square white feedhorns. Right: four events observed by ANITA III Gorham et al. 2018. The anomalous event is A, arriving from 35° below the horizontal. It has the same polarization (horizontal) and polarity as events B and C, which are extensive air showers arriving directly at the detector. It does not show the expected reversed polarity of event D, which also arrives from below, having been reflected off the ice. Image credit: (left) Jeff Kowalski, Peter Gorham, and Christian Miki; (right) Gorham et al. (2018), with permission, copyright 2018 by the American Physical Society.

With the possible exception of the anomalous ANITA events, none of these experiments has identified an ultra-high-energy neutrino. Current limits on the flux, shown in Figure 8.31, are still above predictions, and a next generation of detectors is likely required for further progress.

8.5.2.3 Lunar Detection

The outer layers of the Moon, being dry, are relatively radio transparent, with absorption lengths of order 50λ (Olhoeft & Strangway 1975). This allows lunar orbiters and Earth-based radio telescopes monitoring the Moon to search for the nanosecond-scale signals caused by UHE neutrinos.

While this presents a huge effective detection volume, the Moon is very distant. The most sensitive experiment to date, performed at Parkes, had a detection threshold of 10^{20} eV (Bray et al. 2015b). Furthermore, because the Moon has no atmosphere, it would be difficult to distinguish a lunar event from that of a cosmic ray. However, it may be possible for the next generation of giant radio telescopes such as the Square Kilometre Array (Bray et al. 2015a) or Five-hundred-meter Aperture Spherical Telescope (FAST) (James et al. 2019).

The original motivation for searching for UHE neutrinos at energies above 10^{20} eV came from the apparent lack of a GZK cutoff in the cosmic-ray spectrum observed by the AGASA experiment (Medina-Tanco 1999). Not only would this imply a large number of neutrino-producing interactions between such cosmic rays and the CMB,

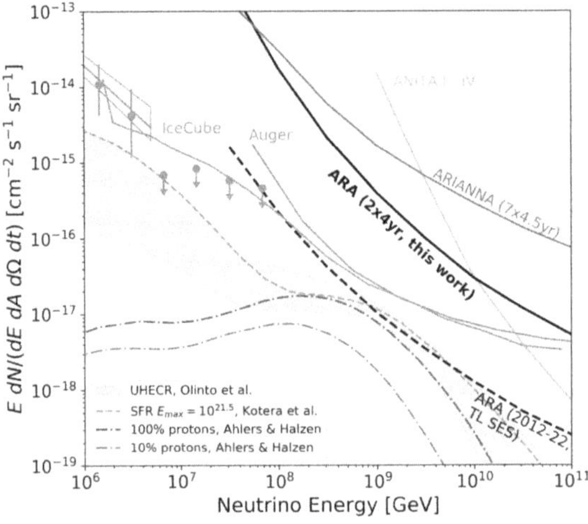

Figure 8.31. Current limits on an ultra-high-energy neutrino flux from the IceCube, Auger, ARA, ANITA, and ARIANNA experiments, compared to predictions from several models of UHE cosmic-ray production and propagation (Allison et al. 2020). At the highest neutrino energies, predictions are less than one per km^2 per year. Image credit: Allison et al. (2020), with permission, copyright 2020 by the American Physical Society.

Figure 8.32. Pierre Auger Observatory. Image courtesy of Steven Saffi.

but theories attempting to explain how such UHE cosmic rays could exist tended to invoke "top-down" mechanisms, such as the decay of as-yet-unknown supermassive particles or topological defects in spacetime (rather than the "bottom-up" acceleration of conventional particles). These models often predicted an associated flux of energetic neutrinos.

Now that the Pierre Auger Observatory (Aab et al. 2017) and Telescope Array (Abbasi et al. 2016; Figure 8.32) have conclusively shown that a cutoff in the cosmic-ray spectrum at GZK energies does exist, the motivation for searching for post-GZK fluxes of UHE neutrinos has largely disappeared. However, as long as the nature of dark matter, the universe's matter–antimatter asymmetry, and a conclusive theory unifying general relativity and quantum mechanics remain open questions, the possibility remains that such extremely rare and energetic neutrinos exist.

References

Aab, A., Abreu, P., Aglietta, M., et al. 2019, JCAP, 2019, 004
Aab, A., Abreu, P., Aglietta, M., et al. 2018a, ApJ, 868, 4
Aab, A., Abreu, P., Aglietta, M., et al. 2018b, ApJ, 853, L29
Aab, A., Abreu, P., Aglietta, M., et al. 2017, JCAP, 2017, 038
Aartsen, M. G., Abbasi, R., Abdou, Y., et al. 2013, PhRvL, 111, 021103
Aartsen, M. G., Abraham, K., Ackermann, M., et al. 2017a, ApJ, 835, 45
Aartsen, M. G., Abraham, K., Ackermann, M., et al. 2015, ApJ, 809, 98
Aartsen, M. G., Ackermann, M., Adams, J., et al. 2019a, PhRvD, 99, 032004
Aartsen, M. G., Ackermann, M., Adams, J., et al. 2020a, PhRvL, 124, 051103
Aartsen, M. G., Ackermann, M., Adams, J., et al. 2020b, PhRvL, 125, 121104
Aartsen, M. G., Ackermann, M., Adams, J., et al. 2017b, JInst, 12, P03012
Aartsen, M. G., Ackermann, M., Adams, J., et al. 2019b, EPJC, 79, 234
Aartsen, M. G., Ackermann, M., Adams, J., et al. 2017c, ApJ, 843, 112
Abbasi, R., Abdou, Y., Abu-Zayyad, T., et al. 2012, APh, 35, 615
Abbasi, R. U., Abe, M., Abu-Zayyad, T., et al. 2016, APh, 80, 131
Abdurashitov, J. N., Gavrin, V. N., Girin, S. V., et al. 1999, PhRvC, 60, 055801
Abe, K., Akutsu, R., Ali, A., et al. 2020, Natur, 580, 339
Abe, K., Haga, Y., Hayato, Y., et al. 2016, PhRvD, 94, 052010
Ackermann, M., Ajello, M., Atwood, W. B., et al. 2015, ApJ, 810, 14
Adrián-Martínez, S., Ageron, M., Aharonian, F., et al. 2016, JPhG, 43, 084001
Adrián-Martínez, S., Al Samarai, I., Albert, A., et al. 2012, APh, 36, 204
Ageron, M., Aguilar, J. A., Al Samarai, I., et al. 2011, NIMPA, 656, 11
Ahmad, Q. R., Allen, R. C., Andersen, T. C., et al. 2002, PhRvL, 89, 011301
Aiello, S., Akrame, S. E., Ameli, F., et al. 2019, APh, 111, 100
Aker, M., Altenmüller, K., Arenz, M., et al. 2019, PhRvL, 123, 221802
Albert, A., André, M., & Anghinolfi, M. 2018, ApJ, 853, L7
Albert, A., André, M., Anghinolfi, M., et al. 2020, ApJ, 892, 92
Albert, A., André, M., Anghinolfi, M., et al. 2017a, PhRvD, 96, 082001
Albert, A., André, M., Anghinolfi, M., et al. 2017b, MNRAS, 469, 906
Albert, A., André, M., Anghinolfi, M., et al. 2017c, PhLB, 769, 249
Albert, A., André, M., Anghinolfi, M., et al. 2018, ApJ, 868, L20
Alexeyev, E. N., Alexeyeva, L. N., Krivosheina, I. V., & Volchenko, V. I. 1988, PhLB, 205, 209
Alimonti, G., Arpesella, C., Back, H., et al. 2009, NIMPA, 600, 568
Allison, P., Archambault, S., Beatty, J. J., et al. 2020, PhRvD, 102, 043021
Altmann, M. F., Mößbauer, R. L., & Oberauer, L. J. N. 2001, RPPh, 64, 97
Álvarez-Muñiz, J., Alves Batista, R., & Balagopal V, A. 2020, SCPMA, 63, 219501
Andres, E., Askebjer, P., Barwick, S. W., et al. 2000, APh, 13, 1
Anker, A., Barwick, S. W., Bernhoff, H., et al. 2019, AdSpR, 64, 2595
Anselmann, P., Hampel, W., Heusser, G., et al. 1992, PhLB, 285, 376
ANTARES Collaboration, Adrián-Martínez, S., Albert, A., et al. 2015, A&A, 576, L8

Antonioli, P., Ghetti, C., Korolkova, E. V., Kudryavtsev, V. A., & Sartorelli, G. 1997, APh, 7, 357
Apel, W. D., Arteaga-Velázquez, J. C., Bekk, K., et al. 2012, APh, 36, 183
Ara Collaboration,, Allison, P., Auffenberg, J., et al. 2012, APh, 35, 457
Askar'yan, G. A. 1962, JETP, 14, 441
Askar'yan, G. A. 1965, JETP, 21, 658
Avrorin, A. D., Avrorin, A. V., Aynutdinov, V. M., et al. 2014, NIMPA, 742, 82
Avrorin, A. D., Avrorin, A. V., Aynutdinov, V. M., et al. 2015, arXiv:1511.02324
Bahcall, J. N., Bahcall, N. A., & Shaviv, G. 1968, PhRvL, 20, 1209
Baumann, D., Beutler, F., Flauger, R., et al. 2019, NatPh, 15, 465
Bechtol, K., Ahlers, M., Di Mauro, M., Ajello, M., & Vandenbroucke, J. 2017, ApJ, 836, 47
Bell, A. R. 1978, MNRAS, 182, 147
Bionta, R. M., Blewitt, G., Bratton, C. B., et al. 1987, PhRvL, 58, 1494
Boger, J., Hahn, R. L., Rowley, J. K., et al. 2000, NIMPA, 449, 172
Borexino Collaboration,, Agostini, M., Altenmüller, K., et al. 2018, Natur, 562, 505
Borexino Collaboration, G., Bellini, J., Benziger, J., et al. 2014, Natur, 512, 383
Bray, J., Alvarez-Muniz, J., Buitink, S., et al. 2015a, in Proc. Advancing Astrophysics with the Square Kilometre Array (AASKA14), 144
Bray, J. D., Ekers, R. D., Roberts, P., et al. 2015b, PhRvD, 91, 063002
Bustamante, M., Baerwald, P., Murase, K., & Winter, W. 2015, NatCo, 6, 6783
Bustamante, M., Baerwald, P., & Winter, W. 2013, ICRC, 33, 795
Cleveland, B. T., Daily, T., & Davis, R. Jr. 1998, ApJ, 496, 505
Cowan, C. L. Jr., Reines, F., Harrison, F. B., Kruse, H. W., & McGuire, A. D. 1956, Sci, 124, 103
Danby, G., Gaillard, J. M., Goulianos, K., et al. 1962, PhRvL, 9, 36
Davis, R., Harmer, D. S., & Hoffman, K. C. 1968, PhRvL, 20, 1205
DONUT Collaboration, Kodama, K., Ushida, N., et al. 2001, PhLB, 504, 218
Fantini, G., Gallo Rosso, A., Zema, V., & Vissani, F. 2018, Introduction to the Formalism of Neutrino Oscillations, 37
Fermi, E. 1949, PhRv, 75, 1169
Follin, B., Knox, L., Millea, M., & Pan, Z. 2015, PhRvL, 115, 091301
Formaggio, J. A., & Zeller, G. P. 2012, RvMP, 84, 1307
Fukuda, S., Fukuda, Y., Hayakawa, T., et al. 2003, NIMPA, 501, 418
Fukuda, Y., Hayakawa, T., Ichihara, E., et al. 1998, PhRvL, 81, 1562
Garrappa, S., Buson, S., Franckowiak, A., et al. 2019, ApJ, 880, 103
Gonzalez-Garcia, M. C., Maltoni, M., & Schwetz, T. 2014, 2014 JHEP, 2014, 52
Gorham, P. W., Allison, P., Baughman, B. M., et al. 2010, PhRvD, 82, 022004
Gorham, P. W., Nam, J., & Romero-Wolf, A. 2016, PhRvL, 117, 071101
Gorham, P. W., Rotter, B., Allison, P., et al. 2018, PhRvL, 121, 161102
Greisen, K. 1966, PhRvL, 16, 748
Grevesse, N., & Sauval, A. J. 1998, SSRv, 85, 161
HESS Collaboration, Abdalla, H., & Aharonian, F. 2018, A&A, 617, A73
Haines, T. J., Bionta, R. M., Blewitt, G., et al. 1986, PhRvL, 57, 1986
Hirata, K., Kajita, T., Koshiba, M., et al. 1987, PhRvL, 58, 1490
Honda, M., Kajita, T., Kasahara, K., Midorikawa, S., & Sanuki, T. 2007, PhRvD, 75, 043006
Huege, T. 2016, PhR, 620, 1
Hümmer, S., Baerwald, P., & Winter, W. 2012, PhRvL, 108, 231101

Hyper-Kamiokande Proto-Collaboration, Abe, K., Abe, K., et al. 2018, arXiv:1805.04163
IceCube Collaboration, Aartsen, M. G., Abraham, K., et al. 2015, arXiv:1510.05223
IceCube Collaboration, 2013, Sci, 342, 1242856
IceCube Collaboration, Aartsen, M. G., Ackermann, M., et al. 2018a, Sci, 361, eaat1378
IceCube Collaboration, Aartsen, M. G., Ackermann, M., et al. 2018b, Sci, 361, 147
IceCube-Gen2 Collaboration, M. G., Aartsen, M., Ackermann, M., et al. 2019, arXiv:1911.06745
James, C. W., Bray, J. D., & Ekers, R. D. 2019, RAA, 19, 019
Janka, H-T. 2017, Neutrino Emission from Supernovae, 1575
Kadler, M., Krau, F., Mannheim, K., et al. 2016, NatPh, 12, 807
Kajita, T., Kearns, E., & Shiozawa, M. 2016, NuPhB, 908, 14
Kamae, T., Karlsson, N., Mizuno, T., Abe, T., & Koi, T. 2006, ApJ, 647, 692
Kato, C., Nagakura, H., Furusawa, S., et al. 2017, ApJ, 848, 48
Keivani, A., Murase, K., Petropoulou, M., et al. 2018, ApJ, 864, 84
Kelner, S. R., Aharonian, F. A., & Bugayov, V. V. 2006, PhRvD, 74, 034018
Krauß, F., Kadler, M., Mannheim, K., et al. 2014, A&A, 566, L7
Kravchenko, I., Cooley, C., Hussain, S., et al. 2006, PhRvD, 73, 082002
Kreisch, C. D., Pisani, A., Carbone, C., et al. 2019, MNRAS, 488, 4413
Kuz'min, V. A. 1966, JETP, 22, 1051
Lehtinen, N. G., Gorham, P. W., Jacobson, A. R., & Roussel-Dupré, R. A. 2004, PhRvD, 69, 013008
Lipari, P., Lusignoli, M., & Meloni, D. 2007, PhRvD, 75, 123005
MacFadyen, A. I., & Woosley, S. E. 1999, ApJ, 524, 262
Marulli, F., Carbone, C., Viel, M., Moscardini, L., & Cimatti, A. 2011, MNRAS, 418, 346
Mathews, G. J., Kusakabe, M., & Kajino, T. 2017, IJMPE, 26, 1741001
Matthiae, G. 2010, NJPh, 12, 075009
Medina-Tanco, G. A. 1999, ApJ, 510, L91
Mikheyev, S. P., & Smirnov, A. Y. 1985, YaFiz, 42, 1441
Mimouni, J. 2015, JPhCS, 593, 012003
Miranda, O. G., & Valle, J. W. F. 2016, NuPhB, 908, 436
Müller, B. 2016, PASA, 33, e048
Munakata, H., Kohyama, Y., & Itoh, N. 1985, ApJ, 296, 197
Murase, K., Oikonomou, F., & Petropoulou, M. 2018, ApJ, 865, 124
Nagashima, Y. (ed) 2014, Beyond the Standard Model of Elementary Particle Physics (Weinheim: Wiley)
Odrzywolek, A., & Heger, A. 2010, AcPPB, 41, 1611
Ohlsson, T. 2016, NuPhB, 908, 1
Olhoeft, G. R., & Strangway, D. W. 1975, E&PSL, 24, 394
Petropoulou, M., Murase, K., Santander, M., et al. 2020, ApJ, 891, 115
Planck Collaboration,, Ade, P. A. R., Aghanim, N., et al. 2016, A&A, 594, A13
Rees, M. J., & Meszaros, P. 1992, MNRAS, 258, 41
Reimann, R. 2019, Galax, 7, 40
Reimer, A., Böttcher, M., & Buson, S. 2019, ApJ, 881, 46
Roberts, A. 1992, RvMP, 64, 259
Schnabel, J. 2015, PhPro, 61, 627
Schröder, F. G. 2017, PrPNP, 93, 1
Smartt, S. J. 2009, ARA&A, 47, 63

Spiering, C. 2019, arXiv:1903.11481
Tanabashi, M., Hagiwara, K., Hikasa, K., et al. 2018, PhRvD, 98, 030001
Urošević, D., Arbutina, B., & Onić, D. 2019, Ap&SS, 364, 185
Vissani, F. 2019, in Solar Neutrinos, ed. M. Meyer, & Z. Zuber, 121
Waxman, E. 1995, ApJ, 452, L1
Wissel, S., Avva, J., Bechtol, K., et al. 2015, ICRC, 34, 1150
Wolfenstein, L. 1979, PhRvD, 20, 2634
Wright, W. P., Gilmer, M. S., Fröhlich, C., & Kneller, J. P. 2017, PhRvD, 96, 103008
Wright, W. P., Nagaraj, G., Kneller, J. P., Scholberg, K., & Seitenzahl, I. R. 2016, PhRvD, 94, 025026
Xue, R., Liu, R-Y., Petropoulou, M., et al. 2019, ApJ, 886, 23
Yoshida, T., Takahashi, K., Umeda, H., & Ishidoshiro, K. 2016, PhRvD, 93, 123012
Zatsepin, G. T., & Kuz'min, V. A. 1966, JETPL, 4, 78

AAS | IOP Astronomy

Multimessenger Astronomy in Practice

Miroslav D Filipović and Nicholas F H Tothill

Chapter 9

Gravitational Wave Astronomy

Paul D Lasky

Gravitational-wave astronomy is a new field of discovery. With the first direct observation of gravitational waves from merging black holes in 2015, this field has expanded significantly with approximately 50 detections in the subsequent five years, including the transformative multimessenger observation of merging neutron stars that was also seen across the electromagnetic spectrum. In this chapter, I briefly review the history of gravitational-wave astronomy, before focusing attention on current observations with ground-based gravitational-wave interferometers. This includes a thorough review of current observations, as well as a look into the potential discovery space of this exciting new field.

9.1 Introduction

A new field of astronomy was born on 2015 September 14 (Abbott et al. 2016e). Almost 100 years in the making, the discovery of gravitational waves is a triumph of modern experimental, theoretical, and computational physics. Approximately 3 $M_{\odot}$ of pure gravitational energy was emitted from the collision of two black holes approximately 400 Mpc from Earth, only to wobble the most precise physics experiment ever built. The distance between mirrors separated by four kilometers changed by about the size of one-thousandth of a proton. And yet, the properties of the merging system such as the progenitor masses and spins can be determined with considerable accuracy.

Since that time, many dozens more detections have been made, including the spectacular observation of a binary neutron star collision in August 2017 (Abbott et al. 2017e). The subsequent observational campaign that saw the event detected across the electromagnetic spectrum signaled the beginning of the new field of multimessenger gravitational-wave astronomy (Abbott et al. 2017f). Among other things, that one event taught us about the speed of gravity (Abbott et al. 2017c), the equation of state of nuclear matter (e.g., Abbott et al. 2017i, 2019m), and the nucleosynthesis of heavy elements, and also gave us a new way of doing cosmology

doi:10.1088/2514-3433/ac2256ch9

independent of the cosmic distance ladder (Abbott et al. 2017g; Hotokezaka et al. 2019).

The future of gravitational-wave astronomy is bright. The two Advanced LIGO detectors (Aasi et al. 2015) in the US and the Advanced Virgo detector (Acernese et al. 2015) in Italy are being joined by the Japanese KAGRA detector (Aso et al. 2013) to create a four-detector network around the globe. All detectors are planned to reach their target design sensitivity by 2022 (Abbott et al. 2020a). The LIGO detectors have guaranteed funding for a significant upgrade around 2025 called A+ (Abbott et al. 2020a), at which point a fifth detector in India is also planned to join the network. Significant planning is also underway for so-called third-generation gravitational-wave detectors, which includes design concepts called the Einstein Telescope (Punturo et al. 2010) and Cosmic Explorer (Abbott et al. 2017h), both slated for operation in the mid-2030s. There is a desire for a third-generation detector in the southern hemisphere, with plans underway for a dedicated high-frequency gravitational-wave detector in Australia to study extremes of nuclear matter (Ackley et al. 2020). Such an instrument would act as a precursor to a fully fledged third-generation Cosmic Explorer South (Bailes et al. 2019)

With their ever-improving sensitivity, ground-based gravitational-wave observatories will eventually be sensitive enough to detect every binary black hole collision in the observable universe, as well as a significant fraction of binary neutron star mergers out to redshifts of a few (Abbott et al. 2017h; Hall & Evans 2019), beyond which one expects very few binary neutron star mergers, if any. But there are numerous other potential sources of interest, including gravitational waves from core-collapse supernovae, “mountains” on rotating neutron stars, and more exotic ones like cosmic strings and primordial gravitational waves from inflation.

Ground-based gravitational-wave experiments operate in the audio frequency band, from tens of Hertz to a few kilohertz. But as optical light is just a tiny fraction of the full electromagnetic spectrum that spans from low-frequency radio waves to very-high-energy gamma rays, so too is there a broad gravitational-wave spectrum, replete with new and fascinating physics waiting to be uncovered. In the millihertz regime, there are space-based gravitational-wave detectors such as the Laser Interferometer Space-Based Antenna (LISA; Armano et al. 2016) and the proposed Chinese detectors TianQin (Luo et al. 2016) and Taiji (e.g., see Xin 2018). The detectors will be so sensitive, they will actually be noise limited by gravitational waves from the foreground of orbiting white dwarf binaries. These missions will also detect the loudest white dwarf binaries, as well as inspiraling binary black hole and double neutron star systems.

Galaxy-sized gravitational-wave detectors operate in the nanohertz regime, attempting to detect gravitational waves from the most massive binary black hole systems in the universe, as well as signatures of the inflationary epoch of the universe, a tiny fraction of a second after the big bang. This is done using precise timing of radio pulsars throughout the Galaxy and correlating the signal, looking for indications of spacetime perturbations between Earth and the pulsars. Three consortia operate individual pulsar timing arrays: the Australian Parkes Pulsar Timing Array (PPTA; Manchester et al. 2013), the European Pulsar Timing Array

(EPTA; Kramer & Champion 2013), and the North American Nanohertz Observatory for Gravitational Waves (NANOGrav; McLaughlin 2013), with other countries looking to join the effort using instruments across the globe, including the SKA. Efforts to combine the data from all three collaborations are ongoing as part of the International Pulsar Timing Array (IPTA; Verbiest et al. 2016).

Finally, gravitational waves produced during the inflationary epoch are also expected to leave an imprint on the cosmic microwave background through so-called B modes. These are detectable with precision polarization measurements of the cosmic microwave background.

This chapter focuses on reviewing gravitational-wave astronomy with ground-based observatories. To date, those are the only gravitational-wave observatories with positive, direct detections of gravitational waves. In the next two sections, we look at a brief history of gravitational-wave astronomy, beginning with the first indirect detection of gravitational waves made by tracking the orbital decay of a binary pulsar system, and then discussing the first direct detection with the LIGO interferometers of two merging black holes. We then move to the bulk of the review, providing an introduction to gravitational-wave interferometers and how they work in Section 9.5; and a quick look at the future of ground-based instruments in Section 9.5.1. We then study different types of gravitational-wave sources, focusing for a large portion on the mergers of compact objects in Section 9.6, looking at detection methods and astrophysical parameter estimation, before concentrating on the astrophysics of various sources, in particular binary black holes (Section 9.7.1), binary neutron stars (Section 9.7.2), and neutron star–black hole binaries (Section 9.7.3). In the remaining sections, we look at alternative sources of gravitational waves for ground-based interferometers, loosely divided into unmodeled, so-called "burst" sources that include supernovae, magnetar flares, pulsar glitches, and cosmic strings (Section 9.8); "continuous-wave" sources primarily from isolated neutron stars; and contributions to the background hum of gravitational waves known as the stochastic gravitational-wave background (Section 9.10).

9.2 Indirect Evidence: The Hulse–Taylor Binary

Despite their existence being predicted in 1916 (Einstein 1916), it took 65 years to indirectly confirm the existence of gravitational waves (Weisberg et al. 1981; Taylor & Weisberg 1982), and 99 years before the first direct detection (Abbott et al. 2016e; also see Book 1, Section 1.3.8 and Chapter 8).

The first confirmation of the existence of gravitational waves was precipitated by the discovery of a highly relativistic double neutron star system PSR 1913+16 Hulse & Taylor 1975; a discovery that was subsequently awarded the 1993 Nobel prize for physics.[1]

As the binary system orbits, the acceleration of the masses cause the emission of gravitational waves, thereby carrying energy away from the system and causing the

[1] https://www.nobelprize.org/prizes/physics/1993/summary/

decay of the orbit. The system is in the relatively weak-field regime, implying a linear post-Newtonian expansion of Einstein's field equation is sufficient to describe the orbital decay. Indeed, the rate of period decrease $\dot{P}_b$ is (Peters & Mathews 1963; Peters 1964)

$$\dot{P}_b = -\frac{192\pi}{5c^5}\left(\frac{2\pi G}{P_b}\right)^{5/3}(1 - e^2)^{-7/2}\left(1 + \frac{73}{24}e^2 + \frac{37}{96}e^4\right)\frac{m_1 m_2}{(m_1 + m_2)^{1/3}}, \tag{9.1}$$

where P_b is the orbital period, e the orbital eccentricity, and $m_{1,\,2}$ the stellar masses. Figure 9.1 from Weisberg & Huang (2016) shows the orbital decay of the Hulse–Taylor binary from observations spanning almost four decades. The observational uncertainty on each data point is so small it is not visible in this figure. The general relativity prediction of the orbital decay shown in Figure 9.1 as the decaying black curve is beautifully consistent with the observational data. This was the first concrete observational evidence for the existence of gravitational waves.

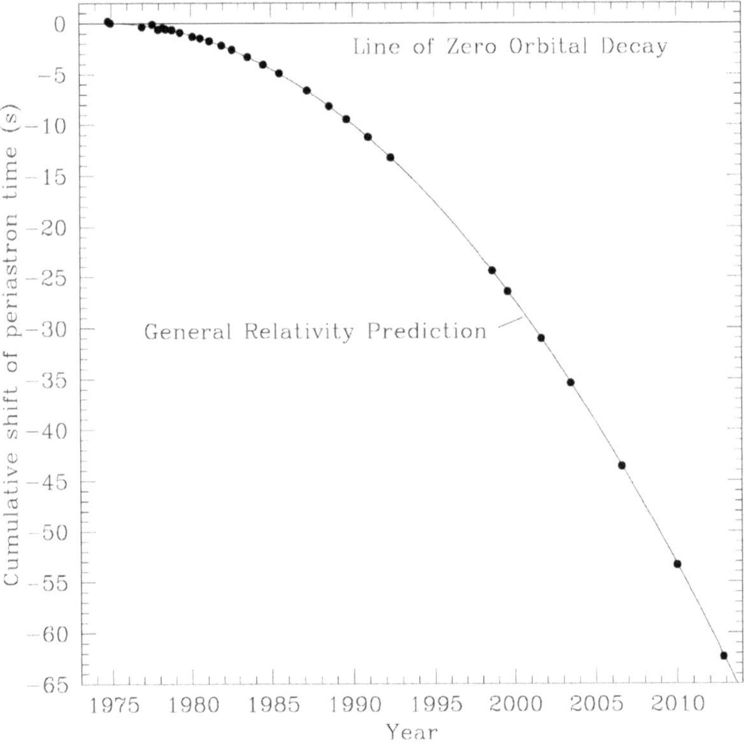

Figure 9.1. Orbital decay of the relativistic Hulse–Taylor double pulsar. The prediction from general relativity that includes orbital angular momentum loss from the emission of gravitational waves is beautifully consistent with the observational data. This was the first observational evidence for the existence of gravitational waves. Image credit: Weisberg & Huang (2016).

9.3 Direct Detection: GW 150914

The initial phase of the LIGO instruments operated six "Science Runs" from 2002 to 2010, iteratively increasing the sensitivity of the instruments, but ultimately not making any detections. Substantial improvements were made from 2010 until 2015, including a factor of about 3 improvement in sensitivity, corresponding to an approximate factor of 30 improvement in sensitive volume—see Section 9.4. In September 2015, Advanced LIGO's first "Observing Run" (O1) began. In fact, on September 14, days before the official start of O1, with the detectors already locked and ready to go, the first ever gravitational-wave measurement was made. Gravitational waves were detected from the collision of two black holes 440^{+160}_{-180} Mpc away, with masses $35^{+5}_{-3}M_{\odot}$ and $30^{+3}_{-4}M_{\odot}$ (Abbott et al. 2016d, 2016e).

The gravitational-wave signal from this event had a matched-filter signal-to-noise ratio (see Section 9.4 for a definition) of 24, corresponding to a false-alarm rate of significantly less than one event per 200,000 years (Abbott et al. 2016e). In other words, one would need to run the two LIGO detectors in the same condition as around the time of the event for more than 200,000 years to get a single noise realization with a comparable matched-filter signal-to-noise ratio. Figure 9.2 shows the filtered and bandpassed gravitational-wave strain of the GW 150914 signal from the first discovery paper (Abbott et al. 2016e). The top-left and top-right columns show the gravitational-wave strain in the Hanford (H1) and Livingston (L1) detectors, respectively. Also shown in the top-right panel in red is the time-shifted and inverted Hanford data that accounts for the gravitational-wave travel time between the detectors and their different orientations. For technical details of the definitions, see Section 9.4. The top-right panel shows that the two data streams have remarkably similar features from approximately 0.32 s to 0.43 s, which include both increasing frequency and amplitude evolution, all the while maintaining similar phase evolution.

The second row shows the gravitational waveforms calculated from a numerical relativity simulation of a binary black hole collision with parameters fit from the observations. These waveforms have also been bandpassed to mimic what one would expect to see in each detector. The third panel shows the residual in each detector, calculated simply as the gravitational-wave strain from the top row subtracted from the strain shown in the second row. This noise is consistent with Gaussian noise. Finally, the bottom row shows the instantaneous gravitational-wave frequency evolution of the signal, showing the characteristic "chirp" signal one expects from colliding black holes.

This first detection of gravitational waves was a momentous occasion, celebrated in part with the 2017 Nobel Prize in Physics being awarded to Rai Weiss, Barry Barish, and Kip Thorne, all awarded "for decisive contributions to the LIGO detector and the observation of gravitational waves."[2] But this first detection was not just a singular discovery; it opened a new observational window on the universe. Since that first discovery, the LIGO gravitational-wave detectors have been joined

[2] https://www.nobelprize.org/prizes/physics/1993/summary/

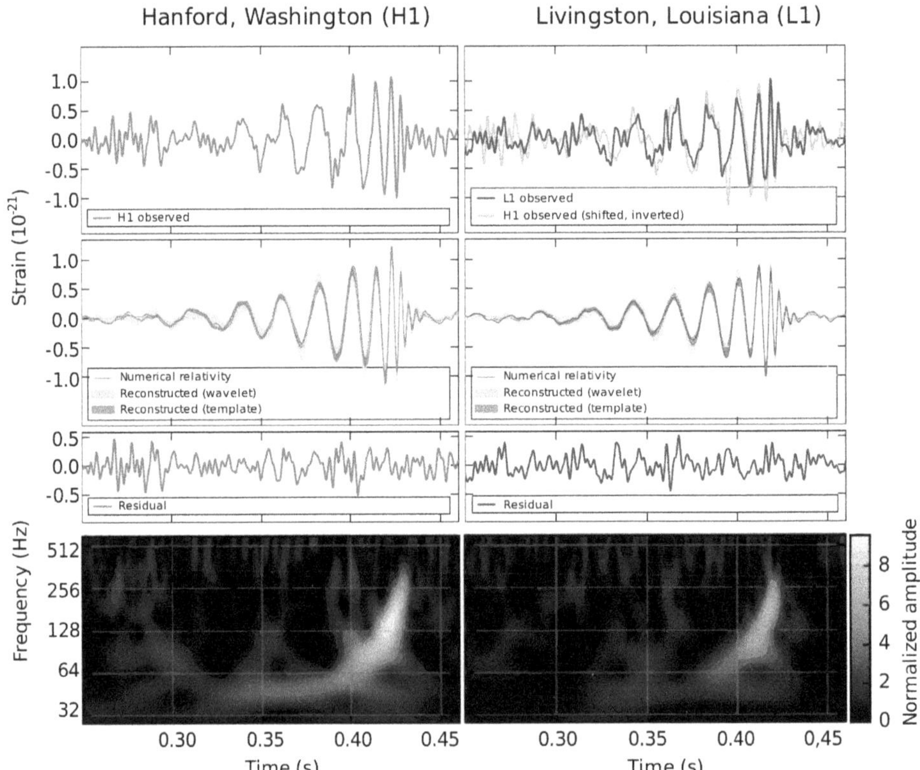

Figure 9.2. Gravitational-wave strain from the first detected binary black hole merger GW 150914 from Abbott et al. (2016e), with permission, copyright 2016 by the American Physical Society. The left and right columns show data from the Hanford and Livingston detectors, respectively. Top panels: gravitational-wave strain in each detector. The top-right panel also shows in red the Hanford data that have been shifted and inverted to account for the gravitational-wave travel time between the detectors and their different orientations. The increasing frequency and amplitude of the signals are characteristic of two compact objects orbiting and coming closer to one another. The relatively sudden decrease in gravitational-wave amplitude indicates the merger of the two black holes to become one. Second row: numerical-relativity-generated gravitational waveforms. The simulations use the best-fit parameters of the colliding system to determine the most likely waveform. Third row: residual signal from subtracting the best-fit numerical relativity waveform from the data; these residuals are consistent with Gaussian noise. Bottom row: time–frequency maps showing the characteristic "chirp" of the increasing amplitude and frequency of the signal.

by Virgo and have collectively detected more than 50 merging systems at the time of writing. This includes the momentous multimessenger observation of colliding neutron stars GW 170817 (see Section 9.7.2 and Abbott et al. 2017e) and potentially some neutron star–black hole mergers (Abbott et al. 2020g, 2021a). The future of gravitational-wave astronomy is indeed bright.

9.4 Ground-based Gravitational-wave Observatories

The first gravitational-wave detectors were resonant bar detectors, also known as Weber bars, which are only sensitive in a narrow frequency range. Controversially,

Weber (1967, 1968, 1969) declared that two of his bar detectors separated by approximately 1000 km had observed multiple coincident signals that "...are neither accidental nor due to seismic or electromagnetic effects," concluding that this "...is good evidence that gravitational radiation has been discovered" (Weber 1969). These claims are commonly disputed, and no subsequent bar detectors have reliably detected gravitational-wave signals. In Australia, a niobium resonant bar detector was built and operated for some time at the University of Western Australia, achieving an incredible strain sensitivity of 7×10^{-19} at close to 1000 Hz (Tobar et al. 1995).

Gravitational waves are quadrupolar, implying they simultaneously stretch and squeeze space in orthogonal directions (see Book 1, Chapter 8). Interferometers therefore provide a natural experimental setting (Gertsenshteĭn & Pustovoĭt 1963; Moss et al. 1971; Weiss 1972), whereby the differential length change or, more correctly, time of flight of photons down orthogonal (or nearly orthogonal) arms can be measured.

In this section, I concentrate only on gravitational-wave interferometers, ignoring for the remainder of this section resonant bar detectors. I begin in Section 9.5 with a discussion of how interferometers work, including the most salient parts of the instruments that make them work. Ultimately, this is not an exhaustive guide to gravitational-wave instrumentation; for such reviews, see Saulson (1994). In Section 9.6, I review basics of gravitational waves from compact binary coalescences, which is a fancy name for mergers that include black holes and/or neutron stars. This includes subsections on detection algorithms and methods used by the LIGO–Virgo–KAGRA collaborations in Section 9.6.3, and astrophysical parameter inference in Section 9.6.4, ultimately leading to a detailed discussion of binary black holes, binary neutron star mergers, and putative neutron star–black hole mergers so far detected in the first few LIGO–Virgo observation runs in Sections 9.7.1, 9.7.2, and 9.7.3, respectively.

While the aforementioned compact binary gravitational-wave sources have all been detected with varying degrees of probability, there are a number of other potential sources that the LIGO–Virgo-KAGRA collaborations actively search for. These are reviewed in the subsequent sections, including various types of unmodeled burst searches such as supernovae, magnetar flares, pulsar glitches, and cosmic strings in Sections 9.8.1, 9.8.2, 9.8.3, and 9.8.4, respectively. Long-lived, nearly monochromatic sources are likely produced from "mountains" on rotating neutron stars; in Section 9.9, I discuss possible emission mechanisms and detection methods for these sources.

Each of the aforementioned sources contributes to the stochastic gravitational-wave background, which is the linear superposition of all individually unresolved gravitational-wave sources throughout the universe. The LIGO–Virgo–KAGRA collaboration actively searches for the stochastic background from various sources including compact binaries, isolated neutron stars, and cosmological sources such as inflationary relics. I review possible sources for the stochastic background in Section 9.10, including derived source amplitudes and various detection methods.

9.5 Introduction to Gravitational-wave Interferometers

While this review is focused on the observational aspects of gravitational-wave astronomy, it is pertinent to provide an overview of the salient features of the experiment that make it possible, including the primary noise sources that affect our ability to detect gravitational-wave signals at different frequencies. This portion of the review is far from complete; for more comprehensive reviews of gravitational-wave interferometer instrumentation, the reader is referred to Saulson (1994), Blair et al. (2012), and Vajente et al. (2019).

The LIGO, Virgo, and KAGRA detectors are modified Michelson interferometers with orthogonal Fabry–Perot cavities; Figure 9.3 shows a schematic of the two LIGO interferometers. Note that the 20 W input power and 100 kW circulating power were the relevant numbers for the initial detection of GW 150914; they have each increased with subsequent observing runs. While the interferometers are significantly more complicated than that shown in Figure 9.3, this schematic is sufficient for the purposes of understanding how the instruments work.

The idea is relatively simple. A coherent light source enters both arms of the interferometer and bounces between the two freely falling test masses in the respective cavities. When a gravitational wave passes through the interferometer, it effectively changes the length of the arms; the quadrupolar nature of gravitational waves implies the relative length between the orthogonal arms changes. When the light source recombines at the beam splitter and passes through to the

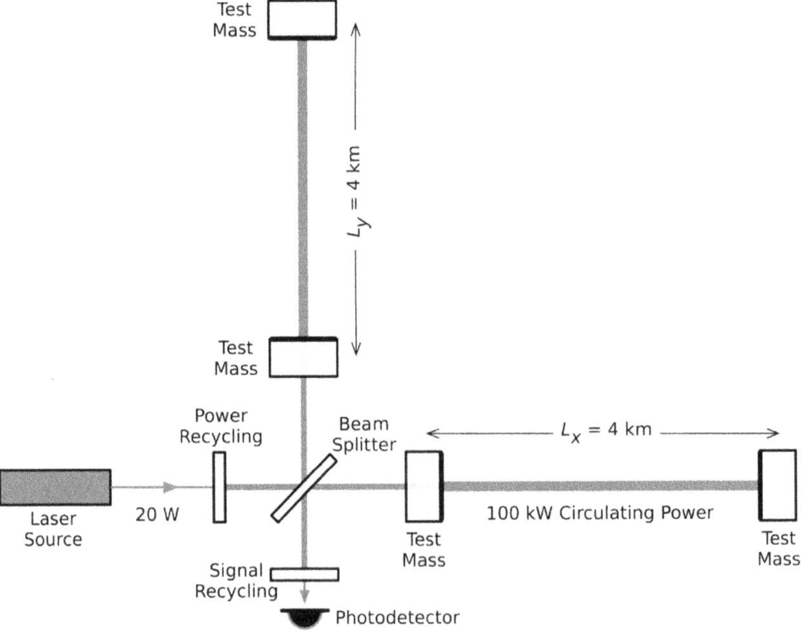

Figure 9.3. Schematic overview of the LIGO interferometers adapted from Abbott et al. (2016e), with permission, copyright 2016 by the American Physical Society. The 20 W input power and 100 kW circulating power have increased since the initial detection of GW 150914 in 2015.

photodetector, relative phase differences between the light from each cavity are measured. This phase change therefore measures the difference in arm length $\Delta L(t) \equiv h(t)L$, where L is the base length of each arm and $h(t)$ is the gravitational-wave strain amplitude projected onto the detector.

One is often asked how light waves can be used to measure gravitational waves when the photons should also be stretched by the gravitational wave. At first blush, this is a worthy critique, but ultimately it comes from thinking about interferometers in the wrong way. The short way to resolve this question is to note that it is the change in light-travel time that is being measured rather than the change in length of the interferometer arms. In this sense, the stretching of the photons due to the gravitational waves themselves becomes irrelevant. A slightly longer answer is that, for gravitational-wave frequencies of less than about 10 kHz, the light-travel time is much shorter than the gravitational-wave period, implying the gravitational-wave strain is effectively constant as the light travels down the kilometer-length arms. For an even more detailed and nuanced answer, including all relevant calculations, the reader is referred to Saulson (1997).

The basic Michelson interferometer setup is modified in a number of ways to mitigate various sources of noise and create an experiment sensitive enough to measure gravitational waves. Many techniques are used to maximize the conversion of gravitational-wave strain to changes in the optical phase at the photodetector. For example, each arm forms a resonant optical cavity, whereby photons bounce between the two test masses, effectively increasing the effect of the gravitational wave on the phase. The power-recycling system at the input acts to significantly amplify the circulating power in the arms, and the quadruple-pendulum systems supported by seismic isolation platforms provide more than 10 orders of magnitude of isolation from ground motion (see Abbott et al. 2016e and references therein).

Ultimately, all sources of noise create a noise budget for the detectors, an example of which is shown in Figure 9.4, which shows the interferometers displacement noise as a function of frequency. The various colored curves show individual noise sources, the thick black curve shows the total noise (the sum of the individual noise sources), and the thin gray curve shows an instance of actual measured noise in the LIGO Livingston interferometer, measured during the second observing run, O2.

If one completely understood all sources of noise in the interferometer, the black and gray curves would lie on top of one another. That they do not at low frequencies is a concern and one that is the topic of ongoing research. The dominant sources of noise at low frequency all constitute various forms of seismic and Newtonian noise. Mitigating these sources of noises and pushing the low-frequency limit of detectors down to and below $\approx$10 Hz is important for significantly improving low-latency detection of gravitational-wave signals with a view to catch, for example, any electromagnetic precursor or prompt emission from binary neutron stars mergers.

The above argument can be seen by looking at how long a typical signal spends in a given frequency band. The frequency evolution of a compact binary coalescence is

$$\frac{df}{dt} = \frac{96\pi^{8/3}G^{5/3}}{5c^5}\mathcal{M}^{5/3}f^{11/3}, \tag{9.2}$$

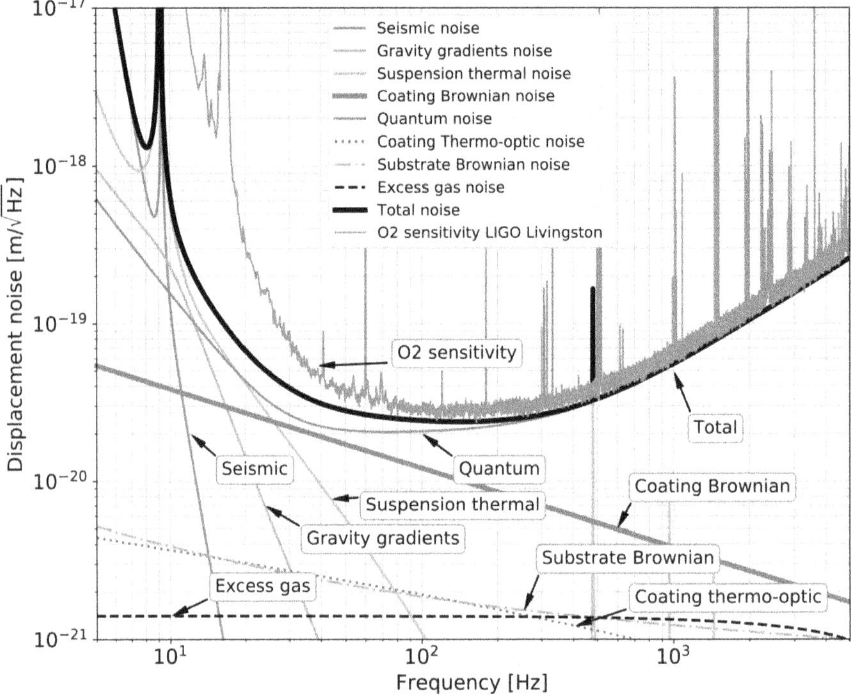

Figure 9.4. Underlying sources of noise for the LIGO Livingston detector during the second observing run. The various colors show the individual contributions to the total noise budget from each noise source, including seismic noise (blue) that dominates at low frequencies and quantum noise (purple) that dominates at higher frequencies. The total noise budget (thick black curve) is simply the sum of all underlying noise sources. The thin gray curve shows an actual noise realization as measured in the interferometer. Image credit: Vajente et al. (2019) copyright 2019, with permission from Elsevier.

where

$$\mathcal{M} = \frac{(m_1 m_2)^{3/5}}{(m_1 + m_2)^{1/5}}, \tag{9.3}$$

is known as the chirp mass, where $m_{1,\,2}$ are the masses of the individual systems in the binary. Integrating Equation (9.2), one can see that a typical binary neutron star system with $m_1 = m_2 = 1.4M_\odot$ spends approximately 55 s between 30 Hz and 1000 Hz, but almost 1000 s between 10 Hz and 30 Hz. That same binary spends more than five days between 1 Hz and 10 Hz!

Although the payoff is obviously significant scientifically, the difficulties with mitigating low-frequency noise are immense (as illustrated by the difference between the thick black and thin gray curves in Figure 9.4) and expensive. The design goal for second-generation detectors such as Advanced LIGO pushes this low-frequency noise wall down to about 10 Hz, while proposed third-generation detectors such as the Einstein Telescope and Cosmic Explorer aim to push sensitivity down to as low as 1 Hz (see Section 9.5.1). There are at least three primary science drivers for pushing to such low frequencies. First, the significant gains one can hope to achieve in the time before detection can trigger electromagnetic telescopes, with an ultimate goal of negative latency (e.g., James et al. 2019). Second, high-mass binary black hole systems merge at lower frequencies, which implies that observing gravitational waves $\lesssim$10 Hz will allow us to probe the relatively high-mass end of stellar-mass binary black holes, a point I discuss in more detail in Section 9.7.1. Third, with these detectors probing mergers at ever greater distances, cosmological redshift can act to move signals to significantly lower frequencies. Increasing the lower-frequency bandwidth of gravitational-wave interferometers will allow us to understand the earliest mergers in the universe.

At the high-frequency end of gravitational-wave interferometers $\gtrsim$50 Hz, sensitivity is almost completely limited by quantum noise. This constitutes two noise sources: photon shot noise and radiation pressure noise, the former being quantum fluctuations in the number of photons providing a random force on the test masses. In principle, shot noise can be mitigated by simply increasing the power in the optical cavities; however, this has the potential to introduce other sources of noise such as parametric instability, which can act to unlock the detector. An alternative approach is to use quantum squeezing, which takes detectors beyond their quantum limits by squeezing the quantum uncertainty on the photons into one of their two quantum conjugate states. Quantum squeezing has been demonstrated with remarkable effect in gravitational-wave interferometers (e.g., Aasi et al. 2013), with a record 5.7 dB of squeezing decreasing the quantum noise by a factor of 2 in the GEO600 interferometer (GEO600 2018); see Figure 9.5.

The smaller the mass of a compact binary merger, the higher the merger frequency. To that end, the merger frequencies of the binary neutron star mergers GW 170817 (Abbott et al. 2017e) and GW 190425 (Abbott et al. 2020b) were too high to be detected by Advanced LIGO and Advanced Virgo. For example, the signal from GW 170817 was lost at approximately 400 Hz, but the merger occurred at $\gtrsim$1000 Hz. Moreover, if a massive neutron star survived the collision, even for a

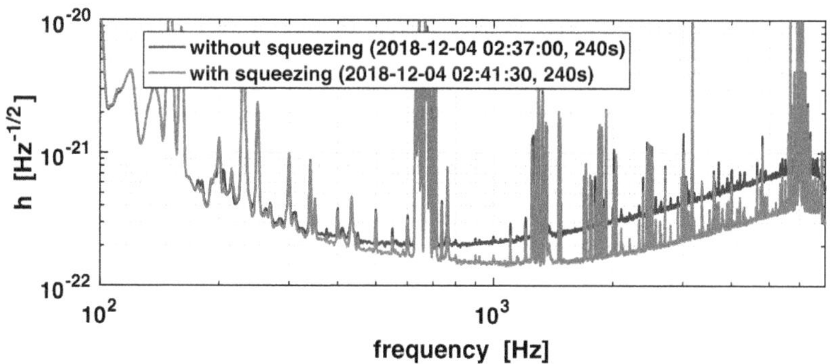

Figure 9.5. The effect of quantum squeezing on gravitational-wave sensitivity. The red curve shows the gravitational-wave strain sensitivity in the GEO600 interferometer without quantum squeezing, and the blue curve shows the effect of 5.7 dB of squeezing, improving the sensitivity at high frequencies by approximately a factor of 2. Image credit: Max Planck Institute for Gravitational Physics and GEO600 (2018).

fraction of a second as one may expect (see Section 9.7.2), then the resultant gravitational-wave frequency of the postmerger remnant would be in the kilohertz regime—far too high for current detectors. This, as well as other potential high-frequency signals from supernovae, isolated neutron stars, and more exotic objects, motivates the need for improved high-frequency sensitivity.

9.5.1 Future Gravitational-wave Interferometers

The Advanced LIGO detectors are due to reach design sensitivity in the next few years, with an already-funded upgrade to the so-called A+ detector due in the middle of the 2020s. Figure 9.6 shows design-sensitivity curves for the current and future generation of ground-based gravitational-wave interferometers. In black, we show the predictions for the design sensitivity of current, second-generation gravitational-wave interferometers: Advanced LIGO (solid black curve; Aasi et al. 2015), Advanced Virgo (dashed black; Acernese et al. 2015), and the Japanese KAGRA detector (dotted–dashed black; Aso et al. 2013). The LIGO-India detector is slated to have the same design specifications as the other two LIGO instruments.

The blue sensitivity curves are the so-called 2.5-generation detectors, which are both scientific discovery machines in their own right but also technology developers for third-generation instruments. The solid blue curve shows A+'s design-sensitivity curve (Miller et al. 2015). The dashed blue curve shows a preliminary design for a dedicated high-frequency detector, NEMO (Ackley et al. 2020) that sacrifices low-frequency sensitivity in a bid to reduce cost, but is competitive with full third-generation detectors

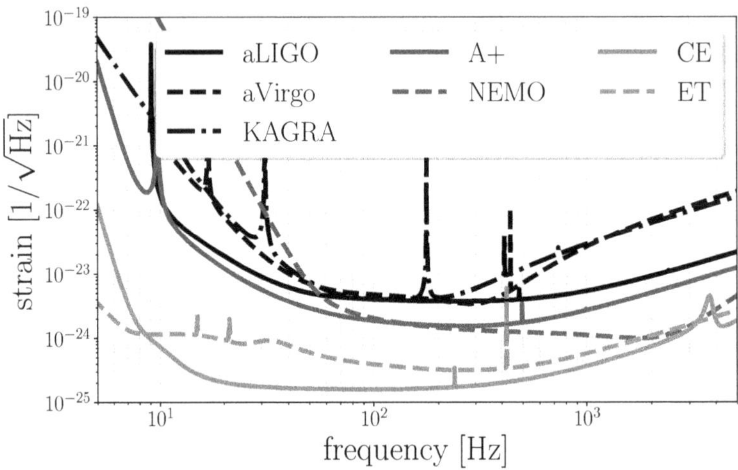

Figure 9.6. Projected current and future-generation gravitational-wave interferometer sensitivity curves. In black are the second-generation detectors at design sensitivity, which includes Advanced LIGO (solid black), aVirgo (dashed black) and KAGRA (dotted–dashed black). The proposed 2.5-generation detectors are in blue, which include the already-funded A+ (solid blue) and the proposed high-frequency detector NEMO (dashed blue). Proposed third-generation instruments include Cosmic Explorer (solid red) and Einstein Telescope (dashed red).

in the kilohertz regime where one can potentially learn about the nuclear equation of state from binary neutron star mergers—see Section 9.7.2.

The red curves show two potential designs for third-generation detectors. The dashed red curve shows the design sensitivity for the proposed European interferometer, the Einstein Telescope (ET; Punturo et al. 2010), which targets a broadband factor of 10 increase in sensitivity over second-generation interferometers, while simultaneously pushing sensitivity at the low-frequency end from $\approx$10 Hz to $\approx$1 Hz. The current design concept envisages an underground facility with three nested detectors in a triangular shape, with 10 km arms.

The US counterpart to the Einstein Telescope is Cosmic Explorer (CE; solid red curve, Abbott et al. 2017h), which also targets a factor of 10 improvement over second-generation detectors. Cosmic Explorer is a more conventional interferometer, in the sense that it is a traditional "L"-shaped observatory, albeit with 40 km arms.

It is worth noting that other design concepts exist, such as a design that includes broadband sensitivity similar to that of Cosmic Explorer and the Einstein Telescope, but with improved sensitivity at high frequencies (Martynov et al. 2019). The original paper includes designs that simply modify the 4 km LIGO interferometer to achieve high-frequency sensitivity, all the way up to new designs with 20 km and 40 km vacuum tunnels. As with the NEMO design study (Ackley et al. 2020), the primary science goal of achieving additional high-frequency sensitivity in the kilohertz regime is to target the equation of state of nuclear matter from binary neutron star mergers.

This review does not go significantly into the science case for full third-generation gravitational-wave interferometers, although it is occasionally and sporadically mentioned. Nor does this review go into detail about required technological advancements to achieve the design sensitivity outlined in Figure 9.6. The Gravitational Wave International Committee (GWIC) commissioned a series of subcommittees to develop reports on the science case, research and development requirements, and computing challenges, as well as governance recommendations and community engagement requirements. These were published in 2019 and can be accessed and read in all their gory detail.[3]

9.6 Compact Binary Coalescences

The LIGO and Virgo collaborations have detected more than 50 gravitational-wave signals since 2015 (Abbott et al. 2021a), each of which has involved the merger of two compact objects, either neutron stars or black holes. These events are collectively described as compact binary coalescences and form the topic of this section.

Compact binary coalescences can typically be described in three distinct phases: the inspiral, merger, and ringdown. An example gravitational waveform is shown as the red curve in Figure 9.7. That waveform is calculated using a numerical relativity

[3] https://gwic.ligo.org/3Gsubcomm/documents.html

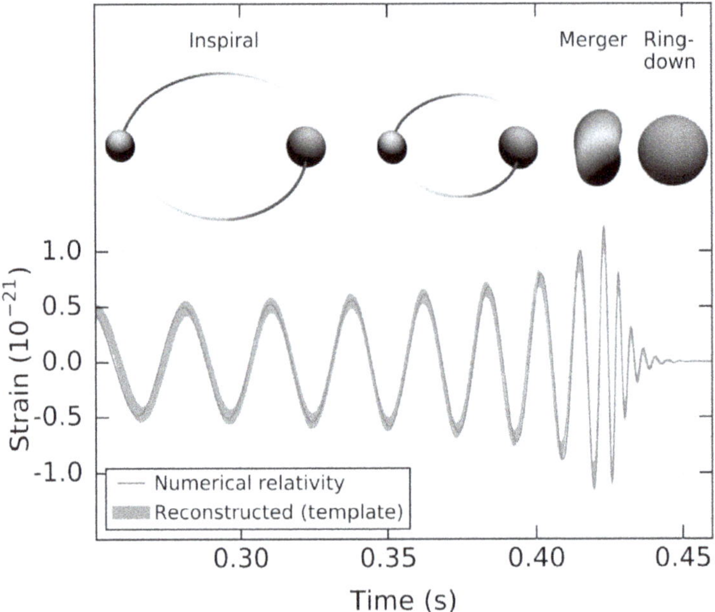

Figure 9.7. Gravitational-wave strain as a function of time for a compact binary coalescence. The red curve shows the strain computed with a numerical relativity simulation for parameters consistent with the first gravitational-wave detection, GW 150914. Highlighted on the top of the plot are the three different phases of a compact binary coalescence—the inspiral, merger, and ringdown phases. The inset images show numerical relativity reconstructions of the black hole apparent horizons throughout the coalescence. Image credit: Abbott et al. (2016e), with permission, copyright 2016 by the American Physical Society.

simulation of a black hole binary coalescence. The parameters of the binary—e.g., the two progenitor masses, the inclination angle, distance, etc. (for definitions of these parameters, see Section 9.7.1)—match those of the parameters for the first detected gravitational-wave signal, GW 150914 (Abbott et al. 2016e). The inset images above the red trace show the black hole apparent horizons during the various stages of the system's evolution, as calculated in the numerical relativity simulation.

The compact binary coalescence shown in Figure 9.7 is that for a binary black hole merger. Compact binary mergers involving one or more neutron stars are significantly more complicated. During the late phase of the inspiral, the waveform changes due to the effect of tidal deformations of the neutron star(s). The ringdown can also be significantly more complicated, particularly in the case of a binary neutron star merger where the remnant is not necessarily a black hole, but may be a short- or long-lived neutron star—we return to these facets of binary neutron star waveforms and physics in Section 9.7.2.

At this point it is worth providing some back-of-the-envelope calculations for the gravitational-wave strain and its evolution. To begin, we can approximate the merger frequency of a system using Keplerian dynamics (of course, one requires general relativity to calculate this properly, but Newtonian gravity gives qualitatively good-enough estimates to understand systems observed and their relevant gravitational-wave frequencies). Consider an equal-mass binary black hole system with total mass $M \equiv m_1 + m_2$. Approximating the horizons of each black hole by their Schwarzschild horizon $R_i = 2Gm_i/c^2$, Keplerian dynamics tells us the gravitational-wave frequency when the two horizons touch is

$$f_{\rm gw} \approx 130\ {\rm Hz}\left(\frac{M}{60M_\odot}\right)^{-1}, \tag{9.4}$$

where $f_{\rm gw} = P/2$, and P is the binary's orbital period. This is roughly consistent with the maximum gravitational-wave frequency observed for the first binary black hole merger, GW 150914 (Abbott et al. 2016e, 2017b). Moreover, such an estimate tells us that the maximum gravitational-wave frequency becomes $f_{\rm gw} \gtrsim 1$ kHz for systems $M \lesssim 10M_\odot$.

The gravitational-wave strain amplitude can be calculated using the quadrupole formula (for a pedagogical derivation, see Flanagan & Hughes 2005). An equal-mass binary neutron star merger with typical progenitor masses $m_1 = m_2 = 1.4M_\odot$ has gravitational-wave strain

$$h \approx 5 \times 10^{-22}\left(\frac{M}{2.8M_\odot}\right)^{5/3}\left(\frac{f_{\rm gw}}{1kHz}\right)^{2/3}\left(\frac{d_L}{40Mpc}\right)^{-1}, \tag{9.5}$$

where d_L is the luminosity distance to the source. Note that the choice of distance normalization here matches that of the distance to the first detected binary neutron star merger, GW 170817 (see Abbott et al. 2017e and Section 9.7.2). Alternatively, we can scale these normalizations to observed binary black hole mergers such as GW 150914 (Abbott et al. 2016e), which gives

$$h \approx 7 \times 10^{-22}\left(\frac{M}{60M_\odot}\right)^{5/3}\left(\frac{f_{\rm gw}}{100\ {\rm Hz}}\right)^{2/3}\left(\frac{d_L}{1\ {\rm Gpc}}\right)^{-1}. \tag{9.6}$$

Note that while GW 150914 was an $M \approx 30M_\odot + 30M_\odot$ system, it was only at a distance of about 500 Mpc, implying the strain in the detector was larger than the above normalization suggests and roughly consistent with the strain shown in Figure 9.7.

9.6.1 Detection Significance

Without context, it is difficult to gauge whether the above gravitational-wave strain estimates render a system detectable by modern interferometers. In principle, we can compare the strain amplitudes with typical sensitivity curves (e.g., Figures 9.4 and 9.6). However, life is more complicated because the signal lasts in the band for a given time, enabling signals that last longer in the band to be "dug out" from the noise. To this end, we define an optimal matched-filter signal-to-noise ratio $\rho_{\rm opt}$ as the noise-weighted inner product

$$\rho_{\rm opt} = \sqrt{\langle h, h\rangle}, \tag{9.7}$$

where $h(t)$ is the strain time series, and the inner product is defined as

$$\langle a, b\rangle = 4Re\int_0^{\infty} df \frac{\tilde{a}(f)\tilde{b}^{\star}(f)}{S_h(f)}. \tag{9.8}$$

Here, $S_h(f)$ is the noise power spectral density of the detector, $\tilde{a}(f)$ is the Fourier transform of $a(t)$ and a $\star$ denotes the complex conjugate. With multiple detectors, a network signal-to-noise ratio is defined as the quadrature sum of signal-to-noise ratios in each detector:

$$\rho_{\rm net}^2 = \sum_{i=1}^{N_{\rm IFO}} \rho_i^2, \tag{9.9}$$

where $N_{\rm IFO}$ is the number of interferometers.

The optimal matched filter assumes one has a waveform model that exactly matches that of the gravitational-wave incident on the detectors. In reality, one performs matched-filter searches and parameter inference with templates that are not exact. To that end, we define a matched-filter signal-to-noise ratio ρ as

$$\rho = \frac{\langle h, u\rangle}{\sqrt{\langle u, u\rangle}}, \tag{9.10}$$

where $u(t)$ is the template. Clearly, when the template matches exactly the time series $u(t) = h(t)$, the matched-filter signal-to-noise defined in the previous equation reduces to the optimal matched-filter signal-to-noise ratio of Equation (9.7), i.e., $\rho = \rho_{\rm opt}$. We discuss waveform templates in more detail in the following section.

The loudest gravitational-wave signal measured to date in terms of signal-to-noise ratio is the binary neutron star merger GW 170817, which has a network signal-to-noise ratio $\rho \approx 32$ (the exact number depends on details of the specific detection algorithm; see Section 9.6.3). The loudest binary black hole merger to date is the first, GW 150914, with $\rho \approx 24$.

The signal-to-noise ratio of a detected signal only tells part of the story, and does not give a sense of the significance of a signal. For this, we turn to the false-alarm rate and false-alarm probability, often abbreviated as FAR and FAP, respectively. The false-alarm probability is the probability that a detected event is generated not by an astrophysical event but by random noise fluctuations in the detectors. This is calculated by time-shifting data between detectors and repeating gravitational-wave searches with this time-shifted data. The false-alarm rate for an event is then the number of events in this time-shifted data with an equal or larger matched-filter signal-to-noise ratio divided by the total time of time-shifted data.

For GW 150914, which was detected only in LIGO Livingston and Hanford (before Virgo came online) the false-alarm rate was less than 1 in 2×10^5 years, corresponding to a false-alarm probability of 2×10^{-7}, or 5.1σ. In other words, one would have to run the two detectors in their same state for more than 2×10^5 years to get a single noise fluctuation with comparable matched-filter signal-to-noise ratio as the event itself! What makes this number even more remarkable is that it is only an upper bound on the false-alarm rate; the calculation is limited by the amount of time-shifted data that could be created with the detectors in an equivalent state. In reality, the false-alarm rate would be significantly smaller: while $\rho_{\text{net}} \approx 24$ for GW 150914, the next highest signal-to-noise ratio in the time-shifted data (excluding the time of signal itself) was $\rho_{\text{net}} \approx 13$. It is not possible to come up with a more nuanced false-alarm rate for GW 150914 given restrictions on the quantity of time-shifted data that can be created, but it is likely that the real false-alarm rate of this event is closer to $1/t_{\text{universe}}$, where t_{universe} is the age of the universe!

In addition to significance calculations, one can quantify the probability that a given event is terrestrial or astrophysical. The latter category can further be divided into probabilities that an astrophysical compact binary coalescence is either a binary black hole merger, a binary neutron star merger, or a neutron star black hole merger. These calculations are important for understanding the underlying rates of mergers throughout the universe and are becoming increasingly important as we begin to use multiple gravitational-wave measurements to understand the properties of the underlying populations. Calculations of whether events are of terrestrial or astrophysical origin involve understanding the statistical distribution of the various events as well as the time-slid data for each individual search pipeline; for full details see Abbott et al. (2019i).

Signals deemed to be of astrophysical origin are further classified into their source type, in particular the probability that they contain zero, one, or two neutron stars. It is worth stressing that this classification is somewhat naive: an object deemed to be a neutron star in this classification is one that has a mass between $1M_{\odot}$ and $3M_{\odot}$. Mergers with at least one of the progenitors in the range between $3M_{\odot}$ and $5M_{\odot}$ are afforded an extra classification known as mass-gap objects (note that this is the so-called lower-mass gap, c.f. the upper-mass gap for objects $\gtrsim 45M_{\odot}$). We can expect these classification schemes to become more nuanced as we develop our understanding of the maximum masses of neutron stars and whether objects can or cannot be formed in the lower-mass gap.

The aforementioned classification schemes and significance calculations are bootstrap methods, in the sense that they rely on empirically measured distributions of the noise properties of the interferometers. Such methods ultimately have limited utility—for critiques, see Capano et al. (2017) and Ashton et al. (2019b). There is, however, an ongoing push for fully Bayesian approaches to calculations of astrophysical odds (Ashton et al. 2019b; Ashton & Thrane 2020; Pratten & Vecchio 2020). These methods rely on the fact that gravitational-wave signals are coherent between any two detectors, whereas stationary and nonstationary noise such as instrument glitches are incoherent (Veitch & Vecchio 2008, 2010; Isi et al. 2018). Ultimately, these calculation methods are independent of the search pipeline as well as the quantity and quality of background data that can be accumulated, while the interferometers are in similar states to when detections were made. Although not widely accepted yet, it is fair to say these Bayesian approaches will become the norm in years to come.

9.6.2 Gravitational Waveforms

In the previous section we used gravitational-wave templates $u(t)$ in the definition of the matched-filter signal-to-noise ratio (Equation (9.10)). The templates will also be required when we look at astrophysical parameter estimation (Section 9.6.4). But what are these templates $u(t)$? In principle, they are the gravitational waveforms derived from solving Einstein's field equations (or your own favorite theory of gravity) that match the compact binary coalescence. For example, suppose we know that two nonspinning, $30M_{\odot}$ black holes merged 1 Gpc from Earth on a quasi-circular orbit (i.e., with zero orbital eccentricity). We can solve Einstein's field equations for the evolution of such a system, calculate the gravitational waveform at a future null infinity using the appropriate inclination angle of the binary, evaluate the antenna-pattern functions of the detectors at the time and sky location of the detected signal, and use the resulting waveform as our template $u(t)$ to calculate the matched-filter signal-to-noise ratio ρ from Equation (9.10). Easy! Of course, in practice we do not know the parameters of the source a priori nor do we know the exact sky location, which means we must repeat the above steps many, many times using different templates. The parameters that produce the highest matched-filter signal-to-noise ratio are then the best-fit parameters. Still easy! Right?

Wrong. It turns out numerical relativity simulations are difficult and extremely computationally expensive. Indeed, the two-body problem in numerical relativity took decades to solve, culminating in the pioneering work of Pretorius (2005) and subsequently Campanelli et al. (2006). But the full, nonlinear, three-dimensional simulations themselves are still too computationally expensive to be used in either detection algorithms or astrophysical parameter inference. In other words, the above "rinse-and-repeat" process of simulating Einstein's field equations, trying it as a template, and then simulating again to find a better signal-to-noise ratio, is not computationally feasible. For this reason, creative algorithms that quickly and reliably generate approximate gravitational waveforms have been created and developed over recent years.

Gravitational-wave "approximants" come in many shapes and sizes. They include, for example, the effective one-body approach (Buonanno & Damour 1999, 2000), the so-called inspiral–merger–ringdown phenomenological (commonly known as IMRPhenom) models (Ajith et al. 2007, 2008; Pan et al. 2008), and numerical relativity surrogate models (Pürrer 2014; Field et al. 2014; Blackman et al. 2015). This has led to a veritable cornucopia of unwieldy acronyms, including the intuitively named `IMRPhenomPv2_NRTidalv2` (Dietrich et al. 2019), the aptly described `SEOBNRv4Tsurrogate` (Lackey et al. 2019), the clear-as-mud `NRSur7dq4`, and its obvious extension `NRSur7dq4Remnant` (Varma et al. 2019). A complete list of all available waveforms, their acronyms, and the papers describing implementation and validations of the approximants in various regimes is available nowhere.[4]

[4] A partial list of waveform approximants used for the most recent LIGO/Virgo catalog of gravitational-wave events is available in Abbott et al. (2021a).

It is worth stating that creating waveform approximants is difficult, but important. There are a myriad of known physical effects that must be taken into account, and no one waveform family currently comes close to including all of these effects. These effects include, for example, spin-induced precession, waveform eccentricity, higher-order modes, and tidal effects. Indeed, while there are waveform families that exist that take into account one or even two of these effects, there are none that include them all. Any astrophysical inference of compact binary coalescences must therefore be interpreted with this limitation in mind. But it is these important families of waveform approximants that allow us to both detect gravitational-wave signals and also identify the physics of their progenitors.

But there is a step in between that we have not yet covered. It turns out that even the fastest waveform families are simply too slow to compute for low-latency gravitational-wave detection of compact binary coalescences. It is to this that we turn presently.

9.6.3 Low-latency Detection

In a book about multimessenger astronomy, it would be remiss to not discuss in some detail low-latency detections and triggers in gravitational-wave astronomy. The primary goal of low-latency detections is to quickly send out alerts to the global array of ground- and space-based telescopes via the Gamma-ray Coordinates Network (GCN).[5] Information in GCNs typically includes sky-map probability distributions (i.e., two-dimensional probability maps of right ascension and declination), as well as probability-based classifications of the signal being astrophysical or terrestrial, and the signal containing zero, one, or two neutron stars. The latter is given to inform astronomers of the probability that a system will be "electromagnetically bright."

For the first multimessenger gravitational-wave detection GW 170817, the initial gravitational-wave sky maps covered some 31 deg^2. This sky region overlapped with the sky localization of the gamma-ray burst observation seen by Fermi's Gamma-ray Burst Monitor (GBM; Meegan et al. 2009), and so began a global search for the multiwavelength counterpart, with the ultimate identification of the host galaxy, kilonova emission, and eventually electromagnetic emission across the band (Abbott et al. 2017f). Figure 9.8 shows these sky maps. Shown in light green is the 90% confidence interval sky map using just the two LIGO instruments, which covers 190 deg^2. In dark green is the 90% sky map for gravitational-wave sky localization using the two LIGO instruments and Virgo. The dark blue contour shows the 90% Fermi-GBM sky localization, and the light blue shows the International Planetary Network triangulation from the time delay between observations of the associated gamma-ray burst between Fermi-GBM and the INTEGRAL spacecraft. The top inset image shows observations of the galaxy NGC 4993 10.9 hr after the gravitational-wave and gamma-ray burst trigger as seen by the 1 m Swope telescope, with the new transient highlighted to the northeast of the Galaxy (Coulter et al. 2017).

[5] https://gcn.gsfc.nasa.gov/

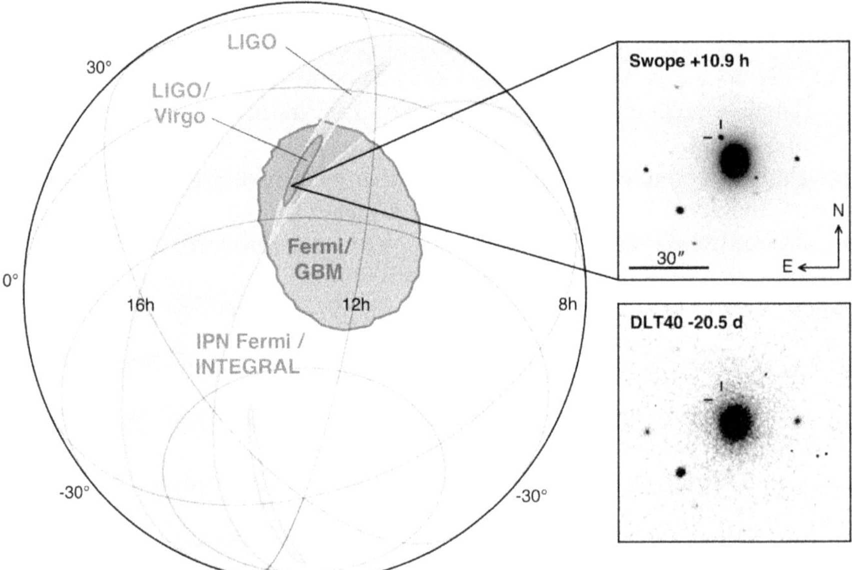

Figure 9.8. Sky location reconstruction of the binary neutron star merger GW 170817. In green (light green) are the gravitational-wave reconstructions from the two LIGO interferometers (three LIGO and Virgo interferometers), and in blue are the different sky localizations from the initial gamma-ray burst detection. The images on the right show galaxy NGC 4993 taken 10.9 hr after the merger with the 1 m Swope telescope (top) and 20.5 days prior to the merger (bottom). Image credit: Abbott et al. (2017f).

The bottom inset image shows a picture of the same galaxy taken some 20.5 days prior to the merger. For a full description of the sequence of events, see Abbott et al. (2017c, 2017f).

The LIGO and Virgo collaborations use five different low-latency pipelines to detect signals in as close to real time as possible. In no particular order, these are the three matched-filter pipelines GstLAL, PyCBC Live, and MBTA Online; the impulse response filter SPIIR; and the burst search Coherent Wave Burst (cWB). Each of these is described in more detail below; results and descriptions of their uses for LIGO and Virgo's second observing run can be found in Abbott et al. (2019b). In general, these low-latency pipelines run in real time as the data are streamed off the detectors. Upon finding a candidate signal that passes some predefined threshold, they are uploaded to the Gravitational wave Candidate Event DataBase[6] (GraceDB). At this point, the signals are analyzed by the Bayestar pipeline (Singer & Price 2016), which creates rapid Bayesian probability sky maps from the trigger information. These sky maps, as well as limited information about the significance of the event, are then transmitted through as a GCN notice. It is worth mentioning that these searches also run in "offline" modes that are more sensitive but slower to run. In such a setting they are able to detect weaker signals, albeit in much longer latency.

[6] https://gracedb.ligo.org/

GstLAL: The GstLAL pipeline (Messick et al. 2017; Sachdev et al. 2019) uses a combination of GStreamer libraries[7] and the LSC algorithm Library[8] to perform time-domain matched filtering of gravitational-wave signals in real time. A gravitational waveform template bank is used for the intrinsic parameters of the source—i.e., the progenitor masses and spins. In general, only models where the individual spins of the compact objects are aligned with the system's orbital angular momentum are used. In this case, the extrinsic parameters—i.e., the distance to the source, the inclination angle of the binaries' orbital angular momentum to the line of sight, etc.—only change the waveform's amplitude and phase at coalescence. Analytic marginalization of this phase at coalescence is used, implying the dimensionality of the template bank is significantly reduced to only the intrinsic source parameters. As the spins are assumed to be aligned with the orbital angular momentum, the number of parameters in the search are the two masses, the two spin magnitudes, as well as the overall scaling amplitude factor. Both point estimates for the gravitational-wave progenitor properties, as well as estimates of the significance of the event, can all be calculated in a matter of seconds; for details, see Sachdev et al. (2019) and references therein.

PyCBC Live: Like GstLAL, the PyCBC Live search pipeline (Usman et al. 2016; Nitz et al. 2018) is a low-latency matched-filter search pipeline that matches detector data against a template bank of precomputed gravitational waveforms. PyCBC Live uses frequency-domain matched filtering performed on finite-sized chunks of data, allowing low- and high-pass filtering to reduce the data size without losing a substantial amount of signal power. As with GstLAL, the template bank consists of quasi-circular, spin-aligned binary black hole waveforms, although the exact placing of the template bank in parameter space is determined based on the noise characteristics of the detectors. This template bank is precomputed, and is only re-calculated if there are significant changes in the power spectral density of the noise (see Usman et al. 2016 for details). Event ranking, significance estimation, and point estimates of the gravitational-wave-event progenitor properties are computed in seconds; for details, see Nitz et al. (2018) and references therein.

MBTA Online: The Multi-Band Template Analysis (MBTA) Online search (Adams et al. 2016) is another low-latency matched-filter search for compact binary coalescences. This pipeline speeds up the search by splitting the matched filter into two or more frequency bands that are analyzed independently, before results are combined to calculate a detection significance. The frequency bands are split so that approximately the same signal-to-noise ratio will be calculated in each band—when two frequency bands are used, this split occurs at approximately 100 Hz with the current sensitivity curve of LIGO/Virgo (of course, the actual split of the signal-to-noise ratio between the bands will depend on the mass of the

[7] https://gstreamer.freedesktop.org/
[8] https://wiki.ligo.org/Computing/LALSuite

merger signal). As with GstLAL and PyCBC, MBTA only uses aligned-spin waveforms in its template bank. For more details about the MBTA Online pipeline see Adams et al. (2016).

SPIIR: The Summed Parallel Infinite Impulse Response (SPIIR; Luan et al. 2012; Hooper et al. 2012; Chu 2017) uses groups of infinite impulse response filters to reconstruct an equivalent matched-filter template for compact binary coalescences. The procedure for creating the template is shown schematically in Figure 9.9. In particular, a series of simple exponentially growing sinusoid functions (panels (a)–(c)) with different growth timescales, frequencies, and cutoff times are summed together linearly with different amplitude prefactors (panel d). As the number of individual components increase, the summed filter tends toward the familiar chirping inspiral waveform, an example of which is shown in panel (e). The process is well suited to parallelization, with considerable speedup gained with the use of graphical processing units (Guo et al. 2018). The SPIIR pipeline has further been used to estimate the near-future potential of negative-latency gravitational-wave detections (James et al. 2019).

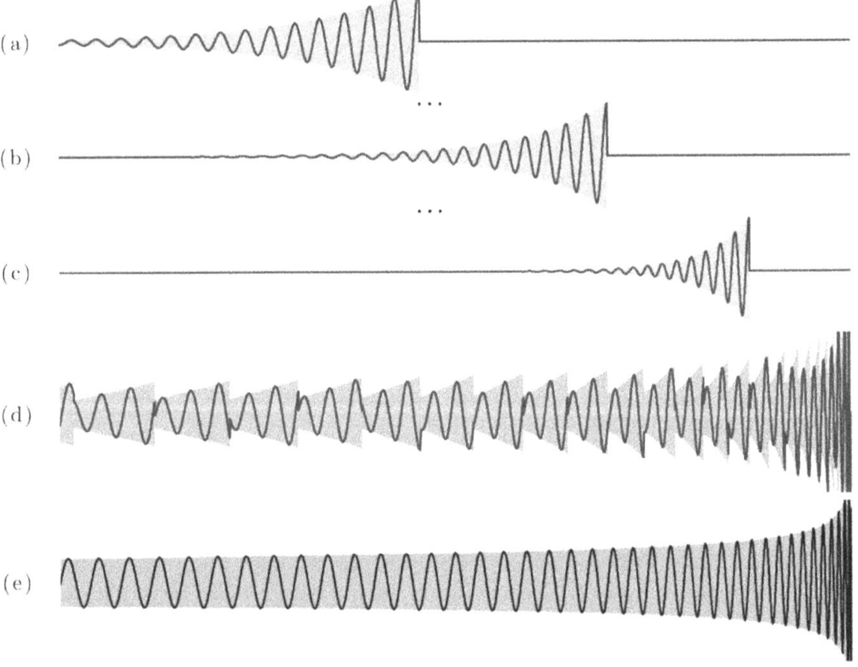

Figure 9.9. Schematic showing the creation of a matched filter for an inspiral signal using the Summed Parallel Infinite Impulse Response Filter (SPIIR). Panels (a)–(c) show single, exponentially growing sinusoid functions that are summed linearly with different amplitudes (panel (d)). In the limit of infinitely many exponentially growing sinusoids, one can exactly replicate a gravitational-wave inspiral template (panel e). Image credit: Hooper et al. (2012), with permission, copyright 2012 by the American Physical Society.

cWB: cWB (Klimenko et al. 2016) is an unmodeled search algorithm, meaning that it does not use a matched-filter template as with the other searches, but instead searches for excess power in the detector's strain data that is coincident between multiple detectors. The search is therefore capable of picking up both compact binary coalescences and also other types of potential sources such as supernovae, cosmic strings, etc. Any type of signal waveform can be reconstructed using a maximum-likelihood method outlined in Klimenko et al. (2016). Moreover, the coherent energy between the multiple detectors can be used to calculate both ranking statistics of various candidates as well as statistical significances.

9.6.4 Astrophysical Inference

The gravitational-wave searches described above provide point estimates for the various parameters of the coalescence. More robust estimates are then inferred using Bayesian parameter estimation pipelines that construct the posterior probability distribution (defined below) for the signal parameters. These include more accurate probability sky maps which can be produced on timescales between a day and a week after the event. These timescales are being enhanced with improvements in sampling techniques, parallelization, and speed of waveform calculation. Here we outline the salient features of Bayesian parameter estimation for gravitational-wave astronomy, but we point the interested reader to Thrane & Talbot (2019) for a more extensive and pedagogical review of the topic.

The main workhorse parameter interference software for the first few years of gravitational-wave discovery was LALInference (Veitch et al. 2015). More recently, a modular Bayesian Inference LiBrarY (Bilby; Ashton et al. 2019a; Romero-Shaw et al. 2020c) has been developed in Python, which is now being used in new LIGO/ Virgo discoveries. Significant speedups for astrophysical parameter inference with slow-to-calculate waveform models has been achieved through a parallel implementation of Bilby, known as Parallel Bilby (pBilby; Smith et al. 2020a). The PyCBC software mentioned in the previous section also has an inference module known, aptly enough, as PyCBC Inference (Biwer & Capano 2019).

Each of the above pieces of software solves Bayes theorem numerically using variations of stochastic samplers to calculate a posterior distribution for the system's parameters given the data. This posterior is denoted $p(\theta|d)$, where d is the data and θ is a vector of system parameters, which, for the case of a quasi-circular, binary black hole inspiral, are typically[9]:

- m_i for $i = 1, 2$: the masses of the two compact objects,
- $\vec{S}_i = (a_i, \theta_i, \phi_i)$: the two spin vectors, where a_i are the dimensionless spin magnitudes, θ_i are the zenith angle between the spin and orbital angular momenta, and ϕ_i are the cosine of that zenith angle,

[9] While this list of parameters fully describes a quasi-circular binary black hole system, they are not always the ones that are used in the stochastic sampling problem. For example, sometimes it is more convenient to calculate combinations of the individual masses such as the chirp mass $\mathcal{M}$ and the mass ratio $q = m_2/m_1$, rather than the individual masses themselves.

- d_L: the luminosity distance to the source,
- RA and Dec: the right ascension and declination of the source, respectively,
- $\theta_{\rm JN}$: the angle between the binary's total angular momentum vector and the line of sight,
- t_c: the time of coalescence,
- ϕ_c: the binary phase at some reference frequency, often considered as the coalescence time, and
- ψ: the polarization angle.

In all, these are 15 parameters that fully describe the binary system. If the system is not on a quasi-circular inspiral, then there are two additional parameters: the eccentricity defined at some reference frequency ϵ and the argument of periapsis ω. Moreover, in the case of two neutron stars, one also has the dimensionless tidal deformabilities of each neutron star Λ_i—for more details see Section 9.7.2.

Bayes theorem is expressed as

$$p(\theta|d) = \frac{\mathcal{L}(d|\theta)\pi(\theta)}{\mathcal{Z}(d)}, \tag{9.11}$$

where $\pi(\theta)$ is the prior probability distribution for the various parameters, $\mathcal{Z}(d)$ is a normalization factor known as the evidence, and $\mathcal{L}(d|\theta)$ is the likelihood function. If we assume the noise in the gravitational-wave detectors is Gaussian (a standard, albeit incorrect assumption due to, among other things, nonstationary artifacts), then the likelihood function is (van der Sluys et al. 2008; van der Sluys 2008; Veitch & Vecchio 2008)

$$\ln \mathcal{L}(d|\theta) = -\frac{1}{2}\sum_k \left\{ \frac{[d_k - u_k(\theta)]^2}{\sigma_k^2} + \ln\left(2\pi\sigma_k^2\right) \right\}, \tag{9.12}$$

where the index k labels the frequency bin, σ is the interferometer noise amplitude spectral density, and $u(\theta)$ is the gravitational-wave template evaluated with parameters θ.

Bayesian inference evaluates Equation (9.11) to determine the posterior probability distribution for each of the source parameters associated with the binary merger. One typically performs simultaneous parameter inference on all model parameters, implying the posterior distribution is a 15-dimensional (depending on the model) distribution. To aid understanding results, one typically then marginalizes (integrates) over parameters to achieve lower-dimensional distributions that are simpler and more convenient to analyze. Of course, covariances between parameters must always be considered when attempting to understand the physics of merger events. Figure 9.10 shows examples of marginalized posterior distributions one achieves from performing parameter estimation on gravitational-wave events. In particular, the left-hand panel shows the posterior reconstruction of the primary m_1 and secondary m_2 masses for all events detected in the first and second observing runs of Advanced LIGO and Virgo, and announced as the first

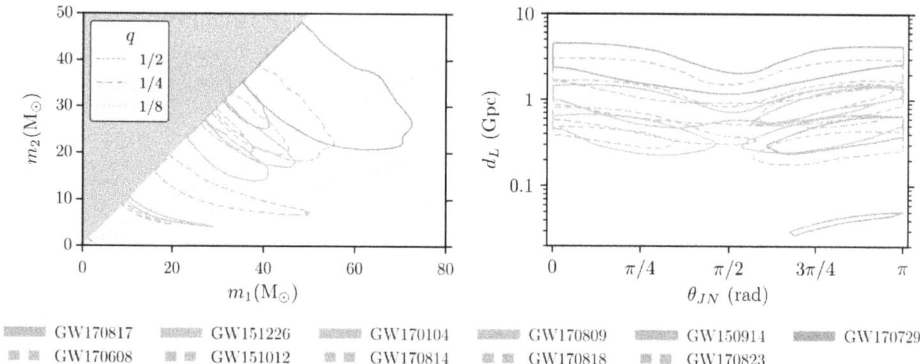

Figure 9.10. Marginalized posterior probability distributions for the parameters of the first binary black hole merger detected, GW 150914. The left-hand panel shows the 90% confidence intervals for the posterior distributions for the systems two source masses. The right-hand panel shows the degeneracy between the luminosity distance to the source D_L and the sources inclination angle with respect to the observer's line of sight $\theta_{\rm JN}$. Image credit: Abbott et al. (2019i), with permission, copyright 2019 by the American Physical Society.

gravitational-wave transient catalog (GWTC-1; Abbott et al. 2019i). The gray region simply acknowledges the fact that one defines $m_1 \geqslant m_2$.

In the left panel of Figure 9.10, one can see that the shape of the contours change from relatively low-mass to relatively high-mass events. This is because at low masses the inspiral part of the signal remains in band for significantly longer, during which the phase evolution is predominantly governed by the chirp mass $\mathcal{M}$ (see Equation (9.3)). On the other hand, at late times in the inspiral, it is the total mass of the system $M = m_1 + m_2$ that is more accurately measured; high-mass systems such as GW 170729 merge at relatively low frequencies, implying they do not have a significant inspiral phase in band, and the total mass is better measured than the chirp mass.

The right-hand panel of Figure 9.10 shows the two-dimensional marginalized distributions for GWTC-1 events for the luminosity distance d_L and the inclination angle $\theta_{\rm JN}$. Here one can see a degeneracy between the two parameters: face-on (face-off) binaries, where the angular momentum axis is pointing toward (away) from the line of sight, have higher amplitude gravitational-wave signals than otherwise identical edge-on binaries. An edge-on binary is therefore inferred to be closer to Earth than a face-on or face-off binary, resulting in the covariances shown in the right-hand panel of Figure 9.10.

In the following sections, we look at what we can learn about the astrophysics and fundamental physics from gravitational-wave detections. From a data-analysis perspective, these inferences rely on understanding the posterior probability distributions as just presented. But at times they also rely on performing model selection. As an example, one might have two different gravitational-wave template banks: one that is created under the assumption that general relativity is the correct theory of gravity, and another under the assumption that the Brans–Dicke scalar-tensor theory is correct, for example. Model selection between the two hypotheses can then be performed to understand which is more likely to have produced the data. To this end, one calculates the evidence $\mathcal{Z}(d)$ by integrating both sides of Equation (9.11) and noting that the posterior probability is normalized. The evidence can then be calculated for each of the models:

$$\begin{aligned} \mathcal{Z}_{\rm GR} &= \int d\theta \mathcal{L}(d|\theta_{\rm GR})\pi(\theta_{\rm GR}), \\ \mathcal{Z}_{\rm BD} &= \int d\theta \mathcal{L}(d|\theta_{\rm BD})\pi(\theta_{\rm BD}), \end{aligned} \tag{9.13}$$

where $\theta_{\rm GR}$ and $\theta_{\rm BD}$ are the model parameters associated with general relativity and Brans–Dicke theory, respectively. One then defines a Bayes Factor as

$$\mathcal{B}_{\rm BD}^{\rm GR} = \frac{\mathcal{Z}_{\rm GR}}{\mathcal{Z}_{\rm BD}}, \tag{9.14}$$

which defines the relative probability that the model for general relativity is preferred over the model for the Brans–Dicke theory of gravity. In principle, one should then calculate an odds ratio $\mathcal{O}_{\rm BD}^{\rm GR} \equiv \Pi_{\rm BD}^{\rm GR}\mathcal{B}_{\rm BD}^{\rm GR}$, where $\Pi_{\rm BD}^{\rm GR}$ is the prior odds that quantifies ones prior belief that one hypothesis is preferred over the other. In practice, that is often difficult to know or quantify, implying it is standard to assume the prior odds is simply unity.

As a matter of convention, one often uses a threshold value of $\ln \mathcal{B} \geqslant 8$ to define "strong evidence" for one hypothesis over another. Some researchers find this contentious, others find it pragmatic.

It is worth closing this section by mentioning that numerous groups are working on developing machine-learning algorithms to significantly increase the speed of gravitational-wave parameter inference, without sacrificing the accuracy of the results (e.g., Gabbard et al. 2019; Chatterjee et al. 2020; Green & Gair 2020). It is also hoped that machine-learning algorithms may be able to provide better characterization of the noise in the interferometer, which would ultimately improve the accuracy of parameter estimation as well. While these methods are certainly promising, they are not widely adopted by the community yet. The coming years may see a shift in this direction, with the ultimate goal of machine-learning algorithms running real-time parameter estimation on live-streamed data, and potentially getting accurate inference results as quickly, if not more quickly, than the online searches discussed in Section 9.6.3. For a review of machine-learning developments in gravitational-wave astronomy including, and beyond, parameter inference, see Cuoco et al. (2020).

9.7 Astrophysics of Compact Binary Coalescences

In the last section we discussed how to detect astrophysical compact mergers and how to extract information about the source properties. We now turn to the question of what we have actually learned in the first five years of gravitational-wave astronomy from, and about, those compact objects. In particular, what astrophysics

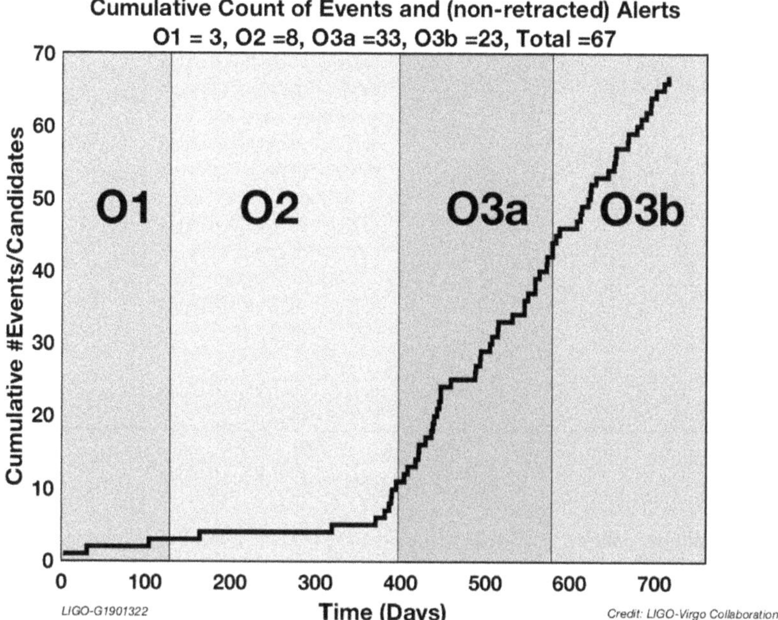

Figure 9.11. Cumulative number of gravitational-wave events as a function of observing time during LIGO/Virgo's observing runs O1, O2, O3a, and O3b. Image credit: LIGO-Virgo (Brian O'Reilly and Dave Reitze) https://dcc.ligo.org/LIGO-G1901322/public.

and fundamental physics have we been able to extract about individual events and the population of events as a whole.

There have been three observing runs of the LIGO/Virgo interferometers, named O1 (2015 September 12–2016 January 19), O2 (2016 November 30–2017 August 25), and O3. The latter run, O3, was split into two runs called O3a (2019 April 1–2019 October 1) and O3b (2019 November 1–2020 March 27).[10] At the time of writing, O3b has finished, although no results have been published. We therefore keep our discussion to that of observing runs O1, O2, and O3a. Figure 9.11 shows the cumulative number of gravitational-wave events seen in all Advanced LIGO/Virgo observing runs. These are events for which detection pipelines triggered, and the trigger was not subsequently retracted. Importantly, some of these events may not be bona fide events that end up being published in catalogs, potentially leading to discrepancies between the numbers in Figure 9.11 and those published numbers.

To date there have been two official gravitational-wave event catalogs published by the LIGO/Virgo collaborations. These are the so-called Gravitational wave Transient Catalogue 1 (GWTC-1; Abbott et al. 2019i) and Gravitational wave Transient Catalogue 2 (GWTC-2; Abbott et al. 2021a). Two other catalog sequences have been published based on non-LIGO/Virgo collaboration work using the publicly available data. One such set of catalogs is known as The First and

[10] Observing run O3b was shortened from the originally planned end date of 2020 April 30, due to the COVID-19 pandemic.

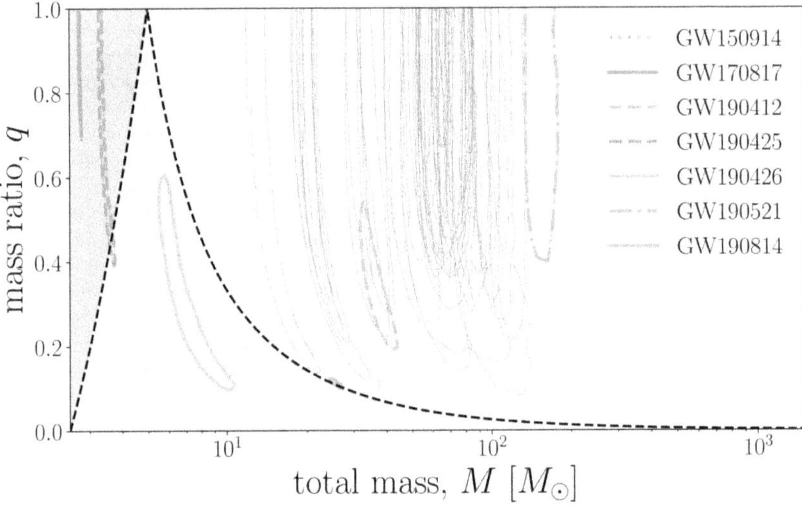

Figure 9.12. Mass ratio $q = m_1/m_2$, as a function of total mass $M = m_1 + m_2$, for all systems published as part of LIGO/Virgo's first and second gravitational-wave transient. Contours represent the 90% credible region for the two-dimensional marginalized posterior distributions. "Special events" that are described independently in the text are shown with different colors and line styles, whereas all other events are shown as gray contours. The parameter space is separated into three regions based on one or both of the compact objects having less than $2.5M_\odot$. The purple region has both $m_{1,\,2} \leqslant 2.5M_\odot$, implying events with majority support in that region are likely binary neutron star mergers. The green pastel region has $m_1 \leqslant 2.5M_\odot$, but $m_2 \geqslant 2.\ 5M_\odot$, implying those events are likely neutron star black hole mergers, and the orange region has both $m_{1,\,2} \geqslant 2.5M_\odot$, implying those events are likely black hole mergers. More nuanced descriptions of these regions, and the events within them, are provided in the text.

Second Open Gravitational wave Catalogues: 1-OGC (Nitz et al. 2019) and 2-OGC (Nitz et al. 2020), respectively. The latter of these announced the discovery of a new marginal black hole merger in addition to LIGO/Virgo's GWTC-2 catalog. The other set of catalogs is from a team at the Institute for Advanced Study at Princeton, whose independently developed detection algorithm (Venumadhav et al. 2019) has claimed a number of new detections in LIGO/Virgo's first and second observing runs (Zackay et al. 2019; Venumadhav et al. 2020; although see Ashton & Thrane 2020 disputing one of these claimed detections). Figure 9.12 shows the 90% confidence intervals for the total mass and mass ratio for all events in the first and second LIGO/Virgo transient catalogs, GWTC-1 and GWTC-2 (Abbott et al. 2019i, 2021a).[11] The plot is divided into three regions based on the expected source properties. In the left-hand region, shaded in purple, both compact objects have masses $m_{1,\,2} \leqslant 2.5M_\odot$, implying both are likely neutron stars. The two detections in these region are GW 170817 (Abbott et al. 2017e) and GW 190425 (Abbott et al. 2020b), both of which we discuss in more detail in Section 9.7.2. The middle section

[11] Note that these catalogs are cumulative, in the sense that any event appearing in GWTC-1 is also included in GWTC-2.

of Figure 9.12 has the secondary compact object $m_2 \leqslant 2.5M_\odot$, but the primary $m_1 \geqslant 2.5M_\odot$, implying any system in this region of the parameter space is likely a neutron star black hole. The only event lying confidently in this region is GW 190426 (Abbott et al. 2021a), although it is worth noting this event has the highest false-alarm rate of all events in both the GWTC-1 and GWTC-2 catalogs—we discuss this event, and neutron star black hole binaries in general, in Section 9.7.3. Finally, the right-hand portion of Figure 9.12, shaded in pale orange, is defined by $m_{1,\,2} \geqslant 2.5M_\odot$, implying those systems are all binary black holes. These are discussed more in Section 9.7.1.

It is important to note that in Figure 9.12 and the following sections, we set the disambiguation between neutron star and black hole as a mass of $2.5M_\odot$. This is different from the $3M_\odot$ often used in LIGO/Virgo publications (e.g., Abbott et al. 2021a)—the correct value for this disambiguation depends on the unknown maximum nonrotating neutron star mass, known as the Tolman–Oppenheimer–Volkoff mass ($M_{\rm TOV}$; Tolman 1939; Oppenheimer & Volkoff 1939). But the distinction is even more subtle; it is possible that low-mass black holes may occupy the same region of parameter space as neutron stars. While the formation scenario for such low-mass black holes is not understood, the confirmed existence of compact objects of $\approx 2.\ 6M_\odot$ (Abbott et al. 2020g) is also not well understood. Ultimately, this means that we need measurements of tidal deformability or electromagnetic counterparts to definitively distinguish between black holes and neutron stars. We discuss this in more detail in the following sections.

In the following sections, we focus primarily on the LIGO/Virgo catalogs and subsequent inferences made from these. In the spirit of acknowledging one's own biases, it is worth mentioning that astrophysical inferences from these first three LIGO/Virgo observing runs are extensive, and the following sections only scratch the surface of what has been achieved in this field. The chosen topics certainly favor my own personal interests; this is not to intentionally downplay other works, but an exhaustive review of all relevant literature would be nigh-on impossible, even though the field itself is only five years old. To quantify this difficulty, at the time of writing this review, the GW 150914 discovery paper (Abbott et al. 2016e), the GW 170817 discovery paper (Abbott et al. 2017e), and the GWTC-1 catalog paper (Abbott et al. 2019i) have approximately 5500, 3500, and 1000 citations, respectively.[12] Each of these papers averages approximately four citations per work day, implying any attempt at an exhaustive review would be outdated before the virtual ink has a chance to dry on the computer screen.

9.7.1 Astrophysics of Black Hole Mergers

The right-hand region of Figure 9.12, shaded in pale orange, shows the 90% confidence intervals for all binary black hole mergers announced in the two LIGO/Virgo transient catalogs (Abbott et al. 2019i, 2021a). Four "special events" are highlighted with different line styles, each of which is discussed separately below.

[12] According to the Astrophysics Data System (ADS) database.

GW 150914—The First Binary Black Hole Detection: The first gravitational-wave detection was also the first time stellar-mass black holes of $\gtrsim 10 M_{\odot}$ were confirmed to exist in the universe, and the first time binary black hole systems in general were confirmed to exist (Abbott et al. 2016e). This event is discussed in some detail in Section 9.3, so we do not go into much detail here. It is worth noting that GW 150914 is still the loudest binary black hole merger detected. The network matched-filter signal-to-noise ratio of $\rho_{\rm mf} \approx 24$ is similar to that of GW 190521_074359 (Abbott et al. 2021a; not to be confused with GW 190521, Abbott et al. 2020f, which is discussed in detail below). GW 150914 is our best black hole merger for performing tests of general relativity—see Section 9.7.4. Other than this, we now know that GW 150914 is quite unremarkable in terms of its masses, spins, etc. For example, the total mass M and mass ratio q for GW 150914 are shown as the dotted pale orange contour in Figure 9.12, which overlays the numerous thin gray contours that describe other unremarkable binary black hole mergers.

GW 190412: The gravitational-wave event GW 190412 (Abbott et al. 2020e) was the first binary black hole merger detected with highly asymmetric masses. Figure 9.12 shows that GW 190412 and GW 190814 are the only binary black hole mergers with posterior distributions that confidently rule them out being equal-mass (i.e., $q = 1$) systems, with component source-frame masses $m_1 = 30.1^{+4.6}_{-5.3} M_{\odot}$ and $m_2 = 8.3^{+1.6}_{-0.9} M_{\odot}$, implying $q = 0.28^{+0.12}_{-0.07}$.

Highly asymmetric systems are also expected to have larger contributions from higher-order multipoles. That is, one can decompose a gravitational waveform into spin-weighted spherical harmonics ${}_{-2}Y_{\ell m}$ as (e.g., Thorne 1980)

$$h_{+} - ih_{\times} = \sum_{\ell \geqslant 2} \sum_{-\ell \leqslant m \leqslant \ell} \frac{h_{\ell m}(t)}{d_L} {}_{-2}Y_{\ell m}(\theta, \phi), \tag{9.15}$$

where θ and ϕ are, respectively, the polar and azimuthal angles of the direction of gravitational-wave propagation from the source to the observer. For most systems, the h_{22} mode is dominant, to the point that astrophysical inferences made with just the h_{22} mode, and with higher-order modes included in the waveform, are indistinguishable (e.g., see Payne et al. 2019). The event GW 190412 was the first to show significant deviations in astrophysical source parameters when analyzed with waveforms that include higher-order modes. For example, the inclusion of higher-order modes was able to break the degeneracy between the system being either face on or face off (i.e., whether the binary's orbital angular momentum vector was pointing toward or away from Earth). A better measure of the inclination angle in this way further leads to tighter constraints on the distance to the source given that these two parameters are highly covariant. Various analyses showed that the presence of higher-order modes was significant, both in terms of Bayes factor calculations comparing waveforms with and without higher-order modes, as well as an increase in matched-filter signal-to-noise ratio when higher-order modes are included (Abbott et al. 2020e). As well as improved parameter estimation, higher-order modes and highly asymmetric-mass systems such as GW 190412 can be used as good tests of general relativity—we return to this in Section 9.7.4.

The event GW 190412 was also interesting because there was evidence for nonzero spins of the progenitor black holes. While the spins of binary black hole systems are described by six parameters (a three-dimensional spin vector for each black hole), effective spin parameters enter waveform calculations that group mass-weighted spin components together. Inference is therefore typically done on these parameters because it is more informative than looking at individual spin components. The two dominant effective spin parameters (Ajith et al. 2011; Schmidt et al. 2015) are the "effective inspiral spin":

$$\chi_{\rm eff} = \frac{(m_1\vec{\chi}_1 + m_2\vec{\chi}_2)\cdot\hat{L}}{m_1 + m_2}, \tag{9.16}$$

which measures the components of the binaries spin vectors that are aligned with the binaries orbital angular momentum vector. Here, $\hat{L}$ is the unit vector of the binary's Newtonian orbital angular momentum, and $\vec{\chi}_i = c\vec{S}_i/(Gm_i^2)$ are the dimensionless spin vectors, where $\vec{S}_i$ are the spin angular momentum vectors of each component. The other spin parameter is the "effective precession":

$$\chi_p = \max\left\{\frac{|\vec{S}_{1\perp}|}{m_1^2}, \kappa\frac{|\vec{S}_{2\perp}|}{m_2^2}\right\}, \tag{9.17}$$

where $\vec{S}_{i\perp} = \vec{S}_i - (\vec{S}_i\cdot\vec{L}_N)\vec{L}_N/|\vec{L}_N|^2$, and $\kappa = q(4q+3)/(4+3q)$. The effective precession parameter measures the binary's in-plane spin components and therefore measure the degree of precession in an orbital system, $\chi_p = 1$ being maximal precession and $\chi_p = 0$ being zero precession.

The effective inspiral spin parameter for GW 190412 confidently rules out zero, with $\chi_{\rm eff} = 0.35^{+0.08}_{-0.11}$. This was not the first purported detection of nonzero spins; both GW 151226 (Abbott et al. 2016b) and GW 170729 (Abbott et al. 2019i) were originally identified as having nonzero $\chi_{\rm eff}$; however, when physically motivated priors are used, both these events become consistent with $\chi_{\rm eff} = 0$ (Miller et al. 2020). The effective precession parameter for GW 190412 also shows evidence for nonzero in-plane spins, although this too is a difficult measurement to make because formally the prior on χ_p rules out zero. Nonetheless, analyses of GW 190412 show $\chi_p = 0.31^{+0.19}_{-0.16}$, implying a 90% lower bound of 0.15. While this is not definitive evidence for precession due to the prior issue, it does imply small values of $\chi_p \lesssim 0.1$ are unlikely. Conversely, the 90% upper bound of $\chi_p < 0.5$ implies strong precession is also ruled out.

While GW 190412 is relatively unique in the catalog of gravitational-wave mergers detected thus far, initial analyses indicate that it is consistent with the low-mass-ratio tail of the distribution of other events and does not necessarily come from a separate population of binary black hole systems (Abbott et al. 2021b). Indeed, while it is the only binary black hole merger other than GW 190814 that

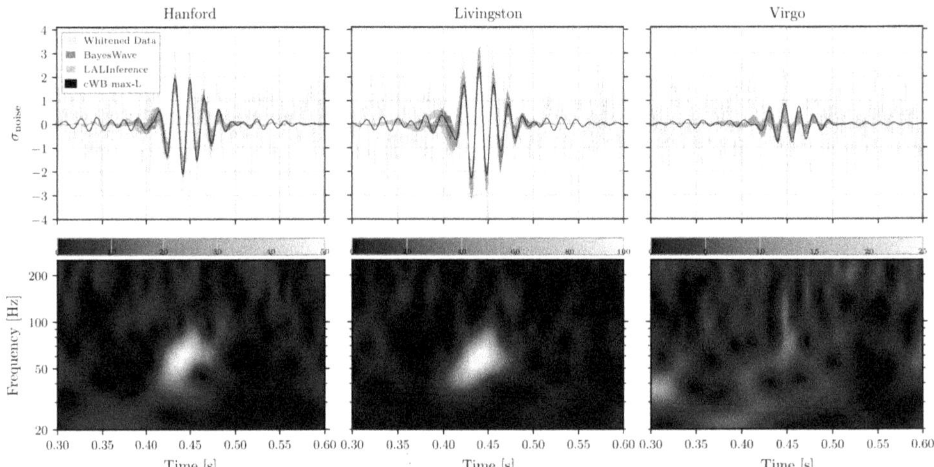

Figure 9.13. Gravitational-wave event GW 190521. The left, middle, and right panels show the data in the LIGO-Hanford, LIGO-Livingston, and Virgo, respectively. The top panel shows the whitened gravitational-wave strain (blue) with various different waveform reconstructions. The bottom panels show the time–frequency map of the whitened data. Image credit: Abbott et al. (2020f).

confidently rules out equal-mass progenitors (i.e., $q = 1$), there is only a 44% chance that it has the smallest mass ratio of black hole mergers among the O3a catalog when GW 190814 is excluded (Abbott et al. 2021a). This fact is because mass ratios are not well constrained for most systems, implying some fraction of the systems shown in the right, orange-shaded section of Figure 9.12 may have a smaller mass ratio than GW 190412. This therefore implies the formation mechanism for GW 190412 is not necessarily different from the other binaries in the catalog—we discuss these formation scenarios in detail below.

GW 190521: The event GW 190521 (Abbott et al. 2020f) stands out in Figure 9.13 as having the highest total mass of all systems thus far detected. Indeed, initial analyses show that this is a system with $m_1 = 85^{+21}_{-14}M_{\odot}$ and $m_2 = 66^{+17}_{-18}M_{\odot}$, with a remnant mass inferred to be $142^{+28}_{-16}M_{\odot}$, making this the first confident inference of an intermediate-mass black hole.[13] More important than this arbitrary classification, the more massive primary black hole (and, to a lesser extent the secondary) lies firmly in a prohibited mass region from stellar evolution and therefore challenges our current understanding of the formation mechanism of systems such as this.

Stellar evolution predicts main-sequence stars with helium core masses above about $30M_{\odot}$ are unstable at the end of their carbon-burning stage (e.g., Woosley et al. 2007). Helium core masses between about $64M_{\odot}$ and $133M_{\odot}$ collapse at this point through a single pulse that completely disrupts the entire star in an event known as a pair-instability supernova (Ober et al. 1983; Bond et al. 1984; Heger & Woosley 2002; Umeda & Nomoto 2002). Such a pair instability is so explosive that no compact

[13] Contrary to other subfields of astronomy, gravitational-wave astronomers often use the classification that intermediate-mass black holes are those between $10^2M_{\odot}$ and $10^5M_{\odot}$.

remnant is left behind, implying one expects a so-called pair-instability supernova mass gap, where no black holes should be created from the explosions of massive stars between $64M_{\odot}$ and $133M_{\odot}$ (see Woosley 2017 and references therein). The primary black hole of GW 190521 is firmly within this mass gap, with only 0.3% posterior probability that $m_1 \leqslant 65M_{\odot}$ (Abbott et al. 2020f). It is therefore an open question as to how GW 190521 formed.

One possibility for the formation of GW 190521 is that it is a second-generation merger. Broadly, there are two classes of formation scenario: isolated, in which the two stars coevolve through the main sequence and their supernovae before eventually coalescing, and dynamical, in which the two black holes evolve independently before becoming dynamically bound through gravitational capture. Second-generation mergers fall into the latter category, where at least one of the progenitors was formed originally through the merger of two black holes. In the case of GW 190521, the proposal is that the primary formed through the merger of two black holes that were below the mass gap. In general, such dynamical captures must occur in dense stellar environments such as globular clusters or active galactic nucleus (AGN) disks, and their recoil velocity must not be sufficiently high that they get kicked from their environment (e.g., Miller & Hamilton 2002; Antonini & Rasio 2016; Mapelli 2016; McKernan et al. 2018; Rodriguez et al. 2019).

For any type of dynamical capture event, one expects no preference for the spins of the black holes to be aligned with the orbital angular momentum of the binary, c.f. isolated evolution, where the spin and orbital angular momentum vectors should be nearly aligned (Belczynski et al. 2002; Campanelli et al. 2006; Stevenson et al. 2017 and references therein). For second-generation mergers, at least one of the binary's components should be spinning rapidly, with dimensionless spin $a \sim 0.7$ (Fishbach et al. 2017). Moreover, for dynamical events, one also expects some fraction of the events to have nonzero orbital eccentricity, c.f. isolated evolution, where the orbits should be quasi-circular (e.g., Rodriguez et al. 2018; Gondán 2019; Zevin et al. 2019). The merger GW 190521 probably exhibits some or all of these properties, indicative of a dynamical capture formation scenario. In particular, it shows evidence of spins misaligned with the orbital angular momentum vector (Abbott et al. 2020f, 2020h) when analyzed with quasi-circular waveforms; the effective precession parameter is measured as $\chi_p \approx 0.7^{+0.2}_{-0.3}$ (waveform systematic uncertainties play a significant role here; Abbott et al. 2020h). However, when analyzed with waveforms that include the effects of eccentricity, these are favored (Romero-Shaw et al. 2020b; Gayathri et al. 2020; Calderón Bustillo et al. 2021). None of the aforementioned analyses include all physics required to fully understand this event. For example, only Gayathri et al. (2020) include eccentricity, precession, and higher-order modes; however, they do so with a very limited number of waveforms, and therefore their analysis method is simply a maximum-likelihood calculation. The creation of fast and reliable waveforms that include all of these physical effects is an ongoing effort.

One intriguing possibility for dynamical capture associated with GW 190521 is that it occurred in the disk of an AGN. Indeed, mass segregation in such disks imply thousands of stellar-mass black holes may exist in the innermost parsec (e.g., Generozov et al. 2018 and references therein). While electromagnetic counterparts are difficult to imagine for binary black hole mergers, the gaseous environment in AGN disks allows for the possibility of an electromagnetic counterpart. One proposed mechanism is that the kicked black hole remnant from the merger may pass through the disk, generating a flare days after the gravitational-wave detection (McKernan et al. 2018). A putative candidate of this sort has been claimed, detected by the Zwicky Transient Facility (Graham et al. 2020), with peak luminosity $\sim$50 days after the gravitational-wave detection. If true, such a result would enable an interesting measurement of Hubble's constant at greater distances than previously done with gravitational waves. However, it is worth being skeptical of the claimed AGN–gravitational-wave association for two primary reasons. First, when analyzed using the most up-to-date gravitational-wave analyses (c.f. the limited information publicly available at the time Graham et al. 2020 was published), the odds for a common source based solely on the sky localization and distances do not satisfy the mantra: "extraordinary claims require extraordinary evidence" (Ashton et al. 2020). Second, the large energy output required to match observations, which comes from accretion onto the merger remnant in the model of Graham et al. (2020), exceeds the merger remnant's Eddington luminosity by many orders of magnitude and will inevitably create feedback, stifling accretion onto the remnant and changing the temperature of the accretion disk in ways that cannot be accounted for in the model (I. Mandel 2020, private communication).

While the previous paragraphs support the notion that GW 190521 formed dynamically, and therefore circumvents the upper-mass gap problem by postulating this is a second-generation merger, there are other options that are being put forward in the literature. For example, it has been proposed that the lower edge of the pair-instability mass gap may actually be significantly higher than previously estimated (e.g., Farrell et al. 2021), or that the progenitor black holes may have formed from Population III stars at high redshift (e.g., Kinugawa et al. 2021; Safarzadeh & Haiman 2020), or the merger of two black holes in colliding ultradwarf galaxies (Palmese & Conselice 2021). Even the individual masses of the black holes that make up GW 190521 are debated; when population-inspired priors are used for the mass of the secondary, its inferred mass falls below the mass gap, while the primary mass is above it (Fishbach & Holz 2020). Ultimately, the true origin of GW 190521 might be difficult to robustly determine, but the near-future accumulation of more events like this is an exciting prospect.

GW 190814: Merger GW 190814 is enigmatic. It has definitively the smallest and best-measured mass ratio with $q = 0.112^{+0.008}_{-0.009}$ (see Figure 9.12) with the primary being a boring $m_1 = 23.2^{+1.1}_{-1.0} M_{\odot}$ (Abbott et al. 2020g), but the secondary mass at a paradoxical $m_2 = 2.59^{+0.08}_{-0.09} M_{\odot}$. Figure 9.14 shows the one-dimensional marginalized posterior distribution for this secondary mass using two different waveform models,

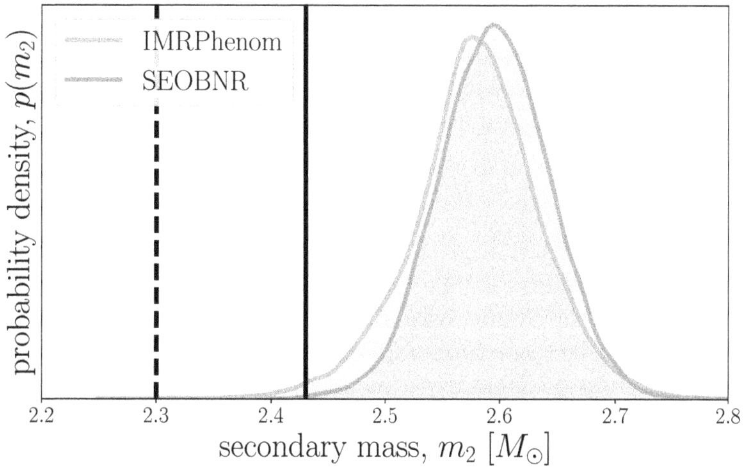

Figure 9.14. Mass of the secondary compact object for gravitational-wave event GW 190814. Shown are two marginalized posterior distributions for the event using different waveforms models to recover the parameters of the system. Both waveform models, IMRPhenomPv3HM and SEOBNRv4HM, include the effects of spin precession and higher-order modes. The vertical dashed black line is the 90% confidence upper limit on the most massive neutron star measured to date (Cromartie et al. 2020), and the vertical solid black line is the 90% confidence upper limit on the inferred maximum neutron star mass from analysis of the binary neutron star merger GW 170817 (Abbott et al. 2018c, 2020g).

IMRPhenomPv3HM (Khan et al. 2019, 2020) and SEOBNRv4PHM (Babak et al. 2017; Ossokine et al. 2020), both of which include precession effects and higher-order modes and show consistent results. Also shown in Figure 9.14 is the 90% upper confidence limit on the most massive neutron star observed in our galaxy, the millisecond pulsar MSP J0740+6620 (Cromartie et al. 2020). No information about the tidal deformability of the secondary is known—more precisely, the posterior distribution for the tidal deformability Λ_2 is equivalent to the prior (Abbott et al. 2020g). Moreover, no electromagnetic counterpart was detected (although at such a high mass ratio one does not necessarily expect an electromagnetic counterpart, even if the secondary was a neutron star—see Section 9.7.3), implying the only information to discern the nature of the secondary is its mass.

The Tolman–Oppenheimer–Volkoff mass (M_{TOV}; Tolman 1939; Oppenheimer & Volkoff 1939) defines the maximum mass a nonrotating neutron star can have without collapsing to a black hole.[14] There are a number of avenues for trying to understand M_{TOV} and to put it in the context of the secondary object in GW 190814.

[14] Upon defining the nuclear physics approximations one wants/needs to make, one can calculate the pressure of nuclear matter at supranuclear densities as a function of density, entropy, temperature, etc. This relation is the equation of state. These calculations are, to say the least, hard, implying many different equations of state exist, all of which have slightly different underlying assumptions that go into the calculations. But given an equation of state, one can solve the general relativistic equations of hydrostatic equilibrium to derive the radius of a given neutron star as a function of its mass and also the maximum mass M_{TOV} allowed by that equation of state. Understanding the maximum mass therefore allows us to understand the equation of state and probe nuclear physics in conditions inaccessible to terrestrial experiments.

From a nuclear-theory perspective, some equations of state can be built that support neutron stars up to and above the $\approx 2.7 M_\odot$ 90% confidence interval upper limit for GW 190814's secondary. However, the binary neutron star merger GW 170817 (see Section 9.7.2) provided informative measurements of the neutron star's tidal deformability, and hence the equation of state, with 90% upper limits on the maximum neutron star mass of $M_{\rm TOV} \leqslant 2.43 M_\odot$ (Abbott et al. 2018c). Comparing the maximum mass posterior with the posterior on the secondary of GW 190814 results in a 3% probability that $m_2 \leqslant M_{\rm TOV}$ (Abbott et al. 2020g). Fattoyev et al. (2020) went one step further, taking a rather generic approach to building equations of state and trying to simultaneously account for the equation-of-state measurements of GW 170817, the secondary of GW 190814 as a neutron star, and the results of terrestrial heavy-ion collision experiments. Their conclusion: the secondary of GW 190814 is "…likely to be the lightest black hole ever discovered" (Fattoyev et al. 2020).

I have shown my bias with the above paragraph and most emphatically by placing GW 190814 in the section on binary black holes rather than the section on neutron star–black hole binaries. I hope to be proven wrong; it would be more exciting from a nuclear physics perspective if $M_{\rm TOV} \gtrsim 2.5 M_\odot$. There are numerous authors clamoring at the door to show that the secondary is compatible with a neutron star. The following is an incomplete list: Godzieba et al. (2021), Dexheimer et al. (2020), Lim et al. (2021), Tsokaros et al. (2020), Huang et al. (2020), Sedrakian et al. (2020), Demircik et al. (2021), and Cao et al. (2020). Each of these papers shows there are important nuclear physics ramifications if one were able to determine that the secondary of GW 190814 is the most massive neutron star observed to date.

There are other ways of inferring the maximum mass of neutron stars that can inform us about the nature of GW 190814. One of the most interesting is to study the galactic population of neutron stars with known masses in a German-tank-like problem, allowing for posterior predictive estimates for the maximum mass of neutron stars. This was originally done in Alsing et al. (2018) and updated in Shao et al. (2020) to account for, among other things, the most massive neutron star measurement (Cromartie et al. 2020). This latter update makes life interesting for the secondary of GW 190814; the posterior predictive distribution for the maximum mass becomes $M_{\rm TOV} = 2.26^{+0.47}_{-0.11} M_\odot$, where the uncertainties are 95% credible intervals (c.f. the 90% upper limit of $M_{\rm TOV} \leqslant 2.6 M_{\rm TOV}$ from Alsing et al. 2018). This long upper-mass tail can accommodate the secondary of GW 190814 being a neutron star without significant tension, although this does not alleviate the tension between this interpretation, terrestrial experiments, and GW 170817 analyses (Fattoyev et al. 2020).

Now that we have failed to resolve the question of the nature of the secondary, we can move on to failing to resolve the formation history of such an exotic system. It is often speculated that there are few, if any, compact objects between $M_{\rm TOV}$ and $\sim 5 M_\odot$ (Bailyn et al. 1998; Özel 2010). This "lower" or "first" mass gap could be due to simple observational biases associated with observing Galactic black holes as X-ray accreting systems (e.g., Kreidberg et al. 2012) or due to details associated with the supernova explosion mechanism (e.g., Belczynski et al. 2012). Much of this latter work claiming that the supernovae cannot create black holes in the lower-mass map relies on post facto interpretations of the observations, rather than deriving or simulating the physics from first principles. As a concrete example, Fryer et al. (2012) propose two related mechanisms—the so-called delayed and rapid explosion mechanisms—one of which produces a significant fraction of black holes in the lower-mass gap while the other does not. "If the observed gap represents the true compact object formation mass distribution, the rapid explosion engine is a better match to the observed gap" (Fryer et al. 2012). In this spirit, if the secondary from GW 190814 is really a first-generation black hole (see below), then these arguments about explosion mechanisms not being able to form mass-gap black holes would simply subside into oblivion.

Even if one can explain the secondary mass through typical stellar evolution, the mass ratio and component masses are difficult to explain through population synthesis simulations, independent of whether the secondary is a black hole or neutron star (e.g., Eldridge & Stanway 2016; Neijssel et al. 2019; Olejak et al. 2020 and references therein). Extreme mass ratios of $q \approx 0.1$ formed through stellar evolution seem only possible at low metallicities (Spera et al. 2019; Giacobbo & Mapelli 2018; Giacobbo et al. 2018). Dynamical capture also tends to favor equal-mass systems (Sigurdsson & Hernquist 1993), particularly in globular clusters (e.g., Rodriguez et al. 2016). This region of parameter space is largely ill understood for dynamical captures in young star clusters (Abbott et al. 2020g). It has been postulated that the secondary from GW 190814 is a second-generation black hole in a globular cluster (Gupta et al. 2020; Abbott et al. 2020g); however, simulations show these seldom merge with significantly more massive black holes (Ye et al. 2020) as is the case here. Finally, it is possible that the secondary of GW 190814 is a primordial black hole, formed through large density fluctuations early in the universe (Clesse & Garcia-Bellido 2020); however, this requires fine-tuning the amount of time that elapses between formation and merger of such objects to explain the merger rate (Vattis et al. 2020).

The LIGO/Virgo Collaboration's conclusion on the formation of GW 190814 is uncharacteristically and unexpectedly sassy: "We conclude that the combination of masses, mass ratio, and inferred rate of GW 190814 is challenging to explain, but potentially consistent with multiple formation scenarios. However, it is not possible to assess the validity of models that produce the right properties but do not make quantitative predictions about formation rates, even at some order-of-magnitude level" (Abbott et al. 2020g).

Population Studies: The previously mentioned "exceptional events" give amazing insight into binary black hole systems, including various things like their formation mechanism, tests of general relativity, and so much more. But we are now truly in the era of regular detections; during LIGO and Virgo's third observing run, binary black hole mergers were being detected at approximately a rate of one per week.

This number of events means we are well into the catalog era of gravitational-wave discovery, where robust population-based inference can be performed.

In general, the tool for population inference is hierarchical Bayesian inference—for an introduction in the context of gravitational waves, see Thrane & Talbot (2019). The LIGO–Virgo collaborations have released their gravitational-wave transient catalogs 1 and 2 (GWTC-1 and GWTC-2) after the second and third observing runs, respectively. With each release, both a "catalog" and "rates and population" paper have been published; the former laid out the details of the individual compact binary merger detections and their parameters, and the latter study their population properties. Clearly, the O3a catalog (Abbott et al. 2021a) and population (Abbott et al. 2021b) papers supersede those of O2 (Abbott et al. 2019g, 2019i), and so in this section we concentrate only on those results.

One of the primary goals of the aforementioned work is to provide updates on the merger rates, with the GWTC-2 analysis giving the binary black hole merger rate as $\mathcal{R}_{\mathrm{BBH}} = 23.9^{+14.9}_{-8.6}\,\mathrm{Gpc}^{-3}\,\mathrm{yr}^{-1}$. Interestingly, there is tentative but not conclusive evidence that the binary black hole merger rate increases as a function of redshift, albeit at a rate that is not faster than the star formation rate.

As discussed above, primarily in the context of GW 190521, an upper-mass gap has been theorized to exist approximately above $45M_{\odot}$ due to pair-instability supernovae. In addition to GW 190521, two other binary black hole mergers have significant posterior support between $45M_{\odot}$ and $100M_{\odot}$, and a merger rate for systems with at least one black hole with mass in this range is $0.71^{+0.65}_{-0.37}\,\mathrm{Gpc}^{-3}\,\mathrm{yr}^{-1}$. A sharp cutoff in the population distribution of primary masses is disfavored, with instead a broken power law, or power law with Gaussian peaking at $33.5^{+4.5}_{-5.5}M_{\odot}$ being the preferred models.

The lower-mass gap is more confusing, with GW 190814 being an outlier to the population distribution of binary black holes. Indeed, if one assumes the secondary of GW 190814 is a low-mass black hole, then there is no mass gap between neutron stars and black holes, in contrast to theoretical expectations (see discussion of GW 190814 above). On the other hand, if the secondary of GW 190814 is a neutron star, then there is support for a dearth of black holes between about $2.6M_{\odot}$ and $6M_{\odot}$, with a peak in the black hole mass spectrum at $7.8^{+2.2}_{-2.1}M_{\odot}$.

Finally, and perhaps most interestingly, despite there being no definitive evidence for precession in any given binary black hole merger, the population does show evidence that some binary black hole systems have component spins misaligned with the orbital angular momentum. Indeed, a significant fraction of systems have their spins more than 90° from their orbital angular momentum. This is difficult to achieve through regular binary stellar evolution, where only the nascent kick from the second supernova can act to misalign the spin from the orbital angular momentum after the common envelope phase. The suggestion is therefore that there is more than one formation channel being seen in the binary black hole mergers so far observed.

The O3a analyses of black hole mergers brought about many interesting observations, both of individual events and populations. There are still approximately 23 events to be analyzed from O3b—see Figure 9.11 and surrounding text.

9.7.2 Astrophysics of Neutron Star Mergers

To date, two binary neutron star systems have been observed in gravitational waves: the celebrated first bona fide multimessenger gravitational-wave event GW 170817 (Abbott et al. 2017e, 2017f) and the relatively unknown GW 190425 (Abbott et al. 2020b)—see the purple regions and contours in Figure 9.12. In all, this gives a binary neutron star merger rate of $\mathcal{R}_{\rm BNS} = 320^{+490}_{-240}\,{\rm Gpc^{-3}yr^{-1}}$, c.f. $\mathcal{R}_{\rm BBH} = 23.9^{+14.9}_{-8.6}\,{\rm Gpc^{-3}\,yr^{-1}}$ for binary black hole systems (Abbott et al. 2021b). One must remember that the gravitational-wave strain from a binary neutron star merger at the same distance as a binary black hole merger is significantly smaller, which is why the detection rate is much lower even though the absolute merger rate is considerably higher.

GW 170817: On 2017 August 17 at about 12:40 UTC, a gravitational-wave signal from coalescing neutron stars began ringing up the LIGO and Virgo interferometers (Abbott et al. 2017e). They finally merged approximately 90 s later, although the merger itself occurred at too high a frequency to be picked up by any of the instruments. Just 1.74 ± 0.05 s later (Abbott et al. 2017c),[15] an excess of gamma-ray photons was detected by both Fermi's Gamma-ray Burst Monitor and INTEGRAL's spectrometer. Those gamma rays were from short gamma-ray burst GRB 170817A, which is now known to be unequivocally associated with the gravitational-wave event. Some 10.9 hr later, the 1 m Swope telescope detected SSS 17a, the first optical photons from a fading source in galaxy NGC 4993 (Coulter et al. 2017) at a luminosity distance $d_L \approx 40$ Mpc (Cantiello et al. 2018; Lee et al. 2018). The system was not detected in X-rays until 9 days later (Troja et al. 2017) and in radio for 16 days (Mooley et al. 2017; Corsi et al. 2017).

The same optical source was subsequently observed by a host of other telescopes over the next tens of days, and also across the full electromagnetic spectrum from gamma rays to radio waves and everything in between. Figure 9.15 shows a timeline of all observations during the first $\approx$30 days, including the gravitational-wave spectrum, gamma-ray counts, optical and UV images of the afterglow, and spectra of the kilonova (Abbott et al. 2017f). More than 940 days after the merger, X-rays from the source were still being observed (Troja et al. 2020).

The gravitational-wave signal from GW 170817 was incredibly loud, with a network matched-filter signal-to-noise ratio of 32.4 (Abbott et al. 2017e). This is primarily due to the two LIGO detectors, with a signal-to-noise ratio of 18.8 and 26.4 in LIGO-Hanford and Livingston, respectively, and the Virgo interferometer only detecting a marginal signal with a signal-to-noise ratio of 2.0. Despite this, Virgo is useful in constraining the sky localization as one is able to use the antenna-pattern function of the detectors to understand that the merger must have occurred in a region of the sky where Virgo was less sensitive at the time. In practical terms, this means that the sky localization using a low-latency pipeline from just the two LIGO detectors was 191 deg^2, which reduces to just 31 deg^2 when Virgo is included—see Figure 9.8.

[15] Interestingly, Abbott et al. (2017f) quote this result as 1.734 ± 0.054 s, and then self-cite their own paper. It is surprising that none of the 3677 authors picked up this error (including myself).

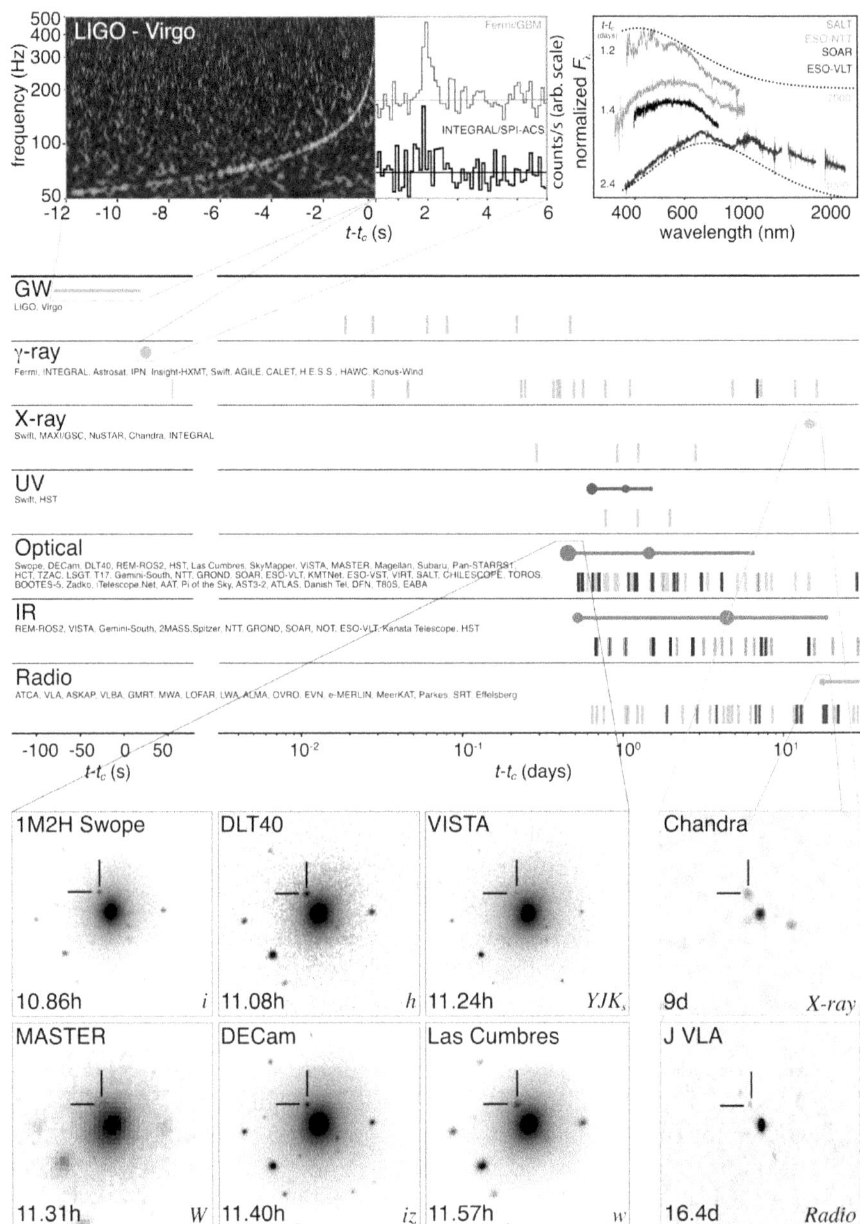

Figure 9.15. Timeline of discovery for the binary neutron star merger GW 170817. The top left is the combined gravitational-wave spectrum from LIGO and Virgo, alongside the gamma-ray observations from Fermi's Gamma-ray Burst Monitor and INTEGRAL's spectrometer. Top right are four spectra taken after 1.2 days (red), 1.4 days (green), 1.4 days (black), and 2.4 days (blue) by four different instruments. The bottom six images on the left (with green borders) are the first six optical images of galaxy NGC 4993, with the event shown in the crosshairs, between 10 hr and 12 hr following the event. The images on the bottom right show a Chandra X-ray image 9 days after the merger and a VLA radio image taken after 16.4 days. The middle panel shows a timeline of all observations taken in the gravitational wave and across the electromagnetic spectrum. Image credit: Abbott et al. (2017f).

High-latency analysis used for electromagnetic follow-up improved the sky localization to 28 $\deg^2$. Subsequent improvements to parameter estimation incorporating recalibrated Virgo data, a reduction in the minimum gravitational-wave frequency used from 30 Hz to 23 Hz, and improved waveform models ultimately reduces the sky localization at the 90% credible interval to 16 $\deg^2$ (Abbott et al. 2019).

Initially published inference of GW 170817 using gravitational-wave data only (i.e., ignoring information about the electromagnetic counterpart) constrained the source luminosity distance to $d_L = 40^{+8}_{-14}$ Mpc (Abbott et al. 2017e) with improved waveform systematics improving this to $d_L = 41^{+6}_{-11}$ Mpc (Abbott et al. 2019l). These distance measures are important: combining them with distance estimates for the electromagnetic counterpart allows one to make a distance-ladder-independent measurement of the Hubble constant that complements other cosmological measurements—we return to these cosmological inferences below.

It is first worth pausing and asking what can, and was, learned from the gravitational-wave data alone, ignoring the electromagnetic counterpart for the moment. As discussed in Section 9.6, the gravitational-wave merger frequency of binary neutron star mergers is well into the kilohertz regime, a region where LIGO/Virgo detectors are not currently sensitive. Indeed, the actual merger of GW 170817 is not seen in gravitational waves because of this lack of sensitivity, and the signal is lost at approximately 400 Hz (see top-left panel of Figure 9.15). No postmerger signal was detected either (Abbott et al. 2017i, 2019m and see discussion below). All information from the gravitational-wave channel is therefore from the inspiral phase.

As neutron stars coalesce, the gravitational-wave phase evolves differently to two point masses (i.e., black holes) due to the tidal deformation of either star. In turn, this provides information about the equation of state and is therefore one of the key pieces of information available from neutron star mergers. The equation-of-state-dependent tidal deformability of a neutron star is a function of the star's radius R and dimensionless Love number k_2 (Hinderer 2008)

$$\lambda = \frac{2}{3G} k_2 R^5. \tag{9.18}$$

It is convenient to undimensionalize the tidal deformability; for the ith neutron star we define

$$\Lambda_i = \frac{c^{2/5}}{G^4} \frac{\lambda_i}{m_i^5}. \tag{9.19}$$

In a binary neutron star merger, Λ_1 and Λ_2 are correlated (e.g., Flanagan & Hinderer 2008 and see the right-hand panel of Figure 9.16). Convenience again dictates a reparameterization (Favata 2014; Wade et al. 2014; Lackey & Wade 2015), where we define two linear combinations of the tidal deformability known as $\tilde{\Lambda}$ and $\delta\tilde{\Lambda}$, which are, respectively,

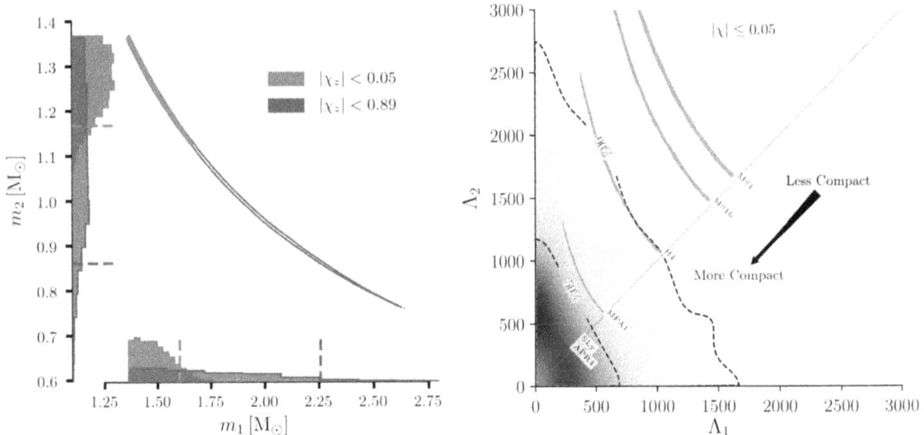

Figure 9.16. Posterior distributions for the masses (left) and tidal deformabilities (right) of the progenitor neutron stars for the GW 170817 signal. The mass posteriors are calculated under two different assumptions, using a low-spin prior such that the magnitude of the in-plane dimensionless spin $|\chi_z| < 0.05$ (blue) and a high-spin prior with $|\chi_z| < 0.89$ (red). The tidal deformability shown on the right are those corresponding to the low-spin prior (see Abbott et al. 2017e for the posterior assuming a high-spin prior). Overplotted on the right-hand panel are some representative equations of state. Image credit: Abbott et al. (2017e), with permission, copyright 2017 by the American Physical Society.

$$\begin{aligned} \tilde{\Lambda} &= \frac{8}{13}\left[(1 + 7\eta - 31\eta^2)(\Lambda_1 + \Lambda_2) + \sqrt{1 - 4\eta}(1 + 9\eta - 11\eta^2)(\Lambda_1 - \Lambda_2)\right], \\ \delta\tilde{\Lambda} &= \frac{1}{2}\left[\sqrt{1 - 4\eta}\left(1 - \frac{13272}{1319}\eta + \frac{8944}{1319}\eta^2\right)(\Lambda_1 + \Lambda_2)\right. \\ &\quad \left. + \left(1 - \frac{15910}{1319}\eta + \frac{32851}{1319}\eta^2 + \frac{3380}{1319}\eta^3\right)(\Lambda_1 - \Lambda_2)\right], \end{aligned} \tag{9.20}$$

where $\eta = m_1 m_2/(m_1 + m_2)$ is known as the symmetric mass ratio. These tidal deformability parameters enter the change in phase of the waveform as

$$\delta\phi(t) = -\frac{39}{2}\tilde{\Lambda}\left(\frac{\pi G\mathcal{M}f}{c^3}\right)^5 + \left(-\frac{3115}{64}\tilde{\Lambda} + \frac{6595}{365}\sqrt{1 - 4\eta}\,\delta\tilde{\Lambda}\right)\left(\frac{\pi G\mathcal{M}f}{c^3}\right)^6. \tag{9.21}$$

The first term on the right-hand side is a fifth-order post-Newtonian term in the waveform expansion, whereas the second term is sixth order (Favata 2014; Wade et al. 2014). The second parameter $\delta\tilde{\Lambda}$ is therefore subdominant to $\tilde{\Lambda}$, and indeed one does not expect to be able to measure $\delta\tilde{\Lambda}$ with Advanced LIGO/Virgo (e.g., Wade et al. 2014). Even for the loud GW 170817 signal, the posteriors for this parameter were uninformative. Figure 9.16 shows the two-dimensional posteriors for the masses m_1 and m_2 (left panel), and the tidal deformabilities Λ_1 and Λ_2 (right panel), of the two objects that coalesced to create the gravitational-wave signal GW 170817 (Abbott et al. 2017e). The mass posteriors show two different sets of contours, corresponding to two different choices of the orbit-aligned dimensionless spin parameter χ_z (the same as $\chi_{\rm eff}$ defined in Equation (9.16)). The so-called low-spin prior $|\chi_z| < 0.05$ can otherwise be known as an "'astrophysical prior," in the sense that the most extreme-spinning Galactic double neutron system is expected to have $\chi_z \lesssim 0.4$ at merger. The high-spin prior $|\chi_z| < 0.89$ is therefore much more agnostic to our astrophysical assumptions, but such large spins cannot be ruled out a priori. The low-spin prior gives masses m_1 between $1.36 M_\odot$ and $1.60 M_\odot$ at 90% confidence, and m_2 between $1.17 M_\odot$ and $1.36 M_\odot$, consistent with the galactic population of double neutron star systems (e.g., Alsing et al. 2018; Farrow et al. 2019). Interestingly, the measured posteriors for the dimensionless effective spin of GW 170817 is consistent with zero, measured as $\chi_{\rm eff} = 0.00^{+0.02}_{-0.01}$ when assuming a low-spin prior and $\chi_{\rm eff} = 0.02^{+0.08}_{-0.02}$ for the high-spin prior (Abbott et al. 2019l).

The right-hand panel of Figure 9.16 shows the tidal deformabilities of the primary and secondary, Λ_1 and Λ_2, respectively, for GW 170817 assuming the low-spin prior. For the equivalent plot for the high-spin prior, see Abbott et al. (2017e). Overplotted on the right-hand panel of Figure 9.16 is a representative set of equation-of-state curves, showing that the stiffest of equations of state in this sample, MS1 and MS1b, are ruled out by the measurements. The two-dimensional posterior distributions for the tidal deformability are complicated, but the LIGO/Virgo analyses can be condensed to just two related salient numbers: $\tilde{\Lambda} \lesssim 800$ and $\Lambda(m = 1.4 M_\odot) \lesssim 800$. It is important to note that $\Lambda_{1,2} = 0$ is not ruled out with these analyses. In other words, using only the gravitational-wave data, one cannot discern whether the merging compact objects were neutron stars or black holes, or one of each.

The above tidal deformability measurements can be improved significantly by utilizing the distance of the electromagnetic counterpart, rather than only the gravitational-wave data to measure the distance, and by assuming both neutron stars have the same equation of state. With these assumptions and assuming a uniform prior on the component masses, the 90% constraints on the tidal deformability is $\tilde{\Lambda} = 222^{+420}_{-138}$, or $\tilde{\Lambda} = 233^{+448}_{-144}$, when the component-mass prior is informed by the measured masses of radio pulsars (De et al. 2018).[16]

Measurements of the neutron star tidal deformability can be converted into more convenient mass–radius constraints that are directly comparable to solutions of the general relativistic equations of hydrostatic equilibrium for different equations of state. These conversions can be done in a number of different ways, including so-called equation-of-state independent methods that rely on universal relations for neutron stars (Yagi & Yunes 2013a, 2013b; Agathos et al. 2015; Yagi & Yunes 2017) and various equation-of-state parameterizations (Read et al. 2009; Lindblom 2010; Hebeler et al. 2013; Riley et al. 2018). The left-hand panel of Figure 9.17 shows one such mass–radius posterior distribution calculated from LIGO/Virgo data using a spectral decomposition approximation for the equation of state (Lindblom 2010), with five different nuclear equation-of-state curves overlaid. In contrast, the right-

[16] De et al. (2018) also derived a posterior for $\tilde{\Lambda}$ with the component-mass prior informed by the masses of Galactic double neutron star systems. While this was a good idea at the time, the measurement of the masses of GW 190425 indicates this is not likely a good proxy for the mass distribution of merging extragalactic neutron stars.

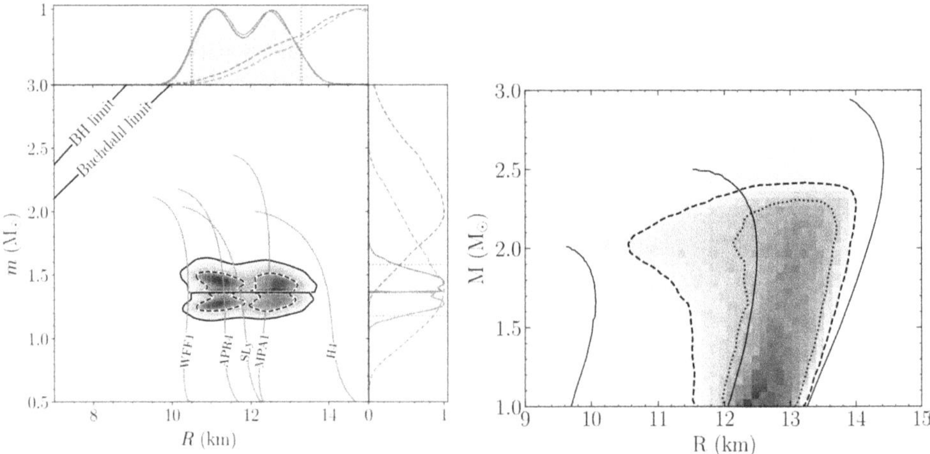

Figure 9.17. Mass–radius posteriors using gravitational-wave data from GW 170817 (left) and X-ray observations of the millisecond pulsar PSR 0300+0451 (right) assuming spectral decomposition and piecewise polytrope parameterizations of the equation of state, respectively. Images credit: left-hand panel from Abbott et al. (2018c), with permission, copyright 2018 by the American Physical Society; right-hand panel from Raaijmakers et al. (2019).

hand panel of Figure 9.17 shows the same parameter space; however, the posterior distribution calculated from pulse-profile modeling of X-ray observations of the millisecond pulsar PSR J0030+0451 (Raaijmakers et al. 2019; Riley et al. 2019) is with the Neutron Star Interior Composition ExploreR (NICER; Gendreau et al. 2016). It is worth mentioning that the right-hand plot in Figure 9.17 assumes a piecewise polytrope equation-of-state parameterization (Read et al. 2009), c.f. a spectral decomposition (Lindblom 2010) for the left-hand plot: other parameterizations are presented in their respective papers, and the results are mostly consistent.

The LIGO/Virgo analysis using a spectral decomposition for the equation of state constrains the two neutron stars to radii $R_{1,2} = 11.9^{+1.4}_{-1.4}$ km (Abbott et al. 2019l), which is largely independent of the neutron star mass in the range given by the low-spin prior. On the other hand, the NICER results prefer a slightly stiffer equation of state, with the equatorial radius for PSR J0030+0451 measured to be $R = 12.71^{+1.14}_{-1.19}$km (Raaijmakers et al. 2019), although this could be larger by up to a kilometer depending on the topology of the hot spot on the neutron star (Riley et al. 2019). Numerous works have explored joint constraints using both LIGO/Virgo and NICER data (e.g., Raaijmakers et al. 2020; Zimmerman et al. 2020; Biswas et al. 2021; Landry et al. 2020; Al-Mamun et al. 2021; Capano et al. 2020), deriving stringent constraints on, for example, the nuclear symmetry energy among other things. Both the tidal deformability measurements of GW 170817 and the NICER measurements of PSR 0300+0451 are driving a mini revolution in our understanding of the equation of state of nuclear matter. Upcoming measurements from both teams, combined with heavy-ion collider experiments and an improved theoretical understanding of building equations of state will continue to drive this exciting field through the coming decade.

One consequence of knowing the equation of state is the opportunity to also understand the maximum, nonrotating neutron star mass—the TOV mass (see Section 9.7.1). Understanding the TOV mass allows, in turn, an understanding of evolutionary pathway for the postmerger remnant of binary neutron star mergers—for a review of postmerger remnants, see Sarin & Lasky (2021). Conversely, if one can measure the TOV mass, this has important input into constraining the equation of state.

There are four possible scenarios that depend on the remnant mass $M_{\rm rem}$ and its relation to $M_{\rm TOV}$. If $M_{\rm rem} \leqslant M_{\rm TOV}$, the remnant will survive as a stable neutron star. If $M_{\rm TOV} < M_{\rm rem} \lesssim 1.2 M_{\rm TOV}$ (Cook et al. 1992, 1994a, 1994b; Breu & Rezzolla 2016), the remnant will survive as a supramassive neutron star, supported by centrifugal support as a result of rapid, albeit solid-body, rotation. In such a case, the remnant is expected to survive for anywhere up to $\approx 10^5$ s before collapsing to form a black hole (Ravi & Lasky 2014). If the remnant is $1.2 M_{\rm TOV} \lesssim M_{\rm rem} \lesssim 1.5 M_{\rm TOV}$ (Baumgarte et al. 2000), the star, known as a hypermassive neutron star, is supported centrifugally by differential rotation. This differential rotation is thought to be quenched on approximately an Alfvén timescale, at which point the star will collapse to form a black hole $\lesssim 1$ s after formation (e.g., Baumgarte et al. 2000; Shapiro 2000). On the other hand, if the remnant has $M_{\rm rem} \gtrsim 1.5 M_{\rm TOV}$, it will promptly collapse to a black hole.

If the postmerger remnant collapses promptly to a black hole, it will emit gravitational waves upward of 6 kHz; even proposed next-generation high-frequency detectors (Martynov et al. 2019; Ackley et al. 2020) will not be able to detect signals from such black holes. On the other hand, if the star survives as either a hypermassive, supramassive, or stable neutron star, it will emit gravitational waves in the first hundreds of milliseconds with comparable amplitude to that of the late inspiral phase (e.g., see Xing et al. 1994; Ruffert et al. 1996; Read et al. 2013). Importantly, the dominant gravitational-wave frequency emitted by the postmerger remnant is strongly correlated with the neutron star compactness and hence the equation of state (e.g., Takami et al. 2015 and references therein). Measurement of gravitational waves from this initial phase will give unprecedented insight into the hot nuclear equation of state, distinct from premerger inspiral measurements and complementary pulsar observations that all measure the cold equation of state. For example, measuring gravitational waves from the postmerger remnant may allow for indirect measurements of phase transitions in the stellar core (e.g., Bauswein et al. 2019; Bauswein & Blacker 2020).

So were gravitational waves detected from the postmerger remnant of GW 1708017? No, and it was not close (Abbott et al. 2017i, 2019m). See Clark et al. (2016) and references therein for expected horizon distances for postmerger remnants. At the sensitivity of the instruments at the time, GW 170817 would have had to merge approximately a factor of 40 times closer for the immediate postmerger remnant (if it existed) to be detected. As strain is inversely proportional to distance, another way of saying this last sentence is that a factor of about 40 improvement in high-frequency (2–4 kHz) sensitivity is required to be able to measure gravitational waves from a postmerger remnant of a GW 170817–like system. It is this that dedicated high-frequency instruments such as NEMO (Ackley et al. 2020), as well as broadband, third-generation instruments such as the Einstein Telescope (Punturo et al. 2010) and CE (Abbott et al. 2017h) hope to achieve.

Gravitational waves from long-lived postmerger remnants may also be highly excited and observable with next-generation gravitational-wave interferometers. Three main gravitational-wave mechanisms include unstable bar modes (e.g., Corsi & Mészáros 2009), magnetic-field-induced ellipticities (e.g., Cutler 2002), and r-mode oscillations (Andersson 1998) that are unstable to the Chandrasekhar–Friedman–Schutz mechanism (Chandrasekhar 1970; Friedman & Schutz 1978). Were gravitational waves from this phase of a putative postmerger remnant detected from GW 170817? Again, the answer is no, and not even close (Abbott et al. 2017i, 2019m). Although one likely does not expect gravitational waves from such an object to be detected in the near future (Sarin et al. 2018), there is some hope one may be able to indirectly infer gravitational-wave emission looking at the spin-down of postmerger remnants via the X-ray afterglow of short gamma-ray bursts (e.g., Fan et al. 2013; Lasky & Glampedakis 2016; Sarin et al. 2020).

Gravitational waves from the postmerger remnant would have been a smoking gun detailing what happened in the aftermath of GW 170817. But hope is not lost without this, and indeed electromagnetic observations can potentially shed some light on the fate of the remnant. For example, some authors (e.g., Granot et al. 2017) attribute some fraction of the 1.74 s delay between the gravitational-wave and gamma-ray observations to the formation and subsequent collapse of a hypermassive neutron star. That is, the gamma-ray burst jet was only launched once the remnant collapsed to form a black hole. It is enticing to speculate that this is true, and indeed if it is, one can make interesting inferences about the equation of state and the TOV mass (Bauswein et al. 2017; Margalit & Metzger 2017; Rezzolla et al. 2018). However, there is ongoing debate about this point—indeed, the existence of long-lived X-ray afterglows has been hypothesized to contain long-lived neutron stars as the central engines (Rowlinson et al. 2013; Lü 2015; Sarin et al. 2019). If true, this would imply the collapse of a neutron star is not required to launch a gamma-ray burst jet. For more conservative constraints on the maximum mass that outlines each possible scenario for the fate of the remnant of GW 170817, see Ai et al. (2020).

The electromagnetic counterpart to GW 170817 provides rich and interesting evidence for the fate of the merger remnant. The kilonova counterpart AT 2017gfo began as a blue thermal spectrum peaking in the optical and ultraviolet, before reddening over the next few days to become dominated by infrared emission (e.g., Evans et al. 2017; Tanvir et al. 2017; Pian et al. 2017). While the late-time red component is attributed to the nucleosynthesis of heavy elements through r-process decay (e.g., Metzger et al. 2010), the early blue component is consistent with lanthanide-poor ejecta with a relatively high-electron fraction, $Y_e \gtrsim 0.25$ (Tanvir et al. 2017; Smartt et al. 2017; Evans et al. 2017; Pian et al. 2017). This blue component may be inconsistent with no neutron star surviving as a remnant (i.e., prompt collapse), because neutrinos from the hot merger remnant are required to irradiate the ejecta and raise the electron fraction (Metzger & Fernández 2014;

Perego et al. 2014; Margalit & Metzger 2017), although this conclusion may be sensitive to the viewing angle (Klion et al. 2021). This does not do justice to the complex and rich physics being understood from the kilonova observations of AT 2017gfo—for a much more detailed overview, see the review by Metzger (2019).

It is worth mentioning that excess X-ray emission 155 days following the merger has been speculatively attributed to a long-lived magnetar remnant (Piro et al. 2019), a scenario that is still consistent with X-ray emission seen at almost 1000 days (Troja et al. 2020).

One of the most striking pieces of information to come from the coincident electromagnetic and gravitational-wave observations is a robust test of the speed of gravity. In principle, the 1.74 s delay between the gravitational waves and gamma rays can be attributed to some combination of the time it took for the jet to launch, jet propagation, and shock formation. However, even conservatively ignoring this, one can constrain the velocity difference between the gravitational waves and the photons, deriving the remarkable result that (Abbott et al. 2017c)

$$-3 \times 10^{-15} \leqslant \frac{\nu_{\rm GW} - \nu_{\rm EM}}{\nu_{\rm EM}} \leqslant 7 \times 10^{-16}, \tag{9.22}$$

where $\nu_{\rm GW}$ and $\nu_{\rm EM}$ are the speed of gravity and the light, respectively. To put these numbers in context, prior to GW 170817, the best fractional bounds from direct measurements of the speed of gravity were between 0.55 and 1.42 (Cornish et al. 2017). In addition to testing the speed of gravity, one can place interesting limits on the Lorentz violation and tests of the equivalence principle—for details, see Abbott et al. (2017c) and references therein.

Gravitational-wave inference of the event GW 170817 goes beyond the masses, tidal deformabilities, and sky localizations. One of the key outcomes was the system's inclination angle with respect to the observer's line of sight. In terms of gamma-ray burst physics, the coincident measurement of the inclination angle with gravitational waves allowed for a first probe of the rich structure of the ultra-relativistic jet emanating from the merger site (Alexander et al. 2018; Mooley et al. 2018b). Such studies, including many subsequent papers on the topic, are helping understand the ultrarelativistic propagation of gamma-ray burst jets and, in turn, multiple other aspects of the complicated physics and astrophysics of gravitational waves.

Further radio observations of the jet showing that it is superluminal (Mooley et al. 2018a) gave tighter constraints on the jet-opening angle, which ultimately allow for a tighter constraint on the Hubble constant of $H_0 = 70.3^{+5.3}_{-5.0}$ km s^{-1} Mpc^{-1} (Hotokezaka et al. 2019). While these constraints are not as strong as current Cepheid-calibrated supernova (Riess et al. 2016, 2019) or cosmic microwave background (Aghanim et al. 2020) constraints, they are independent of the cosmic distance ladder (Schutz 1986) and therefore provide a complementary cosmological probe. Estimates (Hotokezaka et al. 2019) show that approximately 50 to 100 GW 170817–like observations should be sufficient to resolve the Hubble tension (e.g., Verde et al. 2019).

GW 190425: The first binary neutron star merger was a special treat—not all mergers are expected to come with electromagnetic counterparts, and indeed the second neutron star merger detected by LIGO/Virgo had no such thing. Despite this, GW 190425—uncommonly known as "The ANZAC day event"—still threw up some interesting surprises (Abbott et al. 2020b). In some senses, GW 190425 is a single-detector event: LIGO Livingston and Virgo were both locked and observing at the time, but the signal-to-noise ratio was so low in Virgo that it did not contribute to the detection. It did, however, contribute to parameter inference. Regardless, the 90% credible interval for the sky localization for GW 190425 from full Bayesian inference covered a whopping 8284 deg^2 (c.f. GW 170817, which was constrained to 28 deg^2). Moreover, the median-inferred distance to GW 190425 is some four times farther than GW 170817, at $d_L = 159^{+69}_{-71}$ Mpc. At such large distance, and correspondingly low signal-to-noise ratio, the posterior distributions for the tidal deformabilities of the two bodies were uninformative. Moreover, combining the tidal-distribution posteriors with those of GW1 70817 yield posterior distributions indistinguishable to those from the analysis of GW 170817 alone (Abbott et al. 2020b; Hernandez Vivanco et al. 2020).

But perhaps it's unfair to only compare GW 190425 to her older sister. After all, GW 190425 does have one main redeeming feature that makes her scientifically fascinating. Figure 9.12 shows GW 190425 has a different total mass than GW 170817. Indeed, the total mass of the system is $M_{\mathrm{tot}} = 3.3^{+0.1}_{-0.1} M_{\odot}$ when assuming a low-spin prior and $M_{\mathrm{tot}} = 3.4^{+0.3}_{-0.1} M_{\odot}$ with the more conservative high-spin prior. Figure 9.18 shows why this is interesting by comparing this total mass with the masses of galactic double neutron star systems with known masses (Farrow et al. 2019). This total mass is more than five standard deviations from the mean of the galactic population, suggesting a different formation mechanism.

There are a number proposed formation mechanisms for GW 190425 that set it apart from the galactic double neutron star population (e.g., Abbott et al. 2020b; Safarzadeh et al. 2020; Romero-Shaw et al. 2020a). It is worth stressing that the masses of the component objects are consistent with the masses of individual neutron stars seen in our galaxy, which have a generally broader mass distribution than those seen in double neutron star systems (Kiziltan et al. 2013; Alsing et al. 2018). The mass distribution of merging neutron stars may therefore be multimodal with some fraction being drawn from a population consistent with the galactic neutron star population, and another population consistent with the remaining neutron star masses. Further neutron star merger observations will discern this mixing fraction, which has important consequences for determining, for example, the electromagnetic counterparts and remnant properties of mergers.

9.7.3 Astrophysics of Neutron Star–Black Hole Mergers

It is difficult to definitively say if we have detected, to date, the merger of a neutron star and a black hole. The case for event GW 190814 was prosecuted in the section on binary black holes rather than in this section, with the weight of evidence pointing toward the secondary compact object in that system being a low-mass

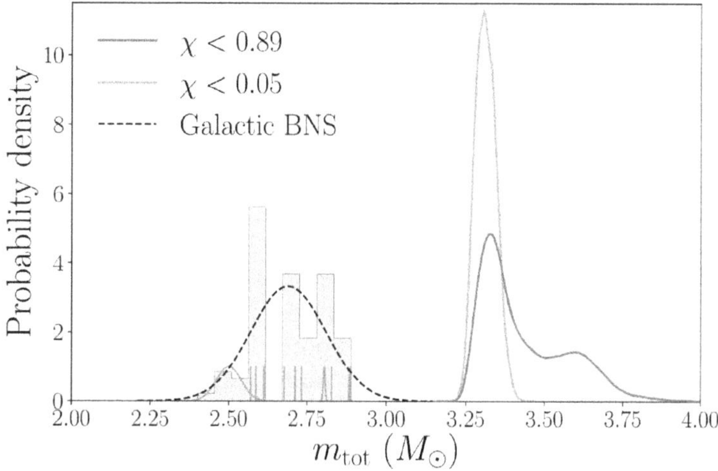

Figure 9.18. Total mass for the binary neutron star merger GW 190425 assuming a low-spin prior (orange) and a high-spin prior (blue curve), compared to the total masses of galactic double neutron star systems with measured masses. The dashed black curve shows a single-Gaussian fit to the distribution of galactic systems. Image credit: Abbott et al. (2020b).

$2.59^{+0.08}_{-0.09} M_\odot$ black hole, rather than the most massive neutron star measured to date. On the other hand, Figure 9.12 shows a singular candidate firmly entrenched in the green pastel shaded region appropriate for neutron star–black hole mergers: GW 190426. Unfortunately, this event has the highest false-alarm rate included in either the first or second of the LIGO/Virgo transient catalogs. With a false-alarm rate of 1.4 per year (Abbott et al. 2021a), it would be folly to espouse GW 190426 as a bona fide neutron star–black hole merger. In fact, because of GW 190426's false-alarm rate, it is not even included in the latest LIGO/Virgo population analyses (Abbott et al. 2021b). The uncertainty associated with the realness of GW 190426 and the nature of GW 190814 imply it is difficult, and somewhat meaningless, to attempt to calculate the merger rate of neutron star–black hole binaries.

With the above in mind, it is still worth discussing the theory of neutron star–black hole mergers. As with many topics, this review is not exhaustive—the interested reader is pointed to the dedicated review by Foucart (2020) for significantly more details.

Loosely, the dynamics of a neutron star–black hole can be divided into two pathways, defined by whether the neutron star gets tidally disrupted by the black hole or not (Lattimer & Schramm 1976). To determine this, we define a disruption radius and the black holes innermost stable circular orbit (ISCO). In Newtonian gravity and the limit of relatively large mass ratios, the disruption radius scales as $R_{\rm dis} \propto (M_{\rm BH}/M_{\rm NS})^{1/3} R_{\rm NS}$, where the constant of proportionality has a mild dependence on the equation of state (Fishbone 1973; Wiggins & Lai 2000). On the other hand, if the black hole's spin $\chi_{\rm BH}$ is aligned with the binary's orbital angular momentum, then the ISCO radius is $R_{\rm ISCO} = f(\chi_{\rm BH}) G M_{\rm BH}/c^2$, where $f(\chi_{\rm BH})$ is a function between one and nine, that is smaller for larger spins (where the spin is in the same direction as the orbital angular momentum; Bardeen et al. 1972). Broadly, if $R_{\rm dis} \gtrsim R_{\rm ISCO}$, the neutron star will be tidally disrupted outside the black hole. Otherwise, the neutron star will effectively be swallowed whole into the black hole.

It is currently believed that most neutron star–black hole binaries will not tidally disrupt (see Foucart 2020 and references therein); such binaries are comparatively uninteresting. The gravitational-wave signal is almost indistinguishable from binary black hole systems of the same mass and spins—i.e., the tidal deformation of the neutron star is not measurable (Lackey et al. 2014). Moreover, without tidal disruption, one does not expect any electromagnetic counterpart.

One the other hand, tidally disrupting binaries provide a wealth of information from both the gravitational-wave signal and potential electromagnetic counterparts. Precise calculations of tidal disruption require general relativistic simulations, which, in the case of disruption, also give estimates for the amount of matter that forms an accretion disk around the black hole after the merger (e.g., Pannarale et al. 2011; Foucart 2012; Kawaguchi et al. 2015; Foucart et al. 2018). Such an accretion disk may wind up a magnetic field and launch a jet that is observed as a short gamma-ray burst (Paschalidis et al. 2015; Siegel & Metzger 2017; Ruiz et al. 2018). The accretion disk itself, which may harbor a few percent of a solar mass of material, is expected to undergo r-process nucleosynthesis, thereby creating observable kilonova signals (e.g., Kasen et al. 2013). Finally, the gravitational0wave signal from tidally disrupted neutron star–black hole mergers potentially provides two complementary pieces of information about the equations of state: the first is the tidal dephasing also seen in binary neutron star mergers, and the second is that the gravitational-wave signal gets cut off at the point of tidal disruption. If measurable, this frequency cut can be used to measure the equation of state (Lackey et al. 2014; Pannarale et al. 2015).

9.7.4 Compact Binary Tests of General Relativity

This final section on compact binary coalescences collects some of the results for using mergers to perform tests of general relativity. Like many other sections, there is an abundance of tests being performed, and this review will only scratch the surface. It is worth emphasizing from the outset that no test of general relativity has yet failed. Actually, this is a remarkable achievement given the orders-of-magnitude difference in both gravitational potential and gravitational curvature being tested with gravitational waves from compact binary mergers compared to previous astronomical and terrestrial tests of general relativity—for reviews, see Psaltis (2008) and Will (2014).

Alongside the second gravitational-wave transient catalog (GWTC-2; Abbott et al. 2021a), the LIGO/Virgo collaboration published a paper with a number of different tests of general relativity (Abbott et al. 2021c), updating constraints on tests of general relativity produced with the first gravitational-wave transient catalog (GWTC-1; Abbott et al. 2019f), and also some of those published with the binary neutron star merger GW 170817 (Abbott et al. 2019k).

One of the cornerstone tests of general relativity is an attempt to verify the famous no-hair conjecture (Israel 1967, 1968; Carter 1971), which states simply that black holes can be described by three parameters—mass, spin, and charge. Astrophysical black holes are not expected to carry significant charge, implying black holes should be described completely by the Kerr spacetime (Kerr 1963) and thus only described by their mass and spin. The remnants of binary black hole mergers provide an excellent environment to test the no-hair theorem: after the merger, the remnant black hole will be in a highly deformed state, but radiate any "hair" to eventually settle to a Kerr black hole. At late times in this radiation process (technically, when the black hole can be considered to be a linear perturbation from the asymptotic state), the gravitational radiation will only carry signatures of the black hole's mass and angular momentum. The damping time and frequencies of all the emitted modes should therefore be related to just these two parameters; measuring any three parameters (e.g., the dominant damping time and frequency and one higher-order mode frequency) allows for a robust test of the no-hair theorem.

Attempts at no-hair tests were performed with the binary merger GW 150914 at the original time of publication (Abbott et al. 2016c); however, only the lowest-order quasi-normal mode frequency and damping time can be measured. However, there are subtleties with these measurements, in particular in how one defines a perturbation as being "small" enough such that one's observations are in the linear regime (Thrane et al. 2017). A partial solution to this using overtones of the fundamental mode has been developed (Giesler et al. 2019), with tests of the no-hair theorem for GW 150914 remaining consistent with the prediction of general relativity (Isi et al. 2019; Calderón Bustillo 2020).

Tests of the no-hair theorem remain difficult simply because one needs to measure higher-order overtones or modes in exponentially damping systems. For this reason, GW 150914 is still the best-measured test of the no-hair theorem. A less difficult, but also less pure test of general relativity is the so-called inspiral–merger–ringdown (IMR) consistency checks. Here, one uses gravitational waveforms of the inspiral portion of the coalescence to calculate an expectation for the dominant ringdown mode. This is considered less pure than the no-hair test because, among other things, waveform systematics and choices in cutoff frequencies between the inspiral, merger, and ringdown phases can bias one's results. Despite this, both IMR analyses of individual events, as well as of the population as a whole, have so far identified no deviations from general relativity (see Abbott et al. 2021c and references therein).

The inspiral part of a waveform can be mathematically described by a post-Newtonian expansion, which is a perturbative expansion method for solving the Einstein field equations (Blanchet et al. 1995; Kidder 1995; Blanchet et al. 2005). At each order in the expansion, one can perturb the gravitational-wave phase away from the Einsteinian value—the coefficients of this expansion can then be measured, with any nonzero value indicating a violation of general relativity at that given post-Newtonian expansion (Arun et al. 2006a, 2006b; Yunes & Pretorius 2009; Mishra et al. 2010; Li et al. 2012). The inspiral and postinspiral (which incorporates both merger and ringdown) expansions are denoted as $\{\varphi_i\}$ and $\{\alpha_i, \beta_i\}$, respectively. Figure 9.19 shows constraints on these various parameters from the second

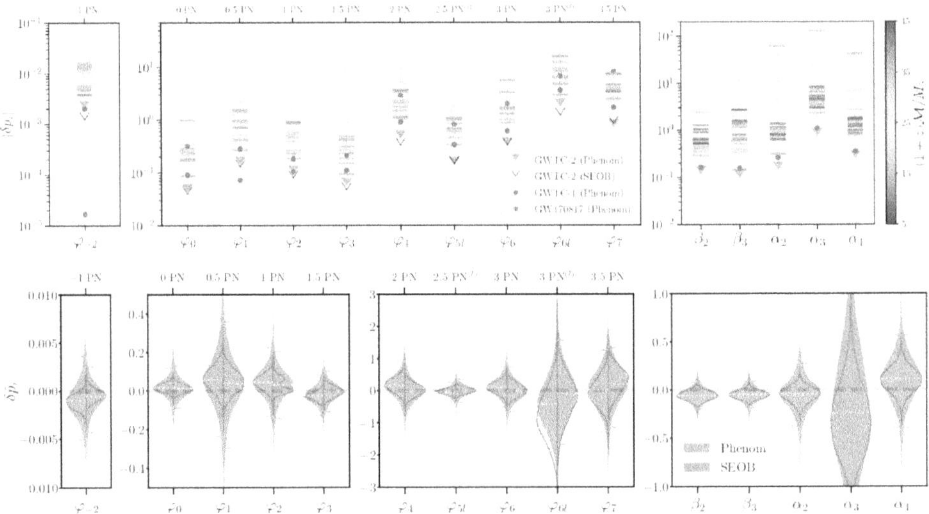

Figure 9.19. Top panel: upper bounds on the size of general-relativity-violating parameters at various post-Newtonian orders for inspiral (left and middle panels) and postinspiral (right panel). Bottom panel: posterior distributions for the same parameters, calculated by combining all binary black hole mergers in the second LIGO/Virgo gravitational-wave transient catalog. General relativity is recovered in the limit that all of these parameters are zero. Image credit: Abbott et al. (2021c), with permission, copyright 2021 by the American Physical Society.

gravitational-wave transient catalog (Abbott et al. 2021c). The top panel shows upper limits on these parameters from individual mergers (assuming different waveform models according to key in the plot). The bottom panel shows population-marginalized posterior distributions for each of these parameters assuming the same waveform families. All parameters are consistent with zero, implying no deviation from general relativity has been detected.

We have hardly scratched the surface of ongoing tests of general relativity with compact binary coalescences—a subject that deserves its own review paper. Ongoing work includes things like attempting to measure extra gravitational-wave polarizations (Eardley et al. 1973a, 1973b; Chatziioannou et al. 2012; Abbott et al. 2017d, 2019k, 2021c; Pang et al. 2020), measurements of the mass of the graviton (Will 1998; Mirshekari et al. 2012; Abbott et al. 2017e, 2019f, 2021c), searches for postmerger echoes from extreme compact objects that differ from black holes (Cardoso et al. 2016a, 2016b; Ashton et al. 2016; Abedi et al. 2017; Lo et al. 2019; Abbott et al. 2021c), and searches for gravitational-wave memory (Braginsky & Thorne 1987; Christodoulou 1991; Lasky et al. 2016; Hübner et al. 2020). While to date these have all come up null, the continued increase in sensitivity and number of detected events make this an interesting field to watch in the coming years.

9.8 Burst Sources

To date, the only gravitational-wave signals that have been detected by LIGO/Virgo are the compact binary coalescences described in the previous sections. However, there is a vast variety of other potential sources of gravitational waves that one may expect to see in the near future as the sensitivity of the instruments continues to increase. These sources can be categorized in a number of ways—herein we use the convention used by the LIGO/Virgo collaboration in defining their working groups: Compact Binary Coalescences, Burst, Continuous Waves, and Stochastic. In this section we look at burst sources of gravitational waves, focusing particularly on gravitational waves from core-collapse supernovae (Section 9.8.1), magnetar flares (Section 9.8.2), pulsar glitches (Section 9.8.3), and cosmic strings (Section 9.8.4).

9.8.1 Core-collapse Supernovae

Core-collapse supernovae were once believed to be the primary target for gravitational-wave detectors such as LIGO/Virgo. Electromagnetic emission from supernovae is not visible until hours after the initial collapse of the star. On the other hand, both gravitational waves and neutrinos are emitted during and in the immediate aftermath of the collapse, and therefore have the potential to provide valuable insight into their inner workings.

Stars with mass $\gtrsim 8 M_\odot$ end their life as core-collapse supernova (e.g., Heger et al. 2003). During the collapse, ~10^{53} erg is released, mostly in neutrinos. On the other hand, the total energy emitted in gravitational waves is comparatively small, $E_{\mathrm{GW}} \sim 10^{47}$ erg; see Abdikamalov et al. (2020), Kalogera et al. (2019), and references therein for reviews of gravitational waves from core-collapse supernovae. At design sensitivity, second-generation gravitational-wave detectors such as Advanced LIGO and Advanced Virgo are not expected to be able to see gravitational-wave signals beyond the Milky Way, while third-generation detectors such as the Einstein Telescope and CE will be able to detect signals out to hundreds of kiloparsec (Gossan et al. 2016). Unfortunately, the rate of core-collapse supernovas in the Milky Way is $1.63 \pm 0.46(100\ \mathrm{yr})^{-1}$ (Rozwadowska et al. 2021), implying prospects of gravitational-wave detection with second-generation interferometers is pessimistic. Rapid rotation of the progenitor may influence the amplitude of fundamental oscillation modes or the so-called low-$T/|W|$ instability, allowing for detections out to ~1 Mpc with third-generation detectors (e.g., Powell & Müller 2020; Shibagaki et al. 2020). The core-collapse supernova rate is much more optimistic at these distances at between 20 and 80 events per 100 years (Nakamura et al. 2016); however only $\sim 1\%$ of progenitors may be rotating fast enough for the low-$T/|W|$ instability to be relevant (Abdikamalov et al. 2020).

The gravitational-wave emission mechanisms for core-collapse supernova are replete with rich and beautiful physics. Four dominant emission mechanisms are expected with a range of characteristic frequencies and timescales—see Müller (2020) for a review of the hydrodynamics, and Abdikamalov et al. (2020) for a review of the gravitational-wave emission mechanisms. These dominant gravitational-wave mechanisms are the signal produced early during the core collapse and bounce phase (e.g., Zwerger & Mueller 1997; Dimmelmeier et al. 2007, 2008; Müller 2013), oscillations of the proto-neutron star (e.g., Torres-Forné et al. 2018; Morozova et al. 2018; Torres-Forné et al. 2019; Radice et al. 2019; Powell & Müller 2020; Shibagaki et al. 2020 and references therein), neutrino-driven convection (e.g., Mueller & Janka 1997; Foglizzo et al. 2006; Marek et al. 2009; Müller 2013; Andresen et al. 2017; Morozova et al. 2018), and the standing accretion shock instability (SASI; e.g., Blondin et al. 2003; Murphy et al. 2009; Kuroda et al. 2016; Andresen et al. 2017).

The above-mentioned emission mechanisms generate relatively short-lived bursts of gravitational waves. Unlike with compact binary coalescences, these signals are not clean, in the sense that attempts at a matched-filter analysis would quickly lose phase of the gravitational-wave signal. For that reason, a majority of gravitational-wave searches for core-collapse supernovae rely on variations of looking for excess power coincident between the two or more detectors. All-sky searches (e.g., Abbott et al. 2017a, 2019a) use the coherent WaveBurst pipeline (Klimenko et al. 2016) described in Section 9.6.3. Signals can also be searched for by combining electromagnetic observations to effectively reduce the trials factor (Abbott et al. 2016a, 2020c). Importantly, with a successful detection, algorithms are in place to perform parameter estimation using a variety of techniques (Heng 2009; Powell et al. 2016, 2017; Suvorova et al. 2019; Afle & Brown 2021; Chan et al. 2020; Bizouard et al. 2021).

9.8.2 Magnetar Flares

Isolated neutron stars are expected to emit gravitational waves through a variety of mechanisms. While some of these are nearly monochromatic signals (discussed in Section 9.9), some are expected to be much shorter lived and difficult to model sufficiently well to perform matched-filter searches. One such example are the gravitational-wave signals expected to be emitted simultaneously with magnetar flares. These highly magnetized neutron stars sometimes emit incredibly large flares, known as giant flares, three of which have occurred in our Galaxy in the last 40 years, with the brightest emitting a total isotropic energy of 2×10^{46} erg of electromagnetic emission (Palmer et al. 2005).

The electromagnetic emission from magnetar flares is powered by the magnetar's magnetic fields, with the giant flares believed to be associated with catastrophic magnetic reconnection events (e.g., Duncan & Thompson 1992; Thompson & Duncan 1995). Gravitational-wave emission must be dominated by large-scale fluid modes in the stellar interior, which can be powered one of two ways (Levin & van Hoven 2011). First, the magnetic field outside the star may catastrophically rearrange, albeit with the magnetic field tied to the stellar crust. This causes the crust to break, launching oscillation modes in the stellar core and crust. Alternatively, the internal magnetic field may rearrange, causing the flare in the first place—this internal rearrangement also launches fluid modes that emit gravitational waves. In addition to the excitement of possibly detecting gravitational waves coincident with these flares, observations of quasi-periodic oscillations in the tails of giant flares (Israel et al. 2005; Strohmayer & Watts 2005) have given hope that such simultaneous observations could in principle allow for the determination of the stellar structure (e.g., Levin 2006; Glampedakis et al. 2006; Gabler et al. 2013; Huppenkothen et al. 2014).

Initial estimates were optimistic that Advanced LIGO/Virgo would be able to detect gravitational waves from magnetar flares. They essentially calculated upper limits on the gravitational-wave energy assuming the internal magnetic field instantaneously rearranged completely, launching fluid modes that coupled to the gravitational-wave emission (Ioka 2001; Corsi & Owen 2011). Such estimates showed that 10^{48}–10^{49} erg could be released, easily detectable from Galactic magnetars. However, these now appear to be optimistic upper limits. A more nuanced calculation of the first mechanism described above, where the external magnetic field causes the crust to crack, showed that there is poor energy conversion into the fundamental *f*-mode, which is the one that couples most strongly to spacetime oscillations (Levin & van Hoven 2011). Similarly, numerical relativity simulations of magnetar interiors showed that the internal rearrangement of the magnetic field is equally disappointing for producing large-amplitude gravitational waves for detection with second-generation instruments (Zink et al. 2012; Ciolfi & Rezzolla 2012). Indeed, these latter works suggest the actual gravitational-wave emission is some 10 orders of magnitude less than the optimistic estimates.

Despite the theoretical pessimism, it is clearly worth searching LIGO/Virgo data for gravitational waves from magnetar flares because the reward for a positive detection would be significant. The best upper limits from the O2 LIGO/Virgo data put an upper energy bound on the magnetar SGR 1806-21, located 8.7 kpc away, of $\approx 3.4 \times 10^{44}$ erg at 150 Hz (lower for higher frequencies relevant for the *f*-mode), which assumes optimal orientation of the source (Abbott et al. 2019d). This is still above the most optimistic estimates from the numerical relativity simulations (Zink et al. 2012; Ciolfi & Rezzolla 2012), and only covered time spans in the data when no giant flares went off, only regular magnetar flares.

While we may have to be lucky to catch a magnetar giant flare, there is also the prospect of enhancing the sensitivity of searches for regular magnetar flares by stacking signals from multiple events (Kalmus et al. 2009). This has been done with data from initial LIGO, searching for gravitational waves at the same frequency in a longer-lasting storm from magnetar SGR 1900+14 (Abbott et al. 2009). As the sensitivity of instruments increase, it may be that creative solutions like this prove more fruitful.

9.8.3 Pulsar Glitches

Another mechanism for isolated neutrons stars to emit potentially detectable gravitational waves is through pulsar glitches. These rapid increases in a neutron star's spin frequency could excite gravitational waves either through the glitch-triggering mechanism or the postglitch relaxation of the interior fluid.

The fractional increase of the star's rotational frequency is $\Delta\nu/\nu \lesssim 10^{-4}$ (e.g., Espinoza et al. 2011 and references therein), with the best timed glitch taking less than 12 s (Palfreyman et al. 2018; Ashton et al. 2020), implying a significant amount of angular momentum is being transferred from the stellar core to crust (e.g., Montoli et al. 2020). One such mechanism for transferring such large quantities of angular momentum is through superfluid vortex avalanches in the core (Anderson &

Itoh 1975). The motion of such vortices could launch gravitational waves through the current quadrupole (Warszawski & Melatos 2012). On the other hand, the recent observation of a null pulse at the time of the 2016 glitch in the Vela pulsar (Palfreyman et al. 2018) has reopened the idea that glitches could be associated with crustquakes of the neutron star (Bransgrove et al. 2020). Although the mechanism that breaks the crust is ill understood, it could conceivably launch a veritable cornucopia of stellar oscillation modes that would emit gravitational waves of varying amplitudes (Keer & Jones 2015).

Spurred on by recent LIGO/Virgo observations of an unusual transient detection (LIGO Scientific Collaboration & Virgo Collaboration 2019; Kaplan et al. 2019) with frequency akin to what one expects from a neutron star fundamental mode (*f*-mode) oscillation, Ho et al. (2020) explored the detectability of gravitational waves coupled to the *f*-mode from nearby glitching pulsars. Assuming the same amount of energy observed in the spin frequency change during the glitch goes into the *f*-mode, they showed that the largest glitches from Vela are observable with Advanced LIGO, while A+ will be able to detect glitches from a handful of other pulsars including the Crab. It is worth mentioning that the *f*-mode oscillation frequency is in the kilohertz regime, where the current generation of detectors has significantly less sensitivity than in their most sensitive region around 100 Hz and are also not well calibrated. Future-generation detectors, either dedicated to kilohertz gravitational waves (Martynov et al. 2019; Ackley et al. 2020) or broadband third-generation telescopes (Punturo et al. 2010; Abbott et al. 2017h) will be considerably more sensitive at these high frequencies.

Despite the physical motivation, the only search for short-lived postglitch gravitational waves published by the LIGO/Virgo collaborations followed the 2006 Vela glitch (Abadie et al. 2011), which was in initial LIGO's fifth science run. Gravitational waves were not detected, and the upper limits were two to three orders of magnitude weaker than predictions for the gravitational-wave energy potentially released during the glitch.

While *f*-mode oscillations described above will damp in $\lesssim$100 ms (Detweiler 1975; Lioutas & Stergioulas 2018), longer-lived gravitational-wave signals could be induced in neutron star glitches. The postglitch spin evolution of pulsars typically shows exponential recovery of the pulsar spin frequency and its time derivative (Lyne et al. 2000; Espinoza et al. 2011), which has been attributed to Ekman flows in the stellar core van (Eysden 2008; Bennett & van Eysden 2010) and the birth of neutron star mountains (Yim & Jones 2020), both of which will emit gravitational waves. Searches for such long-lived gravitational waves postglitch have not detected anything (Keitel et al. 2019).

9.8.4 Other Burst Sources

There are other potential sources of bursts of gravitational waves proposed in the literature that this exciting new field of discovery will potentially unveil. One of the most discussed are oscillations and reconnection events from cosmological cosmic strings, which are topological defects formed in the early universe (e.g., see Damour &

Vilenkin 2000, 2001, 2005 and references therein). The LIGO/Virgo collaborations have put constraints on such putative sources (Abbott et al. 2018a, 2019a) that are complementary to those constraints from, for example, stochastic background searches for gravitational waves (see Section 9.10), pulsar timing observations (Blanco-Pillado 2018), the cosmic microwave background, and big bang nucleosynthesis (Pagano et al. 2016).

Other exciting possibilities include potential gravitational waves coincident with fast radio bursts or other types of short bursts of electromagnetic radiation. This is certainly an exciting prospect because it would allow for an immediate identification of the source of gravitational waves, which, without the electromagnetic counterpart, may be otherwise difficult to characterize. Such coincident detections almost provide complementary information about the source physics. As a concrete example, consider detecting gravitational waves from a magnetar flare or core collapse (as described in the previous sections); while the electromagnetic signature comes from external, relatively low-density regions around the explosion, the gravitational waves originate from the high-density core.

Of course, one of the most exciting prospects for the future of gravitational-wave astronomy is to detect hitherto unpredicted sources, thereby discovering new phenomena in the universe. Detecting such unknown unknowns requires unmodeled gravitational-wave searches. While many techniques have been, and continue to be, developed to search for such gravitational waves, the understanding and characterization of such signals will provide significant challenges (e.g., Divakarla et al. 2020).

9.9 Continuous-wave Sources

Continuous-wave gravitational waves are long lived, monochromatic or nearly monochromatic signals. While they are emitted as nearly monochromatic in the source frame, the signal in the detector frame is Doppler modulated by Earth's motion relative to the source and also the source's motion if, for example, it is in a binary. As the signal is long lived, the amplitude is also modulated by the changing antenna-pattern function of the detectors—for a thorough overview of these issues and more in terms of gravitational-wave detection of continuous-wave sources, see Jaranowski et al. (1998).

The most promising target for continuous gravitational waves are those emitted from neutron stars. Consider an isolated neutron star, modeled as an axisymmetric rigid body with principal moments of inertia $I_1 = I_2 \neq I_3$, such that the star's ellipticity is defined as $\epsilon = (I_3 - I_1)/I_1$. The characteristic gravitational-wave strain is (Zimmermann 1979, 1980)

$$h_0 = \frac{4\pi^2 G}{c^4}\frac{I_1 f_{\rm gw}^2 \epsilon}{d} = 4.2 \times 10^{-26}\left(\frac{\epsilon}{10^{-6}}\right)\left(\frac{I_1}{10^{45}{\rm g\ cm}^2}\right)\left(\frac{P}{10\ {\rm ms}}\right)^{-2}\left(\frac{d}{1{\rm kpc}}\right)^{-1}, \tag{9.23}$$

where $f_{\rm gw} = 2\nu = 2/P$ is the gravitational-wave frequency that is emitted at twice the spin frequency of the star ν, and P is the stellar spin period.

Gravitational waves from isolated neutron stars have not yet been detected. Perhaps the greatest unknown is the size of the neutron star deformation parameterized by the ellipticity ϵ. Such nonaxisymmetries can arise from the internal stellar magnetic field (Bonazzola & Gourgoulhon 1996; Haskell et al. 2008), internal thermal gradients (Bildsten 1998; Ushomirsky et al. 2000), or accretion forming “mountains” on the surface of the star (Melatos & Phinney 2001; Payne & Melatos 2004, 2007)—for reviews of these mechanisms, see Riles (2013), Lasky (2015), and Glampedakis & Gualtieri (2018). As an example, the internal magnetic field of a star creates an ellipticity that scales as the square of the volume-averaged magnetic field (Cutler 2002) for normal matter. The relative strength of the toroidal and poloidal components of the field further affect the ellipticity (Haskell et al. 2008; Colaiuda et al. 2008; Ciolfi et al. 2009). Perhaps more interestingly, but also more speculatively, the scaling of the ellipticity can differ depending on the state of neutron star matter. For example, if the matter in the core is a superconductor, then the ellipticity scales linearly with the magnetic field (Cutler 2002), with exotic states such as crystalline color superconductors, or color-flavor-locked superconductors (Owen 2005; Haskell et al. 2007; Glampedakis et al. 2012) having higher ellipticities for the same magnetic field strengths.

Relevant questions are then, what is a typical ellipticity, and what is the minimum ellipticity, rotating neutron stars can have? The truth is no one knows. Gravitational-wave upper limits provide the most stringent constraints on these ellipticities. Gravitational-wave searches targeting known pulsars are significantly more sensitive than all-sky searches because one knows the likely frequency and frequency evolution of the signal. In this case, searches in Advanced LIGO/Virgo data give upper limits on ellipticity of $\epsilon \approx 10^{-8}$ for pulsars J0437–4715 and J0711–6830 (Abbott et al. 2020d)—see Riles (2017) for a review of LIGO/Virgo searches for continuous gravitational-waves sources. On the other hand, all-sky searches may yield interesting results because only a very small fraction of all the neutron stars in our galaxy have been observed—upper limits here depend sensitively on the assumptions used in the search, but the best results include upper limits of $\epsilon \lesssim 10^{-7}$ for sources within 100 pc and rotating faster than 5 ms (Steltner et al. 2021), and $\epsilon \lesssim 10^{-8}$ within 170 pc, albeit with stricter constraints on the stellar spin evolution (Dergachev & Papa 2020). While there is no true lower bound on the possible ellipticity of a neutron star, the fact that the ellipticity scales with the magnetic field as described above implies such a lower bound may exist anywhere from $\epsilon \sim 10^{-9}$ down to $\epsilon \sim 10^{-12}$. Empirically, there is little one can say about this lower bound, although see Woan et al. (2018) for a speculative lower bound of $\epsilon \sim 10^{-9}$ based on an apparent cutoff in the P–$\dot{P}$ distribution of millisecond pulsars.

Long-lived nonaxisymmetries are not the only potential source of continuous gravitational waves from isolated neutron stars, with another likely mechanism being that from unstable oscillation modes in the star. Probably most relevant to this are the so-called *r*-modes: inertial modes where the restoring force is the Coriolis force (Andersson 1998; Friedman & Morsink 1998; Andersson & Kokkotas 2001). These modes are always unstable to the Chandrasekhar–Friedman–Schutz instability (Chandrasekhar 1970; Friedman & Schutz 1978), whereby the mode that is retrograde in the comoving frame becomes prograde in the inertial frame, implying that as gravitational-wave energy is emitted, the mode paradoxically grows in amplitude. The open question here is therefore the damping mechanism and saturation amplitude of the instability. Certainly viscous dissipation is expected to be the primary mode damper, although the effectiveness of this depends on whether this is shear or bulk viscosity, which in turn depends on the star's temperature—see Glampedakis & Gualtieri (2018) and references therein.

The r-mode instability is particularly interesting for accreting systems, such as with low-mass X-ray binaries like Scorpius X-1. Observationally, accreting millisecond X-ray pulsars and nuclear-powered X-ray pulsars have an abrupt cutoff at about 730 Hz (Chakrabarty et al. 2003; Patruno 2010; Patruno & Watts 2021); however, measured accretion torques suggest at least some of these systems should be spinning faster. One potential explanation is that, as these stars are being spun up through accretion, they are losing angular momentum through the emission of gravitational waves (Papaloizou & Pringle 1978; Wagoner 1984; Bildsten 1998). (It is worth mentioning this is not the only explanation for the cutoff, with other possibilities including magnetosphere/disk interactions, e.g., White & Zhang 1997; Patruno et al. 2012.) From this torque-balance argument one can derive an equation for the gravitational-wave emission amplitude as a function of the star's spin frequency and X-ray luminosity (Wagoner 1984), which is commonly known as the torque-balance limit.

According to the above torque-balance argument, the loudest gravitational-wave sources should also be brightest X-ray sources, implying Scorpius X-1 should be the loudest of them all (Watts et al. 2008). One difficulty with Scorpius X-1 is that the putative neutron star's spin period is not known, implying the torque-balance strain is parameterized over possible values of this spin frequency, and search algorithms must search over this parameter (and others), e.g., see Messenger et al. (2015) and references therein. Gravitational-wave limits on Scorpius X-1 are only just approaching the torque-balance limit in small windows of the possible parameter space (Abbott et al. 2019h, 2019j).

There are other searches for continuous gravitational waves from isolated neutron stars, including from young supernova remnants and targeted searches for neutron stars in the Galactic Center, e.g., see Abbott et al. (2019a, 2019e), Papa et al. (2020) and references therein. A more speculative prospect is the potential detection of continuous gravitational waves from ultralight boson clouds around stellar-mass black holes (Arvanitaki et al. 2010; Arvanitaki & Dubovsky 2011), which can potentially be measured directly (Arvanitaki et al. 2015; Brito et al. 2017; Palomba et al. 2019; Sun et al. 2020; Zhu et al. 2020) or indirectly through the impact they have on, for example, the spin distribution of black holes (Ng et al. 2021).

Continuous gravitational waves have not been detected but are one of the more exciting prospects for this new field of astronomy. A positive detection could allow us to peer directly into the heart of neutron stars, probing physics inaccessible to laboratory experiments or traditional astronomy techniques.

9.10 Stochastic Backgrounds

Our final type of expected source for gravitational-wave interferometers are stochastic backgrounds, linear superpositions of gravitational-wave signals from sources that are not individually resolvable because they are either too weak or too numerous. Such sources include compact binary systems that are more distant than individually resolvable sources described in previous sections, other astrophysical sources such as the superposition of gravitational waves from all isolated neutron stars in the universe, as well as cosmological sources such as those produced during the inflationary epoch of the universe; each of these is described in more detail below. For thorough reviews of stochastic gravitational-wave backgrounds and detection methods, see Romano & Cornish (2017) and Christensen (2019).

In general, one can split stochastic background signal types into two classes: so-called "popcorn" and "continuous" signals, where the former gives a visceral sense for the signal (imagine cooking popcorn on a stove or in a microwave, with the Poissonian distribution of popping corn kernels). As an example, consider the stochastic background from binary black hole mergers: two black holes merge somewhere in the universe every $\sim 250s$, however most signals are only in the band for $\lesssim 1s$, implying the total number of signals in band at any given time is much less than one (Abbott et al. 2018b, 2019c, 2021b). Pop……pop…pop ………pop. On the other hand, consider binary neutron star systems, which merge somewhere in the universe every $\sim 10s$, but are in the band for up to a minute or more. In general, there are ten or more binary neutron star signals in the band at any given time (Abbott et al. 2018b), albeit with most of them being too weak to detect.

Searching for stochastic backgrounds generally relies on cross-correlating signals from multiple detectors, where the signal will add coherently, but the uncorrelated noise between the detectors will add incoherently (Christensen 1992; Allen & Romano 1999).[17] By normalizing the cross-correlation statistic and appropriately defining its spectral shape, the cross-correlation statistic can be defined such that it is equal to the gravitational-wave energy density of the universe at a given frequency (see Abbott et al. 2019c and references therein)

$$\Omega_{\rm gw}(f) = \frac{1}{\rho_c}\frac{d\rho_{\rm GW}}{d\ln f}. \tag{9.24}$$

Here, $\rho_c = 3c^2H_0^2/(8\pi G)$ is the critical closure density of the universe, H_0 is the Hubble expansion rate, and $\rho_{\rm GW}$ is the energy density of gravitational waves in the universe (akin to, for example, ρ_Λ, $\rho_{\rm DM}$, etc., in the standard cosmological model which are the energy densities in dark energy and dark matter respectively). From the above equations, one can also calculate the integrated energy density $\Omega_{\rm GW} = \int \Omega_{\rm GW}(f) d\ln f$. One can then place constraints on the isotropic stochastic background assuming different spectral shapes:

[17] A relatively new method has been proposed that uses Bayesian inference to search each data segment, and statistically determine the existence of a stochastic background (Smith & Thrane 2018; Hernandez Vivanco et al. 2019; Smith et al. 2020b) This method is not yet fully developed, but shows promise for being more sensitive than cross-correlation.

$$\Omega_{\mathrm{GW}}(f) = \Omega_{\mathrm{ref}}\left(\frac{f}{f_{\mathrm{ref}}}\right)^{\alpha}. \tag{9.25}$$

For example, compact binary coalescences will produce a background with $\alpha = 2/3$ (Regimbau 2011), and cosmic string models and gravitational waves from fiducial slow-roll inflation models will produce a flat spectrum ($\alpha = 0$) (Caprini & Figueroa 2018), while other astrophysical sources will produce a variety of other power-law slopes or spectral shapes (Christensen 2019).

No stochastic background of gravitational waves has been observed. The best upper limits come from analysis of the first and second observing runs from LIGO/Virgo (Abbott et al. 2019c; although one can expect better results from the third observing run). For a flat $\alpha = 0$ spectrum, the 95% credible upper limits are $\Omega_{\mathrm{GW}} < 6.0 \times 10^{-8}$, whereas for an $\alpha = 2/3$ power law consistent with compact binary coalescences, the upper limit measured at a fiducial reference frequency of $f_{\mathrm{ref}} = 25$ Hz is $\Omega_{\mathrm{GW}} < 4.8 \times 10^{-8}$. These results represent slices of the two-dimensional posterior distribution in the Ω_{ref}–α plane shown in Figure 9.20. One can compare the above compact binary limit to predictions for the background based on the (relatively well known) local merger rate, which yields for binary black holes and inferred amplitude at 25 Hz of $\Omega_{\mathrm{BBH}} = 5.3^{+4.2}_{-2.5} \times 10^{-10}$ and binary neutron stars of $\Omega_{\mathrm{BNS}} = 3.6^{+8.4}_{-3.1} \times 10^{-10}$ (Abbott et al. 2019c). One can see that the observational upper limits are still almost two orders of magnitude higher than the predicted backgrounds. Of course, it is much more difficult/uncertain to make predictions for cosmological backgrounds, and therefore refer the reader only to the reviews cited above.

The above limits refer to stochastic backgrounds where sources are assumed to be distributed isotropically on the sky, and where the general relativity is the correct theory of gravity, implying only tensorial gravitational-wave modes exist. One can

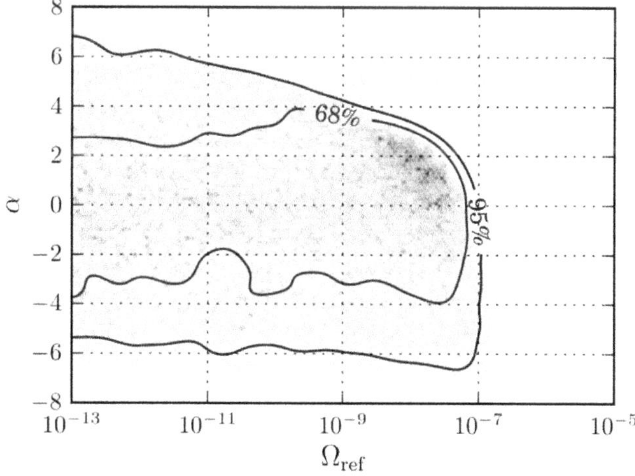

Figure 9.20. Posterior distribution for the isotropic stochastic gravitational-wave background, assuming a power-law spectral slope α and amplitude Ω_{ref}—see Equation (9.25). Image credit: Abbott et al. (2019c), with permission, copyright 2019 by the American Physical Society.

also search for anisotropic sky distributions (Abbott et al. 2019a) and also scalar and/or vector modes (Abbott et al. 2018l) in the stochastic background.

References

Aasi, J., Abadie, J., Abbott, B. P., et al. 2013, NaPho, 7, 613
Aasi, J., Abbott, B. P., Abbott, R., et al. 2015, CQGra, 32, 074001
Abadie, J., Abbott, B. P., Abbott, R., et al. 2011, PhRvD, 83, 042001
Abbott, B. P., Abbott, R., Abbott, T. D., et al. 2016a, PhRvD, 94, 102001
Abbott, B. P., Abbott, R., Abbott, T. D., et al. 2016b, PhRvL, 116, 241103
Abbott, B. P., Abbott, R., Abbott, T. D., et al. 2016c, PhRvL, 116, 221101
Abbott, B. P., Abbott, R., Abbott, T. D., et al. 2017a, PhRvD, 95, 042003
Abbott, B. P., Abbott, R., Abbott, T. D., et al. 2017b, AnP, 529, 1600209
Abbott, B. P., Abbott, R., Abbott, T. D., et al. 2020a, LRR, 23, 3
Abbott, B. P., Abbott, R., Abbott, T. D., et al. 2020b, ApJ, 892, L3
Abbott, B. P., Abbott, R., Abbott, T. D., et al. 2019a, PhRvD, 100, 024017
Abbott, B. P., Abbott, R., Abbott, T. D., et al. 2019b, ApJ, 875, 161
Abbott, B. P., Abbott, R., Abbott, T. D., et al. 2019c, PhRvD, 100, 061101
Abbott, B. P., Abbott, R., Abbott, T. D., et al. 2019d, ApJ, 874, 163
Abbott, B. P., Abbott, R., Abbott, T. D., et al. 2019e, ApJ, 875, 122
Abbott, B. P., Abbott, R., Abbott, T. D., et al. 2019f, PhRvD, 100, 104036
Abbott, B. P., Abbott, R., Abbott, T. D., et al. 2020c, PhRvD, 101, 084002
Abbott, B. P., Abbott, R., Abbott, T. D., et al. 2019g, ApJ, 882, L24
Abbott, B. P., Abbott, R., Abbott, T. D., et al. 2019h, PhRvD, 100, 062001
Abbott, B. P., Abbott, R., Abbott, T. D., et al. 2019i, PhRvX, 9, 031040
Abbott, B. P., Abbott, R., Abbott, T. D., et al. 2019j, PhRvD, 100, 122002
Abbott, B. P., Abbott, R., Abbott, T. D., et al. 2018a, PhRvD, 97, 102002
Abbott, B. P., Abbott, R., Abbott, T. D., et al. 2017c, ApJ, 848, L13
Abbott, B. P., Abbott, R., Abbott, T. D., et al. 2017d, PhRvL, 119, 141101
Abbott, B. P., Abbott, R., Abbott, T. D., et al. 2017e, PhRvL, 119, 161101
Abbott, B. P., Abbott, R., Abbott, T. D., et al. 2017f, ApJ, 848, L12
Abbott, B. P., Abbott, R., Abbott, T. D., et al. 2018b, PhRvL, 120, 091101
Abbott, B. P., Abbott, R., Abbott, T. D., et al. 2018c, PhRvL, 121, 161101
Abbott, B. P., Abbott, R., Abbott, T. D., et al. 2019k, PhRvL, 123, 011102
Abbott, B. P., Abbott, R., Abbott, T. D., et al. 2018l, PhRvL, 120, 201102
Abbott, B. P., Abbott, R., Adhikari, R., et al. 2009, ApJ, 701, L68
Abbott, B. P., Abbott, R., Abbott, T. D., et al. 2016d, PhRvX, 6, 041014
Abbott, B. P., Abbott, R., Abbott, T. D., et al. 2016e, PhRvL, 116, 061102
Abbott, B. P., Abbott, R., Abbott, T. D., et al. 2017g, Natur, 551, 85
Abbott, B. P., Abbott, R., Abbott, T. D., et al. 2017h, CQGra, 34, 044001
Abbott, B. P., Abbott, R., Abbott, T. D., et al. 2017i, ApJ, 851, L16
Abbott, B. P., Abbott, R., Abbott, T. D., et al. 2019l, PhRvX, 9, 011001
Abbott, B. P., Abbott, R., Abbott, T. D., et al. 2019m, ApJ, 875, 160
Abbott, R., Abbott, T. D., Abraham, S., et al. 2020d, ApJ, 902, L21
Abbott, R., Abbott, T. D., Abraham, S., et al. 2020e, PhRvD, 102, 043015
Abbott, R., Abbott, T. D., Abraham, S., et al. 2020f, PhRvL, 125, 101102
Abbott, R., Abbott, T. D., Abraham, S., et al. 2020g, ApJ, 896, L44

Abbott, R., Abbott, T. D., Abraham, S., et al. 2021a, PhRvX, 11, 021053
Abbott, R., Abbott, T. D., Abraham, S., et al. 2021b, ApJL, 913, L7
Abbott, R., Abbott, T. D., Abraham, S., et al. 2020h, ApJL, 900, L13
Abbott, R., Abbott, T. D., Abraham, S., et al. 2021c, PhRvD, 103, 122002
Abdikamalov, E., Pagliaroli, G., & Radice, D. 2020, arXiv:2010.04356
Abedi, J., Dykaar, H., & Afshordi, N. 2017, PhRvD, 96, 082004
Acernese, F., Agathos, M., Agatsuma, K., et al. 2015, CQGra, 32, 024001
Ackley, K., Adya, V. B., Agrawal, P., et al. 2020, PASA, 37, e047
Adams, T., Buskulic, D., Germain, V., et al. 2016, CQGra, 33, 175012
Afle, C., & Brown, D. A. 2021, PhRvD, 103, 023005
Agathos, M., Meidam, J., Pozzo, W. D., et al. 2015, PhRvD, 92, 023012
Aghanim, N., Akrami, Y., Ashdown, M., et al. 2020, A&A, 641, A6
Ai, S., Gao, H., & Zhang, B. 2020, ApJ, 893, 146
Ajith, P., Babak, S., Chen, Y., et al. 2008, PhRvD, 77, 104017
Ajith, P., Babak, S., Chen, Y., et al. 2007, CQGra, 24, S689
Ajith, P., Hannam, M., Husa, S., et al. 2011, PhRvL, 106, 241101
Al-Mamun, M., Steiner, A. W., Nattila, J., et al. 2021, PhRvL, 126, 061101
Alexander, K. D., Margutti, R., Blanchard, P. K., et al. 2018, ApJ, 863, L18
Allen, B., & Romano, J. D. 1999, PhRvD, 59, 102001
Alsing, J., Silva, H. O., & Berti, E. 2018, MNRAS, 478, 1377
Anderson, P. W., & Itoh, N. 1975, Natur, 256, 25
Andersson, N. 1998, ApJ, 502, 708
Andersson, N., & Kokkotas, K. D. 2001, IJMPD, 10, 381
Andresen, H., Müller, B., Müller, E., & Th, H. 2017, MNRAS, 468, 2032
Antonini, F., & Rasio, F. A. 2016, ApJ, 831, 187
Armano, M., Audley, H., Auger, G., et al. 2016, PhRvL, 116, 231101
Arun, K. G., Iyer, B. R., Qusailah, M. S. S., & Sathyaprakash, B. S. 2006a, PhRvD, 74, 024006
Arun, K. G., Iyer, B. R., Qusailah, M. S. S., & Sathyaprakash, B. S. 2006b, CQGra, 23, L37
Arvanitaki, A., Baryakhtar, M., & Huang, X. 2015, PhRvD, 91, 084011
Arvanitaki, A., Dimopoulos, S., Dubovsky, S., Kaloper, N., & March-Russell, J. 2010, PhRvD, 81, 123530
Arvanitaki, A., & Dubovsky, S. 2011, PhRvD, 83, 044026
Ashton, G., Ackley, K., Hernandez, I. M., & Piotrzkowski, B. 2020, arXiv:2009.12346
Ashton, G., Birnholtz, O., & Cabero, M. 2016, arXiv:1612.05625
Ashton, G., Hübner, M., Lasky, P. D., et al. 2019a, ApJS, 241, 27
Ashton, G., Lasky, P. D., Graber, V., & Palfreyman, J. 2020, NatAs, 3, 1143
Ashton, G., & Thrane, E. 2020, MNRAS, 498, 1905
Ashton, G., Thrane, E., & Smith, R. J. E. 2019b, PhRvD, 100, 123018
Aso, Y., Michimura, Y., Somiya, K., et al. 2013, PhRvD, 88, 043007
Babak, S., Taracchini, A., & Buonanno, A. 2017, PhRvD, 95, 024010
Bailes, M., McClelland, D., Thrane, E., et al. 2019, arXiv:1912.06305
Bailyn, C. D., Jain, R. K., Coppi, P., & Orosz, J. A. 1998, ApJ, 499, 367
Bardeen, J. M., Press, W. H., & Teukolsky, S. A. 1972, ApJ, 178, 347
Baumgarte, T. W., Shapiro, S. L., & Shibata, M. 2000, ApJ, 528, L29
Bauswein, A., Bastian, N.-U. F., Blaschke, D. B., et al. 2019, PhRvL, 122, 061102
Bauswein, A., & Blacker, S. 2020, EPJST, 229, 3595

Bauswein, A., Just, O., Janka, H.-T., & Stergioulas, N. 2017, ApJ, 850, L34
Belczynski, K., Kalogera, V., & Bulik, T. 2002, ApJ, 572, 407
Belczynski, K., Wiktorowicz, G., Fryer, C. L., Holz, D. E., & Kalogera, V. 2012, ApJ, 757, 91
Bennett, M. F., & van Eysden, C. A. 2010, MNRAS, 409, 1705
Bildsten, L. 1998, ApJ, 501, L89
Biswas, B., Char, P., Nandi, R., & Bose, S. 2021, PhRvD, 103, 103015
Biwer, C. M., & Capano, C. D. 2019, PASP, 131, 024503
Bizouard, M.-A., Maturana-Russel, P., Torres-Forné, A., et al. 2021, PhRvD, 103, 063006
Blackman, J., Field, S. E., Galley, C. R., et al. 2015, PhRvL, 115, 121102
Blair, D. G., Howell, E. J., Ju, L., & Zhao, C. 2012, Advanced Gravitational Wave Detectors (Cambridge: Cambridge Univ. Press)
Blanchet, L., Damour, T., Esposito-Farèse, G., & Iyer, B. R. 2005, PhRvD, 71, 124004
Blanchet, L., Damour, T., Iyer, B. R., Will, C. M., & Wiseman, A. G. 1995, PhRvL, 74, 3515
Blanco-Pillado, J. J., Olum, K. D., & Siemens, X. 2018, PhLB, 778, 392
Blondin, J. M., Mezzacappa, A., & DeMarino, C. 2003, ApJ, 584, 971
Bonazzola, S., & Gourgoulhon, E. 1996, A&A, 312, 675
Bond, J. R., Arnett, W. D., & Carr, B. J. 1984, ApJ, 280, 825
Braginsky, V. B., & Thorne, K. S. 1987, Natur, 327, 123
Bransgrove, A., Beloborodov, A. M., & Levin, Y. 2020, ApJ, 897, 173
Breu, C., & Rezzolla, L. 2016, MNRAS, 459, 646
Brito, R., Ghosh, S., Barausse, E., et al. 2017, PhRvD, 96, 064050
Buonanno, A., & Damour, T. 1999, PhRvD, 59, 084006
Buonanno, A., & Damour, T. 2000, PhRvD, 62, 064015
Calderón Bustillo, J., Lasky, P. D., & Thrane, E. 2020, arXiv:2010.01857
Calderón Bustillo, J., Sanchis-Gual, N., Torres-Forné, A., & Font, J. A. 2021, PhRvD, 126, 201101
Campanelli, M., Lousto, C. O., Marronetti, P., & Zlochower, Y. 2006, PhRvL, 96, 111101
Cantiello, M., Jensen, J. B., Blakeslee, J. P., et al. 2018, ApJ, 854, L31
Cao, Z., Chen, L.-W., Chu, P.-C., & Zhou, Y. 2020, arXiv:2009.00942
Capano, C., Dent, T., Hanna, C., et al. 2017, PhRvD, 96, 082002
Capano, C. D., Tews, I., Brown, S. M., et al. 2020, NatAs, 4, 625
Caprini, C., & Figueroa, D. G. 2018, CQGra, 35, 163001
Cardoso, V., Franzin, E., & Pani, P. 2016a, PhRvL, 116, 171101
Cardoso, V., Hopper, S., Macedo, C. F. B., Palenzuela, C., & Pani, P. 2016b, PhRvD, 94, 084031
Carter, B. 1971, PhRvL, 26, 331
Chakrabarty, D., Morgan, E. H., Muno, M. P., et al. 2003, Natur, 424, 42
Chan, M. L., Heng, I. S., & Messenger, C. 2020, PhRvD, 102, 043022
Chandrasekhar, S. 1970, PhRvL, 24, 611
Chatterjee, D., Ghosh, S., Brady, P. R., et al. 2020, ApJ, 896, 54
Chatziioannou, K., Yunes, N., & Cornish, N. 2012, PhRvD, 86, 022004
Christensen, N. 1992, PhRvD, 46, 5250
Christensen, N. 2019, RPPh, 82, 016903
Chu, Q. 2017, PhD thesis, The University of Western Australia
Ciolfi, R., Ferrari, V., Gualtieri, L., & Pons, J. A. 2009, MNRAS, 397, 913
Ciolfi, R., & Rezzolla, L. 2012, ApJ, 760, 1
Clark, J. A., Bauswein, A., Stergioulas, N., & Shoemaker, D. 2016, CQGra, 33, 085003

Clesse, S., & Garcia-Bellido, J. 2020, arXiv:2007.06481
Colaiuda, A., Ferrari, V., Gualtieri, L., & Pons, J. A. 2008, MNRAS, 385, 2080
Cook, G. B., Shapiro, S. L., & Teukolsky, S. A. 1992, ApJ, 398, 203
Cook, G. B., Shapiro, S. L., & Teukolsky, S. A. 1994b, ApJ, 424, 823
Cook, G. B., Shapiro, S. L., & Teukolsky, S. A. 1994a, ApJ, 422, 227
Cornish, N., Blas, D., & Nardini, G. 2017, PhRvL, 119, 161102
Corsi, A., Hallinan, G., Mooley, K., et al. 2017, GCN Circ. 21815
Corsi, A., & Mészáros, P. 2009, ApJ, 702, 1171
Corsi, A., & Owen, B. J. 2011, PhRvD, 83, 104014
Coulter, D. A., Foley, R. J., Kilpatrick, C. D., et al. 2017, Sci, 358, 1556
Cromartie, H. T., Fonseca, E., Ransom, S. M., et al. 2020, NatAs, 4, 72
Cuoco, E., Powell, J., Cavaglià, M., et al. 2020, arXiv:2005.03745
Cutler, C. 2002, PhRvD, 66, 084025
Damour, T., & Vilenkin, A. 2000, PhRvL, 85, 3761
Damour, T., & Vilenkin, A. 2001, PhRvD, 64, 064008
Damour, T., & Vilenkin, A. 2005, PhRvD, 71, 063510
De, S., Finstad, D., Lattimer, J. M., et al. 2018, PhRvL, 121, 091102
Christodoulou, D. 1991, PhRvL, 67, 1486
Demircik, T., Ecker, C., & Järvinen, M. 2021, ApJL, 907, L37
Dergachev, V., & Papa, M. A. 2020, PhRvL, 125, 171101
Detweiler, S. L. 1975, ApJ, 197, 203
Dexheimer, V., Gomes, R. O., Klähn, T., Han, S., & Salinas, M. 2021, PhRvC, 103, 025808
Dietrich, T., Samajdar, A., Khan, S., et al. 2019, PhRvD, 100, 044003
Dimmelmeier, H., Ott, C. D., Janka, H. T., Marek, A., & Müller, E. 2007, PhRvL, 98, 251101
Dimmelmeier, H., Ott, C. D., Marek, A., & Janka, H. T. 2008, PhRvD, 78, 064056
Divakarla, A. K., Thrane, E., Lasky, P. D., & Whiting, B. F. 2020, PhRvD, 102, 023010
Duncan, R. C., & Thompson, C. 1992, ApJ, 392, L9
Eardley, D. M., Lee, D. L., & Lightman, A. P. 1973a, PhRvD, 8, 3308
Eardley, D. M., Lee, D. L., Lightman, A. P., Wagoner, R. V., & Will, C. M. 1973b, PhRvL, 30, 884
Einstein, A. 1916, Sitzungsberichte der Königlich Preußischen Akademie der Wissenschaften (Berlin), 688
Eldridge, J. J., & Stanway, E. R. 2016, MNRAS, 462, 3302
Espinoza, C. M., Lyne, A. G., Stappers, B. W., & Kramer, M. 2011, MNRAS, 414, 1679
Evans, P. A., Cenko, S. B., Kennea, J. A., et al. 2017, Sci, 358, 1565
Fan, Y.-Z., Wu, X.-F., & Wei, D.-M. 2013, PhRvD, 88, 067304
Farrell, E. J., Groh, J. H., Hirschi, R., et al. 2021, MNRAS, 502, L40
Farrow, N., Zhu, X.-J., & Thrane, E. 2019, ApJ, 876, 18
Fattoyev, F. J., Horowitz, C. J., Piekarewicz, J., & Reed, B. 2020, PhRvC, 102, 065805
Favata, M. 2014, PhRvL, 112, 101101
Field, S. E., Galley, C. R., Hesthaven, J. S., Kaye, J., & Tiglio, M. 2014, PhRvX, 4, 031006
Fishbach, M., & Holz, D. E. 2020, ApJ, 904, L26
Fishbach, M., Holz, D. E., & Farr, B. 2017, ApJ, 840, L24
Fishbone, L. G. 1973, ApJ, 185, 43
Flanagan, É. É., & Hinderer, T. 2008, PhRvD, 77, 021502
Flanagan, É. É., & Hughes, S. A. 2005, NJPh, 7, 204

Foglizzo, T., Scheck, L., & Janka, H.-Th. 2006, ApJ, 652, 1436
Foucart, F. 2012, PhRvD, 86, 124007
Foucart, F. 2020, FrASS, 7, 46
Foucart, F., Hinderer, T., & Nissanke, S. 2018, PhRvD, 98, 081501
Friedman, J. L., & Schutz, B. F. 1978, ApJ, 222, 281
Friedman, J. L., & Morsink, S. M. 1998, ApJ, 502, 714
Fryer, C. L., Belczynski, K., Wiktorowicz, G., et al. 2012, ApJ, 749, 91
Gabbard, H., Messenger, C., Heng, I. S., Tonolini, F., & Murray-Smith, R. 2019, arXiv:1909.06296
Gabler, M., Cerdá-Durán, P., Stergioulas, N., Font, J. A., & Müller, E. 2013, PhRvL, 111, 211102
Gayathri, V., Healy, J., Lange, J., et al. 2020, arXiv:2009.05461
Gendreau, K. C., Arzoumanian, Z., Adkins, P. W., et al. 2016, Proc. SPIE, 9905, 99051H
Generozov, A., Stone, N. C., Metzger, B. D., & Ostriker, J. P. 2018, MNRAS, 478, 4030
GEO600, 2018, New Squeezing Record at GEO600, https://www.geo600.org/67701/new-squeezing-record-at-geo600
Gertsenshteĭn, M. E., & Pustovoĭt, V. I. 1963, JETP, 16, 433
Giacobbo, N., & Mapelli, M. 2018, MNRAS, 480, 2011
Giacobbo, N., Mapelli, M., & Spera, M. 2018, MNRAS, 474, 2959
Giesler, M., Isi, M., Scheel, M. A., & Teukolsky, S. A. 2019, PhRvX, 9, 041060
Glampedakis, K., Jones, D. I., & Samuelsson, L. 2012, PhRvL, 109, 081103
Glampedakis, K., & Gualtieri, L. 2018, Gravitational Waves from Single Neutron Stars: An Advanced Detector Era Survey Vol. 457, 673
Glampedakis, K., Samuelsson, L., & Andersson, N. 2006, MNRAS, 371, L74
Godzieba, D. A., Radice, D., & Bernuzzi, S. 2021, ApJ, 908, 122
Gondán, L., & Kocsis, B. 2019, ApJ, 871, 178
Gossan, S. E., Sutton, P., Stuver, A., et al. 2016, PhRvD, 93, 042002
Graham, M. J., Ford, K. E. S., McKernan, B., et al. 2020, PhRvL, 124, 251102
Granot, J., Guetta, D., & Gill, R. 2017, ApJ, 850, L24
Green, S. R., & Gair, J. 2020, arXiv:2008.03312
Guo, X., Chu, Q., Chung, S. K., et al. 2018, CoPhC, 231, 62
Gupta, A., Gerosa, D., Arun, K. G., et al. 2020, PhRvD, 101, 103036
Hall, E. D., & Evans, M. 2019, CQGra, 36, 225002
Haskell, B., Andersson, N., Jones, D. I., & Samuelsson, L. 2007, PhRvL, 99, 231101
Haskell, B., Samuelsson, L., Glampedakis, K., & Andersson, N. 2008, MNRAS, 385, 531
Hebeler, K., Lattimer, J. M., Pethick, C. J., & Schwenk, A. 2013, ApJ, 773, 11
Heger, A., Fryer, C. L., Woosley, S. E., Langer, N., & Hartmann, D. H. 2003, ApJ, 591, 288
Heger, A., & Woosley, S. E. 2002, ApJ, 567, 532
Heng, I. S. 2009, CQGra, 26, 105005
Hernandez Vivanco, F., Smith, R., Thrane, E., & Lasky, P. D. 2019, PhRvD, 100, 043023
Hernandez Vivanco, F., Smith, R., Thrane, E., & Lasky, P. D. 2020, MNRAS, 499, 5972
Hinderer, T. 2008, ApJ, 677, 1216
Ho, W. C. G., Jones, D. I., Andersson, N., & Espinoza, C. M. 2020, PhRvD, 101, 103009
Hooper, S., Chung, S. K., Luan, J., et al. 2012, PhRvD, 86, 024012
Hotokezaka, K., Nakar, E., Gottlieb, O., et al. 2019, NatAs, 3, 940
Huang, K., Hu, J., Zhang, Y., & Shen, H. 2020, ApJ, 904, 39

Hübner, M., Talbot, C., Lasky, P.D., & Thrane, E. 2020, PhRvD, 101, 023011
Hulse, R. A., & Taylor, J. H. 1975, ApJ, 195, L51
Huppenkothen, D., Watts, A. L., & Levin, Y. 2014, ApJ, 793, 129
Ioka, K. 2001, MNRAS, 327, 639
Isi, M., Giesler, M., Farr, W. M., Scheel, M. A., & Teukolsky, S. A. 2019, PhRvL, 123, 111102
Isi, M., Smith, R., Vitale, S., et al. 2018, PhRvD, 98, 042007
Israel, G. L., Belloni, T., Stella, L., et al. 2005, ApJ, 628, L53
Israel, W. 1967, PhRv, 164, 1776
Israel, W. 1968, CMaPh, 8, 245
James, C. W., Anderson, G. E., Wen, L., et al. 2019, MNRAS, 489, L75
Jaranowski, P., Królak, A., & Schutz, B. F. 1998, PhRvD, 58, 063001
Kalmus, P., Cannon, K. C., Márka, S., & Owen, B. J. 2009, PhRvD, 80, 042001
Kalogera, V., Bizouard, M.-A., Burrows, A., et al. 2019, BAAS, 51, 239
Kaplan, D., Friedman, J., Read, J., & Growth Collaboration, 2019, GCN Circ. 26243
Kasen, D., Badnell, N. R., & Barnes, J. 2013, ApJ, 774, 25
Kawaguchi, K., Kyutoku, K., Nakano, H., et al. 2015, PhRvD, 92, 024014
Keer, L., & Jones, D. I. 2015, MNRAS, 446, 865
Keitel, D., Woan, G., Pitkin, M., et al. 2019, PhRvD, 100, 064058
Kerr, R. P. 1963, PhRvL, 11, 237
Khan, S., Chatziioannou, K., Hannam, M., & Ohme, F. 2019, PhRvD, 100, 024059
Khan, S., Ohme, F., Chatziioannou, K., & Hannam, M. 2020, PhRvD, 101, 024056
Kidder, L. E. 1995, PhRvD, 52, 821
Kinugawa, T., Nakamura, T., & Nakano, H. 2021, MNRAS, 501, L49
Kiziltan, B., Kottas, A., De Yoreo, M., & Thorsett, S. E. 2013, ApJ, 778, 66
Klimenko, S., Vedovato, G., Drago, M., et al. 2016, PhRvD, 93, 042004
Klion, H., Duffell, P. C., Kasen, D., & Quataert, E. 2021, MNRAS, 502, 865
Kramer, M., & Champion, D. J. 2013, CQGra, 30, 224009
Kreidberg, L., Bailyn, C. D., Farr, W. M., & Kalogera, V. 2012, ApJ, 757, 36
Kuroda, T., Kotake, K., & Takiwaki, T. 2016, ApJ, 829, L14
Lackey, B. D., Kyutoku, K., Shibata, M., Brady, P. R., & Friedman, J. L. 2014, PhRvD, 89, 043009
Lackey, B. D., Pürrer, M., Taracchini, A., & Marsat, S. 2019, PhRvD, 100, 024002
Lackey, B. D., & Wade, L. 2015, PhRvD, 91, 043002
Landry, P., Essick, R., & Chatziioannou, K. 2020, PhRvD, 101, 123007
Lasky, P. D., & Glampedakis, K. 2016, MNRAS, 458, 1660
Lasky, P. D. 2015, PASA, 32, e034
Lasky, P. D., Thrane, E., Levin, Y., Blackman, J., & Chen, Y. 2016, PhRvL, 117, 061102
Lattimer, J. M., & Schramm, D. N. 1976, ApJ, 210, 549
Lee, M. G., Kang, J., & Im, M. 2018, ApJ, 859, L6
Levin, Y. 2006, MNRAS, 368, L35
Levin, Y., & van Hoven, M. 2011, MNRAS, 418, 659
Li, T. G. F., Del Pozzo, W., Vitale, S., et al. 2012, PhRvD, 85, 082003
LIGO Scientific Collaboration & Virgo Collaboration, 2019, GCN Circ. 26222
Lim, Y., Bhattacharya, A., Holt, J. W., & Pati, D. 2021, PhRvC, 104, L032802
Lindblom, L. 2010, PhRvD, 82, 103011
Lioutas, G., & Stergioulas, N. 2018, GReGr, 50, 12

Lo, R. K. L., Li, T. G. F., & Weinstein, A. J. 2019, PhRvD, 99, 084052
Lü, H.-J., Zhang, B., Lei, W.-H., Li, Y., & Lasky, P. D. 2015, ApJ, 805, 89
Luan, J., Hooper, S., Wen, L., & Chen, Y. 2012, PhRvD, 85, 102002
Luo, J., Chen, L.-S., Duan, H.-Z., et al. 2016, CQGra, 33, 035010
Lyne, A. G., Shemar, S. L., & Graham Smith, F. 2000, MNRAS, 315, 534
Manchester, R. N., Hobbs, G., Bailes, M., et al. 2013, PASA, 30, e017
Mapelli, M. 2016, MNRAS, 459, 3432
Marek, A., Janka, H. T., & Müller, E. 2009, A&A, 496, 475
Margalit, B., & Metzger, B. D. 2017, ApJ, 850, L19
Martynov, D., Miao, H., Yang, H., et al. 2019, PhRvD, 99, 102004
McKernan, B., Ford, K. E. S., Bellovary, J., et al. 2018, ApJ, 866, 66
McLaughlin, M. A. 2013, CQGra, 30, 224008
Meegan, C., Lichti, G., Bhat, P. N., et al. 2009, ApJ, 702, 791
Melatos, A., & Phinney, E. S. 2001, PASA, 18, 421
Messenger, C., Bulten, H. J., Crowder, S. G., et al. 2015, PhRvD, 92, 023006
Messick, C., Blackburn, K., Brady, P., et al. 2017, PhRvD, 95, 042001
Metzger, B. D., Martínez-Pinedo, G., Darbha, S., et al. 2010, MNRAS, 406, 2650
Metzger, B. D. 2019, LRR, 23, 1
Metzger, B. D., & Fernández, R. 2014, MNRAS, 441, 3444
Miller, J., Barsotti, L., Vitale, S., et al. 2015, PhRvD, 91, 062005
Miller, M. C., & Hamilton, D. P. 2002, MNRAS, 330, 232
Miller, S., Callister, T. A., & Farr, W. M. 2020, ApJ, 895, 128
Mirshekari, S., Yunes, N., & Will, C. M. 2012, PhRvD, 85, 024041
Mishra, C. K., Arun, K. G., Iyer, B. R., & Sathyaprakash, B. S. 2010, PhRvD, 82, 064010
Montoli, A., Antonelli, M., Magistrelli, F., & Pizzochero, P. M. 2020, A&A, 642, A223
Mooley, K. P., Deller, A. T., Gottlieb, O., et al. 2018a, Natur, 561, 355
Mooley, K. P., Hallinan, G., & Corsi, A. 2017, GCN Circ. 21814
Mooley, K. P., Nakar, E., Hotokezaka, K., et al. 2018b, Natur, 554, 207
Morozova, V., Radice, D., Burrows, A., & Vartanyan, D. 2018, ApJ, 861, 10
Moss, G. E., Miller, L. R., & Forward, R. L. 1971, ApOpt, 10, 2495
Mueller, E., & Janka, H. T. 1997, A&A, 317, 140
Müller, B. 2020, LRCA, 6, 3
Müller, B., Janka, H.-T., & Marek, A. 2013, ApJ, 766, 43
Murphy, J. W., Ott, C. D., & Burrows, A. 2009, ApJ, 707, 1173
Nakamura, K., Horiuchi, S., Tanaka, M., et al. 2016, MNRAS, 461, 3296
Neijssel, C. J., Vigna-Gómez, A., Stevenson, S., et al. 2019, MNRAS, 490, 3740
Ng, K. K. Y., Vitale, S., Hannuksela, O. A., & Li, T. G. F. 2021, PhRvL, 126, 151102
Nitz, A. H., Capano, C., Nielsen, A. B., et al. 2019, ApJ, 872, 195
Nitz, A. H., Dal Canton, T., Davis, D., & Reyes, S. 2018, PhRvD, 98, 024050
Nitz, A. H., Dent, T., Davies, G. S., et al. 2020, ApJ, 891, 123
Ober, W. W., El Eid, M. F., & Fricke, K. J. 1983, A&A, 119, 61
Olejak, A., Fishbach, M., Belczynski, K., et al. 2020, ApJL, 901, L39
Oppenheimer, J. R., & Volkoff, G. M. 1939, PhRv, 55, 374
Ossokine, S., Buonanno, A., Marsat, S., et al. 2020, PhRvD, 102, 044055
Owen, B. J. 2005, PhRvL, 95, 211101
Özel, F., Psaltis, D., Narayan, R., & McClintock, J. E. 2010, ApJ, 725, 1918

Pagano, L., Salvati, L., & Melchiorri, A. 2016, PhLB, 760, 823
Palfreyman, J., Dickey, J. M., Hotan, A., Ellingsen, S., & van Straten, W. 2018, Natur, 556, 219
Palmer, D. M., Barthelmy, S., Gehrels, N., et al. 2005, Natur, 434, 1107
Palmese, A., & Conselice, C. J. 2021, PhRvL, 126, 181103
Palomba, C., D'Antonio, S., Astone, P., et al. 2019, PhRvL, 123, 171101
Pan, Y., Buonanno, A., Baker, J. G., et al. 2008, PhRvD, 77, 024014
Pang, P. T. H., Lo, R. K. L., Wong, I. C. F., Li, T. G. F., & Van Den Broeck, C. 2020, PhRvD, 101, 104055
Pannarale, F., Berti, E., Kyutoku, K., Lackey, B. D., & Shibata, M. 2015, PhRvD, 92, 081504
Pannarale, F., Tonita, A., & Rezzolla, L. 2011, ApJ, 727, 95
Papa, M. A., Ming, J., Gotthelf, E. V., et al. 2020, ApJ, 897, 22
Papaloizou, J., & Pringle, J. E. 1978, MNRAS, 184, 501
Paschalidis, V., Ruiz, M., & Shapiro, S. L. 2015, ApJ, 806, L14
Patruno, A. 2010, ApJ, 722, 909
Patruno, A., Haskell, B., & D'Angelo, C. 2012, ApJ, 746, 9
Patruno, A., & Watts, A. L. 2021, Accreting Millisecond X-ray Pulsars, Vol. 461, 143
Payne, D. J. B., & Melatos, A. 2004, MNRAS, 351, 569
Payne, D. J. B., & Melatos, A. 2007, MNRAS, 376, 609
Payne, E., Talbot, C., & Thrane, E. 2019, PhRvD, 100, 123017
Perego, A., Rosswog, S., Cabezón, R. M., et al. 2014, MNRAS, 443, 3134
Peters, P. C. 1964, PhRv, 136, 1224
Peters, P. C., & Mathews, J. 1963, PhRv, 131, 435
Pian, E., D'Avanzo, P., Benetti, S., et al. 2017, Natur, 551, 67
Piro, L., Troja, E., Zhang, B., et al. 2019, MNRAS, 483, 1912
Powell, J., Gossan, S. E., Logue, J., & Heng, I. S. 2016, PhRvD, 94, 123012
Powell, J., & Müller, B. 2020, MNRAS, 494, 4665
Powell, J., Szczepanczyk, M., & Heng, I. S. 2017, PhRvD, 96, 123013
Pratten, G., & Vecchio, A. 2020, arXiv:2008.00509
Pretorius, F. 2005, PhRvL, 95, 121101
Psaltis, D. 2008, LRR, 11, 9
Punturo, M., Abernathy, M., Acernese, F., et al. 2010, CQGra, 27, 194002
Pürrer, M. 2014, CQGra, 31, 195010
Raaijmakers, G., Greif, S. K., Riley, T. E., et al. 2020, ApJ, 893, L21
Raaijmakers, G., Riley, T. E., Watts, A. L., et al. 2019, ApJ, 887, L22
Radice, D., Morozova, V., Burrows, A., Vartanyan, D., & Nagakura, H. 2019, ApJ, 876, L9
Ravi, V., & Lasky, P. D. 2014, MNRAS, 441, 2433
Read, J. S., Baiotti, L., Creighton, J. D. E., et al. 2013, PhRvD, 88, 044042
Read, J. S., Lackey, B. D., Owen, B. J., & Friedman, J. L. 2009, PhRvD, 79, 124032
Regimbau, T. 2011, RAA, 11, 369
Rezzolla, L., Most, E. R., & Weih, L. R. 2018, ApJ, 852, L25
Riess, A. G., Casertano, S., Yuan, W., Macri, L. M., & Scolnic, D. 2019, ApJ, 876, 85
Riess, A. G., Macri, L. M., Hoffmann, S. L., et al. 2016, ApJ, 826, 56
Riles, K. 2013, PrPNP, 68, 1
Riles, K. 2017, MPLA, 32, 1730035
Riley, T. E., Watts, A. L., Bogdanov, S., et al. 2019, ApJ, 887, L21
Riley, T. E., Raaijmakers, G., & Watts, A. L. 2018, MNRAS, 478, 1093

Rodriguez, C. L., Amaro-Seoane, P., Chatterjee, S., & Rasio, F. A. 2018, PhRvL, 120, 151101
Rodriguez, C. L., Chatterjee, S., & Rasio, F. A. 2016, PhRvD, 93, 084029
Rodriguez, C. L., Zevin, M., Amaro-Seoane, P., et al. 2019, PhRvD, 100, 043027
Romano, J. D., & Cornish, N. J. 2017, LRR, 20, 2
Romero-Shaw, I. M., Talbot, C., Biscoveanu, S., et al. 2020c, MNRAS, 499, 3295
Romero-Shaw, I. M., Farrow, N., Stevenson, S., Thrane, E., & Zhu, X.-J. 2020a, MNRAS, 496, L64
Romero-Shaw, I. M., Lasky, P. D., Thrane, E., & Calderon Bustillo, J. 2020b, ApJL, 903, L5
Rowlinson, A., P T O'brien,, Metzger, B. D., Tanvir, N. R., & Levan, A. J. 2013, MNRAS, 430, 1061
Rozwadowska, K., Vissani, F., & Cappellaro, E. 2021, NewA, 83, 101498
Ruffert, M., Janka, H.-T., & Schaefer, G. 1996, A&A, 311, 532
Ruiz, M., Shapiro, S. L., & Tsokaros, A. 2018, PhRvD, 98, 123017
Sachdev, S., Caudill, S., Fong, H., et al. 2019, arXiv:1901.08580
Safarzadeh, M., & Haiman, Z. 2020, ApJL, 903, L21
Safarzadeh, M., Ramirez-Ruiz, E., & Berger, E. 2020, ApJ, 900, 13
Sarin, N., Lasky, P. D., Sammut, L., & Ashton, G. 2018, PhRvD, 98, 043011
Sarin, N., & Lasky, P. D. 2021, GReGr, 53, 59
Sarin, N., Lasky, P. D., & Ashton, G. 2019, ApJ, 872, 114
Sarin, N., Lasky, P. D., & Ashton, G. 2020, PhRvD, 101, 063021
Saulson, P. R. 1994, Fundamentals of Interferometric Gravitational Wave Detectors (Singapore: World Scientific)
Saulson, P. R. 1997, AmJPh, 65, 501
Schmidt, P., Ohme, F., & Hannam, M. 2015, PhRvD, 91, 024043
Schutz, B. F. 1986, Natur, 323, 310
Sedrakian, A., Weber, F., & Li, J. J. 2020, PhRvD, 102, 041301
Shao, D.-S., Tang, S.-P., Jiang, J.-L., & Fan, Y.-Z. 2020, PhRvD, 102, 063006
Shapiro, S. L. 2000, ApJ, 544, 397
Shibagaki, S., Kuroda, T., Kotake, K., & Takiwaki, T. 2020, MNRAS, 493, L138
Siegel, D. M., & Metzger, B. D. 2017, PhRvL, 119, 231102
Sigurdsson, S., & Hernquist, L. 1993, Natur, 364, 423
Singer, L. P., & Price, L. R. 2016, PhRvD, 93, 024013
Smartt, S. J., Chen, T.-W., Jerkstrand, A., et al. 2017, Natur, 551, 75
Smith, R., Ashton, G., Vajpeyi, A., & Talbot, C. 2020a, MNRAS, 498, 4492
Smith, R., & Thrane, E. 2018, PhRvX, 8, 021019
Smith, R. J. E., Talbot, C., Hernandez Vivanco, F., & Thrane, E. 2020b, MNRAS, 496, 3281
Spera, M., Mapelli, M., Giacobbo, N., et al. 2019, MNRAS, 485, 889
Steltner, B., Papa, M. A., Eggenstein, H. B., et al. 2021, ApJ, 909, 79
Stevenson, S., Berry, C. P. L., & Mandel, I. 2017, MNRAS, 471, 2801
Strohmayer, T. E., & Watts, A. L. 2005, ApJ, 632, L111
Sun, L., Brito, R., & Isi, M. 2020, PhRvD, 101, 063020
Suvorova, S., Powell, J., & Melatos, A. 2019, PhRvD, 99, 123012
Takami, K., Rezzolla, L., & Baiotti, L. 2015, PhRvD, 91, 064001
Tanvir, N. R., Levan, A. J., Gonzalez-Fernandez, C., et al. 2017, ApJ, 848, L27
Taylor, J. H., & Weisberg, J. M. 1982, ApJ, 253, 908
Thompson, C., & Duncan, R. C. 1995, MNRAS, 275, 255

Thorne, K. S. 1980, RvMP, 52, 299
Thrane, E., Lasky, P. D., & Levin, Y. 2017, PhRvD, 96, 102004
Thrane, E., & Talbot, C. 2019, PASA, 36, e010
Tobar, M. E., Blair, D. G., Ivanov, E. N., et al. 1995, AuJPh, 48, 1007
Tolman, R. C. 1939, PhRv, 55, 364
Torres-Forné, A., Cerdá-Durán, P., Obergaulinger, M., Müller, B., & Font, J. A. 2019, PhRvL, 123, 051102
Torres-Forné, A., Cerdá-Durán, P., Passamonti, A., & Font, J. A. 2018, MNRAS, 474, 5272
Troja, E., Piro, L., van Eerten, H., et al. 2017, Natur, 551, 71
Troja, E., van Eerten, H., Zhang, B., et al. 2020, MNRAS, 498, 5643
Tsokaros, A., Ruiz, M., & Shapiro, S. L. 2020, ApJ, 905, 48
Umeda, H., & Nomoto, K. 2002, ApJ, 565, 385
Ushomirsky, G., Cutler, C., & Bildsten, L. 2000, MNRAS, 319, 902
Usman, S. A., Nitz, A. H., Harry, I. W., et al. 2016, CQGra, 33, 215004
Vajente, G., Gustafson, E. K., & Reitze, D. H. 2019, in Advances in Atomic, Molecular, and Optical Physics, Vol. 68, ed. L. F. Dimauro, H. Perrin, & S. F. Yelin (New York: Academic), 75
van der Sluys, M. V., Röver, C., Stroeer, A., et al. 2008, ApJ, 688, L61
van der Sluys, M. 2008, CQGra, 25, 184011
van Eysden, C. A. 2008, CQGra, 25, 225020
Varma, V., Field, S. E., Scheel, M. A., et al. 2019, PhRvR, 1, 033015
Vattis, K., Goldstein, I. S., & Koushiappas, S. M. 2020, PhRvD, 102, 061301
Veitch, J., Raymond, V., Farr, B., et al. 2015, PhRvD, 91, 042003
Veitch, J., & Vecchio, A. 2008, PhRvD, 78, 022001
Veitch, J., & Vecchio, A. 2010, PhRvD, 81, 062003
Venumadhav, T., Zackay, B., Roulet, J., Dai, L., & Zaldarriaga, M. 2019, PhRvD, 100, 023011
Venumadhav, T., Zackay, B., Roulet, J., Dai, L., & Zaldarriaga, M. 2020, PhRvD, 101, 083030
Verbiest, J. P. W., Lentati, L., Hobbs, G., et al. 2016, MNRAS, 458, 1267
Verde, L., Treu, T., & Riess, A. G. 2019, NatAs, 3, 891
Wade, L., Creighton, J. D. E., Ochsner, E., et al. 2014, PhRvD, 89, 103012
Wagoner, R. V. 1984, ApJ, 278, 345
Warszawski, L., & Melatos, A. 2012, MNRAS, 423, 2058
Watts, A. L., Krishnan, B., Bildsten, L., & Schutz, B. F. 2008, MNRAS, 389, 839
Weber, J. 1967, PhRvL, 18, 498
Weber, J. 1968, PhRvL, 20, 1307
Weber, J. 1969, PhRvL, 22, 1320
Weisberg, J. M., & Huang, Y. 2016, ApJ, 829, 55
Weisberg, J. M., Taylor, J. H., & Fowler, L. A. 1981, SciAm, 245, 74
Weiss, R. 1972, Quarterly Progress Report, Research Laboratory of Electronics (MIT) No. 105, 54
White, N. E., & Zhang, W. 1997, ApJ, 490, L87
Wiggins, P., & Lai, D. 2000, ApJ, 532, 530
Will, C. M. 1998, PhRvD, 57, 2061
Will, C. M. 2014, LRR, 17, 4
Woan, G., Pitkin, M. D., Haskell, B., Jones, D. I., & Lasky, P. D. 2018, ApJ, 863, L40
Woosley, S. E. 2017, ApJ, 836, 244
Woosley, S. E., Blinnikov, S., & Heger, A. 2007, Natur, 450, 390

Xin, L. 2018, PhyW, 31, 6
Xing, Z., Centrella, J. M., & McMillan, S. L. W. 1994, PhRvD, 50, 6247
Yagi, K., & Yunes, N. 2013a, PhRvD, 88, 023009
Yagi, K., & Yunes, N. 2013b, Sci, 341, 365
Yagi, K., & Yunes, N. 2017, PhR, 681, 1
Ye, C. S., Fong, W.-f., Kremer, K., et al. 2020, ApJ, 888, L10
Yim, G., & Jones, D. I. 2020, MNRAS, 498, 3138
Yunes, N., & Pretorius, F. 2009, PhRvD, 80, 122003
Zackay, B., Venumadhav, T., Dai, L., Roulet, J., & Zaldarriaga, M. 2019, PhRvD, 100, 023007
Zevin, M., Samsing, J., Rodriguez, C., Haster, C.-J., & Ramirez-Ruiz, E. 2019, ApJ, 871, 91
Zhu, S. J., Baryakhtar, M., Papa, M. A., et al. 2020, PhRvD, 102, 063020
Zimmerman, J., Carson, Z., Schumacher, K., Steiner, A. W., & Yagi, K. 2020, arXiv:2002.03210
Zimmermann, M. 1980, PhRvD, 21, 891
Zimmermann, M., & Szedenits Jr, E. 1979, PhRvD, 20, 351
Zink, B., Lasky, P. D., & Kokkotas, K. D. 2012, PhRvD, 85, 024030
Zwerger, T., & Mueller, E. 1997, A&A, 320, 209

AAS | IOP Astronomy

Multimessenger Astronomy in Practice

Miroslav D Filipović and Nicholas F H Tothill

Chapter 10

Dark Matter

Csaba Balazs

The existence of dark matter has been firmly established at astrophysical and cosmological scales via its gravitational effects. In contrast, we do not know what dark matter is made of. One possibility is that dark matter is composed of novel, yet undetected, fundamental particles. If this is the case, we can look for dark matter particles in various ways, such as indirect and direct detection. This chapter describes the basics of the evidence and the search methods for dark matter.

10.1 Observational Evidence

10.1.1 Knowns and Unknowns

This chapter provides a brief introduction to the dark matter problem. The material in this chapter is mostly qualitative, focusing on introducing the most important concepts that arise when discussing dark matter. The interested reader is encouraged to consult the references and other abundant resources to acquire a more advanced understanding of the problem.

Presently, there are more issues around dark matter that we do not understand than we do. The concept of dark matter emerged in the late 19th century when astrophysicists observed dark voids between clusters of stars and galaxies (Secchi 1877). They debated whether these regions lacked matter or contained some that were not visible to their telescopes (Ranyard et al. 1894). At the beginning of the 20th century, dynamical evidence for dark matter followed from estimates based on the velocity dispersion of stars observed in the Milky Way (Thomson & Kelvin 2010). During the early 20th century, using the virial theorem, the Swiss-American astronomer Fritz Zwicky (1898–1974) estimated the mass of the Coma cluster and found that it contains more dark than visible matter (Zwicky 1933). In the 1970s, emerging rotation curves confirmed that dark matter is abundant in galaxies and considerably affects their circular velocity profile (Rubin & Ford 1970; Freeman 1970; Rogstad & Shostak 1972; Whitehurst & Roberts 1972; Ostriker et al. 1974; Bosma 1978; Rubin et al. 1978). Further evidence of dark matter emerged at all

doi:10.1088/2514-3433/ac2256ch10

astrophysical scales, from Galactic size to the cosmic microwave background (CMB), during the last and this century.

Based on the astrophysical observations, it is possible to deduce some properties of dark matter. Straightforward inference suggests that dark matter interacts with light very weakly. Consequently, it must be composed of electrically neutral or fractionally charged elementary particles. Dark matter must gravitate, otherwise it cannot affect the velocity distribution of visible matter. It must be stable or decay with a lifetime of the order of the observable universe. Dark matter appears to be nearly collisionless, which suggests that it is composed of constituents that are relatively weakly interacting with each other. Finally, it must be cold enough on average to form large-scale astrophysical structures.

Speculation regarding the nature of this mysterious dark matter is still ongoing. Initial assumptions, during the last centuries, were straightforward and simple, conjecturing that dark matter consists of ordinary but low-luminosity or relatively small celestial objects that cannot be resolved by the observer's telescope. Examples of such objects would be burned-out stars, low-density clouds of gas, meteors, and comets. Along the lines of this argument, later, black holes were added to the list. Another line of thought proposed that the measured anomalies in the motion of stars and galaxies originated in non-Newtonian effects modifying gravity or dynamics at low accelerations. Dark astrophysical objects, including black holes, have been searched for but have not been found to contribute enough energy density to account for the gravitational effects of dark matter. Additionally, observationally confirmed Big Bang nucleosynthesis calculations indicate that dark matter is nonbaryonic. Combined with the fact that known nonbaryonic matter that matches the properties of dark matter is not abundant enough to account for dark matter, this strongly suggests that dark matter might be made of one or more types of yet undiscovered elementary particles.

The assumption that dark matter consists of a new type of elementary particles is especially attractive for a particle physicist. This is because most scenarios that solve some of the problems of the Standard Model[1] of elementary particles feature dark matter candidates. An example is the question of why there is more than one fundamental force in the Standard Model. Occam's razor[2] suggests that the ultimate theory describing our universe should have a single force that gives rise to all interactions we observe. On its face, this problem, like several others of the Standard Model, has nothing to do with dark matter. Remarkably, however, one finds that unification of the known fundamental forces implies the existence of new particles, some of which are perfect dark matter candidates. This example is not the only one that delivers to us unexpected dark matter candidates. The strong charge-parity problem, supersymmetry, extra dimensions, string theory— and the list goes on and on—all feature novel particles that fit the known properties of dark matter. This,

[1] The Standard Model is the name given in the 1970s to a theory of fundamental particles and how they interact.

[2] Occam's razor is the principle that when two explanations could account for all the facts, the simpler one is more likely to be correct.

and the rising concordance of this hypothesis and cosmology, is a striking theoretical hint of the particle nature of dark matter.

While we already know a few things about the properties of dark matter particles, we are far from fully understanding them. A particle is defined by its spin, mass, and charge, and we do not have this information for dark matter particles yet. Ultimately, future observations and experiments will confirm these properties. Meanwhile, based on existing data and theoretical arguments, we are able to anticipate the possible values of the defining properties. This is because if dark matter is composed of new fundamental particles, these particles have to fit in the framework of an extension of the Standard Model. These extensions fall into two broad categories: bottom-up and top-down models. The former introduces a set of new particles and interactions without trying to understand the full underlying theory (which might be too early to contemplate), while the latter derives them from new first principles. Adding a dark matter particle to the Standard Model and coupling it to the Higgs boson is an example of the bottom-up approach, and supersymmetry is a top-down theory. Crucially, in both approaches, experiments, well beyond astrophysics, and the theoretical structure of the Standard Model constrain the possibilities for the spin, mass, and charges of the dark matter candidate particles. This drives the quest for the defining properties of dark matter candidates and in turn for the search for dark matter particles.

10.1.2 Galactic Motion

10.1.2.1 Galactic Rotation Curves

Most galaxies are observed to rotate faster than they are expected based on Newtonian gravity and their inferred visible mass. The simplified argument goes as follows. The total mass of a galaxy and the distribution of this mass can be well estimated based on the visible stars. Using this mass distribution and Newtonian gravity and dynamics, which applies well to the nonrelativistic motion at hand, one can calculate the angular velocity of stars within the Galaxy as a function of distance from its center. Such a result, for the spiral galaxy Triangulum (also known as Messier 33, M33), is shown in Figure 10.1.

The figure shows the curve, as a solid line, that fits observations the best. The data in blue, closer to the center of the Galaxy, come from direct observation of starlight, while the data in yellow come from radio observations of the 21 cm spectral transition line of hydrogen. The prediction based on the visible matter distribution is shown by the dashed line. It is clear from the figure that there is a significant discrepancy between theory and observation. The discrepancy is accounted for by the presence of dark matter, which by increasing the mass of a galaxy generates more gravitational pull, thereby boosting the angular velocity of stars. The overlaid visible image of the Galaxy also hints at the spatial extent of the dark matter halo around the Galaxy. The rising rotation curve well beyond the visible edge of the Galaxy suggests that the visible matter is embedded in a much larger halo of dark matter. A similar discrepancy between visible matter and gravitational effects happens in many galaxies, indicating varied amounts of dark matter around them. Galactic rotation

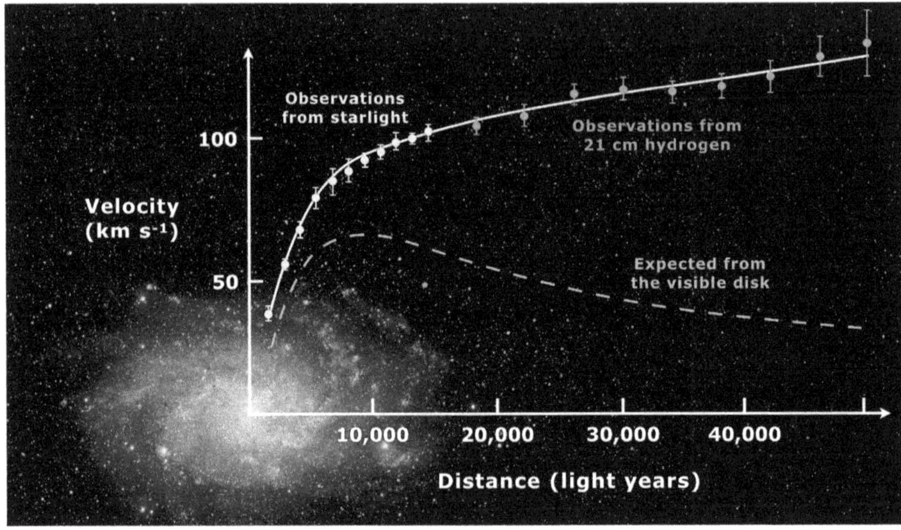

Figure 10.1. Rotation curve of the Triangulum galaxy (M33) overlaid on its visible image. Observation is shown by the yellow and blue points and error bars, while the prediction based on the visible matter distribution is shown by the dashed line. The solid curve that fits the data best significantly differs from the prediction, and the discrepancy is accounted for by the presence of dark matter. Image credit: Wikipedia: Mario De Leo (CC BY-SA 4.0).

curves allow us to calculate the amount and distribution of the invisible gravitating component.

One of the earliest quantitative estimates of the amount of dark matter in galaxies came from applying the virial theorem to them (Zwicky 1933). When applied to a galaxy, the theorem relates the average kinetic energy of its gravitating objects to their average potential energy. Given a velocity distribution of the gravitating ingredients, one can deduce the spatial distribution, and the total mass of the system, from the virial theorem. Once again, using the observed velocity distributions one arrives in a "virial discrepancy," indicating more matter within the Galaxy than the visible component (Zwicky 1937; Penzias 1961; Field 1971; Rood et al. 1970; Rood 1974).

10.1.3 Galaxy Clusters

Dark matter at the scale of galactic clusters is fairly evident because there is more than one way to estimate the mass of these clusters and there is a significant discrepancy between these mass estimates and the amount of visible matter. Surprisingly, about 90% of visible mass in galaxy clusters is in the form of gas (about 10% being in actual galaxies). Thus, mass estimates of galaxy clusters can be based on the X-rays emitted by the gas in them. The temperature of the intergalactic gas can be estimated from their X-ray spectra from which their pressure can be derived. Having the pressure, and assuming that it is balanced by the gravitational force that keeps the cluster together, the mass distribution of the cluster can be obtained.

Figure 10.2. Strong gravitational lensing effect in the group of galaxies called SDSS J0928+2031. The faint arcs are distorted images of galaxies or clusters behind the bright central objects. Image credit: ESA/Hubble & NASA, M. Gladders et al. Acknowledgment: Judy Schmidt.

Gravitational lensing is another way galaxy cluster mass distributions can be estimated. Because spacetime is bent around them gravitating objects affect the path of photons. Because of this, galaxy clusters act like rough lenses by distorting light that passes through or close to them. A specific type of distortion, called strong lensing, appears when the light behind a cluster is bent into shapes of arcs around the strongly gravitating parts of the cluster. This effect is clearly seen in Figure 10.2. Because the amount of distortion is proportional to the mass of the cluster, and its shape is indicative of the mass distribution, this effect can be used to infer the total mass and the mass distribution of a cluster. Lensing indicates a significant amount of dark matter in many galaxy clusters.

Another way to estimate the mass of galaxy clusters is by measuring their radial velocity profile. This technique, in spirit, is similar to the one estimating mass and its distribution in galaxies. There are other techniques that are used to estimate galaxy masses and the distribution of mass within them. For these, and for the caveats in some of the methods, the interested reader should consult the recent literature, such as Ntampaka et al. (2017).

At the spatial scale of galaxy clusters, there is another remarkable phenomenon that provides us with information about dark matter: colliding clusters. The Bullet Cluster, also known as 1E 0657–56, is an example of two galaxy clusters in the process of collision. In Figure 10.3 the colliding galaxies are overlaid in orange and

Figure 10.3. The Bullet Cluster (1E 0657–56): a set of galaxies, shown by white and yellow blobs, on a collision course. The pink and blue areas show most of the visible and dark matter in the clusters. Separation of these indicates a weakly interacting dark matter constituent. Image credits: X-ray: NASA/CXC/CfA/M. Markevitch et al.; optical: NASA/STScI; Magellan/U. Arizona/D. Clowe et al.; lensing map: NASA/STScI; ESO WFI; Magellan/U. Arizona/D. Clowe et al.

white from optical images of Magellan and the Hubble Space Telescope. Hot gas, which makes up most of the visible mass in the clusters, is highlighted by pink as seen by the Chandra X-ray Observatory. Blue areas show the distribution of dark mass in the system inferred from gravitational lensing. Comparison of the X-ray and gravitational lensing observations yield that most of the mass in the cluster is in the form of dark matter. Additionally, the clear separation of dark and visible matter strongly suggests that dark matter has interactions with visible matter and itself that are much weaker than the interaction within the cloud of visible gas.

10.1.4 Structure Formation

So far, we saw that dark matter appears to be abundant on the length scale of galaxies and galaxy clusters. Our understanding of the large-scale structure formation in the observable universe indicates that dark matter is a significant component of it at the largest scales. After inflation, the spatial distribution of matter must have been extremely uniform. As the universe evolved under the influence of gravity, minute density fluctuations became amplified. Regions with slightly higher density attracted more and more matter and gradually gravitationally collapsed into denser and denser clumps. This led to the formation of the largest structures in the observable universe, which host most galaxy clusters.

Early galaxy catalogs, such as the SDSS or the 2dF Galaxy Redshift Survey (2dFGRS) mapped out these largest superstructures of the universe during the last decade. Some of their results are shown in the top and left quadrants of Figure 10.4. One of the internal scales shows these regions extend to billions of light-years in space, which is a considerable portion of our observable universe. In comparison the right and bottom quadrants show the results of numerical calculations by the Millennium Simulation Project (MPA). These calculations simulate the evolution of density fluctuations under the force of gravity during the lifetime of the universe. They are not intended to precisely match the Galaxy surveys because the initial distribution of the density fluctuations is unknown. But they achieve a striking

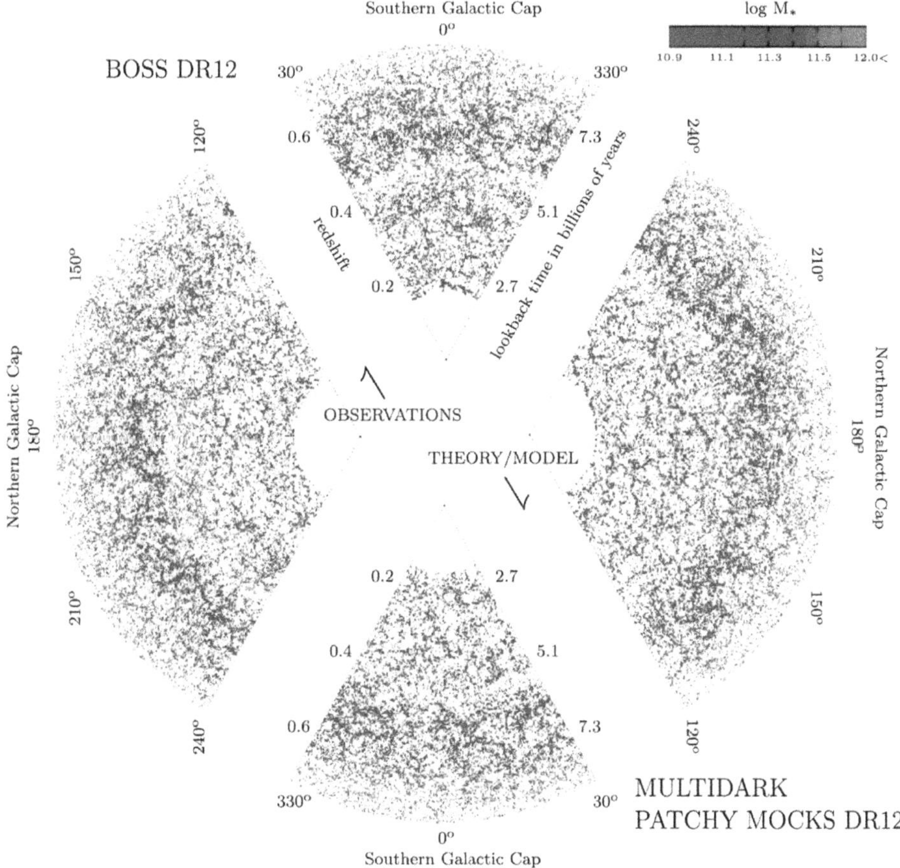

Figure 10.4. Superstructures, containing galaxy clusters, forming the largest-scale structure of our visible universe. Observations from the SDSS-III Baryon Oscillation Spectroscopic Survey are shown in the top and left quadrants. Results of numerical simulation to match are shown in the right and bottom quadrants. Good statistical agreement with observation is only obtained if the correct amount of dark matter is included in the simulations. Image credit: Kitaura (2016) © 2016 The Authors, Published by Oxford University Press on behalf of the Royal Astronomical Society.

statistical similarity to the observations, provided that about 80% of the initial amount of matter in them is dark. If only the visible amount of matter would be included in these simulations, the observed superstructures would never develop because the gravity created by visible matter is not enough to build them up. This is a notable indirect indication of dark matter throughout our whole observable universe.

10.1.5 CMB

Because dark matter significantly affects the density fluctuations of visible matter it leaves its gravitational imprint on the CMB. The CMB is a relic radiation from the time in the early universe when its temperature dropped below about 3000 K and protons and electrons combined into atoms. When this happened, about 380,000 years after the Big Bang, the universe became transparent to electromagnetic radiation. Looking outward from Earth, back in time, the surface of the last scattering can be seen as a curtain blocking electromagnetic radiation from earlier times.

When decoupled from the plasma, the CMB photons were in the visible range (about 0.26 eV) but the expansion of our spacetime redshifted them to microwaves (about 0.2348 meV). This stochastic background of photons appears to be emitted by the surface of the last scattering and permeates the visible universe. The CMB photons are detected, with impressive precision, by the Planck satellite and other means. The precision is impressive because early seeds of minute energy density fluctuations present in the CMB are clearly resolved by PLANCK today as temperature fluctuations of the order of microkelvin. These temperature fluctuations can be mapped onto multipole moments of a spherical harmonic decomposition of the angular variation of the CMB.

The amplitude of the CMB temperature fluctuations as a function of the multipole moment l and the angular scale (on top of the plot) is shown in Figure 10.5.[3] The observed data come from four sources: Planck, its predecessor the Wilkinson Microwave Anisotropy Probe (WMAP), the Atacama Cosmology Telescope (ACT), and the South Pole Telescope (SPT). The solid curve is the best fit of the Lambda-Cold Dark Matter (Lambda-CDM) cosmological model to the data with 26.5% dark matter contribution to the total energy density of the universe.

We are able to infer the amount of dark matter in our universe from CMB temperature fluctuations because, due to their significant gravitational coupling in the early eons, the latter is rather sensitive to the former. Figure 10.6 shows how varying baryon and dark matter contributions to the total energy density modulates the CMB power spectrum. By comparing the data in Figure 10.5 to the predictions in Figure 10.6 we can conclude that in the context of the Lambda-CDM model the unique combination of 4.9% of baryons and 26.5% of dark matter gives the best fit to the observations (Tanabashi et al. 2018).

[3] The inverse of l (which itself can be thought of as the angular “frequency”) is the angular scale or “wavelength” of the temperature fluctuations.

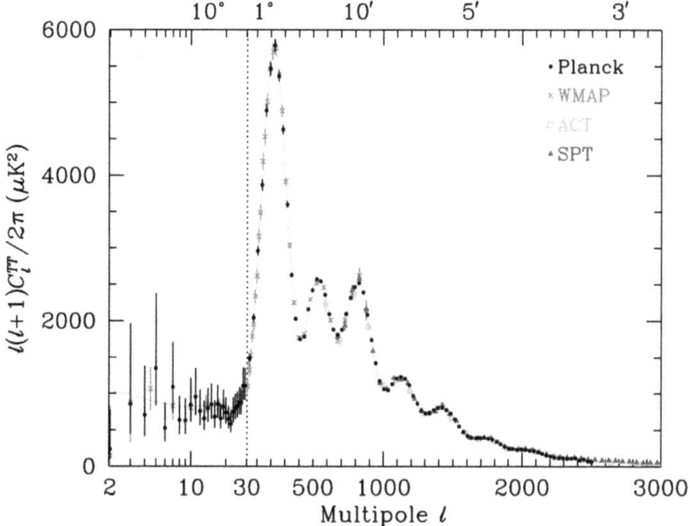

Figure 10.5. CMB temperature fluctuations as a function of the angular multipole l. Observations come from four different experiments, SPT, ACT, WMAP, and Planck. In order to fit these observations well, the theoretical prediction must include the 26.5% dark matter contribution to the total energy density of the universe. Image credit: Tanabashi et al. (2018), with permission, copyright 2018 by the American Physical Society.

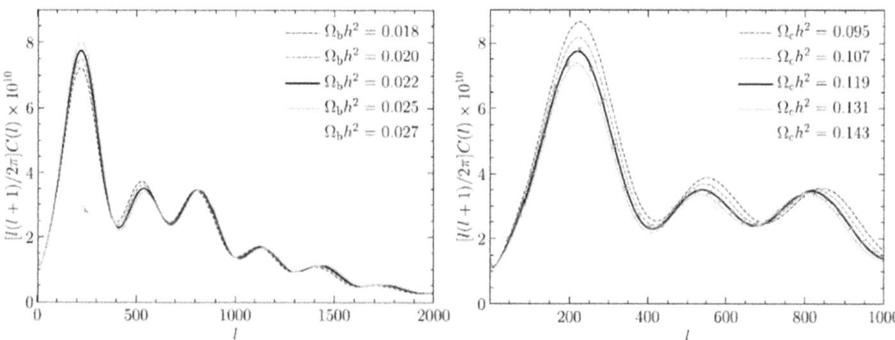

Figure 10.6. The variation of the theoretical prediction of the CMB temperature fluctuations with the amount of baryons and dark matter in the universe. Presently, the unique combination of 5% of baryons and 27% of dark matter gives the best fit to the observations. Image credit: Dodelson & Schmidt (2021), copyright 2021, with permission from Elsevier.

10.1.6 Cosmic Concordance

Dark matter appears to gravitationally exhibit itself from the distance scale of galaxies to the full observable universe. Different galaxies and clusters may, however, contain different amounts of dark matter. How do we know that the different observations consistently indicate the presence of the same agent? We have an answer to this question by combining many observations at the largest scales. Figure 10.7 shows the amount of large-scale structure as a function of distance

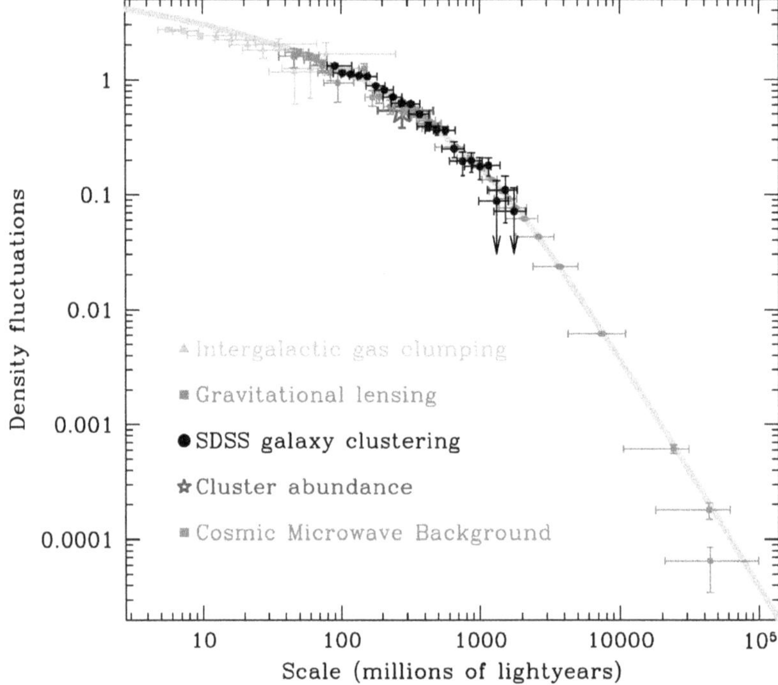

Figure 10.7. Density fluctuations as a function of distance scale from Earth. The observations come from various different means as explained in Tegmark et al. (2004). The solid curve is the prediction of the Lambda-CDM model assuming 25% dark matter in the universe. Image credit: Tegmark et al. (2004) and Max Tegmark (https://space.mit.edu/home/tegmark/sdss.html).

observed by several different means. (This plot is a rescaled version of the one published in Tegmark et al. 2004, where the origin of the data is explained.) The amount of density fluctuations on the vertical axis scales with the amount of structure at the given scale. The solid line represents a prediction of the Lambda-CDM cosmological framework assuming 5% matter and 25% dark matter. The agreement between theory and observations is impressive except at the lowest distance scales. This plot shows that dark matter is able to account for the missing mass throughout several orders of magnitude on the length scale of the observable universe. This represents the joint triumph of modern cosmology and astrophysical observation.

10.2 Indirect Searches

Fueled by clear observational evidence and strong theoretical motivation, dark matter particles are being searched for by various means. These searches typically assume a detectable interaction between dark matter and Standard Model particles. At the moment this assumption is mostly based on simple theoretical considerations. One of these considerations is that dark matter particles, being massive, most likely couple to the Higgs boson as any known massive particles.

This section lists a few indirect ways of searching for dark matter particles. These are called indirect searches because they detect secondary particles that originate from physical processes involving dark matter particles rather than the dark matter particles themselves. Indirect detection, typically based on astrophysical observation, is promising because it looks for dark matter in places where gravitational evidence indicates its presence. As with any astrophysical observation, however, background and foreground originating from standard astrophysical processes inject significant uncertainty, making indirect detection challenging.

10.2.1 Gamma Rays

Somewhat ironically, the strongest indirect constraints on the physical properties of dark matter particles come from observations of gamma rays. This is connected to the fact that in various new physics hypotheses that extend the Standard Model, gamma rays arise from dark matter self-annihilation, semiannihilation, or decay (Bringmann & Weniger 2012; D'Eramo et al. 2013).

The spectrum of these gamma rays is expected to be either relatively broad or fairly sharp depending on the particular annihilation channel. In many particle physics models, being electrically neutral, dark matter particles are their own antiparticles and typically self-annihilate into quarks, leptons, or massive gauge bosons. These standard particles fragment and hadronize, through which neutral pions are created, and these pions then decay to a pair of photons. Although most of the mass of the dark matter particles is converted into photonic energy, because many photons are created in a single annihilation process, each individual photon has about an order of magnitude less energy than the mass of the dark matter particle. Thus, such self-annihilation channels tend to produce relatively broad spectral features.

In contrast, in some models, dark matter particles may dominantly self-annihilate, for example via quantum loops, directly into two photons. This process gives rise to gamma rays with sharply peaked spectra somewhat below the mass of the dark matter particle. These sharp features offer the advantage of easier detection because most astrophysical backgrounds are expected to be smoother than them. For this reason such peaks, boxes, and steps are important targets of dark matter direct detection.

Annihilation or decay signals may arrive from areas of high dark matter density, and one of these promising regions is the center of our own Galaxy. Predictions show that the gamma-ray flux from dark matter annihilation around the center of the Milky Way is among the most detectable (Bringmann & Weniger 2012). Unfortunately, significant systematic uncertainty is injected into the measurements by the diffuse gamma-ray emission foreground along the line of sight between the galactic center and us. This emission comes from charged cosmic-ray particles that propagate in the ISM.

Dwarf spheroidal satellite galaxies within the extended halo of the Milky Way, which are above or below the galactic plane, are also promising targets to observe because of the lower foreground. Additionally, it is believed that they contain an enhanced amount of dark matter. For this reason, the strongest gamma-ray constraint on dark matter particles comes from them. The Fermi Large Area

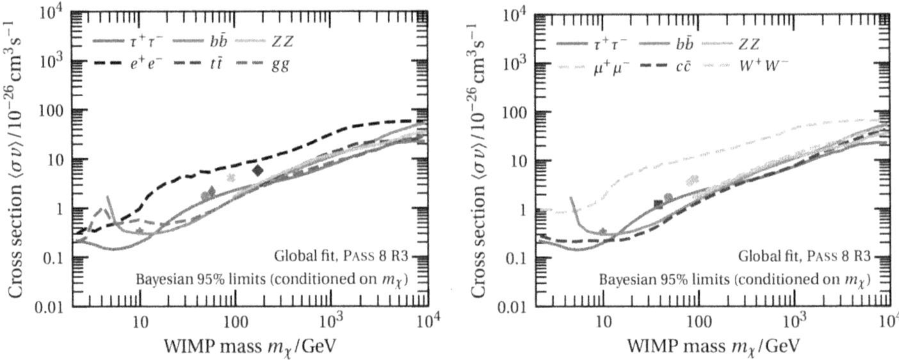

Figure 10.8. Exclusion limits from 27 dwarf spheroidals on dark matter annihilation cross section versus mass. Areas above the colored curves are excluded for the various annihilation channels. For the explanation of the colored symbols, see the text. Images credit: Hoof et al. (2020).

Telescope (LAT) observed over 40 dwarf galaxies and their data are used to put stringent constraints on dark matter annihilation.

The 2019 status of these exclusion limits, from 27 dwarf spheroidal galaxies, in the plane of dark matter annihilation cross section versus dark matter mass, is shown in Figure 10.8. The regions above the colored curves are excluded for the various annihilation channels as indicated by the caption. The various symbol shapes indicate the best-fit points to an excess observed from the Galactic center. These results cast a shadow of doubt on the dark matter explanation of the excess (Anderson & Bechtol et al. 2017; Hoof et al. 2020). The gamma-ray limits impose stringent constraints on dark matter annihilation in various popular new physics models (Athron et al. 2017a, 2017b, 2017c, 2019).

10.2.2 Charged Cosmic Rays

In some hypotheses the annihilation or decay products of dark matter particles can be electron–positron or proton–antiproton pairs. Unlike photons that reach us in a more or less straight line, these charged particles join the intragalactic flux of other charged cosmic rays generated by standard astrophysical processes. Diffusing through the Milky Way, these cosmic rays arrive at Earth and are measured by our instruments. The challenge in this case is to differentiate charged particles originating from dark matter processes from the rest.

An important part of this challenge is to understand the propagation of charged cosmic rays through our Galaxy. The standard Galactic propagation model is the so-called diffusion-convection model (Berezinsky et al. 1990). This model is an improved version of the simplest propagation model called the leaky box model. The diffusion-convection model describes the homogeneous propagation of charged particles in a volume containing the galactic plane. This volume is typically assumed to be cylindrical in shape with the Milky Way in its middle. A coupled set of diffusion equations, called cosmic-ray transport equations, has to be solved, with

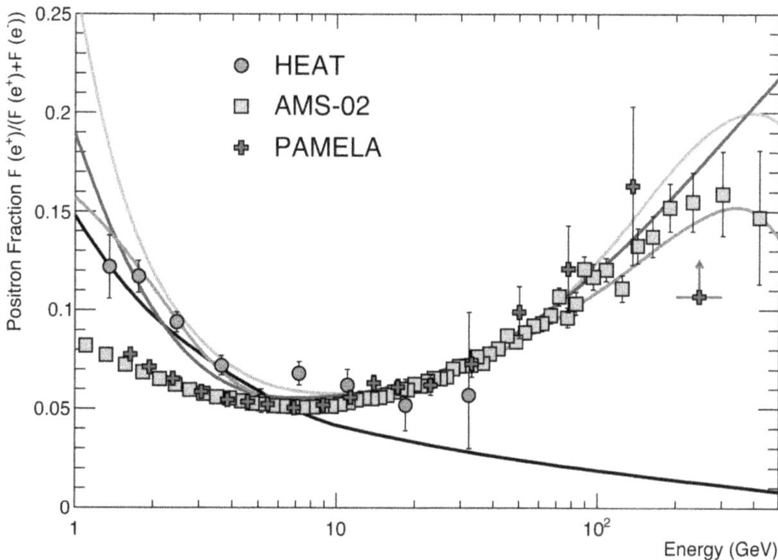

Figure 10.9. Cosmic ray positron fraction spectrum as the function of energy as measured by three different experiments: HEAT, AMS-02 and PAMELA. The black line shows the model prediction of standard astrophysical secondary production. The coloured lines are predictions to model the excess of positrons above 10 GeV. Green: dark matter decay; blue: modified propagation physics; red: production in pulsars. Image credit: Tanabashi et al. (2018), with permission, copyright 2018 by the American Physical Society.

each equation representing the density evolution of a propagating species, within this volume with certain boundary conditions. This task is highly nontrivial and requires a complex numerical code, such as GalProp (Orlando et al. 2018). The main input is the distribution of injected charged particles and the result of interest is the number density of cosmic-ray species near Earth.

The injection spectrum is a crucial input, and it is one of the hardest to pin down. That is because there is considerable uncertainty due to gaps in our knowledge of the distribution of standard astrophysical charged cosmic-ray sources such as supernovae, active galactic nuclei, quasars, and gamma-ray bursts. The present confusing situation is well illustrated by the excess observed in the ratio of the flux of positrons to the total flux of electrons and positrons shown in Figure 10.9.

Over the last decade the positron fraction was measured by several collaborations including the High-Energy Antimatter Telescope (HEAT), the Payload for Antimatter Matter Exploration and Light-nuclei Astrophysics (PAMELA), and most recently the Alpha Magnetic Spectrometer (AMS-02). They all found a systematic rise of the positron fraction in the energy region above 10 GeV. In this region the theoretical prediction based on the diffusion-convection model gives a falling spectrum indicated by the black solid line. The discrepancy may come from dark matter, and one such scenario is shown by the green line on the plot. Equally possibly, however, the discrepancy can be explained by the modification of physics describing the propagation as shown by the blue line. Alternatively, a nearby population of active pulsars could be the culprit as suggested by the red line.

Certainly, we need more data and increased precision theoretical calculations to decide what is happening.[4]

10.2.3 Neutrinos, CMB

Cosmic neutrinos are promising harbingers of annihilating or decaying dark matter particles because they hardly interact with anything while propagating to us from their source. But because neutrinos interact very weakly with matter, and perhaps weakly with dark matter, they might be rarely produced directly by dark matter and are hard to detect. Despite this, cosmic neutrinos from the Galactic center and dwarf galaxies are already used to constrain dark matter annihilation and decay.

The main reason why neutrinos are unique and promising indirect detection tools is that dark matter particles, having reasonably high mass and scattering cross sections, might become gravitationally trapped and accumulate in the core of stars and similar compact objects. If they do then their standard annihilation or decay products interact with the nuclear matter in the core of these objects and eventually produce stable standard particles including high-energy neutrinos. These neutrinos can easily escape the compact object, propagate to Earth and reach a neutrino telescope. Due to its closeness our Sun is the most promising source of such neutrinos (Silk et al. 1985) followed by the core of the Earth (Freese 1986; Krauss et al. 1986; also see Book 1). Neutrino detectors, such as SuperKamiokande, and telescopes, such as AMANDA, ANTARES, and IceCube, already set limits on dark matter annihilation and decay within the Sun (see Chapter 8), although at the time of writing, these limits are weaker than those arising from gamma rays.

Energy injection from dark matter annihilation or decay may have modified the spectrum of the CMB photons during the dark ages of the universe. The cosmic dark ages took place after electrons and protons formed dominantly atomic hydrogen in the early universe. In the lack of a substantial structure, such as stars, the homogeneous and hot gas was immersed in a bath of CMB photons. Because the fundamental physics leading to and governing this era is well known, the amount of ionization and the energy deposited by the gas into CMB photons can be fairly well estimated.

Annihilation or decay of dark matter particles could ionize the hydrogen gas during the dark ages. This additional ionization deposits energy to the photon bath and modifies its density fluctuations. Using the energy spectrum of the annihilation or decay products, an estimate of how these products cool, and how the additional ionization affects the photons, it is possible to calculate the change in the CMB anisotropy. Resulting limits on dark matter annihilation are competitive with gamma-ray constraints (Slatyer 2018).

[4] AMS-02 and PAMELA differ significantly from HEAT and the predictions below 10 GeV because the data were taken at different phases in the solar cycle when solar wind affects charged cosmic rays around Earth differently.

10.3 Direct Searches

The average energy density of dark matter is about five times higher than the average energy of standard matter in the universe. Dark matter is expected to accumulate in deep potential wells, such as the center of galaxies or stars, but in those places it typically overlaps with a large amount of matter, which adds considerable backgrounds and foregrounds to observation. Areas with relatively high dark matter density and relatively low visible matter density, such as dwarf galaxies or bullet-like clusters, are lower density and farther away. As we saw these circumstances make indirect detection of dark matter particles particularly challenging. Compared to collider experiments, for example, an additional difficulty is that the observed system at hand is not fully under our control. Consequently, using indirect observation it is hard to deduce the spin, mass, and charges of particles.

Direct detection of dark matter particles offers a reasonable compromise. These experiments are based on the premise that our galaxy, confined to a substantial potential well, is rotating inside an extended dark matter halo. According to estimates and observations there is about a 0.2–1.5 GeV cm^3 density of dark matter around Earth (Buch et al. 2019). As dark matter is almost collisionless with standard matter, the rotational speed of the dark matter halo around the center of the Galaxy is assumed to be negligible. Considering that the Sun is rotating around our galactic center with a speed of about 230 km s^{-1}, this implies considerable dark matter flux through Earth. Thus, in principle, it is possible to discover dark matter particles with a dark matter detector on Earth.

In such a detector dark matter particles are expected to collide with quarks or gluons (effectively nuclei) or electrons, depositing energy into the detector material. This, in turn, would lead to observable effects such as scintillation, ionization, or release of heat. A dark matter detector, of course, must have extremely good background suppression because, due to the low scattering cross sections, the expected number of dark matter collisions with detector particles is very low. Consequently, these detectors must be very strongly shielded against standard particles, such as cosmic rays, which are typically present on the surface of Earth. For this reason dark matter detectors are placed deep underground and shielded with highly advanced materials and methods.

10.3.1 Modulation Signals

As we saw in the previous section a well-shielded detector on Earth could detect dark matter particles streaming through it. The flux of these particles, however, is not constant for at least two reasons. The first is that the number and velocity distribution of the particles in the dark matter halo does not necessarily have to be constant and homogeneous. But because observations suggest that dark matter interacts very weakly with itself and standard matter, the zeroth-order assumption typically is that these effects can be neglected.

A nonnegligible modulation effect, however, arises due to the rotation of Earth around the Sun. Due to this rotation, the relative velocity of Earth and the dark matter halo is assumed to be

$$v(t) = v_S + v_E \cos(\theta)\cos(\omega(t - t_0)). \tag{10.1}$$

Here v_S is the velocity of Sun around the galactic center, v_E is the velocity of Earth around the Sun, $\theta \approx 60^\circ$ is the angle between Earth's rotational axis and the galactic plane, and $\omega = 2\pi/T$ with $T = 1$ year. The phase is typically fixed to t_0= June 2, when $\vec{v}_S + \vec{v}_E$ is maximal.

Due to the varying speed of Earth through the dark matter halo, a modulation of flux is expected in dark matter detectors. Notably, such a variation was detected by the DAMA/LIBRA Collaboration during its 14 year data collection. The modulation observed by DAMA/LIBRA is statistically significant and closely agrees with the expected phase (Bernabei et al. 2018). Two other modulation experiments, COSINE-100 and ANAIS-112, with similar sensitivity, however, contradict the DAMA/LIBRA results because they do not find a modulation in their signal (Adhikari et al. 2019; Amaré et al. 2019). Additionally, during the last decade, several direct detection experiments ruled out scattering cross sections required by the DAMA/LIBRA signal in the context of the simplest dark matter models, which further weakens the DAMA/LIBRA claim. The general consensus is that further experimental input is needed to clarify the situation.

10.3.2 Crystal Detectors

The first experiment trying to directly detect dark matter particles was using about a kilogram of high purity germanium crystal (Ahlen et al. 1987). Germanium appeared to be particularly suitable as the energy to create an electron-hole pair in this semiconductor is quite low (about 2.9 eV). The idea was that scattering of dark matter particles on the target material would transfer this amount of energy, producing an electron-hole pair, which, in an internal electromagnetic field, would lead to a cascade of such pairs. Because the number of signal events scales with the volume of the detector, increasing the detector volume is vital. Germanium diodes larger than 1 kg, however, are hard to build because of the high capacity of the material electronic noise scales with the size of the crystal. To circumvent this, the latest germanium detectors are built from an array of kilogram-sized elements, increasing the total mass of the detector to the 100 kg range (Jiang et al. 2018).

Another type of high purity crystal that is used in dark matter direct detection is NaI. This material is transparent to light and dark matter particles on the target material; scattering can easily produce photons in it which, in turn, can be detected by photomultipliers placed around the crystal. This simple scintillation detector design does not require a stable external potential as a germanium diode and is capable of operating for long periods. The main challenge with NaI detectors is to decrease the intrinsic backgrounds arising from impurities within the crystal. At the present rate of purification, this background is about one daily count per kilogram. This background forbids the use of NaI crystals as event detectors for dark matter particle scattering on the target material. They can, however, be used as modulation detectors, for example in DAMA/LIBRA, to detect the change of a signal at the top of a supposedly constant background and foreground.

10.3.3 Cryogenic Detectors

Dark matter scattering on a crystalline material may create phonons,[5] which, under suitable circumstances, can be detected. To detect such phonons the internal environment has to be optimally "quiet," in other words the crystal has to be cooled close to zero Kelvin. The produced phonons deposit energy throughout the crystal in the form of heat, and the sensitivity of the detector is quantified by

$$\sigma^2 = c_1\, T\, (C(T)\, T + c_2\, E). \tag{10.2}$$

Here, c_1 and c_2 are constants specific to the given crystal, T is the temperature and $C(T)$ is the heat capacity of the material, and E is the amount of energy deposited into it. The above equation clearly shows that the detector becomes sensitive to a minute amount of energy deposits only if the first term in the parenthesis is comparable to the second. For this reason these detectors are operated below 100 μK. Germanium and silicon are suitable for this detector material because their heat capacity quickly decreases with temperature at low temperatures as $C(T) \sim T^3$.

The small temperature variation of the crystal can be detected by thin wires surrounding it, which are kept just below their critical temperature. Below the critical temperature these wires are superconductive, that is, their resistance is practically zero. When they are slightly heated, however, they become ordinary conductors and their resistance soars. Thus, as long as they are kept around their critical temperature, measuring the resistance of these wires detects minute temperature variations of the crystal very sensitively. As described earlier, ionization is expected to happen simultaneously in Ge crystals, and this second detection channel can be effectively used to reject background events. Consequently, modern Ge-based detectors use a dual-channel detection method.

While offering the advantage of two-channel signal discrimination and precise energy resolution, cryogenic detectors are challenging and expensive to run due to their low operating temperatures. The size of an individual detector is also limited to the kilogram scale, thus requiring many linked modules.

10.3.4 Liquid Noble Gases

Noble gases, such as neon, krypton, argon, or xenon, can easily be ionized and also transparent to light so detecting scintillation in them is a good option. Among the listed noble gases, argon or xenon liquifies at higher temperatures than the others and, because the density of atoms in the detector is an advantage, they are more suitable for dark matter detection. Although argon is about five orders of magnitude more abundant in the atmosphere than xenon, the latter contains only one type of radioactive isotope, which makes it the first choice for a noble-gas-based dark matter detector.

Similarly to germanium-based ones, xenon detectors can utilize dual-channel detection based on scintillation and ionization signals. The scintillation signal is the

[5] A phonon is a definite discrete unit or quantum of vibrational mechanical energy, equivalent to a photon is a quantum of electromagnetic or light energy.

primary one hitting the photomultipliers first. Meanwhile, ionized electrons are driven to the surface of the liquid and generate a secondary scintillation signal scattering on atoms of the gaseous phase of xenon. Detection of these two separate signals allows for determining the location of the interaction vertex between dark matter and xenon atoms and discrimination between nuclear and electronic recoils.

10.3.5 Bubble Chambers

Superheated liquids were one of the earliest particle detectors and were already used to track collisions and decays of standard particles during the 1950s. Due to their early success,[6] they are utilized as dark matter detectors now. Bubble chambers typically contain a superheated liquid, and interaction within this liquid above a certain threshold of momentum transfer triggers a thermal phase transition. This is because when a particle deposits enough energy into the liquid the latter transitions to its gaseous phase around the interaction region.

Phase transition typically proceeds by bubble nucleation, not unlike hot steam bubbles forming within a boiling kettle. Because the probability of bubble nucleation is a function of the energy loss of the incoming particle, nuclear and electron scattering can be discriminated. This greatly helps in reducing the background from standard incoming particles. Bubbles forming in a transparent liquid are photographed by multiple cameras so their number and location can be precisely determined within the volume of interest. Multiple bubbles along a path indicate standard particles, which, due to their stronger interaction with the liquid, trigger bubble formation with a much higher probability than dark matter ones.

Beyond simplicity an advantage of bubble chambers is scalability as their volume can easily be increased into the 100 kg range. A drawback of bubble chambers is that the amount of energy deposited into the liquid cannot be determined because any energy deposition above a certain threshold will create bubbles. The threshold energy for bubble formation can be controlled by adjusting the intensive thermal properties, such as pressure and temperature, of the liquid. Further drawbacks are the complicated operational regime and the substantial dead time of the detector.

10.4 Concluding Thoughts

There are numerous open questions to answer regarding dark matter. For particle physicists perhaps the most important of them is: What is dark matter made of? It will take joint experimental and theoretical effort to answer this question. On the experimental side, to confirm the identity of dark matter particles, we will need a wide range of observations from astrophysics through high energy to precision physics. On the theory side, we will need to come up with new ideas and systematically examine the plausibility of various existing ones. Regardless, this will be an exciting and rewarding journey throughout the next decades.

[6] The 1960 Nobel Prize went to Donald A. Glaser (1926–2013) for the invention of the bubble chamber.

References

Adhikari, G., Adhikari, P., Barbosa de Souza, E., et al. 2019, PhRvL, 123, 031302
Ahlen, S. P., Avignone, F. T., Brodzinski, R. L., et al. 1987, PhL, B195, 603
Albert, A., Anderson, B., Bechtol, K., et al. 2017, ApJ, 834, 110
Amaré, J., Cebrián, S., Coarasa, I., et al. 2019, PhRvL, 123, 031301
Athron, P., Balázs, C., Bringmann, T., et al. 2017a, EPJC, 77, 879
Athron, P., Balázs, C., Bringmann, T., et al. 2017b, EPJC, 77, 824
Athron, P., Balázs, C., Bringmann, T., et al. 2017c, EPJC, 77, 568
Athron, P., Balázs, C., Beniwal, A., et al. 2019, EPJC, 79, 38
Berezinsky, V. S., Bulanov, S. V., Dogiel, V. A., & Ptuskin, V. S. 1990, Astrophysics of Cosmic Rays
Bernabei, R., Belli, P., Bussolotti, A., et al. 2018, Univ, 4, 116
Bernabei, R., Belli, P., Bussolotti, A., et al. 2018, NPAE, 19, 307
Bosma, A., et al. 1978, PhD thesis, Groningen Univ.
Bringmann, T., & Weniger, C. 2012, PDU, 1, 194
Buch, J., Leung, J. S. C., & Fan, J. 2019, JCAP, 2019, 026
D'Eramo, F., McCullough, M., & Thaler, J. 2013, JCAP, 2013, 030
Dodelson, S., & Schmidt, F. 2021, in Modern in Modern Cosmology, ed. S. Dodelson, & F. Schmidt (2nd ed.; New York: Academic), 231
Field, G. B. 1971, ApJ, 170, 199
Freeman, K. C. 1970, ApJ, 160, 811
Freese, K. 1986, PhLB, 167, 295
Hoof, S., Geringer-Sameth, A., & Trotta, R. 2020, JCAP, 2020, 012
Jiang, H., et al. 2018, PhRvL, 120, 241301
Kitaura, F.-S., Rodríguez-Torres, S., Chuang, C. H., et al. 2016, MNRAS, 456, 4156
Krauss, L. M., Srednicki, M., & Wilczek, F. 1986, PhRvD, 33, 2079
Ntampaka, M., Trac, H., Cisewski, J., & Price, L. C. 2017, ApJ, 835, 106
Orlando, E., Johannesson, G., Moskalenko, I. V., Porter, T. A., & Strong, A. 2018, NPPP, 297-299, 129
Ostriker, J. P., Peebles, P. J. E., & Yahil, A. 1974, ApJ, 193, L1
Penzias, A. A. 1961, AJ, 70, 293
Ranyard, A. C., Baden-Powell, B. F. S., Grew, E. S., Webb, W. M., & Proctor, R. A. 1894, Structure of the Milky Way, Knowledge: A Monthly Record of Science, Vol. 17 (Wyman & Sons)
Rogstad, D. H., & Shostak, G. S. 1972, ApJ, 176, 315
Rood, H. J., Rothman, V. C. A., & Turnrose, B. E. 1970, ApJ, 162, 411
Rood, H. J. II 1974, ApJ, 188, 451
Rubin, V. C., & Ford, W. K. Jr 1970, ApJ, 159, 379
Rubin, V. C., Ford, W. K. Jr+, & Thonnard, N. 1978, ApJ, 225, L107
Secchi, A. 1877, L'Astronomia in Roma nel pontificato di Pio IX: memoria (Tipografia della Pace)
Silk, J., Olive, K. A., & Srednicki, M. 1985, PhRvL, 55, 257
Slatyer, T. R. 2016, in Proc. Theoretical Advanced Study Institute in Elementary Particle Physics: Anticipating the Next Discoveries in Particle Physics (TASI 2016), 297
Tanabashi, M., Hagiwara, K., Hikasa, K., et al. 2018, PhRvD, 98, 030001
Tegmark, M., Blanton, M. R., Strauss, M. A., et al. 2004, ApJ, 606, 702

Thomson, W., & Kelvin, B. 2010, Cambridge Library Collection—Physical Sciences (Cambridge: Cambridge Univ. Press)
Whitehurst, R. N., & Roberts, M. S. 1972, ApJ, 175, 347
Zwicky, F. 1933, AcHPh, 6, 110
Zwicky, F. 2009, GReGr, 41, 207
Zwicky, F. 1937, ApJ, 86, 217

Chapter 11

Multimessenger SETI and Techniques

Branislav Vukotić, Milan M Ćirković and Miroslav D Filipović

The recent developments in observational techniques across the whole electromagnetic domain have paved the way for rethinking old and developing new SETI approaches, often under new labels such as the "search for technosignatures." In addition to describing the various specific SETI efforts, this chapter investigates the methodological grounds of SETI studies in general, in order to facilitate further development of search techniques and models. The synthesis of various SETI approaches is discussed, in particular how different aspects of the entire enterprise relate to contemporary astrobiology, the search for biosignatures, the issues of galactic habitability, and our knowledge on the evolution of life in its widest cosmic context.

> It may be objected that the above is nothing more than a series of imperfectly proved hypotheses. But granting its improbability, it suffices that this explanation is not impossible. For then I have shown that the problem is not insoluble, and nature will have found a better solution than mine.
>
> —Ludwig Boltzmann (1895)

Life thrives only when there is a biosphere of mutually interacting life. Interactions make the foundation for life evolve and adapt in order to be sustainable and survive. Life would naturally spread, diversify, and evolve in order to better adapt and interact with its environment. It is hard to imagine a lush biosphere consisting only of a single species. Does this transcend to intelligent and aware life? Are there any other advanced species in a space with numerous Earth-like planets? As in countless other instances all over the realm of science, Boltzmann's wise words quoted above demonstrate how we should think about solving the astrobiological puzzle. We are investigating the outcomes of an evolutionary process, in the domains of both biology and culture, with a myriad entanglements and contingencies which we cannot follow in detail. A simplified understanding may be the best we can aspire to—but it also may be enough.

doi:10.1088/2514-3433/ac2256ch11

11.1 Motivation to Do SETI

Life is destined to seek other life. That is the root of our need to socialize, and it goes much deeper than the scope of the social sciences, into the very foundations of the evolution that made us be. This is the main drive that makes us reach for deep space and connect to other life in order to spread our evolution experiment spacewide. Of course, in order to sustain our technology addiction over longer periods of time in harsh environments, it would be a good starting point to have it done here on Earth first. But for the most part, it is not a matter of escaping the possible rage of "Mother Earth" owing to our ecological negligence but rather a matter of integration into a spacewide network of life. Hence, likely, our fascination with the clear night sky, shown for instance in Kant's famous dictum on the starry heavens and the moral law. There must be something or someone out there.

Understanding life even here in our own backyard is extraordinarily complicated, let alone across vast distances of space. This is probably the reason why at the beginning of the space age, which roughly coincided with the foundation of SETI as a scientific endeavor, the SETI methodology was largely abstract. However, the discovery of exoplanets has made a significant impact on our understanding of life in space with the advent of astrobiology, the hunt for biomarkers, and other empirical insights that are closer to our everyday experience. This recent astrobiological revolution has initiated the rethinking of traditional SETI strategies and is establishing SETI in the mainstream of scientific research as a more active and experience-based enterprise, although traditional methods are still present in practice.

To give just one among many examples, the study of Zuckerman (2019) points out that *in situ* detection is superior over listening-oriented and other passive SETI efforts. Indeed, reaching outside Earth would be beneficial to SETI even if we happen to not discover new life. It is, therefore, of paramount importance to attempt to put the entire field into as wide a context as possible, especially as appropriate to the current space age.

Settling humans on Mars is likely to happen very soon. With no obvious theoretical obstacles to travel and life-sustaining processes (Zubrin 2011; Muscatello et al. 2005), it appears to be just a matter of perfecting existing technology. Surely, the best time to start is immediately. Colonizing outside our own planet would give chances to alien SETI projects to detect our interplanetary activity and possibly try to contact the solar system. And, colonizing the solar system is likely to have a profound influence on our SETI strategies. It should not be considered as leisure-time stargazing into the vast space separated from us by a glassy sheet, as one would do in an underwater sea tunnel with transparent walls. Moving outside of Earth would mean that we are present on the "radar" of extraterrestrials and that we are not passive observes, but that our actions can make a difference to our galactic neighbors. Ashworth (2019) recently confirmed the old conjecture of Ronald Newbold Bracewell (1921–2007). Viktor Borisovich Shklovsky (1893–1984) and other pioneers state that expansionist civilizations would quickly become dominant in the Galaxy, given the required interstellar travel timescales. Because traditional SETI has appeared together with the dawn of the space age, it does go hand in hand with the ability of civilization to expand. In the absence of the results of traditional radio SETI, novel approaches that are based on the possibility of expanding beyond the home planet should be considered and the resulting prospects for detection examined.

In his pioneering work, Drake (1961) highlights the importance of a complete astronomical picture of life when the search for other life is concerned. Therefore, the assessment of SETI phenomena would inevitably include our knowledge of the development and evolution of life in space. The literature on SETI shows steady growth together with our increased knowledge of life in general.

In an initiative to build a comprehensive bibliography of SETI-related works, Reyes & Wright (2019) report an ADS bibliography group with 2908 SETI-related entries (accessed on 2020 February 21). After the initial few dozens of papers in the 1960s and 197s, the 1980 s witnessed a sharp rise in SETI research that has remained approximately constant until the present, with an average of $\approx$50 papers per year. In the rest of the introductory section, we give a brief insight into the literature in order to highlight the methodological background of the main SETI paradigms and approaches. One should keep in mind at all times that SETI literature might not be immediately recognized as such, for various reasons, due mainly to the controversial reputation the field has always carried like a burden, especially in the aftermath of the NASA HRMS cancellation in 1993 (Garber 1999; NASA 2018). This "winter of discontent" has had an effect even on people who have actually been engaged in SETI research, being reluctant to openly proclaim so; the modern impetus to try to subsume SETI under the label of "search for technosignatures" is based, in part, on the same motivation. While the history of science will undoubtedly have the last word in the decades—or perhaps centuries—to come, there is an unfortunate air of adversity (and, sometimes, ridicule) often surrounding the conventional label.

11.1.1 Methodological Grounds

In the last couple of years, roughly since about 2015, we have witnessed a renaissance of our efforts (in the SETI), understood in the most general sense. In order to understand (detect, discern from natural) a particular message or effect ("SETI activity"), we should first try to predict and conceptualize the expected ways that intelligence and its activity could be manifested. One concern is how human civilization is seen from afar, which is connected to both the development of our observational capacities and to the so-called "METI controversy" (from Messaging to ExtraTerrestrial Intelligence) that unfolded in the recent decade or so. To rule out the absence of detection, first we must try and rethink all possibilities. However, it might not be so simple and may require a unified approach across multiple viable modes of detection. Figure 11.1 demonstrates how SETI is embedded in wider astrobiological and, ultimately, cosmological contexts. Modern cosmology has made huge strides in accounting for the initial cosmological conditions around the Big Bang and the subsequent processes of structure formation. The latter is nowadays successfully developed and modeled within the predominant λCDM (Lambda cold dark matter) paradigm, and the level of detail captured by the

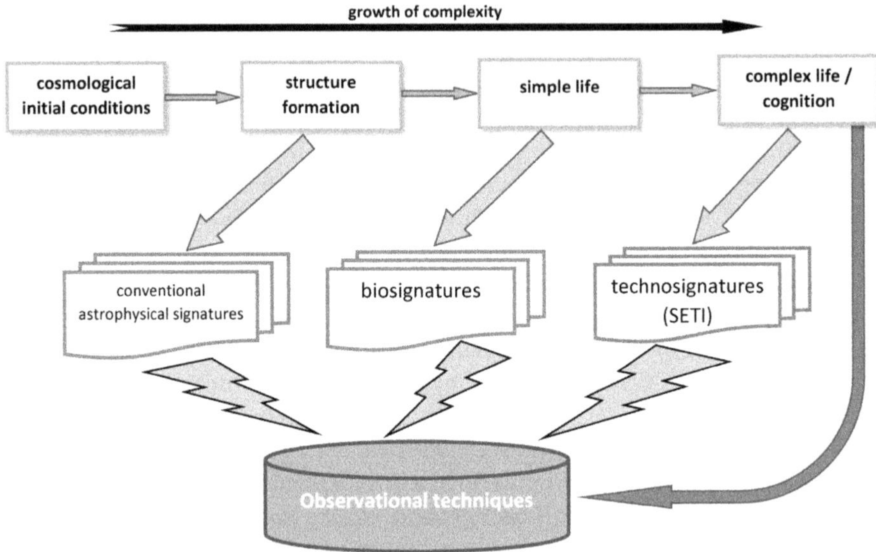

Figure 11.1. Schematic description of SETI areas and their relation to our research efforts.

contemporary massive numerical simulations is astonishing and rises exponentially. At least some of the cosmological structure supports habitable regions, and at least some of these habitable regions are reasonably expected to develop simple life forms. On Earth, this step happened quickly, in contrast to the much longer timescale for the development of complex life and, ultimately, intelligence capable of reflecting upon these issues. To what extent is this situation we perceive in our past average or typical is unclear, but there is no reason to doubt that the causal sequence occurs everywhere in the manner schematically shown in the diagram.

Of course, apart from the inferred growth of complexity, there is an additional asymmetric process at play, namely: observation selection. There is only a limited part of the total parameter space that can contain histories consistent with the dynamical laws leading to the present and to us as specific kinds of intelligent observers. The problem here—and it is indeed a key problem for SETI studies—is that there is simply no way of establishing to what extent we are typical intelligent/cognitive/self-aware systems among all such environments produced, in the fullness of space and time, by the universal cosmic evolution. Even a vague answer to this question (not to mention what is indeed desirable, namely a precise quantitative account) seems to be far beyond our present understanding, so we have to use some simplifying assumptions like Copernicanism: the assumption that we are fairly typical within our reference class. This is the ultimate grounding of all SETI activities, because all astronomical data are, in the final analysis, obtained by observational techniques devised by our particular form of cognition, including peculiarities of our perception and inference. There is no way around it, and while it may not be particularly important when we look at conventional signatures of purely (astro)physical processes like star formation or cosmic-ray acceleration, it becomes of paramount importance when dealing with bio- and especially technosignatures.

Some proposed SETI models, such as the "Zoo" hypothesis (Forgan 2011; Ball 1973) or directed panspermia (Crick & Orgel 1973; Sleator & Smith 2021), imply the causality between the categories presented in Figure 11.1. Also, the class of daisy-world models investigates the influence of life on its surroundings and consequent astrophysical signatures. For a fruitful approach, this should be kept in mind and the described effects clearly separated. Another practical example in this manner is the search for excess infrared heat from stars and galaxies that might be the result of energy dissipation by advanced civilizations (Wright 2020; Wright et al. 2014a, 2014b, 2019; Zackrisson et al. 2015).

Most SETI arguments concern the radio-frequency domain. This is the direct consequence of attempts to redesign traditional SETI, which relies on radio detection and transmission. This should be expanded to other multimessenger domains including particle emission (Carstairs 2002; Tough 2000).

In what follows, we give a description of some possibilities for SETI activity in other domains. With some of them being quite exotic, given the current state of our technology, it might be a viable path for future generations of SETI astronomers. While reviewing several-decades-long SETI activity in Russia, Gindilis & Gurvits (2019) highlighted that future SETI should be based on a wealth of experience and material from numerous past attempts. Apart from the purely practical benefits, understanding details of previous empirical searches underlines the essential methodology at the very core of the field: in an area lacking solid theoretical foundations, it is of foremost importance to sample different parts of the huge parameter space and to obtain as many negative results as possible. The chapter is concluded with a synthesis of other possibly relevant phenomena and by considering possible benefits of a unified approach, connected with the recently suggested Dysonian SETI, in particular how different aspects of SETI relate to galactic habitability and to our knowledge of life in general.

11.2 Various Approaches to SETI

What is more abundant, simple or intelligent life in the universe? The answer to this crucial question might also change from epoch to epoch, reflecting the evolutionary nature of the subject matter. So far, we will assume that there is no reason to prefer one over the other, and a detailed discussion of this topic is beyond the scope of this chapter. One characteristic is common to both types of life, simple and intelligent: the ability to interact and change the environment. Apart from SETI based on messages, it is these kinds of deviations that should be searched for in order to detect ETI or simple exolife. In the case where the above assumption on expansion-oriented ETIs holds, they are likely to be easier to detect than biosignatures from simple life forms This is due either to the signatures of their interplanetary and interstellar movement or their feats of microengineering.

The history of modern SETI is entangled with the appearance of modern technology, which has enabled us to creep out into the vastness of space. At the beginning of the 20th century, the founders of radio technology immediately advertised the potential of radio waves in communicating with extraterrestrial civilizations. From Nikola Tesla's (1856–1943) display of a radio-controlled boat at the Madison Square Garden (New York, USA) exhibition in 1898, it took more than half a century for this technology to mature and enable efficient communication with our space vehicles, effectively kick-starting the space age. Parenthetically, both Tesla and his main rival Guglielmo Marconi (1874–1937) were enthusiastic about using radio waves for interplanetary communication, which was appropriate for the time when most people believed Mars to be not only habitable but actually inhabited by intelligent beings (e.g., Dick 1996). In parallel, the light from electrically powered light bulbs spread like raging fire across the shadowed side of our planet. If there is any other similar civilization out there, within the 30 pc radius from Earth, it could have become aware of our presence in this manner, as well as in number of other ways and techniques, the most poignant being the detection of anthropogenic air pollution in Earth's atmosphere.

The Arecibo message (Staff at the National Astronomy & Ionosphere Center 1975) will reach its intended designation, the globular cluster M13, in a little more than 20,000 years. In terms of the human civilization development rate, this is a substantial time interval. It might be that our understanding and conduction of SETI will significantly change during this time period, making this effort obsolete and only of historical significance. Obviously, the Arecibo message was not intended to be replied to; rather, its purpose was to demonstrate the technological ability to send an intentional technogenic message over distances comparable to the spatial scale of the Galaxy. The message was fairly simple yet highly informative—and experimenting with the message design was one of the purposes of this experiment. Our further messaging actions should be tailored to smaller required time intervals, i.e., to closer targets, in order to incorporate the epistemological time frame.

This might also invoke the possibility of initiating contact. Slim probabilities of radio-SETI detection are reported by (Ashworth 2016). The author further argues that quantifying these assumptions is crucial for estimating the feasibility of radio SETI. Zubrin (2017) underscores problems with radio communication, considering it an inefficient means of communication and suggesting a biological material-based communication. The analysis of Haqq-Misra et al. (2013) found that the radio background emission from Earth, consisting of radio communication that is not intended for messaging extraterrestrials, has a greater potential than METI attempts to be detected. Messerschmitt & Morrison (2012) underscore the difficulties in designing and operating two-way communication on SETI scales, concluding that they result in serious implications for sending and receiving such messages.

In the footsteps of modern SETI pioneers, Billingham & Benford (2014) discard messaging activity in favor of lower cost and more efficient (in terms of energy and time requirements) traditional SETI "listening." Despite the negative SETI results so far, Brin (2014) discusses ways to rethink traditional SETI in the light of scientific discoveries that have occurred in the meantime.

One of the approaches to analyzing large volumes of data was pioneered by the SETI community. According to Werthimer et al. (2001) the SETI@home project utilizes desktop computers of over a million volunteers, effectively becoming the largest supercomputer on the planet, averaging 20 teraflops at a time. The project searches for narrowband signals in Arecibo telescope radio data. In a similar fashion, optical data are searched for short pulses.

11.2.1 Intentional Messages/Beacons

Although some of the arguments raised against METI (Messaging Extraterrestrial Intelligence) rely on the possible dangers of messaging extraterrestrials, the results of game theory based on a model from de Vladar (2013) largely discard these dangers in favor of pursuing active SETI. Haqq-Misra (2019) states that METI might be conducted in addition to existing SETI efforts but also highlights that the possible dangers of conducting METI are indeterminable at present. In addition, Zaitsev (2008) points out that the power used for METI is about a million times smaller than the power transmitted in order to obtain radar images of asteroids and planets.

On the other hand, Shostak (2011a) argues that omnidirectional beacons are very inefficient and costly. This was first pointed out by the Northern Irish physicist Sir David Bates (1916–1994) in the 1970s, where the naive assumptions of early SETI pioneers were criticized (Bates 1974, 1978). Seth Shostak (1943–) proposed the use of directed low-power emissions. In Shostak (2011b) he also proposed the use of short pulses as an efficient means of getting their attention. Once there is an alert, the attention can be directed to constructing a low-power message receiver that can integrate the signal from the target over a longer time. A similar approach to scanning large spatial angles for short-term pulses is presented in Benford et al. (2010) and also in Scheffer (2005).

From the above we can see that the practical use of SETI is a two-way street. When thinking about a particular search strategy, one needs to take into account not just the means of detecting a specific signal but also the means of transmitting such a signal over vast distances of space and making it detectable. At present, we've seemed to have gone far from the traditional SETI. Instead of searching for encrypted messages from space, we seem to understand the difficulties imposed by a large volume of space and time even on simple beacon-based emissions. The cost-optimized beacon of search strategies (Benford et al. 2010) is likely to be the main characteristic of the SETI for years to come, especially given the rapid development of our data-processing capabilities that rely more heavily on machine-learning techniques.

11.2.2 Parasitic Searches

The amount of SETI search space completed to date present is very small in comparison to the size of the total available search space (Wright et al. 2018a). Davies & Wagner (2013) state that existing databases should be searched for indirect ETI evidence, such as footprints of nonhuman technology, e.g., the analysis of photographs from the Moon's surface. As part of ongoing planet search programs, optical narrowband data are searched for SETI signals (Werthimer et al. 2001).

Heller (2019) demonstrates the substantial success of en masse SETI efforts, conducted through social media, that have demonstrated our collective ability to understand and code SETI messages.

The following might be an example of a practical deployment of novel SETI approaches. Borra & Trottier (2016) reported the detection of periodic spectral modulations in SDSS data for 234 stars, consistent with the existence of ETI (elaborated in Borra 2012), but underscoring the need for further investigations and tests of their results. Another example of improving upon methods of the traditional SETI is the autocorrelation search on Allen Telescope Array observations (Harp et al. 2018). Although a negative SETI result is reported, they propose expanding their efforts to interferometer databases worldwide, estimating an upper limit of a 5% chance of success.

The energetic nature of X-ray photons enables communication only with temporal modulation, unlike the wave modulations possible at lower energies (Corbet 1997). The author also summarizes the possibilities of deploying parasitic searches for a very short pulse modulation that a sufficiently advanced civilization may induce in X-ray binaries. It is emphasized that the available X-ray satellite technology at the time could scan down to microsecond resolution. In addition, he also proposes that a similar approach is possible in the radio and other wavebands.

Cabrol (2016) investigates connections between biomarkers and SETI and discusses how the application of artificial intelligence (AI) and other modern tools to big data can influence SETI research in the near future. Her study clearly demonstrates what is still regarded as suspicious in some conservative circles: that SETI is a legitimate, full-fledged sector of the overall astrobiological effort and that in modern times there is no justification whatsoever for its marginalization or systemic doubts about it.

11.2.3 In Situ

More than a half-century ago, Bracewell (1960) argued about the possibility that ETI could send probes to our solar system that, upon arrival, would try to draw our attention with radio emissions within the solar system. Gertz (2016) argues that alien interstellar probes, as an efficient way of conducting SETI searches, are more likely to be found than interstellar beacons.

Building up the case for in situ SETI over messaging-based SETI is also done in the work of Gertz (2017), which introduces the system of interstellar nodes that act as hubs and root the space probes. Galera et al. (2019) investigate percolation as a means of settling the present nondetection of extraterrestrial civilizations (ETCs). On the other hand de Magalhães (2016) and Zaitsev (2006) propose that despite slim chances of success, message-based SETI could increase the chances of us being detected by ETCs. This in turn might result in their initiative to contact us and therefore increase the chances for SETI success.

In addition, Gertz (2019) argues that sending Oumuamua-like objects[1] as SETI probes is more efficient than sending out messages. Also, Rose & Wright (2004)

[1] One of the first interstellar asteroids, Oumuamua is Hawaiian for "a messenger from afar arriving first."

propose the sending of an inscribed matter package instead of traditional messaging. Their considerations are based on the energy expenses and the density of information that could, using present-day (e.g., quantum lithography) or near-future technologies, be achieved. However, this might be read from a slightly different point of view as well: communication via inscribed matter packages will be rather slow by wall-clock time, although perhaps not by the average bits/unit of time, indicating a long-living and extremely stable interstellar society.

11.3 Radio SETI

In the era of modern radio telescopes, such as the Five hundred meter Aperture Spherical Telescope (FAST), Low Frequency Array (LOFAR), ASKAP, Parkes, SKA, etc., one of the biggest challenges for radio SETI is the automatic processing of large data sets in order to remove the radio-frequency interference (RFI)[2] and to classify and label the detected signals for an efficient subsequent analysis (also see Chapter 12). Using a data spectrogram as a 2D image, Harp et al. (2019) argue that their machine-learning algorithm has a high level of discrimination of source types, achieving a very low rate of false positives. In an earlier work, Cox et al. (2018) also employed a spectrogram as a favorable way of presenting radio data when utilizing neural networks to classify signals for the purpose of SETI.

To address the future challenges of radio astronomy, Ekers & Bell (2001) summarize the importance of improving the methods for RFI isolation. Dana (2018) analyzed the trajectories of satellites in order to predict their locations in the sky and reduce their RFI signature in radio observations.

The recent work of Pinchuk & Margot (2020) investigates correlation-based RFI removal as an alternative to widely used direction-based approaches (e.g., Harp 2005). They also describe their usage of machine-learning techniques in this field, which is of obvious methodological relevance to future expanded searches, including those of Breakthrough Listen (Isaacson et al. 2017).

The recent collaboration between the Chinese Academy of Sciences and the University of California at Berkeley resulted in the application of Berkeley's SERENDIP VI (Search for Extraterrestrial Radio Emissions from Nearby Developed Intelligent Populations)—a real-time SETI spectrometer—to the largest single-aperture radio telescope on the planet, FAST, located in China. In Zhang et al. (2020), the authors describe their robust RFI rejection filter that combines narrowband RFI removal with a multibeam approach and machine learning.

11.3.1 Limits of Radio SETI

The recent work by Sheikh et al. (2020), as part of the Breakthrough Listen search surveyed the directions from which Earth's transits across the Sun are apparent to possible extraterrestrials. The selected targets have distances of up to $\approx$150 pc. The survey is performed with the Green Bank Telescope and reported no SETI detections with a stated sensitivity of up to a thousand times fainter than the Arecibo telescope

[2] For a description of RFI and its effects on SETI, see the report by McNally (2000).

radar transmitter. Zhang et al. (2020) estimated that a source at 50 pc distance should radiate at least be 3.53×10^{13} W in order to be detected by FAST. The total solar luminosity is in excess of 10^{26} W. A simple calculation reveals that the FAST SETI survey of the Andromeda galaxy (Gray & Mooley 2017), located at a distance of $\approx$750 kpc, would require detectable sources to have a luminosity of at least $(750/0.05)^2 \times 3.53 \times 10^{13} 10^{-4}\ L_{\odot}$, which is clearly in the domain of civilizations that are capable of utilizing energy at the stellar emission levels. However, this does not take into account limitations to signal-to-noise ratio caused by, among other things, scintillations in the ISM. This might limit the effective range of radio SETI to sources at distances of the order of magnitude of only 100 pc (Cordes et al. 1997).

For comparison, the total energy consumption of humanity amounts to $<10^{-8}\ L_{\odot}$. If this amount of energy would be radiated in an omnidirectional beacon, it would be detectable at a distance of 7.5 kpc, which is within the Milky Way confinements. Even with present state-of-the-art technology, we are still unable to detect alien civilizations that are at our level of energy consumption, at least outside our immediate interstellar neighborhood. At greater distances, this type of SETI observation could plausibly detect only civilizations that are developed enough to harness much larger energies from their surroundings (civilizations in excess of Kardashev's Type 2). Garrett (2018) describes the potential of very long baseline interferometry (VLBI) to isolate the RFI but at the same time argues that a well-dedicated array of transmitters can detect civilizations even below the 2.0 level of Kardashev's[3] scale (Figure 11.2). Gravitational lensing is also proposed as a way to reduce the source power required to overcome the detection threshold (Maccone 2010).

Further, concerns about the possibilities of interstellar communication with ETI add another layer of difficulties. Assuming an omnidirectional emission of 10^{12} W,

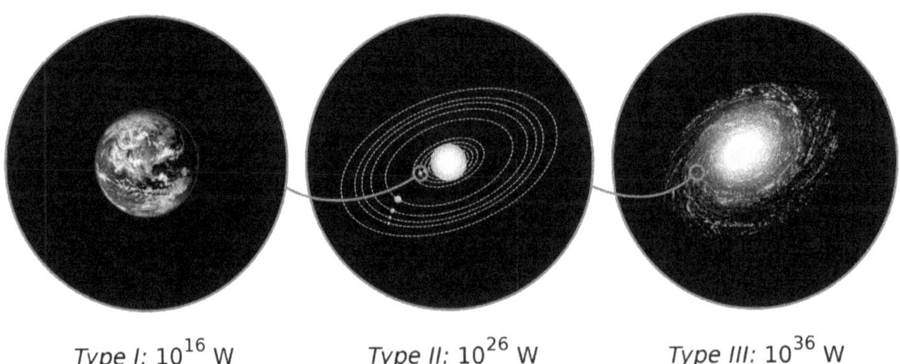

Figure 11.2. Three milestone values of Kardashev's continuous scale for levels of civilization development. A Type 1.0 (planetary) civilization can use and store all of the energy that reaches its planet from its parent star. A Type 2.0 civilization (stellar civilization) could harness the total energy of its planet's parent star, and a Type 3.0 (galactic) civilization would control energy on the scale of its entire host galaxy. Humans are currently in a Type 0 civilization. Image credit: Wikipedia: Indif (CC BY-SA 3.0).

[3] Nikolai Semenovich Kardashev was a Russian astrophysicist (1932–2019).

Fridman (2011) estimates that the detectable transmission (with instruments such as Arecibo or SKA) of information the size of the human genome, to a distance of 1000 lt-yr, would require more than 10^5 yr (see also Bates 1978). Such call-response times seem enormous and excessive from our present-day and anthropocentric point of view, but one should keep in mind that this might be due solely to our parochialism. The fact that we have difficulty imagining social structures stable over, for example, 10^5 yr should not be overrated because those timescales are still negligible in comparison to the (Charley) Lineweaver timescale indicating the age of most planetary habitats in the Milky Way (Lineweaver 2001; Lineweaver et al. 2004). The Copernican principle suggests that, lacking deeper insights into cultural dynamics, the ages of extraterrestrial civilizations should roughly follow the age distribution of planets. Therefore, it is conceivable that the Milky Way contains a population of very old localized (i.e., of Kardashev's Type 2 or less) and extremely stable civilizations; the communication on timescales similar to 10^5 yr would not present a challenge for such civilizations as it does for us.

Given the possibility that SETI targets are transitory and rare, Garrett et al. (2017) discuss the advancements for SETI research that are made possible by the new generation of radio telescopes that have wide fields of view and are able to scan the whole sky simultaneously, such as MWA, ASKAP, and others. There is also a possibility to observe a bit wider part of the radio domain, just outside of the standard "radio window," below the 5 MHz cutoff frequency of the ionospheric plasma (Fields et al. 2013). A significant improvement in this direction might come with the project of setting up radio telescopes outside of Earth's atmosphere. The work of DeBiase (2005) describes the concept of utilizing the antennae and other equipment aboard existing satellites in geosynchronous orbits to serve as a very large SETI telescope. As another visionary project, Gorgolewski (2002) proposes the construction of arrays on the far side of the Moon or VLBI elements in the Moon's orbit as a solution to avoid the RFI generated on Earth (see also Book 1, Section 1). The possible benefits of such undertakings will also depend on human's ability to preserve the RFI contamination of the Moon's far side, despite the significant technological activity and logistics that are required to operate such instruments.

Clearly, the advancements in modern-day SETI equipment are impressive (Garrett 2015), both in the instrumental and data-processing domains. This will surely enable us to reach deeper into unexplored aspects of SETI and further constrain the prospects for detection of alien intelligence in conjunction with the modern studies of galactic habitability and the general evolution of life in the universe.

11.4 Optical and Infrared SETI

While radio telescopes can detect radiation at extremely low energies, at shorter wavelengths the radiation has somewhat higher energy, which would require larger emitting power in order for such radiation to be detected across large distances. To achieve higher emitting power, even the radio SETI uses directed emission. At optical and IR frequencies, the directed emission is instead the rule—if any chance of detection is to be expected. However, this might not be a disadvantage. If there is no other possibility than to detect directed emissions, especially from larger distances, then it is more likely that such signals are of technological, rather than natural, origin. This relies on Olbers' paradox (Heinrich Wilhelm Olbers; 1758–1840) of the dark night sky, which states that despite the "countless" number of stars up there, the night sky is still dark. It follows that the more distant an emitting alien civilization is, the lower the chances are that such an emission would reach Earth by mistake.

Almost contemporary with the modern development of radio SETI, Schwartz & Townes (1961) suggested the use of optical masers (later referred to as lasers) for communication over long distances. Fundamentally, lasers can operate in two modes, continuous and pulsed. Later on, it became apparent that the pulsed mode gives better prospects for detection because, unlike the continuous mode, the pulses can be emitted at a wide wavelength range on nanosecond (ns) timescales or even shorter. Instead of surveying the sky for a particular unknown frequency, it is much easier to look for short pulses at a wider frequency range. In addition, we do not know of any natural phenomena that emit light pulses on such a short timescale (Lampton 2000). The detection of a pulsed signal can be performed simultaneously with three detectors within the same time interval of the pulse. This is the way to rule out the significant number of false positives, e.g., source scintillation caused by turbulent air cells in Earth's atmosphere, when compared to two detector systems (Wright et al. 2001). Another argument in favor of pulsating sources is presented by Borra (2012), who discusses the possibility that sources pulsating on short timescales show a periodic signature in their spectra (spectral comb). This would make the detection of such sources even easier.

The energetic argument might be considered from yet another point of view. Because the optical regime permits squeezing more information into a given amount of time, Shvartsman et al. (1993) argue that civilizations that are more developed than ours might use it as a preferable means of communication. A civilization that has reached a higher rank on Kardashev's scale is also able to utilize more energy and further improve the range of its optical transmissions. In addition, radio waves have enormous wavelengths, when compared with optical radiation. This makes them much easier to handle and manipulate with existing technology. That is the reason why achievements in optical interferometry still haven't reached the level of radio interferometry. However, with sufficient advancements in technology, optical instruments may become far superior to their radio counterparts. With basic calculations, Shuch (2001) showed that because the collector area, in terms of radiation wavelength, is much larger for optical equipment, the power gain is much larger at optical wavelengths. In this manner, the human eye has a power gain comparable to the Arecibo radio telescope.

In addition to the SETI projects that use optical equipment to detect pulsed SETI signals (Stone et al. 2005; Drake et al. 2010; Wright et al. 2018c), gamma-ray collecting equipment may also be used (Eichler & Beskin 2001; Holder et al. 2005; Cherenkov Telescope Array Consortium 2019). Ground-based gamma-ray telescopes collect optical flashes of Čerenkov light that is produced by gamma-ray cascades in Earth's atmosphere. They have a large area for collecting optical

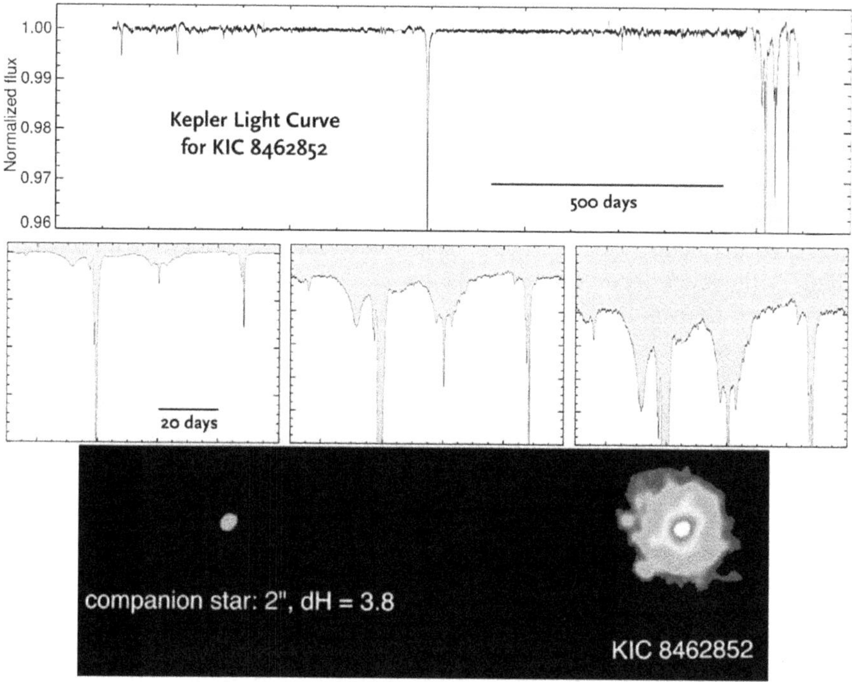

Figure 11.3. The curious case of KIC 8462852. The top and middle panels show four years of Kepler observations. The bottom panel is from the KECK AO *H* band. Image credit: Boyajian et al. (2016). © 2016 The Authors Published by Oxford University Press on behalf of the Royal Astronomical Society.

photons and are optimized to detect pulsed signals. The performance of such instruments for optical photometry and spectroscopy is discussed in Lacki (2011). Later work by the same author (Lacki 2014) discusses the potential of this method to detect transients of small bodies as far as the Oort Cloud. Abeysekara et al. (2016) used the VERITAS (Very Energetic Radiation Imaging Telescope Array System) gamma-ray instrument to observe objects from the Kepler mission database (KIC 8462852; Figure 11.3), which is hypothesized to have a swarm of artificial objects in order to explain the peculiar light curve. They report that the telescope has a wide field of view (more than 3°) and pointing accuracy better than 20″ with sensitivity of $\approx$1 photon per m^2 for the 12 ns pulse integration time. The same object was observed from Panama's Boquete Optical SETI observatory by Schuetz et al. (2016), but with a much higher threshold of 67 photons per m^2.

Despite the progress made in the detection of pulsed signals, in the last few years there has also been growing interest in the detection of continuous signals. Stewart et al. (2018) argue that after recent advancements in technology and the results of the Kepler mission, continuous signal searches are plausible. They suggest searching for directed signals from M31. Hippke (2018b) argues that at optimal optical frequency, a laser beam can be detected up to a 3 kpc distance and that characteristics of optical SETI lasers are below ones envisioned for a laser-powered interstellar flight. Lately, Benton (2019) reported the possible continuous laser source detection threshold of less than a photon per m^2 per second. A possible middle-ground approach between pulsed and continuous signals is discussed in Stanton (2019): to detect the time variation of the sources, estimating that the required power levels would be of the order of 10^8 W.

Numerous objects have so far been observed in searches for optical SETI signatures. Reines & Marcy (2002) report the observations of several hundred main-sequence stars with Keck Echelle spectrometer in order to detect narrow lines indicative of artificial laser emission. Extremely narrow emission lines were also searched with Keck spectroscopy (Tellis & Marcy 2015). Their observations included more than a thousand Kepler objects of interest that have at least one exoplanet. They state that the lasers with kilowatt power can be detected from distances of 1000 lt-yr. From the Lick observatory, Stone et al. (2005) observed several thousand nearby stars, up to 200 pc distance. Together with the Arecibo radio SETI SERENDIP survey (Werthimer et al. 2001), the SERENDIP survey is presented. It searches nanosecond visual wavelength pulses targeting nearby F–M-class stars, globular clusters, and galaxies. The all-sky optical SETI at Harvard is presented in Howard et al. (2007). Clark & Cahoy (2018) propose a laser detection and communication system based on the megawatt (MW) class laser, which is detectable at the TRAPPIST-1 (Gillon et al. 2017) distance ($\approx$12 pc) across the whole habitable zone of that planetary system (0.045 au). The three systems with planets in their habitable zones (Trappist-1, GJ 422, and Wolf 1061) were observed with the South African Large Telescope (SALT; Welsh et al. 2018).

Because of the larger required energies, optical and infrared SETI are conducive at somewhat shorter distances than the radio SETI projects, but there should be no such limitations in our immediate Galactic surroundings. Howard & Horowitz (2001) stated that even a modest kilowatt power infrared laser could be detectable with the Terrestrial Planet Finder at a distance of 15 pc. On the other hand, a simple far-end analysis of Lampton (2000) highlights the fact that up to 1000 lt-yr of interstellar space is transparent in all directions. Due to lower extinction, the distances are larger and outside of the Galactic Plane; there should be no obstacles in reaching even other galaxies. Reines & Marcy (2002) argue that their survey of nearby stars would have detected a >50 kW laser emission.

Kingsley (1993; 2001) argues that the future of SETI will be more oriented toward photon-based instruments and that the majority of SETI funding will go to optical projects. He further argues that if by the year 2020 optical SETI does not yield positive detections, it would probably mean that we had not accumulated sufficient observing time, rather than there being no other alien civilizations out there beaming their lasers toward Earth.

11.5 SETI at High Energies

The development of optical SETI has followed the footsteps of its radio counterpart. However, despite the use of Čerenkov telescopes for optical SETI purposes and, e.g., atmospheric X-ray detectors, which have been around almost from the dawn of modern SETI, high-energy SETI did not develop correspondingly. At short wavelengths, the energy requirements are even larger than for the optical SETI.

The largest nuclear probes, as the most energetic man-made events, released around 10^{17}J of energy, with a few percent in the form of X-ray and gamma radiation. For comparison, gamma-ray bursts (GRBs), believed to be beamed emissions from exploding massive stars, releases around 10^{44} J of gamma radiation (comparable with the energy release of a supernova) and are detectable across cosmological distances.

The energetic argument has more strength in this case than for the optical SETI. Very advanced alien civilizations, capable of harvesting the required energies, can possibly engage in such forms of SETI messaging activity. Corbet (1997) suggests that sufficiently advanced civilizations can possibly modulate the emission of an X-ray binary system in order to message other civilizations.

The property of neutrinos to weakly interact (with anything) surely adds to their robustness as information carriers. On the other hand it makes them much more difficult to detect than electromagnetic waves, and they require high-energy processes in order to be created. For a discussion of the possibility of using neutrinos for communication, their production, and detection (including stellar gravitational lensing), see Hippke (2018a) and references therein. The suggested usage for communication might be direct, as information-carrying beams, or indirect, as energy carriers shot into the core of variable stars in order to modulate their oscillation periods. Gravitational lensing is further examined in Jackson (2019), where the author argues that using a neutron star or a black hole as a transmitting lens is likely to require a Kardashev Type 2 civilization.

At the technology level required to modulate the emission of a binary star or to utilize large energy budgets, it is very likely that such a civilization would be capable of interstellar travel and have macroengineering capabilities. Such civilizations might be easier to detect by looking for technosignatures of such engineering feats, including high-energy radiation from the propulsion of their spaceships, macro-processing objects such as Freeman Dyson (1923–2020) spheres, and other swarm-like objects.[4]

11.6 Natural Phenomena?

Before the space age, the universe was largely perceived as a collection of stars and gas organized into galaxies. Modern technologies, which significantly improved our ground-based detectors and also enabled their spread across the entire solar system, aside from opening new vistas for SETI, have significantly contributed to a more detailed understanding of the universe. We have just scratched the surface of enormous underlying complexity of the newly discovered exotic natural phenomena such as quasars, pulsars, the cosmic microwave background, GRBs, and galactic

[4] For a discussion of the possibility of detecting the transit signatures of such structures, see Wright et al. (2016) and Wright & Sigurdsson (2016).

shocks. Most of these phenomena are not completely understood and are still the subject of very active research.

Throughout the history of our species, attempts to explain natural phenomena that were beyond our comprehension at the time have resulted in various constructions, including mythology, grand design, visiting aliens, and so on. None of them satisfied the basic criteria of a valid scientific hypothesis. However, the application of the Copernican principle to the discovery of numerous possibly habitable Earth-like exoplanets implies that simple life is likely widespread in the universe. Together with extensive work on SETI development, this gave the context for which some of the phenomena from nature might be explained by the ETI hypothesis, satisfying the standards of scientific method application. Below we outline some of these cases.

11.6.1 Fast Radio Bursts

A discovery of a 30 Jy radio burst that lasted less than 5 ms was presented in Lorimer et al. (2007). Their high dispersion measure suggested a highly energetic, extragalactic event, such as a neutron star or pulsar-related phenomena (Thornton et al. 2013; Lyutikov et al. 2016; Cordes & Chatterjee 2019; Metzger et al. 2019). There were also plausible explanations through Galactic origins, such as stellar flares (Loeb et al. 2014; Maoz et al. 2015) or a magnetar, but they also advised caution because similar events can be of terrestrial origin (Burke-Spolaor et al. 2011).

Originally, they were considered as one-time events, similar to GRBs, but Spitler et al. (2016) announced the discovery of a repeating fast radio burst (FRB). The bursts repeated on timescales of $\sim$1 s but the work reported no statistically significant periodicity, while detections in Fonseca et al. (2020) repeat on timescales of several days. In a recent work, Connor et al. (2020) argue that nonrepeating bursts might be the consequence of a geometrical effect. The relativistic effects cause the short-lasting burst to have a very compact opening angle that sweeps across Earth too quickly to be detected.

Lingam & Loeb (2017a) examine the possibility that FRBs are the result of alien ships' light-sail propulsion systems. They show that the parameters of FRBs are consistent with artificial beams. Assuming an artificial origin of FRBs, they even estimated the upper limit for the number of Galactic ETIs that produce FRBs to 10^4. While there is still a possibility FRBs are periodic, their sporadic timing is not compatible with communication beacons. Anyway, a plausible ETI-based hypothesis that explains this phenomenon will surely benefit from future searches that will constrain the parameters of FRBs.

11.6.2 Changes in Stellar Brightness

Whether the occultation effect will give rise to a revolution in SETI searches as it did with the exoplanets is an open challenge for future SETI attempts. The Kepler photometry survey discovered numerous exoplanets and exoplanet candidates, but an F-type star, KIC 8462852 (Boyajian's star), at an estimated photometric distance of 450 pc, shows up to 20% brightness dips that occur irregularly on timescales between 5 and 80 days (Figure 11.3; Boyajian et al. 2016).

Further examinations concluded that KIC 8462852's brightness changes are difficult to explain with stellar variability (Montet & Simon 2016; Simon et al. 2018). Plausible explanations included small bodies and their fragments (Marengo et al. 2015; Boyajian et al. 2016; Bodman & Quillen 2016; Neslušan & Budaj 2017) or ISM (Lacki 2016). Abeysekara et al. (2016) reported no evidence of pulsed optical beacons. Harp et al. (2016) have found no radio signals while Thompson et al. (2016) found no significant emission in the submillimeter domain. Further infrared observation analysis from Lisse & Sitko (2015) suggests an ongoing late heavy bombardment but otherwise no other significant features, such as dust clouds or accreting disks, are found. Lipman et al. (2019) calculated that a 24 MW laser could be detected from the distance of KIC 8462852 but found no such emission in the spectra of this object. Also, an earlier work by Schuetz et al. (2016) did not find periodic laser pulses.

Wright & Sigurdsson (2016) reviewed and analyzed the explanations for the brightness behavior of KIC 8462852. They argued that the most likely explanation is interstellar matter or an intervening dim object (black hole, neutron star, or a white dwarf) with a disk-like structure around it. The possibility of a swarm-like Dyson sphere is also examined, noting that an increased reddening in some future brightness dip (likely caused by dust) would disfavor such explanation. On the other hand, Arnold (2005) simulated photometric transit curves of various geometrical shapes and stated that a Jupiter-sized body would produce transit curve residuals of 10^{-4}. A multitude of such objects could improve the detection prospects by an order of magnitude. A possibility that stellar transits can be used for interstellar transmissions is put forward in the same work but also examined in Forgan (2019a) as a Galactic-wide communication network. Unlike recent transit photometry studies and discovered FRBs, variable stars have been studied from the very beginnings of modern astronomy. A possibility for a galactic communication scheme that uses modulated signals from variable stars, instead of transits, is presented in Learned (2008) and Hippke et al. (2015).

11.7 Other Phenomena

11.7.1 Gravitational Waves

Gravitational waves were predicted at the beginning of the past century and was very soon suggested as a possible communication means (Kudrin 1973). These ripples in the spacetime continuum propagate unobstructed through space and reach detectors without being distorted on their way, with their intensity scaling inversely with the distance squared from the source (see Chapter 9). A gravitational-wave detector can consist of multiple stations. Each station uses laser interferometry to detect a local change in the distance between two points on the station caused by the passage of the gravitational wave (see also Book 1, Chapter 8). Combining the data from different stations triangulates the direction of the wave propagation, due to the difference in wave arrival times at each station, which is essentially the same principle used in satellite navigation systems.

The only problem is that huge masses, much larger than standard stellar ones, are involved in making such waves. Objects like this are not readily found in our proximity and the first detection of gravitational waves (LIGO detector) was from the merging of binary black holes at a 400 Mpc distance (Abbott et al. 2016). Such an event was barely detected with our present state-of-the-art detectors that can detect 10^{-18} m of local space disturbances.

Unlike FRBs, we have a pretty good idea of how gravitational waves are formed because their existence was predicted more than a hundred years ago (see also Book 1, Chapter 1). However, given their stability and potential to travel across cosmological distances, it is hypothesized that they might be selected as a rule-of-thumb communication means for an ETI that is sufficiently advanced to manipulate them. Is the intention of such ETIs to communicate with equally developed ETIs, or are they aware that even relatively low developed civilizations, such as humanity, is able to detect such waves is a matter for further speculation. If gravitational-wave detectors continue to develop at the current pace, the likely answer to such a question would probably be found by the end of this century. We will find out whether advanced abilities to detect gravitational waves adds significantly to the possibility of their manipulation.

For a point-mass lens, Deguchi & Watson (1986) state that for the wavelength λ, the diffraction parameter is

$$y = \frac{2\pi}{\lambda} r_s, \tag{11.1}$$

where r_s is the Schwarzschild radius of the lens. A simple calculation reveals that for a lens mass M at frequency ν, the lensing effect is possible if, approximately, the product $\nu\, M > 10^{35}$ kg s^{-1}. The solar-mass lens could therefore be efficient at frequencies above 10^5 Hz. Hippke (2018a) also discusses the possibility of focusing the gravitational waves and states that no natural sources are expected to emit gravitational waves at $>10^3$Hz. The highest frequency of a black hole merger from Abbott et al. (2016) was 150 Hz. An ETI using its host star as a gravitational-wave lens is therefore expected to broadcast at frequencies exceedingly higher than the ones resulting from natural phenomena.

Maccone (1993) proposes a space mission that would place a detector in the solar focal point to exploit the lensing effect for SETI purposes. The focus of the Sun's gravitational lens is located at about 550 au; for comparison, at the end of October 2020, the most distant human-made probe, NASA's Voyager 1, is located at the heliocentric distance of about 152 au, which it reached 43 years and 3 months into the mission. Therefore, a solar focus mission is, in principle, feasible with current technology. An approach to SETI that can exploit the ability of gravitational waves to trigger the detectors irrespective of their propagation direction is coordinated signaling (discussed in Seto 2019). Predicting a gravitational-wave-emitting event, ETI can send a coordinated SETI signal that a potential receiver would pick up from the same direction as the gravitational wave at around the same time. Abramowicz and collaborators argue in Abramowicz et al. (2020) that advanced ETIs are likely to converge toward the center of the Milky Way in order to study the central black hole. They could sustain a Jupiter-size mass in the orbit around the black hole and emit continuous gravitational waves for long periods of time. They highlighted that such a scenario could easily be tested with our present abilities to detect gravitational waves within a one-year observation window.

11.7.2 Inscribed Matter Packages

Apart from using wave or particle radiation, a possible method of ETI communication might be the use of inscribed matter packages (Rose & Wright 2004; Sullivan 2004; Hippke et al. 2018). When compared to electromagnetic-wave SETI, an analogy would be like contacting someone by sending a letter, instead of using a mobile phone. The advantage of this method is that it does not require synchronization between sender and receiver in order to convey a message. It is evidently a slower process of communication but another advantage is that it has no sensitivity limit like radiation receivers.

The intention to send such packages to a specific location in the Milky Way would surely require construction in a vehicle-like manner in order to enable navigation (which might be expensive), while sending them out from Earth like leaflets from an airplane would require a larger number of packages to achieve any effect, possibly resulting in similar costs for both approaches. A more efficient possibility is that a multitude of such small packages can be launched piggybacked on a vehicle with a Voyager-like purpose and released into interstellar space once the vehicle has left the solar system. The free-floating packages could then be picked up by passing stars or drift farther away into the depths of the Milky Way. Once inside a different stellar system, packages could announce their presence with miniature radio beacons (Bracewell 1960).

Such a scenario would resemble an interstellar panspermia process. A similarity might be even more striking when considering the possibility that genetic codes might be used as an information-carrying medium (shCherbak & Makukov 2013; Davies 2012). Zubrin (2017) suggests using microbes and their genetic code as the inscribed matter packages. This further illustrates how the SETI might involve various aspects of our research, not just limited to intelligent life per se. Various aspects of astrobiology that are related to retrieving simple alien life could also be of SETI importance.

11.8 Technosignatures

In the footprints of alien technology, Davies (2012) suggests that no intentional communication might be attempted by ETIs until they are aware of our presence. Even with extensive beacon or other emitting attempts, there is still some considerable time before the signatures of existence reach faraway alien worlds. A search for unintentional signatures of ETI existence, such as the products of their technological activity, might be a better strategy for a rookie in the Milky Way SETI scene.

Harris (2002) argues that a signature of antiproton annihilation is likely to be a sign of intelligent activity, rather than antimatter's natural existence. Also, signal variability would favor intelligent origin. High-energy processes, such as matter–antimatter annihilation or nuclear explosions, which are envisioned for spaceships' propulsion systems, are a likely source of high-energy waste radiation in the form of gamma rays. Harris (1986) examines the detection of gamma-ray emission from an antimatter propulsion spaceship. While high proper motion is outlined as the most distinctive signature of a traveling spaceship, it was estimated that a space-borne gamma-ray detector could detect such a ship to a distance of up to 300 pc. A later work by Teodorani (2014) also outlines high proper motion as the most important characteristic. While emissions of such sources in the radio and optical are examined it is also mentioned that X-ray and gamma-ray transient outbursts should be searched for. Such an approach is certainly more in line with the multimessenger studies and is more robust in the sense of detecting diverse propulsion systems.

Another promising approach in the detection of technosignatures might be the search for RFI or some other forms of waste radiation. While the RFI from Earth limits the detection capability of our equipment, that fact might be turned around and worked in favor of SETI by scanning space for ETI-generated RFI. The detection thresholds of radio surveys (Norris 2017) are lower than 0.01 mJy. A simple calculation reveals that at this level of sensitivity, an omnidirectional emission from a 1 kW home appliance, e.g., a microwave oven, would be detected at a distance of 0.3 pc. With the successful deployment of SKA pathfinders, such as MWA (Figure 11.4) and ASKAP, it is realistic to expect that envisioned sensitivity

Figure 11.4. A pathfinder for the next generation of high-sensitivity aperture array radio telescopes. Antenna tile of the Murchison Widefield Array in the Australian desert. Image credit: Natasha Hurley-Walker.

levels, of the order of microjanskys, will be achieved in the next generation of radio telescopes. This will enable the detection of microwave-oven-like appliances of "Alpha Centaurians." An observable side of such a planet might as well look similar to the night side of Earth, spotted with numerous lights, when observed with such a device. Assuming a century-long radio-communication window for ETIs and SKA detection limit out to a distance of 100 pc, Forgan & Nichol (2011) conclude that such SETI is likely to be less efficient than the dedicated communication techniques. However, humanity has done very little in this direction so far, and playing the intentional emission card might not be ripe for future SETI. If ETIs are similar to us, we should expect that they will be more dedicated to making noise with the extensive usage of mobile phones and other gadgets.

On the other hand, any form of waste-product detection resulting from ETI activity should be considered in the light of sustainability arguments (Schmidt & Frank 2019). Apart from global natural and technological extinction hazards, the longevity of ETIs would also depend on their sustainability. The more sustainable ETIs are likely to have a lower waste footprint. This might produce an interesting observational selection effect that long-lived civilizations are harder to detect than short-lived ones, at least at the beginning of one's SETI efforts.

11.8.1 Dyson Spheres

Apart from engaging in interstellar travel, advanced civilizations are likely to practice extensive forms of processing activity in order to harness the energy of the host star for their technological needs (Figure 11.5). In an extensive review on Dyson spheres, Jason Wright (2020) calculates that such constructions are most

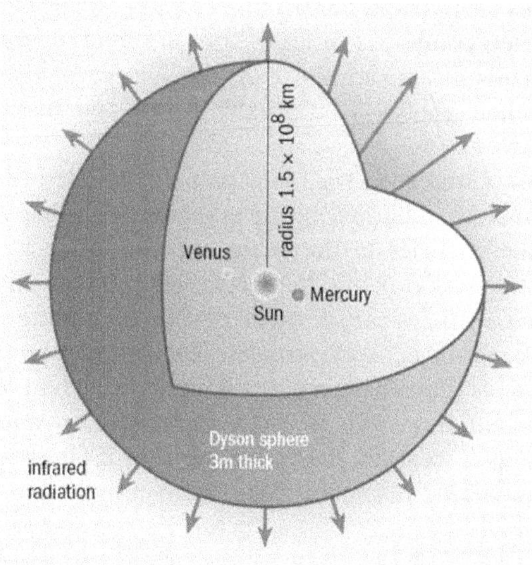

Figure 11.5. A sketch of our own "future" Dyson sphere. Image credit: © IOP Publishing. Reproduced with permission. All rights reserved. https://physicsworld.com/a/the-search-for-astroengineers/.

efficient at temperatures of $\approx$100 K. Much more colder than that and they are not likely to serve any meaningful purpose. On the other hand, the author also concludes that somewhat higher temperatures are not unreasonable and should be also considered for these kinds of SETI searches. Clearly, the 100 K emissions are in the domain of radio and infrared SETI, but the possibility for detection at shorter wavelengths arises from the influence that the Dyson sphere radiation might have on the host star, possibly causing its expansion. Wright (2020) outlines that this effect has not been studied thus far but it could have strong observable consequences for the detection of Dyson spheres.

Although the emission of Dyson spheres is most likely to appear in the transitional region between the radio and infrared domains, for a positive distinction of a Dyson sphere from a stellar object, it is best to observe it in multiple domains (Wright 2020). The prospects for Dyson sphere detection in infrared are examined by Zackrisson et al. (2018). This work considers the influence of plausible, excess IR stellar emissions, on Gaia[5] spectroscopy and the resulting stellar parameters (such as distance). At larger distance scales, a somewhat earlier work (Zackrisson et al. 2016) studies the IR excess for whole galaxies and the possibility that it can be explained by the technological activity of galaxywide civilization(s). Lacki (2019a) considers a similar argument, detecting the Dyson spheres cloaking color signatures in the spectrum of their host galaxies.

11.8.2 Collimated Particle Beams

Hippke (2018a) noted that the scientific examination of the possibility of using neutrinos to communicate was predicted in science fiction. Collimated particle beams appear in science fiction as the ultimate destruction weapons, but also have everyday applications (Wikipedia contributors 2020a, 2020b). They have a shorter collimation range when compared to lasers, but are more energetic and efficient in target disruption. While lasers only heat up the surface of the target, the particle beams can deposit their energy at a greater depth and disrupt the target more efficiently.

The Large Hadron Collider (LHC) is capable of accelerating charged particles close to the speed of light. Charged particle beams are subject to electromagnetic repulsion between the particles in the beam and the influence of electromagnetic fields on their path, such as Galactic magnetic fields. Upon exiting the accelerator, the positively charged particles can be made neutral by adding electrons in order to make a high-speed beam of neutral particles. The production of highly collimated neutral particle beams is plausible even at humanity's current level of technology.

In a similar way that humanity is using explosives for construction purposes, apart from their use as a weapon, it is plausible that ETIs would also use particle beams for macroengineering purposes. Such beams might also be used to propel a spaceship.

---

[5] An astrometric space observatory of the ESA.

The detection of such beams would require a gamma-ray-like detector design because at high energies, photons behave similarly to particles. Also, there is the possibility of indirect detections. When particle beams hit particles of other matter, gamma rays are produced. Lacki (2015) argues about the possibility that ETIs might use Planck energy accelerators simply for the purposes of scientific experimentation and that such devices would surely have a large amount of waste radiation that could be detected at slightly higher thermal radiation energies than Dyson spheres, apart from producing extremely high-energy radiation such as particles, photons, and neutrinos.

11.8.3 Other Possibilities

Besides Dyson spheres, an interesting possibility for large-scale engineering, but possibly on a smaller length scale, are shell worlds (Roy et al. 2013)—Moon-sized bodies that have an enclosed surface in order to sustain a comfortable atmosphere over long periods of time. Such objects could be even utilized as ships that support a self-sufficient traveling community, for long-duration space travel. Unlike Dyson spheres, their smaller size would certainly make them harder to detect in terms of reflected stellar light or a blackbody-like spectra from their predominantly featureless surfaces. On the other hand, their smaller size would make them easier to produce and thus have a number of advantages over stellar-sized artificial objects.

Our present megastructure buildings, for the purpose of exploration, resource exploitation, and transport, among other things, might indeed extrapolate to our Kardashev level 2 future—including Dyson spheres and other gigantic structures, such as the possibility of building swarm-like gamma-ray shields (Ćirković & Vukotić 2016). At least those are the manifestations that one would be most likely to detect, given our present experience of perceiving such achievements of humans, for example, the Great Wall in China being visible from Earth's orbit. However, present and new technologies that will open the door to our future in space might also leave subtle footprints. Apart from enabling further prospects from ETI detection on their own, such footprints can be of great use for follow-up observations and confirmation of detections that are made in more conventional SETI modes.

The present attempts of humanity to move to renewable energy sources have not met our current needs. A possible way of sustaining sufficient energy production and reducing pollution, at our current technology level, might be to go for nuclear fission power generation on a larger scale. One of the best ways to dispose of larger amounts of nuclear waste from such activities is to dump them into the Sun. Whitmire & Wright (1980) argued that spectral signatures of stellar photospheres that are used for nuclear waste disposal would be detectable for A5–F2-type stars. F-type stars, which are the closest in terms of stellar parameters to our Sun, thus might present viable targets for this kind of SETI search. Another possibility would be to detect industrial pollution in planetary atmospheres. Lin, Abad, and Loeb (Lin et al. 2014) calculated that a James Webb Space Telescope would be able to detect chlorofluorocarbons, currently found in the Earth's atmosphere as a product of industrial pollution, at 10 times the current terrestrial level, for an integration time of up to 2 days.

Another prospect is to detect stellar light reflected from flat surfaces of artificial structures that are in a geometrically favored position with respect to Earth for such effect to occur, similar to reflections of sunlight from satellites in the night sky. Natural objects are likely to be spherical and lack larger flat areas on their surfaces. Finding glints of reflected stellar light in the survey data would be a likely sign of an artificial object (Lacki 2019b). Their occurrence might indeed be rare, given the favorable geometrical requirements, especially at larger distances. The detection rate will not grow with the sensitivity of deployed detectors, which is a likely case for natural phenomena. Such glints and other possible anomalies in astrophysical data are a likely sign of ETI activity.

In the scenario where an ETI extensively uses photovoltaic arrays on the planetary surface to harness the host star energy, Lingam & Loeb (2017b) note that such activity may leave a spectral edge signature in the host star's light that is reflected from the surface of the planet. They also noted that such ETIs may redistribute light and heat (e.g., onto the night side of the tidally locked planet), which offers even better chances for detection than finding the spectral edge directly. Loeb & Turner (2012) state that our current observational capabilities can detect the light signal that corresponds to a major city on the night side of Earth from a Kuiper Belt object. A change in the distance from the Sun will cause a different light-curve slope for artificial and solar-illuminated objects. They also proposed that spectral phase modulation on the night-side light of exoplanets, caused by their orbital motion, could be detectable with future generations of telescopes.

Detection of power-beaming leakage (Benford & Benford 2016) is yet another possible SETI activity. The laser power beams could be used by ETIs to launch satellites for interplanetary travel or interstellar launches. Such laser beams are likely to be more powerful but less collimated than the ones envisioned for interstellar communication. Benford & Benford (2016) even suggest that ETIs could use this leakage for intentional communication by emitting messages with these power beams.

Apart from detecting proof of ETI's existence from space, Davies (2012) suggests various possibilities for astroforensics. The work states that finding even small plutonium deposits on Earth would strongly suggest past alien technological activity because plutonium is an excellent nuclear fuel and its half-life is much smaller than the age of Earth.

11.9 A Unified Approach: Future Prospects

The chances of detecting technosignatures of a civilization should correlate with the ability of that civilization to survive, expand, and leave more detectable traces of its activity. This is in direct connection with Galactic treatments for evolving and supporting the development of such civilizations in particular and life in general (Filipovic et al. 2013). Olson (2015) argues that expansionist civilizations can have a significant impact on the evolution of the universe. Their activity releases radiation from otherwise pressureless matter, and this is likely to affect the evolution of models on cosmological scales. This connection, between the detectable manifestations of advanced civilizations and parameters of their home galaxies, is a stepping stone for rethinking the existing and developing novel SETI approaches, in particular, by developing a holistic SETI approach across multiple electromagnetic wavelengths, which is similar to the multimessenger approach that is readily used in studies of astronomical objects and phenomena. Forgan (2019b) extensively reviewed radio and optical SETI while also discussing its relation to solving Fermi's paradox. In another work that utilizes Fermi's paradox as the boundary condition for SETI, Olson (2017) assumes that some fraction of Kardashev's scale type 3 civilizations are expanding aggressively and concludes that, given the adopted model of expansion, such civilizations can hardly be observed, or observable, at cosmological distances. In addition, clusters of such civilizations in regions with a high number density of galaxies are more likely to be observed than civilizations in an isolated galaxy. This would make rich clusters of galaxies, like the Comma or the Fornax clusters, natural targets for extragalactic SETI observations. The downside, however, would be that these objects have long lookback times in any realistic cosmological model, so we may be observing them simply before Type 3 civilizations emerged there.

Studies of Galactic habitability also rely on Fermi's paradox as a crucial modeling constraint and the ability to resolve such paradox is in direct connection with our possibility to communicate or detect other civilizations in space. All studies that estimate prospects of ETI detection have some model of ETI development that must conform to Galactic loneliness, presently perceived by humanity. These models are usually based on our assumptions and extrapolations of the development of technological civilization and their aspirations in terms of spreading and utilizing resources available across space. We extrapolate from our present knowledge of civilization development here on Earth and induced life-friendly conditions of habitats that are available across the Galaxy.

While there are (and will be) many debates on the origin and evolution of life, the life-friendly part of the parameter space of surface conditions on exoplanets is unambiguous, at least the position of the habitability peak for life as we know it. While the possibility should not be discarded that biospheres based on biochemistry different from the one on Earth exist out there, which also implies the possibility of habitats other than planetary surfaces, the search for Earth-like planets with habitable surfaces seems more and more promising in forming the solid foundations for the next generation of SETI projects. Three decades ago, the existence of Earth-like planets in our Galaxy was only in the domain of speculation. Today, we are certain that there are numerous Earth-like habitats across the Galaxy. This certainly adds weight to the Copernican assumption of present astrobiological studies (Ćirković & Vukotić 2013) that we are most likely a typical case of intelligent life in space. The next step is to answer how much is typical in space and time. This is one of the key questions in modern studies of Galactic habitability. The answer to such a question would have a significant impact on our SETI models and future SETI approaches, particularly for target selection. This is especially relevant for detecting alien civilizations at close range, where, in terms of detection capabilities, it is most likely to detect an ETI signal. It is therefore of great importance to estimate the epoch in Galactic history when our neighborhood enters the period of rapid ETI development or is colonized by ETIs from other, more life-friendly parts of the Galaxy.

Olson (2020) argues that the likelihood for ETI detection, such as the ones derived from the Copernicanism-based self-sampling assumption, should be revised assuming self-indication. In a nutshell, this would eliminate the need to assume an ETI appearance rate, because it relies on the typicality of favorable conditions for the emergence of ETIs, rather than the typicality of ETI emergence itself. Studies on Galactic habitability incorporate knowledge from most fields of astronomy, forming an astrobiological synthesis with other disciplines (Ćirković 2012), such as biology or chemistry. In addition to SETI surveys at a multitude of wavelengths, a multimessenger approach is therefore readily applied in studies of Galactic habitability that provide a modeling foundation for SETI studies. Together with questions on the nature and development of life, SETI is becoming one of the key convergence points for the scientific enterprise of our civilization.[6]

Present SETI visions will define our future in space (Wright & Oman-Reagan 2018). The achievements we make today are an important part of our cosmic identity through which other ETIs will rate us, likely in the not-so-distant future.

Acknowledgments

This work was financed by the Ministry of Education, Science, and Technological Development of the Republic of Serbia through the contract number 451-03-9/2021-14/200002. We also acknowledge all our colleagues who have (in)directly contributed to the content of this chapter with numerous discussions: Ana Vudragović, Anders Sandberg, Amedeo Balbi, Slobodan Perović, and Richard Gordon.

References

Abbott, B. P., Abbott, R., Abbott, T. D., et al. 2016, PhRvL, 116, 061102

Abeysekara, A. U., Archambault, S., Archer, A., et al. 2016, ApJ, 818, L33

Abramowicz, M., Bejger, M., Gourgoulhon, E., & Straub, O. 2020, NatSR, 10, 704

Arnold, L. F. A. 2005, ApJ, 627, 534

Ashworth, S. 2016, JBIS, 69, 419

Ashworth, S. 2019, Astronautical Evolution, issue 150, https://astronauticalevolution.wordpress.com/2019/11/02/the-destiny-of-civilisations-a-problem-for-seti/ (accessed: 24-Feb-2020)

Ball, J. A. 1973, Icar, 19, 347

Bates, D. R. 1974, Natur, 252, 432

Bates, D. R. 1978, Ap&SS, 55, 7

Benford, G., Benford, J., & Benford, D. 2010, AsBio, 10, 491

Benford, J. N., & Benford, D. J. 2016, ApJ, 825, 101

Benton, D. M. 2019, PASP, 131, 074501

[6] For some of the studies on galactic habitability that are, for the most part, motivated by SETI, see Ćirković & Bradbury (2006), Ćirković & Vukotić (2008), Forgan (2009), Morrison & Gowanlock (2015), Zackrisson et al. (2016), Đošović et al. (2019), and Stojković et al. (2019).

Billingham, J., & Benford, J. 2014, JBIS, 67, 17
Bodman, E. H. L., & Quillen, A. 2016, ApJ, 819, L34
Borra, E. F. 2012, AJ, 144, 181
Borra, E. F., & Trottier, E. 2016, PASP, 128, 114201
Boyajian, T. S., LaCourse, D. M., Rappaport, S. A., et al. 2016, MNRAS, 457, 3988
Bracewell, R. N. 1960, Natur, 186, 670
Brin, D. 2014, JBIS, 67, 8
Burke-Spolaor, S., Bailes, M., Ekers, R., Macquart, J.-P., & Crawford III, F. 2011, ApJ, 727, 18
Cabrol, N. A. 2016, AsBio, 16, 661
Carstairs, I. R. 2002, A&G, 43, 6.26
Cherenkov Telescope Array Consortium, Acharya, B. S., Agudo, I. et al 2019, Science with the Cherenkov Telescope Array (Singapore: World Scientific)
Ćirković, M. M. 2012, The Astrobiological Landscape (Cambridge: Cambridge Univ. Press)
Ćirković, M. M., & Bradbury, R. J. 2006, NewA, 11, 628
Ćirković, M. M., & Vukotić, B. 2008, OLEB, 38, 535
Ćirković, M. M., & Vukotić, B. 2013, IJAsB, 12, 87
Ćirković, M. M., & Vukotić, B. 2016, AcAau, 129, 438
Clark, J. R., & Cahoy, K. 2018, ApJ, 867, 97
Connor, L., Miller, M. C., & Gardenier, D. W. 2020, MNRAS, 497, 3076
Corbet, R. H. D. 1997, JBIS, 50, 253
Cordes, J. M., & Chatterjee, S. 2019, ARA&A, 57, 417
Cordes, J. M., Lazio, J. W., & Sagan, C. 1997, ApJ, 487, 782
Cox, G. A., Egly, S., Harp, G. R., et al. 2018, arXiv:1803.08624
Crick, F. H. C., & Orgel, L. E. 1973, Icar, 19, 341
Dana, R. 2018, AAS Meeting 231, 253
Davies, P. C. W. 2012, AcAau, 73, 250
Davies, P. C. W., & Wagner, R. V. 2013, AcAau, 89, 261
de Magalhães, J. P. 2016, SpPol, 38, 22
de Vladar, H. P. 2013, IJAsB, 12, 53
DeBiase, R. L. 2005, JBIS, 58, 357
Deguchi, S., & Watson, W. D. 1986, PhRvD, 34, 1708
Dick, S. J. 1996, The Biological Universe: The Twentieth-Century Extraterrestrial Life Debate and the Limits of Science (Cambridge: Cambridge Univ. Press)
Došovic, V., Vukotic, B., & Ćirkovic, M. M. 2019, A&A, 625, A98
Drake, F. D. 1961, PhT, 14, 40
Drake, F. D., Stone, R. P. S., Werthimer, D., & Wright, S. A. 2010, in Astrobiology Science Conference 2010: Evolution and Life: Surviving Catastrophes and Extremes on Earth and Beyond, Vol. 1538, 5211
Eichler, D., & Beskin, G. 2001, AsBio, 1, 489
Ekers, R. D., & Bell, J. F. 2001, in IAU Symp. 196, Preserving the Astronomical Sky, ed. R. J. Cohen, & W. T. Sullivan (Cambridge: Cambridge Univ. Press), 199
Fields, D. E., Kennedy, R. G., Roy, K. I., & Vacaliuc, B. 2013, AcAau, 82, 251
Filipovic, M. D., Horner, J., Crawford, E. J., Tothill, N. F. H., & White, G. L. 2013, SerAJ, 187, 43
Fonseca, E., Andersen, B. C., Bhardwaj, M., et al. 2020, ApJ, 891, L6
Forgan, D. H. 2009, IJAsB, 8, 121
Forgan, D. H., & Nichol, R. C. 2011, IJAsB, 10, 77

Forgan, D. H. 2011, IJAsB, 10, 341
Forgan, D. H. 2019a, IJAsB, 18, 189
Forgan, D. H. 2019b, Solving Fermi's Paradox, Cambridge Astrobiology (Cambridge: Cambridge Univ. Press)
Fridman, P. A. 2011, AcAau, 69, 777
Galera, E., Galanti, G. R., & Kinouchi, O. 2019, IJAsB, 18, 316
Garber, S. J. 1999, JBIS, 52, 3
Garrett, M. A. 2018, arXiv:1810.07235
Garrett, M., Siemion, A., & van Cappellen, W. 2017, arXiv:1709.01338
Garrett, M. A. 2015, AcAau, 113, 8
Gertz, J. 2016, JBIS, 69, 88
Gertz, J. 2017, JBIS, 70, 454
Gertz, J. 2019, JBIS, 72, 282
Gillon, M., Triaud, A. H. M. J., Demory, B.-O., et al. 2017, Natur, 542, 456
Gindilis, L. M., & Gurvits, L. I. 2019, AcAau, 162, 1
Gorgolewski, S. 2002, in 34th COSPAR Scientific Assembly, Vol. 34, 2492
Gray, R. H., & Mooley, K. 2017, AJ, 153, 110
Haqq-Misra, J. 2019, Fut, 106, 33
Haqq-Misra, J., Busch, M. W., Som, S. M., & Baum, S. D. 2013, SpPol, 29, 40
Harp, G. R. 2005, RaSc, 40, RS5S18
Harp, G. R., Ackermann, R. F., Astorga, A., et al. 2018, ApJ, 869, 66
Harp, G. R., Richards, J., Shostak, S., et al. 2016, ApJ, 825, 155
Harp, G. R., Richards, J., Tarter, S. S. J. C., et al. 2019, arXiv:1902.02426
Harris, M. J. 1986, Ap&SS, 123, 297
Harris, M. J. 2002, JBIS, 55, 383
Heller, R. 2019, IJAsB, 18, 296
Hippke, M. 2018a, AcAau, 151, 53
Hippke, M. 2018b, JApA, 39, 73
Hippke, M., Learned, J. G., Zee, A., et al. 2015, ApJ, 798, 42
Hippke, M., Leyland, P., & Learned, J. G. 2018, AcAau, 151, 32
Holder, J., Ashworth, P., LeBohec, S., Rose, H. J., & Weekes, T. C. 2005, ICRC, 5, 387
Howard, A., & Horowitz, P. 2001, Icar, 150, 163
Howard, A., Horowitz, P., Mead, C., et al. 2007, AcAau, 61, 78
Isaacson, H., Siemion, A. P. V., Marcy, G. W., et al. 2017, PASP, 129, 054501
Jackson, A. A. 2019, arXiv:1905.05184
Kingsley, S. A. 1993, Proc. SPIE, 1867, 75
Kingsley, S. A. 2001, Proc. SPIE, 4273, 72
Kudrin, V. B. 1973, Prir, 6, 98
Lacki, B. C. 2011, MNRAS, 416, 3075
Lacki, B. C. 2014, MNRAS, 445, 1858
Lacki, B. C. 2015, arXiv:1503.01509
Lacki, B. C. 2016, arXiv:1610.03219
Lacki, B. C. 2019b, PASP, 131, 084401
Lacki, B. C. 2019a, PASP, 131, 024102
Lampton, M. 2000, in ASP Conf. Ser. 213, Optical SETI: The Next Search Frontier (San Francisco, CA: ASP), 565

Learned, J. G., Kudritzki, R.-P., Pakvasa, S., & Zee, A. 2008, arXiv:0809.0339
Lin, H. W., Gonzalez Abad, G., & Loeb, A. 2014, ApJ, 792, L7
Lineweaver, C. H. 2001, Icar, 151, 307
Lineweaver, C. H., Fenner, Y., & Gibson, B. K. 2004, Sci, 303, 59
Lingam, M., & Loeb, A. 2017a, ApJ, 837, L23
Lingam, M., & Loeb, A. 2017b, MNRAS, 470, L82
Lipman, D., Isaacson, H., Siemion, A. P. V., et al. 2019, PASP, 131, 034202
Lisse, C. M., Sitko, M. L., & Marengo, M. 2015, ApJ, 815, L27
Loeb, A., Shvartzvald, Y., & Maoz, D. 2014, MNRAS, 439, L46
Loeb, A., & Turner, E. L. 2012, AsBio, 12, 290
Lorimer, D. R., Bailes, M., McLaughlin, M. A., Narkevic, D. J., & Crawford, F. 2007, Sci, 318, 777
Lyutikov, M., Burzawa, L., & Popov, S. B. 2016, MNRAS, 462, 941
Maccone, C. 2010, in Astrobiology Science Conference 2010: Evolution and Life: Surviving Catastrophes and Extremes on Earth and Beyond, Vol. 1538, 93-iaf-710
Maccone, C. 1993, in 44th International Astronautical Federation Congress, Graz, 5016
Maoz, D., Loeb, A., Shvartzvald, Y., et al. 2015, MNRAS, 454, 2183
Marengo, M., Hulsebus, A., & Willis, S. 2015, ApJ, 814, L15
McNally, D. 2000, AdSpR, 26, 347
Messerschmitt, D. G., & Morrison, I. S. 2012, AcAau, 78, 80
Metzger, B. D., Margalit, B., & Sironi, L. 2019, MNRAS, 485, 4091
Montet, B. T., & Simon, J. D. 2016, ApJ, 830, L39
Morrison, I. S., & Gowanlock, M. G. 2015, AsBio, 15, 683
Muscatello, A., Zubrin, R., Ohman, C., & Booth, S. 2005, in Space Resources Roundtable VII: LEAG Conference on Lunar Exploration, Vol. 1287, 68
NASA, 2018, arXiv:1812.08681
Neslušan, L., & Budaj, J. 2017, A&A, 600, A86
Norris, R. P. 2017, NatAs, 1, 671
Olson, S. J. 2015, CQGra, 32, 215025
Olson, S. J. 2017, IJAsB, 16, 176
Olson, S. J. 2020, arXiv:2002.08194
Pinchuk, P., & Margot, J. 2020, AAS Meeting 235, 227.07
Reines, A. E., & Marcy, G. W. 2002, PASP, 114, 416
Reyes, A., & Wright, J. T. 2019, arXiv:1908.02587
Rose, C., & Wright, G. 2004, Natur, 431, 47
Roy, K. I., Kennedy, R. G. III, & Fields, D. E. 2013, AcAau, 82, 238
Scheffer, L. K. 2005, RaSc, 40, RS5012
Schmidt, G. A., & Frank, A. 2019, IJAsB, 18, 142
Schuetz, M., Vakoch, D. A., Shostak, S., & Richards, J. 2016, ApJ, 825, L5
Schwartz, R. N., & Townes, C. H. 1961, Natur, 190, 205
Seto, N. 2019, ApJ, 875, L10
shCherbak, V. I., & Makukov, M. A. 2013, Icar, 224, 228
Sheikh, S. Z., Siemion, A., Enriquez, J. E., et al. 2020, AJ, 160, 29
Shostak, S. 2011a, AcAau, 68, 347 (SETI Special Edition)
Shostak, S. 2011b, AcAau, 68, 362
Shuch, H. P. 2001, Proc. SPIE, 4273, 128

Shvartsman, V., Beskin, G., Mitronova, S., et al. 1993, in ASP Conf. Ser. 47 (San Francisco, CA: ASP), 381
Simon, J. D., Shappee, B. J., Pojmański, G., et al. 2018, ApJ, 853, 77
Sleator, R., & Smith, N. 2021, in Planet Formation and Panspermia: New Prospects for the Movement of Life through Space Astrobiology Perspectives on Life of the Universe, ed. J. Seckbach, B. Vukotić, & R. Gordon (Beverly, MA: Wiley-Scrivener), 119
Spitler, L. G., Scholz, P., Hessels, J. W. T., et al. 2016, Natur, 531, 202
Staff at the National Astronomy & Ionosphere Center, 1975, Icar, 26, 462
Stanton, R. H. 2019, AcAau, 156, 92
Stewart, A., Lubin, P., Zhang, Q., & Taylor, S. 2018, in 42nd COSPAR Scientific Assembly, Vol. 42, F3.8–5–18
Stojković, N., Vukotić, B., & Ćirković, M. M. 2019, SerAJ, 198, 25
Stone, R. P. S., Wright, S. A., Drake, F., et al. 2005, AsBio, 5, 604
Sullivan, W. T. 2004, Natur, 431, 27
Tellis, N. K., & Marcy, G. W. 2015, PASP, 127, 540
Teodorani, M. 2014, AcAau, 105, 547
Thompson, M. A., Scicluna, P., Kemper, F., et al. 2016, MNRAS, 458, L39
Thornton, D., Stappers, B., Bailes, M., et al. 2013, Sci, 341, 53
Tough, A. 2000, in ASP Conf. Ser. 213 (San Francisco, CA: ASP), 445
Welsh, B., Vallerga, J., Kotze, M., & Wheatley, J. 2018, AAS Meeting 231, 104.01
Werthimer, D., Anderson, D., Bowyer, S., et al. 2001, Proc. SPIE, 4273, 104
Whitmire, D. P., & Wright, D. P. 1980, Icar, 42, 149
Wikipedia contributors, 2020a, Particle beam—Wikipedia, the free encyclopedia, https://en.wikipedia.org/w/index.php?title=Particle_beam&oldid=944874033 (accessed 23-April-2020)
Wikipedia contributors, 2020b, Particle-beam weapon—Wikipedia, the free encyclopedia, https://en.wikipedia.org/w/index.php?title=Particle-beam_weapon&oldid=948239123 (accessed 23-April-2020)
Wright, J. T., Griffith, R. L., Sigurdsson, S., Povich, M. S., & Mullan, B. 2014b, ApJ, 792, 27
Wright, J. T., Mullan, B., Sigurdsson, S., & Povich, M. S. 2014a, ApJ, 792, 26
Wright, J. T. 2020, SerAJ, 200, 1
Wright, J. T., Cartier, K. M. S., Zhao, M., Jontof-Hutter, D., & Ford, E. B. 2016, ApJ, 816, 17
Wright, J. T., Kanodia, S., & Lubar, E. 2018a, AJ, 156, 260
Wright, J. T., & Oman-Reagan, M. P. 2018b, IJAsB, 17, 177
Wright, J. T., & Sigurdsson, S. 2016, ApJ, 829, L3
Wright, J. T., Zackrisson, E., & Lisse, C. 2019, BAAS, 51, 366
Wright, S. A., Drake, F., Stone, R. P., Treffers, D., & Werthimer, D. 2001, Proc. SPIE, 4273, 173
Wright, S. A., Horowitz, P., Maire, J., et al. 2018c, Proc. SPIE, 10702, 107025I
Zackrisson, E., Calissendorff, P., Asadi, S., & Nyholm, A. 2015, ApJ, 810, 23
Zackrisson, E., Calissendorff, P., González, J., et al. 2016, ApJ, 833, 214
Zackrisson, E., Korn, A. J., Wehrhahn, A., & Reiter, J. 2018, ApJ, 862, 21
Zaitsev, A. 2006, RPPh, 69, 3101
Zaitsev, A. L. 2008, arXiv:0804.2754
Zhang, Z.-S., Werthimer, D., Zhang, T.-J., et al. 2020, ApJ, 891, 174
Zubrin, R. 2011, The Case for Mars (New York: Simon & Schuster)
Zubrin, R. 2017, JBIS, 70, 163
Zuckerman, B. 2019, arXiv:1912.08386

Chapter 12

Data Science (Mining and Processing) with Machine Learning in the Era of Multimessenger Astronomy

Ray Norris

The role of data science in astronomy is to extract scientific knowledge from astronomical data, and that role has become increasingly challenging as data volumes from large telescopes continue to increase inexorably. To this must be added the increasing challenge from extracting and combining the information available from multimessenger data sets. This increasing data diversity presents new challenges to data science for which our old techniques are ill equipped. Here, I describe the tools and techniques that are available now and the developments that will lead to new tools in the future.

12.1 Introduction: Why Multimessenger Science Requires New Approaches

Astronomy and astrophysics are in the midst of a revolution, driven by instrumentation and techniques that would have been unthinkable just a few years ago. One obvious change is the volume of data: just a few years ago, a night's observation on one of the world's major telescopes might result in a few megabytes of data, whereas now gigabytes are normal, terabytes are common, and petabytes are closer than they may appear. A less obvious but equally important driver is that a few years ago, an optical astronomer or a radio astronomer might observe their galaxy and publish a paper on it using data almost exclusively based on their own data, whereas now a referee is unlikely to accept that paper unless they compare it with data at other wavelengths. This very act of comparison would have been very difficult a few years ago, whereas now we have vast interoperable databases available online and the tools with which to compare them.

doi:10.1088/2514-3433/ac2256ch12

Comparing data from different wavelengths (or with non-electromagnetic messengers) presents its own challenges. Comparing an optical and near-infrared image is often straightforward as both images largely show the same objects. Comparing optical and radio is much harder because some of the brightest optical sources are stars that may be undetectable in the radio, and the brightest radio objects may be high-redshift galaxies that are extremely faint or invisible in the optical. Even worse, the strong radio sources may consist of a pair of synchrotron lobes separated by arcminutes, and the optical counterpart may be a faint galaxy somewhere between them. But even that seems easy when trying to compare the optical with a gamma-ray image with a resolution of arcminutes, or a gravitational-wave detection with an error ellipse covering tens of square degrees. Comparison, or even simultaneous analysis, of conventional electromagnetic data with high-energy or non-electromagnetic data may require new approaches and new infrastructure (Allen et al. 2018; Shawhan et al. 2019).

Multimessenger astronomy using gravitational waves, cosmic rays, neutrinos (Aiello et al. 2020a, 2020b) poses individual challenges that surpass those of conventional electromagnetic observations.

These multimessenger comparisons seem very ambitious compared to the simple cross-matching of two nearby wavelengths a few years ago, but the task is not as hard as it may seem because we are also developing techniques more sophisticated than those in use a few years ago, many of them drawing from machine learning or artificial intelligence, which for the purposes of this chapter we will call collectively "machine learning" or ML. However, most ML algorithms used in astronomy are currently optimized for relatively small volumes of data. Current and forthcoming instrumentation typically produces terabytes or even petabytes of data, and so ML approaches also need upgrading (Allen et al. 2019).

In Section 12.2, I will first discuss the volumes of data encountered of multimessenger astronomy and how data science can contribute to that, and then discuss in subsequent sections source extraction, cross-matching, redshift determination, classification, searches for the unexpected, and visualization. Most of this discussion is focused on continuum measurements, as spectroscopic techniques are beyond the scope of this chapter.

12.2 Data

12.2.1 The Growth in Data Volumes

Table 12.1 shows the growth in typical telescope data rates over the past few decades. Radio telescopes generally have the highest data volumes, closely followed by large optical telescopes; other multimessenger telescopes have more manageable data volumes. Thus, the data challenges are primarily driven by radio and optical telescopes. For example, the computing requirements for the SKA increase as shown in Figure 12.1.

Until about 1981, astronomical data were stored and transmitted in a large variety of proprietary formats, making a comparison of data from different wavelengths very difficult. 1981 saw the introduction of the FITS (Flexible Image Transport System; Wells et al. 1981), which revolutionized astronomy by making

Table 12.1. Illustrative Data Rates Associated with a Sample of Telescopes

Year Start	Name	Raw Data (Gbit/s)	Processed data (Gb/day)	References
1978	VLA[a]	691	12.2	Thompson et al. (1980)
1990	*HST*[b]	0.000 2	27.4	
2010	Pan-STARRS[c]	?	440	Flewelling et al. (2016)
2015	LIGO[d]	0.09	2×10^3	
2018	MeerKAT[e]	5120	3×10^4	
2019	ASKAP[f]	68,400	2×10^5	Johnston et al. (2008)
2022	LSST[g]	0.21	2×10^4	LSST Science Collaboration & Abell (2009)
2025?	SKA1[h]	10^8?	2×10^6	
2025?	KM3NeT[i]	31	2726	Hofestädt (2019)
2025?	CTA[j]	20	3×10^5	Paz Arribas (2016)
2035?	SKA2[k]	10^9?	10^9?	

Notes:
[a] https://www.nasa.gov/mission_pages/hubble/main/index.html
[b] https://www.nasa.gov/mission_pages/hubble/main/index.html
[c] https://www.ligo.caltech.edu/
[d] https://www.ligo.caltech.edu/
[e] http://www.ast.uct.ac.za/
[f] https://doi.org/10.1098/rsta.2019.0060
[g] https://doi.org/10.1098/rsta.2019.0060
[h] https://doi.org/10.1098/rsta.2019.0060
[i] https://www.skatelescope.org/technology/
[j] https://www.skatelescope.org/technology/
[k] https://www.skatelescope.org/technology/

it easy to compare or combine data from different telescopes. Over the past 40 years, FITS has become an almost universal standard in astronomy, and almost all astronomical data are now available in FITS format. Initially, the FITS standard contained all data in 80-character lines, but the option of including binary tables was later introduced, making it more efficient for larger images, or other non-image data, such as visibility data from radio interferometers.

With even larger data volumes now becoming commonplace in astronomy, FITS is no longer the optimum solution, and so other data formats such as the Hierarchical Data Format (HDF; Fortner 1998) are increasingly being used for large data sets.

A further format that is increasingly used is the Hierarchical Progressive Survey (HiPS) format (Fernique et al. 2015), which is based on hierarchical tiling of sky regions, using the HEALPix tessellation of the sky at a variety of spatial resolutions. HiPS enables multiresolution zooming and panning of an image (e.g., http://emu-survey.org/pilot).

12.2.2 The Virtual Observatory

The International Virtual Observatory Alliance (IVOA) was formed in 2002 June with a mission to "facilitate the international coordination and collaboration

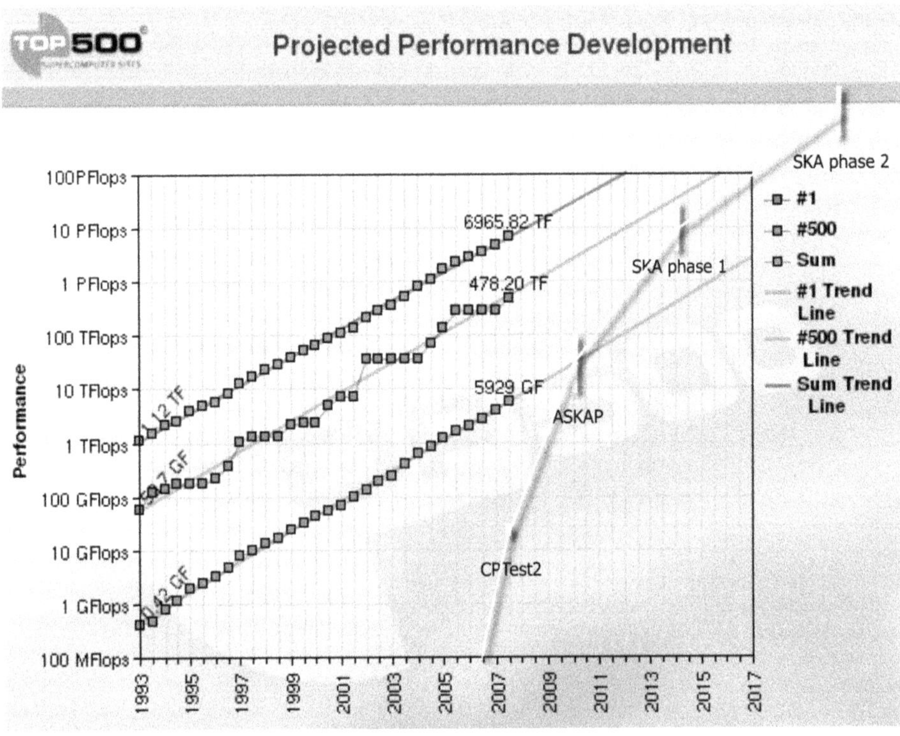

Figure 12.1. The projected computer requirements for the Square Kilometre Array, which approaches the performance of the world's fastest computer. Image credit: Tim Cornwell.

necessary for the development and deployment of the tools, systems and organizational structures necessary to enable the international utilization of astronomical archives as an integrated and inter-operating virtual observatory (VO)."[1]

The development of those tools and standards has been immensely successful, and almost every astronomer uses them daily, although many are not aware of it. Before the VO, it was extremely difficult to compare data from different wavelengths, whereas tools such as TopCat, Aladin, SkyView, and NED now make it simple. We all routinely access data at all wavelengths from online databases such as NED, SIMBAD, IPAC, and Skyview, as well as the many online databases hosted by individual telescopes, projects, or observatories, not only for conventional electromagnetic observations but also cosmic rays (Maurin et al. 2014; Pizzolotto et al. 2017).

Virtually all of these tools and resources rely on the underlying standards and protocols developed by the IVOA. But just as many people routinely use the web without being aware of the protocols and technology that make it possible, the underlying protocols and technology of the VO are invisible to most astronomers.

[1] Text taken from http://IVOA.net in 2020 September.

12.2.3 Large-n Astronomy

The availability of large databases and the tools to handle the data also opens up new opportunities. Specifically, there is a growing movement toward what I call "large-n astronomy," where instead of studying the astrophysics of individual objects, we study the statistical properties of populations of objects. This approach is complementary to the traditional approach of detailed individual study but depends critically on the ability to combine large volumes of data from very different techniques.

Such "large-n" astronomy also enables new approaches to traditional questions. For example, to study the properties of a population of galaxies discovered in the radio, it may not be necessary to know the redshift of every galaxy. Instead, in some circumstances, the properties of the population may be determined if the redshift distribution of the sample is accurately known.

The current generation of astronomical surveys will vastly increase the number of known astronomical objects in just a few years. For example, radio surveys driven by SKA precursor radio telescopes will increase the number of known radio sources from about 2.5 million to 100 million (Norris 2017b). Such massive changes trigger a qualitative change in the way we do astronomy. Rather than studying individual objects, or classes of objects, these millions of sources enable fine-grained statistical studies. The large numbers enable the sample to be chopped up along many different axes, so even subtle correlations can be detected, which enables new approaches to traditional questions. For example, to study the properties of a population of galaxies, we do not need to know the redshift of every galaxy. Instead, the properties of the population may be modeled if the redshift distribution of the sample is known.

"Large-n astronomy," in which the statistical properties of a large sample of galaxies are studied, is likely to become an increasingly important part of modern astronomy.

12.2.4 Sexagesimal or Degrees?

Traditionally, the position of an astronomical object was specified by its right ascension in hours, minutes, and seconds, and its declination in degrees, minutes, and seconds. While charming in their antiquity, these units make calculation difficult and invite a variety of incompatible formats when entering them into a computer interface. Starting in the 1980s, a few space missions started measuring right ascension and declination in degrees, and this has slowly spread to all areas of astronomy and astrophysics. It is now common to see astronomical positions listed only in degrees, and it is perhaps only a matter of time before the old sexagesimal units vanish from astrophysical publications and become a historical curiosity.

12.3 Citizen Science

Citizen Science is scientific research conducted by members of the public. It is particularly relevant to astronomy because of the large data volumes and the difficulty of identifying or classifying objects in images—a task that is easy for the human brain, but difficult for algorithms.

SETI@home (Anderson et al. 2002), first released to the public in 1999, used citizens' computers, rather than their brains, to search for signals from extraterrestrial civilizations. This was followed by the Galaxy Zoo project, later renamed Zooniverse, discussed further below.

Another major astronomical citizen science project was the launch in 2006 of Stardust@home, which encouraged users to search images from the Stardust spacecraft for tiny interstellar dust impacts (Westphal et al. 2005).

The importance of these projects is that ML algorithms still fall well short of the abilities of the human brain to classify or interpret images. There may come a time when this is no longer true, but at present even human brains with little training can easily beat a sophisticated convolutional neural net. However, human brains are much slower, hence the need to enlist large numbers of citizen scientists.

Recently, Wright et al. (2017) have demonstrated that a combination of citizen science and deep learning methods can outperform the capabilities of either method individually.

12.3.1 Zooniverse

Zooniverse started as the Galaxy Zoo project (Lintott et al. 2008), which is probably the most successful astronomical citizen science project to date. Galaxy Zoo invited participants to classify galaxies in images from the SDSS. It achieved over 4×10^7 classifications by about 10^5 participants.

The success of Galaxy Zoo encouraged other fields to set up similar projects, but the first implementation of Galaxy Zoo used purpose-built code that was difficult to use for other projects. The Galaxy Zoo group therefore built a second-generation code, which provided a general-purpose infrastructure platform on which other citizen science projects could be built. Tens of citizen science projects, collectively known as Zooniverse projects, now use this platform, and over 100 journal papers have now been written based on Zooniverse projects.

Most astronomical citizen science projects are based on optical and radio wavelength data, with few multimessenger projects, with one notable exception: the Gravity Spy project (Zevin et al. 2017), which enlists the help of citizen scientists to search for gravitational waves in data from the LIGO and VIRGO gravitational-wave observatories.

12.3.2 Radio Galaxy Zoo

Traditionally, radio sources found in a survey are cross-matched with optical/IR host galaxies using visual inspection. This process would clearly not be feasible for the many galaxies in the evolutionary map of the universe (EMU) project (Norris et al. 2011), and so EMU initiated the Radio Galaxy Zoo (RGZ; Banfield et al. 2015) project, in collaboration with the Galaxy Zoo project, in which citizen scientists cross-match radio sources to IR sources (Figure 12.2).

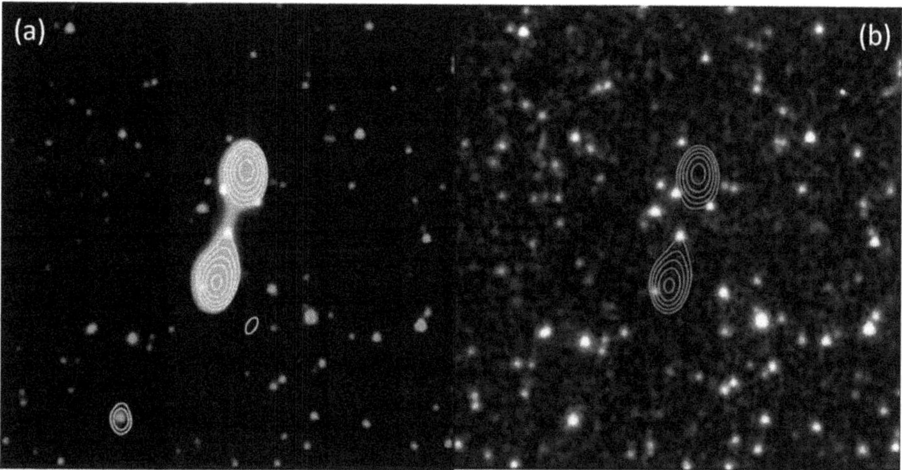

Figure 12.2. Two screenshots from RGZ showing radio contours overlaid on IR. The position of the slider at the bottom determines whether radio or infrared is most prominent. In (a) the radio is most prominent, and in (b) the infrared is most prominent. Image courtesy of Ivy Wong and http://radio.galaxyzoo.org.

To develop the technique, radio data were imported from the ATLAS project (Norris et al. 2006; Middelberg et al. 2008) and the FIRST survey (White et al. 1997; Becker et al. 1995), and infrared data were imported from the SWIRE survey (Lonsdale et al. 2003) and WISE (Wright et al. 2010).

Citizen scientists were then shown an image in which radio contours were overlaid on infrared images and participants were asked to identify the radio components and the infrared host galaxy, for each source. Thus the participants both classified the morphology of the radio source and performed a cross-identification to the infrared.

RGZ was very successful, with about 10,000 citizen scientists performing cross-matches, resulting in over 1.6 million cross-matches. As expected, the resulting database generated a number of astrophysical papers (Banfield et al. 2016; Kapińska et al. 2017; Garon et al. 2019; Rodman et al. 2019; Tang et al. 2020). What was not predicted is that, perhaps even more importantly, RGZ has been invaluable as a tool for training and testing ML algorithms for cross-matching (Alger et al. 2018; Lukic et al. 2018; Galvin et al. 2019; Ralph et al. 2019; Wu et al. 2019).

Although the original RGZ has now closed down, having achieved its goal, a new project, "Radio Galaxy Zoo: LOFAR," has started to classify radio sources discovered with the LOFAR radio telescope (van Haarlem et al. 2013). A new project, "Radio Galaxy Zoo: EMU" is also under construction.

12.4 Source Extraction and Event Detection

12.4.1 Optical/Infrared

Most optical images are obtained from CCD arrays mounted on telescopes, but historical images may be obtained from scanning photographic plates. The digitization of scanned plates is a specialized topic involving many more steps and corrections than extracting sources from CCD arrays and will not be discussed here.

The most widely used source extraction tool in optical astronomy is SExtractor (Bertin & Arnouts 1996), but many others are also in use. The generic description below is applicable to most of these and omits most of the fine detail.

The vast majority of pixels of an optical image belong to the dark background, upon which are a relatively small number of pixels containing objects. The first job of a source extractor is to characterize the background by measuring its mean and rms noise, both of which may vary across the field. The next step is to define a threshold for the background and subtract the background from the image to segment the background from the objects. After optional filtering, islands of pixels above the threshold are identified and then deblended to separate out adjacent objects (e.g., two nearby stars) that may be in the same island, using a decision tree process.

An important step is an "aperture correction," which corrects the measured intensities in the pixels for the wings of the point-spread function, which may spread outside the measured pixels. The task of the aperture correction is to estimate this missing flux. Several different aperture corrections are available such as ISOCOR in SExtractor (which assumes a Gaussian profile), Kron (1980; which models the emission from an object with an elliptical profile), and other more sophisticated aperture corrections.

The choice of aperture correction is a complex issue, depending on the instrument, the object being studied, and the presence of nearby sources. For example, the best aperture correction for a star is likely to differ from that for a galaxy, even in the same image. Many packages include an "AUTO" option, where the software attempts to use the optimum correction for each object, and catalogs often list a selection of measured photometry made with different aperture corrections, so that the end user can choose the aperture correction best suited to a particular measurement.

12.4.2 Radio

The discussion here is confined to the most widespread application, which is the extraction of sources from images from aperture synthesis radio telescopes.

Radio astronomy differs from optical astronomy in that there is no single widely adopted source extractor, but there are many different source extractors each with its own adherents. It also differs in that a radio source hosted by a single galaxy often consists of several well-separated components. To avoid confusion, we therefore use the terminology that a "source" refers to all the emission from one physical object such as a galaxy. That source may consist of several "islands" of emission, each of which consists of an area of emission in which the individual pixels are above some threshold. Each island may consist of several "components," each of which consists of a group of pixels surrounding a maximum. Often a component will be well fitted by a Gaussian.

Another difference between radio images and optical/IR images is that the mean value of a radio image is zero. In most cases, this means that the mean value of non-source pixels is also close to zero, except in the case of strong sources, which can cause the surrounding pixels to have a mean negative value, which appears as the so-called "CLEAN bowl" around the strong sources.

The earliest source finders such as SAD ("Search-and-Destroy") in AIPS (Wells 1985) and "IMSAD" in Miriad (Sault et al. 1995) worked by finding the peak pixel in an image, fitting a Gaussian to the surrounding region, subtracting that from the data, and then looking for the next highest pixel, and so on. Such source finders work effectively on images consisting mainly of well-separated compact sources but perform less well in more complex images.

Most recent major radio surveys instead use tools such as BLOBCAT (Hales et al. 2012), Aegean (Hancock et al. 2018), Pybdsf (Mohan & Rafferty 2015), and Selavy (Whiting & Humphreys 2012). All these tools use flood-fill algorithms, in which islands of emission are identified, and then Gaussians are fitted to individual components within each island. Comparisons of these tools (Hopkins et al. 2015) show that these tools perform reasonably well on compact objects, but perform less well on extended objects.

To overcome this limitation, two source finders better optimized for extended sources have recently been published: PROFOUND (Hale et al. 2019) and CAESAR (Riggi et al. 2019). These are currently being evaluated for use in the EMU (Norris et al. 2011) and VLASS (Lacy et al. 2020) surveys.

For spectral-line data cubes, any of the above source finders may be used in individual spectral planes, but because spectral lines generally occupy several planes, this will be less efficient than a source finder that operates in the cube rather than in individual planes. Such a source finder, named SOFIA (Serra et al. 2015), has been developed for the WALLABY (Koribalski et al. 2020) and other major spectral-line surveys.

12.4.3 Multimessenger

X-ray telescopes produce images with similar properties to optical telescopes, and so optical source extractors such as SExtractor are widely used in X-ray images.

Many other multimessenger telescopes have smaller data rates than the optical/infrared and radio telescopes, and so the source-finding challenge is quite different.

For example, detecting gravitational waves relies on searching for a particular signature in time series data (Figure 12.3). ML techniques have turned out to be particularly valuable for extracting gravitation-wave signals from the massive and complex LIGO databases (Allen et al. 2019; Wang et al. 2020) and pulsar timing arrays (Chen et al. 2020), in many cases using astrophysical modeling to inform the algorithm. They also are important for the extraction of neutrino events (Aiello et al. 2020c).

Detection of gamma rays by the Čerenkov Telescope Array (Paz Arribas 2016) involves the detection of a pulse that is extended in both space and time so that a wavefront detection system is required (Zoli et al. 2015).

12.5 Cross-matching

When using two data sets from two different telescopes, or taken at two different wavelengths, it is necessary to cross-match the images so we know we are comparing like with like. Fifty years ago this might have involved measuring the positions of

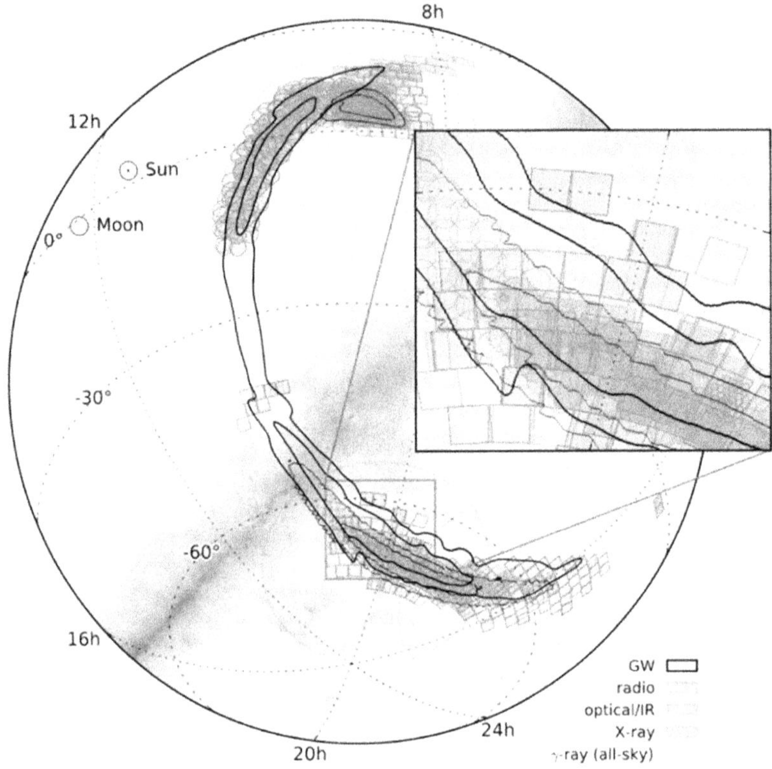

Figure 12.3. The footprint of a LIGO detection of a gravitational-wave source (contours) with the footprints of the optical, radio, and X-ray observations attempting to locate the source. Image credit: Abbott et al. (2016).

guide stars on a photographic plate. Nowadays the images are more generally made available as FITS files with coordinate scales derived at the observatory, which are generally fairly reliable.

The sources of astrometric error when comparing objects found in two images may be subdivided into:

- Uncertainties in reference frames,
- Uncertainty in the measured coordinates of the observation,
- Uncertainty as to whether the centroid of the object is in the same absolute place in the sky at all wavelengths.

12.5.1 Reference Frames and Coordinate Systems

A decade ago, reference frames were poorly established, and there was often confusion over which reference frame was being used. A common problem was that the frame established for a telescope would be calibrated in the northern hemisphere and then extrapolated to the southern hemisphere, resulting in an offset of a few arcseconds compared to one calibrated in the southern hemisphere. Thankfully, the demands of large surveys, especially spacecraft-based surveys, with their need for accurate reference frames over the whole sky, have driven a solution to that problem. Starting in about 2000, the IAU FITS group developed a set of standards for defining world coordinate systems (WCSs), which have now been almost universally adopted by observatories and astronomical software packages (Greisen & Calabretta 2002).

At the same time as the WCS was being developed, radio and optical observational programs defined the "International Celestial Reference System (ICRS)," which is referred to the equinox and epoch of Julian Date 2000.0, and is often written as (J2000). The ICRS has been adopted by the International Astronomical Union (IAU) as the standard celestial reference system and is now used by virtually all modern astronomical observations. Its origin is at the barycenter of the solar system, with axes that are intended to be an inertial frame that is assumed to be stationary with respect to fixed frames of reference such as distant quasars and the cosmic microwave background.

The ICRS is defined in space by the International Celestial Reference Frame (ICRF), which is based on several hundred extragalactic radio sources, mostly quasars, which lie at sufficiently large distances that they appear to be stationary with current technology. However, their positions can be measured to milliarcsecond accuracy using the technique of Very Long Baseline Interferometry (VLBI). At optical wavelengths, the ICRS is realized by the Hipparcos Celestial Reference Frame (HCRF), a collection of about 100,000 stars from the Hipparcos Catalogue. This includes a collection of radio stars, which, together with quasars, are used to tie the radio and optical references frames together. With the advent of Gaia (Perryman 2002), the optical realization of the ICRS will be tied to the radio-based ICRF using only the (apparently stationary) extragalactic sources (Gaia Collaboration et al. 2018; Lindegren et al. 2018).

Earlier observations may be found in the literature that are referred to the equinox and epoch of the Besselian years B1900.0 and B1950.0. Fortunately, the WCSs have provided accurate algorithms, now implemented in software such as astropy[2] and in most online databases, so that this coordinate transformation is now trivial for the user.

Fifty years ago, errors in reference frames and coordinate systems often dominated positional uncertainties. Because of the advances discussed above, this source of uncertainty is now negligible for almost all astronomical purposes.

12.5.2 Uncertainty in the Measured Coordinates of the Observation

The positions of the radio and optical sources in the ICRS are known to milliarcsecond accuracy, and so any observation should ideally include these sources in the observation so that the coordinate system of the observation is then known to a similar accuracy.

For large surveys, this may be often achievable, but more often the ICRS sources are too faint, or too widely separated, to be included in an observation. In this case,

[2] https://www.astropy.org/

another survey observation is used as an intermediary. For example, most space-based large surveys include ICRS sources, and their positional calibration is constant across the sky, and so source positions are generally accurate to a small fraction of their spatial resolution. Thus many ground-based observations will base their positional calibration on cross-matches with space-based surveys

The position of an object on an astronomical image ideally reflects the projected position in the sky but may deviate from it by instrumental effects, such as refraction by the atmosphere or ionosphere, aberration by lenses, or distortion of the receiving element (which was especially important in the days of photographic plates). In practice, such errors are small in modern optical surveys using CCD cameras (Bernstein et al. 2017).

Such scaling distortions are usually minimal in interferometric radio observations as the reference frame of the image depends on the fundamental physics and locations of the antennas, and so only one source of known position needs to be observed as a calibrator. Exceptions occur at low frequency, when ionospheric refraction becomes important and must be modeled, and at high frequency, where tropospheric refraction becomes important.

12.5.3 Challenges of Multimessenger Cross-matching

When comparing two images at nearby optical wavelengths, at similar resolution, cross-matching is simple because both images mostly show the same objects, albeit at different brightnesses. A simple cross-correlation will suffice to check, and if necessary adjust, the relative astrometry of the two images, and then sources can be directly compared.

As the wavelengths become more separated, and scales differ, the problem becomes harder. One image may have many more objects per square arcsecond than the other, and the brightest objects at one wavelength may be insignificant at the other, as shown in Figure 12.4.

An obvious solution, still widely used, is just to take the nearest object ("the nearest-neighbor algorithm"), which can be quite effective and is still used in tools like TopCat. More elegant solutions are discussed below.

Whatever algorithm is adopted, multimessenger cross-matching presents special challenges such as

- Images at different wavelengths may contain completely different objects,
- Images at different wavelengths may have very different scales and positional uncertainties.

In some cases, the order of cross-matching images makes a large difference to the final outcome. Figure 12.4 shows a deep optical image overlaid on a radio image. Experience has shown that the brightest optical objects (stars and nearby galaxies) tend to be weak at radio wavelengths, while the strongest radio sources are often high-redshift galaxies that are faint at optical wavelengths. A direct cross-match between radio and optical therefore often produces a high false-identification rate.

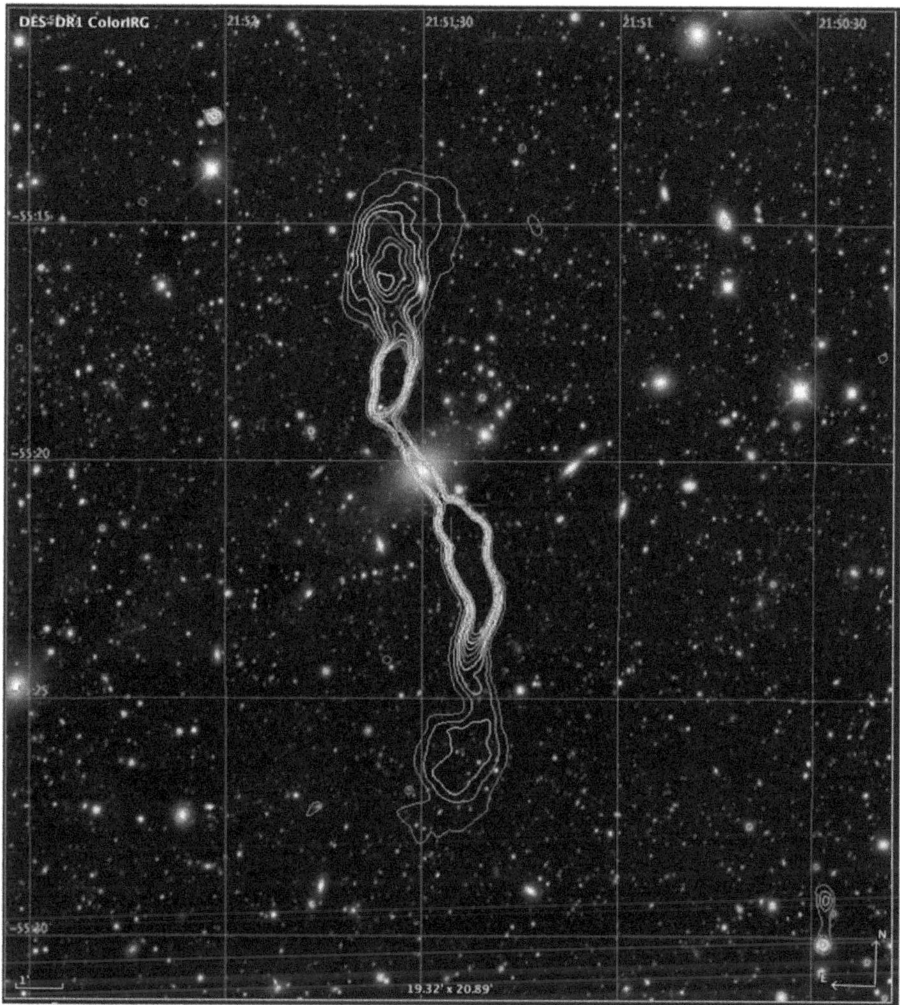

Figure 12.4. The radio galaxy PKS 2148–555, showing the disparity in scale sizes and identifications between different wavelengths. Contours are from the EMU Pilot Survey (Norris et al. 2021b) with a resolution of ~12″. Color is the composite color image from the Dark Energy Survey DR1 (Abbott et al. 2018) with a resolution of ~1″. The image was generated using the Aladin tool (Bonnarel et al. 2000).

Experience has shown (Norris et al. 2006) that the radio sources tend to be relatively brighter in the near-infrared (NIR) than in the optical. Therefore (at least with a radio resolution of several arcseconds or more), cross-matching the radio to the NIR, and then the NIR to the optical, will produce a lower radio-optical false-ID rate than doing it directly.

This problem becomes most acute when the positional uncertainties are very different, such as trying to cross-match gamma-ray or gravitational-wave events, with error ellipses of tens of degrees, against an optical catalog, or lists of event triggers generated by a telescope such as LSST (LSST Science Collaboration et al. 2009).

It may be more efficient to match an extreme event against a wavelength dominated by high-energy processes, such as radio or X-ray, and then cross-match those candidates against the optical, rather than try to cross-match the extreme event directly to optical. In any case, the algorithms for cross-matching the large databases involved will need to be considerably faster and more sophisticated than those currently in use (Allen et al. 2019).

This last scenario raises another problem in that the field of view of the comparison wavelengths needs to be at least as large as the error ellipse of the extreme event, which is difficult when studying a time-variable event.

12.5.4 Nearest Neighbor

The most obvious way of cross-matching two catalogs of sources is simply to find the nearest object in one catalog to an object in the other catalog. This "nearest-neighbor" (NN) technique can be very effective and gives a low false-ID rate in situations where the same objects are likely to appear in both catalogs, such as a catalog of stars at two different optical wavelengths, and the sources are not so closely packed that ambiguities arise. It is probably the most widely used method of cross-matching sources and is implemented in astropy and tools such as TopCat (Taylor 2005). It should be noted that these tools generally use a much faster algorithm, the k-d tree, rather than a simple list sort, but the overall result is identical.

An extension of this simple technique, used by the Gaia project (Marrese et al. 2017), is the "Best neighbor" technique where the probability of a potential neighbor being correct is evaluated taking into account positional uncertainties and source densities in the two catalogs.

12.5.5 Likelihood Ratio

In cases where there are likely to be several candidate matches within a reasonable search radius, it is necessary to use a more sophisticated approach than NN. The "likelihood ratio" (LR) algorithm was first introduced by de Ruiter et al. (1977) and subsequently improved and developed for different applications by Wolstencroft et al. (1986), Sutherland & Saunders (1992), Pineau et al. (2011), and Weston et al. (2018). The key point of the LR technique is to estimate a prior likelihood that different sources might be associated with a target source. For example, it might be expected that brighter sources in one band are likely to be brighter sources in another band, and so a match between two bright sources may be given a higher prior likelihood that a match between a bright and a faint source.

The other key innovation of the LR algorithm is that for every potential match, a likelihood is calculated of the match being correct, and of the match being false. The ratio of these two likelihoods is then known as the likelihood ratio, and the match is chosen that has the highest likelihood ratio.

The LR technique is often used for observations where the source density is so high that an NN match is likely to give a high false-ID rate. However, it still assumes that a source in one catalog has a single identification in another catalog. This is not the case for radio data, where a single source may consist of several different components, in which case even more sophisticated approaches are needed.

12.5.6 Multiple-component Radio Sources

A single radio galaxy often has several distinct components corresponding to its core, jets, and lobes.

The lobes are often brighter than the core, making cross-identification using NN or LR doomed to fail. As a result, radio-optical matches were usually done by eye. which is feasible with catalogs containing up to a few thousand members. However, modern surveys can include millions of sources, placing a high priority on the development of automated techniques that recognize the existence of several radio components for each optical galaxy (Figure 12.5).

The first attempts to construct algorithms to cross-match complex radio sources (Proctor 2006; Kimball & Ivezić 2008; van Velzen et al. 2015) used simple algorithms that only identified specific cases. Recently, a number of authors have attempted more general algorithms, described in the next section, which either take a Bayesian approach or else use ML techniques.

Whichever technique is used, the problem of cross-identification becomes inseparable from the problem of source classification. For example, a pair of nearby compact radio components might either be a pair of nearby star-forming galaxies, or

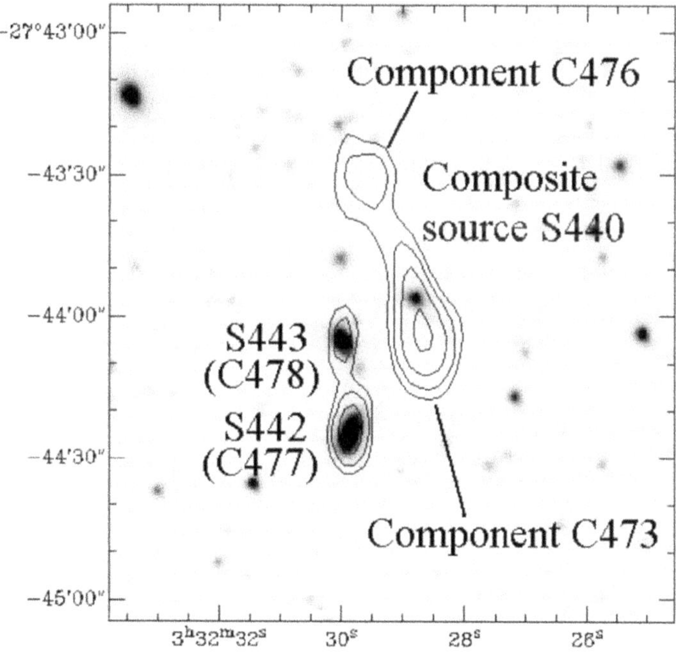

Figure 12.5. Figure showing the difficulty of cross-matching radio and optical/infrared. Contours are radio (Norris et al. 2006) and grayscale is infrared (Lonsdale et al. 2003). The double radio source on the left is two separate star-forming galaxies, while the double radio source on the right is a double-lobed AGN.

they may be the two lobes of a double-lobed radio galaxy. The key to distinguishing between these two options is to look at the optical or infrared counterparts. If the radio objects do not have optical counterparts, but there is an elliptical galaxy at their midpoint, they are almost certainly a pair of lobes associated with the elliptical galaxy host. Conversely, if each of the radio components is coincident with a star-forming galaxy, then they are almost certainly the radio emission of a pair of star-forming host galaxies.

In practice, as discussed above in Section 12.5.3, experience has shown (Norris et al. 2006) that a much higher cross-match success rate is achieved if the cross-identification is initially done with infrared rather than optical catalogs. The infrared can then be cross-matched with optical to obtain the final, optical-radio cross-match.

12.5.7 Bayesian Approaches

The first attempt to develop a more sophisticated technique for multiple-component radio sources was by Fan et al. (2015), who developed a Bayesian framework that assigned prior probabilities to various configurations of the optical host galaxy. For example, if three radio components are in a straight line, associating the optical host galaxy with the central component has a much higher probability than assigning it to one of the outer components.

This simple technique, shown in Figure 12.6 was surprisingly successful and correctly matched 86% of sources.

However, it has two key drawbacks. First, it assigned a low likelihood when the jet was significantly bent, and it was difficult to insert further refinements to the Bayesian priors. Second, it was a "greedy" algorithm, meaning that once a radio component was assigned to an optical source, it would not be subsequently considered for a (possibly better) match to another source.

To overcome these two limitations, Fan et al. (2020) developed the "integer linear programming" (ILP) approach. ILP allows the construction of more complex hypotheses. Having constructed a number of hypotheses for each source, ILP then chooses a solution that maximizes the likelihood over the whole field, rather than focusing on the likelihood for each source separately. ILP is currently being evaluated on data from the EMU survey (Norris et al. 2011).

12.5.8 Machine-learning Techniques

Convolutional neural nets (CNNs), first developed for computer vision (Krizhevsky et al. 2012), are ideal for classifying multiple-component radio sources and can be used to cross-match them. The first attempt to use a CNN for radio source classification was Aniyan & Thorat (2017), followed closely by increasingly sophisticated treatments by Lukic et al. (2018) and Alger et al. (2018), who each trained and tested their code using RGZ. The most complex CNN for this purpose was CLARAN (Wu et al. 2019), which used a computer vision code, again using RGZ data, to both classify radio sources and cross-match them with infrared sources. However, each of these approaches was experimental, and there is not yet a

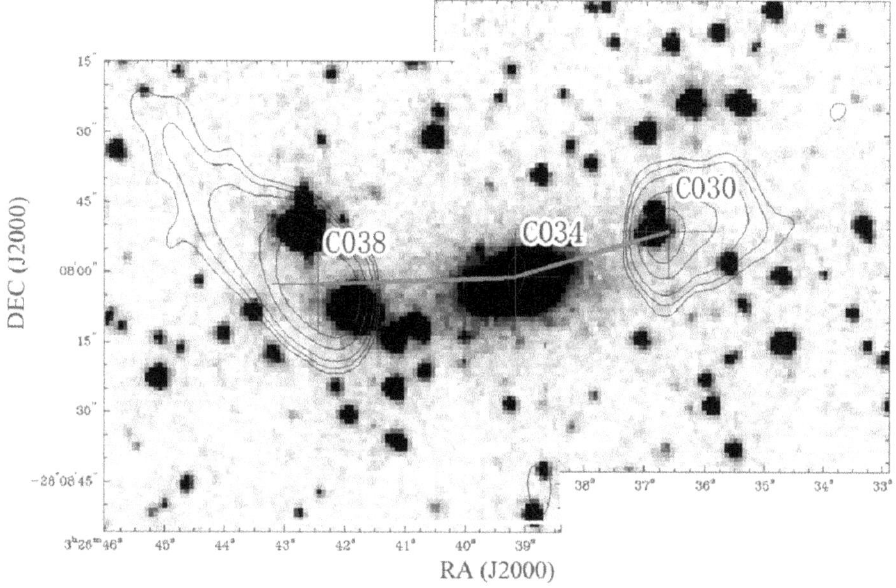

Figure 12.6. An example (contours: radio, grayscale: infrared) of a source identified using a Bayesian approach. This triple source was incorrectly classified as a double in Fan et al. (2015) because the three components are not in a perfectly straight line, but later work (Fan et al. 2020) is able to cope with bent radio galaxies.

generally available CNN for routine classification and cross-identification of radio sources.

An alternative approach uses a Self-Organized Map (SOM). SOMs had been pioneered by Geach (2012) for galaxy classification, and by Polsterer et al. (2015) for radio source classification. It was first applied to the problem of cross-matching radio to IR sources by Galvin et al. (2019, 2020). The technique shows great promise and is currently being applied to the EMU Pilot Survey (Norris et al. 2021b).

These sophisticated classification and cross-matching algorithms tend to be computationally expensive. One approach is to speed up the SOM using an autoencoder (Ralph et al. 2019). Another approach is to reduce the number of sources that need classifying. Only about 10% of sources in modern radio surveys consist of multiple components, with most consisting of a single radio component, which in most cases is the synchrotron radio emission from a single host galaxy, due either to AGN or star-forming activity. Therefore, a significant performance improvement is achieved by preclassifying radio sources into “simple” (i.e., a single radio component corresponding to one host galaxy) or “complex” (multiple radio components from one host galaxy). Segal et al. (2019) showed that this can be achieved simply using a complexity measure based on a compression algorithm, while Park et al. (in preparation) evaluated a number of algorithms for this purpose and found that even simple algorithms are effective in separating simple from complex sources.

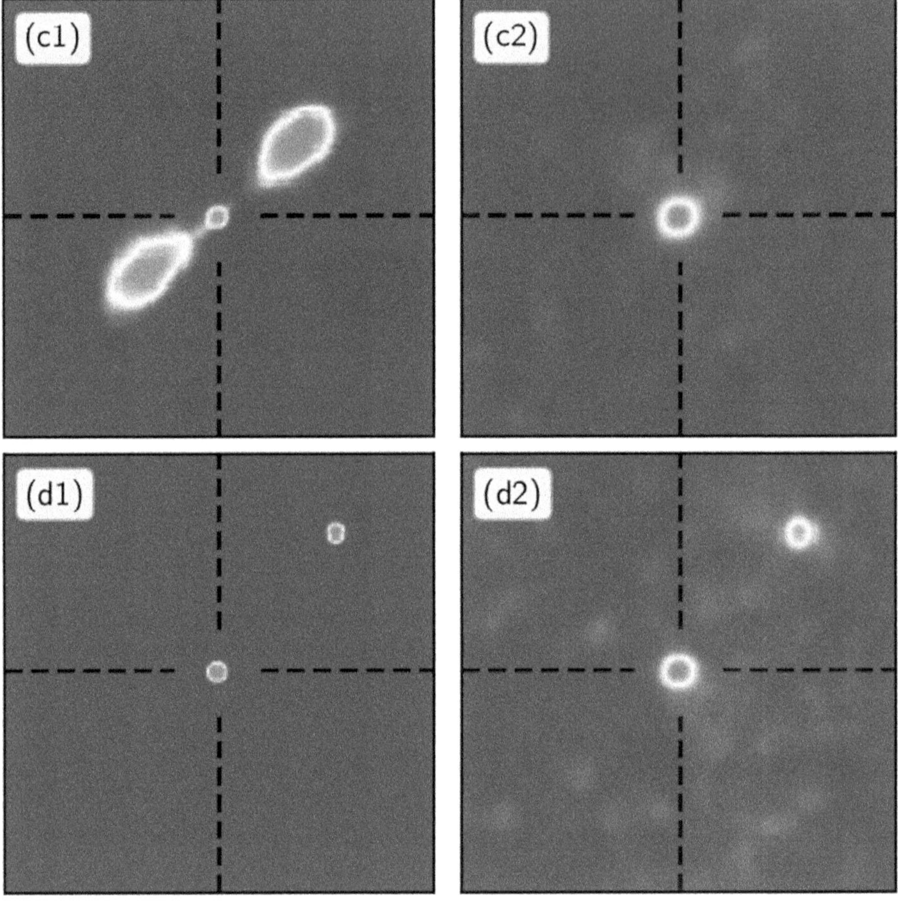

Figure 12.7. A figure from Galvin et al. (2020) in which the SOM has correctly identified the infrared host galaxy (C2) at the center of the radio galaxy (C1), and correctly identified the two infrared host galaxies (d2) corresponding to the two radio components (d1).

12.6 Redshifts

12.6.1 Spectroscopic Redshifts

It is difficult to understand the astrophysics of any astronomical object unless its distance is known. Traditionally, the gold standard of distance measurement has been optical spectroscopy. However, conventional single-slit spectrometers limit the number to a few galaxies per night. Since the start of the millennium, multiobject spectrographs have steadily increased the number of spectra possible (Sharp et al. 2006; de Jong et al. 2012). However, even when a spectrograph can take thousands of spectra per night, there is still not enough telescope time to measure redshifts of the millions of galaxies discovered by surveys and so alternative techniques become important.

Spectroscopy is not confined to optical wavelengths. Neutral hydrogen spectroscopic surveys have been very successful (Meyer et al. 2004; Huang et al. 2012; Koribalski et al. 2020) although the weakness of the neutral hydrogen line, and the limited frequency range of telescopes, limits them to redshifts less than $z \sim 0.5$.

Spectroscopy of the CO molecule has also been used to measure redshifts both of galaxies (Combes 2018; Weiß, 2013; Scoville et al. 2016) and clusters (Riechers et al. 2014; Emonts et al. 2016). The limited bandwidth of conventional millimeter telescopes made it hard to do a survey for redshifts except where an indication of the redshift was already known. However, the wide bandwidth of ALMA, aided by the various transitions of CO, made it possible to do a spectral scan of a region of galaxies, because, whatever the redshift, at least one transition of CO would fall within the bandpass (Blain et al. 2002). This technique of millimeter spectroscopic surveys has become even more important as [C II] was also found to be a valuable tracer of molecular gas at high redshifts (Maiolino et al. 2015).

12.6.2 Photometric Redshifts

Because spectroscopic surveys are too slow to measure redshifts for millions of galaxies, photometric redshifts are becoming increasingly important (Salvato et al. 2019).

The first pioneering work on photometric redshifts (Baum 1962; Butchins 1981; Loh & Spillar 1986) estimated redshift by matching the emission received through standard optical photometric filters to that expected from a selection of standard galaxy templates. This assumed a priori knowledge of the spectral energy distribution (SED) of the galaxies and also assumed that the templates were appropriate for the galaxies under investigation. This latter assumption might fail if either the target galaxies were at a significantly different redshift from the template galaxies, as galaxies evolve over cosmic time, or if the galaxies are chosen from a parent sample that differs (perhaps in a way that is not evident) from the parent sample of the template galaxies. An example of the latter case is that optical counterparts of radio sources fit templates worse than optically selected samples (Norris et al. 2019).

There are two broad categories of photometric redshift estimation: template fitting and ML techniques. We now consider each of these separately.

12.6.3 Template Fitting

Template-fitting techniques (Arnouts et al. 1999; Benítez 2000; Duncan et al. 2018a) use a library of template galaxy spectra that are shifted to different redshifts. The total energy measured in each photometric filter band is then estimated and fitted to the observed photometric data.

The accuracy of the resulting photo-z's will improve with the depth of the data, which results in a smaller photometric error, the number of the bands available, and decreasing spacing between the filters (Budavári & Szalay 2008; Benítez et al. 2009). In the case of the COSMOS field (Scoville et al. 2007), some 30 bands are used, so that the photometric measurements can almost be described as low-resolution spectroscopy, resulting in high-quality redshifts (Salvato et al. 2011).

However, in template fitting, the choice of the template library is extremely important. A standard template library will suffice for most normal galaxies at moderate redshifts, but additional templates must be included if the sample contains significant numbers of radio or X-ray-selected sources, or contains sources at high redshift. For X-ray sources, the template library must also change with the depth of the X-ray survey (Salvato et al. 2011; Hsu et al. 2014).

12.6.4 Machine Learning

ML-based techniques were first introduced by Firth et al. (2002) and Tagliaferri et al. 2003). They differ from template fitting in that no a priori knowledge of templates or astrophysics is assumed. Instead, a training/validation set is constructed consisting of galaxies for which measured spectroscopic redshifts are available, together with the same photometric or other data that is available for the target objects. Once trained, the algorithm can be applied to a set of data with unknown redshifts. However, it is important that the photometric data for both sets are closely matched.

Many different algorithms have been used (e.g., random forest, neural networks, nearest neighbors, support vector machines, Gaussian process regression, and self-organized maps) although comparisons (Norris et al. 2019; Luken et al. 2019; K. J. Luken 2020, in preparation) show that the choice of algorithm is less important than data conditioning and the choice of features. Data conditioning includes algorithms to estimate missing data and, most importantly, identifying which features in the data are most important to a successful estimation, in a feature selection phase (Brescia et al. 2013; D'Isanto et al. 2018). In particular, Polsterer et al. (2014) and Cavuoti et al. (2014) have shown that feature selection should be data driven.

An important goal of modern algorithms is, rather than generating a best guess at a redshift, to generate a probability distribution function (Duncan et al. 2018b).

12.6.5 ML versus Template Fitting, and Hybrid Approaches

Several studies have compared the performance of ML and template approaches (Hildebrandt et al. 2008; Dahlen et al. 2013; Abdalla et al. 2011; Norris et al. 2019). In uniform samples that are well represented by templates, templates can outperform ML techniques, while with good training sets that accurately match the target data, ML methods can be more accurate than template fitting.

An enormous advantage of ML methods is that all types of data are grist to the mill, and they can use nonphotometric information such as morphology, radio polarization, or photometric gradients (Gieseke et al. 2011; Norris et al. 2013). Cavuoti et al. (2017a) have even used the spectroscopic type provided by template fitting. Clearly, ML techniques have enormous power, but the technique is in development, and that enormous power to use many types of data has not yet been successfully harnessed.

The two approaches are complementary, and this complementarity has inspired hybrid approaches that use both template fitting and ML in a single algorithm to improve the accuracy of photo-z estimates (Cavuoti et al. 2017b; Duncan et al. 2018b) as shown in Figure 12.8. Duncan et al. (2018b) divide their data into several

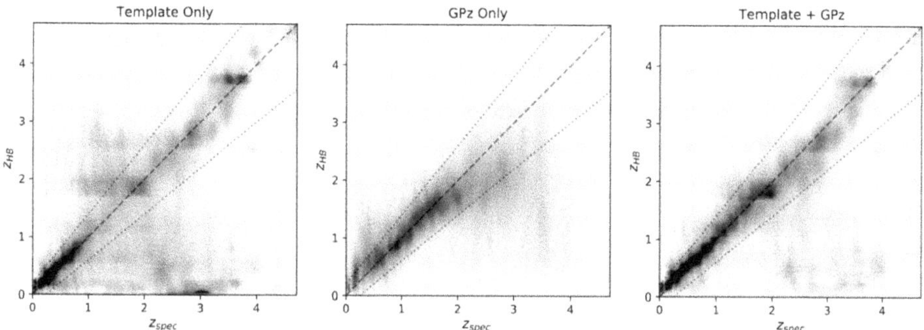

Figure 12.8. Stacked probability distributions ($Z_{\rm spec}$ versus $Z_{\rm H\beta}$) for photometric redshift estimates derived from template fitting (left), Gaussian processes ML technique (center), and combined template + Gaussian processes (right), as a function of spectroscopic redshift. The dashed gray line corresponds to the 1:1 relation (Duncan et al. 2018b). © 2018 The Author(s), Published by Oxford University Press on behalf of the Royal Astronomical Society.

astrophysical classes and train their algorithms separately on these classes. They then combine the results from the different sets, together with template-fitting results, using a Bayesian estimator.

12.6.6 Multimessenger Photometric Redshifts

Nearly all traditional photometric redshift template fitting codes are limited to the use of optical or near-infrared data, because (a) the templates are not well defined at other wavelengths, and (b) the quality of the photometric data at other wavelengths has, in the past, generally been inferior to optical/IR data. This is also largely true of ML techniques.

Exceptions include Rowan-Robinson et al. (2008), who used IRAC mid-infrared data from the SWIRE survey (Lonsdale et al. 2003), Pearson et al. (2013) who used Herschel and submillimeter data, and several authors who have used radio data together with optical/IR data (Aretxaga et al. 2007; Norris et al. 2019; Luken et al. 2019).

Similarly, most studies comparing the performance of photo-z methods have also been based on optically selected samples, which can be suboptimum. For example, the median redshift of radio sources in modern radio surveys is $z \sim 1.2$, compared to the median redshift $z < 0.2$ of most wide optical surveys (Norris et al. 2013), so optical templates are poorly matched to most samples of radio sources.

Furthermore, such approaches miss the potential power of ML algorithms, which (at least in principle) are capable of using many forms of multimessenger data to inform the estimation of redshift. The use of nonphotometric information such as morphology, radio polarization, or photometric gradients has already been discussed above, and to these, we might add other multimessenger data. However, the use of other multimessenger data raises other challenges.

An additional challenge when computing photo-z for large multimessenger surveys is that the photometry will often be heterogeneous, using different photometric catalogs of different depths, with different aperture corrections.

12.7 Classification

After a survey has been completed, and the multiple wavelengths and messengers have been cross-identified as discussed above, the next most important step is to classify the objects in the survey.

One of the most powerful tools for this purpose is to examine the infrared colors, such as those measured by WISE (Wright et al. 2010). For example, the measured WISE colors can be used to identify the type of object as shown in Figure 12 of Wright et al. (2010). However, this diagram is only correct for low-redshift galaxies and becomes more complex for high-redshift objects (Shobhana et al. in preparation).

Optical colors can also be used to classify objects, with red/yellow colors typically indicating a passive early-type galaxy, with bluer colors typically indicating a star-forming galaxy (Masters et al. 2011; Schawinski et al. 2014). If spectroscopy is available, line ratios are a powerful tool for diagnosing the galaxy type (Kewley et al. 2019).

If the object is a radio source, then the radio morphology can be an important indicator. Galactic objects such as H II regions and SNRs are often obvious from their morphology (Bozzetto et al. 2017; Yew et al. 2021), while double-lobed sources usually indicate an AGN. However, about 90% of radio sources are compact or unresolved at arcsecond scales, and so morphology is unhelpful here. Most galaxies, whether star-forming or AGN, tend to have a spectral index[3] $\alpha \sim -0.7$, although a galaxy with a spectral index significantly flatter or steeper than −0.7 generally indicates an AGN. Instead, the presence of an AGN may be diagnosed either by the radio–far-infrared relation (Norris et al. 2006) or by the presence of polarization—any extragalactic polarized radio source is either an AGN or a very low-redshift star-forming galaxy.

If the object is detected in x-rays, then the hardness can be used to classify the source. For example, Filipovic et al. (1998) and Rosati et al. (2002) use the hardness ratio HR, defined as HR= (H − S)/(H + S), where H and S are the net count rates in the Chandra 2–7 keV and 0.5–2 keV bands, respectively. Sources with HR > 0.2 are generally Type 2 AGNs, whereas sources with HR $\leqslant$ 0.2 are generally either Type 1 AGNs or star-forming galaxies.

From all these tests, we can assemble some idea of the nature of the object that we are studying. The process of classification is sometimes likened to placing the object into a category, but this is probably an oversimplification that can be counter-productive. For example, a single galaxy might be variously described as a star-forming galaxy, a Seyfert galaxy, an AGN, or a luminous infrared galaxy (LIRG). So, into which category is it inserted? Instead of inserting objects into categories, Rudnick (2021) has argued that it is better to add labels to an object, recognizing that one object may have many different labels. It is the collection of those labels, rather than the presence in any one category, that accurately describes the object.

[3] Here we adopt the convention that the spectral index α is defined by the flux density S and frequency ν as $S(\nu) \propto \nu^{\alpha}$.

12.8 Searching for the Unexpected

Norris (2017a) has shown that over half the major discoveries in astronomy are unplanned or unexpected and generally result from observing the universe in an unexplored part of observational parameter space, rather than from testing a hypothesis.

In the past, these unexpected discoveries were often made serendipitously by an astronomer examining their data carefully and noticing an anomaly. For example, pulsars were discovered when a young Ph.D. student, Jocelyn Bell, noticed some "scruff" on the chart recorder and refused to accept her supervisor's explanation that it was probably caused by radio-frequency interference (Bell Burnell 2009).

Could Jocelyn Bell's discovery take place with a modern telescope? Possibly not, as the large volumes of data prevent researchers from inspecting their data by eye to look for anything odd. Of course, anything significantly wrong with the data will hopefully be picked up by validation techniques, but anything as subtle as a bit of "scruff" will probably be missed, or, worse, removed by radio-frequency-interference mitigation techniques.

In an imaging survey, anything that appears in output images may well be spotted, as happened with the discovery of "odd radio circles" by the EMU team in 2020 (Norris et al. 2021a). But anything in that EMU data that did not appear in an image (such as a spectral line or time-varying signal) would probably be missed. Even in the image domain, an object that was too different from the objects under study (e.g., too large, or too diffuse) would probably be missed. It is therefore essential to design new telescopes or projects explicitly to enable them to discover the unexpected.

Therefore, telescopes must be designed explicitly to maximize their ability to discover the unknown, which is potentially more important than the stated science goals. Specifically, we need to design ML algorithms that will search the data for the unexpected.

One example of such an algorithm was by a team that trained a random forest algorithm to search SDSS spectroscopy for spectra, unlike the majority. The result (Baron & Poznanski 2017) was impressive, with several "weird" objects discovered out of more than 2 million SDSS spectra.

12.9 Conclusion

Astronomy is in the middle of a revolution. Gone are the days when an astronomer might make a single observation of an object at one wavelength and publish a paper on it. Nowadays, we routinely compare our data with data at other wavelengths or from other astronomical "messengers," and we have the tools and data readily available to do so.

Next-generation electromagnetic telescopes now produce such enormous volumes of data that specialized knowledge is needed to make effective use of them. Even more challenging, for the electromagnetic astronomer, is to make effective use of the rapidly growing body of non-electromagnetic data from the emerging generation of telescopes operating on high-energy particles and gravitational waves.

To make effective use of those multimessenger and multiwavelength data requires an astronomer to be conversant with the techniques and tools available at all wavelengths and for all messengers. It is no longer sufficient to be an expert in one wavelength and rely on the skill of your collaborators to deal with the other wavelengths. In this chapter, I have briefly outlined some of those tools and techniques and hopefully provided sufficient citations to guide the reader to more detailed descriptions.

References

Abbott, B. P., Abbott, R., Abbott, T. D., et al. 2016, ApJ, 826, L13
Abbott, T. M. C., Abdalla, F. B., Allam, S., et al. 2018, ApJS, 239, 18
Abdalla, F. B., Banerji, M., Lahav, O., & Rashkov, V. 2011, MNRAS, 417, 1891
Aiello, S., Albert, A., Garre, S. A., et al. 2020a, CoPhC, 256, 107477
Aiello, S., Ameli, F., Andre, M., et al. 2020b, CoPhC, 256, 107433
Aiello, S., Albert, A., Alves Garre, S., et al. 2020c, JInst, 15, P10005
Alger, M. J., Banfield, J. K., Ong, C. S., et al. 2018, MNRAS, 478, 5547
Allen, G., Anderson, W., Blaufuss, E., et al. 2018, arXiv:1807.04780
Allen, G., Andreoni, I., Bachelet, E., et al. 2019, arXiv:1902.00522
Anderson, D. P., Cobb, J., Korpela, E., Lebofsky, M., & Werthimer, D. 2002, Commun. ACM, 45, 56
Aniyan, A. K., & Thorat, K. 2017, ApJS, 230, 20
Aretxaga, I., Hughes, D. H., Coppin, K., et al. 2007, MNRAS, 379, 1571
Arnouts, S., Cristiani, S., Moscardini, L., et al. 1999, MNRAS, 310, 540
Banfield, J. K., Andernach, H., Kapińska, A. D., et al. 2016, MNRAS, 460, 2376
Banfield, J. K., Wong, O. I., Willett, K. W., et al. 2015, MNRAS, 453, 2326
Baron, D., & Poznanski, D. 2017, MNRAS, 465, 4530
Baum, W. A. 1962, in IAU Symp. 15, Problems of Extra-Galactic Research, ed. G. C. McVittie (New York: MacMillan), 390
Becker, R. H., White, R. L., & Helfand, D. J. 1995, ApJ, 450, 559
Bell Burnell, J. 2009, in 27th IAU General Assembly, Accelerating the Rate of Astronomical Discovery, ed. R. P. Norris, & C. L. Ruggles, PoS(sps5)014
Benítez, N., Moles, M., Aguerri, J. A. L., et al. 2009, ApJ, 692, L5
Benítez, N. 2000, ApJ, 536, 571
Bernstein, G. M., Armstrong, R., Plazas, A. A., et al. 2017, PASP, 129, 074503
Bertin, E., & Arnouts, S. 1996, A&AS, 117, 393
Blain, A. W., Smail, I., Ivison, R. J., Kneib, J. P., & Frayer, D. T. 2002, PhR, 369, 111
Bonnarel, F., Fernique, P., Bienaymé, O., et al. 2000, A&AS, 143, 33
Bozzetto, L. M., Filipović, M. D., Vukotić, B., et al. 2017, ApJS, 230, 2
Brescia, M., Cavuoti, S., D'Abrusco, R., Longo, G., & Mercurio, A. 2013, ApJ, 772, 140
Budavári, T., & Szalay, A. S. 2008, ApJ, 679, 301
Butchins, S. A. 1981, A&A, 97, 407
Cavuoti, S., Amaro, V., Brescia, M., et al. 2017a, MNRAS, 465, 1959
Cavuoti, S., Tortora, C., Brescia, M., et al. 2017b, MNRAS, 466, 2039
Cavuoti, S., Brescia, M., & Longo, G. 2014, in IAU Symp. 306, Statistical Challenges in 21st Century Cosmology, ed. A. Heavens, J.-L. Starck, & A. Krone-Martins (Cambridge: Cambridge Univ. Press), 307

Chen, M., Zhong, Y., Feng, Y., Li, D., & Li, J. 2020, arXiv:2003.13928
Combes, F. 2018, A&ARv, 26, 5
Dahlen, T., Mobasher, B., Faber, S. M., et al. 2013, ApJ, 775, 93
de Jong, R. S., Bellido-Tirado, O., Chiappini, C., et al. 2012, Proc. SPIE, 8846, 84460T
de Ruiter, H. R., Willis, A. G., & Arp, H. C. 1977, A&AS, 28, 211
D'Isanto, A., Cavuoti, S., Gieseke, F., & Polsterer, K. L. 2018, A&A, 616, A97
Duncan, K. J., Brown, M. J. I., Williams, W. L., et al. 2018a, MNRAS, 473, 2655
Duncan, K. J., Jarvis, M. J., Brown, M. J. I., & Röttgering, H. J. A. 2018b, MNRAS, 477, 5177
Emonts, B. H. C., Lehnert, M. D., Villar-Martín, M., et al. 2016, Sci, 354, 1128
Fan, D., Budavári, T., Norris, R. P., & Basu, A. 2020, MNRAS, 498, 565
Fan, D., Budavári, T., Norris, R. P., & Hopkins, A. M. 2015, MNRAS, 451, 1299
Fernique, P., Allen, M. G., Boch, T., et al. 2015, A&A, 578, A114
Filipovic, M. D., Pietsch, W., Haynes, R. F., et al. 1998, A&AS, 127, 119
Firth, A. E., Somerville, R. S., McMahon, R. G., et al. 2002, MNRAS, 332, 617
Flewelling, H. A., Magnier, E. A., Chambers, K. C., et al. 2016, arXiv:1612.05243
Fortner, B. 1998, Dr Dobb's J Software Tools, 23, 42
Gaia Collaboration, Mignard, F., Klioner, S. A., et al. 2018, A&A, 616, A14
Galvin, T. J., Huynh, M., Norris, R. P., et al. 2019, PASP, 131, 108009
Galvin, T. J., Huynh, M. T., Norris, R. P., et al. 2020, MNRAS, 497, 2730
Garon, A. F., Rudnick, L., Wong, O. I., et al. 2019, AJ, 157, 126
Geach, J. E. 2012, MNRAS, 419, 2633
Gieseke, F., Polsterer, K. L., Thom, A., et al. 2011, arXiv:1108.4696
Greisen, E. W., & Calabretta, M. R. 2002, A&A, 395, 1061
Hale, C. L., Robotham, A. S. G., Davies, L. J. M., et al. 2019, MNRAS, 487, 3971
Hales, C. A., Murphy, T., Curran, J. R., et al. 2012, MNRAS, 425, 979
Hancock, P. J., Trott, C. M., & Hurley-Walker, N. 2018, PASA, 35, e011
Hildebrandt, H., Wolf, C., & Benítez, N. 2008, A&A, 480, 703
Hofestädt, J. 2019, EPJJWC, 207, 08001
Hopkins, A. M., Whiting, M. T., Seymour, N., et al. 2015, PASA, 32, e037
Hsu, L.-T., Salvato, M., Nandra, K., et al. 2014, ApJ, 796, 60
Huang, S., Haynes, M. P., Giovanelli, R., & Brinchmann, J. 2012, ApJ, 756, 113
Johnston, S., Taylor, R., Bailes, M., et al. 2008, ExA, 22, 151
Kapińska, A. D., Terentev, I., Wong, O. I., et al. 2017, AJ, 154, 253
Kewley, L. J., Nicholls, D. C., & Sutherland, R. S. 2019, ARA&A, 57, 511
Kimball, A. E., & Ivezić, Ž. 2008, AJ, 136, 684
Koribalski, B. S., Staveley-Smith, L., Westmeier, T., et al. 2020, Ap&SS, 365, 118
Krizhevsky, A., Sutskever, I., & Hinton, G. E. 2012, in Advances in Neural Information Processing 25 (Burlington, MA: Morgan Kaufmann Publishers), 1097
Kron, R. G. 1980, ApJS, 43, 305
Lacy, M., Baum, S. A., Chandler, C. J., et al. 2020, PASP, 132, 035001
Lindegren, L., Hernández, J., Bombrun, A., et al. 2018, A&A, 616, A2
Lintott, C. J., Schawinski, K., Slosar, A., et al. 2008, MNRAS, 389, 1179
Loh, E. D., & Spillar, E. J. 1986, ApJ, 303, 154
Lonsdale, C. J., Smith, H. E., Rowan-Robinson, M., et al. 2003, PASP, 115, 897
LSST Science Collaboration, Abell, P. A., Julius, A., et al. 2009, arXiv:0912.0201
Luken, K. J., Norris, R. P., & Park, L. A. F. 2019, PASP, 131, 108003

Lukic, V., Brüggen, M., Banfield, J. K., et al. 2018, MNRAS, 476, 246
Maiolino, R., Carniani, S., Fontana, A., et al. 2015, MNRAS, 452, 54
Marrese, P. M., Marinoni, S., Fabrizio, M., & Giuffrida, G. 2017, A&A, 607, A105
Masters, K. L., Maraston, C., Nichol, R. C., et al. 2011, MNRAS, 418, 1055
Maurin, D., Melot, F., & Taillet, R. 2014, A&A, 569, A32
Meyer, M. J., Zwaan, M. A., Webster, R. L., et al. 2004, MNRAS, 350, 1195
Middelberg, E., Norris, R. P., Cornwell, T. J., et al. 2008, AJ, 135, 1276
Mohan, N., & Rafferty, D. 2015, PyBDSF: Python Blob Detection and Source Finder, Astrophysics Source Code Library, ascl:1502.007
Norris, R. P. 2017a, PASA, 34, e007
Norris, R. P. 2017b, NatAs, 1, 671
Norris, R. P., Afonso, J., Bacon, D., et al. 2013, PASA, 30, e020
Norris, R. P., Afonso, J., Appleton, P. N., et al. 2006, AJ, 132, 2409
Norris, R. P., Hopkins, A. M., Afonso, J., et al. 2011, PASA, 28, 215
Norris, R. P., Intema, H. T., Kapińska, A. D., et al. 2021a, PASA, 38, e003
Norris, R. P., Marvil, J., Collier, J. D., et al. 2021b, PASA, 38, e046
Norris, R. P., Salvato, M., Longo, G., et al. 2019, PASP, 131, 108004
Paz Arribas, M. 2016, PhD thesis, Humboldt University, Berlin
Pearson, E. A., Eales, S., Dunne, L., et al. 2013, MNRAS, 435, 2753
Perryman, M. A. C. 2002, Ap&SS, 280, 1
Pineau, F. X., Motch, C., Carrera, F., et al. 2011, A&A, 527, A126
Pizzolotto, C., Di Felice, V., D'Urso, D., et al. 2017, ICRC, 301, 227
Polsterer, K. L., Gieseke, F., & Igel, C. 2015, in ASP Conf. Ser. 495, Astronomical Data Analysis Software and Systems XXIV (ADASS XXIV), ed. A. R. Taylor, & E. Rosolowsky (San Francisco, CA: ASP), 81
Polsterer, K. L., Gieseke, F., Igel, C., & Goto, T. 2014, in ASP Conf. Ser. 485, Astronomical Data Analysis Software and Systems XXIII, ed. N. Manset, & P. Forshay (San Francisco, CA: ASP), 425
Proctor, D. D. 2006, ApJS, 165, 95
Ralph, N. O., Norris, R. P., Fang, G., et al. 2019, PASP, 131, 108011
Riechers, D. A., Carilli, C. L., Capak, P. L., et al. 2014, ApJ, 796, 84
Riggi, S., Vitello, F., Becciani, U., et al. 2019, PASA, 36, e037
Rodman, P. E., Turner, R. J., Shabala, S. S., et al. 2019, MNRAS, 482, 5625
Rosati, P., Tozzi, P., Giacconi, R., et al. 2002, ApJ, 566, 667
Rowan-Robinson, M., Babbedge, T., Oliver, S., et al. 2008, MNRAS, 386, 697
Rudnick, L. 2021, Galax, 9, 85
Salvato, M., Ilbert, O., Hasinger, G., et al. 2011, ApJ, 742, 61
Salvato, M., Ilbert, O., & Hoyle, B. 2019, NatAs, 3, 212
Sault, R. J., Teuben, P. J., & Wright, M. C. H. 1995, in ASP Conf. Ser. 77, Astronomical Data Analysis Software and Systems IV, ed. R. A. Shaw, H. E. Payne, & J. J. E. Hayes (San Francisco, CA: ASP), 433
Scaife, A. M. M. 2020, RSTPA, 378, 20190060
Schawinski, K., Urry, C. M., Simmons, B. D., et al. 2014, MNRAS, 440, 889
Scoville, N., Aussel, H., Brusa, M., et al. 2007, ApJS, 172, 1
Scoville, N., Sheth, K., Aussel, H., et al. 2016, ApJ, 820, 83
Segal, G., Parkinson, D., Norris, R. P., & Swan, J. 2019, PASP, 131, 108007

Serra, P., Westmeier, T., Giese, N., et al. 2015, MNRAS, 448, 1922
Sharp, R., Saunders, W., Smith, G., et al. 2006, Proc. SPIE, 6269, 62690G
Shawhan, P. S., Brady, P. R., Brazier, A., et al. 2019, in ASP Conf. Ser. 523, Astronomical Data Analysis Software and Systems XXVII, ed. P. J. Teuben, M. W. Pound, B. A. Thomas, & E. M. Warner (San Francisco, CA: ASP), 705
Sutherland, W., & Saunders, W. 1992, MNRAS, 259, 413
Tagliaferri, R., Longo, G., Andreon, S., et al. 2003, in Lecture Notes In Computer Science, Vol. 2859 (Stuttgart: Springer), 226
Tang, H., Scaife, A. M. M., Wong, O. I., et al. 2020, MNRAS, 499, 68
Taylor, M. B. 2005, in ASP Conf. Ser. 347, Astronomical Data Analysis Software and Systems XIV, ed. P. Shopbell, M. Britton, & R. Ebert (San Francisco, CA: ASP), 29
Thompson, A. R., Clark, B. G., Wade, C. M., & Napier, P. J. 1980, ApJS, 44, 151
van Haarlem, M. P., Wise, M. W., Gunst, A. W., et al. 2013, A&A, 556, A2
van Velzen, S., Falcke, H., & Körding, E. 2015, MNRAS, 446, 2985
Wang, H., Wu, S., Cao, Z., Liu, X., & Zhu, J.-Y. 2020, PhRvD, 101, 104003
Weiß, A., De Breuck, C., Marrone, D. P., et al. 2013, ApJ, 767, 88
Wells, D. C. 1985, Nrao'S Astronomical Image Processing System (AIPS) (Boston, MA: Springer), 195
Wells, D. C., Greisen, E. W., & Harten, R. H. 1981, A&AS, 44, 363
Weston, S. D., Seymour, N., Gulyaev, S., et al. 2018, MNRAS, 473, 4523
Westphal, A. J., Butterworth, A. L., Snead, C.J., et al. 2005, LPSC, 36, 1908
White, R. L., Becker, R. H., Helfand, D. J., & Gregg, M. D. 1997, ApJ, 475, 479
Whiting, M., & Humphreys, B. 2012, PASA, 29, 371
Wolstencroft, R. D., Savage, A., Clowes, R. G., et al. 1986, MNRAS, 223, 279
Wright, D. E., Lintott, C. J., Smartt, S. J., et al. 2017, MNRAS, 472, 1315
Wright, E. L., Eisenhardt, P. R. M., Mainzer, A. K., et al. 2010, AJ, 140, 1868
Wu, C., Wong, O. I., Rudnick, L., et al. 2019, MNRAS, 482, 1211
Yew, M., Filipović, M. D., Stupar, M., et al. 2021, MNRAS, 500, 2336
Zevin, M., Coughlin, S., Bahaadini, S., et al. 2017, CQGra, 34, 064003
Zoli, A., Bulgarelli, A., Rosa, A. De., et al. 2015, ICRC, 34, 944

Lightning Source UK Ltd.
Milton Keynes UK
UKHW031037260122
397711UK00005B/91

9 780750 323451